5 € Bot

Corn:
Chemistry and Technology

Edited by
Stanley A. Watson
Paul E. Ramstad

Published by the
American Association of Cereal Chemists, Inc.
St. Paul, Minnesota, USA

Cover photograph courtesy of the Council for Agricultural
  Science and Technology, Ames, Iowa

Library of Congress Catalog Card Number: 87-070831
International Standard Book Number: 0-913250-48-1

©1987 by the American Association of Cereal Chemists, Inc.

All rights reserved.
No part of this book may be reproduced in any form
by photocopy, microfilm, retrieval system, or any other means,
without written permission from the publisher.

Copyright is not claimed in any portion of this work written by United States
Government employees as a part of their official duties.

Reference in this volume to a company or product name by personnel of the U.S.
Department of Agriculture is intended for explicit description only and does
not imply approval or recommendation of the product by the U.S. Department of
Agriculture to the exclusion of others that may be suitable.

Printed in the United States of America

American Association of Cereal Chemists, Inc.
3340 Pilot Knob Road
St. Paul, Minnesota 55121, USA

CONTRIBUTORS

R. J. Alexander, Penick & Ford, Ltd., Cedar Rapids, Iowa

M. O. Bagby, Northern Regional Research Center, U.S. Department of Agriculture, Agricultural Research Service, Peoria, Illinois

G. O. Benson, Department of Agronomy, Iowa State University, Ames, Iowa

C. D. Boyer, Department of Horticulture, The Pennsylvania State University, University Park, Pennsylvania

L. L. Darrah, U.S. Department of Agriculture, Agricultural Research Service, Department of Agronomy, University of Missouri, Columbia, Missouri

R. E. Hebeda, Enzyme Bio-Systems Ltd., Bedford Park, Illinois

F. L. Herum, Department of Agricultural Engineering, Ohio Agricultural Research and Development Center, The Ohio State University, Columbus, Ohio

L. D. Hill, Department of Agricultural Economics, University of Illinois at Urbana-Champaign, Urbana, Illinois

M. N. Leath, U.S. Department of Agriculture, Economic Research Service, National Economics Division, Washington, DC

W. F. Maisch, Brown Forman Corporation, Louisville, Kentucky

S. W. Marshall, Crookham Company, Caldwell, Idaho

J. B. May (*retired*), A. E. Staley Manufacturing Company, Decatur, Illinois

F. T. Orthoefer, A. E. Staley Manufacturing Company, Decatur, Illinois

R. B. Pearce, Department of Agronomy, Iowa State University, Ames, Iowa

L. W. Rooney, Cereal Quality Laboratory, Soil and Crop Sciences Department, Texas A&M University, College Station, Texas

S. O. Serna-Saldivar, Centro de Investigaciones en Alimentos, University of Sonora, Hermosillo, Sonora, Mexico

J. C. Shannon, Department of Horticulture, The Pennsylvania State University, University Park, Pennsylvania

R. D. Sinram, A. E. Staley Manufacturing Company, Decatur, Illinois

C. L. Storey (*retired*), U.S. Grain Marketing Research Laboratory, U.S. Department of Agriculture, Agricultural Research Service, Manhattan, Kansas

S. A. Watson (*retired*), Director's Office, Ohio Research and Development Center, The Ohio State University, Wooster, Ohio

E. J. Weber, U.S. Department of Agriculture, Agricultural Research Service, Department of Agronomy, University of Illinois, Urbana, Illinois

N. W. Widstrom, Coastal Plains Experiment Station, U.S. Department of Agriculture, Agricultural Research Service, Tifton, Georgia

C. M. Wilson, U.S. Department of Agriculture, Agricultural Research Service, Department of Agronomy, University of Illinois, Urbana, Illinois

K. N. Wright, Wright Nutrition Service, Decatur, Illinois

M. S. Zuber (*retired*), Department of Agronomy, University of Missouri, Columbia, Missouri

PREFACE

This is a book about a most remarkable cereal grain, *Zea mays* L., maize, or, as it is better known in the United States and Canada, corn. We realize that it is called "maize" in most other countries, but for uniformity, we have used the term "corn" throughout this book except in a few instances where "maize" seemed more appropriate. Since this is a technical work of worldwide interest, we have preferred to use metric and IS units throughout, with the English equivalents or conversion factors given for those readers not accustomed to using metric units. The symbol "t" is used to denote metric tons (tonnes).

Corn is indigenous to North America. It was developed by Central American natives many centuries before Columbus saw it. Corn was the foundation of the extensive North and South American ancient civilizations and was important in the agriculture of more recent Indian populations. Although Columbus carried corn seed to Europe, where it became established as an important crop in southern latitudes, it was the opening of the fertile plains of the midwestern United States that accelerated the development of modern corn culture. These developments were brought about, first, through work of ingenious farmers and, later, through the efforts of many research scientists.

In countries where corn is an important crop, it is the principal component of livestock feeds, and most of it is fed to farm animals. In only a few countries is corn a major constituent of human diets. In developed countries, corn is consumed mainly as popcorn, sweet corn, corn snacks, and occasionally as corn bread. However, most consumers are not aware that corn plays an important role in the production of meat, milk, and eggs or that corn is an important source of the sweeteners, starches, oil, and alcohol used in many foods, beverages, and numerous other products.

Much of the importance of corn is derived from its low cost, which is the result of corn's high productivity. Although, at this time, the major corn-growing areas seem to have a large surplus of corn in storage, a burgeoning population will need more food, and a drought or other natural disaster could lead to rapid utilization of the surplus. Therefore, research on improvement in productive efficiency and new uses of corn will continue. Progress in utilization of corn is dependent on new information on the physical, chemical, and biological properties of corn resulting from research by chemists, engineers, food scientists, nutritionists, and other researchers.

Corn is important in international trade. In recent years, it has been a major agricultural export of the United States, Argentina, Brazil, South Africa, Thailand, and Rumania. The United States, as the world's largest producer, has also been the largest exporter. Most corn gluten feed, a by-product of the growing corn wet-milling industry, is exported from the United States. Recent replacement of sucrose by high-fructose corn syrup in the United States has had a major impact on the economies of countries exporting sugar to the United States, but this development, together with development of a fuel ethanol business in the United States, has had a favorable effect on utilization of the U.S. corn crop.

Although other books on the chemistry and technology of corn have been published, none is recent. Much new knowledge is contained in this book. Authorities in many fields have contributed chapters or have reviewed manuscripts. Among them are agronomists, geneticists, entomologists, mycologists, food scientists, chemists, engineers, and economists who hold responsible positions in universities, government, and industry. Anyone interested in any aspect of corn research and development, marketing, utilization, etc., should find this volume useful. The editors, as representatives of The American Association of Cereal Chemists, are grateful to each of the authors and reviewers. We also wish to thank the AACC staff for their advice, help, and technical editing. We also owe thanks and appreciation to the Agricultural Research and Development Center, The Ohio State University, Wooster, Ohio, for use of facilities and to secretaries Georgia Miller and Gwen Alaura, photographer Ken Chamberlain, and draftsman, Newell Hartrum.

Stanley A. Watson
Paul E. Ramstad

CONTENTS

CHAPTER	PAGE

1. Corn Perspective and Culture. G. O. BENSON and R. B. PEARCE 1

    I. Introduction 1

    II. Description and Adaptation 1

    III. Origin and History 3

    IV. Production and Use 4

    V. Corn Culture 7
       A. Growth and Development—B. Climatic Requirements—C. Soil and Crop Management—D. Pest Management

    VI. Summary 28

2. Breeding, Genetics, and Seed Corn Production. M. S. ZUBER and L. L. DARRAH .. 31

    I. Introduction 31
       A. Reproduction of the Corn Plant—B. Kernel Structure—C. Endosperm Dosage Effects—D. Endosperm Variation

    II. Progress in Corn Improvement 37
       A. Early Inbreeding and Hybridization Experiments—B. Use of Open-Pollinated Varieties, 1910–1936—C. Introduction of Hybrids, 1937–1965—D. Single-Cross and Modified Single-Cross Hybrids, 1966 to the Present—E. Host-Plant Resistance

    III. Breeding Techniques for Selection of Improved Genotypes 39
       A. Mass Selection—B. Random Selfed Lines—C. S_1 and S_2 Line Selection—D. Recurrent Selection—E. Extraction of Inbred Lines from Populations Improved by Recurrent Selection—F. Inbred Line Development by Recycling Existing Lines—G. Evaluation of Experimental Material

    IV. Kernel Modification Through Breeding 44
       A. Single-Mutant Endosperm Genes—B. Altering Kernel Composition and Integrity by Selection

    V. Seed Production 47
       A. Crossing Techniques—B. Isolation Requirements—C. Seed Conditioning—D. Seed Classification—E. Contract Production

3. Structure and Composition. STANLEY A. WATSON 53

    I. Introduction 53

    II. Gross Structural Features 53

    III. Physical Properties 59

   IV. Structural Details 61
       A. Germ—B. Endosperm—C. Pericarp

   V. Composition 69
       A. Kernel Development—B. Distribution in the Mature Kernel—C. Minerals—D. Vitamins—E. Nitrogenous Components—F. Lipids—G. Carbohydrates—H. Miscellaneous Substances

   VI. Summary 78

4. Harvesting and Postharvest Management. FLOYD L. HERUM 83

   I. Introduction 83
       A. Evolution in Corn Systems—B. Labor Requirements in Production—C. Mechanization of Handling and Transportation

   II. Harvesting Methods 86
       A. Ear Corn Harvest—B. Field Shelling—C. Future Trends in Harvest Mechanization

   III. Corn Drying 89
       A. Moisture Reduction and Shrink—B. Drying Mechanisms—C. Corn Drying Systems

   IV. Handling Systems 103
       A. Ear Handling—B. Shelled Corn Handling

   V. Storage and Storage Management 112
       A. Ear Corn Storage—B. Shelled Corn Storage

5. Measurement and Maintenance of Quality. STANLEY A. WATSON 125

   I. Introduction 125

   II. Grades and Standards 126
       A. History—B. Sample Collections—C. Classes and Grades—D. Grading Procedures—E. Accuracy of Grade Determinations—F. Importance of Grades in Corn Marketing—G. Grades and Standards for Export Trade

   III. Broken Corn, Foreign Material, and Dust 134
       A. BCFM—B. Dusts

   IV. Moisture 136
       A. Importance of Grain Moisture—B. Accounting for Moisture in Marketing—C. Moisture Measurement

   V. Density 143
       A. Specific Density—B. Test Weight (Bulk Density)—C. Corn Maturity and Test Weight—D. Feeding Value of Low-Test-Weight Corn—E. Test Weight in Milling Applications

   VI. Hardness 149
       A. Significance—B. Measurements

   VII. Physical Quality Changes 152
       A. Mechanical Damage—B. Stress Cracks—C. Breakage Susceptibility

VIII. Fungal Invasion 156
   A. "Damage" Grade Factor—B. Role of Microorganisms in Quality Deterioration—C. Field Fungi—D. Storage Fungi—E. Detection of Fungal Invasion—F. Incidence of Fungi in Stored Corn—G. Loss Prevention—H. Mycotoxins

IX. Quality in Corn Utilization 168
   A. Seed Corn—B. Beverage Alcohol—C. Dry Milling and Snack Foods—D. Wet Milling—E. Feed Manufacturing

X. Conclusions and Trends 173

6. Effect and Control of Insects Affecting Corn Quality. CHARLES L. STOREY 185

I. Introduction 185

II. Important Insects Affecting Corn Quality 185

III. Development of Insects in Stored Grain 188
   A. Developmental Stages—B. Factors Affecting Development

IV. Sampling and Measurement of Insect Populations in Corn 189

V. Protective Measures and Control of Insects in Corn Storage 190
   A. Environmental Protective Measures—B. Prevention by Chemical Treatment of the Corn—C. Remedial Control Measures for Existing Populations

VI. Summary 197

7. Economics of Production, Marketing, and Utilization. MACK N. LEATH and LOWELL D. HILL 201

I. Introduction 201

II. Trends in Supply and Production Costs 202
   A. Acreage, Yield, and Production—B. Production Costs and Returns

III. Trends in Utilization 210
   A. Domestic Livestock Feed—B. Food and Industrial Products—C. Seed Use—D. Exports

IV. The Marketing System 219
   A. Overview of Marketing Flows—B. Farm Drying and Conditioning—C. Farm Storage and Marketing—D. Commercial Handling and Storage

V. Domestic Transportation and Flow Patterns 225
   A. Modal Shares—B. Movements to Domestic Markets—C. Movements to Export Regions—D. Developments in Recent Years

VI. Trends in World Corn Markets 230
   A. Major Coarse-Grain Markets—B. Corn Supply and Demand—C. Major Competing Exporters—D. Policies of Major Trading Countries—E. Implications for U.S. Exports

VII. Corn Pricing System 241
   A. Organized Grain Exchanges—B. Country Elevators

VIII. Government Programs Affecting the Industry 243
  A. Price-Support Programs—B. Production Adjustment Programs—C. Commodity Storage Programs—D. Commodity Disposal Programs

IX. Summary 249

8. Carbohydrates of the Kernel. CHARLES D. BOYER and JACK C. SHANNON 253

  I. Introduction 253

  II. General Considerations 254
    A. Transfer into the Kernel—B. Distribution Within the Kernel—C. Endosperm Carbohydrates

  III. Simple Carbohydrates 257
    A. Monosaccharides—B. Disaccharides and Trisaccharides—C. Sugar Alcohols—D. Phytate—E. Metabolic Intermediates

  IV. Complex Carbohydrates—Structural 258
    A. Cell Walls—B. Cellulose—C. Pentosans—D. Other Cell Wall Components

  V. Complex Carbohydrates—Storage 260
    A. Water-Soluble Polysaccharides—B. Starch

  VI. Summary 269

9. Proteins of the Kernel. CURTIS M. WILSON 273

  I. Introduction 273

  II. Protein Fractionation Overview 274

  III. Endosperm Proteins 277
    A. Prolamins and Protein Body Components—B. Nonprolamin Proteins

  IV. Embryo Proteins 293

  V. Protein Bodies 296

  VI. Protein Assay 298
    A. Total Nitrogen and Amino Acids—B. Assays for Improved Nutritional Value

  VII. Genetics 301

  VIII. Uses of Corn Protein 305
    A. Food and Feed—B. Industrial Uses and Potential

10. Lipids of the Kernel. EVELYN J. WEBER 311

  I. Introduction 311

  II. Oil Content 312
    A. Effect of Agronomic Practices on Oil Content—B. Genetic Control of Oil Content—C. Value of High-Oil Corn in Animal Feeding

III. Fatty Acid Composition 316
   A. Effect of Agronomic Practices on Fatty Acid Composition—B. Genetic Control of Fatty Acid Composition

IV. Lipid Class Composition 320
   A. Triacylglycerols—B. Phosphoglycerides—C. Glycosylglycerides—D. Glycosylsphingolipids—E. Sterols—F. Hydrocarbons—G. Polyisoprenoid Alcohols—H. Waxes—I. Cutin—J. Carotenoids—K. Tocols

V. Distribution of Lipids in the Corn Kernel 337
   A. Dissected Grain—B. Starch Lipids

VI. Enzymes and Hormones 341

VII. The Future 342

11. Corn Dry Milling: Processes, Products, and Applications. RICHARD J. ALEXANDER 351

I. Introduction 351
   A. History of Corn Dry Milling—B. Present Milling Capacity in the United States

II. Tempering-Degerming Systems 353
   A. Process with the Beall Degerminator—B. Alternative Milling Systems

III. Dry-Milled Products—Types, Volumes, and Composition 355

IV. Industrial Applications 357
   A. Current Market Volumes—B. Brewing—C. General Food Uses—D. Corn-Based Fortified Foods—E. Nonfood Uses—F. By-Products and Animal Feed

V. Future of Corn Dry Milling 371

12. Wet Milling: Process and Products. JAMES B. MAY 377

I. Introduction 377

II. The Process 377
   A. Steeping—B. Separation of Kernel Components—C. Corn Varieties Processed Commercially—D. Feed Production Process—E. Germ Processing

III. Yields, Production, and Marketing of Products 390

IV. Finishing of Wet-Milling Products 392
   A. Starch—B. Ethanol—C. Feed By-Products and Miscellaneous

V. Pollution Control 393
   A. Waste—B. Air Quality

VI. Trends 395
   A. Automation—B. Utilities

13. Food Uses of Whole Corn and Dry-Milled Fractions. LLOYD W. ROONEY and SERGIO O. SERNA-SALDIVAR 399

I. Introduction 399

II. Uses of Corn in the United States 399
   A. Sweet Corn—B. Degerminated Corn Products—C. Fermentation

III. Corn-Based Ready-To-Eat Breakfast Cereals 402
   A. Corn Flakes—B. Other Corn-Based Breakfast Cereals

IV. Maize Processing and Food Use Around the World 405
   A. Traditional Milling and Grinding—B. Alcoholic Beverages—C. *Arepa*—D. Porridges

V. Alkaline-Cooked Products 410
   A. Traditional Process for Tortillas—B. Modern Methods of Preparing Alkaline-Cooked Corn Products—C. Nixtamalized Dry Corn Flours—D. Preparation of Dry Nixtamalized Corn Flours—E. Experimental Preparation of Dry Masa Flours—F. Hominy

VI. Snacks from Corn 418
   A. Corn and Tortilla Chips—B. Extruded Snacks—C. Popcorn—D. Parching

VII. Factors Affecting Food Corn Quality 422
   A. Corn Properties for Alkaline Processing—B. Cooking Properties of Corn—C. Changes in Corn Kernel Structure During Cooking—D. Nutritional Value and Protein Fortification of Corn Products

VIII. Future of Corn Foods 426

14. Sweet Corn. STEPHEN W. MARSHALL 431

I. Introduction 431

II. Origin and History 432

III. Elevated-Sugar Endosperm Mutant Types 432
   A. Sugary Mutants—B. Other Mutants

IV. Sweet Corn Breeding—Past, Present, and Future 437
   A. Insect Resistance—B. Disease Resistance—C. General Hybrid Sweet Corn Improvement

V. Production and Processing of Sweet Corn Seed and Sweet Corn 440
   A. Seed—B. Sweet Corn for Processing

VI. Marketing Sweet Corn 444

15. Nutritional Properties and Feeding Values of Corn and Its By-Products. KENNETH N. WRIGHT 447

I. Introduction 447

II. Nutritional Value of Corn Proteins 449
   A. Normal and High-Protein Corn—B. High-Lysine Corn and Synthetic Lysine

III. Nutritional Value of Corn Lipids 452
   A. Corn Oil—B. Carotenoids

IV. Vitamins 454

V. Minerals 455

VI. Antinutrients 456

VII. Corn Feed By-Products from Food Processes 456

VIII. Corn Wet-Milling Feed Products 457
A. Corn Gluten Feed—B. Corn Gluten Meal—C. Condensed Fermented Corn Extractives (Steep Liquor)

IX. The Distilling Industry 467
A. Corn Distiller's Dry Grains—B. Dried Distiller's Solubles

X. Corn Dry-Milling 472
A. Process—B. Hominy Feed

16. Corn Starch Modification and Uses. F. T. ORTHOEFER 479

I. Introduction 479

II. Properties Involved in Modification 479
A. Molecular Characteristics—B. Granule Structure and Character—C. Viscosity

III. Objectives of Modification 481

IV. Characteristics of Modified Starches 483

V. Manufacture of Derivatives 484

VI. Modification 485
A. Acid Thinning—B. Bleaching or Oxidation—C. Cross-Linking—D. Derivatization or Substitution—E. Other Derivatives—F. Starch as a Source of Chemicals

VII. Corn Starch Utilization 495
A. In Foods—B. In Paper—C. Future for Starch Utilization

17. Corn Sweeteners. RONALD E. HEBEDA 501

I. Introduction 501

II. History 502

III. Chemistry 504
A. Starch Hydrolysis—B. Acid Hydrolysis—C. Enzyme Hydrolysis—D. Isomerization

IV. Manufacturing Processes 508
A. Dextrose—B. High-Fructose Corn Syrup—C. Corn Syrups—D. Maltodextrins

V. Properties 522
A. Dextrose—B. High-Fructose Corn Syrup—C. Corn Syrup—D. Maltodextrin

VI. Applications 525
    A. Dextrose—B. High-Fructose Corn Syrup—C. Corn Syrup—D. Maltodextrins

18. Corn Oil: Composition, Processing, and Utilization.
F. T. ORTHOEFER and R. D. SINRAM 535

    I. Introduction 535

    II. Production and Markets 535

    III. Composition and Characteristics of Corn Oil 536

    IV. Recovery of Crude Oil 538

    V. Corn Oil Processing 538
        A. Crude Corn Oil—B. Crude Oil Filtration and Degumming—C. Caustic Refining—D. Physical Refining by Steam—E. Bleaching—F. Winterization—G. Hydrogenation—H. Blending of Hydrogenated Oil—I. Deodorization—J. Quality Analyses for Corn Oil

    VI. Processing of Corn Oil Coproducts 545
        A. Corn Lecithin—B. Soapstock—C. Vegetable Oil Distillate

    VII. Handling 545
        A. Storage—B. Loadout and Shipping—C. Prevention of Oxidation

    VIII. Nutrition 546

    IX. Uses and Applications 547
        A. Margarines—B. Frying or Salad Oils—C. Other Uses

    X. Trends 549

19. Fermentation Processes and Products. WELDON F. MAISCH 553

    I. Introduction 553

    II. Fermentation Substrate 554
        A. Starch-Bearing Substrates—B. Hydrolyzed Substrates—C. Analytical Methods

    III. Products of Conversion and Fermentation 557
        A. Ethanol—B. Other Fermentation Products—C. Beer Production

    IV. Substrate Preparations 560
        A. Cooking—B. Starch Conversion

    V. Fermentation Process 563
        A. Organisms—B. Fermentation Mode

    VI. Fermentation Losses 568

    VII. Product Finishing 570
        A. Distillation—B. Absolute Ethanol—C. Aging

    VIII. Economic Importance 572

20. Biomass Uses and Conversions.
MARVIN O. BAGBY and NEIL W. WIDSTROM 575

    I. Introduction 575

   II. Corncob Residue 575
      A. Cob Collection—B. Processing, Conversion, and Uses

  III. Stalk Residue 579
      A. Stalk Collection and Storage—B. Processing, Conversion, and Uses

  IV. Sugar and Ethanol from the Corn Plant 582
      A. Background—Cornstalk Juice Sugar—B. Plant-Sugar Relationship—C. Measurements for Comparisons—D. Corn Compared with Other Crops as Ethanol Sources

   V. Summary 587

    Index 591

Corn:
Chemistry and Technology

CHAPTER 1

CORN PERSPECTIVE AND CULTURE

G. O. BENSON
R. B. PEARCE
Department of Agronomy
Iowa State University
Ames, Iowa

I. INTRODUCTION

Corn (*Zea mays* L.) is the only important cereal crop indigenous to the Americas. In the United States, corn production is more than double that of any other crop. Worldwide, corn, wheat (*Triticum aestivum* L.), and rice (*Oryza sativa* L.) greatly exceed the production of other crops.

Corn is also called *maize* and *Indian corn*. The term *corn* is also used in some countries to designate other crops: in England, wheat; in Scotland and Ireland, oats; and in some parts of Africa, grain sorghum.

II. DESCRIPTION AND ADAPTATION

Corn is a tall annual plant belonging to the grass family (Gramineae). It has a fibrous root system (Fig. 1) and an erect stalk with a single leaf at each node and leaves in two opposite ranks. Each leaf consists of a sheath surrounding the stalk and an expanded leaf blade connected to the sheath by a blade joint (collar). Corn is a cross-pollinated species and is monoecious, i.e., it has separate male (tassel) and female (ear) flowers located on the same plant. Normally the tassel is located at the top of the main stalk and the ears are located at the end of short branches (shanks) that develop from nodes lower on the stalk. Commercial cultivars in some cases do develop elongated branches (tillers or suckers), but they normally develop only one or two ears per plant. The ears grow to contain 300–1,000 developed kernels arranged in rows along a rachis (cob).

Corn is a warm-season crop requiring warmer growing temperatures than the small grains. It shows little growth at temperatures below 10°C or above 45°C. Highest yields are found in areas where corn takes 130–140 days to mature.

Corn requires abundant sunlight for optimum yields and does not grow well in shade. Due to its long growing season, it requires abundant moisture. Its deep root system allows it to utilize moisture down to 2 m in deep soils. The rate of

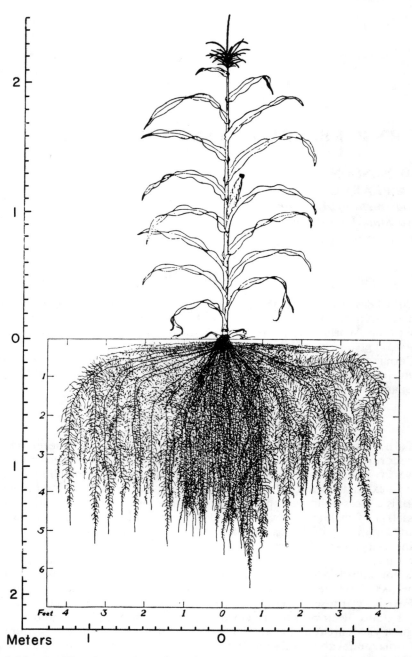

Fig. 1. Mature stalk and root system of a typical corn plant (cv. Krug) grown on the Nebraska Experiment Station farm. (Reprinted, with permission, from Kiesselbach, 1949)

maturity is affected by day length, with short photoperiods favoring early flowering.

Even with these environmental restraints, corn has proved to be as adaptable and variable as any other crop. It has cultivars adapted to climates from the tropics to temperate regions (0–55° latitude), to altitudes from sea level to 12,000 ft, and to growing seasons ranging from 42 to 400 days.

III. ORIGIN AND HISTORY

Corn apparently originated in Mexico and spread northward to Canada and southward to Argentina. Although secondary centers of origin in South America are possible, the oldest archaeological corn (7,000 years) was found in Mexico's valley of Tehuacan. The female inflorescence of this 5000 B.C. corn had reached a degree of specialization that precluded the possibility of natural seed dissemination. Thus, the oldest corn of record was dependent upon humans for its survival.

Numerous theories of origin have been offered over the years, only two of which receive serious consideration today. One is that teosinte (*Z. mexicana*) is the wild progenitor of corn; the other is that a wild pod corn, now extinct, was the ancestor of domesticated corn. Although perhaps more students of corn seem to accept the first theory, others are equally convinced of the second (Brown et al, 1984).

Aside from its possible role in the origin of corn, teosinte has had a major impact on its evolution. In Mexico particularly, introgression between corn and teosinte has probably occurred for centuries and continues to this day. The effects are evident in the morphology and cytology of both species. In addition, genes for resistance to certain viruses may have reached corn through its introgression with teosinte (Brown et al, 1984).

Following European discovery of the Americas, corn moved quickly to Europe, Africa, and Asia. From Spain, it spread northward to the short-growing-season areas of France, Germany, Austria, and eastern Europe. In Africa, much of the corn is derived from later introductions from the southern United States, Mexico, and parts of eastern South America. Corn of tropical lowland Africa is similar to the lowland and tropical corns of Central and South America. In Asia, the most widely used and productive corns are derived from Caribbean-type flints introduced in relatively recent times. However, older and distinct types of corn can also be found.

Whereas most of the modern races of corn are derived from prototypes developed by early native agriculturists of Mexico and Central and South America, one outstanding exception is solely the product of postcolonial North America. It is the yellow dent corn that dominates the U.S. Corn Belt, Canada, and much of Europe today. The origin and evolution of this remarkable race of corn have been clearly documented and confirmed.

In the early 1800s the late-maturing Virginia Gourdseed and the early-maturity Northeastern Flints were crossed, and the superiority of the hybrid was recognized and described. The cross was repeated many times during the western migration of settlers, and out of these mixtures eventually emerged the Corn Belt dents, the most productive race of corn found anywhere in the world.

The highly selected cultivars of Corn Belt dents formed the basis of hybrid

4 / *Corn: Chemistry and Technology*

corn and were the source of the first inbred lines used to produce hybrids. Germ plasm from some of these cultivars (Reid, Lancaster, Krug, etc.) figure prominently in the ancestry of hybrids used in the Corn Belt even to this day (Brown et al, 1984).

IV. PRODUCTION AND USE

World corn production has increased since 1930, with a dramatic increase occurring in the last 35 years (Fig. 2). This increase has been due both to

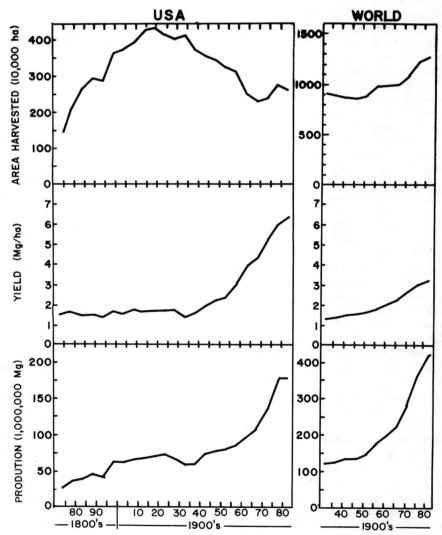

Fig. 2. Harvested area, yield, and production of corn, using five-year averages for the United States (1870–1984) and the world (1930–1982). (Data from USDA, 1930, 1983)

increases in land area used in corn production and to increased yield per unit of land area. In the United States, corn production has had a similar trend (Fig. 2). However, before 1920, the increased corn production was primarily due to increased land area, whereas after 1935, land area devoted to corn production declined and the increase in corn production was due to increased yield per unit of this area.

Every continent except Antarctica produces corn. North America produces over 50% of the world's total, and Oceania hardly produces a significant amount (Table I). The largest single area of production is the U.S. Corn Belt, which is made up of all or part of 12 states (Iowa, Illinois, Nebraska, Minnesota, Indiana, Ohio, Wisconsin, Michigan, Missouri, South Dakota, Kentucky, and Kansas, in order of production). The U.S. Corn Belt produces 40% of the world's total, with

TABLE I
Annual World Corn Area, Yield, and Production for Continents, Areas, and Countries of Major Production for 1980–1981 to 1982–1983[a,b]

	Harvested Area (1,000,000 ha)	Yield (t/ha)	Production (1,000,000 t)
World total	130.4	3.3	429.1
North America	40.3	5.4	217.6
United States	29.8	6.6	196.6
Mexico	7.4	1.4	10.4
Canada	1.1	5.8	6.4
Central America	1.7	1.4	2.4
Asia (minus USSR)	36.8	2.3	85.9
China	19.7	3.1	61.4
India	5.9	1.1	6.7
Indonesia	2.8	1.5	4.1
Thailand	1.6	2.2	3.5
Philippines	3.3	1.0	3.3
Europe (minus USSR)	11.4	4.7	54.0
Romania	3.1	3.3	10.2
Yugoslavia	2.3	4.5	10.1
France	1.6	5.7	9.3
Hungary	1.2	6.0	7.1
Italy	1.0	7.0	6.8
South America	18.4	2.0	36.6
Brazil	13.1	1.8	23.0
Argentina	3.1	3.2	10.0
Africa	19.9	1.3	26.1
S. Africa	4.6	2.0	9.2
Kenya	1.6	1.3	2.1
Zimbabwe	1.2	1.6	2.0
Nigeria	2.0	0.9	1.8
Malawi	1.2	1.1	1.3
Zambia	1.0	1.0	1.0
Tanzania	1.3	0.6	0.8
Soviet Union (USSR)	3.6	2.9	10.3
Oceania	0.1	4.6	0.4

[a] Source: USDA (1983).
[b] Conversions: 1 hectare (ha) = 2.471 acres; 1 metric ton (t) = 2,205 lb; 1 t/ha = 892 lb/acre = 15.93 bu (56 lb)/acre.

other significant areas being China (14%), the Balkan area (8%), Western Europe (5%), southern Brazil (5%), and Mexico, the USSR, Argentina, and South Africa (around 2% each).

Corn grown in subsistence agriculture has been and is currently being used as a basic food crop. But as countries become more developed and shift to a more urban population, with a concurrent increase in demand for wheat flour and animal-derived foods, the primary use for corn has shifted to feed for animals. As a result, in developed countries, over 90% of the corn produced or imported is used for animal feed (Watson, 1977).

Worldwide, the three crops with the greatest production are wheat, corn, and rice (Table II). Of these, the greatest land area by far is devoted to wheat, but because of lower average yield per unit of land area, the total production of wheat is not much more than that of corn or rice. Barley (*Hordeum vulgare* L.) is a distant fourth in total production. Note that the four crops produced the most are in the grass family: two are primarily food crops (wheat and rice), and two are primarily feed crops (corn and barley). For the most-produced nongrass crop, soybeans (*Glycine max* (L.) Merrill), production is still less than 25% as much as corn, wheat, or rice (Table II).

Corn, a warm-season grass, has the potential for high yields in any area with the proper environment, good management, and a history of corn breeding for cultivar development. Of the major crops, corn's yield per unit of land area is second only to that of potatoes (*Solanum tuberosum* L.) (Table II). Very few crops produce over 2 t/ha (1,784 lb/acre or 32 bu/acre) on a worldwide basis; potatoes, corn, and rice are the primary exceptions. Not all corn-producing areas have corn yields that exceed 2 t/ha. In fact, the United States, Canada,

TABLE II
Area Harvested, Yield, and Production Data of Major Crops for the 1982–1983 Production Season[a,b]

	United States			World		
	Area Harvested (100,000 ha)	Yield (t/ha)	Production (1,000,000 t)	Area Harvested (1,000,000 ha)	Yield (t/ha)	Production (1,000,000 t)
Corn	29.6	7.2	213.1	129	3.5	443
Wheat	31.9	2.4	76.5	236	2.0	473
Rice	1.3	5.3	7.0	145	2.9	409
Barley	3.7	3.1	11.4	80	2.1	163
Soybeans	29.1	2.2	63.4	53	1.8	95
Sorghum	6.5	3.7	21.4	48	1.5	72
Potatoes[c]	0.1	6.7	0.7	15	4.3	64
Oats	4.3	2.1	9.0	27	1.8	48
Rye	0.3	1.8	0.5	17	1.8	30
Peanuts	0.6	2.8	1.6	18	1.0	18
Sunflower	2.1	1.3	2.8	14	1.2	16
Beans (dry)	0.7	1.3	0.9	25	0.6	14

[a] Source: USDA (1983).
[b] Conversions: 1 hectare (ha) = 2.471 acres; 1 metric ton (t) = 2,205 lb; 1 t/ha = 892 lb/acre = 15.93 bu (56 lb)/acre = 14.87 bu (60 lb)/acre.
[c] Grain weight equivalent, i.e., moisture of potatoes adjusted from its average (76%) to the average for grain (15%).

Europe, Argentina, the USSR, and China are the only areas to significantly yield more than this amount (Table I).

V. CORN CULTURE

A. Growth and Development

Corn develops from a small seed to a plant typically 2–3.5 m (6.5–11.5 ft) tall in a few weeks. All normal corn plants grow and develop in a similar manner; however, plant size, length of growth period, and yield potential vary greatly depending on the production location.

The plant and the production methods described here relate most closely to the U.S. Corn Belt, but the actual plant and production principles do not vary greatly from those of other major corn-growing areas of the world. Time of appearance aboveground and percent of final dry weight of various plant parts is shown in Fig. 3 (Ritchie and Hanway, 1982). Only by understanding how a corn plant develops can a grower make the best corn management decisions or deal with problems that may develop during the growing season.

GERMINATION AND VEGETATIVE DEVELOPMENT

The major structures of the corn kernel consist of the pericarp (seed coat), endosperm, and embryo (germ) (Chapter 3). The pericarp is a thin outer layer

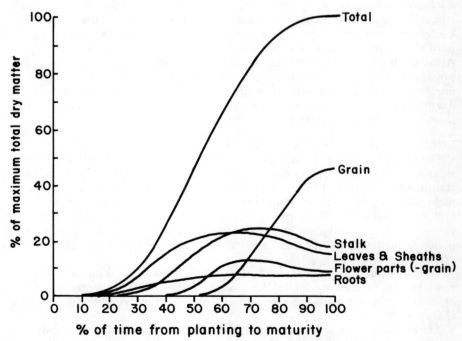

Fig. 3. Percent of total dry weight at various stages of growth and distribution of weight among plant parts. (Adapted from Ritchie and Hanway, 1982)

that protects the enclosed endosperm and embryo. The endosperm consists mainly of starch, which serves as the major energy source for the germinating seed and small seedling. The embryo consists of the initial parts of the young plant.

If the seed is planted in warm, moist soil, it germinates and emerges typically in one to two weeks (Benson and Reetz, 1984). The actual time varies greatly depending on soil temperature.

The first structure to break through the seed coat is the radicle, followed by the coleoptile (pointed structure that aids in emergence), and then the seminal roots. Growth of the coleoptile and elongation of the mesocotyl (first internode) cause the plant to emerge from the soil surface. When the coleoptile reaches light, the tip splits and the first true leaves emerge.

Figure 4 is a corn seedling about one week after emergence. In addition to roots emanating from the seed, the nodal (crown) root system has begun to develop. Additional whorls of nodal roots will develop in future weeks as this root system rapidly takes over the nutrient- and water-absorption roles. These roots and new leaves are developing from the growing point (apical meristem), which is still 2.5–3.8 cm below the soil surface.

Ritchie and Hanway (1982) explain corn plant development in great detail. They describe a plant that develops 20–21 leaves, produces silk in about 65 days, and reaches physiological maturity in 125 days. All corn plants have the same development pattern; however, the number of leaves and time between growth stages vary with hybrid maturity, planting date, location, etc. By the time five or six leaves have fully emerged (the leaf collar is visible) from the whorl, stalk elongation has been sufficient to elevate the growing point above the soil surface. Under good growing temperatures, a new leaf emerges every three to four days until the tassel emerges, indicating that the plant has achieved full height. Although the plant described may produce 20 or more leaves, only 14 or 15 are present at the completion of the vegetative stage. This is because the lower leaves are torn from the plant by nodal root growth and lower stalk expansion. Growth during the latter part of the vegetative stage shows very rapid leaf-area formation and reproductive development.

POLLINATION AND KERNEL DEVELOPMENT

The emergence of the tassel from the whorl and of silks from the ear shoots signals a critical stage in plant development. Corn is monoecious—the tassels are staminate flowers and the ear shoots are pistillate flowers. This separation of male and female parts into separate structures is important not only for understanding how pollination occurs, but as the key to the improvement of corn by geneticists and plant breeders. Although now visible in the nondissected plant for the first time, the tassel and ear shoot were both initiated early in the plant's life.

The individual stamens are not unlike other grasses. At anthesis, anthers are extruded; they soon open and pollen escapes. Pollen is carried by the wind and most of it falls on the silks of other corn plants—thus, cross pollination occurs (Kiesselbach, 1949).

The key to the crucial pollination stage is the female flowering structure or ear. Each potential kernel (ovule) produces a tubelike structure called a silk that grows until it emerges from the husks surrounding the ear tip. Pollen shed by tassels falls on these silks. A single pollen grain germinates on each silk and produces a pollen tube, and in about one day each ovule is fertilized (Chapter 2 provides details of this process). A key factor in the pollination process is that silks must have emerged at the time pollen is shed (Aldrich et al, 1975). It is not uncommon for environmental stresses, especially dry weather, to delay silking more than pollen shed, which can result in poor pollination and reduced yields.

After fertilization is accomplished, the main function of the plant is to develop the corn ears. The cob is the major area of growth for the first 10–14 days. This is followed by a period of rapid and essentially linear deposition of dry matter in

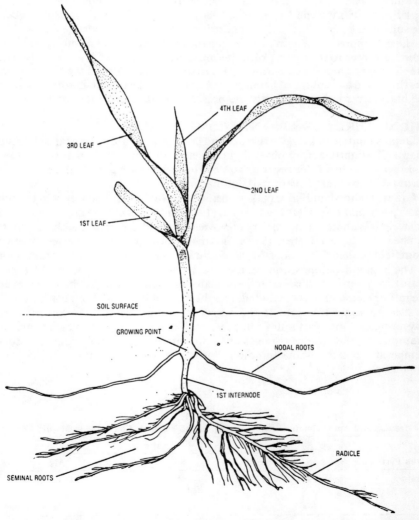

Fig. 4. Establishment of the newly emerged corn seedling. (Reprinted, with permission, from Benson and Reetz, 1984)

the kernels. Ritchie and Hanway (1982) identify postsilking stages as blister, milk, dough, dent, and physiological maturity. The root system has not been described in detail, but it probably achieves its greatest depth sometime in the middle of the reproductive stage. Although rooting depth varies greatly due to genotype and environment, depths of 1–2 m are considered normal.

Physiological maturity is reached when kernels on the ear have achieved maximum dry matter accumulation. It is important to know when corn has reached physiological maturity, and work by Daynard and Duncan (1969) confirmed that the so-called "black layer" is a good visual indicator of maturity. This dark closing layer develops between the basal endosperm and the vascular area in the pedicel (kernel tip). Another useful maturity indicator is the kernel milk line (Afuakwa and Crookston, 1984). The milk line is a boundary between the solid and liquid phases of the maturing endosperm. It is visible on the side of the kernel opposite the germ. Following the dent stage, this line moves gradually from the crown to the tip of a kernel as maturity approaches. It is typically visible over at least a three-week period and is easier to observe than the black layer. It has the added advantage of helping one to estimate the time to maturity, whereas the black layer indicates strictly the presence or absence of maturity.

THE MATURE CORN PLANT—COMPOSITION

Details relative to kernel structure and composition are covered in subsequent chapters. Slightly before physiological maturity, dry weight reaches its zenith and roots and lower leaves start to deteriorate. Figure 3 gives the relative dry matter distribution of plant parts at maturity.

Grain, as shown in Fig. 3, makes up about 42% of the dry matter, and the other plant parts make up 58%. Larson et al (1978) indicate that the grain to nongrain dry weight is about equal. In very high-yielding fields, over half of the dry matter can be in the grain (Barber, 1977). This distribution obviously varies somewhat from field to field. The root system weight is not included in these comparisons.

The mineral composition of the grain and stover fractions (11.3 t/ha of grain) are given in Table III. Only the major nutrients are listed; however, even if the minor elements were included, less than 5% of the total dry matter would be in the inorganic fraction. If the stover is left in the field following kernel physiological maturity, its composition changes due to weathering. For example, rainfall leaches significant amounts of potassium from the stover. Temperature and relative humidity primarily determine the rate of kernel dry-down.

TABLE III
Amount and Approximate Distribution (kg/ha) of Nutrients Contained in Mature Corn[a,b]

Crop Part	Nutrient					
	N	P	K	Ca	Mg	S
Grain	190	34	45	4	18	16
Stover	78	15	179	39	38	18
Total	268	49	224	43	56	34

[a] Data from Barber (1977).
[b] Based on grain yield of 11.29 t/ha and stover yield of 9.86 t/ha.

B. Climatic Requirements

Corn is grown over a wide range of climatic conditions, although most production is in the middle latitudes, between 30° and 47°. Some corn in northern Europe is produced farther north, but much of this is used for forage (mainly silage) and not grain. Temperature and moisture are the key factors that determine whether or not corn is adapted to an area. Growing season length also is a limiting factor. Solar radiation is ultimately the key environmental factor.

Climate is key in determining major production areas. Specific weather factors greatly influence the year-to-year production potential, so management practices are determined by the expected weather and its variability. For example, in the U.S. Corn Belt, annual yield varies by 20% or more due to weather. For individual farmers, the weather-induced yield variability is even greater.

GROWING SEASON

Growing season is normally defined as the length of the freeze-free period. However, in some cases, planting starts before the average freeze-free date has passed. This is possible because of the time delay between planting and emergence and because the growing point remains below the soil surface for some time following plant emergence. Thus, although short periods of freezing temperatures may damage exposed leaves, the plant usually recovers.

It is important to know not only the average dates of the last spring and first fall freeze (northern hemisphere), but also the odds associated with later-than-normal spring and earlier-than-normal fall freeze dates. This helps in risk assessment and in choosing hybrid maturity. Length of growing season can also be defined according to moisture relationships. In parts of the world, the "rainy season" sets limits on the length of the growing season.

TEMPERATURE

Although temperature and moisture are discussed separately, they interact and rarely operate independently. The developmental aspects rather than the physiological processes are emphasized here. Blacklow (1972) describes the influence of temperature on corn germination and emergence. Radicle and shoot elongation is greatest at about 30°C (86°F) and almost nonexistent at 9°C (48°F) and 40°C (104°F). Soil temperature at this early stage is the dominant factor. Time to emergence in the field ranges from less than a week at soil temperatures of around 25°C (77°F) to more than three weeks in the 10–13°C (50–55°F) range.

The relationship between temperature and development through the later vegetative stages appears to be similar to that of early stages. Temperatures of 21°C (daily minimum) and 32°C (daily maximum) give the fastest rate of development (Brown, 1977). However, development during the reproductive stages is less temperature-sensitive than development during the vegetative stages. A curvilinear relationship exists between temperature (15–30°C) and the time from emergence to tassel initiation (Brown, 1977). Photoperiod and day-night temperature range can also influence when tassels are initiated.

In addition, early-season temperatures influence both the corn plant and potential plant pests. Growth rates of plant pests are not always the same as growth rates for corn. This is especially true under cool early-season conditions, in which corn is often at a relative disadvantage to some plant pathogens, insects, and weed species.

Numerous systems that relate air temperature to rate of corn development have been tried (Shaw, 1977a). The one most used in the United States is the growing-degree unit (GDU) system, which assumes a linear response to temperature from 10 to 30°C. The formula used is:

$$[(\text{Daily max temp} + \text{Daily min temp})/2] - 10°C = GDU.$$

Daily GDUs for a particular period are then summed and related to crop development. Temperatures below 10°C and above 30°C are used in the equation as 10°C and 30°C, respectively. This system is discussed again later as it relates to hybrid maturity rating systems. Another somewhat different system developed in Canada is referred to as the Ontario or Canadian heat unit system (Brown, 1977).

Optimum temperatures for corn development may not be optimum for maximum grain yields. This is especially true for the later parts of the vegetative period and through most of the reproductive period. Temperatures considered optimum for development are too high for maximum yield, especially if water is limiting.

Highest corn yields are associated with daytime maximums of 24–30°C (75–86°F) unless rainfall is optimum at all times (Aldrich et al, 1975). Cool nights and sunny days with moderate temperature are considered ideal. Even in the U.S. central Corn Belt, temperatures are not considered ideal. Above-normal temperatures from planting through June and slightly below-normal temperatures in July and August are considered ideal for that region. High temperature stress during the pollination period can result in poor kernel set (Herrero and Johnson, 1980).

MOISTURE

In most important corn-growing areas, a shortage of water, especially at critical crop stages, is often a serious yield-limiting factor. Moisture availability to the corn plant involves the amount in the soil, the soil texture, and the atmospheric demand for water (Shaw, 1977a). Atmospheric demand for water is a function of solar radiation, wind, humidity, and temperature of the air. The amount of stress a plant is under depends on the relationship between atmospheric demand and soil moisture availability. Shaw (1977b) reports that when actual evapotranspiration drops below potential evapotranspiration, stress occurs. The amount of yield loss varies depending on factors such as the amount that actual evapotranspiration falls below potential evapotranspiration, crop stage, and soil fertility. Asghari and Hanson (1984) showed that when N was adequate, July precipitation was the single most important factor relating to corn yields.

The amount of moisture in the soil when dry periods develop can delay moisture stress. Good soils can store more than 5 cm of plant-available moisture per 30 cm of soil depth. This means that, if the rooting depth is 1.5–2 m, 40–50% of the total moisture needed for a crop of corn may be present as soil moisture. Water use by corn depends on several factors, the two most important being atmospheric demand and crop stage. Seasonal totals typically are in the 40–65 cm (16–26 in.) range. On an average midsummer day in the middle latitudes, typical usage is approximately 0.5 cm (0.2 in.). The actual amount varies greatly due to the factors previously discussed.

The reduction in yield due to moisture stress varies with crop stage. Shaw (1977b) has summarized the results of several researchers in Fig. 5. The upper and lower limits and an average value are indicated. Identical stress is difficult to simulate, but this diagram indicates that greater yield reductions occur around the tasseling-silking stage. Plant nutrient shortages are hard to prevent when moisture stress occurs, due to the drying of surface soil layers where most of the nutrients are located.

Excess moisture can also reduce corn yields. This is obvious where plants are killed due to flooding; however, the greatest reductions are due to saturated soil conditions. Poor soil aeration because of ponding can influence root growth and nutrient availability and cause denitrification. Excess water often delays planting, which reduces yields. Tilling wet soils results in soil compaction.

C. Soil and Crop Management

To this point we have considered items basically outside the control of the individual corn producer. We now consider those items that are influenced to varying degrees by the producer's management decisions. For clarity, most management decisions are discussed separately; however, numerous

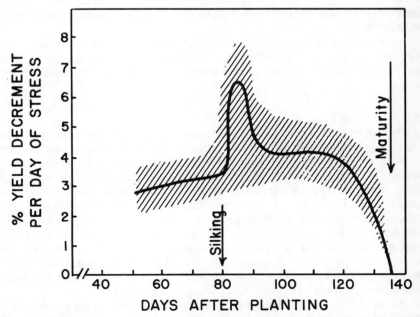

Fig. 5. Relationship between age of crop and percentage yield decrement due to one day of moisture stress. (Reprinted, with permission, from Shaw, 1977a)

interactions exist. It is impossible to say which practices are best to use, especially across environments. However, the factors to consider and, for some inputs, recommendations within a range are given. This discussion applies to the major corn-growing areas; in areas with more extreme conditions, recommendations would vary considerably.

ESTABLISHING YIELD GOALS

One does not just decide what yield will be produced and then add the needed inputs. Soils and weather are the environmental factors that form the limits for any realistic yield goal. One must know not only the average weather during the growing season but the degree, frequency, and timing of deviations from the average as well.

Where they exist, the best sources for evaluating soils are modern soil surveys. Such surveys give detailed descriptions of the physical and chemical characteristics of soils, which in turn influence the adaptability of soils for corn production. Often, average yield expectations with a specified management level are also included in the survey data. Because management level is difficult to specify, the yields produced by different farmers on the same soil type may vary considerably, but the relative yield relationships between soil types could still be valid.

After the soils have been characterized and weather expectations for a given location are known, research can define more precisely the management practices needed to produce maximum profit. One needs to also ask whether the practices used for short-term profit are compatible with the long-term productivity of a given soil. For example, continuous corn production may give maximum returns in the short run, but, if soil erosion is excessive, the long-term productivity of the soil may be reduced.

In summary, a yield goal of 5 t/ha (80 bu/acre) may be too high at some locations, whereas 10 t/ha (160 bu/acre) may be low for others. Only by having a good understanding of the factors contributing to yield can one make realistic plans.

CROPPING SYSTEMS

Corn can be grown under a wide range of soil and climatic conditions and is therefore grown as part of many different cropping systems. It can be grown continuously or in rotation with other crops. Some of the more common crops grown in rotation with corn are oats, wheat, barley, soybeans, grain sorghum, alfalfa, and numerous clovers and grasses. Various forms of multiple cropping include corn, especially in more tropical areas. Corn production is most intensive in areas where it has the greatest competitive advantage. In dry areas, considerable corn is grown with the aid of irrigation, but the majority of corn crops grown in the world must depend on natural precipitation. Numerous factors can influence which cropping system is best for a particular location. Ultimately, the key items will be profitability, soil resource, pest control, climatic risk, and the role of livestock (Aldrich et al, 1975).

Since the 1950s, the trend in major corn-growing areas has been to reduce the amount of small grains and forage legumes grown in rotation with corn. For example, in the U.S. Corn Belt, the most common rotational crop has become soybeans. The reason behind such changes has been economic. Greater profits, lower labor requirements, and new technology have all contributed to this trend.

Availability of commercial N at a reasonable price has allowed the purchased N to substitute for N produced by forage legumes. Insecticides that control corn rootworms reduce the impact of this insect when corn is grown after corn. Herbicides allow an intensification of row crop production as they, in effect, substitute for labor. Weeds can be controlled by mechanical cultivation or by hand, but the amount and cost of such operations limit the number of hectares that can be handled in such fashion. Improvement of corn tolerance to major diseases and insect pests has given stability to corn production under more intense cropping systems than was possible in the past.

The economic optimum rate of N needed for corn varies with many factors; however, one of the most important is the prior crop. Also, corn yields are often higher where corn is not grown after corn, even when the N rates applied should be adequate (Higgs et al, 1976; Voss and Shrader, 1979). One of the reasons soybeans have become a popular rotation partner with corn is obvious from Table IV. Soybeans, a legume, provide some N benefit, although less than alfalfa, to the corn that follows. Other rotational benefits are also present, as corn yields following other crops, including soybeans, frequently average about 10% higher than when corn follows corn.

The greatest negative consequence of cropping systems involving mainly row crops is soil erosion. Because such systems do not give adequate ground cover for part of the year, the erosion potential is greater than when solid-seeded small grains, forage legumes, or grasses are grown. Whether the erosion potential is enough to be of concern depends on many factors, including soil slope. On level to moderately sloping soils, erosion can be held fairly well in check with the use of conservation tillage. Erosion becomes severe if row crops are grown on steep slopes or even on moderate slopes if the tillage systems bury or remove too much of the crop residue.

Growing corn year after year in the same field may also result in "soil tilth" problems. The excessive cultivation of soils that are row-cropped can create problems with soil structure, especially when soils are tilled when they are too wet.

Concern over the long-term effect of continuous corn production and corn residue management is addressed by Barber (1979). The soil organic matter level reaches an equilibrium depending on the cropping system used over a long

TABLE IV
Impact of Prior Crop and N Rates on Corn Yields (Mg/ha)
in Northern Iowa (1971–1978)[a]

N (kg/ha)	Prior Crop Grown		
	Continuous Corn	Soybeans	Alfalfa Hay
0	3.07	6.52	8.28
67	5.77	7.40	8.03
134	7.09	8.03	8.28
202	7.40	8.15	8.09

[a] Data from Voss and Shrader (1979).

period of time. Most researchers feel that rotations including forage legumes come to equilibrium at a higher level of soil organic matter than that reached by continuous cropping with nonleguminous row crops. However, how a crop like corn is managed is still a factor. Where corn yields are high and the crop residues are returned to the soil, organic matter is higher than where the reverse is true. Larson et al (1972) estimated that 5.4 t/ha (2.5 ton/acre) of corn residue was necessary to maintain organic carbon levels in a southwest Iowa soil.

Corn is sometimes grown in a multiple cropping system. Such systems involve the growing of two or more crops on the same field in a year. Sequential cropping (growing one crop after another in the same year) and intercropping are types of multiple cropping and include numerous different systems. Sequential systems are most common where the growing season is long, and intercropping systems are most popular where labor is readily available and inexpensive. The type of livestock produced has a great influence on cropping systems. If beef or dairy cattle are produced, forages are needed, unlike the conditions in a cash-grain operation. More of the corn probably will be used for silage and greater amounts of hay and pasture will be needed. Ideally, such operations are well fitted to areas with sloping soils where intensive row crops would result in excess erosion. Poultry and swine are less able to utilize the forage crops well suited to more sloping land.

SOIL CONSERVATION AND TILLAGE

Historically, tillage and soil conservation have been treated as separate topics. However, in recent years, the topics have become closely allied as various conservation tillage methods have come to be looked upon as a major type of soil erosion control. Tillage was previously thought of almost exclusively in terms of its soil loosening, seedbed preparation, and weed control effects.

To plan needed conservation methods or evaluate the seriousness of potential erosion at a particular location, a soil loss prediction mechanism is needed. Development of equations to calculate soil loss due to water erosion started in approximately 1940 in the U.S. Corn Belt. Following many revisions, the one in present use is the universal soil loss equation (Wischmeier and Smith, 1978).

By using the soil loss equation, a technician can predict soil loss and compare this with what is known as "soil loss tolerance." This is the maximum amount of soil erosion that will permit a high level of productivity to be sustained both indefinitely and economically. The effect on erosion reduction of various conservation practices can then be tested by adjustment of the appropriate factors in the equation.

Cropping and tillage systems can have major impacts on potential soil erosion, as illustrated in Table V. Less row cropping, delayed tillage, and greater crop residues on the soil surface all reduced potential soil losses. One of the major reasons conservation tillage methods are becoming so popular is that they permit increased intensity of row crop production, while keeping soil loss below established tolerance levels, on soils where this would be impossible with conventional tillage methods.

Techniques for conserving soil are numerous. Cover crops, strip cropping, contour strip cropping, wind breaks, grass waterways, terracing, choice of crops grown, and some form of conservation tillage are the most common. No "best system" exists, as whatever method or methods are selected must be adaptable to the local situation and be economically feasible.

The objective of a tillage system is to provide a good seedbed, help control weeds, and modify the soil environment to make it as favorable as possible for corn growth while minimizing soil erosion. Components of the soil environment include structure, texture, temperature, moisture, air, and nutrient availability. As the importance of the components changes with soil and climate, the optimum tillage system varies greatly depending upon location (Larson and Hanway, 1977). It was once rather easy to separate tillage into primary (the first and often the major tillage operation) and secondary (usually harrowing, firming, and smoothing) components. However, with some newer systems, the distinction is rather blurred.

No attempt is made here to describe the multitude of tillage systems used for corn. Suffice it to say that the most common systems are based on a moldboard plow, chisel plow, disk plow, rotary tiller, or large disk harrow for primary tillage. This is followed by various operations involving the disk harrow, spring-tooth harrow, and spike-tooth harrow to further prepare the soil for planting. Such systems are often referred to as "conventional tillage." With the greater emphasis on conservation, any system that reduces loss of soil or water relative to conventional tillage is referred to as "conservation tillage" (Manning and Fenster, 1983). Such systems emphasize leaving varying amounts of crop residue on the soil surface. Conservation tillage has many forms, ranging from those in

TABLE V
Crop Management Factors Developed for Wisconsin—
Ratio of Soil Loss from Cropping Systems Relative
to Continuous Fallow Set at 1.0[a]

Cropping System[b]	Residues Plowed Under		Residues Left on Surface; Wheel-Track Planting
	In Fall	In Spring	
Continuous row crop	0.43	0.38	0.26
Rotation			
RRROM[c]	0.21	0.17	0.12
RROMM	0.12	0.10	0.07
ROMMM	0.06	0.04	0.03

[a] Data from Thompson and Troeh (1978).
[b] Corn yields 5+ t/ha, hay yields 8–12 t/ha.
[c] R = Row crop, O = oats, M = grass-legume meadow.

TABLE VI
Crop and Tillage Effects on Surface Cover and Soil Loss, 4% Slope[a]

Tillage	Percent Cover		Soil Loss (t/ha) After	
	Soybeans	Corn	Soybeans	Corn
No-till	26	69	12.2	2.2
Chisel (up and down slope)	12	25	27.5	13.6
Moldboard plow	1	7	36.3	19.8

[a] Data from Moldenhauer et al (1983).

which only a small slit is opened in the soil ahead of the planter unit to full-width tillage that could even involve a moldboard plow. In the latter case, the soil surface could likely be left very rough, with crop residue present. Thus, any definition of "conservation tillage" depends on what is considered to be "conventional tillage" for a given area. The type of tillage can have a great influence on surface cover and soil loss, as shown in Table VI.

Success of a tillage system depends greatly on how well it is adapted to the soils and climate of a given location. For example, temperate region soils do respond differently to tillage; however, these differences can be even more extreme when tropical soils are considered (Agboola, 1981).

WATER MANAGEMENT

The importance of water in corn production was discussed earlier. Aldrich et al (1975) indicate that on nonirrigated farms, assuming that good corn management practices are used, lack of water often is the greatest limiting factor to higher yields. Rainfall from planting to maturity in most corn-growing areas is seldom sufficient to supply the 40–65 cm (16–26 in.) of water needed for high yields. One must remember that part of rainfall is lost to runoff, drainage, and evaporation. Also, as was indicated earlier, the timing of rainfall is very important. Utilizing the annual precipitation as efficiently as possible increases yields. Excess water is also a problem for many corn producers. In the more humid areas, water excesses and shortages may occur in the same field at different times of the year.

The major factors in water availability, other than the precipitation level itself, is how much water the soil can store per unit of depth and to what depth the subsoil is favorable for water storage and root growth. Soil storage capability determines how much moisture can be in reserve to supplement crop season rainfall.

Management practices that decrease evaporation and runoff and increase infiltration and storage capacity are desirable in most cases. Little can be done in the short run to increase the water-holding capacity of soils; however, in tillage systems that leave the surface rough and with more crop residues at or near the surface, less water runs off. Farming on the contour and using terraces on sloping soils also helps. By maintaining high soil fertility and producing high yields, these practices return greater amounts of crop residue to the soil and increase the yield produced per unit of available water. Some adjustments in planting rates, hybrid selection, and tillage and fertilization practices may be in order when soil moisture reserves are below normal before the crop season begins.

In more arid regions, corn yields often are extremely low without irrigation, and water availability determines whether corn can or cannot be grown successfully. In more humid regions, irrigation is often profitable on sandy soils because they have such low water-holding capacity. On fine-textured soils in humid regions, most corn is not irrigated as the extra costs involved are greater than the value of the extra production. Obviously a "gray area" exists between these examples, where the irrigation versus nonirrigation choice is a difficult one.

Excess water, at least for part of the year, can be a problem with fine-textured soils so that drainage becomes an important management input. Irrigation may also require certain drainage needs. Drainage is a very old practice; field drains were reported in use as early as 400 B.C. in the Nile Valley. Corn yield increases due to drainage were reported to have been 150% in the 19th century for the U.S. Corn Belt (Sundquist et al, 1982).

Most drainage systems are either random or regular, and either surface or subsurface (Thompson and Troeh, 1978). Random systems are used to drain scattered wet spots, whereas regular systems feature uniformly spaced equipment to drain an entire area. Surface drains remove excess water before it infiltrates into the soil, and subsurface drains remove water that has infiltrated into the soil. Where the water infiltration rate of a soil is slow, surface drainage tends to be best. Where infiltration is more rapid or artesian water exists, subsurface methods are more popular. Ditches are the main method of surface drainage, whereas tiles are the main method used in subsurface systems.

NUTRIENT REQUIREMENTS

No simple model can be used to define the nutrient additions needed for corn. Native soil fertility varies greatly from soil to soil, and the way these soils have been managed over the years greatly influences present soil fertility levels. Yield potential, soil type, and production inputs must be considered when determining nutrient needs. However, when crop removal and nutrient loss due to erosion and leaching exceed the natural additions from the weathering of soil minerals, soil fertility is depleted. This is the case with corn production, especially when the entire plant is harvested for silage.

In spite of the complexity of this topic, a good understanding of a few key factors can serve as a basis for corn fertilization decisions. What nutrients are most likely to limit corn production? How does one determine this and decide how much of what nutrient should be added to the soil? These are economic decisions, and farmers who use plant nutrients efficiently have a distinct competitive advantage. The great increase in the use of fertilizers in the United States since World War II is shown in Table VII.

Essential Elements. Plants acquire carbon, oxygen, and hydrogen from the air and water. The other elements come from the soil. The six that are used in the greatest quantities are nitrogen, phosphorus, potassium, calcium, magnesium,

TABLE VII
Estimated Use of Fertilizer N, P_2O_5, and K_2O on Corn and Percent
of Harvested Acres Fertilized for Various Years in the United States[a]

Year	Average Application Rate (kg/ha) on Acreage Receiving			Percent of Harvested Acres Receiving		
	N	P_2O_5	K_2O	N	P_2O_5	K_2O
1947	11	26	13	44	44	44
1950	17	26	17	48	48	48
1954	30	31	28	60	60	60
1959	46	41	41	61	60	60
1965	82	56	54	87	82	77
1970	125	80	81	94	90	85
1975	118	65	75	94	86	82
1980	147	74	96	96	87	81

[a] Data from Sundquist et al (1982).

and sulfur. These are referred to as macronutrients and are sometimes further subdivided into primary (N, P, and K) and secondary (Ca, Mg, and S) nutrients. N, P, K, and sometimes S are applied as fertilizers to the soil as manure and/or commercial fertilizers. Ca and Mg are supplied as part of liming programs. Figure 6 gives the typical rates of uptake of N, P, and K for a 120-day growing season.

The nutrients iron, manganese, copper, zinc, boron, molybdenum, chlorine, and cobalt are used by plants in very small amounts and are called micronutrients or sometimes trace or minor elements.

Identifying Nutrient Needs. For most corn producers, the goal is to add nutrients up to the level that gives maximum net income. This rate is somewhat less than the amount needed for maximum yield. What methods are used to determine how much fertilizer is needed?

Historically, many applications have been made by trial and error. By observation and some rough measurements, the producer determined whether the fertilized area was more profitable than the unfertilized area. The next step was to compare rates or methods of fertilizer application. Whether such tests were on-farm observations or sophisticated field experiments, the question ultimately was why so much difference existed from soil to soil and field to field. The main reason found was that the soils varied in the availability of the various essential elements.

Such research led to the development of soil testing as a method of evaluating the nutrient status of soils and determining how this related to crop response to possible nutrient additions. Such tests do not measure the total amount of a nutrient in a soil sample. The goal is to develop an "index of nutrient availability" based on chemical analysis. Numerous laboratory tests exist, but correlation studies must be conducted to see how well the soil tests relate to actual nutrient availability to plants. These correlation tests are typically greenhouse studies. Calibration tests, that provide the basis for fertilizer recommendations, must then be conducted in the field. Such experiments are

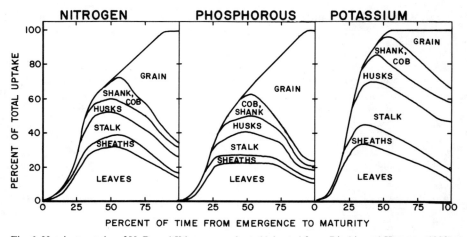

Fig. 6. Nutrient uptake of N, P, and K by a corn plant. (Adapted from Ritchie and Hanway, 1982)

designed to study the response to various rates of a nutrient on soils with different test levels.

Olson et al (1982) report that nutrient sufficiency levels must come from long-term calibration of soil tests with field yield responses. Such research establishes response probabilities ranging from "almost certain" to "unlikely," depending on test levels. Such test levels are often expressed as very low, low, medium, high, etc. Some tests for P, K, and lime needs are quite good, but for most other nutrients the reliability ranges from fair to poor (Aldrich et al, 1975). Summarizing the work on corn fertilization in Europe, Raillard (1980) agrees by saying P and K needs can be predicted by soil analysis, but N needs are more difficult to predict.

Plant analyses can also be useful in determining nutrient deficiencies; however, they are often not useful for correcting the problem in the year they are taken. They also require considerable skill in interpretation. Deficiency symptoms, like plant analysis, have their greatest value for obvious growth problems. The two most distinctive deficiency symptoms on corn are those for N and K. N deficiency early in the growing season is typified by spindly pale-green to yellow plants. As plants get larger the lower leaves show a V-shaped yellowing from the leaf tips. This is in contrast to K deficiency, in which yellowing and then dying of leaf margins occurs.

Cropping history is valuable in determining N needs for corn, as discussed earlier and shown in Table IV. Based on this and several other studies in Iowa (Voss and Shrader, 1979), the amount of N needed for maximum corn yield under the conditions of these experiments can be determined (Table VIII). Based on a study of crop rotation by N rate in Wisconsin, Higgs et al (1976) suggested slightly lower rates. Tests in England with low-yielding corn for forage (Pain et al, 1980) indicated that the optimum N rates were 70–80 kg of N per hectare (62–71 lb/acre). Ebelhar et al (1984) showed that in climates mild enough to grow winter annual legumes as a cover crop, hairy vetch supplied N equivalent to 90–100 kg/ha (80–89 lb/acre) of fertilizer N to the following corn crop.

N in soils is mainly in the organic matter. Part is in plant residues that are decaying and part in the more stable humus fraction. The N cycle is very

TABLE VIII
Nitrogen Rates Needed for Maximum Corn Yield[a]

Crop Rotations[b]	Nitrogen (kg/ha) per Year
Continuous corn	168–224
Rotation	
CSb	112–168
CSbCOM	22–56
CSbCOM	56–90
CCCOM	22–56
CCCOM	56–112
CCCOM	112–168
COMM	0–22

[a] Data from Voss and Shrader (1979).
[b] C = Corn, Sb = soybeans, O = oats, M = alfalfa meadow. Year of corn reported on is underlined. Note that in some cases the rotation is identical but the corn reported on is in different places in the sequence.

complex, but conditions that determine whether mineralization (inorganic ions released) or immobilization (inorganic ions converted to organic form) takes place are key to the availability of N for corn growth. Mineralization results in the production of ammonium (NH_4^+), which is often converted to nitrate (NO_3^-). Key to what occurs in the soil is the ratio of carbon to nitrogen (C:N) (Thompson and Troeh, 1978). As the C-N ratio increases, immobilization is favored, with the reverse being true as the ratio narrows.

Although both ammonium and nitrate can be utilized by corn, most of the ammonium applied to or produced in soils is converted to nitrate by microbial action. Therefore, most of the N taken up by corn is in the NO_3^- form, which is very mobile. It can be leached with water or lost by denitrification. Denitrification, which occurs under anaerobic conditions, converts the NO_3^- form of N to gaseous forms.

Both the mineral and the organic fractions of soils supply the other essential nutrients to plants. The chemistry in each case is different, and where good soil tests exist, they are the best method to assess whether nutrient supplementation is needed.

Nutrient Sources and Application. In addition to the nutrients from the soil or prior crop residues, animal manure is a valuable source of nutrients when available (Ketcheson and Beauchamp, 1978). Before the development of commercial fertilizer technology, manure was the only transportable nutrient source for corn production. Manures vary greatly in nutrient content and effectiveness due to source, bedding content, storage method, application method, etc. The greatest variation is relative to N content and losses because manure often contains a large NH_3 content that is easily lost due to volatilization. A common estimate of the nutrient content for 900 kg of fresh cattle manure with some bedding is 4.5 kg of N, 1 kg of P, and 3.8 kg of K (Aldrich et al, 1975). Not all the nutrients in manure are available to plants in the year applied. With the rapid expansion of livestock confinement systems, manure in liquid form has become readily available. Analysis of N, P, and K content of manure from such systems must be done so that rational application rates can be determined. In some cases, municipal sludge or wastes from other sources may be available.

The base material for most of the fertilizer N industry is anhydrous ammonia (NH_3^+). Ammonia is produced by the German-developed Haber-Bosch process by the reaction of N and H under high pressure and temperature. Natural gas is the main H source used, and the N comes from the air. Ammonia (82% N) can be applied directly by injection into the soil or serve as the base material from which other N fertilizers are made. Other widely used N sources and typical N contents include ammonium sulfate (21% N), ammonium nitrate (33.5% N), and urea (46% N). The most popular nonpressure liquid N sources typically contain 28–32% N and are mixtures of ammonium nitrate and urea. There is no "best" N source; each has its unique characteristics that require it to be managed accordingly. In Europe, the dry forms of N are most widely used (Cooke, 1982), whereas ammonia and other liquid forms are the most popular in the United States. Various ammonium phosphate fertilizers also supply significant amounts of N.

Although less transient than N, soil P is nevertheless very complicated chemically. It is found in soils in both organic and inorganic forms; inorganic P comes from the mineral, apatite. Which of the solution P ions occurs in a given soil depends mainly on the pH. The orthophosphate ion is the most common form absorbed by plants (Thompson and Troeh, 1978). As pointed out earlier, no universally acceptable soil test for available P exists. The choice of an extracting solution will vary with soils, with pH being a key factor.

Phosphorus fertilizers in the United States are labeled not according to their total P content, but according to the percent "available P," as determined by solubility in a neutral ammonium citrate solution. The exact system used varies in many countries. The percent available P may be as percent P, percent P_2O_5, or both, as given in the countries belonging to the European Economic Community (Cooke, 1982). Although some ground rock phosphate is sold for direct application, the major P fertilizers are those that have been processed from rock phosphate into superphosphate, triple superphosphate, monoammonium phosphate, diammonium phosphate, or ammonium polyphosphate. The choice of P fertilizer source for corn depends on price, application methods best suited to local soils and climate, sources available, equipment available, etc.

Plants absorb large amounts of potassium in the form of the K^+ ion. Soils contain large amounts of K in the micas and feldspars; however, these sources are unavailable to plants except over many years due to weathering. Potassium in the soil solution and that in the exchangeable form are the direct sources of K. Soil tests used to predict K requirements typically measure the level of exchangeable K in soils. The most widely used fertilizer materials are potassium chloride and potassium sulfate. The fertilizer analysis is typically reported as percent water-soluble K_2O or K or both.

Sulfur and micronutrient deficiencies in corn are not nearly as widespread as deficiencies for N, P, and K. However, deficiencies of any nutrient can cause yield reductions. Most problems with these nutrients relate to specific soils. Although calcium and magnesium deficiencies may occur, the major reason for adding Ca and Mg is as part of a liming program to raise soil pH. Ca and Mg carbonates are the most common liming materials used. In addition to the chemical nature of lime, the fineness of grind also influences how quickly lime becomes effective. The optimum pH for soils depends on which crops are grown; however, for corn, the most favorable range is pH 6.0–7.0 in most soils.

The choice of fertilizer sources and application methods can be complicated by the choice of tillage system. Such questions are more frequent as corn growers do less tillage and incorporation of fertilizer materials into the soil is more difficult. Fox and Hoffman (1981) report that some N is lost in a no-till system when urea or urea-ammonium nitrate solutions are not incorporated. They also report that the amount of loss is closely associated with how long the urea fertilizer is on the soil surface before a significant rain. Losses are insignificant if rains come within two days of when the urea was applied. Mengel et al (1982) report that subsurface placement of N fertilizer in a no-till system produces higher corn grain yields than those obtained with surface applications. The greatest practical application from this work is in the case of urea-ammonium nitrate solutions, which have often been applied to the surface. Injecting these materials in a no-till system appears to be a better way to improve N use efficiency.

SEED SELECTION

An important item in any corn management program is deciding what cultivars to plant. In the developed countries, the seed planted will be a hybrid. In underdeveloped nations, it may be a hybrid, but more likely it will be an open-pollinated cultivar. Important factors to consider when selecting cultivars include seed quality, maturity, yield, standability, and ear retention. Tolerance to locally important disease and insect problems is also a consideration in cultivar selection.

Most purchased seed is of high quality, has been tested for both warm and cold germination, and is properly labeled according to existing seed laws. In spite of this, growers should not assume that all corn seed is of high germination and should check the label carefully. Seed quality is of greater concern in those areas where seed production, conditioning, and storage are not done by professionals.

Knowing the maturity of cultivars is very important. Growing cultivars that have a high probability of reaching physiological maturity before the first fall freeze is of special importance in areas with the short growing seasons of Canada, Europe, and the northern United States. Conversely, growing cultivars that are too early for an area will mean lower yields. In some areas, maturity selection means fitting critical periods (silking, for example) to times when weather conditions are expected to be most favorable.

No universally used cultivar maturity rating system exists. The U.S. growing degree unit and Ontario heat unit systems, discussed earlier, are based on air temperature and are sometimes used to classify cultivar maturity. Commonly used are some form of "relative maturity system," such as the Minnesota Agricultural Experiment Station or FAO systems. Because no system can indicate the exact number of days from planting to maturity, the value of any system is to give an idea of the comparative maturity of different cultivars (Aldrich et al, 1975). Many European countries require strict testing of cultivars for maturity, yield, and other performance traits before they can be placed on an approved list for farmer purchase.

Assuming that the seed will grow and is of a known maturity, the most important trait then becomes harvestable yield. Ability to stand until time of machine grain harvest is of obvious importance to the commercial corn farmer. If the corn is harvested for forage, by hand or directly by livestock, standability is not as important. Cultivar selection criteria for forage vary somewhat from those for corn harvested for grain. One cultivar may outyield another by 1–2 t/ha of grain even when both are commonly grown in an area. Where good scientifically run tests exist, they are excellent sources of information for cultivar selection. In some areas, such tests are conducted by public institutions and serve as good comparisons of cultivars. Private seed companies normally have good information comparing cultivars they market.

PLANTING DATES AND DEPTH, RATES, AND ROW SPACING

These are corn-growing practices that require experimentation in the region of ultimate application. This is especially true for planting dates and rates, which are sensitive to climatic differences. For example, Larson and Hanway (1977) report that optimum planting dates in the United States range from February to May as one goes from south to north. Berger (1962) indicates that planting dates range from March in southern Italy to May in the northern corn-growing areas of Europe. In the southern hemisphere, planting is typically during September to November. In locations where corn production is not restricted by the length of the freeze-free growing season, planting dates are often based on periods of expected moisture stress or high temperature stress, as these relate to crop stage.

Much of the corn grown in the world is in temperate areas where optimum planting dates are often slightly before the average last-freeze date. Thus, corn producers often start planting, assuming that soil temperatures are not too low or the soil too wet, about two weeks before the average date of the last 0° C air temperature. Figure 7 gives relative yield expectations according to planting dates based on tests at several Iowa locations. The typical date of last 0°C temperature for these locations is May 5. At these locations, early-planted corn may have its leaves damaged by a later-than-normal freeze. In most cases, the growing point, which is still below the soil surface, survives and produces new leaf tissue, and these fields will outyield replanted fields. Again, planting date decisions are best determined by long-term studies that represent a given corn-growing region.

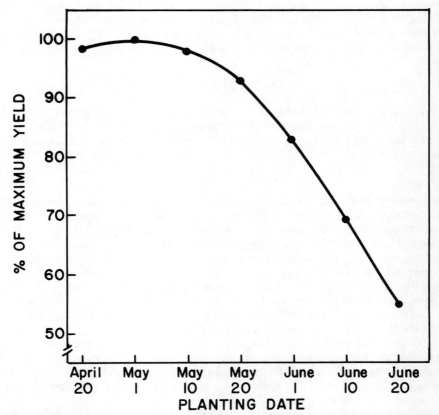

Fig. 7. Effect of planting date on corn yield response in Iowa. (Reprinted, with permission, from Benson, 1982)

Corn is typically planted 4–5 cm deep. For early planting, especially when the soil is cool and wet, the ideal depth is slightly less. When the surface soil is dry, especially when planting has been delayed and soils are warm, the optimum depth is slightly deeper.

Planting rates have greatly increased over the last 40 years, with a subsequent increase in yield. The switch to hybrids, improvement in the tolerance of hybrids to increased stand densities, increased fertilizer use, better plant distribution within the row, narrowing of row widths, earlier planting, and other management changes all have contributed to increased planting rates. The difference between planting rates and plant population (stand density) is important. Planting rate is the number of seeds planted per unit of land area, whereas stand density is the number of plants present in a given area. Most producers plant 10–15% more seeds than their stand density goal contains.

The optimum stand density depends on factors such as hybrid, moisture stress, soil fertility, and yield goal. Typical stand densities range from 50,000 to 65,000 plants per hectare (20,000–26,000 plants per acre). This may be too high or too low depending on local conditions.

On a worldwide basis, corn is usually grown in rows 50–100 cm (20–40 in.) apart (Berger, 1962). Where cultivation is done by hand, the spacing may be even closer; however, in more arid locations, rows may be 2–3 m (6.5–10 ft) apart. Where cultivation and harvesting are mechanized, row spacing of about 76 cm (30 in.) appears to be becoming standard.

D. Pest Management

Weeds, diseases, and insects are pests that must receive attention in any corn management program. Klassen (1979) defines pest management as "a discipline for the selection, integration, and implementation of pest control actions on the basis of predicted economic, ecological, and sociological consequences."

WEEDS

Weeds compete with corn for moisture, nutrients. and light. An inverse relationship exists between weed growth and corn yield. Weeds can cause harvest problems and foster insects and diseases. Although the list of weed species troublesome in corn is long and varies greatly by location, the genera judged to cause the most problems on a worldwide basis are *Setaria, Echinochloa, Amaranthus,* and *Cyperus* (Behrens, 1979). Grass and broadleaved weeds of both annual and perennial life cycles cause large corn yield losses. The first step in any weed control system should be the identification of weeds present, especially those that are most serious. Management strategies to best control the problem can then be devised.

Corn management techniques that can aid in weed control include: 1) crop rotation; 2) delayed planting, which permits weeds to germinate and then be killed by preplant tillage; 3) uniform and high corn stand densities; 4) soil drainage; 5) cultivation; and 6) hand weeding. With the advent of herbicides, a powerful new tool was added to the list of weed control techniques.

Sundquist et al (1982) reported that 98% of all U.S. corn land received some herbicide treatment in 1980. Modern weed control programs involve less tillage to kill weeds and fewer row cultivations. With the use of herbicides, the use of a harrow or rotary hoe to control small annual weeds has been reduced. Earlier corn plantings have put an added strain on weed control because preplant tillage has been reduced and early corn growth is slowed more than that of many weeds when temperatures are cool (Ammon and Heri, 1979).

Herbicide applications are classified into three application periods relative to corn planting. Preplant applications are made before corn is planted, preemergence applications after planting but before the weeds and corn emerge, and postemergence applications after weeds and corn emerge. Any herbicide that is used must be used in a manner consistent with the label and meet all legal requirements set forth by the governmental agencies with such authority in each country.

DISEASES

The severity and incidence of an infectious corn disease depends on the presence of a virulent pathogen, a susceptible host, and a favorable environment for the disease. For some diseases, viruses for example, a vector such as a certain insect is necessary to spread the disease. Fungal diseases are the most important in corn; however, bacteria, viruses, and nematodes can also be important (Cassini and Cotti, 1979).

Stuckey et al (1984) present a very logical sequence to corn disease management. They first suggest determining the need for disease control. Steps required include disease identification, determining disease severity, and estimating loss potential versus the cost of control. After the need for control is assessed, possible disease control methods include resistant hybrids, crop rotation, tillage systems, chemical treatments (mainly seed treatment), stress protection, and in some rare cases exclusion or quarantine methods.

Corn diseases are typically grouped into six disease categories: 1) leaf diseases, 2) seed and seedling diseases, 3) stalk rots, 4) ear rots, 5) viral diseases, and 6) damage by nematodes. Shurtleff (1980) is an outstanding source of information on corn disease identification and control.

INSECTS

Many of the same principles discussed for diseases relative to infestation severity and management, with some modification, also apply to insects that attack corn. The principles of pest management that include a heavy emphasis on scouting to see when insect populations or plant injury approaches economic threshold levels have application for several corn insects.

As in other corn production or corn protection practices, corn has varying insect problems depending upon location. According to Allemann (1979), corn rootworms are the corn pest of greatest economic concern in the United States but can be controlled by crop rotation, whereas stem borers and foliar pests are often of greater concern in other areas. Cutworms are a problem to some degree in most corn-growing areas.

In addition to attacking the plant directly, insects are often a problem in stored grain (Chapter 6). Storing clean grain in a bin that is properly managed relative to grain temperature and moisture keeps insect infestation problems to a minimum.

VI. SUMMARY

Corn is one of the three great grain crops of the world. It has been intensively studied by many disciplines. The information gained has led to dramatic yield increases, especially in the last 35 years. Because this crop is grown in such diverse environments and under many different management systems, any attempt to describe its adaptation and suggested production methods cannot take into account all local conditions.

Yields continue to increase and can be expected to continue in the future. The source of these future yield increases will depend, to an even greater extent than in the past, on genetic improvements. Grain quality and composition will receive greater attention in future research efforts.

LITERATURE CITED

AFUAKWA, J. J., and CROOKSTON, R. K. 1984. Using the kernel milk line to visually monitor grain maturity in maize. Crop Sci. 24:687-691.

AGBOOLA, A. A. 1981. The effects of different soil tillage and management practices on the physical and chemical properties of soil and maize yield in a rain forest zone of western Nigeria. Agron. J. 73:247-251.

ALDRICH, S. R., SCOTT, W. O., and LENG, E. R. 1975. Modern corn production. A and L Publications, Champaign, IL.

ALLEMANN, D. V. 1979. Maize pests in the USA. Pages 58-63 in: Maize—Technical Monograph. E. Hafliger, ed. CIBA-GEIGY Ltd., Basle, Switzerland.

AMMON, H. U., and HERI, W. J. 1979. Weed control in European maize growing. Pages 46-50 in: Maiz—Technical Monograph. E. Hafliger, ed. CIBA-GEIGY Ltd., Basle, Switzerland.

ASGHARI, M., and HANSON, R. G. 1984. Nitrogen, climate, and previous crop effect on corn yield and grain N. Agron. J. 76:536-542.

BARBER, S. A. 1977. Plant nutrient needs. Pages 41-51 in: Our Land and Its Care. The Fertilizer Institute, Washington, DC.

BARBER, S. A. 1979. Corn residue management and soil organic matter. Agron. J. 71:625-627.

BEHRENS, R. 1979. Weed control in U.S. maize. Pages 38-45 in: Maize—Technical Monograph. E. Hafliger, ed. CIBA-GEIGY Ltd., Basle, Switzerland.

BENSON, G. D. 1982. Corn production in Iowa. Pm-409. Iowa Coop. Ext. Serv., Ames.

BENSON, G. O., and REETZ, H. F. 1984. Corn Plant Growth—From Seed to Seedling. NCH-3, National Corn Handbook. Purdue University Cooperative Extension Service, West Lafayette, IN.

BERGER, J. 1962. Maize Production and the Manuring of Maize. Centre d'Etude de L'Azote, Geneva, Switzerland.

BLACKLOW, W. M. 1972. Influence of temperature on germination and elongation of the radicle and shoot of corn (*Zea mays* L.). Crop Sci. 12:647-650.

BROWN, M. D. 1977. Response of maize to environmental temperatures—A review. Pages 15-26 in: Agrometeorology of the Maize (Corn) Crop. World Meteorological Org., Geneva, Switzerland.

BROWN, W. L., ZUBER, M. S., DARRAH, L. L., and GLOVER, D. Y. 1984. Origin, Adaptation, and Types of Corn. NCH-10, National Corn Handbook. Purdue University Cooperative Extension Service, West Lafayette, IN.

CASSINI, R., and COTTI, T. 1979. Parasitic diseases of maize. Pages 72-81 in: Maize—Technical Monograph. E. Hafliger, ed. CIBA-GEIGY Ltd., Basle, Switzerland.

COOKE, G. W. 1982. Fertilizing for maximum yield. The Macmillan Co., New York.

DAYNARD, T. B., and DUNCAN, W. G. 1969. The black layer and grain maturity in corn. Crop Sci. 9:473-476.

EBELHAR, S. A., FRYE, W. W., and BLEVINS, R. L. 1984. Nitrogen from legume cover crops for no-tillage corn. Agron. J. 76:51-55.

FOX, R. H., and HOFFMAN, L. D. 1981. The effect of N fertilizer source on grain yield, N uptake, soil pH, and lime requirement in no-till corn. Agron. J. 73:891-895.

HERRERO, M. P., and JOHNSON, R. R. 1980. High temperature stress and pollen viability of maize. Crop Sci. 20:796-800.

HIGGS, R. L., PAULSON, W. H., PENDELTON, J. W., PETERSON, A. F., JACKOBS, J. A., and SHRADER, W. D.

1976. Crop rotations and nitrogen. Res. Bull. R2761. Univ. of Wisconsin, Madison.

KETCHESON, J. W., and BEAUCHAMP, E. G. 1978. Effects of corn stover, manure, and N on soil properties and crop yield. Agron. J. 70:792-797.

KIESSELBACH, T. A. 1949. The structure and reproduction of corn. Res. Bull. 161. Nebraska Agric. Exp. Stn., Lincoln.

KLASSEN, W. 1979. Concepts of pest management. Pages 403-419 in: Introduction to Crop Protection. E. B. Ennis, Jr., ed. Am. Soc. Agron., Crop Sci. Soc. Am., Madison, WI.

LARSON, W. E., and HANWAY, J. J. 1977. Corn production. Pages 625-669 in: Corn and Corn Improvement. G. F. Sprague, ed. Am. Soc. Agron., Madison, WI.

LARSON, W. E., CLAPP, C. E., PIERRE, W. H., and MORACHAN, Y. B. 1972. Effects of increasing amount of organic residues on continuous corn: II. Organic carbon, nitrogen, phosphorus, and sulfur. Agron. J. 64:204-208.

LARSON, W. E., HOLT, R. F., and CARLSON, C. W. 1978. Residues for soil conservation. Pages 1-15 in: Crop Residue Management Systems—Proceedings of the Symposium. W. R. Oschwald, ed. Spec. Pub. 31. Am. Soc. Agron., Madison, WI.

MANNING, J. V., and FENSTER, C. R. 1983. What is conservation tillage? J. Soil Water Cons. 38:141-143.

MENGEL, D. B., NELSON, D. W., and HUBER, D. M. 1982. Placement of N fertilizers for no-till and conventional till corn. Agron. J. 74:515-518.

MOLDENHAUER, W. C., LONGDALE, G. W., FRYE, W., McCOOL, D. K., PAPENDICK, R. I., SMIKA, D. E., and FRYREAR, D. W. 1983. Conservation tillage for erosion control. J. Soil Water Cons. 38:144-151.

OLSON, R. A., FRANK, K. D., GRABOUSKI, P. H., and REHM, G. W. 1982. Economic and agronomic impacts of varied philosophies of soil testing. Agron. J. 74:492-499.

PAIN, B. F., PHIPPS, R. H., and GIBB, R. W. 1980. Effect of inorganic fertilizers and livestock slurry on yield and composition of forage maize. Pages 95-97 in: Production and Utilization of the Maize Crop. E. S. Bunting, ed. The Hereward and Stourdale Press, Ely, Cambridgeshire, England.

RAILLARD, D. 1980. Aspects agronomiques d'une production rationnelle de mais en Europe. Pages 15-20 in: Production and Utilization of the Maize Crop—Proceedings of the Symposium. E. S. Bunting, ed. The Hereward and Stourdale Press, Ely, Cambridgeshire, England.

RITCHIE, S. W., and HANWAY, J. J. 1982. How a corn plant develops. Spec. Rep. 48. Iowa Coop. Ext. Serv., Ames.

SHAW, R. H. 1977a. Climatic requirement. Pages 591-623 in: Corn and Corn Improvement. G. F. Sprague, ed. Am. Soc. Agron., Madison, WI.

SHAW, R. H. 1977b. Water use and requirements of maize—A review. Pages 119-134 in: Agrometeorology of the Maize (Corn) Crop. World Meteorological Org., Geneva, Switzerland.

SHURTLEFF, M. C. 1980. Compendium of Corn Diseases. The American Phytopathological Society, St. Paul, MN.

STUCKEY, R. E., NYVALL, R. F., KRAUSZ, J. P., and HORNE, C. W. 1984. Corn disease management. NCH-4, National Corn Handbook. Purdue University Cooperative Extension Service, West Lafayette, IN.

SUNDQUIST, W. B., MENZ, K. M., and NEUMEYER, C. F. 1982. A technology assessment of commercial corn production in the United States. Stn. Bull. 546. Minnesota Agric. Exp. Stn., St. Paul, MN.

THOMPSON, L. M., and TROEH, F. R. 1978. Soils and Soil Fertility. McGraw-Hill, New York.

USDA. 1930. Stat. Bull. 28. U.S. Dept. Agric., Washington, DC.

USDA. 1983. Agricultural Statistics. U.S. Dept. Agric., Washington, DC.

VOSS, R. D., and SHRADER, W. D. 1979. Crop rotations—Effect on yields and response to nitrogen. Pm-905. Iowa State Univ. Coop. Ext. Serv., Ames.

WATSON, S. A. 1977. Industrial utilization of corn. Pages 721-763 in: Corn and Corn Improvement. G. F. Sprague, ed. Am. Soc. Agron., Madison, WI.

WISCHMEIER, W. H., and SMITH, D. D. 1978. Predicting rainfall erosion losses. Agric. Handbook 537. U.S. Dep. Agric., Washington, DC.

CHAPTER 2

BREEDING, GENETICS, AND SEED CORN PRODUCTION

M. S. ZUBER (*retired*)
Department of Agronomy
University of Missouri
Columbia, Missouri

L. L. DARRAH
U.S. Department of Agriculture
Agricultural Research Service
Department of Agronomy
University of Missouri
Columbia, Missouri

I. INTRODUCTION

A. Reproduction of the Corn Plant

Corn (*Zea mays* L.) has a monoecious flowering habit. The male and female flowers are separate but on the same plant. The staminate (male) flowers are borne in the tassel and the pistillate (female) flowers are borne on the ears. Corn is predominantly cross-pollinated; pollen from any tassel can randomly pollinate the silks on the ears of adjacent plants or even its own silks. The average corn tassel produces 25 million pollen grains, and most ears have 500–1,200 kernels. If an ear had 1,000 kernels, about 25,000 pollen grains would be available for each kernel.

Cells of a typical corn plant have nuclei with 20 chromosomes. Ten of these chromosomes are derived from the egg cell (premature kernel) and 10 from the sperm nucleus (pollen grain) that fertilizes the egg. In the tassel, each spore mother cell (microsporocyte) divides once reductionally and once equationally (meiosis), forming four spores, with 10 chromosomes each. Each spore's nucleus divides equationally (mitosis), forming a vegetative (tube) nucleus and a generative nucleus. The latter again divides equationally to form two sperm cells, so that a mature pollen grain has three haploid nuclei with 10 chromosomes each. Stepwise development of the male gamete is shown in Fig. 1.

In the pistillate inflorescence (ear), a single spore mother cell (megasporocyte) of each functioning flower goes through meiosis and forms four spores, with each nucleus having 10 chromosomes. Three of these spores abort. The remaining spore undergoes three mitotic divisions to give an eight-nucleate embryo sac. The nuclei are separated into four at each end (steps 1–11 in Fig. 2). Next, the eight nuclei become organized in the embryo sac with three nuclei at each end and two in the center. These latter two polar nuclei fuse and are fertilized to eventually become the endosperm (step 13 in Fig. 2). One of the three nuclei at the basal end of the embryo sac enlarges and becomes the egg cell. The mature embryo sac, when ready for fertilization, contains a one-nucleate egg and two fused polar nuclei. The fertilized egg nucleus becomes the diploid zygote and the fertilized polar nuclei become the triploid endosperm nucleus (step 14 in Fig. 2).

Pollination takes place when silks protruding from the ear shoot intercept pollen grains from the tassel. Pollen grains germinate on the silk and send out a tube that grows down the center of an individual silk, from which it finally enters the embryo sac, ruptures, and releases two sperms, each carrying 10 chromosomes. The nucleus of one sperm fuses with the egg to form the zygote, which has 20 chromosomes. This number persists in the somatic cells of the plant and the nucleus of all new cells that appear during growth have the 2n number of 20 chromosomes. The other sperm fuses with one of the two polar nuclei, which then fuses with the other polar nucleus, forming a primary endosperm nucleus

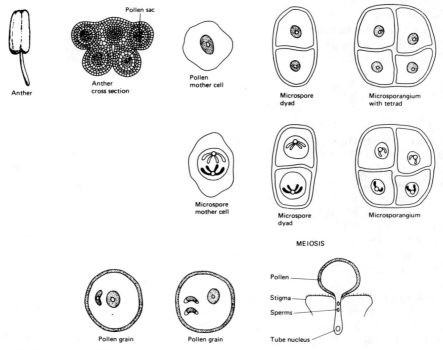

Fig. 1. Development of the male gametophyte of higher plants. (Reprinted, with permission, from Kiesselbach, 1949)

Breeding, Genetics, Seed / 33

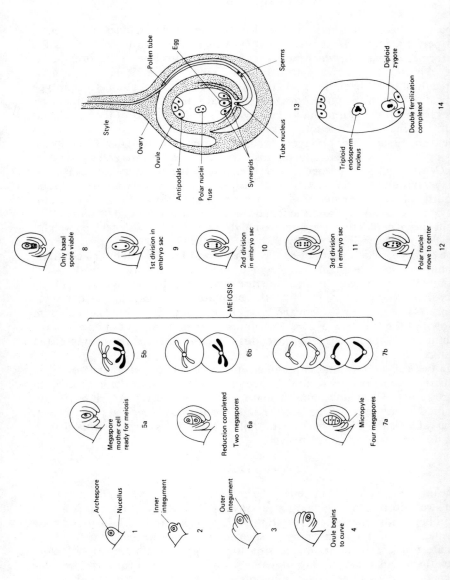

Fig. 2. The development of the female gametophyte of higher plants (stages numbered in order). a, Anatomical structure; b, the nuclear cytology at the same stage. (Reprinted, with permission, from Rédei, 1982; © Macmillan Publishing Co.)

with 30 chromosomes. This phenomenon is referred to as double fertilization (step 14 in Fig. 2).

B. Kernel Structure

The corn seed is a single fruit called the kernel. It includes an embryo, endosperm, aleurone, and pericarp (Fig. 3). The pericarp (seed coat) is the transformed ovary wall, which covers the kernel and furnishes protection for the interior parts. The pericarp is maternal tissue. Hence, it has no genetic contribution from the pollen grain that fertilized the ovule, but has a genotype identical to that of the plant on which the seed was borne. Pericarp thickness ranges from 25 to 140 μm among genotypes. The dry weight of the pericarp is usually less than 2% of the total kernel weight.

Aleurone tissue is a single layer of cells lying immediately under the pericarp. The single layer is characteristic of most U.S. Corn Belt strains and exotic corn from Central and South America. One exception is the multiple layering (two to five layers) found in some selections of the race Coroico, an exotic strain found in South America. The nuclei of aleurone cells have three sets of 10 chromosomes, two sets coming from the female and one set from the male parent.

Endosperm makes up the greater part of the kernel, usually 80–85% by weight. Endosperm cells are filled mainly with starch grains.

The embryo, which is about 8–10% of the total kernel weight, is actually a dormant young corn plant that becomes active and initiates growth under certain exterior temperature and moisture conditions. In a cross section of the embryo, the rudimentary parts of the ultimate corn plant are evident, for instance, the plumule with stem and leaves and the primary root. The scutellum is known as the first leaf. However, it functions not as a leaf but as a storage organ; it also helps in digesting the endosperm during the early growth of the seedling.

The pedicel serves as the attachment point of the kernel to the cob. The conductive tissue located within the pedicel is referred to as the hilar layer in mature kernels. When physiological maturity of the kernel is reached, no further movement of nutrients occurs. At this time, the hilar layer becomes dark brown or nearly black and is referred to as the "black layer." It is commonly used as an indicator of physiological maturity.

C. Endosperm Dosage Effects

The immediate effect of pollen on the kernel is known as xenia. For example, when sweet corn with the sugary gene (su) is pollinated with normal (starchy) dent pollen (with gene Su), the resulting endosperm is starchy ($Sususu$). Because endosperm tissue has 30 chromosomes, 20 coming from the female and 10 from the male, four combinations are possible for a single gene. For example, a close association exists between vitamin A and yellow endosperm (Table I). The color of an endosperm with genes yyy is completely white; with yyY it is light yellow, with yYY moderate yellow, and with YYY intense yellow. These various combinations are obtained by making reciprocal crosses. If yy (white) is used as the female and YY (yellow) as the male, the result is an endosperm with yyY. With YY as the female and yy as the male, the resulting endosperm has YYy.

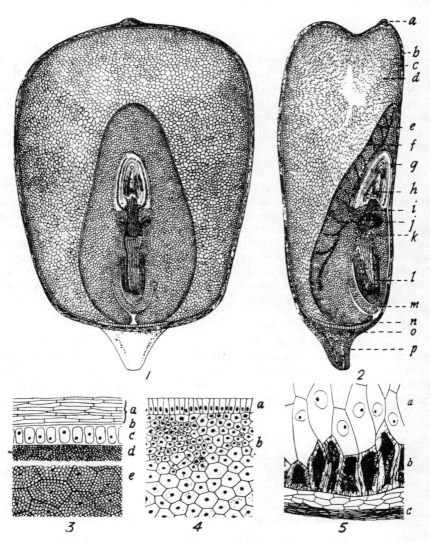

Fig. 3. The mature kernel. 1 and 2, Vertical sections in two planes of a mature kernel of dent corn, showing arrangement of organs and tissues: a, silk scar; b, pericarp; c, aleurone; d, endosperm; e, scutellum; f, glandular layer of scutellum; g, coleoptile; h, plumule with stem and leaves; i, first internode; j, lateral seminal root; k, scutellar node; l, primary root; m, coleorhiza; n, basal conducting cells; o, brown abscission layer; p, pedicel or flower stalk (×7). 3, Enlarged section through pericarp and endosperm: a, pericarp; b, nucellar membrane; c, aleurone; d, marginal cells of endosperm; e, interior cells of endosperm (×70). 4, Enlarged section of scutellum: a, glandular layer; b, interior cells (×70). 5, Vertical section of the basal region of endosperm: a, ordinary endosperm cells; b, thick-walled conducting cells of endosperm; c, abscission layer (×350). (Reprinted, with permission, from Rédei, 1982; © Macmillan Publishing Co.)

This phenomenon is known as endosperm dosage. Researchers have shown endosperm-dosage effects for several other endosperm genes such as waxy (gene symbol wx), and high amylose (gene symbol ae).

D. Endosperm Variation

Kernel types can be grouped as dent, flint, flour, sweet, pop, and pod. Endosperm composition can be changed by a single gene difference such as those between floury (fl) versus flint (Fl), sugary (su) versus starchy (Su), or waxy (wx) versus nonwaxy (Wx) (see Fig. 2, Chapter 3). However, the volume of endosperm, which determines the size of the kernel, is polygenic in its inheritance. In pod corn, a single-gene trait (Tu) produces long glumes enclosing each kernel. This type of corn has no practical use except as an ornament.

DENT CORN

Modern dent corn originated from the hybridization of a late-flowering southern dent called Gourdseed with early-flowering northern flints. Dent corns are characterized by the presence of vitreous, horny endosperm at the sides and back of the kernel, while the central core extending to the crown of the kernel is soft and floury. Upon drying, the center part collapses to give a distinct indentation. The degree of denting varies with the genetic background. Most of the corn grown in the U.S. Corn Belt is dent corn with a yellow endosperm. The acreage of white is much less, usually less than 3%. White maize is largely used in the manufacture of food products (Chapters 11 and 12).

FLINT CORN

Flint corns have a thick, hard, vitreous endosperm surrounding a small granular center. The relative amounts of soft and vitreous starch may vary among corn strains. Flint kernels are smooth and rounded with no denting. Very little flint corn is grown commercially in the United States, but flints are grown extensively in Argentina, other areas of South America and Latin America, and Northern Europe.

FLOURY CORN

Floury corn is one of the oldest types of corn grown in the world; it can be traced back to the ancient Aztecs and Incas. Since the endosperm is composed

TABLE I
Relation Between Number of Alleles for Yellow Endosperm and Units of Vitamin A[a]

Number of Alleles for Yellow	Genetic Composition of Endosperm	Endosperm Color	Average Units of Vitamin A per Gram of Material
0	yyy	White	0.05
1	yyY	Light Yellow	2.25
2	yYY	Moderate yellow	5.00
3	YYY	Intense yellow	7.50

[a]Source: Mangelsdorf and Fraps (1931), used by permission.

entirely of soft starch with hardly any hard, vitreous material, it is easily ground and was used extensively by the American Indians. Upon drying, the kernels tend to shrink uniformly, and as a consequence, very little or no denting occurs. Very little floury corn is grown in the United States, but it is grown widely in the Andean region of South America.

SWEET CORN

Several mutant genes condition the endosperm to make it sweet when consumed about 18-20 days postpollination. Sweet corn genetics and breeding are discussed in Chapter 14.

In the United States, sweet corn is an important crop for human food, being consumed directly in fresh form or as canned and frozen products. Generally, all of the mutants used for sweet corn involve the prevention of or reduction in the rate of conversion of sugar into starch. Most mutants affecting sweet corn are recessive, and plants must be isolated from foreign pollen sources that would have a xenia effect resulting in starchy endosperm.

II. PROGRESS IN CORN IMPROVEMENT

A. Early Inbreeding and Hybridization Experiments

G. H. Shull (1909), who worked at the Carnegie Institute, has been given credit for suggesting the development of pure inbred lines in corn. The method involved self-fertilization (selfing) and selection of homozygous biotypes. In the inbreeding process, the effects of many deleterious recessive genes are expressed; these genes survive in a open-pollinated variety because of cross-pollination and maintenance in a heterozygous state. In 1910 and 1911, Shull urged various experiment stations to try pure-line development. He envisioned using inbred lines as parents in single crosses to develop hybrids. But, due to the reduced vigor of the inbred lines, the seed yield was extremely low, making it impractical to use single crosses as commercial hybrids. These problems were solved when Jones and Mangelsdorf (1926) suggested the use of double crosses. Double-cross hybrids, which result from crossing two single crosses, benefit from hybrid vigor in seed production.

B. Use of Open-Pollinated Varieties, 1910-1936

Early varieties such as Reid Yellow Dent and others were developed from crosses between early, northern flints and late, southern dents such as Gourdseed. Similarly, the Lancaster open-pollinated variety was developed from the cross of an early flint with a large, late, rather rough-grained variety. During this period, farmers selected within open-pollinated varieties for maturity, ear and kernel type, and adaptation. As a consequence, many different open-pollinated varieties were developed, some of which were unique for a specific character. The large number of open-pollinated varieties stimulated corn shows, where exhibitors vied for the best 10-ear sample. Unfortunately, seed from the best 10-ear sample would not yield any better than seed from random ears from the storage crib from which the 10-ear sample was picked.

This was due to effects of uncontrolled pollination. Many of the better open-pollinated varieties later served as sources for inbred line development.

C. Introduction of Hybrids, 1937–1965

In 1933, less than 1% of the corn acreage in the United States was planted with hybrids. By 1945, the amount was approximately 90%. The use of double-cross hybrids allowed seed producers to take advantage of heterosis in the production of seed of the final cross, which gave a great improvement over the relatively poor seed yield of an inbred line. The number of double-cross combinations that can be made from a given number of lines can reach astronomical numbers. For example, from 10 inbred lines, 45 single crosses and 630 double crosses can be made. From 20 inbred lines, 14,535 double crosses are possible. Obviously, seed could not be made nor yield trials conducted on this large a number of experimental hybrids. Studies conducted by Jenkins (1934) resulted in a procedure for predicting double-cross yield performance from single-cross yield trials. Only seed of the top predicted double-cross hybrids was made and evaluated. Many of the best-performing hybrids of that era, such as US13 and US523W, were developed by this method.

D. Single-Cross and Modified Single-Cross Hybrids, 1966 to the Present

The development of the early inbred lines mainly involved inbreeding in open-pollinated varieties, and the main emphasis was to develop lines that could endure the inbreeding process. These lines were generally not vigorous, and yields were low. As time progressed, breeders improved the elite older lines by recycling. Also, new lines were developed from synthetics and populations that had been improved by some form of recurrent selection. The resulting inbred lines from these sources had more vigor, could tolerate greater stress by increased plant densities, and had increased grain yield. Increased inbred yields made possible the use of single crosses in commercial production. The change from double-cross to single-cross hybrids was quite rapid because farmers liked the increased plant and ear uniformity. The cost of single-cross hybrid seed was greater than for double-cross hybrid seed, but farmers were willing to pay extra for uniformity and, frequently, greater yields. In some instances, seed of superior-performing single crosses was very difficult and costly to produce, due to the low yield of the seed parent of the cross. To alleviate this problem, seed producers used sister-line crosses for the female parent. This method increased seed production by 10–50%. For example, a single cross of inbred parent A by inbred parent B would be formed as $(A_1 \times A_2) \times B$, where A_1 and A_2 were closely related sister lines. The greater the difference in relationship between A_1 and A_2, the larger the expected seed parent yield. These crosses are known in the trade as modified single crosses and usually are not as uniform as true single crosses.

E. Host-Plant Resistance

Many of the major diseases of corn have been controlled by breeding for resistance. After a source of resistance is located, transferring it to adapted material has been rather routine. Resistance to most of the diseases in corn is polygenic, but single-gene resistance has been successful in controlling the leaf blight *Helminthosporium turcicum* and common rust (*Puccinia sorghi*). Some diseases, such as stalk and ear rots, are under genetic control, but when subjected to the stresses of environments, even the most resistant strains show stalk rot symptoms. Although corn breeders have had success in lowering economic losses caused by the many pathogens that infect corn, new diseases, such as anthracnose (*Colletotrichum graminicola*) and new isolates and races of older diseases do occur, making breeding for disease resistance a continuous task.

Breeding for resistance to economically important insects has been less productive than breeding for disease resistance. The lag in progress has been partly due to the need to develop methods for rearing insects for hand infestation, as depending upon natural infestation gave unpredictable results. For example, practically no progress was made in breeding for resistance to the European corn borer (*Ostrinia nubilalis*) until rearing techniques were developed so that either egg masses or the larvae could be placed on plants in large numbers. Progress has been made in breeding for resistance to the first generation of the European corn borer, but because of the added effort required and possibly fewer sources of resistance, not much progress has been made in breeding for resistance to the second generation. Other corn-damaging insects for which varying degrees of resistance have been found are the corn earworm (*Heliothis zea*), fall armyworm (*Spodoptera frugiperda*), southwestern corn borer (*Diatraea grandiosella*), and western and northern corn rootworm (*Diabrotica* sp.). Differences have been found among genotypes for resistance to stored grain insects, but host-plant resistance has not been widely used as a means of control. Resistance to insects is either chemical or morphological or both. An example of morphological resistance is the long, tight husks found in corns adapted to the southern United States, which give protection against damage by the corn earworm.

III. BREEDING TECHNIQUES FOR SELECTION OF IMPROVED GENOTYPES

A. Mass Selection

Early improvement and development of the hundreds of open-pollinated varieties in the early 1900s was realized by mass selection. The technique involves no pollen control; each ear is pollinated by a random mixture of pollen from neighboring plants in the field. Therefore, selection progress was slow, but it was effective for the more simply inherited characters such as ear and plant height, ear number, adaptation, maturity, and kernel and ear characteristics. Grain yield improvement by mass selection was more difficult because of random pollination from both good- and poor-yielding plants. Also, the effects of genotype and environment could not be separated. Gardner (1961) modified the mass selection methodology by dividing the field into blocks and selecting the highest-yielding plants in each. With separation of genotypic and environmental effects, Gardner was able to realize significant yield increases.

B. Random Selfed Lines

When corn breeders first began inbreeding, they selfed a large number of plants. Seed from individual ears was planted ear-to-row (i.e., seed from one ear planted in one row), and the inbreeding process was continued for five to seven generations. Since the source materials were often open-pollinated varieties, many deleterious traits emerged that were not uncovered in the random mating of open pollination. Therefore, the main objective at that time was development of lines that could survive the inbreeding process. Selection for specific traits was practically nil. The different characteristics associated with inbred lines came about largely at random and not necessarily by selection through early testing or by challenging plants with pathogens or insect infestations.

C. S_1 and S_2 Line Selection

During the past two decades, many new techniques for inbred line development have emerged. Among these are selection among and within S_1 or S_2 progenies, derived from one or two generations of selfing, respectively. Some breeders prefer the S_2 generation for testing because more seed is available. The usual procedure is to evaluate large numbers of S_1 or S_2 progenies grown at three or four locations with two or three replications per location. Visual selection and responses to pathogens and insects are used to eliminate progenies from further consideration at the S_1 stage of testing. Usually, grain yield data are not considered important among S_1 progenies. Progenies that survive can be inbred further or intermated to form a new breeding population (recombination) and the same S_1 or S_2 selection scheme can be repeated. This latter procedure, generally termed recurrent selection, is considered a very efficient way for culling out those progenies that show undesirable traits before proceeding to the more expensive stage of measuring combining ability for yield. Usually when S_1 and S_2 progenies are evaluated, no pollinations are made and the selected progenies are recombined or further selfed using remnant seed.

D. Recurrent Selection

Recurrent selection (Sprague and Eberhart, 1977) is used for traits that are polygenically inherited. Polygenic inheritance occurs where many genes, each with a small effect, control expression of a trait. Recurrent selection increases or decreases the frequency of alleles (an allele is one of a pair or series of forms of a gene at a specific locus on a chromosome) by selecting within a normal distribution of genotypes. The effect of recurrent selection—a change in the mean of a breeding population—is depicted in Fig. 4. Note that all possible genotypes exist in the infinite population, but only a finite number can be sampled in a selection experiment.

Many versions of recurrent selection have been developed: S_0, S_1, S_2, full-sib, half-sib, ear-to-row, reciprocal recurrent, and reciprocal full-sib recurrent selection. These methods differ in the progeny that are evaluated in replicated field trials. In the methods involving selfing, selfed progenies are evaluated. In full-sib selection, full-sib progenies (having the same male and female parents) from paired-plant crosses within a population are evaluated. Half-sib (progenies having one parent in common) selection utilizes a test cross to a common tester, which might be a population, an inbred line, or a hybrid. The test-cross progenies are then grown for evaluation. Ear-to-row selection is a half-sib method in which one location's yield trial is grown in isolation and is pollinated by selected male plants grown as border rows. Reciprocal recurrent selection and reciprocal full-sib recurrent selection are interpopulation improvement procedures that require two breeding populations. Reciprocal recurrent selection includes two half-sib selection experiments: in the first experiment, population B is the tester for population A and, in the reciprocal experiment, population A is the tester for population B. Inbred lines from the respective populations may also be used as testers. Reciprocal full-sib recurrent selection evaluates the full-sib crosses from pairs of plants from different populations, e.g., a plant from population A crossed reciprocally with a plant from population B.

The steps completing a cycle of selection include the generation of progeny, the evaluation of progeny, and the recombination (intermating) of selected entries using remnant seed. Evaluation trials are usually grown at two or three sites, with two or three replications at each site. The season-by-season activities for the various methods of population improvement are shown in Table II.

Gains from recurrent selection for the population's yield improvement alone range from negative to over 10% per cycle, with an average in the 2–5% range (Hallauer and Miranda, 1981). Gains from reciprocal recurrent selection range from 1 to 7% per cycle, with most researchers reporting 3–7%. To make

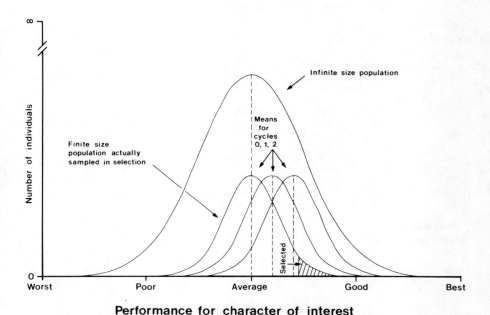

Fig. 4. Response to recurrent selection: distribution of genotypes in a breeding population of finite size versus that in a conceptually infinite-size population.

comparisons, rates of gain per cycle must be adjusted for seasons per cycle to obtain rates per year. For instance, a rate of 2% gain per cycle using mass selection compares favorably with 5% per cycle using S_1 selection if no winter nurseries are used, because mass selection requires one season per cycle and S_1 selection requires three seasons per cycle.

E. Extraction of Inbred Lines from Populations Improved by Recurrent Selection

Populations improved by recurrent selection have been excellent sources for the development of inbred lines. Hallauer et al (1983) compared inbred lines developed from cycle 0, cycle 5, and cycle 7 in Iowa Stiff Stalk Synthetic (BS13)

TABLE II
Summary of Seasons and Activities Required for Various Recurrent Selection Methods

Selection Method	Season 1	Season 2	Season 3	Season 4
Mass	Yield trials, selection and recombination			
Ear-to-row	Yield trials, selection and recombination			
S_0[a]	Selection and selfing of individual plants	Recombination of selected entries		
S_1	Self-selected plants	Yield trials	Recombination of selected entries	
S_2	Self-selected plants (S_1)	Self-selected plants (S_2)	Yield trials	Recombination of selected entries
Full-sib	Make full-sib progenies by recombination	Yield trials		
	or Make full-sib progenies	Yield trials	Recombination of selected entries	
Half-sib	Make test-cross progenies	Yield trials	Recombination of selected entries	
Reciprocal recurrent	Make reciprocal test-cross progenies	Yield trials	Recombination of selected entries	
Reciprocal full-sib recurrent	Make reciprocal full-sib progenies	Yield trials	Recombination of selected entries	

[a] S_0 plant selection would generally not be used for yield improvement. Because selection is based on a single plant, the trait being selected must be simply inherited and not greatly influenced by environment.

that had been selected for yield through recurrent selection. All lines were crossed to the inbred line Mo17 as a common tester. Yield trials were conducted in Iowa at four locations for five years. B37 (cycle 0) × Mo17 yielded 7.8% more than B14 (cycle 0) × Mo17; B73 (cycle 5) × Mo17 yielded 8.5% more than B37 × Mo17; and B84 (cycle 7) × Mo17 yielded 11.5% more than B73 × Mo17. Comparison of B14 (cycle 0) × Mo17 and B84 (cycle 7) × Mo17 showed an increase of 33.2%. These results demonstrate the potential for new line development from populations improved by recurrent selection.

F. Inbred Line Development by Recycling Existing Lines

Most development and improvement of inbred lines by commercial breeders involve the recycling of old elite inbred lines. Inbred line families such as B14A, B37, B73, C103/Mo17, Oh43, and WF9 have been used for development of many second-cycle lines. In the recycling process, breeders have added genes for disease and insect resistance, stalk and root quality, earlier maturity, and increased yield. The process crosses the elite line by another line that has the attributes desired for improvement of the older line. Selfing may begin in the F_2 generation or after one or two backcrosses to the recurrent parent. Further inbreeding with intensive selection for the desired trait completes the process.

G. Evaluation of Experimental Material

One of the most important tasks corn breeders have is to evaluate newly developed experimental materials. The difficulty in this task is to separate genetic and environmental effects. The usual procedure is to evaluate the material in performance trials conducted over two or three years at a minimum of six locations. More locations are preferred, but the number is determined by the resources available. Testing sites should be located in areas where the newly developed material is likely to be marketed.

SELECTION TRIALS

Most breeders use a screening type of trial for eliminating genotypes that are very obviously poor. Usually, large numbers of many genotypes are observed at a few locations. Sometimes, inoculations with prevalent leaf diseases and stalk rot pathogens are included. Sprague (1952) reported that combining ability (yield) is determined early in the inbreeding process. If a group of S_1 lines were top-crossed to a common tester and their S_5 derivatives were also top-crossed to the same tester, the relative yield rankings between the S_1 and S_5 progeny would be close. Although early testing has proven to be a useful method, most breeders generally do not use this approach, as they prefer to screen S_1 families visually and discard poor genotypes that would otherwise have gone through yield trials under the early-testing procedure.

EXPERIMENTAL HYBRIDS

Newly developed hybrids are usually placed in screening trials conducted at four to six locations. The elite performers are tested further at a larger number of locations over several years. Performance testing over a number of environments is likely to identify the weaknesses of a new hybrid. Those hybrids that survive rigorous testing are usually grown in larger strip tests for evaluation by farmers. If reaction is favorable, the selected hybrids are put into pilot seed production and also entered in state variety trials before being placed into large-scale production.

STATE HYBRID PERFORMANCE TRIALS

Almost every state in the United States where corn is grown conducts performance trials. The objective is to provide farmers, seed producers, and extension staff with an unbiased estimate of a hybrid's genetic potential for yield and other agronomic characteristics. Most states charge a "per entry" fee to cover expenses of conducting the trial and publishing the data. A company usually enters a hybrid after it has survived rigorous screening and is considered close to being commercially released. Most states also include the most widely grown hybrids in the trials to provide comparative information on popular hybrids for which seed is available. Often, the newer hybrids entered by companies have not reached the stage of sizeable seed availability.

IV. KERNEL MODIFICATION THROUGH BREEDING

A. Single-Mutant Endosperm Genes

Corn kernels can be altered by genetic means to give modifications in starch, protein, oil, and other aspects such as pericarp thickness or kernel hardness. During the last four decades, much interest has been generated in specialty corns that are used by corn processors for specialty products.

STARCH MODIFICATION

Most of the genes affecting endosperm composition are recessive. One exception is the floury gene *Fl1*, which is partially dominant.

Starch from normal dent or flint corn is composed of 73% amylopectin (the starch fraction with branched molecules) and 27% amylose (the fraction with linear molecules). Waxy corn (having the *wx* gene) was first found in China, but waxy mutations have also been found in American dent strains. Starch from this mutant is 100% amylopectin. Waxy corn is easy to identify. Its starch and its pollen grains stain a reddish brown when subjected to a dilute iodine-potassium iodide solution, whereas the starch and pollen grains from normal dent corn stain blue. Yields of the first waxy hybrids were somewhat less than those for their normal dent counterparts, but the newer waxy hybrids are comparable to the better dents.

The endosperm mutant amylose-extender (*ae*) was found by R. P. Bear in 1950 (Vineyard et al, 1958). It increased the amylose fraction of the starch to 50% and above. The kernel of this corn is characterized by a tarnished, translucent, and partially full appearance. The *ae* gene, plus modifiers, gives a range in amylose content of 50–80%, but the amylose content can be stabilized at various intermediate levels. For example, in class-5 high-amylose corn, the amylose content is 50–60% and in class 7, it is 70–80%.

Several other mutant genes, either alone or in combination, affect starch composition by changing the amylose-amylopectin ratio (i.e., the ratio of straight-chain versus branched molecules). Among these are dull (du) and sugary-2 ($su2$). Both of these genes have been studied in combination with wx and ae (Alexander and Creech, 1977), but commercial usage of these gene combinations has not developed.

Both waxy and high-amylose hybrids are grown under contract for corn wet-milling. Since both genes are recessive, the fields in which they are produced must be isolated from normal dent corn. Limited acreages of waxy corn are also grown as feed for cattle and other livestock. Discussion of the products made from waxy and high-amylose corn starches is covered in Chapter 16.

PROTEIN MODIFICATION

Several endosperm mutants that alter the balance of amino acids have been identified. The most important of these is opaque-2 (o_2). Mertz et al (1964) reported that o_2 reduced zein in the endosperm and increased lysine. Other mutant genes with similar effects are floury-2 (fl_2) and opaque-7 (o_7).

Kernels with the o_2 gene are characterized by a soft, chalky, nontransparent appearance, with very little hard vitreous or horny endosperm. This type of kernel is more prone to damage by kernel rots, insects, rodents, and harvesting machinery. Much improvement has been made in increasing resistance to ear and kernel rots by selection for more vitreous o_2 types (CIMMYT, 1982). However, lysine levels tend to decrease, so selection must be accompanied by chemical endosperm analysis to retain high levels of lysine.

Yields of the first o_2 hybrids were 85–90% of those of the normal dents. However, through selection, the yields of modified o_2 material have been improved during the last decade.

High-lysine corn can be an important source for a balanced protein in the diets of nonruminants. Several nutritional studies have shown the potential value of high-lysine corn in helping meet human food needs in the less developed countries. In the United States, the use of high-lysine corn has been restricted because it yields less than normal dent corn and because a nutritionally balanced protein from corn is not needed when soybean (*Glycine max*) meal is readily available. The trade-off in growing high-lysine corn is that of losing calories per hectare in exchange for a gain in higher quality protein.

B. Altering Kernel Composition and Integrity by Selection

OIL

The oil content of hybrids from the U.S. Corn Belt ranges from 3.5 to 6.0%, with an average of about 4.5%. The long-time Illinois selection experiment (Dudley and Lambert, 1974) showed that oil content can go from as low as 0.1% to as high as 19.6%. Development of wide-line nuclear magnetic resonance spectroscopy has given researchers a nondestructive, rapid method of determining the oil content of samples as small as an individual kernel (Bauman et al, 1965). High-oil hybrids, with greater than 6% oil content, are lower in yield than hybrids with less than 6% oil. Increasing oil content genetically is not a difficult task, because variation occurs in existing germ plasm and most of it is heritable (Alexander and Creech, 1977).

Oil quality is a function of the relative amounts of unsaturated and saturated fatty acids. Oils with a high proportion of linoleic acid and lesser amount of oleic, palmitic, and stearic acids are desired and often recommended for human diets. The linoleic acid content of oil in commercial corn produced in the United States before 1964 averaged 58.7%, but by 1968, it had increased to 61.9%. Corn inbreds vary significantly in linoleic acid. Alexander and Creech (1977) found that 169 inbred lines from the northern U.S. Corn Belt averaged 58%, whereas 63 inbred lines from the southern United States averaged 48%. The amount of fatty acids is under genetic control and can be altered through breeding (see Chapter 10).

PROTEIN QUANTITY

Protein quantity in corn grain is a function of cultural practices and heredity. The current average protein content of U.S. hybrids ranges between 9 and 11% (moisture-free basis). Through selection, protein content can be altered. The long-time Illinois selection experiment covering 70 generations of selection for protein has produced corn with a low of 4.4% and corn with a high of 26.6% (Dudley and Lambert, 1974). Currently, not much interest exists in developing hybrids with higher protein potential because economically available soybean protein can produce a ration that is balanced with respect to the essential amino acids.

KERNEL INTEGRITY

Damage to kernels during harvesting, drying, elevating, and moving grain through commercial channels has become of great concern, especially to the export trade. Contributing to the problem has been the change from harvesting on the ear to using field picker-shellers. No artificial drying was needed for corn harvested on the ear, as it dried naturally in a storage crib. Because one of the advantages of combine harvesters is relatively earlier harvesting to reduce field losses, the grain usually has a high moisture content and requires artificial drying. Due to limited drying capacity, most farmers dry grain rapidly at high temperatures, often in excess of 80°C. Excessively rapid removal of moisture causes stress cracks to occur in the kernels. When the grain is moved through market channels, kernels hit hard objects that cause them to break, resulting in fine particles that lower the value of the product.

Research focused on solving the corn breakage problem is being coordinated through regional project NC-151.[1] Methods of determining breakage susceptibility have been developed (Chapter 5), which have been used to show that many kernel characteristics are related to the breakage problem. Some of these characteristics are 1) the ratio of vitreous to nonvitreous endosperm, 2) kernel density, 3) average kernel weight, 4) pericarp quantity and quality, 5) test weight, and 6) kernel size and shape. Most of these characteristics are heritable, but at this time, corn breeders have not given high priority to selection for kernel breakage reduction, as no reward is given the farmer for improved grain quality.

Kernel integrity can be measured by subjecting a sample of the corn lot to be tested to mechanical breakage in a machine designed for that purpose (Chapter 5). The Stein Breakage Tester (SBT) has been widely used in testing corn for

---

[1]North Central Regional Research Project NC-151, Marketing and delivery of quality cereals and oilseeds in domestic and foreign markets. Ohio Agricultural Research and Development Center, Wooster.

breakage susceptibility (Watson et al, 1986). Other breakage testers, developed in various research programs and claimed as improvements over the SBT, were compared with it in a collaborative study (Watson and Herum, 1986). The study showed that a tester developed at the University of Wisconsin (Singh and Finner, 1983) had the least error, best repeatability, and, unlike the SBT, was adaptable to automated operation.

Research (L. F. Bauman, Purdue University, Lafayette, IN, personal communication) has shown that differences exist among genotypes for kernel fracturing caused by fast, high-temperature drying. Thus, selection for resistance to this kind of kernel fracturing should be possible. Marked differences have been shown in rates of artificial drying among corn genotypes (Stroshine et al, 1981).

Another solution to the problem would be to let corn dry in the field to a moisture content that would require less artificial drying and use lower drying temperatures. Hybrids within the same maturity groups differ in rates of drying in the field and are known in the trade as slow and fast dryers. Development of fast-drying hybrids has been by chance, as corn breeders are not sure what selection criteria to use. Until farmers are rewarded for high-quality grain that endures processing with little breakage, the likelihood of corn breeders developing hybrids with high kernel integrity is rather remote.

V. SEED PRODUCTION

Seed production requires special care. Hybrid plants with the same pedigree but from seed produced by two individuals can vary as much as two different hybrids, just because of careless production and/or conditioning techniques.

After the seed producer chooses the desired hybrid, parental seed stocks must be located. Many large companies produce their own seed stocks, but smaller producers may buy seed from foundation seed stock companies that specialize in developing inbred lines and producing seed of public inbred lines. Seed of inbred line parents is sold on a thousand-viable-kernel basis. For example, if the germination is 90%, the number of seeds is adjusted to 1,111 kernels to provide 1,000 viable kernels.

A. Crossing Techniques

The planting pattern is sometimes a function of the planting equipment, where the number of male rows may be one to two and the number of female rows may be four to eight. Tassels are removed from the female rows by hand, with workers either walking or riding a special apparatus that carries six or eight workers through the field at one time. Mechanical cutters and pullers are also used but usually require a follow-up by a walking crew to remove tassels that were missed.

A method alternative to tassel removal is the use of male sterility, which can either be genetic or cytoplasmic. Genetic male sterility has not been used to any great extent. Before 1970, over 90% of the corn seed in the United States was produced in Texas male-sterile cytoplasm. But after the 1970 catastrophe, when *Helminthosporium maydis* race T caused damage to hybrids made using Texas male-sterile cytoplasm, cytoplasmic male sterility was not used for a few years.

In 1971, male-sterile cytoplasms were found that did not confer susceptibility to any foliar diseases. Since that time, the smaller producers have gradually returned to the use of cytoplasmic male sterility. Some of the cytoplasmic male-sterile sources now used belong to the "C" and "S" groups (Beckett, 1971). The larger producers have not returned to the use of cytoplasmic male sterility.

One of the problems associated with seed production via the cytoplasmic male-sterile method is the need to restore pollen fertility in the farmers' fields. This is accomplished by breeding into the male parent a restorer gene (Rf) or by blending (usually 50-50) seed produced by the cytoplasmic male-sterile method with seed of the same hybrid produced in normal cytoplasm on detasseled plants. Most seed producers use the latter method.

B. Isolation Requirements

Usually, a seed-production field should be isolated from other corn by a distance of not less than 200 m (660 ft). However, this distance can be reduced by planting border rows of the male parent. A rule of thumb is that each border row is the equivalent of 5 m (16.5 ft) of isolation, reducing the distance required to not less than 135 m (440 ft). For example, a 40-ha (100-acre) field could have 13 border rows and 135 m (440 ft) of isolation. Other factors, such as natural barriers (trees) or time of planting, can affect the isolation distance (Jugenheimer, 1976).

C. Seed Conditioning

HARVESTING

All seed corn is harvested on the ear. Following harvest, ears are run over an endless belt where off-type and moldy ears and husks are manually removed before drying.

DRYING

Most dryers use heated, forced air that moves through a bin of ear corn. Temperatures range from 26 to not over 40°C. Fast drying at higher temperatures may reduce seed quality.

SHELLING

After drying, the ear corn is shelled, usually at 12–14% moisture to reduce mechanical damage to the kernel. Shellers for seed corn are operated at slower speeds than those for market corn.

CLEANING AND SIZING

Shelled corn is cleaned by an air-screen machine that removes cob pieces, very light and moldy kernels, and foreign debris. Sizing and grading is necessary for plate planters, but plateless planters can plant either sized or unsized seed. However, most corn seed sold in the United States is sized, because it is marketed based on the number of kernels per 22.6-kg (50-lb) bag. Seed is separated by mechanical means according to width, thickness, and length of kernel. The first division is round vs. flat kernels, and both of these divisions are subsequently divided into large, medium, and small fractions.

SEED TREATMENT

Seed is usually treated with a fungicide by the slurry method, providing a film around the seed that helps retard growth of soilborne fungi after the seed is planted. Sometimes an insecticide is added to protect the seed from storage insects.

GERMINATION TESTS

Both "warm" and "cold" tests are performed. The warm test is conducted at a temperature of about 27°C at a high relative humidity. The cold test is conducted at low temperatures (10°C), somewhat similar to soil temperatures experienced at planting time, especially under adverse conditions. Soil, with soilborne fungi, is applied over the seed at the start of the test. Usually, the results of both warm and cold germination tests appear on the tag attached to each bag of seed.

MARKETING

Large seed companies have well-organized sales departments with regional, district, and local salespeople (usually farmers). Smaller seed companies have a seed sales manager but rely more on local salespeople. Corn seed sales in the United States are very competitive.

D. Seed Classification

BREEDER SEED

Seed of inbred lines developed by the plant breeder and initially increased is termed "breeder seed." The breeder and/or his employer, be it a public or private firm, becomes responsible for increasing and maintaining the seed. Breeder seed is the source for foundation seed.

FOUNDATION SEED

This seed is the next step in the production of commercial seed that seed companies produce to sell farmers for commercial production. In the United States, several foundation seed companies specialize in producing inbred lines and sister-line crosses that are sold to seed producers for producing F_1 hybrids.

REGISTERED SEED

Registered seed is produced from either breeder or foundation seed under the approval and requirements of some state certifying agency. It is the seed used to produce certified seed.

CERTIFIED SEED

Certified seed is the progeny of foundation or registered seed used to produce commercial hybrid seed under the requirements of a certifying agency. These requirements may include specification of: 1) origin of parental seed; 2) the strain; 3) evidence of performance; 4) inspection of the growing crop for off-type plants, isolation, and detasseling; 5) germination percentage; and 6) grading and sizing. Very little hybrid seed corn is grown under certification. Occasionally, contracts between seed companies and U.S. seed producers require certification. By this means, an absentee contractor relies on a seed certification agency to make sure the seed was produced under the arrangements agreed upon.

E. Contract Production

SPECIAL ENDOSPERM TYPES

Waxy and high-amylose corn are usually grown under contract to ensure that the processor receives a pure product, as both of these mutants are recessive, and contamination with normal dent pollen gives off-type kernels. Another advantage for the producer is that a market for the crop is assured. In addition, the crop does not pass through elevators and terminals but is delivered by the contractee to the contractor's designated location.

WHITE CORN

Some of the white corn produced in the United States is grown under contract for many of the reasons previously mentioned. By contracting production, a contractor is usually ensured a higher-quality product at an agreed-upon price.

LITERATURE CITED

ALEXANDER, D. E., and CREECH, R. G. 1977. Breeding special industrial and nutritional types. Pages 363-390 in: Corn and Corn Improvement. G. F. Sprague, ed. Am. Soc. Agron., Madison, WI.

BAUMAN, L. F., CONWAY, T. F., and WATSON, S. A. 1965. Inheritance of variation in oil content of individual corn kernels. Crop Sci. 5:137-138.

BECKETT, J. B. 1971. Classification of male-sterile cytoplasms in maize (*Zea mays* L.). Crop Sci. 11:724-727.

CIMMYT. 1982. CIMMYT Review 1982. Centro Internacional de Mejoramiento de Maiz y Trigo, El Batan, Mexico.

DUDLEY, J. W., and LAMBERT, R. J. 1974. Seventy generations of selection for oil and protein concentration in the maize kernel. Pages 181-212 in: Seventy Generations of Selection for Oil and Protein Concentration in Maize. J. W. Dudley, ed. Spec. Pub. Crop Sci. Soc. Am., Madison, WI.

GARDNER, C. O. 1961. An evaluation of effects of mass selection and seed irradiation with thermal neutrons on yield of corn. Crop Sci. 1:241-245.

HALLAUER, A. R., and MIRANDA Fo., J. B. 1981. Quantitative Genetics in Maize Breeding. Iowa State University Press, Ames.

HALLAUER, A. R., RUSSELL, W. A., and SMITH, O. S. 1983. Quantitative analysis of Iowa Stiff Stalk Synthetic. Proc. Stadler Symp., 15th, Univ. of Missouri, Columbia. Univ. Missouri Printing Services, Washington Univ., St. Louis.

JENKINS, M. T. 1934. Methods of estimating the performance of double crosses in corn. J. Am. Soc. Agron. 26:199-204.

JONES, D. F., and MANGELSDORF, P. C. 1926. Crossed corn. Conn. Agric. Exp. Stn. Bull. 273.

JUGENHEIMER, R. W. 1976. Corn Improvement, Seed Production, and Uses. John Wiley & Sons, Inc., New York.

KIESSELBACH, T. A. 1949. The structure and reproduction of corn. Neb. Agric. Exp. Stn. Res. Bull. 161.

MANGELSDORF, P. C., and FRAPS, G. S. 1931. A direct quantitative relationship between vitamin A in corn and the number of genes for yellow endosperm. Science 73:241-242.

MERTZ, E. T., BATES, L. S., and NELSON, O. E. 1964. Mutant gene that changes protein composition and increases lysine content of maize endosperm. Science 145:279-280.

RÉDEI, G. P. 1982. Genetics. Macmillan Publishing Co., New York.

SHULL, G. H. 1909. A pure line method of corn breeding. Am. Breeders Assoc. Annu. Rep. 5:51-59.

SINGH, S. S., and FINNER, M. F. 1983. A centrifugal impactor for damage evaluation of shelled corn. Trans. ASAE 26:1858-1863.

SPRAGUE, G. F. 1952. Early testing and recurrent selection. Pages 400-417 in: Heterosis. J. W. Gowen, ed. Iowa State College Press, Ames.

SPRAGUE, G. F., and EBERHART, S. A. 1977. Pages 330-339 in: Corn and Corn Improvement. G. F. Sprague, ed. Am. Soc. Agron., Madison, WI.

STROSHINE, R. L., EMAN, A., TUITE, J., CANTONE, F., KIRLEIS, A., BAUMAN, L. F., and OKOS, M. R. 1981. Comparison of drying rates and quality parameters for selected corn inbreds/hybrids. Paper 81-3529. Am. Soc. Agric. Eng., St. Joseph, MI.

VINEYARD, M. L., BEAR, R. P., MacMASTERS, M. M., and DEATHERAGE, W. L. 1958. Development of "Amylomaize"—Corn hybrids with high amylose starch: I. Genetic considerations. Agron. J. 50:595-598.

WATSON, S. A., and HERUM, F. L. 1986. Comparison of eight devices for measuring breakage susceptibility of shelled corn. Cereal Chem. 63:139-142.

WATSON, S. A., DARRAH, L. L., and HERUM, F. L. 1986. Measurement of corn breakage susceptibility with the Stein breakage tester: A collaborative study. Cereal Foods World 31:366-367, 370, 372.

CHAPTER 3

STRUCTURE AND COMPOSITION

STANLEY A. WATSON (*retired*)
Director's Office
Ohio Agricultural Research and Development Center
The Ohio State University
Wooster, Ohio

I. INTRODUCTION

An intimate and precise knowledge of the structure and composition of the corn plant is necessary for understanding and optimizing plant growth to achieve the highest level of grain production. Likewise, knowledge of the structure and composition of the mature kernel is necessary to development of improved methods of storage and handling to preserve initial quality. Such knowledge is also valuable for most efficient utilization, whether for seed production, for animal feeding, or for industrial or food product development.

For the purposes of this book, the corn plant has been adequately described in Chapters 1, 2, and 20 and in other publications (Kiesselbach, 1949; Sprague, 1977). This chapter's discussion concerns properties of the developing and mature corn kernel.

II. GROSS STRUCTURAL FEATURES

Corn kernels are produced on the female inflorescence called the ear. About 800 kernels are produced on a properly developed ear and must be removed from the inner cylinder of the ear, known as the cob, by the process of shelling (Chapter 4). If corn is to be used for silage, the entire ear is ground up with the leaves and stalk. In some rations, whole, mature ear-corn is ground for cattle feeding, since these animals can utilize some of the cellulosic components of the cob. A discussion of cob structure and composition is given in Chapter 20.

The corn kernel is classified botanically as a caryopsis, i.e., a dry, indehiscent, single-seeded fruit (Fig. 1, also Fig. 3 in Chapter 2). This kind of fruit, in which the mature ovary wall (pericarp) does not separate naturally from the seed, is characteristic of all cereal grains. The kernel is attached to the cob by the pedicel. During grain development, conducting elements in the pedicel carry the products of photosynthesis into the developing kernel (see Chapter 8 for a description of this process). When the kernel is removed from the ear, the pedicel is broken randomly, leaving a jagged end. The conical structure that remains attached to the kernel is called the tip cap.

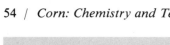

Fig. 1. Dent corn kernel longitudinal section, 40 μm thick, perpendicular to face of kernel. P, pericarp; Al, aleurone layer; FE, floury endosperm; HE, horny endosperm; HL, hilar layer; TC, tip cap; SA, silk attachment; Sc, scutellum attachment. The embryo includes the following parts: Cp, coleoptile; Pl, plumule; VB, vascular bundle; RB, adventitious root bud; FI, first internode; ScN, scutellar node; VC, vascular cylinder; Cr, coleorhiza; PrR, primary root; RC, root cap (a mass of cells indistinguishable from the coleorhiza at this magnification). Parts Sc through RC are collectively termed "germ." (×7.6) (Reprinted, with permission, from Wolf et al, 1952a)

Differences in size and shape of corn kernels are due to differences in genetic background and also to placement on the ear. Kernels at the shank (butt) end are large and rounded, and those at the tip end are small and round. Kernels in between are usually flattened due to pressure from adjacent kernels during growth. When adjacent kernels are missing, central kernels are round. An average dent corn kernel weighs 250–300 mg, with a range of 100–600 mg; an average kernel from the center of an ear measures about 4 mm thick, 8 mm wide, and 12 mm long. Variety and environment produce differences in shape and size. Inherited differences in number of rows and endosperm characteristics are additional factors. Pomeranz et al (1985), in measuring size classes of three dent corn hybrids by sieving, showed that 38–42% of the kernels would pass through 8.33-mm (21/64-in.) round holes and 4.76-mm (12/64-in.) slots. These kernels had a relative sphericity of 0.62–0.65 (round = 1.0) and had average kernel weights of 248–252 mg. The remainder were the small kernels (average weight, 169.7 mg), which passed through a 6.75-mm (17/64-in.) round hole sieve, or the largest kernels (average weight, 366.8 mg), which passed through a 11.1-mm (28/64-in.) round hole but were retained on a 5.95-mm (15/64-in.) slot. The seed corn industry separates seeds into eight size classes through screens with round or slotted holes and classifies sizes as: extra large (over 9.5 mm or 24/64 in.), large round, large flat, medium flat, medium round, small round, small flat, and extra small (through 6.7 mm or 17/64 in.). The seed corn industry discards the extra large and extra small seeds to animal feed but packages and labels the other sizes for specified sizes of corn planter plates (Craig, 1977). In an average lot of dent corn, about 75% of the kernels would be classified as flat, 20% as round, and 5% as very small.

The five general classes of corn—flint corn, popcorn, flour corn, dent corn, and sweet corn (Fig. 2)—are based on kernel characteristics. Flint corn has a rounded crown and the hardest kernels due to the presence of a large and continuous volume of horny (also called corneous) endosperm (Fig. 2b). Flint corn varieties are most popular in Argentina, some parts of Italy, and Africa. Popcorn (Fig. 2c) is a small flint corn type (see Chapter 13). Flour corn generally also has a rounded or flat crown but contains virtually all floury or soft endosperm (Fig. 2d). Flour corn varieties are only grown by certain indigenous populations in Latin America for direct consumption as food. Dent corn (2a), as the name implies, has a depressed crown that forms as the maturing kernel dehydrates. Apparently, the rigidity of the cylinder of horny endosperm prevents the central core of floury endosperm from shrinking uniformly during dry-down. This causes the crown to be sucked in, resulting in a dent and a central fissure. Many races of each of these types were endemic to various parts of North and South America, but their origin is lost in antiquity (Goodman, 1978). Corn-Belt dents of the United States apparently developed in precolonial North America from natural hybridization between northern flint and southern flour corns (Chapter 1).

Since dent corn is a derivative of flint-flour crosses, it can show significant differences in the ratio of horny to floury (H/F ratio) endosperm caused by heritable and environmental influences (Hamilton et al, 1951). Ten different dent corn commercial hybrids were evaluated for relative hardness by Pomeranz et al (1986b), using several different methods of hardness evaluation. Stenvert hardness values (see Chapter 5) ranged from 20.4 sec (hard) to 12.0 sec (soft),

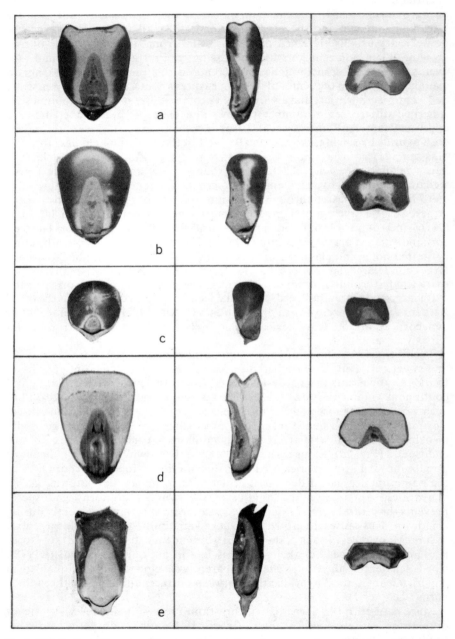

Fig. 2. Distribution of horny and floury endosperm in various types of corn kernels: a, dent corn; b, flint corn; c, popcorn; d, high-lysine flour corn ($o_2\ o_2$); e, sweet corn ($su\ su$). Left column, longitudinal section parallel to germ front; center column, longitudinal section perpendicular to germ front; right column, cross section. All sections were cut approximately through the median line. (×2)

whereas near-infrared (1,680 λ) values ranged from 433.2 (hard) to 306.4 units (soft). Among other physical measurements, test weight and density values showed the highest correlation with hardness values. All hybrids were dried on the ear at low temperatures. Data from A. W. Kirleis[1] has shown that these differences in hardness are correlated with differences in the H/F ratios. Furthermore, individual kernels within each lot differ significantly in the H/F ratio observed. This can be seen in the cross sections of the three kernels (Fig. 3) that represent the H/F ratio ranges from each of three separate commercial hybrids. These hybrids had been classified as hard, intermediate, and soft by the Stenvert hardness test (15.6, 13.8, and 10.5 sec., respectively) and by Purdue milling evaluation factor (see Chapter 5) values of 49.5, 43.4, and 30.4,

[1] A. W. Kirleis, Purdue University, West Lafayette, IN, private communication, 1985.

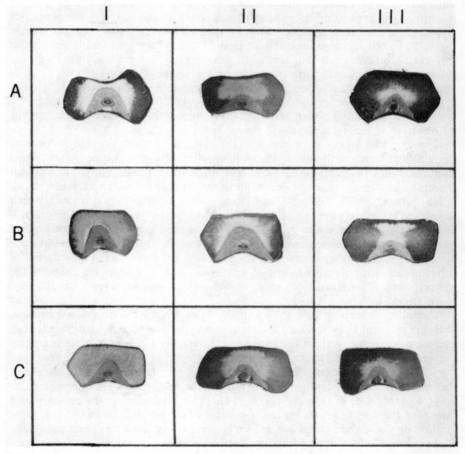

Fig. 3. Cross sections of kernels from corn hybrids classified as hard (A), intermediate (B), and soft (C), showing maximum ranges of horny and floury endosperm areas within kernels of each hybrid. I, most floury; III, most horny. Kernels (except popcorn) were mounted for cutting as described by Kirleis et al (1984). (×3)

respectively. Thus, considerable overlap exists in the H/F ratios between and within hybrids. Differences are caused by environment on the ear, within fields, and between fields due to moisture, temperature, and soil nitrogen supply and uptake. High soil nitrogen has a significant influence in producing a higher H/F ratio (Hamilton et al, 1951). When corn is being bred for increased hardness, it is probably more valid to evaluate hardness on a bulk sample using actual hardness measurements, such as grinding resistance, than to measure differences in H/F ratio.

Five recessive genes were initially identified as producing floury endosperm. They are: fl_1, fl_2, o_1, o_2, and h (Neuffer et al, 1968); additional floury (fl) and opaque (o) mutants have been reported (Nelson, 1978). The o_2 and fl_2 genes, when double recessive, also produce increased lysine content in endosperm protein (Nelson et al, 1965). Ortega and Bates (1983) present evidence that differences in hardness among opaque-2 populations is heritable and is related to differences in two "zeinlike" protein components. The o_2 and fl_2 genes restrict synthesis of the low-lysine protein, zein, thus increasing lysine on a dry weight basis. Presently, no commercial high-lysine hybrids have acceptable grain yield or kernel properties,[2] but research on population improvement at the International Center for Improvement of Maize and Wheat (CIMMYT) at El Batán, Mexico, has produced modified opaque-2 varieties called "quality protein maize" (Vasel et al, 1980) that have harder kernels due to the presence of significant amounts of horny endosperm of higher than normal lysine content (Robutti et al, 1974a; Ortega and Bates, 1983).

The dent and flint corn endosperms carry dominant genes and normal amounts of starch with normal starch properties. However, numerous recessive mutant genes (such as waxy [wx] and amylose extender [ae]) alter starch composition. Some recessive genes interfere with starch synthesis (Neuffer et al, 1968; Nelson, 1978). The most well known of the latter is sugary-1 (su_1), which produces standard sweet corn (Fig. 2e). However, as described in Chapter 14, many of the other recessive genes are now being used to produce sweet corn that is sweeter than the normal, sugary-1 sweet corn. These recessive genes produce extremely small or shrunken kernels at maturity, in proportion to their degree of interference with starch or protein synthesis. In the rest of this chapter, the descriptions of structural, physical, and chemical properties refer only to dent corn, unless otherwise noted.

Corn kernels can differ significantly in color from white to yellow, orange, red, purple, and brown. Color differences may be due to genetic differences in pericarp, aleurone, germ, and endosperm (Neuffer et al, 1968). The pericarp can be colorless, orange, cherry red, red, dark red, brown, or variegated; the aleurone layer can be colorless, red, red-purple, purple, or brown; the germ can be colorless, yellow, orange red, or purple; the endosperm is either colorless, yellow, orange, or orange-red (Wolf et al, 1952a). Obviously, the pericarp and aleurone must be colorless for the true color of the endosperm to be seen. Only yellow or white dent corns are grown commercially. Some hybrids may have light tan and light orange pericarps. Flint corn grown in Argentina has orange-red endosperm.

---

[2]A. F. Troyer. Breeding for improved amino acid content in corn. Presented at the National Feed Ingredient Association Meeting, Chicago, IL, May 2, 1984. (Copies available from author, DeKalb-Pfizer Seed Co., DeKalb, IL.)

III. PHYSICAL PROPERTIES

Table I gives a tabulation of known physical properties of kernel and bulk dent corn. Compared with other cereal grains, such as wheat and sorghum, dent corn has larger kernel size and lower specific gravity, but in many other physical and chemical properties these grains are similar. The values given in this table are for mature "average-quality" grain samples at 15% moisture content (MC) unless otherwise specified.

Most of the properties vary with differences in moisture content (Chung and Converse, 1971; Nelson, 1980) and from variety to variety, year to year, and region to region of production. For example, moisture deficiency during growth may produce smaller kernels; an early frost may produce kernels of lower density. Flint and popcorn kernels naturally are of greater density than dent corn due to the greater proportion of the denser horny endosperm. Different levels of soil nitrogen or genetic differences may result in differences in endosperm H/F ratio and in density. Kernel size and shape, bulk density, true density, and porosity (void volume) are parameters used in studying hydrodynamic, aerodynamic, and heat and mass transfer problems in grain (Chapters 4 and 5). These properties are important to take into account in the design of handling and storage equipment. The void volume is a measure of the space between kernels in a bulk. Generally, the lower the bulk density, the higher

TABLE I
Physical Properties of Dent Corn[a]

Property	English Units	International Units
Kernels per ear	800	...
Range	(500–1,200)	
Kernels per unit weight, lb; kg	1,300	2,900
Range	(900–1,800)	(2,000–4,000)
Bulk density,[b] lb/bu	58.3	...
Range	(52–60)	
Bulk density,[c] lb/ft^3; kg/m^3	45.4	727
Range	(40–46)	(641–737)
Specific gravity	1.26[d]	...
Void volume,[e] %	42.3	...
Thermal conductivity, BTU/(hr)(ft^2)(F/in)	1.22	6.33[f]
Specific heat[c]	0.486	...
Heat energy content,[g] BTU/lb; KJ/kg	8,075	18,740
Angle of repose, degrees	35	...
Range	(34/43.5)[h]	...

[a] All values are for mature grain at 15% MC, except as noted. Single values are the most probable median value at 15% MC.
[b] Anonymous (1983), 80–88° F.
[c] Brooker et al (1974).
[d] Nelson (1980).
[e] Thompson and Isaacs (1967).
[f] $J/cm^2 \cdot hr^{-1} \cdot °C^{-1}$.
[g] Keener et al (1985).
[h] Moisture range of 7.5–23.1%. Brooker et al (1974).

the void volume. Corns have an average void volume of 42.3% (Thompson and Isaacs, 1967), which is intermediate compared with void volumes of other cereals: oats, 50%; barley, 47%; wheat, 40%; and sorghum and soybeans, 36% (Anonymous, 1983). Void volume influences the rate of passage of air or fumigants through the grain in a bin.

When measuring physical properties, the moisture content of the grain must be accurately determined. At the same moisture content, properties differ depending on whether the corn reached its moisture content by desorption or absorption of water vapor because of the hysteresis effect (Chapter 4). Chung and Converse (1971) measured the changes in several physical properties at different moisture levels in corn kernels of different size ranges. Differences in bulk density, specific gravity, and kernel weight were found. They showed that the test weight of mixed kernels was lower than for lots having kernels of uniform size and shape, and that the rate of weight change with respect to moisture content change was almost twice as much for the mixed corn as for the samples of uniform size and shape. Large flat kernels had greater true density than small kernels.

These changes in kernel properties with changes in moisture content are due largely to absorption or desorption of water vapor, which produces changes in density and volume. Such changes are even more dramatic when the kernels are put in contact with liquid water, such as is involved when corn is steeped in water for starch production. Fan et al (1962) showed that volume increases of popcorn and dent corn at temperatures of 0–100°C were linearly related to weight increase. Floury kernels of opaque-2 corn swell to a greater volume than kernels of dent corn (Watson and Yahl, 1967; Ratkovic et al, 1982). Like most dehydrated plant materials, corn absorbs water at an initially rapid rate to fill capillaries. This is followed by a gradually falling rate as the water diffuses through the tissue and is absorbed into cell components; the rate levels off at 4–8 hr at temperatures up to 71°C. Above that temperature, starch gelatinization causes faster swelling. Water first enters the kernel through the tip cap, moves quickly through voids in the pericarp by capillary action, and enters the endosperm through the crown (Cox et al, 1944). From there on, wetting of the cellular contents of endosperm and germ is by the slower diffusion process, which increases in rate as the temperature of the water increases (Fan et al, 1962). The germ absorbs three to five times more water than the endosperm (Ratkovic et al, 1982). Syarief et al (1984), in evaluating the moisture-absorbing properties of the individual kernel parts, found that diffusion coefficients increased with increasing moisture contents over a range of 5–30% MC (dry basis). The diffusion coefficient of germ was 3.5–3.8 times that of floury endosperm and 4.7–5.3 times that of horny endosperm. Data for the pericarp indicated that the diffusion characteristics varied with location on the kernel.

Data in Table I showing thermal conductivity of corn as $6.33 \text{ J}/\text{cm}^2 \cdot \text{hr}^{-1} \cdot {}^\circ\text{C}^{-1}$ indicate that corn is a poor heat conductor (for comparison, thermal conductivity of concrete is 36–47 J units). This means that development of heat within a bin of corn dissipates only slowly from the heated area. In the case of heat caused by the action of microorganisms or insects, continual generation of heat eventually causes the temperature to exceed the combustion temperature of the corn, resulting in "spontaneous" combustion. Alternatively, a grain mass cooled by aeration during winter remains cooler than the outside temperature for many weeks into a summer storage period.

The heat energy content, also called heat of combustion, of corn—18,740 KJ/kg (8,075 BTU/lb), as given in Table I—is sufficient to make corn grain useful as fuel if circumstances warrant. Such cases include corn that has been damaged to the extent that it is no longer useful for animal feed, or periods when the price of corn is depressed below that of common fuels. Keener et al (1985) burned corn in a fluidized sand-bed burner (designed for burning ground corncobs) with 75% efficiency, about the same as that for burning ground cobs (heat energy content, 18,800 KJ/kg or 8,088 BTU/lb). By comparison, corn priced at \$2.85/bu (15.5% MC) would be equivalent to liquid fuels, such as propane at \$0.80/gallon (100% conversion). Burning corncobs would be even more cost competitive with propane as fuel.

IV. STRUCTURAL DETAILS

The corn kernel is a seed and therefore contains a complete embryo and all of the structural, nutritional, and enzymatic apparatus required to initiate embryo growth and development. Figure 1 shows the internal relationships of the various components in a one-dimensional section. Reference to the several sections in Fig. 2 provides an understanding of the three-dimensional relationships.

A. Germ

The germ is composed of the embryo and the scutellum (Fig. 1). The scutellum, which functions as a nutritive organ for the embryo, makes up 10-12% of the kernel dry weight (Table II). Pomeranz et al (1986a) found that

TABLE II
Weight and Composition of Component Parts of Dent Corn Kernels from Seven Midwest Hybrids[a]

Part	Percent Dry Weight of Whole Kernel	Composition of Kernel Parts (% db)					
		Starch	Fat	Protein	Ash	Sugar	Unaccounted For
Endosperm							
Mean	82.9	87.6	0.80	8.0	0.30	0.62	2.7
Range	81.8–83.5	86.4–88.9	0.7–1.0	6.9–10.4	0.2–0.5	0.5–0.8	
Germ							
Mean	11.1	8.3	33.2	18.4	10.5	10.8	8.8
Range	10.2–11.9	5.1–10.0	31.1–35.1	17.3–19.0	9.9–11.3	10.0–12.5	
Pericarp (bran)							
Mean	5.3	7.3	1.0	3.7	0.8	0.34	86.7
Range	5.1–5.7	3.5–10.4	0.7–1.2	2.9–3.9	0.4–1.0	0.2–0.4	
Tip cap							
Mean	0.8	5.3[b]	3.8	9.1	1.6	1.6	78.6
Range	0.8–1.1	...	3.7–3.9	9.1–10.7	1.4–2.0	...	
Whole kernels							
Mean	100	73.4	4.4	9.1	1.4	1.9	9.8
Range	...	67.8–74.0	3.9–5.8	8.1–11.5	1.37–1.5	1.61–2.22	

[a] Data of samples 1–6 and 8 in Earle et al (1946).
[b] Composite.

germs isolated from kernels classified into seven size ranges weighed a relatively constant 11% of their kernel weights. Dry matter per 10 germs increased on average for three hybrids from 0.167 to 0.342 g as kernel weights increased from 170 to 366 mg per kernel. The germ stores nutrients and hormones, which are mobilized by enzymes elaborated during the initial stages of germination. All cells of the embryo and scutellum are potentially metabolically active upon hydration. Figure 4 shows how the conducting network branches from the embryo into all parts of the scutellum and also shows other details of the cellular makeup of the embryo and scutellum. The landmark publications of Wolf et al (1952a, 1952b, 1962c, 1952d), from which Figs. 4 and 5 were taken, provide much specific information about the fine structure of the kernel.

Additional details of the scutellum in Fig. 5 show that these parenchyma-type cells contain a nucleus, dense cytoplasm, and clear objects that contain liquid oil. These clear objects are specific organelles known as oil bodies, or "spherosomes," and are found in all oilseeds (Gurr, 1980) as well as in corn scutellum[3] (see Chapter 10). Semadeni (1967) isolated spherosomes from maize seedlings and found that they contained lipid hydrolytic enzymes as well as all necessary enzymes for lipid synthesis. Walls of the scutellum cells are thick and contain numerous pits and intercellular spaces that facilitate movement of material among cells. On the outer margin of scutellum is a single layer of secretory cells, which forms the primary contact between germ and endosperm. This layer also folds inward in the scutellum, forming "glands," which may provide a greater surface for secretion (Wolf et al, 1952d). These epithelial cells are cylindrical in shape and attached laterally only near their bases. The cell walls are composed almost entirely of an araban type of hemicellulose and very little cellulose (Seckinger et al, 1960). During germination, the outer ends of the cells swell and the cell walls becomes thin. Hydrolytic enzymes are secreted and diffuse into the endosperm to digest starch and protein. The resulting sugars and amino acids are then translocated through the scutellum to the embryo.

In the ungerminated kernel, a dark amorphous layer, called the cementing layer, is located between the epithelial layer and the endosperm (Fig. 5). It is not digested by protease or carbohydrate enzymes (Wolf et al, 1958), but analysis shows that it contains protein, hemicellulose, and probably cellulose (Seckinger et al, 1960). The cementing layer is of endosperm origin (apparently degraded endosperm cellular material) and forms a tight bond between germ and endosperm (Wolf et al, 1952c). This accounts for the difficulty in separating purified germ fractions by dry milling. In wet milling, the cementing layer probably accounts for the long steeping of 25 hr or more required to obtain satisfactory germ recovery. The reaction of sulfur dioxide (SO_2) in producing good starch recovery requires only about 4 hr (Watson and Sanders, 1961). A fairly clean separation of germ from endosperm after only a 4-hr water steep can be made by an unconventional technique. When kernels of 38–40% MC are passed through a smooth roller mill with a gap slightly smaller than the kernel width, the resulting pressure produces a shearing force that tears the endosperm away from the germ with very little damage to the germ (Stewart and Watson, 1969).

---

[3]L. Yatsu, Southern Regional Research Center, USDA/ARS, New Orleans, LA, unpublished data, 1976.

B. Endosperm

The endosperm constitutes 82–84% of the kernel dry weight and is 86–89% starch by weight (Table II). It is comprised of elongate cells (Fig. 1) packed with

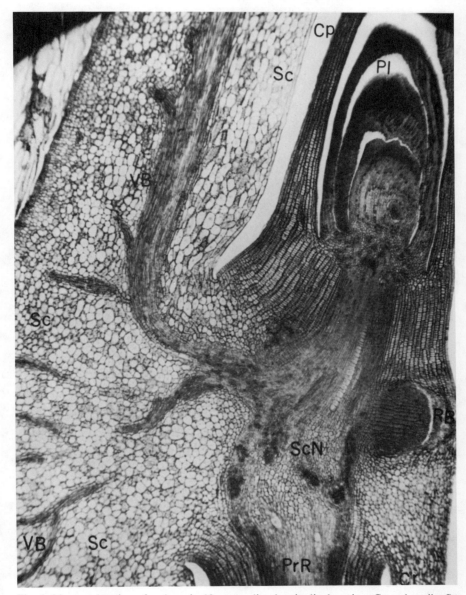

Fig. 4. Mesocotyl region of embryo in 10-μm median longitudinal section. Cp, coleoptile; Sc, scutellum; Pl, plumule; ScN, scutellar node; RB, root bud; VB, vascular bundle; PrR, primary root. (×18) (Reprinted, with permission, from Wolf et al, 1952d)

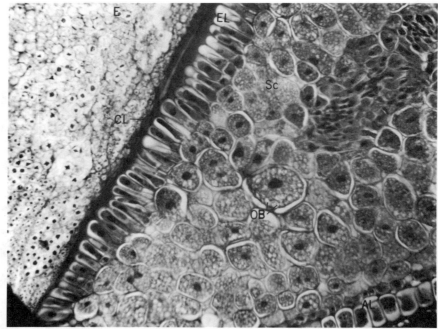

Fig. 5. Section of scutellum-endosperm interface (10 μm). E, endosperm; Sc, scutellum; EL, epithelial layer; CL, cementing layer; OB, oil bodies; AL, aleurone layer. (×90) (Reprinted, with permission, from Wolf et al, 1952d)

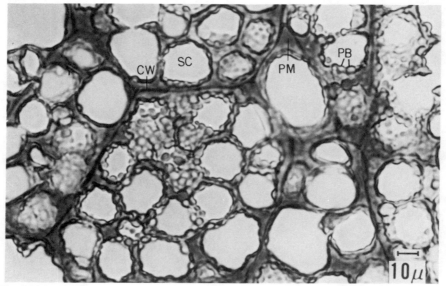

Fig. 6. Light micrograph through 10-μm section of horny endosperm destarched with amylase. PM, protein matrix; PB, protein bodies; SC, starch granule cavity; CW, endosperm cell wall. (×520) (Reprinted, with permission, from Christianson et al, 1969)

starch granules of 5–30 μm embedded in a continuous protein matrix (Fig. 6). As mentioned earlier, starchy endosperm is of two types, floury and horny. The floury endosperm surrounds the central fissure and is opaque to transmitted light. Duvick (1961) has explained the opacity of floury endosperm as being due to light refraction from minute air pockets around starch granules, which result from tearing of the thin protein matrix as it shrinks during drying. The matrix no longer completely surrounds the starch granules, which assume a round shape. Figure 7 shows the great difference in the amount of protein matrix in cells of horny and floury endosperm tissue from a dent corn kernel. The endosperms of flour corns, such as opaque-2 (Fig. 8) have a thin protein matrix throughout; hence, air pockets may be found around most starch granules that have a round shape (Robutti et al, 1974b). In the horny endosperm, the protein matrix is thicker and remains intact on drying. During drying, the plastic starch granules in the horny endosperm are compressed into polyhedral shapes (Fig. 7). Endosperm cells become progressively smaller from the central fissure to the outer endosperm. In most dent and flint corn kernels, a layer of cells lies just beneath the aleurone. This is called the subaleurone or peripheral endosperm (Figs. 7 and 9; also Fig. 2 in Chapter 8). It contains the smallest cells and may

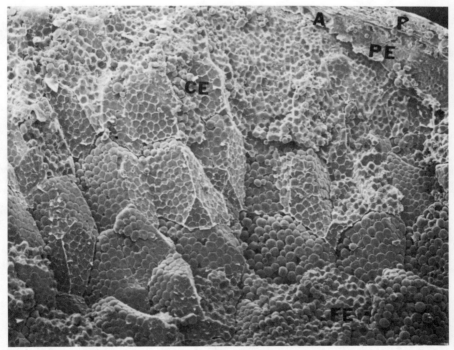

Fig. 7. Scanning electron micrograph of fractured endosperm showing how horny (corneous) endosperm (CE) breaks along cell walls because of the strength of the protein matrix, whereas floury endosperm (FE) fractures across cells, exposing round starch granules not completely surrounded by protein matrix. PE, peripheral endosperm; A, aleurone layer; P, pericarp. (×165) (Courtesy C. Earp and L. Rooney, Texas A&M University)

contain very small starch granules surrounded by a very thick protein matrix (Watson et al, 1955; Wolf et al, 1969). A flint corn variety dissected by Hinton (1953) contained 3.9% of the most dense peripheral endosperm, which showed 27.7% protein upon analysis. Christianson et al (1969) showed that the endosperms of the corn types differ not only in protein matrix thickness but also in H/F relationships, thickness of the subaleurone layers, cell sizes, and protein components.

The protein matrix is composed of an amorphous protein material in which are embedded discrete protein bodies, as shown in Fig. 6 (Christianson et al, 1969; see Chapter 9 for a discussion of the protein components of these structures). Protein bodies are comprised almost entirely of zein, a protein fraction extremely low in lysine. Opaque-2 kernels have a higher content of lysine because the protein bodies are either few and very small (Wolf et al, 1969) or absent (Robutti et al, 1974b).

To recover starch by wet milling, the granules must be released from the matrix. This can be accomplished by treating corn (or endosperm) with alkali, which disintegrates and disperses the matrix. The more traditional method is to treat the corn with a reducing agent (preferably SO_2) in a steeping process at 45–50°C (Watson, 1984). The SO_2 causes the matrix structure to weaken, apparently because it breaks disulfide cross-links and forms soluble

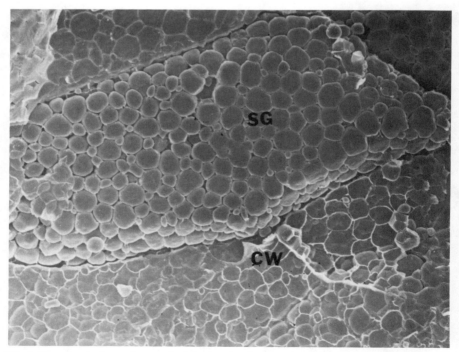

Fig. 8. Scanning electron micrograph of fractured endosperm from high-lysine (opaque-2) hybrid corn. SG, starch granule; CW, cell wall. (×95) (Courtesy C. Earp and L. Rooney, Texas A&M University)

Structure and Composition / 67

S-sulfoproteins, which prevent disulfide reformation (Boundy et al, 1967a). When the SO_2-treated endosperm is pulverized in water, the starch granules are recovered free of adhering matrix and can be separated by a gravitational procedure such as centrifugation (Chapter 12). The matrix fragments are

Fig. 9. Transection through pericarp, TC through E, and outer endosperm, AL through SE (10 μm). TC, tube cells; CC, cross cells; M, mesocarp; E, epidermis; AL, aleurone; SA, subaleurone; SE, starch endosperm. Swollen by steeping at 8°C for three days. (×325) (Reprinted, with permission, from Wolfe et al, 1952b)

recovered as a high-protein fraction called "gluten," which also contains endosperm cell walls and small starch granules. The walls of endosperm cells are thin, cellulosic membranes containing many pits that are greatly enlarged during the steeping process (Earp et al, 1985). The protein portion of the gluten can be isolated as a nearly "pure" protein (15.7% N) suitable for use in foods (Reiners et al, 1972). However, corn gluten does not have the viscoelastic properties so characteristic of wheat gluten. The lipid extract of corn gluten is rich in yellow carotenoid pigments (Chapter 10).

In the dry-milling process, the primary product is isolated pieces of endosperm, which are recovered by progressive grinding, sieving, and aspiration (Chapter 11). Dry grinding of floury endosperm causes breakage across the cell contents, releasing some free starch granules and producing a rough surface with many exposed starch granules and very little starch granule damage. Horny endosperm breaks more along cell wall lines but also across cells, with little release of starch granules but with much granule damage.

The outer layer of endosperm, the aleurone layer, is a single layer of cells of an entirely different appearance (Figs. 1, 5, and 9). This layer, which covers the entire starchy endosperm and the germ, is interrupted only at the hilar layer at the kernel tip. It is thinnest over the germ. Hinton (1953) estimated that the aleurone layer in a flint corn variety made up 2.2% of the kernel dry substance and contained 19.2% protein. The contents of aleurone cells are granular in appearance and contain protein granules but no starch (Fig. 2 in Chapter 8). Pomeranz (1973) has excellent photomicrographs of cereal aleurone cells, and although corn aleurone was not studied, the structure of aleurone cells in all cereal grains appear to be quite similar. These cells are also rich in minerals and protein, which are of high quality but probably not available nutritionally to digestive enzymes unless the cells are opened by grinding (Saunders et al, 1969).

C. Pericarp

The outermost structure of the seed is a thin, hyaline, almost invisible membrane termed the seed coat (Wolf et al, 1952b). It adheres tightly to the outer surface of the aleurone layer and is thought to impart semipermeable properties to the corn kernel. (However, the aleurone layer may be just as important in this function.) The presence of a semipermeable membrane can be demonstrated in intact, fully hydrated corn kernels, which show osmotic pressure effects on transfer between pure water and salt solutions.

All tissues exterior to the seed coat are the true fruit coat and are collectively named the pericarp or hull (Fig. 9). Pericarp makes up 5–6% of the kernel dry weight (Table II). All parts of the pericarp are composed of dead cells that are cellulosic tubes (Wolf et al, 1952b). The innermost tube-cell layer is a row of longitudinal tubes pressed tightly against the aleurone layer. Next is a very loose and open area called the cross-cell layer, which has a great deal of intercellular space. This layer is covered by a thick and rather compact layer, known as the mesocarp (Fig. 9), composed of closely packed, empty, elongate cells with numerous pits (Fig. 10). These pits and open areas in the cross cell layer provide capillary interconnections between all cells, which facilitates water absorption. An outer layer of cells, the epidermis, is covered by a waxy cutin layer that probably retards moisture exchange. Helm and Zuber (1969) found pericarp thickness to vary from 62 to 160 μm in 33 corn inbred lines and showed that breeding for either thick or thin pericarps should be successful (Helm and Zuber, 1972). Although the word "bran" is sometimes used as a synonym for pericarp, its use should be restricted to the pericarp-containing product of a dry-milling or wet-milling process that includes tip cap, aleurone layer, and adhering pieces of starchy endosperm. In the wet-milling process, it is termed "fiber."

The pericarp extends to the base of the kernel, uniting with the tip cap. Inside the tip cap are spongy, star-shaped cells connected only by the ends of the branches, thus forming an open structure that is continuous with the cross-cell layer (Wolf et al, 1952b). Vascular bundles from the cob pass through the tip cap and terminate in the basal portion of the pericarp. The tip cap constitutes about 1% of kernel dry weight (Table II). Pulling the tip cap off exposes a dark brown circular layer, known as the hilar or black layer, that lies against the base of the germ and endosperm. The hilar layer appears to be an abscission layer, with a probable function of sealing the tip of the kernel. The appearance of the black layer coincides approximately with kernel maturation and cessation of dry substance accumulation (Daynard and Duncan, 1969; see also Chapter 1).

V. COMPOSITION

A. Kernel Development

Chapter 2 describes the double fertilization process in corn, in which one set of fused nuclei leads to embryo development and another set to endosperm

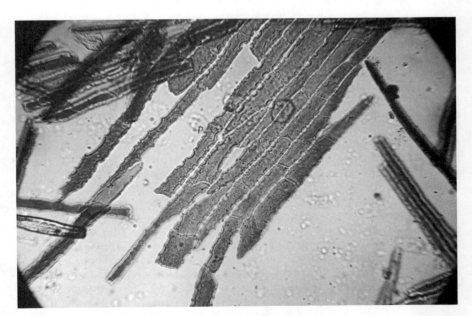

Fig. 10. Corn pericarp cell types, showing interlocking nature and numerous pits (P). Separated by soaking in 10% potassium hydroxide for 5 hr at room temperature. (×175)

development. Kiesselbach (1949) says that the endosperm starts to develop first (before the embryo) within 3–5 hr after fertilization by nuclear divisions only. About 50 hr after fertilization, when there are 128–256 free nuclei, cell walls form between nuclei; then cell division continues until the entire endosperm vacuole is filled with cells. Kiesselbach (1949) states:

> For some time after the endosperm becomes cellular, cell divisions occur throughout but soon cease, or become rare on the interior, divisions then being confined to cells of the peripheral zone, several cells deep.

Growth then continues by division from this peripheral meristem layer. Cell division ceases about the 28th day after fertilization (DAF), as judged by maximum DNA content in the endosperm (Fig. 11, part 6). Endosperm growth then continues to maturity by cell enlargement to accommodate the starch and storage protein synthesized. The meristem layer later differentiates into the mature aleurone layer. Kiesselbach (1949) observed that

> starch formation begins two weeks or less after fertilization. The first cells to show starch are in the upper crown part of the kernel and this formation progresses toward the basal part of the endosperm.

Starch content of the endosperm, expressed either as percent of total dry matter (Evans, 1941; Wolf et al, 1948; Chapter 8) or on a kernel weight basis (Fig. 11-1) (Cerning and Guilbot, 1971), increases very rapidly in the next 14 days and levels off at about 30 DAF. Sugar content (Fig. 11-7) accumulates to 15 DAF and then declines as starch deposition accelerates. Synthesis of proteins in the endosperm (Fig. 11-3) is initiated about the same time (Ingle et al, 1965), starts out fast, slows down as starch synthesis levels off, but accelerates in the later stage of kernel development (Fig. 11-3), coincident with the rapid decline of free amino acids (Fig. 11-8). Bressani and Conde (1961) and Watson (1949) reported that total protein content declines to about 37 DAF, due to faster synthesis of starch than protein, and levels off or increases slightly. The peak in RNA content at about 25 DAF (Fig. 11-5) suggests that enzyme protein synthesis has been largely completed. The deposition of high-molecular-weight alkali-soluble proteins known as glutelins, which make up the continuous phase in the protein matrix surrounding starch granules, is completed in 30–35 DAF (Watson 1949; Bressani and Conde, 1961). Matrix formation in corn endosperm is probably similar to that observed for wheat (Bechtel and Barnett, 1986), in which (storage) protein gradually surrounds starch granules and remnants of cytoplasm in the later stages of endosperm maturity. Although Bressani and Conde (1961) show that the alcohol-soluble storage protein, zein, accumulates at a steady rate throughout kernel development, other data (Watson, 1949) indicate that the major deposition occurs after 30 DAF. Zein synthesis may account for the second spurt in protein synthesis shown in Fig. 11-3.

The corn embryo is much slower to develop than is the endosperm (Kiesselbach, 1949). At 13 DAF, the embryo is an undifferentiated mass of cells showing faint outlines of future subdivisions. At 21 DAF, the embryo is still very small in relation to the rapidly growing endosperm, but the plumule and radical (root) are definite and distinct. Kiesselbach (1949) claims that "seed carefully dried at this stage may germinate." From this stage on, development is rapid to maturity. Deposition of lipids begins at about this time (Fig. 11-4) and continues at a steady state to maturity (Evans, 1941). Lipid deposition may continue after starch and protein have stopped and until the plant dies. Note that RNA content of the embryo (Fig. 11-5) increases to maturity in the embryo and is available for rapid enzyme production during germination.

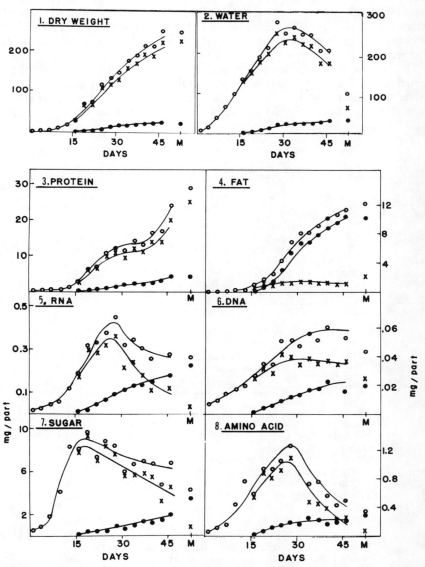

Fig. 11. Changes in dry weight, water, and components of the embryo and endosperm over a 46-day development period after pollination. M; comparable data from the analysis of mature grains; o-o, whole grains; X-X, endosperms; •-•, embryo. (Adapted from Ingle et al, 1965)

B. Distribution in the Mature Kernel

As already described, the mature corn kernel has four easily separable parts: tip cap, pericarp, endosperm, and germ. Table II gives data on the average composition of each of these parts with respect to starch, fat, protein, ash, and sugar for seven U.S. dent hybrid corns. Table III expresses the percentage of each constituent in the kernel parts. The major component of corn is starch, of which 98% is in the endosperm (Earle et al, 1946). On the whole-kernel basis, starch content is 72–73% (Tables II and IV). The endosperm also contains 74% of the kernel protein, of which the majority is insoluble storage proteins (see Chapter 9).

The germ is the major depository of lipids, which amount to 83% of the total kernel lipids. The greater part of the germ lipids are triacylglycerides, which, on extraction, give the well-known corn oil of commerce (Chapters 10 and 18). The germ, being potentially metabolically active tissue, contains 70% of the kernel

TABLE III
Percentage of Total Indicated Constituent in Specified Fraction[a]

Part	Starch	Fat	Protein	Ash	Sugar	Unaccounted For
Endosperm	98.1	15.4	73.8	17.9	28.9	26
	97.8–98.7	13.3–17.4	69.5–78.9	12.6–23.3	23–37.3	
Germ	1.5	82.6	26.2	78.4	69.3	12
	0.7–1.7	80.9–85.0	18.4–27.8	72.4–83.3	60.8–75.1	
Bran	0.6	1.3	2.6	2.9	1.2	54
	0.4–0.7	0.8–1.7	1.4–2.6	0.9–3.6	0.7–1.7	
Tip cap	0.1	0.8	0.9	1.0	0.8	7.0
	...	0.4–1.0	0.5–1.2	0.7–1.6	0.6–1.1	

[a] Averages of samples 1–6 and 8 in Earle et al (1946).

TABLE IV
Proximate Analysis of Corn Grain

Characteristic	Range[a]	Average[b]
Moisture (%, wet basis)	7–23	16.0
Starch (%, dry basis)	61–78	71.7
Protein[c] (%, dry basis)	6–12	9.5
Fat (%, dry basis)	3.1–5.7	4.3
Ash (oxide) (%, dry basis)	1.1–3.9	1.4
Pentosans (as xylose) (%, dry basis)	5.8–6.6	6.2
Fiber (neutral detergent residue) (%, dry basis)	8.3–11.9	9.5
Cellulose + lignin (acid detergent residue) (%, dry basis)	3.3–4.3	3.3
Sugars, total (as glucose) (%, dry basis)	1.0–3.0	2.6
Total carotenoids (mg/kg)	5–40	30

[a] Numerous sources, including Miller (1958) and unpublished data.
[b] Moisture, starch, protein, and fat values are averages of corn purchased on the open market during 1980–1984 in Illinois, Iowa, and Indiana. Unpublished data.
[c] N × 6.25.

sugar and 26% of the kernel protein. Most of the germ proteins are albumins or globulins and probably are components of the enzymatic apparatus of the cells.

C. Minerals

Corn germ is rich in mineral elements (ash), as shown in Table II; it contains 78% of the kernel minerals (Table III), probably because they are essential for early growth of the embryo. The most abundant inorganic component is phosphorus (Table V). It is largely present as the potassium-magnesium salt of phytic acid—the hexaphosphate ester of inositol. Phytin is an important storage form of phosphorus (Hamilton et al, 1951; O'Dell et al, 1972), which is liberated by phytase enzymes to initiate embryo development. Sulfur, the fourth most abundant element in corn, is largely present in organic form as a constituent of the amino acids methionine and cystine. The toxic heavy metals are present at levels far below a level causing animal toxicity unless the corn is grown on land amended with sewage sludge high in such metals (Hinesley et al, 1978). Corn is an important source of the essential element selenium in animal rations (Chapter 15).

D. Vitamins

Corn contains two fat-soluble vitamins, A (β-carotene) and E, and most of the water-soluble vitamins (Table VI; see also Chapter 15). The β-carotene content

TABLE V
Inorganic Components of Dent Corn Grain (dry basis)[a]

Component	Range	Average
Total ash (oxide), %	1.1–3.9	1.42
Phosphorus[b]		
Total, %	0.26–0.75	0.29
Inorganic, %	0.01–0.2	0.08
Potassium, %	0.32–0.72	0.37
Magnesium, %	0.09–1.0	0.14
Sulfur, %	0.01–0.22	0.12
Chlorine, %	...	0.05
Calcium, %	0.01–0.1	0.03
Sodium, %	0.0–0.15	0.03
Iodine, mg/kg	73–810	385.0
Iron, mg/kg	1–100	30.0
Zinc, mg/kg	12–30	14.0
Fluorine, mg/kg	...	5.4
Manganese, mg/kg	0.7–54	5.0
Copper, mg/kg	0.9–10	4.0
Lead, mg/kg	0.2–0.3	0.27
Cadmium, mg/kg	0.04–0.15	0.07
Chromium, mg/kg	0.06–0.16	0.07
Selenium, mg/kg	0.01–1.00	0.08
Cobalt, mg/kg	0.003–0.34	0.05
Mercury, mg/kg	0.002–0.006	0.003

[a] From Patrias and Olson (1969), Miller (1958), Anonymous (1969, 1982), Schneider et al (1953).
[b] Largely phytin phosphorus.

of corn is genetically variable in inbreds and hybrids. It is gradually destroyed by oxidation, along with the other carotenoid pigments, during prolonged storage (Watson, 1962).

The water-soluble vitamins thiamine (B_1) and pyridoxine are present in concentrations sufficient to be important in animal rations. Niacin is present at high concentrations in the bound form, which is mostly unavailable to monogastric animals. However, treatment with alkali, such as in masa preparation (Chapter 13), makes it available (McDaniel and Hundley, 1958).

Other minor components, which usually are not classed as vitamins but which have a major role in embryo growth, are the heterocyclic compounds adenine (5.4 mg/kg), adenosine (9.6 mg/kg), uridine (18.3 mg/kg), and uracil (2.8 mg/kg) (Christianson et al, 1965). Corn is a fairly rich source of the plant growth hormone indoleacetic acid (IAA, or auxin), which is present mainly as sugar esters and peptides. Bandurski and Schultze (1977) found free IAA at 0.5–1.0 mg/kg and ester IAA at 71.6–78.6 mg/kg. Quaternary compounds (betaine, trigonelline, and choline) are present in corn at microgram levels, mainly in the germ (Christianson et al, 1965).

E. Nitrogenous Components

Table IV gives single values for the composition of whole corn; these can be used as a general average. However, well-documented evidence shows that the protein content is influenced by the available soil nitrogen (Hamilton et al, 1951; Schneider et al, 1953; Pierre et al, 1977) and by genetics (Alexander and Creech, 1977). Dudley et al (1974) have shown that the total protein content of corn can vary from 4.4 to 26.6% by selection for protein content without regard to grain yield. Application of fertilizer to normal dent corn can increase the total protein content to about 11.5%, but growing corn in soil depleted of nitrogen can produce kernels with as low as 6% protein (Pierre et al, 1977). Differences in kernel protein and yield response to soil nitrogen levels may also be influenced by genetic differences in the capacity of genotypes to take up nitrogen from the

TABLE VI
Vitamin Content of Dent Corn (dry basis)[a]

Vitamin	Range	Average[b]
Vitamin A, mg/kg	...	2.5
Vitamin E, IU/kg[c]	17–47	30
Thiamine, mg/kg	3.0–8.6	3.8
Riboflavin, mg/kg	0.25–5.6	1.4
Pantothenic acid, mg/kg	3.5–14	6.6
Biotin, mg/kg	...	0.08
Folic acid, mg/kg	...	0.3
Choline, mg/kg	...	567
Niacin, mg/kg	9.3–70	28
Pyridoxine, mg/kg	...	5.3

[a] From Miller (1958), Anonymous (1969, 1982).
[b] Best values, Anonymous (1982).
[c] Calculated by multiplying 1.49 × (mg of α-tocopherol/μg + 0.1 γ-tocopherol/kg); Combs and Combs (1985). One IU = 1 mg of standard DL-α-tocopherol.

soil and translocate it to the sink kernels (Pollner et al, 1979). Nitrogen availability to the plant can also be reduced by competition from weeds and by drought. Table IV gives the entire range of protein contents found in all kinds of corn, but in the United States, the range for bulked, normal dent corn is probably not much more than ± 1.0% from the average value. For example, corn from the eastern Nebraska-western Iowa area generally averages 0.5% higher protein contents than corn grown in Illinois. Differences from farm to farm show a wider range, probably amounting to ± 2% protein, due to different amounts of soil nitrogen available to the plant. Speculations that the protein of corn has been declining since the introduction of hybrid corn were refuted by Earle (1977), who found no significant trend in protein from 1907 to 1972 except for a period of six years between 1930 and 1936 when protein content was high, probably due to drought that prevented full kernel growth. Other data obtained from corn millers in the central Corn Belt indicate that no change occurred from 1976 to 1984. The single-value data in Table IV for protein, oil, and starch represent average analyses from millions of bushels of corn used by several corn wet-milling companies in the Midwest over the last five years.

Changes in total protein content are primarily changes in endosperm protein content, mainly zein (Hamilton et al, 1951). As the protein content increases, the amount of horny endosperm increases. Hinton (1953) reported that horny endosperm was 1.5–2.0% higher in protein percentage than floury endosperm in a flint corn of European origin, but the amount is closer to 1.0% higher in dent corn. The dense subaleurone endosperm region contained 28% protein in Hinton's (1953) material and constituted 3.9% of the endosperm. These cells contain small starch granules and a very thick protein matrix that can cause difficulty in starch purification (Watson et al, 1955). The total protein of corn is a complex of numerous protein molecules, each of which are polymers of amino acids. The protein and amino acid compositions are described in Chapter 9.

F. Lipids

The lipid content of corn is influenced mainly by genetics (Dudley et al, 1974; see Chapter 10) but not by fertility unless nutrients are severely restricted. Corn that does not mature properly may contain less oil than normal, but if the growing season has been ideal during kernel development and full maturation is attained with no killing frost until late in the season, the fully mature kernels can have up to 0.5% higher oil content than normal. Although good-yielding hybrids with as high as 7–8% oil content have been produced (Watson and Freeman, 1975; Alexander and Creech, 1977), grain yields are not optimum. Higher-oil hybrids are produced for special purposes, such as a superior swine feed, or for use by corn starch manufacturers reaching for higher oil yield. Consequently, the goal in development of commercial corn hybrids has been for maximum grain yield, with the result that oil content genes have been allowed to randomly segregate. As the grain yield of Corn-Belt hybrids has increased, the oil percentage has declined (Fig. 12). During the years 1954–1982, the average yield of corn in Illinois increased from 2.5 t/ha (41 bu/acre) to 7.28 t/ha (116 bu/acre). A plot of oil content vs bushels per acre has the same slope and follows close to the same pattern as the oil per year plot. Two of the three sets of data presented by Earle (1977) show trends similar to the data in Fig. 12. During these same years, the linoleic acid content of corn oil steadily increased from about 55–56% in 1954 to 60–62% in 1982.

G. Carbohydrates

Starch is the major carbohydrate in corn, making up 72–73% of the kernel. Sugars, present mainly as sucrose, glucose, and fructose, amount to only 1–3% of the kernel (Table II). The polymeric carbohydrates are described in Chapter 8.

The fiber content of corn kernels given in Table IV was determined by the neutral detergent extraction method of Goering and Van Soest (1970) amended by inclusion of an amylase to obtain better starch removal (AACC, 1983). The resulting value is termed neutral detergent fiber (NDF), insoluble fiber, or "dietary fiber." This method now replaces the crude fiber method that was in use for so many years. The crude fiber method has been discredited because it underestimates cell wall content due to hydrolysis of hemicellulose and some cellulose. For comparison, the crude fiber value for whole corn is about 2.5%. Some investigators contend that NDF is not equivalent to dietary fiber because soluble, indigestible polysaccharides, which may be present, also have important intestinal bulking properties. Unlike whole wheat, in which soluble fiber amounts to 11% of total fiber, the soluble fiber content of corn is negligible. However, in corn meal (endosperm), soluble fiber is about 0.5%, or 12% of the 4.1% total fiber present (Table VII; Nyman et al, 1984). The many methods that have been proposed for analysis of both soluble and insoluble fiber are reviewed by Wisker et al (1985).

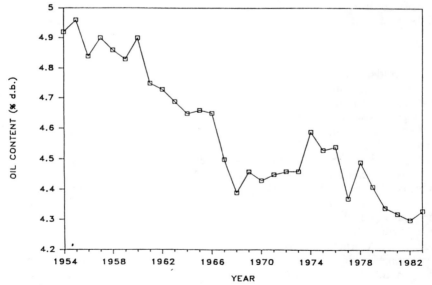

Fig. 12. Decline in total oil content of corn from the upper central Corn Belt (northern Illinois, southern Wisconsin, western Indiana, and northeast Iowa) from 1954 to 1982. (Courtesy C.P.C. International, Inc., Argo, IL)

The acid detergent fiber procedure solubilizes all pentosans and other hemicelluloses, leaving only cellulose and lignin. Sandstead et al (1978) reported 0.1% lignin in corn bran by the acid detergent fiber procedure, which suggests that a value of 1.4% for whole corn by the Keason chemical method (Nyman et al, 1984) is erroneous. Van Soest et al (1979) found 0.2% lignin in whole corn, suggesting a small lignin content in germ. Lignified cells in the scutellum are seen as spiral conducting cells in Fig. 4.

Table VII gives the best available estimates for the distribution of NDF and fiber components among kernel parts. The pericarp and tip cap make up 80% of the kernel total fiber. The material unaccounted for in Table II is largely accounted for by the NDF values. New information on the importance of fiber in proper bowel function in humans and its relation to prevention of several diseases (Kies et al, 1982) has brought forth a plethora of food products containing added fiber. As a result, a food-grade corn bran is being offered by some dry millers at an NDF level of 50–65%, and at least one wet miller offers a purified corn bran containing 88–92% NDF.

Extraction of pericarp with hot, dilute alkali produces a 53% yield of a rather homogeneous xylan-araban polysaccharide (Wolf et al, 1953). This corn hull gum, as it has been called (Watson, 1959), is readily purified by precipitation in methanol. This substance is structurally a polymer composed of 50–54% D-xylose units linked by α-(1→3) bonds into an elongated molecule having short side chains containing some α-D-xylose, L-arabinose (33–35%), β-DL-galactose (8–11%), and α-D-glucuronic acid (3–6%). This xylan is apparently linked to the cellulose skeleton in the cell wall by ester linkage cross-bonding through ferulic acid (4-hydroxy-3-methycinnamic acid) and diferulic acid (Markwalder and

TABLE VII
Fiber Components of Corn Kernel Parts

Part	Percent of Kernel Dry Substance[a]	Insoluble Neutral Detergent Fiber	Hemi-cellulose	Cellulose[b]	Lignin[b]	Soluble Fiber	Total Fiber[c]	Percentage of Kernel Fiber[a]
Whole kernel	100	9.5[d]	6.7	3.0	0.2	0.1	9.5	100
Starchy endosperm	87[e]	1.0[b]	...	...	...	0.5[b]	1.5	12
Aleurone endosperm	2.0[f]	50.0[g]	...	...	...	25[g]	75	15
Germ	11.0	11.0[d]	18[d]	7[d]	1.0[g]	3.0[g]	14	16
Pericarp (bran)	5.3	90	67[d]	23[d]	0.1[h]	0.2	90.7	51
Tip cap	0.8	95[g]	70[g]	...	2.0[g]	...	95	0.1

[a] Based on data in Table II.
[b] Data of Van Soest et al (1979) and Nyman et al (1984).
[c] Insoluble neutral detergent fiber plus soluble fiber.
[d] Unpublished data, S. A. Watson.
[e] Data from Table II minus aleurone estimate of 2.0%.
[f] Hinton (1953).
[g] Estimate by author.
[h] Sandstead et al (1978).

Neukom, 1976). Boundy et al (1967b) isolated a small amount of a mucopolysaccharide or glycopeptide from corn pericarp that was extractable by rigorous treatment with alkali or trichloroacetic acid. It is composed of a peptide containing 17 amino acids, a hexosamine, and hydroxyproline. The latter bifunctional molecule has been suggested as a cross-linking agent between the glycopeptide, the xylan, and the cellulose framework of cell walls. The mucopolysaccharide is apparently an integral part of the cell wall matrix (see Chapter 8). Investigations on the structure of plant cell walls is currently a very active field of research that is revealing a very complex structure (Fincher and Stone, 1986).

H. Miscellaneous Substances

In addition to the substances already mentioned, corn contains many other chemical compounds and materials in very low concentrations. Many of these substances, such as enzymes or their precursors, are vital to the growth of the embryo. The roles of other substances, such as a low level of a trypsin inhibitor (Melville and Scandalios, 1972; Halim et al, 1973) are not understood. Mature, dry corn does have a low level of α-amylase activity, which increases on germination (Chapman, 1976). α-Amylase is absent in the dry grain but develops during germination exclusively in the scutellum (Dure, 1960) as well as in deembryonated kernels at a much lower level than in barley and without the assistance of an embryo factor. Lectins, which are carbohydrate binding proteins with hemagglutination activity, have been shown to be present in corn endosperm and germ (Newberg and Concon, 1985).

VI. SUMMARY

The corn kernel, like other seeds, is a uniquely packaged storage organ that contains all of the necessary components required to propagate the species. Fortunately for the human race, it also contains high levels of starch, protein, oil, and other nutritionally valuable substances. Scientific investigation of the macro and micro structural and chemical properties of corn has improved our ability to utilize it in many food and industrial products. Furthermore, the knowledge of structural and chemical component relationships continues to assist those who seek to improve corn for its numerous uses through plant breeding. Corn's primary use is as a source of energy in human foods and animal feeds because of its high content of starch. This property sustains a large animal industry in the United States and in many other parts of the world and also makes corn a basic food in Latin North America, South America, and parts of Africa and Asia.

LITERATURE CITED

AACC. 1983. Approved Methods of the American Association of Cereal Chemists, 8th ed. Method 32-20, first approval October 1977; final approval October 1982. The Association, St. Paul, MN.

ALEXANDER, D. E., and CREECH, R. G. 1977. Breeding special industrial and nutritional types. Pages 363-390 in: Corn and Corn Improvement, 2nd ed. G. W. Sprague, ed. Am. Soc. Agronomy, Madison, WI.

ANONYMOUS. 1969. United States-Canadian Tables of Feed Composition. Natl. Acad. Sci., Washington, DC.

ANONYMOUS. 1982. US-Canadian Tables of

Feed Composition, 3rd ed. Natl. Academy Press, Washington, DC. 148 pp.
ANONYMOUS. 1983. ASAE data on grain storage. Pages 294-313 in: Agricultural Engineers Yearbook. Am. Soc. Agric. Eng., St. Joseph, MI.
BANDURSKI, R. S., and SCHULTZE, A. 1977. Concentration of indole-3 acetic acid and its derivatives in plants. Plant Physiol. 60:211-213.
BECHTEL, D. B., and BARNETT, D. B. 1986. A freeze fracture study of storage protein accumulation in unfixed wheat starchy endosperm. Cereal Chem. 63:232-240.
BRESSANI, R., and CONDE, R. 1961. Changes in the chemical composition and in the distribution of nitrogen of maize at different stages of development. Cereal Chem. 38:76-84.
BOUNDY, J. A., TURNER, J. E., WALL, J. S., and DIMLER, R. J. 1967a. Influence of commercial processing on composition and properties of corn zein. Cereal Chem. 44:281-287.
BOUNDY, J. A., WALL, J. S., TURNER, J. E., WOYCHICK, J. H., and DIMLER, R. J. 1967b. A muco-polysaccharide containing hydroxy-proline from corn pericarp: isolation and composition. J. Biol. Chem. 242:2410-2415.
BROOKER, D. H., BAKKER-ARKEMA, F. W., and HALL, C. W. 1974. Drying Cereal Grains. AVI Publishing Co., Westport, CT.
CERNING, J., and GUILBOT, A. 1971. Die Entwicklung hochmolekularer Kohlenhydrate im Maizkorn wahrend des Reifeprozesses. Staerke 23:238-244.
CHAPMAN, A. J. 1976. Investigations of the nature of amylase enzymes from incubated, deembryonated *Zea mays* kernels. M.S. thesis, Univ. of Massachusetts, Amherst.
CHRISTIANSON, D. D., WALL, J. S., and CAVINS, J. R. 1965. Location of nonprotein nitrogenous substances in corn grain. J. Agric. Food Chem. 13:272-280.
CHRISTIANSON, D. D., NIELSEN, H. C., KHOO, U., WOLF, M. J., and WALL, J. S. 1969. Isolation and chemical composition of protein bodies and matrix protein in corn endosperm. Cereal Chem. 46:372-381.
CHUNG, D. S., and CONVERSE, H. H. 1971. Effect of moisture on some physical properties of grains. Trans. ASAE 14:612-614, 620.
COMBS, S. B., and COMBS, G. P. 1985. Varietal differences in the vitamin E content of corn. J. Agric. Food Chem. 33:815-817.
COX, M. J., MacMASTERS, M. M., and HILBERT, G. E. 1944. Effect of the sulfurous acid steep in corn wet milling. Cereal Chem. 21:447-465.
CRAIG, W. F. 1977. Production of hybrid corn seed. Pages 700-722 in: Corn and Corn Improvement. G. F. Sprague, ed. Am. Soc. Agron., Madison, WI.
DAYNARD, T. B., and DUNCAN, W. G. 1969. The black layer and grain maturity. Crop Sci. 9:473-476.
DUDLEY, J. W., LAMBERT, R. J., and ALEXANDER, D. E. 1974. Seventy generations of selection for oil and protein concentration in the maize kernel. Pages 181-212 in: Seventy Generations of Selection for Oil and Protein in Maize. J. W. Dudley, ed. Crop Sci. Soc. Am., Madison, WI.
DURE, L. S. 1960. Site of origin and extent of activity of amylase in maize germination. Plant Physiol. 35:925-934.
DUVICK, D. N. 1961. Protein granules in maize endosperm cells. Cereal Chem. 38:374-385.
EARLE, F. R. 1977. Protein and oil in corn: Variation by crop years from 1907 to 1972. Cereal Chem. 54:70-79.
EARLE, F. R., CURTIS, J. J., and HUBBARD, J. E. 1946. Composition of the component parts of the corn kernel. Cereal Chem. 23:504-511.
EARP, C. F., McDONOUGH, C. M., and ROONEY, L. W. 1985. Changes in the protein matrix and endosperm cell walls of corn during lactic and sulfurous acid steeping. (Abstr.) Cereal Foods World 30:545.
EVANS, J. W. 1941. Changes in the biochemical composition of the corn kernel during development. Cereal Chem. 18:468-473.
FAN, L. T., CHU, P. S., and SHELLENBERGER, J. A. 1962. Volume increase of kernels of corn and sorghum accompanying absorption of liquid water. Biotechnol. Bioeng. 4:311-322.
FINCHER, G. B., and STONE, B. A. 1986. Cell walls and their components in cereal grains. Pages 207-295 in: Advances in Cereal Science and Technology, Vol. 8. Y. Pomeranz, ed. Am. Assoc. Cereal Chem., St. Paul, MN.
GOERING, H. K., and VAN SOEST, P. J. 1970. Forage Fiber Analysis. Agriculture Handbook 379. U.S. Dept. Agric., Agric. Res. Serv., Washington, DC. 20 pp.
GOODMAN, M. M. 1978. A brief survey of the races of maize and current attempts to infer racial relationships. Pages 143-158 in: Maize Breeding and Genetics. D. B. Walden, ed. John Wiley & Sons, New York.
GURR, M. I. 1980. The biogenesis of triglycerides. Pages 205-248 in: The Biochemistry of Plants. A Comprehensive Treatise. Vol 4, Lipids: Structure and Function. P. K. Stumpf

and E. C. Conn, eds. Academic Press, New York.

HALIM, A. H., WASSOM, C. E., and MITCHELL, H. L. 1973. Trypsin inhibitor in corn (Zea mays L.) as influenced by genotype and moisture stress. Crop Sci. 13:405-406.

HAMILTON, T. S., HAMILTON, B. C., JOHNSON, B. C., and MITCHELL, H. H. 1951. The dependence of the physical and chemical composition of the corn kernel on soil fertility and cropping systems. Cereal Chem. 28:163-176.

HELM, J. L., and ZUBER, M. S. 1969. Pericarp thickness of dent corn inbred lines. Crop Sci. 9:803-804.

HELM, J. L., and ZUBER, M. S. 1972. Inheritance of pericarp thickness in corn belt maize. Crop Sci. 12:428-430.

HINESLEY, T. D., ALEXANDER, D. E., ZIEGLER, E. L., and BARRETT, G. L. 1978. Zinc and cadmium accumulation by corn inbreds grown on sludge amended soil. Agron. J. 70:425-428.

HINTON, J. J. C. 1953. The distribution of protein in the maize kernel in comparison with that of wheat. Cereal Chem. 30:441-445.

INGLE, J., BIETZ, D., and HAGEMAN, R. H. 1965. Changes in composition during development and maturation of maize seeds. Plant Physiol. 40:832-835.

KEENER, H. M., HENRY, J. E., SCHONAUER, S. L., and ANDERSON, R. J. 1985. Burning shelled corn as an alternate fuel. Ohio Report, July-August, pp. 60-62.

KIES, C., BALTERS, S., KAN, S., LO, B., WESTRING, M. E., and FOX, H. M. 1982. Impact of corn bran on nutritional status and on gastrointestinal tract function of humans. Pages 33-44 in: Maize: Recent Progress in Chemistry and Technology. G. E. Inglett, ed. Academic Press, New York.

KIESSELBACH, T. A. 1949. The structure and reproduction in corn. Neb. Agric. Exp. Stn. Res. Bull. 161:3-96.

KIRLEIS, A. W., CROSBY, K. D., and HOUSLEY, T. L. 1984. A method for quantitatively measuring vitreous endosperm area in sectioned sorghum grain. Cereal Chem. 61:556-558.

MARKWALDER, H. U., and NEUKOM, H. 1976. Diferulic acid as a possible cross link in hemicelluloses from wheat germ. Phytochemistry 15:836-837.

McDANIEL, E. G., and HUNDLEY, J. M. 1958. Alkali treated corn and niacin deficiency. (Abstr.) Fed. Proc., Fed. Am. Soc. Exp. Biol. 17:1897.

MELVILLE, J. C., and SCANDALIOS, J. G. 1972. Maize endopeptidase: Genetic control, chemical characterization, and relationship to an endogenous trypsin inhibitor. Biochem. Genet. 7:15-31.

MILLER, D. F. 1958. Composition of cereal grains and forages. Publ. 585. Natl. Acad. Sci., Washington, DC.

NELSON, O. E. 1978. Gene action and endosperm development in maize. Pages 389-401 in: Maize Breeding and Genetics. D. B. Walden, ed. John Wiley & Sons, New York.

NELSON, O. E., MERTZ, E. T., and BATES, L. S. 1965. Second mutant gene affecting the amino acid pattern of maize endosperm proteins. Science 150:1469-1470.

NELSON, S. O. 1980. Moisture-dependent kernel- and bulk density for wheat and corn. Trans. ASAE 23:139-143.

NEUFFER, M. G., JONES, L., and ZUBER, M. S. 1968. The Mutants of Maize. Crop Sci. Soc. Am., Madison, WI. 74 pp.

NEWBERG, D. S., and CONCON, J. M. 1985. Lectins in rice and corn endosperm. J. Agric. Food Chem. 33:685-687.

NYMAN, M., SILJESTRÖM, M., PEDERSEN, B., BACH KNUDSEN, K. E., ASP, N.-G., JOHANSSON, C.-G., and EGGUM, B. O. 1984. Dietary fiber content and composition in six cereals at different extraction rates. Cereal Chem. 61:14-19.

O'DELL, B. L., de BOLAND, A. R., and KOIRTYOHANN, S. R. 1972. Distribution of phytate and nutritionally important elements among morphological components of cereal grains. J. Agric. Food Chem. 20:718-721.

ORTEGA, E. I., and BATES, L. S. 1983. Biochemical and agronomic studies of two modified hard-endosperm opaque-2 maize (Zea mays L.) populations. Cereal Chem. 60:107-111.

PATRIAS, G., and OLSON, O. E. 1969. Selenium contents of samples of corn from midwestern states. Feedstuffs, Oct. 25, pp. 32-33.

PIERRE, W. H., DUMENIL, L., JOLLEY, V. D., WEBB, J. R., and SHRADER, W. D. 1977. Relationships between corn yield, expressed as percentage of maximum, and the N percentage in the grain. 1. Various N-rate experiments. Agron. J. 69:215-220.

POLLNER, W. G., EBERHARD, D., KLEIN, D., and DHILLON, B. S. 1979. Genetic control of nitrogen uptake and translocation in maize. Crop Sci. 19:82-86.

POMERANZ, Y. 1973. Structure and mineral composition of cereal aleurone cells as shown by scanning electron microscopy. Cereal Chem. 50:504-511.

POMERANZ, Y., CZUCHAJOWSKA, Z., MARTIN, C. R., and LAI, F. S. 1985. Determination of corn hardness by the Stenvert hardness tester. Cereal Chem. 62:108-112.

POMERANZ, Y., CZUCHAJOWSKA, Z., and LAI, F. S. 1986a. Gross composition of coarse and fine fractions of small corn samples ground on the Stenvert hardness tester. Cereal Chem. 63:22-26.

POMERANZ, Y., HALL, G. E., CZUCHAJOWSKA, Z., and LAI, F. S. 1986b. Test weight, hardness, and breakage susceptibility of yellow dent corn hybrids. Cereal Chem. 63:349-351.

RATKOVIC, S., DENIC, M., and LAHAJNAR, G. 1982. Kinetics of water imbibition by seed: Why normal and opaque-2 maize kernels differ in their hydration properties. Period. Biol. 84:180-182.

REINERS, R. A., PRESSICK, J. C., URQUIDI, R. L., and WARNECKE, M. O. 1972. Corn protein concentrates for food use. Proc. Int. Congr. of Nutrition, 9th, Mexico City, Mexico. Sept. 3-9.

ROBUTTI, J. L., HOSENEY, R. C., and DEYOE, C. W. 1974a. Modified opaque-2 corn endosperms. I. Protein distribution and amino acid composition. Cereal Chem. 51:163-172.

ROBUTTI, J. L., HOSENEY, R. C., and WASSOM, C. E. 1974b. Modified opaque-2 corn endosperms. II. Structure viewed with a scanning electron microscope. Cereal Chem. 51:173-180.

SANDSTEAD, H. H., MUNOZ, J. M., JACOB, R. A., KLEVAY, L. M., RECK, S. J., LOGAN, G. M., Jr., DINTZIS, F. R., INGLETT, G. E., and SHUEY, W. C. 1978. Influence of dietary fiber on trace element balance. Am. J. Clin. Nutr. 31:S180-S184.

SAUNDERS, R. M., WALKER, H. G., Jr., and KOHLER, G. O. 1969. Aleurone cells and the digestibility of wheat millfeeds. Poult. Sci. 48:1497-1503.

SCHNEIDER, B. H., LUCAS, H. L., and BEESON, K. C. 1953. Nutrient composition of corn in the United States. J. Agric. Food Chem. 1:172-177.

SECKINGER, H. L., WOLF, M. J., and MacMASTERS, M. M. 1960. Hemicelluloses of the cementing layer and of some cell walls of the corn kernel. Cereal Chem. 37:121-128.

SEMADENI, E. G. 1967. Enzymatische Characterisierung der Lysosomquivalente (Sphaerosome) von Maiskeimlingen. Planta 72:91-118.

SPRAGUE, G. F. 1977. Corn and Corn Improvement. Am. Soc. Agron., Madison, WI. 774 pp.

STEWART, C. W., and WATSON, S. A. 1969. Corn degermination process. U.S. patent 3,474,722, assigned to Corn Products Co., New York.

SYARIEF, A. M., GUSTAFSON, R. J., and MOREY, R. V. 1984. Moisture diffusion coefficients for yellow-dent corn components. Paper 84-3551. Am. Soc. Agric. Eng., St. Joseph, MI.

THOMPSON, R. A., and ISAACS, G. W. 1967. Porosity determination of grains and seeds with an air comparison pycnometer. Trans. ASAE 10:693-696.

VAN SOEST, P. J., FADEL, J., and SNIFFEN, C. J. 1979. Discount factors for energy and protein in ruminant feeds. Pages 63-75 in: 1979 Cornell Nutrition Conf. Cornell Univ., Ithaca, NY.

VASAL, S. K., VILLEGAS, E., BJARNSON, M., GELEW, B., and GOERTZ, P. 1980. Genetic modifiers and breeding strategies in developing hard endosperm opaque-2 materials. Pages 37-73 in: Improvement of Quality Traits of Maize for Grain and Silage Use. W. G. Pollmer and R. H. Phipps, eds. Martinus Nijhoff Publishers, Amsterdam.

WATSON, S. A. 1949. An agronomic and biochemical comparison of four strains of corn which differ widely in total grain nitrogen. Ph.D. thesis, Univ. of Illinois, Urbana.

WATSON, S. A. 1959. Corn hull gum. Pages 299-306 in: Industrial Gums. R. L. Whistler, ed. Academic Press, New York.

WATSON, S. A. 1962. Yellow carotenoid pigments of corn. Pages 92-100 in: Proc. Annu. Hybrid Corn Industry Res. Conf., 17th. Am. Seed Trade Assoc., Washington, DC.

WATSON, S. A. 1984. Corn and sorghum starches: Production. Pages 417-468 in: Starch: Chemistry and Technology, 2nd ed. R. L. Whistler, J. N. BeMiller, and E. F. Paschall, eds. Academic Press, Orlando, FL.

WATSON, S. A., and FREEMAN, J. E. 1975. Breeding corn for increased oil content. Pages 251-275 in: Proc. Annu. Corn and Sorghum Res. Conf., 30th. Am. Seed Trade Assoc., Washington, DC.

WATSON, S. A., and SANDERS, E. H. 1961. Steeping studies with corn endosperm sections. Cereal Chem. 38:22-33.

WATSON, S. A., and YAHL, K. R. 1967. Comparison of the wet-milling properties of opaque-2 high-lysine corn and normal corn. Cereal Chem. 44:488-498.

WATSON, S. A., SANDERS, E. H., WAKELY, R. D., and WILLIAMS, C. B. 1955. Peripheral cells of the endosperms of grain

sorghum and corn and their influence on starch purification. Cereal Chem. 32:165-182.
WISKER, E., FELDHEIM, W., POMERANZ, Y., and MEUSER, F. 1985. Dietary fiber in cereals. Pages 169-238 in: Advances in Cereal Science and Technology, Vol. 7. Y. Pomeranz, ed. Am. Assoc. Cereal Chem., St. Paul, MN.
WOLF, M. J., MacMASTERS, M. M., HUBBARD, J. E., and RIST, C. E. 1948. Comparison of corn starches at various stages of kernel maturity. Cereal Chem. 25:312-325.
WOLF, M. J., BUZAN, C. L., MacMASTERS, M. M., and RIST, C. E. 1952a. Structure of the mature corn kernel. I. Gross anatomy and structural relationships. Cereal Chem. 29:321-333.
WOLF, M. J., BUZAN, C. L., MacMASTERS, M. M., and RIST, C. E. 1952b. Structure of the mature corn kernel. II. Microscopic structure of pericarp, seed coat, and hilar layer of dent corn. Cereal Chem. 29:334-348.
WOLF, M. J., BUZAN, C. L., MacMASTERS, M. M., and RIST, C. E. 1952c. Structure of the mature corn kernel. III. Microscopic structure of the endosperm of dent corn. Cereal Chem. 29:349-361.
WOLF, M. J., BUZAN, C. L., MacMASTERS, M. M., and RIST, C. E. 1952d. Structure of the mature corn kernel. IV. Microscopic structure of the germ of dent corn. Cereal Chem. 29:362-382.
WOLF, M. J., MacMASTERS, M. M., CANNON, J. A., ROSEWALL, E. C., and RIST, C. E. 1953. Preparation and some properties of hemicelluloses from corn hulls. Cereal Chem. 30:451-470.
WOLF, M. J., MacMASTERS, M. M., and SECKINGER, H. L. 1958. Composition of the cementing layer and adjacent tissues as related to germ-endosperm separation in corn. Cereal Chem. 35:127-136.
WOLF, M. J., KHOO, U., and SECKINGER, H. L. 1969. Distribution and subcellular structure of endosperm protein in varieties of ordinary and high-lysine maize. Cereal Chem. 46:253-263.

CHAPTER 4

HARVESTING AND POSTHARVEST MANAGEMENT

FLOYD L. HERUM
*Department of Agricultural Engineering
Ohio Agricultural Research and Development Center
The Ohio State University
Columbus, Ohio*

I. INTRODUCTION

This chapter describes overall systems for harvesting, drying, handling, and storage of market corn, illustrating relationships that exist between these various components, while integrating and interpreting recent technological developments. Although production and handling of corn is global, this report emphasizes systems that have been developed in the United States. Mechanization in most of the other countries in which corn in commercially produced has tended to follow the same capital-intensive, low-labor pattern. A brief history of harvesting methods is included, as many less-developed countries have neither the technological infrastructure nor the resources to emulate the degree of mechanization of those more developed. One of the more elementary systems described may be the most suitable, at least at this time.

The specialized topics of seed harvest and storage, with special emphasis on quality rather than cost (Justice and Bass, 1978) are not included here.

A. Evolution in Corn Systems

Today's corn production and marketing systems have evolved to compete worldwide with other corn-exporting nations and with alternative feed grains. With per-unit profit margins exceedingly narrow, fewer operators produce and handle greater quantities with larger and more sophisticated equipment and facilities. Figure 1 illustrates a present-day field sheller, designed to harvest up to 30 t/hr (1,200 bu/hr) under ideal conditions; a modern farm crop drying, handling, and storage facility is shown in Fig. 2. These specialized systems have reduced costs per unit of production, but in such a dispersed system, "improvements" adopted at one place in the system often lead to problems at another. Practices that might improve or even maintain original corn quality, regardless of how laudable, generally have not been economically justifiable.

Fig. 1. Six-row field sheller, rated at a capacity of 30 t/hr under ideal harvesting conditions. (Courtesy New Idea Farm Equipment Corp., Coldwater, OH)

Fig. 2. Modern farm crop drying, handling, and storage facility. (Courtesy Meyer Morton Co., Morton, IL)

As documented by many researchers, field shelling and the subsequent rapid drying of corn cause stress cracks and greater breakage susceptibility of corn kernels compared to that resulting from ear harvest and drying (Thompson and Foster, 1963). High-moisture field shelling physically damages kernels, increasing their rate of storage deterioration from two to five times that of gently shelled corn (USDA, 1969). Greater amounts of broken kernels, called fines, are generated in the mechanized handling systems, which were not designed for fragile kernels. Fines contribute to dust explosions and fires, aeration and storage difficulties, reduced yields in milling, screening discounts, and respiratory hazards for workers.

Concentration of marketing into fewer but larger facilities, necessary for unit train, barge, and ship transport, increases the magnitude of dust-related problems and leads to closer scrutiny by regulatory agencies (such as the Environmental Protection Agency and Occupational Safety and Health Administration in the United States) in an environment continually more urban and socially sensitive. Large corn bins and silos, even if adequately designed, require more care in filling, aeration, fumigation, and emptying than the smaller containers they replace.

Finally, the market reflects growing consumer apprehensions about diverse quality factors, such as the presence of live or dead insects, levels of toxins and toxin-forming microflora, pesticide residues, and nutritive characteristics (see Chapters 5 and 6). New techniques and equipment for inspecting, testing, cleaning, segregating, protecting, and detoxifying are being developed to ease these fears, placing additional stresses on management skills and financial resources throughout the production and marketing system.

B. Labor Requirements in Production

Corn harvesting in the United States has become fully mechanized over the past half-century. Labor requirements for corn production diminished slowly from 49 man-hours per metric ton in 1910 to 39 in 1939 but have dropped drastically since then (Johnson and Lamp, 1966). One operator of a present-day six-row field sheller, harvesting as much as 20 t/hr (800 bu/hr), equals the output of 50 workers hand-picking ear corn from rows of standing stalks. Drying capacity appears to be the major impediment to further increases in field harvesting rates and corresponding reductions in labor requirements.

C. Mechanization of Handling and Transportation

Bag handling and storage of cereal crops, once prevalent throughout much of the world, is being replaced by a universal system of bulk handling equipment and storage silos. Most designs, however, were developed before the abrupt increase in corn breakage resulted from adoption of field shelling and artificial drying. At the farmstead, 3.5% broken corn and foreign material (BCFM) resulted from drying and handling corn that was field-shelled at 19% moisture content (MC); the amount increased to 4.4% when corn was harvested at 24% MC (Pierce and Hanna, 1985). The alarming increase in both the number and severity of elevator explosions and fires (National Materials Advisory Board, 1982a) over the past two decades has led to the establishment of industry task groups, such as the Fire and Explosion Research Council, formed by the National Grain and Feed Association in 1978. A national conference on grain elevator design was conducted the following year by the same organization, which also published a guide to elevator design (Natl. Grain Feed Assoc., 1979), and a national symposium was conducted on management and utilization of grain dust (Miller and Pomeranz, 1979). These efforts, building upon studies of actual kernel damage in grain-handling facilities (Fiscus et al, 1971; Foster and Holman, 1973), are leading to the redesigning of structures and handling equipment, identification of ignition sources, better techniques of housekeeping and dust control, and the introduction of risk management techniques.

Reports of increased breakage in overseas shipments were confirmed by Hill et al (1979). In one shipload of corn, BCFM, initially 4.9% upon leaving a Great Lakes port, had increased to 20.3% on a barge at Rotterdam. However, a portion of this increase may have been due to segregation at loading and unloading.

II. HARVESTING METHODS

Present-day harvesting methods reflect developments in machinery design and power availability over the past century, coupled with advances in corn breeding and agronomic practices. Quick and Buchele (1978) provide a detailed and illustrated history of the evolution of corn harvesting techniques and equipment, from simple utensils to today's multirow combines. Field shelling necessitates integration of the producer's combine, transport wagons, and dryer into an orchestrated operation, as drying capacity must equal harvest rates with high-moisture shelled corn. The techniques of operations research are being extended to production agriculture, using current-generation table-top computers that easily handle optimizing procedures too complex for hand calculations. Benock and Loewer (1981) describe a program to model on-farm harvesting, hauling, drying, and storage systems. Such analyses should soon replace trial-and-error selection of machine sizes and numbers.

A. Ear Corn Harvest

Corn was harvested, dried, and stored on the ear until equipment for field shelling and artificial drying was developed less than three decades ago. Even with as much as 24% MC, ear corn generally dries naturally when stored in slatted cribs of recommended widths. Systems for mechanization of ear corn harvest were proposed (Richey and Peart, 1973), but only after field shelling had become well established. Unless drying costs increase substantially, or cobs gain much greater value (such as becoming a source of drying energy), producers presently have little incentive to revert to ear harvesting and storage.

HAND HARVEST
Until the end of the 1800s, corn was harvested entirely by hand, with the help of a variety of tools. Where the entire plants were collected for animal feed, stalks were cut, tied into bundles, and shocked in the field, with ears husked out when dry. Hand-held knives evolved into horse-drawn cutting sleds, which in

turn were replaced by mechanical corn binders and binder-shockers. In drier climates where natural drying in the field proceeded more rapidly, ears were husked from standing stalks in the row and thrown directly into a wagon drawn alongside. The loads of ear corn were then shoveled or elevated into corn cribs for storage and natural drying.

In northern regions, the husking "peg," a metal hook held in the palm of the hand (Quick and Buchele, 1978), greatly speeded husk removal where clean husking was desired. Husks were retained on ears grown in southern regions, reportedly as a shield against insect attack in storage. An experienced husker could hand pick 2.5 t (100 bu) per day from rows standing in the field, but hand husking was considered by most to be an arduous, unpleasant task.

MACHINE HARVEST

The first patent on a mechanical corn snapper was issued in 1850 (Quick and Buchele, 1978), but development of commercially successful snapping or husking machines depended on convenient tractor power. Davidson (1931) states that "power from the tractor motor through the power take-off has greatly added to the usefulness of the machine." By 1928, two-row corn pickers were being sold in considerable numbers. Young (1931) cited problems of field losses and poor machine reliability, as did Shedd (1933), who reported field losses of 9.1–19.3% in 1931 and 1932 field tests, due principally to the inability to recover ears from lodged (fallen) stalks. He stressed the need for genetically improved varieties for mechanical harvesting. Wileman (1933) concurred, reporting an inverse relationship between shank diameter and picker losses. The introduction shortly thereafter of hybrids tailored to meet specific needs greatly eased this problem of field losses.

Pairs of inclined, counter-rotating snapping rolls became the customary devices for detaching ears from the stalks. Although designed to be very aggressive, these often plugged when operating in dry, lodged stalks. In spite of all warnings, many farmers lost fingers, hands, arms, and feet attempting to clear clogged snapping rolls. With 70,000 corn pickers in use, 434 corn picker accidents were reported in Iowa in 1948 (Scranton, 1952). The addition of stripper bars above the snapping rolls, which reduced shelling losses at the rolls, diminished but has not eliminated the risk of such injuries.

Corn pickers and snappers had identical gathering mechanisms. Pickers included additional sets of rolls to remove the husks. Snappers, without such husking beds, were lighter and less expensive.

EAR CORN SHELLING

To reduce its bulk by half and make it relatively free-flowing, so that it can be handled in modern marketing facilities, corn must be shelled from the cob. Cylinder shellers rub and separate the kernels from the cob, usually with aspiration for cleaning. As yet relatively unsophisticated in design, shellers must be rugged and durable. Modern shellers often include low-pressure pneumatic conveyors to dispose of husks and cobs. They are presently manufactured in sizes up to 25 t/hr (1,000 bu/hr), but actual capacities are substantially reduced by higher moisture contents and/or larger proportions of husks.

B. Field Shelling

Although an Australian field sheller was demonstrated in the 1920s (Quick and Buchele, 1978) in a climate where postharvest drying was unnecessary, development of systems for field shelling in the humid regions of the world, which are dependent upon adequate systems for artificial drying, proceeded slowly. McKibben (1929) reported "not very satisfactory" results when he harvested 0.93 ha (2.3 acres) of corn with a cutter-bar combine designed for small grains. With an experimental stalk gathering and cutting assembly on his combine, Logan (1931) experienced no special problems and noted that corn combining success depended only upon economical methods of drying. Whole-plant harvesters placed an extra burden on the separation mechanisms, however, and never became commercially successful.

Retarded by the great depression of the 1930s, field shelling efforts were renewed as World War II approached. A two-row cornpicker in which the husking bed had been replaced by a cylinder sheller was found to be economically feasible (Skelton and Bateman, 1942), even though the price of the wet corn was discounted when the corn was sold directly from the field. Commercial development of picker-shellers proceeded rapidly after 1945, and performance tests of production machines were reported in the early 1950s (Burrough and Harbage, 1953; Herum and Barnes, 1954). Optimum harvesting periods were found to be shorter than those available for ear harvest; gathering

Fig. 3. Combine sieves. These leave substantial proportions of broken kernels in the field.

losses increased while sheller losses diminished as the harvest season progressed.

Pickard (1955) conducted laboratory trials in 1950 showing that the standard rasp-bar cylinder of a grain combine could shell ears of corn. This led to many field studies, including development of better devices to gather the corn, with and without severing the stalks. Production self-propelled combines, fitted with newly developed corn heads, were successfully field-tested by Goss et al (1955) and Hurlbut (1955) and greatly increased the potential usage of the expensive combine.

Although Morrison (1955) had reported effects of combine cylinder speed and adjustment on shelling losses and kernel damage, farm applications of grain combines for corn harvest were not completely successful, due to excessive separation losses and kernel breakage. These were attributed to improper adjustments, excessive cylinder and forward travel speeds, and high moisture contents (Byg and Hall, 1968). As operators were often unaware of either the magnitude or causes of these losses, such as fines left in the field by combine sieves (Fig. 3), educational programs were initiated through state extension services.

Many subsequent studies, such as that by Chowdhury and Buchele (1978), further documented the effects of combine cylinder speed, and concave number and spacing, on damage to corn kernels. Finally developed to provide more gentle threshing action, "rotary" field harvesters are becoming established as "state-of-the-art" for corn and other fragile crops. However, Paulsen and Nave (1980) found little difference in kernel damage between properly adjusted conventional and rotary combines, except for a breakage reduction in the rotary unit at lower moisture contents.

C. Future Trends in Harvest Mechanization

If current world overproduction of corn and low market prices continue, economics will probably exert the greatest influence on changes in harvest mechanization. Field harvesters may have reached a plateau in size, complexity, and cost. Producers prefer machines with ever-larger capacity to reduce their weather risks, but such machines are unwieldy and difficult to utilize effectively in smaller or undulating fields. Rather, single-purpose corn harvesters are envisioned; these would require fewer and simpler adjustments and controls than those needed for multipurpose combines. Such specialized machines, utilizing present-day monitors and microprocessors to provide more precise control than possible by an operator (Brizgis et al, 1980), could reduce kernel damage and separation losses (Wood and Kerr, 1980) over a broad range of corn moisture contents.

When hybrids are developed that maintain yield with earlier maturities, or that naturally dry more rapidly in the field, producers may elect to harvest at lower moisture contents, reducing drying costs substantially, increasing dryer capacities, and improving the quality of the product.

III. CORN DRYING

Although drying studies had been initiated in the late 1920s for wheat harvested under adverse weather conditions (Hurst, 1927), systems for artificial drying of corn received little attention until almost two decades later. Hukill (1948) described an experimental "all-crop mechanical drier" consisting of a furnace connected by a fabric duct to a portable wooden drying box. He discussed the classic paradox of cross-flow dryers, that high air velocities increase drying rates and reduce the moisture content difference across the bed, but also diminish drying efficiencies. Manufacturers responded quickly to the needs resulting from adoption of field shelling, and self-contained batch dryers were commercially available by the early 1950s.

Dryer applications, however, were not an instant success. Producers did not understand the complexities of grain drying. Few used moisture meters and, if they did, had not learned that moisture content could not be accurately measured on samples immediately upon removal from the dryer. The author probed a small bin of "dried" corn for an apprehensive farmer in 1953; when withdrawn, the probe was too hot to hold bare-handed.

More detailed descriptions of drying fundamentals and equipment are provided by Brooker et al (1974) and Foster (1982).

A. Moisture Reduction and Shrink

As a living entity composed mostly of carbohydrates and liberally inoculated with a variety of microflora, the corn kernel is vulnerable to attack and deterioration. Its respiration rate depends upon its moisture content, temperature, and degree of injury. As described in Chapter 5, the viabilities of attacking organisms, including insects, depend upon availability of food substances and the temperature, composition, relative humidity (RH), and oxygen content of the immediately surrounding (interstitial) gases.

All cereal crops, including corn, have historically been marketed at the maximum moisture content associated with an acceptable rate of deterioration. A satisfactory drying system economically reduces the moisture content to this level, with a loss of kernel quality that the ultimate user perceives as acceptable.

MOISTURE CONTENT AT HARVEST

Corn producers tend to begin harvest early, often at moisture contents well above those recommended by the harvester manufacturers, usually below 26%, wet basis. In addition to reducing gathering losses, early commencement of harvest offers better weather conditions and a longer harvest season. Although corn topping (Allen et al, 1982) and other procedures have been tested to speed natural drying in the field, operators continue to face strong incentives to begin well before the optimal harvest moisture is reached for maximum energy efficiency of the entire system, as described by Loewer et al (1984a).

SAFE STORAGE MOISTURES

A hygroscopic material, the corn kernel loses or gains moisture depending on whether its vapor pressure is greater or less than that of the immediate surroundings. Equilibrium moisture content is that kernel moisture content at which interior and external vapor pressures are equal and no net moisture exchange takes place. Although the equilibrium moisture content varies somewhat among corn varieties and is affected by prior treatments, results of many studies have been consolidated into the commonly accepted equilibrium isotherms of Fig. 4, which are based upon the Chung equation (Chung and Pfost, 1967). Each isotherm shown is presumed to be an average of data from tests of both adsorption and desorption; a hysteresis loop results (Ngoddy and Bakker-Arkema, 1975) if a single point is not reached from both directions. Data are not included below 20% RH, because they are irrelevant for market corn, nor above 90% RH, at which kernel decomposition disrupts measurements before equilibration is reached.

Many references list "safe" storage moisture contents for grains, but these are inadequate and often misleading. Storage stability depends upon the relative humidity of the interstitial gases, a function of both moisture content and temperature. *Aspergillus halophilicus* is viable at the lowest relative humidity, 65–70% (Christensen and Sauer, 1982), but seldom affects stored corn. Microflora activity is prevented when interstitial relative humidity is maintained below this range, as indicated in Fig. 4.

A reduction in temperature thus exerts two independent effects on storage

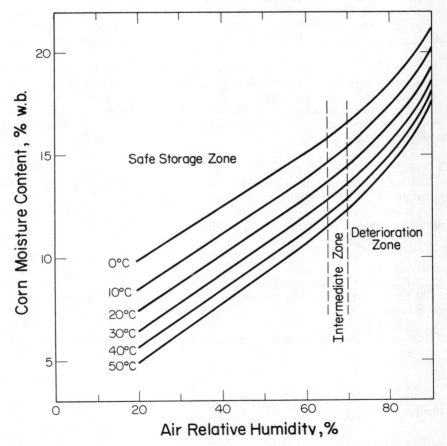

Fig. 4. Equilibrium moisture isotherms for corn, showing zones of storage stability. (Adapted from Chung and Pfost, 1967)

stability. The first is the well-known temperature effect on biological activity, halving the rate with each reduction of approximately 10°C in temperature within the range of temperatures in which each microorganism is viable. The second, assuming that a small change in air relative humidity in a bed of corn entails no measurable change in corn moisture content, is that a drop in temperature may prevent further microbiological activity entirely. Cooling without aeration from 10 to 0°C in 15.5% MC corn, for example, reduces the interstitial relative humidity from approximately 72 to 64%, preventing further microbiological activity. At relative humidities above the range shown in Fig. 4, grain deterioration will result from microflora activity over time, as described in Chapter 5.

Two or more batches of corn at different moisture contents are commonly blended to obtain a desired average value. When grain was commingled to give an average moisture content of 15–18%, even with as much as 15% difference in initial moisture contents, Sauer and Burroughs (1980) reported that equilibrium was more than 80% complete in 24 hr. They concluded from their inoculated samples of freshly harvested wet corn that different moisture contents of corn could be blended with minimal risk of mold damage or aflatoxin development if the average moisture content was at or below that recommended for normal storage.

Practical experiences with blending have shown, however, that serious spoilage can result from blending lots of corn with large differences in moisture content, probably due to deterioration that has already occurred in the wet component before blending. Unexplained spoilage problems in bins and holds of ships may have been due to injudicious blending, a means of disposing of deteriorating lots of wet corn.

LOSS OF WEIGHT IN DRYING

Drying weight loss, termed shrink, is a simple calculation, but it can be no more precise than our ability to measure moisture contents accurately. Such calculations are complicated by the wet-basis definition of moisture content used in commerce and are derived from the dry matter (nonwater) portion:

$$DM = W_1(1 - M_1) = W_2(1 - M_2)$$

in which:

DM is the nonwater mass of the batch,

W_1 and W_2 are initial and final batch weights, respectively, and

M_1 and M_2 are initial and final moisture contents respectively, given as decimals.

From this, one calculates any new weight based on a change in moisture content:

$$W_2 = W_1 \frac{1 - M_1}{1 - M_2} .$$

Further, we can calculate the loss of water, ΔW;

$$\Delta W = W_1 \frac{M_2 - M_1}{1 - M_2} ,$$

so that

$$\frac{\Delta W}{W_1} = \frac{M_2 - M_1}{1 - M_2}.$$

From this it is clear that the percent moisture loss depends upon the final moisture content only, as

$$\frac{\Delta W}{W_1} = \frac{1}{1 - M_2}$$

for each 1% change in moisture content. At a final moisture content of 15.5%, this equals 1.183% loss in weight per percent loss in moisture content. Table I includes representative values of moisture shrink based on both initial wet weight, W_1, and final dry weight, W_2.

When corn is dried and handled at the farmstead, total shrink averages 1.24% per 1% moisture removed (Hurburgh et al, 1983). Most commercial dryer operators increase the calculated drying shrink slightly above those of Table I to include screening and handling losses, which average about 0.9% in drying, regardless of the amount of moisture removed (Hurburgh and Moechnig, 1984).

B. Drying Mechanisms

Even if crushed kernels were not objectionable, the water in a mature corn kernel is held too tightly to be removed by squeezing or expression, such as is done with sugar cane. Instead, the moisture to be removed is vaporized and migrates from the kernel to a lower external vapor pressure. It must then be absorbed or carried away by a moisture sink. About 2.26 MJ/kg (970 BTU/lb) of energy is required to vaporize free water, such as that held in wet kernels of 30% MC or greater. Heats of sorption, which increase as drying progresses, must be added to this heat of vaporization. Drying by vaporization is a complex combination of heat and moisture transfer; at any point in the drying process, the moisture content is greatest and the temperature lowest at the center of the kernel, and both are continuously changing as drying proceeds.

The rate of drying is limited by resistances to both heat and moisture transfer, both within the kernel and at its surface. Mechanisms involved are not yet well understood, as moisture is presumed to transit by a combination of capillarity, liquid diffusion, and vapor diffusion. Syarief et al (1984) reported moisture

TABLE I
Weight Shrink Due to Moisture Loss in Drying to 15.5%, wb

Initial Moisture Content (%)	Percent Shrink Based Upon	
	Wet Weight	Dry Weight
35	23.1	30.0
30	12.3	20.7
25	8.0	12.7
20	3.8	5.6
18	2.1	3.0

diffusion coefficients for the different components of the corn kernel. Thermal conductivities diminish as drying proceeds, and they vary with both position and time. As total resistances to both heat and moisture flow increase with larger particle sizes, smaller or flat kernels can be expected to exhibit greater drying rates. Further, since the pericarp is considered to provide a major portion of the total resistance to moisture flow, fracturing the pericarp increases kernel drying rates.

Because air-to-kernel heat transfer rates are low, alternate mechanisms for increasing these rates have been studied. Using heated sand, Tessier and Raghavan (1984) removed 4% MC with a contact time of only 40 sec, reporting improved fuel efficiency. Headley and Hall (1963) vibrated corn under an infrared source, drying a 5-cm layer from 35 to 15% MC in 0.75 hr with no apparent damage. Development efforts appear to be continuing on microwave-heated drying, generating the heat within the moisture of the kernel, but costs of energy conversion and application have thus far exceeded those of current drying systems.

HEAT SOURCES

Conventional drying requires massive amounts of thermal energy. A 12.7 t/hr (500 bu/hr) dryer consumes perhaps 700 MJ/hr (660,000 BTU/hr) per percentage point of moisture removed. Therefore, even small dryers use large amounts of energy, which constitutes the major operating cost.

Natural gas is a preferred fuel, as it is low in cost and requires only simple burners. Since it is clean burning, the products of its combustion can be mixed with the drying air for greatest efficiency. Manufactured gases, such as butane and propane, are equally desirable but more costly.

Fig. 5. Passive solar collection in shed roof and wall, adjacent to drying bin, facilitates use of solar energy for space heating as well as for crop drying. (Courtesy Dept. of Agricultural Engineering, University of Tennessee)

Fuel oils and most nonconventional fuels, such as biomass, require heat exchangers to avoid fouling of the corn by combustion particulates (Sizemore et al, 1984). This increases equipment costs and reduces thermal efficiencies (Morey et al, 1984), except where gasification occurs outside the dryer (Payne et al, 1983). Recent developments of small fluidized-bed direct combustion, such as that by Keener et al (1983), in which the heated air was described as totally clean, provide the potential to fully exploit biomass as a heat source.

Solar energy can reduce fuel costs for low-temperature drying (Fraser and Muir, 1980) and has been collected from both fixed and portable passive collectors (Eno and Felderman, 1980). Drying would appear to be an ideal application of solar energy, as energy is used as collected and no storage is necessary. But high fixed costs per unit of energy collected, undependability of supply in autumn months, and short seasons of use restrict its economic justification. Authors of a recent report on on-farm demonstrations of solar energy for crop drying in nine states (USDA, 1983) were cautiously optimistic about the future of solar drying if collectors can be used for other heating purposes, as in Fig. 5.

As a source of drying energy, electricity is severely limited by cost and by the capacity of distribution systems. Its use has been restricted to supplemental heating in natural air systems to assure continued drying during periods of adverse weather conditions.

MOISTURE SINKS

Current practical dryers use air both to apply the drying heat to the kernel and to absorb and carry away the water vapor. As this humid air is at least partially exhausted, at temperatures above ambient, some heat is lost. Desiccants such as silica gel (Danzinger et al, 1972) or bentonite (Graham et al, 1983) have been tested for drying air dehumidification in a closed recirculation system, but the energy savings have not justified the additional cost and complexity of the systems.

DRYING RATES

Drying rates of ears of corn (Sharaf-Eldeen et al, 1980) and fully exposed corn kernels (Hustrulid and Flikke, 1959; Misra and Brooker, 1980; Li and Morey, 1984) have been extensively studied. In "thin-layer" drying tests, the kernel is exposed to a constant drying environment and its weight loss recorded over time. Typical results are shown in Fig. 6a, with weight diminishing with time and approaching an equilibrium weight, W_e, as an asymptote.

The curve is usually normalized, however, by plotting the moisture ratio (MR) as the ordinate as in Fig. 6b, so that the value of the ordinate begins at 1.0 and diminishes toward 0. If the data fit an exponential curve, of the form $MR = e^{-\kappa\Theta}$, in which Θ represents time, the curve of Fig. 6b is linear in semilog coordinates as in Fig. 6c. Here, the slope of the curve, k, is defined as the drying constant. Values of k depend not only upon kernel initial moisture content and properties of the drying air, but they tend to diminish as the kernel dries. Note that the slope of the initial portion of the curve, k_1, is typically greater than that of the latter portion, k_2. The simple logarithmic model is adequate for designing practical drying processes, however, which generally take place in the upper range of moisture ratios. For example, if corn is dried from 35 to 17.6%, dry basis (26 to

15%, wet basis) with heated air, the moisture ratio drops from 1.0 to about 0.5, a range in which the drying constant may be treated as constant.

Moisture ratios are generally calculated from moisture contents on the dry basis, as net weights may be substituted directly for moisture contents:

$$MR = \frac{M_\Theta - M_e}{M_i - M_e} = \frac{W_\Theta - W_e}{W_i - W_e} ,$$

in which the subscript Θ designates a specific time after drying begins, e refers to equilibrium, and i represents initial conditions.

DEEP-BED DRYING

Thin-layer drying tests measure the drying rate of an individual kernel or ear, but the drying efficiency approaches zero. In practical drying systems, a bed or layer is sufficiently deep to extract a reasonable portion of the energy from the drying air. The layer nearest the incoming air begins to dry immediately, at its greatest rate, acting as a thin layer, as shown in Fig. 6a. Successive layers start losing moisture only after the preceding ones no longer absorb all the drying potential from the air. Hukill (1947) defined these layers as depth factors and developed a procedure for calculating the average moisture content of each as a function of dimensionless time units, initial moisture content, and drying air conditions. Introduction of computers, both analog (Hamdy and Barre, 1970) and digital (Thompson et al, 1968) permitted modeling of drying to quickly evaluate effects of changes in corn properties and drying parameters on overall effects, capacities, and efficiencies of different dryer configurations.

DRYING EFFECTS ON QUALITY

As noted previously, the transition to field shelling and heated air drying has been accompanied by an increase in breakage susceptibility, the tendency for kernels to disintegrate in handling (Kline, 1972). Gunasekaran and Paulsen (1985) reported that the generation of BCFM in breakage testers increased by as much as 10 times when drying air temperatures were increased from 20 to 65° C

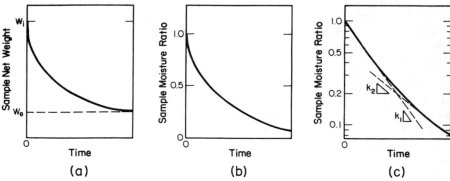

Fig. 6. Typical thin-layer drying curves: a, sample weight vs. drying duration (W_i = initial weight, W_e = equilibrium weight); b, sample weight converted to moisture ratio; c, sample moisture ratio plotted on log scale, giving drying constants k_1 for initial drying rate and k_2 for final drying rate.

(68 to 149°F). Generation of internal stresses, which often cause visible stress cracks in the endosperm (Chapter 5), is proportional to the rate of drying, itself a function of the temperature and relative humidity of the drying air (Gustafson and Morey, 1979; Fortes and Okos, 1980; Gunasekaran and Paulsen, 1985). Further, if kernel temperatures exceed critical values, puffiness reduces test weight. Even higher temperatures discolor the kernels, reducing the grade as described in grain standards. Excessive screenings, composed of fines that pass the official 4.76-mm (12/64-in.) round hole dockage sieve, must be removed as BCFM and sold at discount to maintain the grade (Chapter 5).

AIRFLOW THROUGH CORN

As air is the usual medium for providing drying heat and removing vapor, the mechanics and costs of providing its flow must be considered. Some elementary relationships are included here; considerably more detail is provided by Brooker et al (1974).

Resistances to airflow through common clean and dry grains and seeds were measured and reported by Shedd (1953). Curves for ear and shelled corn are extracted in Fig. 7, showing airflow per unit area resulting from pressure differences per unit depth. Expected airflow rates are reduced if finer particles are present but are increased if larger particles, such as cob pieces, are present in shelled corn.

The log-log coordinate system of Fig. 7 facilitates placing a broad range of data on a single graph but tends to mask the relationships between flow and pressure. At very low velocities, the curve for shelled corn has a slope of unity, meaning that air velocity is directly proportional to pressure difference. This flow condition, known as laminar flow, fits the equation:

$$F = 0.080\ P,$$

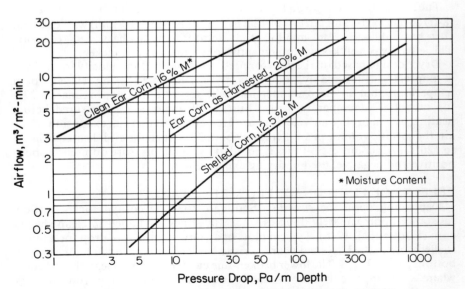

Fig. 7. Airflow per unit area vs. pressure drop for corn. (Adapted from Shedd, 1953)

in which F is m³/m²-min and P is Pa/m, and is typical of aeration systems for shelled corn.

Airflow rates for drying, however, generally fall in the range of 3–15 m/min (10–50 ft³/ft²-min). Here the average slope is 0.67 and the flow, termed transitional, is between laminar and turbulent. Flow rates can be approximated by the equation:

$$F = 0.21 P^{0.67}.$$

To maintain the same rate of airflow per unit of contents, which is necessary to meet drying time constraints, the theoretical air power requirements increase by 58% for each 20% increase in the depth of a drying bed. Peterson (1982) shows that the additional cost of using an 11-m (36-ft) diameter bin for natural air drying rather than one of 9 m (30-ft) for the same mass of corn can be recovered in 1.1 years due to reductions in fan fixed and operating costs.

The clean ear corn curve of Fig. 7 has a slope of 1/2, meaning that airflow is completely turbulent and fits the equation:

$$F = 3.0 \ P^{0.50}$$

with the same units as above. Here an increase in depth of 20% requires 73% more power for the same airflow per unit of volume.

Since pressure in ear corn storages is usually determined by wind velocity, the effects of husks and shelled corn need to be taken into account. The curve for "ear corn as harvested, 16% MC" is approximated by the equation:

$$F = 0.85 \ P^{0.58}.$$

The static pressure drop due to a wind of 32.6 km/hr (20 mph) across a 2.4-m (8-ft) corncrib gives an airflow of 12.5 m³/m²-min (41 ft³/ft²-min) with clean ear corn but only 4.9 m³/m²-min (16 ft³/ft²-min) for the "as harvested" ear corn. This effect was well known to perceptive operators of corn pickers.

Dryer design based on the data for pressure drop vs. flow velocity from Fig. 7 is imprecise. Presumably, the tests were conducted using air at room temperature, but air viscosity increases with temperature, approximately 27% for each 100°C temperature rise. Further, if airflows are measured before heating, actual drying-air velocities are greater due to thermal expansion of the air; the specific volume of air increases about 34% when the air is heated from 20 to 120°C. Air compressibility is ignored at the small pressure differences, both positive and negative, used in drying and aeration.

EFFECTS OF FINES ON AIRFLOW RESISTANCE

Depending upon the size distribution of the fine corn particles, airflow resistance approximately doubles when 10% fines are added to clean shelled corn (Haque et al, 1978). Typical plots of airflow vs. pressure drop for a bin of shelled corn, called "system characteristic curves," pass through (0,0) as in the bottom part of Fig. 8. The lower broken-line curve represents clean shelled corn, whereas the upper curve shows the doubled pressure resulting from the 10% fines. Grama et al (1984) estimate that 3% BCFM affects aeration power requirements only slightly but increases air power by 147–364% in corn dryers because of higher airflow requirements. Although individual particles of fines dry more rapidly than whole kernels, their presence retards drying by restricting airflow rates.

C. Corn Drying Systems

A variety of drying systems has evolved, from the simplest fan connected to the plenum chamber of a drying-storage bin, to today's continuous-flow, recirculating commercial dryers. Each is a compromise between the optimum and the possible for a number of factors: air temperatures and velocities, rate of drying, drying efficiencies, kernel quality, air power, fuel source, fixed costs, and management expertise required. Holmes et al (1985) compare the cost effectiveness of five different drying systems, using recent cost data.

FAN CHARACTERISTICS

Fans provide airflow for drying and aeration. To select a fan for a given task, one must understand the differences among many types and sizes of fans. Vaneaxial or tubeaxial fans are generally less expensive but are limited to lower pressures. In larger sizes, tip speeds may approach sonic velocities, making fans unpleasantly noisy. Centrifugal fans are preferred when higher pressures are required. Those with backward-curved blades are generally recommended, as they are not susceptible to overloading due to reductions in pressure, a problem with forward-curved blades. Fig. 8 also illustrates the differences between the performance curves of two 7.5-kW fans produced by one manufacturer.

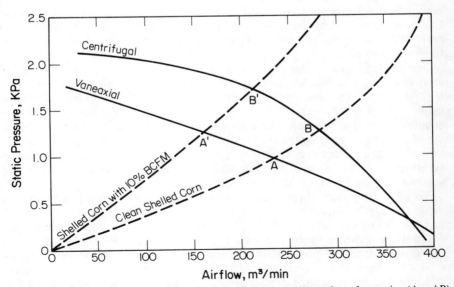

Fig. 8. Performance curves for two 7.5-kW fans (solid lines), showing points of operation (A and B) with a typical system characteristic curve (dashed line) for clean shelled corn and points of operation (A' and B') with a system characteristic curve for shelled corn with 10% broken corn and foreign material (BCFM).

Although not specified, the centrifugal fan presumably had backward-curved blades.

FAN SELECTION

A fan can be selected only after the system characteristic curve has been plotted, such as that shown for clean shelled corn in Fig. 8. When fan curves are superimposed, the vaneaxial fan is shown to operate at point A and the centrifugal at point B. In the same system but with 10% fines, operating points shift to A' and B', respectively. Note the substantial reductions in airflow due to fines.

Fans should be chosen not only to provide the needed airflow, but also to operate somewhere near their maximum efficiencies. These are usually in the range of 50–70% of maximum airflow. As fluid power is the product of pressure and flow rate, maximum efficiency can be estimated by locating the point on the fan performance curve at which the largest rectangle can be constructed beneath the curve. Performance curves for several fans often must be evaluated before one is identified that provides the needed airflow within its range of maximum efficiency.

IN-STORAGE DRYING

Before development of specialized drying equipment, moisture contents of both ear and shelled corn were reduced in their storages. Drying of ear corn, historically accomplished naturally in slatted cribs, could be speeded by distributing large airflows through central ducts in modified cribs (Holman, 1948). Heat was added cautiously due to risks of fire in dried husks. Artificial drying of ear corn never became a common practice, since drying usually proceeded naturally, and has disappeared with the emergence of field shelling. Ear corn drying instructions issued by the U.S. Department of Agriculture in the early 1950s, using natural air (USDA, 1952a) and heated air (USDA, 1952b), have not been revised.

Being relatively free-flowing, shelled corn permits a greater variety of in-storage drying systems. All are systems that use natural air or low heat (up to a 6°C rise) to avoid serious overdrying at the bottom, which would result from the use of heated air. Perforated floors assure linear airflow if fine particles are evenly dispersed by mechanical grain spreaders, but such dispersion may double the air power requirements for the entire batch (Chang et al, 1983). Drying is typically begun when the filling depth exceeds 1 m (3 ft). Because successive layers increase resistance to airflow and reduce the rate of drying, they must be added carefully (Morey et al, 1979).

To keep air power requirements as low as possible, drying is intended to progress only fast enough to avoid spoilage or toxin development in the upper, last-to-dry layer (Ross et al, 1979). Pierce and Thompson (1982) describe their computer simulation, based on 10 years of weather data, to select and manage an energy-efficient bin drying system requiring as little added energy as 675 kJ/kg (290 BTU/lb) of water removed. Although in-bin drying systems are energy-efficient and produce corn quality approaching that of corn dried on the ear, they are vulnerable to adverse weather conditions and require more management than heated air drying systems.

Grain stirrers are marketed to aid drying in in-storage systems. One or more vertical augers rotate in a pattern throughout the bin, loosening and mixing the corn as drying proceeds. Loewer et al (1984b) and Wilcke and Bern (1986) reported that their use was only marginally economical.

To reduce the risks of spoilage in in-storage drying systems, mold inhibitors, termed mycostats, have been metered into the natural air. Sulphur dioxide (Eckhoff et al, 1984), ammonia, formaldehyde, and methylene-bis-propionate (Nofsinger et al, 1979) have controlled spoilage in natural-air systems with airflows as low as 3 $m^3/min \cdot t$ (2.6 cfm/bu). Application costs, coupled with safety considerations, corrosion problems, and residual odors, restrict the advantages of these mold inhibitors.

BATCH-IN-BIN SYSTEMS

As the name of this method implies, a shallow bed of corn is dried to the desired average moisture content in a perforated-floored bin, with subsequent removal to storage. Drying is completed in hours rather than many days as in natural-air systems. Use of bins became feasible when convenient means of emptying, such as sweep augers, were developed. Corn depths are limited to about 0.9 m (3 ft) without grain circulation, to avoid overdrying at the bottom. Such overdrying can be controlled by the use of recirculation, in which the sweep auger removes the grain from the bottom as it dries and places it back on the surface (Roberts and Brooker, 1975). Drying is typically finished upon completion of one cycle of recirculation.

SPECIALIZED DRYER DESIGNS

Equipment specifically designed to dry agricultural grains combines high air temperatures and flow rates with relatively thin drying layers to complete drying in only a few hours. Dryer system sophistication varies widely, from small batch cross-flow units for farm use to fully automated concurrent-countercurrent continuous-flow dryers for commercial operations. Features of the more common of these are described briefly.

Cross-Flow Batch Dryers. As specialized versions of the batch-in-bin concept, cross-flow batch dryers increase the area of exposure by placing the drying bed vertically between perforated sheets around the hot air plenum. High air temperatures with large airflow rates, coupled with slender drying columns of 0.3–0.4 m (12–16 in.), provide rapid batch drying but reduce energy efficiencies. Substantial differences in moisture content develop from inner to outer layers of the column (Gustafson and Morey, 1981). Heat input is halted when the average moisture content is about 2% above the final value desired, and this portion is removed during the cooling portion of the cycle.

Differences in moisture content across the column are reduced in a recirculating batch dryer, which returns the discharge to the top of the drying column, blending its various moisture contents. Recirculation tends to generate fines unless it is very gentle.

Continuous-Flow Cross-Flow Dryers. These dryers complete the drying process as the grain makes a single pass through the column. Heated air is provided for drying in the upper portion and unheated air for cooling in the lower portion. The same moisture differential across the column results as in batch dryers unless vanes are installed in the column to mix the grain as it passes. With sensors, the rate of discharge can be automatically regulated to deliver corn at the desired final average moisture content (Zachariah and Isaacs, 1966), but control system stability can be a problem if the wet corn that is input varies widely in moisture content.

A continuous-flow cross-flow dryer was briefly marketed in which a layer of wet corn dried as it was conveyed horizontally on a perforated bed. Heated air was blown upward through the layer. This was described, incorrectly, as a fluidized bed dryer. Air velocities in a true fluidized bed equal the terminal velocity of the particle, causing flotation and natural mixing within the bed. Corn kernels, with high terminal velocities, require excessive air power to be fluidized in large beds. Uniform particle size, shape, and density are also required, as particles with lower terminal velocities are discharged with the drying air. For these reasons, fluidized bed drying of whole corn has not been found to be practical.

Dryeration. In their pioneering study, Thompson and Foster (1963) noted the effects of both drying and cooling rates on stress crack formation, especially below 19% MC. To permit redistribution of moisture and relief of stresses between final drying stages, "dryeration" was conceived (McKenzie et al, 1966); the hot corn is withdrawn from the dryer and held in a separate bin for several hours before being cooled with natural air. Using the sensible heat in the corn to remove the final 2–4% MC, this procedure improves kernel quality, raises drying efficiency, and increases the capacity of the dryer by eliminating its cooling function.

Efforts continue to improve both efficiency and product quality in cross-flow dryers. Shrouds installed to recirculate part of the exhaust air show promise on larger commercial units; automatic damper controls will probably be needed to achieve maximum efficiencies. A system of gates was proposed to periodically reverse the direction of airflow in one cross-flow dryer marketed, but the promised benefits did not appear to justify the mechanical complexity of the device. Gustafson et al (1983) showed that short pauses between drying cycles, called tempering, reduce breakage susceptibility of dried corn while increasing drying efficiencies.

To help select the best combination of tempering and drying, Brook and Bakker-Arkema (1980) developed a computer optimization technique, considering both maximum kernel temperature and breakage susceptibility. Bakker-Arkema and Schisler (1984) reported results with a differential grain-speed cross-flow dryer, which included tempering between the two drying stages, improving both drying uniformity and corn quality.

Concurrent-Countercurrent Dryers. Hall and Anderson (1980) reported results with their latest concurrent-flow heating and counter-flow cooling dryer, rated at 250 t/hr (10,000 bu/hr) based on a reduction of 5% MC. With the heated air and wet corn moving in the same direction, all corn is subjected to identical conditions, and exceptional temperatures of as high as 399° C (750° F) could be used. Cooling shock was minimized in the counter-flow cooling section. Excellent energy efficiencies, as low as 3,330 kJ/Kg (1,430 BTU/lb) of water removed, were reported, coupled with less breakage susceptibility than for corn from cross-flow dryers.

FUTURE DRYER DESIGN AND CONTROL

As initial drying of high-moisture corn must be accomplished within a few hours of harvest, this drying must be performed at or near the production site. Dryers designed for farm use will increase in size as producers continue to specialize, permitting increases in efficiency and improved control. Although overly optimistic claims abound, newly developed "computerized grain management" control systems have the potential to monitor both ambient air and drying grain conditions to regulate airflows in natural air systems, plus air temperatures in dryers, to optimize the drying process. Success depends upon development of appropriate instructions programmed into the microprocessor, which doubtless will require time and experience.

If fuel costs continue to increase more rapidly than the value of corn, greater drying efficiencies must be found. To this end, U.S. manufacturers might increase drying column thicknesses and reduce airflow rates, approaching the more energy-efficient designs of European dryers. Drying at higher air temperatures, with inherent increases in thermal efficiency, will be attempted for markets for which quality may be less restrictive, but newly developed quality monitors, such as density and breakage testers (Singh and Finner, 1983), will be applied to assure the quality demanded for specialized markets.

Although it increases the physical size of continuous-flow dryers, the tempering of sections between applications of heated air improves drying efficiencies (Harnoy and Radajewski, 1982) and reduces the breakage susceptibility of dried corn (Gustafson et al, 1983).

For commercial-sized operations, the inherent advantages of concurrent-countercurrent-flow dryers, contrasted to cross-flow dryers, make further development inevitable. Modifications, such as equalizing the residence time for all paths through the dryer, and innovative concepts for energy reuse, will further improve both corn quality and drying efficiencies.

IV. HANDLING SYSTEMS

Most corn-handling equipment was developed when corn was still harvested and dried on the ear, and kernels were much more tolerant of rough handling. As increasing damage levels became apparent, conferences first emphasized causes and solutions at the production level in a Grain Damage Symposium (1972) and a Corn Quality Conference (1977). The disastrous terminal elevator explosions of late 1977 prompted an International Symposium on Grain Dust (Miller and Pomeranz, 1979), coupled with intensive research on causes, control, and containment of grain dust explosions and fires.

The National Materials Advisory Board documented the number of explosions in U.S. grain-handling facilities (1982a), showing an abrupt increase from the mid-1960s through the end of 1977, although numbers reported vary considerably between sources. Recent reductions may be attributed to risk management evaluations, such as that of Kameyama et al (1982), and industry-wide campaigns conducted by major grain-handling and processing associations, based upon recommendations such as those of the National Materials Advisory Board (1982b).

The primary recommendation, improved housekeeping, has led to the development of a variety of dust collection and disposition plans. Large vacuum systems are being installed to reduce dust layers below arbitrary levels. Scrubbers, precipitators, centrifugal separators, and filters are used for final collection of dust. One large elevator reported the collection of 13,000 t of dust in one year, even though this amount was only 0.29% of its throughput. Studies continue on economical utilization of grain dust (Miller and Pomeranz, 1979).

Recent epidemics of *Aspergillus flavus* in field corn, especially in the southeastern part of the United States, have led to studies of hazards to farmers and grain handlers of airborne dusts that contain aflatoxin. Burg and Shotwell (1984) measured as much as 13,000 ng of aflatoxin per cubic meter of air near an elevator conveyor belt.

As a means of retaining grain dust within the batch during handling, the addition of as little as 0.02% mineral or soybean oil, sprayed onto the corn, drastically reduces dustiness for as much as six months after application (Lai et al, 1981). The U.S. Food and Drug Administration has approved white mineral oil as a dust suppressant on commodity grain and food for human consumption (FDA, 1983), and equipment for application of the oil is now commercially available. Added water, at rates too small to increase measured moisture content, also reduce dustiness, but the water is absorbed and must be reapplied at each handling. Extensive tests on various additives to control dust in grains are described in a U.S. patent (Barham and Barham, 1980).

Fiscus et al (1971) studied corn breakage occurring at specific locations in a grain-handling facility, especially in free fall, spouting, and bucket elevators. They concluded that free fall from heights greater than 12 m (40 ft) caused more breakage than any other process, and that breakage increased as both moisture content and temperature diminished.

A persistent problem in commercial grain elevators, often of greater operational importance than kernel damage, is erosion or wear of equipment at points where grain slides or impinges upon it. Ceramic materials cemented at critical locations resist wear better than alloy steels, but neither cushions the impacts. Polyurethanes and specially composed plastics with ultra-high molecular weight have been developed to provide both wear resistance and reduced impact.

A. Ear Handling

With few exceptions, ear corn is shelled from the crib before marketing. Thus, ear corn is handled only at the farmstead. Hydraulic or mechanical lifts were common to raise the front of a wagon when unloading at open inclined flight elevators. Large double-crib granaries with overhead bins sometimes contained slow-moving vertical elevators with large buckets that could accommodate corn ears. A mechanical drag, inserted into the crib as it was emptied, greatly speeded the removal of ear corn for shelling. Controls were developed to automatically withdraw ear corn from the crib into a grinder (Forth et al, 1951), which could have evolved into complete ear-handling systems if field shelling had not intervened.

B. Shelled Corn Handling

Much of the material-handling equipment at the farmstead is equivalent in design to that found at grain elevators and terminals, but smaller and less sophisticated. One exception is the hopper-bottomed wagon used to transport shelled corn from the field to the drying-storage site. Discharge was by gravity, in early models, but most recent designs incorporate short, high-capacity, inclined augers for unloading. Gravity discharge restricts wagon capacity and increases overall wagon height, reducing stability. Capacities exceed 20 t in recent designs, and large flotation tires are employed to permit operation in muddy fields and to reduce soil compaction.

FLOW THROUGH ORIFICES

Bins and tanks, both flat- and hopper-bottomed, empty through orifices into conveying equipment. Chang et al (1984) determined the relationships between flow rate and orifice area (Fig. 9) for dry shelled corn, noting that higher moisture contents reduced flow rates very little except in larger orifices.

Bridging at bin discharges is usually broken by externally mounted vibrators. Vibration, however, tends to consolidate some hard-to-flow materials. Devices are marketed to be installed inside the throat of the orifice, where they can assure flow and, in some cases, control delivery rates.

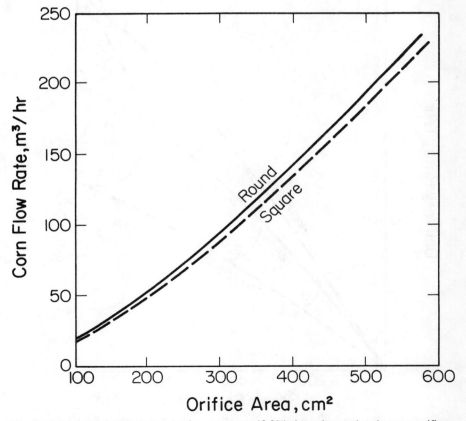

Fig. 9. Flow rates of shelled corn (moisture content 12.3%) through round and square orifices. (Adapted from Chang et al, 1984)

SCREW CONVEYORS

A screw conveyor consists of an Archimedes' screw rotating in an enclosing tube or open "U-tube." With only one moving component in contact with the corn, screw conveyors are used extensively in low-cost conveying systems. Commonly called augers, their efficiencies, durabilities, and effects on product vary widely. In a well-designed system, the desired capacity is obtained with a large-diameter screw rotating relatively slowly, essentially full. Conversely, partially filled conveyors, rotating rapidly, can cause considerable kernel damage (Sands and Hall, 1971). Hanger suspensions, arranged so that the screw is not supported by the tube, reduce kernel pinching and breakage but make cleanout difficult unless the conveyor slopes sufficiently. A screw conveyor is occasionally used to mix insecticides or moisture into a mass of corn as it is being conveyed.

Capacities and power requirements depend upon the size and rotational speed of the screw, angle of inclination, and moisture content of the corn. Figure 10

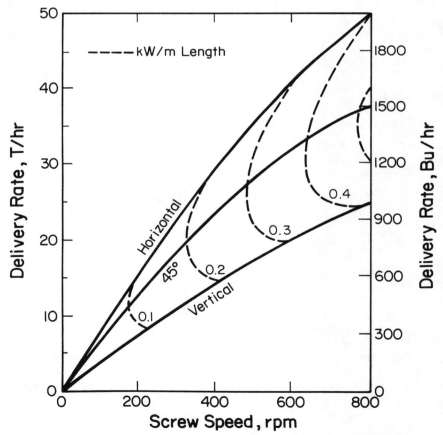

Fig. 10. Capacity and power requirements of a 15-cm (6-in.) auger conveyor handling dry shelled corn. (Adapted from White et al, 1962)

illustrates the relationships between these factors for dry corn, whereas Fig. 11 shows the substantial capacity reductions and increases in power requirements for wet corn (White et al, 1962). McFate and George (1971) provide equivalent data for 20-cm (8-in.) screw conveyors. In larger sizes, power demands may exceed that available from farm electric motors, and power-take-off shafts are provided; a 25-cm (10-in.) diameter auger 24 m (80 ft) in length requires 30 kW (40 HP), according to its manufacturer. Industrial sizes up to 61 cm (24 in.) in diameter can convey several hundred tons per hour. Capacity is independent of length, whereas power requirements are considered to be proportional to length.

Relying heavily upon friction for their operation, screw conveyors tend to wear rapidly, especially their flighting, and they are inefficient in use of energy. Since torque on the screw can be substantial, less-expensive units with lightweight shafts should be driven from the input end, especially if the flighting is not attached continuously along the shaft. This wraps the flighting more tightly about the shaft due to torsional deflection, increasing greatly its torsional rigidity. If driven from the discharge end, shaft deflection compresses the flighting, which may lead to its buckling and, in extreme cases, to breaking of the welds and crumpling of sections of flighting inside the tube.

When the auger is operating at smaller slopes, up to about 30° from horizontal, the corn slides by gravity down the continuously advancing face of the flighting as well as along the bottom and sides of the tube. The limiting angle depends upon the coefficient of sliding friction of the corn on steel and the pitch-to-diameter ratio of the screw. Capacity decreases as slope increases.

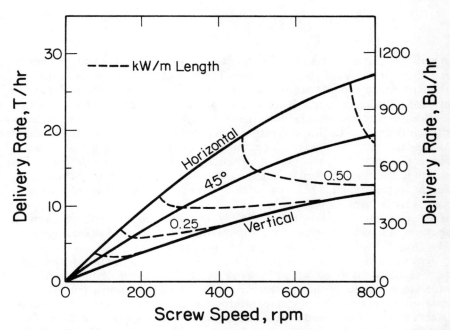

Fig. 11. Capacity and power requirements of a 15-cm (6-in.) auger conveying shelled corn of 25% moisture content. (Adapted from White et al, 1962)

Larger-capacity screw conveyors, when used at elevators and terminals, are typically installed horizontally or at small slopes and are usually of the U-tube configuration. This permits inspection of the entire screw in place, if needed. Little data have been published on operating characteristics of such installations. Because of previously cited corn damage problems, screw conveyors are being replaced by chain and flight conveyors in larger commercial elevators.

At slopes greater than about 30°, the corn no longer slides by gravity down the advancing flighting face. The auger speed must then be sufficient to throw the corn radially against the tube wall, which retards its rotation and causes it to be advanced by the screw (Rademacher, 1981). Power requirements increase rapidly and capacity diminishes very rapidly with increasing slope, especially with high-moisture corn as shown in Fig. 11.

BUCKET ELEVATORS

Bucket elevators, called "legs," convey the product vertically and are the heart of a handling-storage system. Being almost frictionless, they elevate with efficiencies exceeding 90%. Capacity is simply the product of belt speed, bucket capacity, and spacing. But the belt speed selected for a centrifugal-discharge leg greatly influences durability and discharge performance.

Figure 12 shows the general arrangement of a head section. As a bucket with contents reaches the head pulley, position A, the average velocity of its center of gravity must increase abruptly from that of the belt, V_b, to some greater velocity, V_{cg}, which is determined by the relative sizes of the radii:

$$V_{cg} = V_b \frac{R_{cg}}{R_b},$$

in which R_{cg} is the radius to the center of gravity of the bucket and its contents and R_b is the radius of the head pulley. The bucket naturally pivots backward about its attachment at the onset of this abrupt acceleration, pitching forward a fraction of a second later as it comes around the top of the head pulley, position B. This latter action, coupled with the centrifugal force on the bucket contents, efficiently empties the buckets into the elevator discharge with a minimum of "down-legging" if the system is operating at optimum belt speed. Changes in bucket size and shape, or contents, may change this optimum speed slightly. The abrupt jerk and subsequent oscillation of the bucket help to empty its contents, but the forces that must be transmitted by the attachments to the belt will cause premature failures if excessive.

A continuous-discharge leg relies only on gravity to empty the buckets. It operates at lower belt speeds, and each triangularly shaped bucket directs the flow from the following bucket into the discharge.

Because bucket elevators typically have the largest concentrations of grain dust, as measured by Wade and Hawk (1980), they are being designed with vent panels to minimize pressure increases due to enclosed explosions. Sensors can be installed to monitor bearing temperatures, a cause of dust ignition if excessive. Other devices are designed to note belt speeds to warn if slippage, a cause of belt fires, is occurring. Elevators are also being located outside the storage complex to isolate explosion effects.

BELT CONVEYORS

Although limited to inclinations of about 20° from horizontal, belt conveyors are universally installed for lateral conveying in larger elevators and grain terminals. In widths of up to 2 m (6 ft) and with few constraints on velocity, current designs handle as much as 1,400 t/hr (55,000 bu/hr) with a minimum of loss or damage. Troughed idler pulleys greatly increase belt capacity, especially over greater distances. The surcharge, the equivalent material that would be carried on a flat belt of the same width, is limited to a maximum slope of about 20° at its edges, due to undulations of the belt as it passes over the idler pulleys. Trippers on overhead belts permit discharge into selected tanks or silos.

For smaller applications, a belt-in-a-tube conveyor—using, for example, a belt 30 cm (12 in.) wide sliding in a 25-cm (10-in.) tube—appears to have the same advantages as the larger units, but friction between belt and tube can greatly increase power requirements per unit conveyed. This friction can be

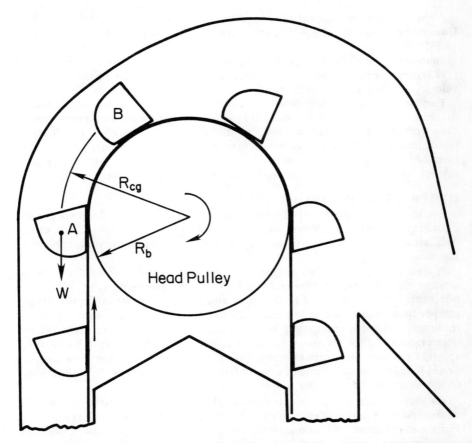

Fig. 12. Arrangement of head section of a bucket elevator, including head pulley, belt with buckets, and discharge hood. A, bucket at head pulley; B, bucket at top of head pulley; R_{cg}, radius to center of gravity of bucket and contents; R_b, radius of head pulley.

reduced by special coatings or by provision of an air cushion beneath the belt, which also diminishes the rate of wear on belt and tube.

PNEUMATIC CONVEYING

Pneumatic vacuum-pressure systems, using flexible suction tubes, offer special convenience for unloading and cleanup in otherwise difficult locations. They are commonly used for ship unloading in European ports. Rotary blowers provide pressure differentials of about 50 kPa (7.5 lb/in.^{2}). To avoid damaging the kernels in the blower, the suction side connects to the input of a centrifugal separator, from which the product drops through a rotary air lock to enter the positive pressure delivery stream. Another separator must be installed at the discharge end. Portable units with tubes up to 25 cm (10 in.) in diameter and advertised capacities of up to 165 t/hr (6,500 bu/hr), depending on rise and length, are commercially available but require power sources of as large as 112 kW (150 hp).

Tests of a commercially produced unit indicated that power requirements for the conveying were proportional to flow rates (Susai and Gustafson, 1982). Conveying rates for 22% MC corn were 8–10% less than those for corn at 16.5% MC.

In their pressure conveying system having a 10-cm (4-in.) diameter, Baker et al (1985) measured minimum specific energy consumptions at an uncompressed air velocity of 20 m/sec (3,940 ft/min).

Except at high air velocities, with uncompressed air flowing at around 30 m/sec (5,900 ft/min), Baker et al (1986) found that pneumatic conveying in their pressure system caused relatively little damage to corn. When batches of brittle and nonbrittle corn were conveyed at 30 m/sec, fines were generated at rates of 1.735 and 0.66%, respectively, but this dropped to 0.83 and 0.37% at 20 m/sec. Damage rates were essentially independent of feed rate.

OTHER HANDLING COMPONENTS

Static cleaners, composed of sloping screens, are often installed beneath the discharge of the leg to remove BCFM by gravity flow. Powered sieves, either oscillating or rotary, may be placed at locations convenient for monitoring their function.

A system of gravity tubes or spouts conducts corn from the leg discharge to storage bins, dryers, rail cars and trucks, or boats. Slopes must be >30° from horizontal for gravity flow of dry corn and 40–45° for wet corn. But slopes steeper than needed, which are often unavoidable, produce excessive velocities in longer sections, increasing kernel damage and equipment wear. Grain decelerators or flow retarders, installed at intervals along the length of a steep spout, have been no more effective for reducing grain damage than cushion boxes (deadheads) at the end of a run (Stephens and Foster, 1977), while adding to the weight and cost of the spout.

As shelled corn containing fine materials is discharged into a storage bin, coarser particles roll and slide to the outer portion of the bin and finer segments remain at the center. This concentration of fines is termed the "spout line." To avoid problems due to spout lines, grain spreaders are installed to more evenly distribute the fine materials within the bin. In smaller bins, unpowered spreaders operate from the kinetic energy of the falling corn. Larger units are powered by electric motors. Without the use of grain spreaders, fines can be distributed by flow regulators to give choke-flow, improving the distribution of fines over that from simple spouts (Chang et al, 1986). However, as previously indicated, fine materials reduce storage stability and should be removed rather than distributed throughout a bin of corn.

Inadequate storage management may result in hang-ups and bridging of spoiling corn. In large silos, several hundred tons of corn may be involved, constituting a serious hazard if workmen attempt to clear the problem by hand. A number of mechanical devices have been marketed to be lowered into the silo for dislodging hang-ups.

En-mass chain and flight conveyors, capable of elevating corn to slopes of 60°, provide greater control of flow rates than is provided by screw or belt conveyors. They are used for metering corn from hopper-bottomed bins, for horizontal transfer, and for other specialized handling needs. Designed for commercial use, many having plastic scrapers to reduce wear, their capacities are 1,000 t/hr (40,000 bu/hr) and greater. In longer units, provision is required to accommodate chain stretch under load.

Dust and segregation problems have been greatly reduced by the use of flow regulators at discharges into storages. Chang et al (1986) reported that dust

Fig. 13. Contained-flow loading spout designed to retain dust within the grain mass and reduce particle size segregation at a barge-loading facility. (Courtesy Dust Control and Loading Systems, Charlevoix, MI)

concentrations were reduced by 80% compared to the level from open spouts, coupled with a broader distribution of fine materials. Application of one of these choked-flow or contained-flow regulators is shown in Fig. 13.

FUTURE TRENDS IN HANDLING EQUIPMENT

New designs of equipment for handling corn will reflect recent research emphasis on causes and prevention of grain dust generation and fires. Conveyors that handle the product more gently, such as large-bucket, slow-speed legs, will diminish grain breakage and dust generation while reducing downtime and maintenance expenses (Fishel, 1983).

V. STORAGE AND STORAGE MANAGEMENT

The types and sizes of corn storage structures required depend upon the form and quality of corn to be held, the anticipated duration of storage, and the degree of mechanical handling desired. Lindblad and Druben (1976) recommend design details for small farm storages in developing countries.

A. Ear Corn Storage

Thousands of smaller corn producers, especially in developing countries, doubtless find that ear harvest and storage is appropriate to their needs. Typical African farmstead corn cribs are described by Ampratwum and Bockhop (1975). Corn cribs, rectangular or cylindrical slatted structures constructed of wood or welded wire, are sited in open areas where wind currents will naturally dry the contents. Cribs should be oriented with the smaller dimension—typically 1.5–2.5 m (5–8 ft) in the United States (Neubauer and Walker, 1961)—in the direction of the prevailing wind.

Anticipating the use of corn sheller drags, producers often construct larger cribs with channels beneath the center of the floor, into which the sheller drags are inserted. Loose boards above the drag are lifted as emptying proceeds.

Ear corn in cribs cannot be economically protected from losses. Although "rodent-proof" cribs have been designed, rodents and squirrels find entry, leaving their droppings and damaged kernels. Pilferage is a problem in some locations. Birds and domestic fowl litter and consume, and, in warmer climates, insects cannot be controlled. Although handfuls of dead weevils were found in malathion-treated corn in Brazil, essentially 100% of the kernels on ears with and without husks were infested and damaged in test cribs (Triplehorn et al, 1966).

B. Shelled Corn Storage

Shelled corn is stored in a large variety of styles and sizes of bins, tanks, silos, and pits. Described by Hyde and Burrell (1982), underground storage would appear to be desirable, as no structure need be purchased. But problems of hermetic sealing, especially in areas with higher water tables, appear to restrict the potential of pit storage as a common practice for market corn.

BIN AND SILO DESIGNS

Many small farms still use 83-t (3,000-bu) flat-bottomed corregated steel bins, although sizes up to 3,000 t (110,000 bu) are marketed for farm use. Commercial elevators, where fully mechanical handling is required, use combinations of silos, cylindrical tanks, and flat storages. Each of these storages facilitates a different degree of mechanical handling. Silos are typically constructed of reinforced concrete and are filled and emptied by belt conveyors with little or no setup involved. Metal hopper-bottomed bins provide the same convenience but are used in smaller enterprises. Somewhat less convenient for emptying are large, cylindrical, welded tanks (sometimes surplus oil tanks) holding 30,000 t (1,000,000 bu) or more, which provide intermediate-length storage.

Flat storages are those that are sufficiently shallow that structural strength is relatively unimportant. These provide large storage volumes at low cost, and, when empty, may be useful for other purposes; however, they require considerably more work and equipment for both filling and emptying. Several smaller units, rather than a single larger one, are required to provide the necessary flexibility for segregation and subsequent blending of different lots.

The engineering and design of larger silos and tanks has become highly specialized, requiring detailed analyses of foundations, wall loadings, and roof forces (Gaylord and Gaylord, 1984). Internal pressures depend upon the geometry and size of the structure, as well as on corn bulk density, angle of repose, and coefficient of friction on the enclosing surfaces. As these corn properties depend upon moisture content, which may change during storage, precise values cannot be predicted and safety factors must be conservatively applied. Normal hoop stresses may be estimated with considerable precision, based on modifications of the classical Janssen's equations, but vertical loads on walls may be greatly increased due to spoilage, freezing, or other instances in which the contents are bonded to the walls. Bucklin et al (1985) have evaluated the buckling resistance of thin-walled bins as a function of internal pressures and vertical wall loads.

Spectacular bin failures have been attributed to sometimes-overlooked sources of exceptional loads in storage structures. Nonsymmetric filling and unloading, including dynamic forces as corn masses flow intermittently, cause highly variable lateral wall and floor pressures (Thompson et al, 1986) and tipping forces. Springtime aeration, intended either to warm or to rehumidify overdried corn, can add sufficient moisture to produce swelling and critical hoop stresses (Risch and Herum, 1982).

Diurnal and seasonal temperature changes, especially abrupt in continental climates, cause expansion and contraction in metal storage walls. The temperature of the enclosed corn, however, changes very slowly. An expanding structure permits its contents to settle but, as the corn is only semifluid, it resists the contraction of the cooling tank. Manbeck (1984) describes these thermally induced pressures, which can be converted to hoop stresses.

Finally, the inherent rigidity of cylindrical bin walls diminishes as their diameter increases; large-diameter walls designed to withstand tension hoop stresses have buckled inward due to high winds. If they are not constructed with deeper corrugations to increase their wall rigidity, such bins may have to be kept at least partially filled during periods of anticipated high winds.

Temporary corn storage in outside piles is becoming common at country elevators when enclosed storage capacity is exceeded. Corn settled in piles has a slope angle (angle of repose) of approximately 22°; thus, piles have heights of about 20% of their diameters. Based upon the equation for the volume of a cone, the capacity of such a pile can be estimated from the equation:

$$T = 0.0381 \, D^3,$$

in which T is the metric tons of corn and D is the diameter in meters, or

$$BU = 0.0425 \, D^3,$$

where BU is bushels and D is diameter in feet. Intended for provisional storage only, such piles suffer weather losses if uncovered or if inadequately sited and

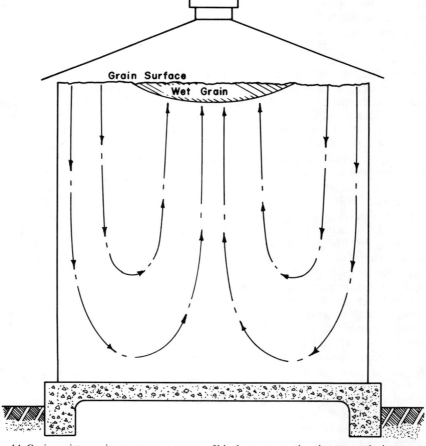

Fig. 14. Grain moisture migrates to upper center of bin due to convective air currents during autumn and early winter. (Reprinted from Anonymous, 1975, by permission of the American Society of Agronomy)

drained. Plastic or canvas covers, held in place either by aeration vacuum or as an air-supported enclosure with positive-pressure aeration, improve protection but add to storage costs.

Operators of all shelled corn storage units must be alert to the hazards of suffocation due to immersion in flowing grain masses, called entrapment, which can occur as grain storages are emptied. Schwab et al (1985), citing 27 such deaths in Nebraska during a decade, measured average vertical pulls on mannequins of as much as 6,150 N (1,384 lb) in flowing corn.

MOISTURE MIGRATION

Moisture migration is the gradual movement of a small proportion of grain moisture from throughout a bin to some specific location, leading to subsequent spoilage if not corrected. It occurs when the temperature of a mass of corn is substantially different from that outside over a period of time. The temperature of the corn near the bin wall tends to follow external temperatures, whereas that at the center remains relatively unchanged. This is due to the low thermal conductivities of grain.

As ambient temperatures fall during autumn and early winter months, the interstitial air in the cooler, outer portions of the bin tends to move downward, displacing and forcing upward the warmer air at the center. As this air contacts the cold corn at the surface, it cools and deposits part of its moisture at the upper surface (Fig. 14). This continues as long as the temperature difference exists, and it can result in spoilage in the affected area even at temperatures as low as 0° C. In small storages, in which moisture migration is usually relatively minor, the wet layer can be removed or remixed into the grain mass by turning.

AERATION

Aeration is as yet the only practical method of controlling moisture migration in larger storages. Aeration, the blowing or drawing of ambient air through the corn at low rates, can control moisture migration by a combination of two mechanisms. It slowly equilibrates temperatures between the center and the exterior of the bin, eliminating the cause of air currents (Foster and Tuite, 1982). But moisture deposition can also be avoided by interception, through a vacuum system that draws the warmer air outside without passing it through a layer of cooler corn (Fig. 15). Of these, temperature equilibration has the greater merit, due to the benefits of lower temperature on storage stability, and, once the contents are cooled to near ambient temperatures, aeration can be discontinued. Most aeration systems are operated as in Fig. 16, which provides both temperature equilibration and warm air interception. Airflows of as little as $1/45$ m^3/min·t ($1/50$ ft^3/min-bu) have been found to be adequate, but systems are often operated with airflows two to five times as great.

Unless the danger of spoilage exists, aeration systems should be operated intermittently as needed, as 0.25–0.5% MC can be removed from the entire bin by continuous operation through the winter months. Multon et al (1980) devised a guide for recommended aeration based on air relative humidity and temperature differences between grain and outside air; its use led to reductions in aeration durations of one third to one half that of previous storage seasons.

Operators of aeration systems need not fear that repeated cooling-heating of corn might increase its susceptibility to breakage. Jindal et al (1978) showed that as many as 16 freezing-thawing cycles had no detrimental effect on kernel strength.

Duct design and layout for aeration should provide uniform pressure, or vacuum, throughout the system. Uniformity of flow near the exterior walls is not required, but extra ducting may be helpful beneath spout lines and other dense areas that divert air streams from where they are most needed. In tall storages, care must be taken to provide sufficient perforated duct area, as air velocities in layers of corn near the ducts is much larger than elsewhere, causing exceptional pressure drops and energy losses immediately adjacent to the ducts. The greater energy efficiency of larger ducts must be balanced against their additional cost and their inconvenience at bin emptying.

Foster and Stahl (1959) noted that pressure systems provided more uniform

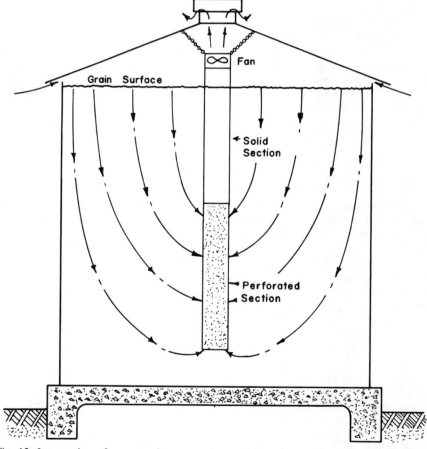

Fig. 15. Interception of warmer air currents in small bins, by aeration, to prevent moisture deposition. (Reprinted from Anonymous, 1975, by permission of the American Society of Agronomy)

aeration than did vacuum in flat storages when duct velocities exceeded 3–4 m/sec (600–800 ft/min).

CONTROLLED-ENVIRONMENT STORAGE

Storage of high-moisture corn under controlled environments that restrict or inhibit biological degradation avoid the costs of drying, but market corn must ultimately be dried to move safely through marketing channels. Applications of organic acids, such as propionic acid or methylene bis propionate, provide brief protection but require uniformity of mixing that is difficult to achieve (Sauer et al, 1975). They also corrode metal bins. Oxygen-free storage produces a fermented product suitable for animal feed. Refrigerated storage, done specifically to extend the time available for drying and using the exhaust heat of

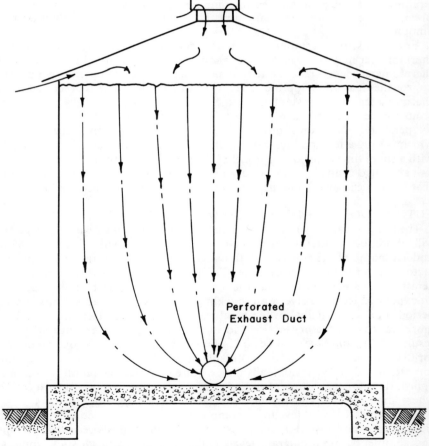

Fig. 16. Common aeration system for larger bins and silos, providing both temperature equilibrium and warm air interception. (Reprinted from Anonymous, 1975, by permission of the American Society of Agronomy)

the heat pump condenser for drying, has been demonstrated (Shove, 1968), but large fixed costs for the required heat pump capacity and insulated structure apparently discouraged commercial development. Burrell (1982) discusses both mechanical and biological considerations in refrigerated storage and drying.

MONITORING STORAGE CONDITIONS

Condition or quality of stored corn is usually evaluated by its moisture content and temperature. Because a simple means for remote measurement of moisture content has not been developed, probing is necessary to determine this factor. Except near the surface, probing is awkward and tedious in large bins. The condition is usually monitored by measuring grain temperature at intervals along thermocouple cables installed vertically in the bin or pushed down into the grain mass from the surface. These cables can be linked to automatic recording instruments that periodically "sweep" all thermocouples. Potential trouble spots are usually identified by abrupt temperature increases. Some recording equipment can be programmed to compare each new reading with that at the previous sweep, noting specially any that have changed more than a preset amount.

For occasional temperature measurements in smaller bins, a bulb thermometer may be inserted, attached to a wooden probe. In theory, a bulb thermometer never reaches the final environmental temperature but only approaches it as an asymptote. The rate at which this occurs is described by the thermometer time constant, the time required to register 62% of the total temperature change when inserted into an environment with a different temperature. The error of measurement is less than 1% in five time constants, which is only a few minutes with smaller bulb thermometers. A thermometer with a small time constant approaches the corn temperature more quickly, but its value also changes more rapidly upon withdrawal, reducing the assurance that the value shown upon inspection represents that sensed in the bin.

FUTURE TRENDS IN CORN STORAGE STRUCTURES

The trend toward design and construction of ever-larger silos and storage bins will doubtless continue, but their successful use will require a better understanding of the effects of both management practices and environmental factors on structural performance. The potential hazards due to springtime aeration have been noted, but the specific extent of the risks is unknown. Unexplained compressive wall buckling has occurred, apparently during periods of exceptionally low temperatures, in corrugated steel bins after 10 or more years of satisfactory service. Moisture migration to cooler outer walls, with resultant corn spoilage, still occurs in bins in which current recommendations for drying and aeration have presumably been followed. The classic equations for lateral and vertical wall loads may require considerable modification to be appropriate for the design of today's larger structures, especially in locations where large and rapid environmental temperature changes occur.

Employing recently developed computer-based systems, commercial grain elevators can now install tracking systems that continuously record all receipts and shipments, monitor grain levels and temperatures in all silos, compute amounts owed, and issue checks. Automated blending systems, using continuous moisture measurement, should soon be available to improve precision in delivering corn to specific customer demands. Clearly, as technology surges ahead, elevator superintendents must trade their shovels and coveralls for M.B.A. diplomas and grey flannel suits.

LITERATURE CITED

ALLEN, R. T., MUSICK, L. D., and HOLLINGSWORTH, L. D. 1982. Topping corn and delaying harvest for field shelling. Trans. ASAE 25:1529-1532.

AMPRATWUM, D. B., and BOCKHOP, C. W. 1975. Storage of grains in the humid tropics. Paper 75-4055. Am. Soc. Agric. Eng., St. Joseph, MI.

ANONYMOUS. 1972. Proc. Grain Damage Symp. April 5-7. Am. Soc. Agric. Eng. and Dept. Agric. Eng., The Ohio State Univ., Columbus.

ANONYMOUS. 1975. Better farm crops. Crops Soils 28(2):17-20.

ANONYMOUS. 1977. Proc. 1977 Corn Quality Conference. Univ. Ill. Agric. Exp. Stn., Urbana-Champaign.

BAKER, K. D., STROSHINE, R. L., FOSTER, G. H., and MAGEE, K. J. 1985. Performance of a pressure pneumatic grain conveying system. Appl. Eng. Agric. 1:72-78.

BAKER, K. D., STROSHINE, R. L., MAGEE, K. J., FOSTER, G. H., and JACKO, R. B. 1986. Grain damage and dust generation in a pressure pneumatic conveying system. Trans. ASAE 29:840-847.

BAKKER-ARKEMA, F. W., and SCHISLER, I. P. 1984. Differential grain-speed crossflow grain dryer. Paper 84-3522. Am. Soc. Agric. Eng., St. Joseph, MI.

BARHAM, H. N., Jr., and BARHAM, H. N. 1980. Methods for the adsorption of solids by whole seeds. U.S. patent 4,208,433.

BENOCK, G., and LOEWER, D. H. 1981. Grain flow restrictions in harvesting-delivery-drying systems. Trans. ASAE 24:1151-1161.

BRIZGIS, L. J., NAVE, W. R., and PAULSEN, M. R. 1980. Automatic cylinder-speed control for combines. Trans. ASAE 23:1066-1071, 1075.

BROOK, R. C., and BAKKER-ARKEMA, F. W. 1980. Design of multi-stage corn dryers using computer optimization. Trans. ASAE 23:200-203.

BROOKER, D. B., BAKKER-ARKEMA, F. W., and HALL, C. W. 1974. Drying Cereal Grains. Avi Publishing Co., Westport, CT. 265 pp.

BUCKLIN, R. A., ROSS, I. J., and WHITE, G. M. 1985. The influence of grain pressure on the buckling load of thin walled bins. Trans.

ASAE 28:2011-2020.

BURG, W. R., and SHOTWELL, O. L. 1984. Aflatoxin levels in airborne dust generated from contaminated corn during harvest and at an elevator in 1980. J. Assoc. Off. Anal. Chem. 67:309-312.

BURRELL, N. J. 1982. Refrigeration. Pages 407-441 in: Storage of Cereal Grains and Their Products, 3rd ed. C. M. Christensen, ed. Am. Assoc. Cereal Chem., St. Paul, MN.

BURROUGH, D. E., and HARBAGE, R. P. 1953. Performance of a corn picker-sheller. Agric. Eng. 34:21-22.

BYG, D. M., and HALL, G. E. 1968. Corn losses and kernel damage in field shelling of corn. Trans. ASAE 11:164-166.

CHANG, C. S., CONVERSE, H. H., and MARTIN, C. R. 1983. Bulk properties of grain as affected by self-propelled rotational type grain spreaders. Trans. ASAE 25:1543-1550.

CHANG, C. S., CONVERSE, H. H., and LAI, F. S. 1984. Flow rate of corn through orifices as affected by moisture content. Trans. ASAE 27:1586-1589.

CHANG, C. S., CONVERSE, H. H., and LAI, F. S. 1986. Distribution of fines and bulk density of corn as affected by choke-flow, spout-flow, and drop height. Trans. ASAE 29:618-620.

CHOWDHURY, M. F., and BUCHELE, W. F. 1978. The nature of corn kernel damage inflicted in the shelling crescent of grain combines. Trans. ASAE 21:610-615.

CHRISTENSEN, C. M., and SAUER, D. B. 1982. Microflora. Pages 219-240 in: Storage of Cereal Grains and Their Products, 3rd. ed. C. M. Christensen, ed. Am. Assoc. Cereal Chem., St. Paul, MN.

CHUNG, D. S., and PFOST, H. B. 1967. Adsorption and desorption of water vapor by cereal grains and their products, Part II. Trans. ASAE 10:552-555.

DANZINGER, M. T., STEINBERG, M. P., and NELSON, A. I. 1972. Drying of field corn with silica gel. Trans. ASAE 15:1071-1074.

DAVIDSON, J. B. 1931. Agricultural Machinery. John Wiley & Sons, New York.

ECKHOFF, S. R., TUITE, J., FOSTER, G. H., ANDERSON, R. A., and OKOS, M. R. 1984. Inhibition of microbial growth during ambient air corn drying using sulphur dioxide. Trans. ASAE 27:907-914.

ENO, B. E., and FELDERMAN, E. J. 1980. Supplemental heat for grain drying with a transportable solar heater. Trans. ASAE 23:959-963.

FDA. 1983. Food additives permitted for direct addition to food for human consumption; White mineral oil. U.S.P. Fed. Regist. 48(242):55727 (Dec. 15).

FISCUS, D. E., FOSTER, G. H., and KAUFMANN, H. H. 1971. Physical damage of grain caused by various handling techniques. Trans. ASAE 14:480-485, 491.

FISHEL, S. 1983. Cargill's slow-speed leg gains industry acceptance. Grain J. 11:6.

FORTES, M., and OKOS, M. R. 1980. Changes in physical properties of corn during drying. Trans. ASAE 23:1004-1008.

FORTH, M. W., MOWERY, R. W., and FOOTE, L. S. 1951. Automatic feed grinding and handling. Agric. Eng. 23:601-605.

FOSTER, G. H. 1982. Drying cereal grains. Pages 79-116 in: Storage of Cereal Grains and Their Products, 3rd ed. C. M. Christensen, ed. Am. Assoc. Cereal Chem., St. Paul, MN.

FOSTER, G. H., and HOLMAN, L. E. 1973. Grain breakage caused by commercial handling methods. U.S. Dep. Agric. Marketing Res. Rep. 968.

FOSTER, G. H., and STAHL, B. M. 1959. Operating grain aeration systems in the corn belt. U.S. Dep. Agric. Marketing Res. Rep. 337.

FOSTER, G. H., and TUITE, J. 1982. Aeration and stored grain management. Pages 117-143 in: Storage of Cereal Grains and Their Products, 3rd ed. C. M. Christensen, ed. Am. Assoc. Cereal Chem., St. Paul, MN.

FRASER, B. M., and MUIR, W. E. 1980. Energy consumptions predicted for drying with ambient and solar-heated air in Canada. J. Agric. Eng. Res. 25:325-331.

GAYLORD, E. H., and GAYLORD, C. N. 1984. Design of Steel Bins for Storage of Bulk Solids. Prentice-Hall, Englewood Cliffs, NJ.

GOSS, R. R., BAINER, R. G., and SMELTZER, D. G. 1955. Field tests of combines in corn. Agric. Eng. 36:794-796.

GRAHAM, V. A., BILANSKI, W. K., and MENZIES, D. R. 1983. Absorption grain drying using bentonite. Trans. ASAE 26:1512-1515.

GRAMA, S. N., BERN, C. J., and HURBURGH, C. R. 1984. Airflow resistance of mixtures of shelled corn and fines. Trans. ASAE 27:268-272.

GUNASEKARAN, S., and PAULSEN, M. R. 1985. Breakage resistance of corn as a function of drying rates. Trans. ASAE 28:2071-2076.

GUSTAFSON, R. J., and MOREY, R. V. 1979. Study of factors affecting quality changes during drying. Trans. ASAE 22:926-932.

GUSTAFSON, R. J., and MOREY, R. V. 1981. Moisture and quality variations across the column of a crossflow grain drier. Trans. ASAE 24:1621-1625.

GUSTAFSON, R. J., MAHMOUD, A. Y., and HALL, G. E. 1983. Breakage susceptibility reduction by short-term tempering of corn. Trans. ASAE 26:918-922.

HALL, G. E., and ANDERSON, R. J. 1980. Batch internal recycling dryer (BIRD). Paper 80-3515. Am. Soc. Agric. Eng., St. Joseph, MI.

HAMDY, M. Y., and BARRE, H. J. 1970. Analysis and hybrid simulation of deep bed drying of grain. Trans. ASAE 13:752-757.

HAQUE, E., FOSTER, G. H., CHUNG, D. S., and LAI, F. S. 1978. Static pressure drop across a bed of corn mixed with fines. Trans. ASAE 21:997-1000.

HARNOY, A., and RADAJEWSKI, W. 1982. Optimization of grain drying—With rest periods. J. Agric. Eng. Res. 27:291-307.

HEADLEY, V. E., and HALL, C. W. 1963. Drying of shelled corn vibrated in an infrared source. Trans. ASAE 6:148-150.

HERUM, F. L., and BARNES, K. K. 1954. What's the best way to harvest corn? Iowa Farm Sci. 9:7-8.

HILL, L. D., PAULSEN, M. R., and EARLY, M. 1979. Corn quality: Changes during export. Spec. Publ. 58. Coll. Agric., Univ. Ill., Urbana.

HOLMAN, L. E. 1948. Adapting cribs for corn drying. Agric. Eng. 29:149-151.

HOLMES, B. J., KLEMME, R. M., and LINDHOLM, J. 1985. Economic evaluation of corn drying systems in the Upper Midwest. Trans. ASAE 28:907-914.

HUKILL, W. V. 1947. Basic principles in drying corn and grain sorghum. Agric. Eng. 28:335-338, 340.

HUKILL, W. V. 1948. Types and performance of farm grain driers. Agric. Eng. 29:53-54, 59.

HURBURGH, C. R., and MOECHNIG, B. W. 1984. Shrinkage and other corn-quality changes from drying at commercial elevators. Trans. ASAE 27:1176-1180.

HURBURGH, C. R., BERN, C. J., WILCKE, W. F., and ANDERSON, M. E. 1983. Shrinkage and corn quality changes in on-farm handling operations. Trans. ASAE 26:1854-1857.

HURLBUT, L. W. 1955. Harvesting corn by combine. Agric. Eng. 36:791-792.

HURST, W. M. 1927. Bulk drying of wheat by forced ventilation with heated air. Agric. Eng. 8:201-203.

HUSTRULID, A., and FLIKKE, A. M. 1959. Theoretical drying curve for shelled corn. Trans. ASAE 2:112-114.

HYDE, M. B., and BURRELL, N. J. 1982. Controlled atmosphere storage. Pages 443-478 in: Storage of Cereal Grains and Their Products, 3rd ed. C. M. Christensen, ed. Am. Assoc. Cereal Chem., St. Paul, MN.

JINDAL, V. K., HERUM, F. L., and MENSAH, J. K. A. 1978. Effects of repeated freezing-thawing cycles on the mechanical strength of corn kernels. Trans. ASAE 21:367-370, 374.

JOHNSON, W. H., and LAMP, B. J. 1966. Principles, Equipment and Systems for Corn Harvesting. Agric. Consulting Assoc., Inc., Wooster, OH.

JUSTICE, D. L., and BASS, L. N. 1978. Principles and Practices of Seed Storage. Agric. Handb. 506. U.S. Dep. Agric., Washington, DC.

KAMEYAMA, Y., LAI, F. S., SAYAMA, H., and FAN, L. T. 1982. The risk of dust explosions in grain processing and handling facilities. J. Agric. Eng. Res. 27:253-259.

KEENER, H. M., HENRY, J. E., and ANDERSON, R. J. 1983. Controllable fluidized-bed direct combustor produces clean high temperature air. Paper 83-3037. Am. Soc. Agric. Eng., St. Joseph, MI.

KLINE, G. L. 1972. Damage to corn during harvest and drying. Pages 79-82 in: Proc. Grain Damage Symp. Am. Soc. Agric. Eng. and Dept. Agric. Eng., The Ohio State Univ., Columbus.

LAI, F. S., MILLER, B. S., MARTIN, C. R., STOREY, C. L., and BOLTE, L. 1981. Reducing grain dust with oil additives. Trans. ASAE 24:1626-1631.

LI, H., and MOREY, R. V. 1984. Thin-layer drying of yellow dent corn. Trans. ASAE 27:581-585.

LINDBLAD, C., and DRUBEN, L. 1976. Small farm grain storage. Action/Peace Corps J. Manual Ser. 2 and VITA Pub. Manual Ser. 35E. ACTION, Washington, DC.

LOEWER, O. J., BRIDGES, T. C., WHITE, G. M., and ROZOR, R. B. 1984a. Optimum moisture content to begin harvesting corn as influenced by energy cost. Trans. ASAE 27:362-365.

LOEWER, O. J., BRIDGES, T. C., COLLIER, D. G., and WHITE, G. M. 1984b. Economics of stirring devices in grain drying. Trans. ASAE 27:603-608.

LOGAN, C. A. 1931. The development of a corn combine. Agric. Eng. 12:277-278.

MAGEE, K. J., STROSHINE, R. L., FOSTER, G. H., and BAKER, K. D. 1983. Nature and extent of grain damage caused by pneumatic conveying systems. Paper 83-3508. Am. Soc. Agric. Eng., St. Joseph, MI.

MANBECK, H. B. 1984. Predicting thermally-induced pressures in grain bins. Trans. ASAE 27:482-486.

McFATE, K. L., and GEORGE, R. M. 1971. Power-capacity relationships of nominal 8-inch screw conveyors when handling shelled corn. Trans. ASAE 14:121-123, 126.

McKENZIE, B. A., FOSTER, G. H., NOYES, R. T., and THOMPSON, R. A. 1966. Dryeration—Better corn quality with high speed drying. Coop. Ext. Serv. AE-72. Purdue Univ., W. Lafayette, IN.

McKIBBEN, E. G. 1929. Harvesting corn with a combine. Agric. Eng. 10:231-232.

MILLER, B. S., and POMERANZ, Y., eds. 1979. Proceedings of the International Symposium on Grain Dust. Div. Continuing Educ., Kansas State Univ., Manhattan. 508 pp.

MISRA, M. R., and BROOKER, D. B. 1980. Thin-layer drying and rewetting equations for shelled yellow corn. Trans. ASAE 23:1254-1260.

MOREY, R. V., CLOUD, H. A., GUSTAFSON, R. J., and PETERSEN, D. W. 1979. Management of ambient air drying systems. Trans. ASAE 22:1418-1425.

MOREY, R. V., THIMSEN, D. P., LANG, J. P., and HANSEN, D. J. 1984. A corncob-fueled drying system. Trans. ASAE 27:556-560.

MORRISON, C. S. 1955. Attachments for combining corn. Agric. Eng. 36:796-799.

MULTON, J.-L., CAHAGNIER, B., and BRYON, G. 1980. Heat and water vapor transfers in a corn (maize) bin and improvement of ventilation management. Cereal Foods World 25:16-20.

NATIONAL GRAIN AND FEED ASSOCIATION. 1979. Practical Guide to Elevator Design. The Association, Washington, DC.

NATIONAL MATERIALS ADVISORY BOARD. 1982a. Prevention of grain elevator and mill explosions. NMAB Rep. 367-2, Natl. Tech. Inf. Serv., Springfield, VA.

NATIONAL MATERIALS ADVISORY BOARD. 1982b. Pneumatic dust control in grain elevators. NMAB Rep. 367-8. National Academy Press, Washington, DC.

NEUBAUER, L. W., and WALKER, H. B. 1961. Farm building design. Prentice-Hall, Englewood Cliffs, NJ.

NGODDY, P. O., and BAKKER-ARKEMA, F. W. 1975. A theory of sorption hysteresis in biological materials. J. Agric. Eng. Res.

20:109-121.

NOFSINGER, G. W., BOTHAST, R. J., and ANDERSON, R. A. 1979. Field trials using extenders for ambient-conditioning high-moisture corn. Trans. ASAE 22:1208-1213.

PAULSEN, M. R., and NAVE, W. R. 1980. Corn damage from conventional and rotary combines. Trans. ASAE 23:1110-1116.

PAYNE, F. A., ROSS, I. J., WALKER, J. N., and BRASHEAR, R. S. 1983. Exhaust analysis from gasification-combustion of corn cobs. Trans. ASAE 26:246-249.

PETERSON, W. H. 1982. "Low-profile" bins save energy and money. Ill. Farm Electrification Counc. Fact Sheet. Dep. Agric. Eng., Univ. Illinois, Urbana.

PICKARD, G. E. 1955. Laboratory studies of corn combining. Agric. Eng. 36:792-794.

PIERCE, R. O., and HANNA, M. A. 1985. Corn kernel damage during on-farm handling. Trans. ASAE 28:239-241, 245.

PIERCE, R. O., and THOMPSON, T. L. 1982. Drying scheduling—A procedure for layer filling low-temperature corn drying systems. Trans. ASAE 25:469-474.

QUICK, G. R., and BUCHELE, W. F. 1978. The Grain Harvesters. Am. Soc. Agric. Eng., St. Joseph, MI. 269 pp.

RADEMACHER, F. J. C. 1981. On possible flow back in vertical screw conveyors for cohesionless granular materials. J. Agric. Eng. Res. 26:225-250.

RICHEY, C. B., and PEART, R. M. 1973. Pallet system of harvest, handling, and storing ear corn. Trans. ASAE 16:40-43.

RISCH, E., and HERUM, F. L. 1982. Bin wall stresses due to aeration of stored shelled corn. Paper 82-4072. Am. Soc. Agric. Eng., St. Joseph, MI.

ROBERTS, D. E., and BROOKER, D. B. 1975. Grain drying with a recirculator. Trans. ASAE 18:181-184.

ROSS, I. J., LOEWER, O. J., and WHITE, G. M. 1979. Potential for aflatoxin development in low temperature drying systems. Trans. ASAE 22:1439-1443.

SANDS, L. D., and HALL, G. E. 1971. Damage to shelled corn during transport in a screw conveyor. Trans. ASAE 14:584-585, 589.

SAUER, D. B., and BURROUGHS, R. 1980. Fungal growth, aflatoxin production, and moisture equilibration in mixtures of wet and dry corn. Phytopathology 70:516-521.

SAUER, D. B., HODGES, T., BURROUGHS, R., and CONVERSE, H. 1975. Comparison of propionic acid and methylene bis propionate as grain preservatives. Trans. ASAE 18:1162-1164.

SCHWAB, C. V., ROSS, I. J., PIERCY, L. R., and McKENZIE, B. A. 1985. Vertical pull and immersion velocity of mannequins trapped in enveloping grain flow. Trans. ASAE 28:1997-2002.

SCRANTON, C. J. 1952. Safety and the mechanical corn picker. Agric. Eng. 30:140-142.

SHARAF-ELDEEN, Y. I., BLAISDELL, J. L., and HAMDY, M. Y. 1980. A model for ear corn drying. Trans. ASAE 23:1261-1265, 1271.

SHEDD, C. K. 1933. A study of mechanical corn pickers. Agric. Eng. 14:123-125.

SHEDD, C. K. 1953. Resistance of grains and seeds to air flow. Agric. Eng. 34:616-619.

SHOVE, G. C. 1968. Application of dehydrorefrigeration to shelled corn conditioning. Trans. ASAE 11:312-317.

SINGH, S. S., and FINNER, M. F. 1983. A centrifugal impactor for damage susceptibility evaluation of shelled corn. Trans. ASAE 26:1858-1863.

SIZEMORE, S. W., LOEWER, O. J., TARABA, J. L., ROSS, I. J., and WHITE, G. M. 1984. Retention of gasification-combustion products by corn. Trans. ASAE 27:1546-1548.

SKELTON, R. F., and BATEMAN, H. P. 1942. Field shelling of corn. Agric. Eng. 23:131-133.

STEPHENS, L. E., and FOSTER, G. 1977. Reducing damage to corn handled through gravity spouts. Trans. ASAE 20:367-371.

SUSAI, A. D., and GUSTAFSON, R. J. 1982. Power requirements and quality changes of material for a pneumatic conveying system. Paper 82-3559. Am. Soc. Agric. Eng., St. Joseph, MI.

SYARIEF, A. M., GUSTAFSON, R. J., and MOREY, R. V. 1984. Moisture diffusion coefficient for yellow-dent corn components. Paper 84-3551. Am. Soc. Agric. Eng., St. Joseph, MI.

TESSIER, S., and RAGHAVAN, G. S. V. 1984. Performance of a sand medium drier for shelled corn. Trans. ASAE 27:1227-1232.

THOMPSON, R. A., and FOSTER, G. H. 1963. Stress cracks and breakage in artificially-dried corn. U.S. Dep. Agric. Mark. Res. Rep. 631.

THOMPSON, S. A., USRY, J. L., and LEGG, J. A. 1986. Loads in a model grain bin as affected by various unloading techniques. Trans. ASAE 29:556-561.

THOMPSON, T. L., PEART, R. M., and FOSTER, G. H. 1968. Mathematical simulation of corn drying—A new model. Trans. ASAE 11:582-586.

TRIPLEHORN, C. A., HERUM, F. L., PIGATTI, D., GIANNOTTI, O., and

PIGATTI, A. 1966. O paiol de tela para armazemento de milho. O. Biologico (Brazil) 32:38-42.

USDA. 1952a. Drying ear corn with unheated air. U.S. Dep. Agric. Leafl. 334.

USDA. 1952b. Drying ear corn with heated air. U.S. Dep. Agric. Leafl. 333.

USDA. 1969. Guidelines for mold control in high-moisture corn. U.S. Dep. Agric. Farmers Bull. 2238.

USDA. 1983. On-farm demonstrations of solar drying of crops and grains. U.S. Dep. Agric. Ext. Serv., Washington, DC.

WADE, F. J., and HAWK, A. L. 1980. Dust measurement inside grain conveying equipment. Paper 80-3560. Am. Soc. Agric. Eng., St. Joseph, MI.

WHITE, G. M., SCHAPER, L. A., ROSS, I. J., and ISAACS, G. W. 1962. Performance characteristics of enclosed screw conveyors handling shelled corn and soybeans. Res. Bull. 740. Agric. Exp. Stn., Purdue Univ., W. Lafayette, IN.

WILCKE, W. F., and BERN, C. J. 1986. Natural-air corn drying with stirring: II. Dryer performance. Trans. ASAE 29:860-867.

WILEMAN, R. H. 1933. The effect of corn plant characteristics on mechanical corn picker loss. Agric. Eng. 14:125-126.

WOOD, J. E., and KERR, R. P., Jr. 1980. Evaluation of grain loss monitor performances. Paper 80-1451. Am. Soc. Agric. Eng., St. Joseph, MI.

YOUNG, R. L. 1931. Present status of mechanical corn picking. Agric. Eng. 12:267-270.

ZACHARIAH, G. L., and ISAACS, G. W. 1966. Simulating a moisture-control system for a continuous flow drier. Trans. ASAE 9:297-302.

CHAPTER 5

MEASUREMENT AND MAINTENANCE OF QUALITY

STANLEY A. WATSON (*retired*)
Director's Office
Ohio Agricultural Research and Development Center
The Ohio State University
Wooster, Ohio

I. INTRODUCTION

It has been said, "Beauty is in the eye of the beholder"; so it is with corn quality. Its definition is different for every commercial user of corn. Corn quality includes a spectrum from perfect, unblemished kernels used for seed to severely damaged kernels useful only as fuel. The quality differences in the products offered to the user start with the hybrid that is planted and are affected by cultural and climatic conditions and by harvesting, drying, storage, handling, and transportation methods (Freeman, 1980). The corn kernel is a living organism and must be treated as such if the highest quality is to be preserved. Corn is not only food for humans and domestic animals but, if care is not taken to control pests, it can also become food for rodents, birds, insects, and fungi. In the process of consuming, contaminating, or decomposing the corn, these pests have adverse effects on quality.

The quality of unblemished kernels varies because of the influences of inheritance and environment on the structure and composition of the kernel components. For example, cultivars may differ in the ratio of horny to floury endosperm, relative size of germ, and thickness of pericarp and may have subtle influences on resistance to pest invasion. Differing percentages of total protein and ratios of protein components (and of amino acids) can be modified by breeding and nitrogen fertility of the soil. Likewise, the amount and composition of oil in the germ can be modified by breeding and by climatic conditions. Changes in any one component obviously affect the percentages of other components. The molecular components of starch granules can be genetically modified. Some intrinsic quality differences are discussed in this chapter, but differences in structure and composition are discussed primarily in Chapter 3.

Quality can be measured by a variety of tests, from simple visual appearance to the more complicated laboratory milling tests. The most commonly used tests are those included in the national grades and standards of corn-growing nations; however, these tests may be superficial and inadequate to all but the least critical user. Each industry, and each individual user within an industry, has developed tests that give the best index of performance in particular processes and products. Research conducted by public and private institutions has developed quick tests of raw corn to measure its end-use attributes. However, the ultimate test is the performance of corn in the actual use in individual manufacturing plants or in consumption. All quick tests must be calibrated against actual commercial experience. Fortunately, this type of testing can be accomplished by a laboratory simulation of the process or use, such as laboratory wet- or dry-milling procedures or feeding trials. Some tests on the raw corn may be imposed by government regulations to monitor situations where public health must be considered. These include tests for the presence of mycotoxins, chemical additives, or insects. This chapter surveys all aspects of the corn quality problem, including cause, recognition, and end-use consequences. The influence of insects on corn quality and the control of insects on corn are described in Chapter 6.

II. GRADES AND STANDARDS

A. History

To facilitate grain marketing for both domestic and export grain, the U.S. Department of Agriculture (USDA), after nearly 10 years of discussion with the grain trade, developed in 1914 a set of standard rules and definitions for uniform marketing. These rules were adopted as official standards in 1916 in the United States Grain Standards Act (Hoffman and Hill, 1976). In spite of great changes in the way corn is harvested, handled, and marketed, few changes have been

TABLE I
United States, Grading Standards for Corn[a]

Grade	Minimum Test Weight per Bushel (lb)	Maximum Limits		
		Broken Corn and Foreign Material (%)	Damaged Kernels	
			Total (%)	Heat-Damaged Kernels (%)
1	56	2.0	3.0	0.1
2	54	3.0	5.0	0.2
3	52	4.0	7.0	0.5
4	49	5.0	10.0	1.0
5	46	7.0	15.0	3.0
Sample Grade[b]	...	...	...	...

[a] Grades and grade requirements for the classes Yellow Corn, White Corn, and Mixed Corn.
[b] U.S. Sample Grade is corn that a) does not meet the requirements for grades No. 1–5; or b) in a 1,000-g sample, contains eight or more stones with aggregate weight in excess of 0.20% of the sample weight, two or more pieces of glass, three or more crotalaria seeds, two or more castor beans, eight or more cockleburs, four or more particles of an unknown substance(s) or a commonly recognized harmful or toxic substance(s), or animal filth in excess of 0.20%; or c) has a musty, sour, or commercially objectionable foreign odor; or d) is heating or otherwise of distinctly low quality (FGIS, 1984a).

made in the grades and standards for corn since 1914. Some changes were made in 1977 (FGIS, 1980a) and others in 1985. The grain standards for corn (Table I) are based on selected physical attributes of the whole kernels. These attributes were chosen to define and measure grain properties because experience had indicated that they provided relatively simple ways to differentiate among lots of corn differing in quality. The U.S. grain grades and standards are administered by the Federal Grain Inspection Service (FGIS, formerly the Grain Division, Agricultural Marketing Service) a unit of the USDA.

B. Sample Collections

Before the quality of a lot of grain can be evaluated, a sample must be obtained that is as representative as possible of the entire lot. A survey of USDA-approved sampling devices and procedures used for FGIS official inspections is given by Parker et al (1982). Other devices may be used for nonofficial inspections if results are equal to those of the approved devices. The trier or probe is the basic tool for obtaining grain samples from trucks, boxcars, hopper cars, or barges if a grade is needed before the lot is unloaded. The trier consists of two concentric metal tubes that can be turned to open or close coordinated slots to collect the grain. Most country elevators receiving grain directly from farmers rely on mechanically probed samples for initial value determination. Because of the need for faster sampling, many elevators are employing automatic probes that use vacuum to pull the grain into the probe and deliver it to a container. Hurburgh and Bern (1983) tested sampling accuracy of commercially available mechanical probes. They found the probes that collect grain by suction overestimated broken corn and foreign matter (BCFM) in corn by 1.5 percentage points, whereas the USDA-approved compartmented hand probe overestimated it by 1.05 percentage points. In lots with up to 5% BCFM, the amount of overestimation increases with the square of the BCFM content. The core-type automatic probe samples BCFM more accurately than the hand probe, and it is now approved for official inspections.

The other class of samplers are those designed to sample moving streams of grain either by hand or automatically. The preferred device for hand sampling is the pelican, a narrow, elongated leather cup that is swung through the grain stream. The Ellis cup is used to collect hand samples from a belt. Diverter-type mechanical samplers, installed at the end of a belt or within a spout are preferred for sampling moving grain streams. They are not only labor-saving devices, but they provide a more representative sample than the hand samplers (FGIS, 1980b). Diverter-type samplers vary in design. In one type, a pelican moves back and forth through the grain stream at regular intervals; in others, a small portion of the stream continuously diverts a primary sample through a splitting device, which delivers a secondary sample for use by the inspector (Watson et al, 1970). A Woodside sampler continuously collects samples from a belt by means of a small bucket conveyor mounted directly over the grain stream.

C. Classes and Grades

The standards specify three classes of corn (FGIS, 1984a): Yellow Corn, White Corn, and Mixed Corn. Yellow corn must contain no more than 5% white corn kernels, but white corn can contain no more than 2% yellow corn. Corn that contains more than 10% of other grains is classed as Mixed Grain. There are also five numerical grades, No. 1 to No. 5, and a lower grade known as Sample Grade. Currently, grade is determined by four factors: test weight, a measure of bulk density; BCFM; damaged kernels; and "heat damaged" kernels. Before September 1985, moisture was also a grade-determining factor, but it is now only a standard, the value of which must be shown on the invoice (FGIS, 1984b). The grade number is determined by whichever of the four factors indicates highest grade number (lowest grade). For example, if BCFM, total damage, and heat damage of a lot of yellow corn are all below the No. 1 grade level of 2, 3, and 0.1%, respectively, but test weight is 53 lb/bu (68.2 kg/hl), the corn is graded U.S. No. 3Y.

As indicated in Table I, the presence of stones, mustiness, sourness, heating, or an objectionable odor requires a designation of Sample Grade. Popcorn or sweet corn kernels in dent corn are graded as foreign material. Three special grades are also recognized: Flint Corn (95% or more of flint kernels); Flint and Dent Corn (a mixture of flint and dent containing more than 50% but less than 95% flint corn); Waxy Corn (95% or more of waxy kernels). "Infested corn" (previously termed "weevily") includes the presence of two or more live weevils or one weevil and five other insects injurious to corn or 15 or more other live insects injurious to stored grain (FGIS, 1980c). The term "distinctly low quality" indicates the presence of broken glass, large stones, or pieces of concrete, animal filth, or an unknown foreign substance such as a seed treatment chemical. Sample Grade was revised in 1985 to include a maximum level of harmful seeds of cocklebur (*Agrostemma githago*), crotalaria (*Crotalaria* spp.), or castor bean (*Ricinus communis*), as shown in Table I.

D. Grading Procedures

Inspection (grading) of a lot of corn begins by reduction of the size of the field sample to at least 2,500 g for an official FGIS inspection. This representative sample must be randomly divided to produce a 1,000-g "work sample." Official division must be made with one of several commercially available dividers, including the Boerner, Gamet rotary, or Cargo dividers. The work sample is used for measuring test weight and BCFM.

Test weight is a measure of bulk density expressed in pounds per bushel. The work sample is dropped 2 in. (5.1 mm) from a funnel into a 1-qt (1.1 L) container that is leveled and weighed. The net weight is multiplied by 32 to give bushel weight (1 bu = 0.3523 hl). The United States is now the only country still using the bushel measure. Use of the bushel for grain trading is confusing because grain is measured and used on a weight basis. In the United States, most other grains are also traded on the bushel basis, except rice and sorghum, which are traded on a hundredweight basis. Conversion to the metric system by the United States would standardize units on a world basis.

The BCFM factor is the sum of all material passing through a 4.8-mm (12/64 in.) round-hole sieve plus all noncorn material remaining on the sieve. Official determination is made using the work sample in the Carter dockage tester, an automated screening device. The cleaned work sample is next split into 250-g portions for determination of damage. The U.S. grade standards (FGIS, 1984a) state that

> Damaged kernels shall be kernels and pieces of kernels of corn which are heat damaged, sprouted, frosted, badly ground damaged, badly weather damaged, moldy, diseased, or otherwise materially damaged.

Most of the listed categories of damage are due to mold invasion. However, the "otherwise materially damaged" category includes insect-bored kernels and dryer damage. Damage caused by excessive temperature during artificial drying includes kernels that are discolored, wrinkled, and blistered, puffed, or swollen from excessive heat or that are obviously crazed or stress cracked (FGIS, 1980b). The damaged category normally can be observed as visible mold on the germ face (called "blue-eye") or a noticeable darkening of the germ. Heat damage is discoloration of the entire kernel due to spontaneous heating resulting from excessive microbial growth in an unfavorable storage situation. Heat damage also includes blackening due to burning from any excessive heat source. Only 0.2% of this category is allowed in U.S. No. 2 corn. Color slides of representative kernels of specific damage categories are available in grading stations for comparison to samples being graded.

E. Accuracy of Grade Determinations

Since only a small sample, subject to bias and variability, represents a large corn shipment, the results of a grade test are really an estimate of the properties of the whole lot. Two types of variation exist: random and nonrandom. Nonrandom variation has several sources: 1) uneven distribution of grain or impurities, 2) improper or inadequate sampling procedures, and 3) inaccurate measurements. Random variation is natural and unavoidable. Every kernel is an individual and differs from other kernels. Likewise ears, fields, farms, and areas also differ. If a load of corn is thoroughly mixed, the closeness of the report to the actual condition of the load is governed by the laws of probability. One can calculate the sample size required to produce a result that has a given desired probability of representing the entire lot. Since most statisticians prefer a probability of at least 0.1 (90% confidence), properties that occur infrequently such as heat damage or aflatoxin content require a very large sample size to achieve acceptable confidence limits. Elam and Hill (1977) have criticized the U.S. grade standards for expecting to find 0.1 or 0.2% heat-damaged corn kernels in 1,000-g samples with any accuracy, as required by U.S. Grades No. 1 and 2. Johnson et al (1969) recommended that 6 lb (2.7 kg) of corn are required to adequately analyze corn lots for aflatoxin. As described later, aflatoxin is usually present in concentrations of several hundred thousand parts per billion in a few kernels, and yet U.S. Food and Drug Administration rules specify that loads containing over 20 ppb cannot be traded (see section on mycotoxins).

Among the nonrandom errors, uneven distribution is more of a problem with some factors than with others. Although test weight can vary within a load, the spread is generally not very great. However, nonuniform distribution of BCFM and moisture in lots of corn pose serious sampling problems. Moisture variation within a lot of corn can result from inadequate mixing of sublots of differing moisture contents (Hill et al, 1986) and differential flow characteristics of dry and damp corn. In the case of BCFM, fine material does not flow as readily as whole kernels and tends to build up in pockets or cores, known as "spout lines," beneath the loading spout. Also the unethical practice of "plugging" is sometimes used, i.e., intentionally placing portions of low-quality grain or BCFM in locations within a load not apt to be sampled.

Because of these variations, the FGIS has minimum patterns for probe sampling all types of containers (FGIS, 1980b; Parker et al, 1982). Lai (1978) has described a mathematical procedure for determining the number and location of probe samples required to adequately describe the fines distribution in a bin of corn. Most country elevators take only one probe per load. Hurburgh (1984) has recommended a two-probe sample pattern that reduces variability in truck sampling. Sampling for the presence of live insects is even more difficult because of their mobility within the mass (Barak and Harein, 1981).

Besides inadequate sampling, nonrandom errors due to inaccurate measurements can result from incorrectly adjusted or calibrated instruments, e.g., an inaccurate thermometer for moisture determination. Also, the analyst may use the wrong procedure, make recording errors, etc.

Some inaccuracies may result because of the speed of grain movement in elevators. This is being helped by introduction of automated measurement of test weight, BCFM, and moisture, which may be used with a computer to calculate and record results. The subjective tests, which include damage, odors, other grain, weed seeds, and foreign objects, will be very difficult to automate.

F. Importance of Grades in Corn Marketing

Hall and Rosenfeld (1982) showed that buyers use discounts based on grade factors to attract the quality of grain they believe best fits their use. Or, inversely, high discounts discourage sellers from offering quality that is deemed unsuited for a particular use. These researchers found that the higher the end-use value of the products, the higher the discounts made on corn of substandard quality. Furthermore, firms with high market share tended to have the highest discount schedules.

The quality of corn produced by farmers varies greatly in all quality factors because of differences in soils, climate, insects, disease, hybrids, and management practices with respect to harvesting, drying, storing, etc. Maintaining all grade levels throughout the marketplace would be confusing and costly. Therefore, blending lots of different qualities at all steps in the marketing chain from farmers, country elevators, terminals, and river elevators, to export elevators is necessary to provide uniform quality. No. 2 grade dominates the U.S. market, whereas No. 3 dominates the U.S. export market. Thus, the best grade (U.S. No. 1) and the poorer grades disappear as the corn proceeds through marketing channels. Farmers have limited options with respect to the quality of corn they can deliver except as corn is blended from different fields or bins as it is moved around the farm or is delivered to the elevator. This is because blending facilities are limited and because the average farmer has only a small pool of grain to work with. The country elevator has a much larger pool of corn having wider differences in quality. Blending the diverse lots to fit the limitations of individual elevator bin capacity, or to fulfill contracts for a specific quality, provides a significant portion of operating profit. Blending profits are achieved by discounting the price paid for the corn based on grade factors that exceed the base grade of U.S. No. 2Y but generally paying no premium for quality better than the U.S. No. 2Y base. Hill (1982a) contends that this practice, especially with respect to moisture, causes the seller supplying the higher-quality grain to subsidize the services received by the seller supplying the lower-quality grain. Competition for farmer's grain is prompting some elevators to develop purchasing strategies that more fairly compensate sellers for actual quality delivered.

The use of discounts in grain purchasing provides a signal by the buyer as to the quality of grain preferred. Thus, with the base set at U.S. No. 2Y corn, the purchaser is telling the seller that the highest price will be paid for delivery of corn exactly at 54.0 lb/bu, 3.0% BCFM, 5.0% damage, 0.2% heat damage, and (formerly) at 15.5% moisture content (MC). Although moisture is no longer a grade-determining factor, the price of the corn must be adjusted to compensate for excess moisture. It is likely that the 15.5% MC base will continue to be used because corn can be stored for a few months at this moisture level with no damage that results in a change in grade (Saul, 1968). Since farmers are generally not rewarded for delivery of quality better than No. 2Y, harvesting and storing practices are generally adjusted to producing corn of less than optimum quality, especially with respect to BCFM and damage. Test weights cannot be manipulated because they are primarily determined by climatic and cultural factors, although drying conditions exert some influence (Hall, 1972). Moisture content can be controlled, but the corn has to be dried to below 16% for storage even of short duration. Farmers who dry their own corn do so for longer-term storage and therefore must dry it to 14% MC or lower. About 50% of the corn is dried on the farm and 50% by elevators (Hill, 1985a).

Grain standards should include only those factors that supply the most useful information to the average prospective buyer. Some FGIS officials (FGIS, 1979; Anonymous, 1981) have stated they do not believe existing standards provide adequate measures of end-use properties. This opinion is echoed by other observers (Elam and Hill, 1977; Hall and Rosenfeld, 1982).

Grain grades were established to allow buyer and seller to be able to agree on a price, using a universal set of rules that both accept and understand. However, the use of grades in marketing is conditioned by the pricing policies of individual marketing organizations. Bermingham and Hill (1978) state,

> The greater the distance between buyer and seller, and the more complex the marketing system, the more that buyer and seller must rely on grading standards and standardized terminology for a description of the characteristics of the grain.

The greatest distance between buyer and seller is in the export market. The closest is in dealings between local people, such as farmer and local elevator owner or farmer and farmer. In the latter case, grade often is not a basis of the transaction.

Specialty corns such as seed corn, waxy corn, popcorn, high-amylose corn, high-lysine corn, and most white corn are generally grown under an identity-preserved contract specifying delivery of strict quality characteristics specified by the buyer. Some manufacturers of specialty corn food products also use the contract method. In the case of sweet corn, especially that intended for canning or freezing, the buyer usually has agents in the field who tell the producer when to harvest to achieve the desired quality.

G. Grades and Standards for Export Trade

Most corn exported from the United States is Grade No. 3 with respect to BCFM, but some buyers specify a maximum of 14.0% MC in order to prevent mold development during shipment. Sale is made on the basis of "origin certificate final," which is a federal white certificate issued on inspection of a sample collected at the time a vessel is loaded. No allowance is made for any change in properties during transit such as additional breakage or mold damage. Other grain-exporting countries (Argentina, South Africa, Thailand, and most European countries) sell on a different system called "fair average quality" (FAQ) (Bermingham and Hill, 1978). Under this system, a representative sample of corn is obtained as a vessel is being unloaded at the destination. Samples from all shipments arriving under an FAQ contract from a particular origin are sent to a central office, where a committee made up of representatives of both buyers and sellers compares all samples for the month from each origin port by mostly visual inspection to determine average quality. When agreement is reached, the lots of median or representative quality may be further tested, then pooled, retested, and used as a quality standard. This becomes the FAQ sample for the month. Lots judged to be of better or poorer quality are priced accordingly. Thus, the FAQ arriving at a particular port differs by country of origin and from month to month. Bermingham and Hill (1978) evaluated FAQ samples of French and Argentine corn by the U.S. grading system and found some above and some below U.S. Grade No. 2.

The advantage of the FAQ system for the seller is that it is easier to provide acceptable shipments in years when crop quality is low. The advantage to the buyer is that payment is made for quality received that month, including any changes occurring during transport or handling. However, the month-to-month quality changes introduces uncertainty about future quality. With the certificate-final system, based on a numerical grade, a buyer knows that the quality shipped was within the limits for the grade ordered. However, the buyer may be disappointed if hot spots have developed in transit, a great deal of breakage has occurred during unloading, or the BCFM has segregated to give portions of the load that are higher in BCFM than the average.

In order to satisfy customers requesting to buy on a numerical grade, most exporting countries have also developed a grading system similar to the U.S. standards. Table II gives the grade standards for South Africa and Table III for Argentina. Both countries agree with the United States in using the lowest factor for grade determination. The definition of damaged kernels is the same for all three countries, being essentially mold and insect damage. They are also similar in that one factor is based on material passing through a sieve. Grading equipment and procedures are similar.

Differences among the three countries are important. Both Argentina and South Africa keep broken corn and foreign material as separate factors. The

Argentines use a 3.17-mm (8/64-in.) triangular-hole sieve to separate broken corn. The South Africans combine broken corn and whole kernels that pass a 6.35-mm (16/64-in.) sieve with mold-damaged kernels and call this fraction "defective maize kernels."

Only South Africa retains moisture as a grade-determining factor, and only the United States has a grade factor for test weight. Only the Argentine system separates damaged kernels, broken kernels, and foreign material as separate factors. Market pricing structures also affect grades. For example, Argentine elevators pay a premium for No. 1 corn, causing farmers to screen corn before sale, but most lots have low BCFM because they are of flint corn, field dried and ear harvested. Therefore, Argentine exporters have not had the breakage problem. As Argentine farmers change to dent corn (for higher yields) and use combine harvesting plus artificial drying, they are observing stress cracking and greater breakage in both dent and flint classes (Paulsen and Hill, 1985b).

Shellenberger (1982) has suggested the establishment of international standards for grain, including a few basic determinations of grain characteristics used universally. An international standard should not prevent the continued use of prevailing standards within each country. International standards would not be required for all export or import transactions. They would merely be available when requested and would provide a uniform standard of comparison worldwide.

TABLE II
South African Grading Standards for Yellow Maize[a]

	Maximum Percent Allowable			
Grade	A Defective Kernels[b]	B Other-Colored Maize Kernels	C Foreign Material	D Collective Deviations in (A) + (B) + (C)[c]
1	9.0	2.0	0.3	9.0
2	20.0	5.0	0.5	20.0
3	30.0	5.0	0.75	30.0

[a] Source: Hill (1982c); used by permission.
[b] Kernels showing mold, insect damage, or mechanical damage, plus pieces of kernels that will pass through a 6.35-mm (16/64-in.) round-hold sieve.
[c] Deviations in A, B, or C must be within specified limits.

TABLE III
Argentine Grading Standards for Maize[a]

	Maximum Percent Allowable		
Grade	Damage[b]	Broken Kernels[c]	Foreign Materials
1	3	2	1
2	5	3	1.5
3	8	5	2

[a] Source: Hill (1982c); used by permission.
[b] Fermented, moldy, sprouted, etc.
[c] Pieces of kernels that will pass through a 3.17-mm triangular-hole sieve.

III. BROKEN CORN, FOREIGN MATERIAL, AND DUST

A. BCFM

When the U.S. grade standards were established, very little broken corn was encountered because ear-dried kernels are tough. It was logical to combine broken corn and foreign material in 1914 when foreign material predominated, but in the 1980s, the situation is reversed. BCFM in corn was first defined as all material passing a 5.6-mm (14/64-in.) round-hole sieve, but in 1921 the screen was changed to 4.8-mm (12/64-in.). The arbitrary nature of either screen size has been questioned, especially by Hill et al (1982). These authors examined fractions of BCFM obtained from over 1,000 Illinois corn samples screened on a succession of sieve sizes from 5.9-mm (15/64-in.) to 1.8-mm (4.5/64-in.). Fractions were analyzed for physical and chemical composition. They concluded that the 3.2-mm (8/64-in.) and the 2.4-mm (6/64-in.) sieves showed the greatest differences between corn and screenings. The material remaining on the 4.8-mm (12/64-in.) and 4.0-mm (10/64-in.) sieves was closer to corn in composition, but none of the differences was sharp. As particle size decreased, protein, ash, and fiber contents increased gradually, whereas fat and digestible energy values decreased.

Although a good case can be made for separate classifications of broken corn and foreign material in grading corn, the separation of pieces of corn from noncorn particles cannot be made by screening alone. However, because the finest material is highest in noncorn material such as dirt (ash), this fraction should be regarded as "foreign material." Large pieces of foreign material would have to be hand-picked, as is presently done. Classifying material passing a fine screen, e.g., 2.3-mm (6/64-in.), as dockage would result in removal of dust from the grain early in the marketing chain, thus diminishing the danger of dust explosions. In the present U.S. grading system, Grade No. 2 permits a maximum of 3% BCFM without a discount, which sends a message to the market that the first 3% of BCFM, whether it is all broken corn or all foreign material, has the same value as corn. Therefore, BCFM cleaned out of lots above the 3.0% limit may be added back to lots below that level. But this practice has been criticized as detrimental to the U.S. export market and may lead to separation of broken corn and foreign material in grading standards.

Hill et al (1982) also found that as the corn passed through the marketing chain, the percentage of broken corn in BCFM increased and the foreign material decreased (Table IV). This primarily reflects the tendency of corn to

TABLE IV
Physical Properties of Materials Passing Through a 4.8-mm (12/64-in.) Sieve, Illinois, 1976–1977[a]

	Country Elevators		Terminal Elevators	
	Receipts	Shipments	Receipts	Shipments
Corn, %	83.4 ± 14.4	87.9 ± 9.6	88.9 ± 10.1	90.0 ± 8.0
Dust and inert materials, %	0.7 ± 2.1	0.4 ± 1.4	0.1	0.2 ± 1.4
Weed seeds, %	3.4 ± 7.6	1.4 ± 3.2	1.2 ± 5.0	1.6 ± 1.6
Corn by-products,[b] %	12.6 ± 11.9	10.3 ± 9.4	9.8 ± 9.2	9.2 ± 7.6

[a] Source: Hill et al (1982); used by permission.
[b] Cobs, chaff, etc.

break up in handling, but elevator blending practices also contribute to the increase. Table V shows how the percentage of samples graded U.S. No. 1 for BCFM declined, whereas No. 2 and No. 3 grades increased as the corn proceeded through the marketing system. The efforts of grain handlers to blend all corn to No. 2 grade is frustrated by the breakage of corn at each handling, which adds to the broken corn content. Most corn lots arriving at export elevators must be screened before loading even to achieve No. 3 grade.

B. Dusts

The material in corn that is of such small particle size as to be of unrecognized origin to the naked eye is termed dust. The particle sizes range from 1,000 μm down to <5 μm (Martin, 1981). Excessive breakage of today's corn has resulted from artificial drying and more rapid handling with more frequent impacts. As a result, dust explosions in grain elevators have become a serious problem (Fugatt, 1978; Kameyana et al, 1982). Explosions occur when finely divided combustible material is suspended in air and a heat source such as a spark is applied. The sudden release of energy can be disastrous (Chapter 4). The dust that is most hazardous is the fine dust that will pass through a 120-mesh sieve and is below 125 μm in size. Fine dust accounts for about 60% of the material in grain elevator dust collection systems (Martin and Lai, 1978). The combustible material in fine dust is mostly starch granules, but some fine protein and fiber is also present.

Although regulatory agencies have suggested that all dust be removed from grain to reduce dust explosion hazards, analysis of the total amount of dust in a corn lot is difficult. The fine dust clings tightly to the grain through electrostatic and other forces but is dislodged during handling and becomes airborne. Martin and Lai (1978) called the adhering dust "residual dust" and developed a quantitative method of measuring it.

Since residual dust can be knocked off the corn during handling and more dust is generated by progressive breakage (Martin, 1985), what is the ultimate solution to the problem of grain elevator explosions and fires? The National Grain and Feed Association, an association of grain merchants and elevator operators, has had a vigorous research program to determine the causes and

TABLE V
Distribution of Corn Samples Among Grade Designations on BCFM[a] Limits, Illinois[b]

		Percent Material Passing a 4.8-mm (12/64-in.) Sieve			
	Maximum BCFM Content	Country Elevators		Terminal Elevators	
Grade		Receipts	Shipments	Receipts	Shipments
1	2.0	81.4	64.7	67.6	40.1
2	3.0	11.7	19.7	17.8	33.5
3	4.0	4.8	9.7	6.8	17.3
4	5.0	1.1	4.2	4.0	5.4
5	7.0	0.3	1.4	2.8	1.2
Sample	>7.0	0.7	0.3	2.0	2.5

[a] Broken corn and foreign material.
[b] Source: Hill et al (1982); used by permission.

remedies (Graham, 1981). They have developed designs for new elevator construction (NGFA, 1979) and have recommended changes in existing installations. They have studied the separation and utilization of dust. A combination of dust removal and improved housekeeping appears to be the first alternative. Government authorities in the United States and other countries are developing dust control rules for operation of grain-handling facilities. Since commercial grain dust has good nutritional value, some of the companies have established central collection locations to pellet dust for incorporation into animal feed. Feeding studies indicate it can replace up to 25% of the corn in animal rations (Schnake, 1981). However, the most basic solution to the dust problem is to provide marketing incentives to remove dust at origin by classifying it as dockage and by development and adoption of methods for measuring breakage susceptibility to discriminate against excessively brittle lots. An intriguing solution to the dust problem is the addition of small amount (0.02%) of mineral or soybean oil (Barham and Barham, 1980; Lai et al, 1981, 1982). Studies by the FGIS have shown that this amount of oil cannot be detected and does not affect grade yet is effective for as long as one year. Additions to wheat in up to three successive doses of 0.02% each did not affect flour yield or bread quality (Youngs et al, 1984). Its use on grain has been approved by U.S. government agencies. Thus, many grain-handling companies are installing equipment for oil addition.

IV. MOISTURE

A. Importance of Grain Moisture

The moisture or water content of corn kernels is a natural component. It is not an instrinsic quality factor, but it has significant influence on quality changes, on processing properties, and on economics. The water in corn is a diluent of the organic substances that make corn useful. At high moisture contents of 28–32% (physiologically mature corn) down to about 20% MC, the kernels have a soft texture that is easily cut and punctured by harvesting and handling machinery. At lower moisture contents, kernels are tougher and can withstand rough handling. Below 12% MC, kernels become especially brittle. (All moisture values in this chapter are given on a wet basis, i.e., weight of water divided by wet weight of corn).

The optimum moisture content of 23–24% for combine harvesting (Chapter 4) is not optimum for transportation, storage, or many end uses. Combine-harvested corn must be dried soon after harvesting to prevent mold, bacterial, and sprout deterioration. The way moisture is removed can also have deteriorative effects on quality. Although biological damage is prevented, physical damage may result. Stress crack formation (discussed under Physical Damage) is due to rapid removal of water. Corn harvested on the ear at about 18–20% MC and dried at ambient or low temperatures has few stress cracks and low breakage susceptibility. This probably is because the higher-moisture cob has a wicking effect that prevents undue moisture differentials within the kernel. Corn dried off the ear in almost any way will show some stress cracks, but the slower it is dried, the fewer there will be. The moisture content of corn also affects its test weight or bulk density values (Hall, 1972; Hall and Hill, 1974). Because water has a specific gravity of 1.0 and corn solids of about 1.3, test weight increases as moisture content decreases.

B. Accounting for Moisture in Marketing

Moisture is important economically because its removal requires energy and therefore increases cost. Furthermore, the user of corn converts only the solid material to end products, whether cornstarch or beefsteaks. In current practice, the buyer discounts the price for moisture content over 15.5% MC, the most commonly traded moisture level. Before September 1985, a moisture content below 14.0% designated the lot as U.S. Grade No. 1 (FGIS, 1984a). The moisture contents at the other U.S. Grade levels were No. 2, 14.1–15.5%; No. 3, 15.6–17.5%; No. 4, 17.6–20.0%; No. 5, 20.1–23%; sample grade, over 23%. Now, moisture is no longer a grade-determining factor, but moisture content must be listed on the inspection certificates for grade (FGIS, 1984b). This change treats moisture as a condition of grain and not as a grade-determining factor, as has been the rule for wheat, barley, oats, triticale, and rye for many years. Hill (1985a) concluded that deletion of moisture as a grade-determining factor for corn (also for soybeans and sorghum) would have no significant impact on corn marketing.

Discounting the price paid for high-moisture corn takes into account not only the actual water content but also the direct and indirect costs of drying to the standard moisture value of 15.5%. The loss in weight on drying is termed "shrink" and includes the known loss of weight of water plus an "invisible" loss of dry matter. The weight loss for each percentage point of moisture removed is predictable and can be calculated with the equations given in Chapter 4 (Foster, 1982). The "invisible" loss is due to loss of dust, glumes (beeswing), small kernels, and possibly dry matter loss during slow drying. These losses occur during drying of corn either on the farm or in an elevator (Hurburgh et al, 1982; Hurburgh and Moechnig, 1982).

The practice of discounting the price paid for low quality while paying no premium for high quality grain has become controversial in recent years in the United States. Some farmers with dry grain have been adding water just before marketing to adjust it to just under the 15.5% MC for corn in order to obtain the optimum price (Bloome et al, 1982; Hill, 1982a). Adding water back to dry grain is not a sound practice because the corn may become more vulnerable to mold attack, especially if water addition is not done carefully (J. E. Tuite, Purdue University, personal communication). Hill (1982b) has proposed that changes in trading practices are needed that more accurately reflect the actual dry matter in a lot of grain. He proposes to retain the 15.5% MC as the basis for trading. At this moisture level, dry matter content is 84.5% and one 56-lb (25.4-kg) bushel of corn contains 47.32 lb (21.5 kg) of dry matter. This would be defined as a "standard bushel." Using this basis, elevator personnel would determine the dry matter content of a load of corn (100 minus moisture content), calculate the number of standard bushels it contained, and pay the quoted price. Thus a 10,000-bu lot at 17.5% MC (46.216 lb of dry matter per bushel) would contain 9,763.3 standard bushels (10,000 × [46.20 ÷ 47.32]) and the seller would be paid accordingly. A person delivering 10,000 bu of corn at 13% MC would be paid for 10,296.8 standard bushels. Hill (1982b) points out that this is not a new idea because many elevators are currently using the shrink factor to adjust wet corn to a 15.5% MC level in order to calculate drying charges. Adoption of this type of pricing would be especially valuable to farmers who artificially dry and store shelled corn, or to farmers in areas, such as the West and Southwest, where corn (also wheat, sorghum, and soybeans) equilibrates with low-humidity air during bin drying to moisture levels below 15%.

C. Moisture Measurement

Accurate measurement of true water content in cereal grains and products is a most difficult task for analytical chemists. Water is ubiquitous and an integral part of grain constituents. It exists in three broad types of interaction with hygroscopic constituents of the kernel, especially carbohydrates and protein. These types are absorbed (free water), adsorbed (bound water), and constituent (water chemically bound by covalent or ionic bonds). Hunt and Pixton (1974) review the fundamentals of moisture determination.

Methods for moisture measurement in cereal grains are of three general types: 1) fundamental or basic reference methods; 2) practical reference methods; and 3) empirical routine commercial methods. The basic reference methods are those that are designed to measure free and bound water most accurately by direct and specific measurement of these two forms of water, avoiding interference from water of constitution or other grain constituents. Three methods are included in this category. The Karl Fischer (KF) method (Hart and Neustadt, 1957) employs a KF reagent that reacts specifically with water, quantitatively releasing free iodine, which is determined by titration. There are many sources of error in the method, but when performed properly it is thought to be accurate (Jones and Brickenkamp, 1981). A second basic method is that of distillation of a ground sample of corn over boiling toluene (110°C) and direct measurement of the volume of water recovered (CRA, 1983). A third method, used in Europe, involves drying a ground sample of corn under vacuum in the presence of a strong desiccating agent, phosphorus pentoxide (P_2O_5) at 45–50°C. About 10 days is required by this method for the sample to reach constant weight. All three of these methods are time-consuming and tedious but are as close to the exact measurement of water as is possible with today's technology. Recent data show that a KF method in which the corn is ground in a sealed ball mill in the presence of just enough methanol to give a homogeneous, weighable paste duplicates results of the toluene distillation method and is highly repeatable (Watson, 1986b).

Although a universally accepted practical reference method for moisture in corn would be desirable, the large number of methods in use (Table VI) attests to the difficulty of knowing precisely when only the physically held water in corn has been completely removed. Parker et al (1982) show that, for wheat, corn, and barley, seven different air-oven and two vacuum-oven methods are used officially around the world, whereas Hunt and Pixton (1974) list 27 methods for cereals, oilseeds, and their products. Most currently used practical reference methods are drying methods that may use different temperatures, different holding times, and either whole or ground corn (Table VI). When wet corn, which is difficult to grind, must be analyzed, several methods use a two-step procedure in which whole kernels are first quantitatively dried to a moisture level that is convenient for grinding and then a weighed portion is dried to measure residual moisture. Methods that utilize whole corn do so for simplicity but also to avoid the error associated with grinding of moisture and with dry substance lost as dust. Nevertheless, grinding speeds the drying rate and requires less heat, hence reduces possible errors due to loss of volatiles. Some methods require that drying time be continued until the sample reaches constant weight. Other methods use a fixed drying time, assuming that all samples analyzed will be completely dry in the same length of time, which was initially determined by constant weighing experiments.

"Drying" methods for determining moisture content do not discriminate among the three classes of water: free, bound, and constituent. In an effort to remove the last traces of the most tightly bound water, one may have to supply enough energy to release some constituent water by chemical decomposition of especially labile bonds, or to volatilize small amounts of organic constituents,

TABLE VI
Referee Oven Methods Used for Determination of Moisture Content of Corn[a]

Group or Country	Method Number	Sample Preparation	Size (g)	Temperature (°C)	Time (hr)
United States	USDA 103° air oven (FGIS, 1984a)	Whole	15[b]	103	72
		Whole	100[b]	103	72
AACC[c]	AACC-44-15A	Whole	15	103	72
	AACC-44-19	Ground	2–3	135	1
	AACC-44-15A (Two-stage)[d]	Whole	20–30	Not specified	14–16
		Ground	2–3	130	1
ASAE[e]	AACC-44-15A	Whole	15	103	72
CRA[f]	...	Ground	5	115[g]	6
Canada	AACC-44-15A	Whole	50	103	72
South Africa	AACC-44-15A	Whole	15	103	72
Argentina	AACC-44-19	Ground	2–3	135	1
EEC[h]	...	Ground	8	130	4
Japan					
Food use	...	Ground	5	135	5
Feed use	AACC-44-19	Ground[i]	2	135	2
Taiwan	...	Ground	3–5	105–110	CW[j]
China (PRC)[k]	...	Ground	3–5	105	CW
USDA proposed	...	Ground	8	135	2

[a] Adapted from Hill (1985b).
[b] 15 g if moisture is below 16%; 100 g if above.
[c] American Association of Cereal Chemists (AACC, 1983).
[d] Two-stage method for samples over 16% moisture content. Not specified for corn.
[e] American Society of Agricultural Engineers (ASAE, 1983).
[f] Corn Refiners Association.
[g] Vacuum oven.
[h] European Economic Community.
[i] Using larger (30–40 g) sample for grinding.
[j] Constant weight.
[k] People's Republic of China.

which are then measured as water loss. Hart and Neustadt (1957) hypothesized that it was impossible to remove all bound water without causing some dry matter loss. Conditions for the USDA Official Air Oven Method (15 g of whole corn, 72 hr, 103°C) were selected to compensate for loss of nonaqueous volatiles by not removing all water. However, Hunt and Pixton (1974) state that loss of nonaqueous volatiles can occur due to a burst of enzymatic activity when the internal corn temperature reaches about 80°C. This could be the result of activation of glutamic acid decarboxylase, an enzyme in dry grain that releases carbon dioxide under certain conditions.

Nevertheless, the USDA adopted the 103°C, 72-hr air-oven method in 1959 based on data of Hart and Neustadt (1957), who demonstrated concurrence with their KF method. In 1961, the American Association of Cereal Chemists adopted this procedure for corn (now Method No. 44-15A [AACC, 1983]). Recent research by chemists from universities, industry, and the FGIS has shown that the USDA method (AACC-44-15A) is about 0.7% MC lower than the best available basic methods (Hill, 1985b; Watson, 1986b). Current research with the KF method suggests that the original calibration data of Hart and Neustadt (1957) was apparently incorrect by about 0.7%.

Most European countries have accepted a method known as ICC-101/1, which was proposed by the International Association for Cereal Chemistry as an international standard method. In this method, corn is ground to specific particle size specifications, then dried 4 hr at 130°C in an air oven. Again, results show a 0.7% MC difference with AACC-44-15A. FGIS is conducting research to develop an oven reference method that would correct this discrepancy. However, Hill (1985b) points out that changing the USDA reference method could have significant economic consequences and would not achieve international uniformity of moisture determination since no major corn-exporting country has adopted this so-called international standard.

The practical empirical methods of measuring moisture currently in use include four main types: 1) wide-line nuclear magnetic resonance (NMR) spectroscopy; 2) rapid automated drying ovens; 3) near-infrared (NIR) spectroscopy; and 4) electronic meters that measure differences in dielectric properties of grain.

NMR very accurately and rapidly measures a physical property of hydrogen atoms in water and oil. NMR is not widely used because of high cost, complexity of operation, and interference of oil content on water signals. In recent years, low-cost, simply operated models have been introduced. The automated drying equipment, represented by the Brabender moisture tester and the Cenco tester, subject small weighed samples to controlled heat on a timed rotating tray. Neither of these instruments has been approved for use with grain because of insufficient accuracy.

The NIR method subjects a ground (unweighed) sample of grain to near-infrared radiation in narrow bands (1.0–2.6 μm) and measures reflected absorption characteristic of specific chemical bonds (Butler, 1983). By use of complicated circuitry and microprocessors, the signals, after calibration against reference methods, provide very accurate analyses of water, protein, oil, and starch in all cereal grains. Although NIR instruments are expensive and complicated compared with the dielectric meters, they are becoming widely used in elevators that handle wheat because of the need to quickly measure protein content as well as moisture (Williams and Norris, 1983). Improved design of instrumentation and intense competition in the last few years has greatly improved the accuracy and ease of operation of NIR instruments. The FGIS is currently testing several models for moisture measurement in corn and oil and moisture measurement in soybeans and has already approved NIR use for protein in wheat (FGIS, 1985).

Of the rapid, practical methods, the "electronic" dielectric moisture meters are most widely used in grain merchandising because of speed, simplicity of operation, and relatively low cost. This method measures differences in dielectric properties of grain samples differing in moisture content (Nelson, 1981). However, the accuracy and precision of measurement are poorer than for the other practical methods, especially above 25% moisture (Fig. 1). Nevertheless, these meters provide the moisture values on which grain merchandising is conducted in the United States and most other countries.

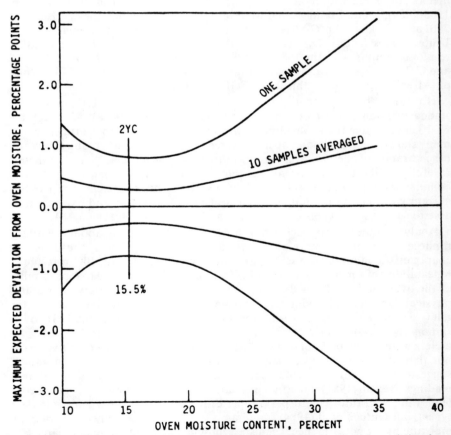

Fig. 1. Random variation of electronic moisture meters with corn. (Reprinted, with permission, from Hurburgh and Hazen, 1983-1984, by courtesy of Marcel Dekker, Inc.)

Several simplified dielectric meters selling for just several hundred dollars with more limited accuracy are now available for use on farms.

A number of different electronic moisture meters are manufactured and sold worldwide, but only the Motomco 919 is recognized as official by the USDA. All other meters must conform to the Motomco results, which are in turn calibrated against the official air-oven method. Hurburgh et al (1985) compared four different brands of moisture meters against each other and against the USDA air-oven method, using 900 samples of 1979 and 1980 Iowa corn. They found a spread of 1.5–3.5 percentage points among the four meters. In 1980, the test data were corroborated by Illinois investigators (Paulsen et al, 1983a). The Iowa and Illinois legislatures, thereupon, enacted legislation requiring recalibration of all meters used in the grain trade. After recalibration at moisture levels below and above 23–24%, all brands of meters agreed with each other but had an average bias of ±0.5% at the lower moisture levels (11–23%) and about ±1.0% above 23% MC. Figure 1 shows the expected limits of random variation obtained from an equation that was derived from data with the 900 samples using four brands of meters compared with data from the air oven method (Hurburgh and Hazen, 1983-1984). Meter brands tested were Motomco 919, Steinlite S5250, Burrows 700, and Dickey-john GACII, which are the most widely used meters in the United States. The USDA evaluated and adopted the Iowa-Illinois calibration results beginning with the 1984 crop by replacing charts C-1 (below 21% MC) and C-2 (21–29% MC) and eliminating the C-13 chart (above 29% MC) because of extreme inaccuracy at this level (FGIS, 1981).

Hurburgh believes that more effort should be made for nationwide standardization of electronic moisture meters (Watson, 1986a). Although the FGIS has adopted the Iowa-Illinois calibration for the Motomco official meter, many state regulating bodies do not enforce similar calibration requirements on other brands of meters. This would require several official "referee" meters carefully calibrated against an accepted oven method. Then all meters in commercial use would be calibrated against the standard meters. The FGIS is currently evaluating these other brands of moisture meters and expects to be able to certify some as official meters in addition to the Motomco 919.

Analysis of the Iowa data (Hurburgh et al, 1985) showed that about 85% of the random errors in moisture values resulted from variation in the dielectric characteristics among corn samples of equal moisture content, 10% was due to repeatability of a meter test on a specific sample, and 5% was due to repeatability of the oven method. Hameda et al (1982) found that corn variety, amount of physical damage, and drying temperature all have significant influence on the relative accuracy of meters with respect to the oven. Following the 1981 Iowa-Illinois recalibration, Hurburgh et al (1985) found that, up to 25% MC, all meters were within 0.1% of oven values on 3,000 samples tested over five years, but that all meters showed large year-to-year deviations from oven values, suggesting that environmental conditions cause the corn to give different readings. Nelson (1981) states that dielectric properties vary among types of corn (e.g., flinty and floury endosperm types differ from each other and from dent types), but Gutheil et al (1984) found an insignificant effect of corn variety on moisture meter performance when six yellow-dent hybrids were compared.

There has been recent interest in the measurement of moisture content in individual corn kernels because of spoilage of corn exported from the United States during the fall of 1984. Although the average moisture content was at an apparent "safe" level, adequate blending of lots of corn of high and low moisture contents may not have been achieved. Hill et al (1985), using an oven technique to determine moisture in single kernels, showed that an export shipment with a bulk average of 15.1% MC has a spread of 12.5–17.5% MC when sampled at origin and destination. The finding of the same moisture range at destination, if verified by further studies, indicates that the expected moisture equilibration was not achieved. This finding, if confirmed, could have significant influence on corn storage, marketing, and transportation. A low-cost method of detecting blends of high- and low-moisture corn has been reported. It utilizes the high-moisture inaccuracy of the Tag-Heppenstall conductance moisture meter in comparison with oven-moisture methods or dielectric meters (Martin et al, 1986). The greater the standard deviation of Tag-Heppenstall data, the greater the spread in moisture content of blends.

V. DENSITY

A. Specific Density

Density, a measure of weight per unit of volume, can be expressed in two ways: bulk density and true density. Determination of true density is made by measuring the weight of a given volume of liquid such as water, toluene, or ethanol displaced by a known weight of the test material (Mohsenin, 1970). Density is usually expressed as specific gravity and compared with water, which has a specific gravity of 1.0. Measuring density of corn in water would give a direct reading, but corn kernels tenaciously trap air bubbles when placed in water, although not in ethanol or toluene. Many agronomists have used ethanol displacement as a useful, convenient way to estimate kernel density. By weighing 100 kernels for the ethanol displacement, one can also have a value for average kernel weight, sometimes expressed as thousand-kernel weight. Although this property is a function of both seed size and density, it is useful in some comparisons.

Density of corn may also be determined by comparison against air or helium, using commercially available instruments that also provide data on porosity or interseed space (Thompson and Isaacs, 1967; Gustafson and Hall, 1970). It is a convenient, nondestructive method but requires a 200-g sample.

Another useful method of comparing densities of different lots of corn in a relative manner is called the flotation test (Wichser, 1961; H. Anderson, The Quaker Oats Company, private communication). In this test, 100 kernels are put into a cylinder containing a mixture of deodorized kerosene and tetrachloroethylene adjusted to a density of 1.275. The number of kernels that float are counted, and relative hardness is read from a chart relating kernel moisture content to percent floaters (Fig. 2). The chart was developed by determining dry-milling factors of different lots of corn against percent floaters. This method is not useful with corn that has been dried at excessive temperatures because large internal void spaces tend to skew the results.

All density tests are affected by moisture content (Nelson, 1980) since water has a lower density than the grain. Tests must be made on corn samples of uniform moisture content or a moisture correction factor must be developed. The density of an average dent corn at 12% MC is 1.2, of flour corn 1.1, and of popcorn or flint corn up to 1.3. The density of a corn kernel is the sum of the density of its components: 63% starch, density 1.5; 8.5% protein, density 1.1; 4% oil, density, 0.9; 12% water, density 1.0; internal voids filled with air, density near zero. Differences in chemical composition among individual kernels of dent corn are probably too small to account for the wide differences in density observed. However, kernels do differ in the amount of void space within them and in ratios of horny to floury endosperm. Horny endosperm is very dense, whereas floury endosperm is full of microfissures or void spaces.

B. Test Weight (Bulk Density)

The most widely used measure of density is test weight, a measure of bulk density obtained by weighing a specific volume of the grain. Measurement of test weight is important in storing and transporting corn because it determines the size of container required for a given lot of corn. In the United States, test weight is measured in pounds per bushel, but in the metric system, it is measured in kilograms per hectoliter. The minimum test weight for U.S. Grade No. 1 corn is 56 lb/bu (1 bu = 32 qt, 1.244 ft^3, or 0.3524 hl), which is 45.02 lb/ft^3 or 72.08 kg/hl.

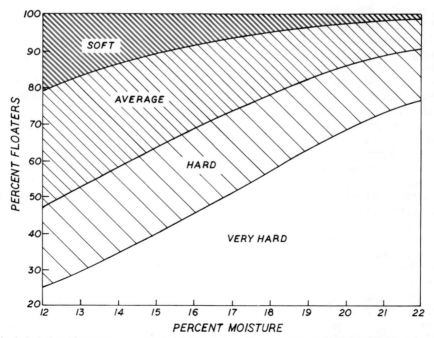

Fig. 2. Relationship between corn kernel moisture content, percent floaters in liquid (1.275 sp. gr.), and relative kernel hardness. (Courtesy Ben Grogg, Quaker Oats Company, Barrington, IL)

Test weight is criticized as a useful measurement because it is a combination of the densities of the kernels and the way they pack in a container. It is relatively independent of kernel size, as illustrated by the fact that many agricultural products including clover seed, wheat, soybeans, and potatoes all have nearly the same test weights (Hlynka and Bushuk, 1959). Pomeranz et al (1985) found that flat corn kernels had about 1.93 kg/hl (1.5 lb/bu) lower test weight than round kernels, but within the two shape classes, size had an insignificant influence on test weight. Moisture content affects test weight values. Nelson (1980) showed that bulk density of corn at 10% MC is 58% of kernel density, whereas at 25% MC it is 54%. It is common practice at country elevators to measure test weight at incoming moisture content and to discount the price both for low test weight and high moisture. To correct this situation, Hall (1972) measured the test weight changes of numerous standard hybrids during drying (Fig. 3) and developed a table with which to correct the test weight for moisture (Hall and Hill, 1974). These authors state that

> The amount of increase in test weight during drying is affected by the (a) initial moisture content of the corn, (b) amount of kernel damage, (c) drying temperature, (d) final moisture content, and (e) variety.

Since local elevators have no way to determine variety, initial corn moisture, or drying temperatures for farm-dried corn, the only two criteria that can be used to adjust test weights are final moisture content and percentage of kernels showing visible mechanical damage (Hall and Hill, 1974). Nelson (1980) developed an

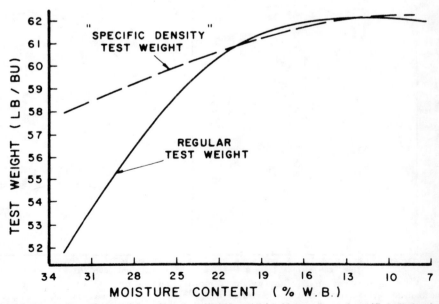

Fig. 3. Comparison of test weights of hand-shelled corn obtained by regular and specific bulk density measures. (Reprinted, with permission, from Hall, 1972)

equation relating bulk density (ρ_b) to moisture (M, in percent, wet basis) at a correlation coefficient of $r = 0.998$:

$$\rho_b = 0.6829 + 0.01422M - 0.0009843M^2 + 0.00001548M^3.$$

Nelson (1980) also developed a similar equation for kernel density, using an air comparison pycnometer to measure density. Hall and Hill (1973, 1974) showed that mechanical damage to kernels during combine harvesting had a pronounced effect on both wet and dry test weights (Fig. 4). They found that high drying temperature reduced dry (15.5% MC) test weight and that this effect was more pronounced when the corn was harvested at 31% MC compared with 24% MC. Specific hybrids differed significantly in the degree of change of test weight on drying and in their response to drying air temperature (Hall and Hill, 1973). At an average of 15.5% MC, corn dried at 21.1° C (70° F) had a test weight about 1.93 kg/hl (1.5 lb/bu) higher than corn dried with air at 104.4° C (220° F). All of the several hundred corn samples evaluated in these tests (made in 1969 and 1972) were at full maturity.

C. Corn Maturity and Test Weight

Most users of corn believe that test weight is a measure of the maturity or plumpness of kernels, but the data of Hall and Hill (1973, 1974) bring this assumption into question. It is true that corn harvested before maturity is low in test weight. Maturity in corn is defined as the cessation of dry weight

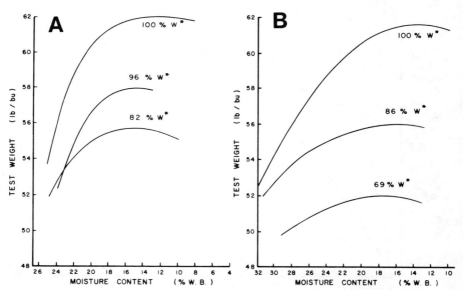

Fig. 4. Drying response curves of test weight vs. moisture content for corn with three levels of mechanical kernel damage that was harvested at 31% moisture content (A) and 25% moisture content (B) and dried in both cases at 77.2° C (170° F) air temperature. (Reprinted, with permission, from Hall and Hill, 1974)

accumulation by the kernels and is identified as attainment of maximum moisture-free weight per thousand kernels and, therefore, maximum yield (Frey, 1981). Maturation is also associated with gradual loss of moisture (Hallauer and Russell, 1966; Schmidt and Hallauer, 1966) and with changes in density and test weight (Hill, 1975; Nelson, 1980). For an estimation of the stage of maturity, the corn producer usually relies on the extent of dent formation or attainment of about 30% MC. However, some fully dented ears may not have reached maximum dry weight. The accuracy of moisture determination with electronic moisture meters is quite poor above 25% MC, and hybrids differ significantly in the moisture content at which maximum kernel weight is obtained (Carter and Poneleit, 1973).

Daynard and Duncan (1969) found that visual observation of a black closing layer across the base of the kernel, called the hylar layer, appeared to provide a quick estimate of maturity. However, subsequent studies have shown the black layer can form at moisture contents ranging from 15.4 to 35% (Carter and Poneleit, 1973). A more accurate quick test of maturity is observation of the "milk line" on the abgerminal (back) side of the kernel. This characteristic is more reliably associated with percent kernel moisture than is the black layer (Afuakwa and Crookston, 1984). Soon after full denting of the kernel, a solid-liquid (milk) interface can be observed on the abgerminal face of the kernel. For both early- and late-maturing hybrids and in every environment, kernels at the half-milk stage (milk line observable on kernels half way down the ear) had a the moisture content that was consistently near 40%, equivalent to 94% of normal yield. Approximately 200 growing-degree days (see Chapter 1) are needed to reduce moisture to 28% for harvesting (16 calendar days in Iowa). The proper maturation of corn is more greatly affected by temperature in the last two weeks of August and most of September than by rainfall except in very dry years (Shaw, 1977). Prolonged periods of cloudiness, alone or combined with abnormally low temperatures, can slow the corn maturation process; this occurred in some parts of the North American Corn Belt in 1967, 1972, and 1974. At the other extreme, moisture stress and high temperatures during the grain-filling period can reduce the maturation rate. In 1970, a severe epidemic of Southern corn leaf blight due to infection by *Helminthosporium maydis* caused premature death of corn plants that stopped grain maturation and resulted in grain with low test weight and reduced thousand-kernel weight (Brekke et al, 1972).

Any growing condition that retards the rate of kernel filling or causes late initiation of kernel development (e.g., late planting) makes the crop vulnerable to early freezing. Corn frozen before maturation is called "soft corn" and is characterized by high moisture content and low test weight. It may also be damaged by mold if harvest is delayed by wet fields, as happened in 1972 in the U.S. Corn Belt.

Slowed or late kernel maturation can have serious effects on grain quality if an early frost kills the corn plant. This may occur in years of an early frost on farms where late-maturing hybrids have been planted in the hope of achieving highest yields in normal years or years with late fall frosts. When the plant is killed but the grain not frozen, the kernels will subsequently dry and be sound but of low test weight. If the kernels are frozen, the soft corn usually must be harvested and dried artificially. Physical damage due to harvesting machinery may be high, and dryer operators may cause additional damage by use of excessive temperatures to meet the drying demand. If harvest is delayed due to wet weather, extensive mold invasion may occur. In any case, the grain will be severely discounted in the marketplace.

D. Feeding Value of Low-Test-Weight Corn

Hicks (1975) reported that corn with a test weight of 40 lb/bu (51.4 kg/hl) has only 93% of the total digestible nutrients of corn with 56 lb/bu (72.0 kg/hl) test weight. He showed that a standard discount schedule of one cent per bushel fairly discounts this loss in value but that graduated discount schedules that deduct five cents per bushel for test weights under 50 lb/bu (64.3 kg/hl) depart widely from the loss in actual value. On the other hand, Hill (1975) reported that three sets of compositional and feeding data show that corn with a test weight of 59.2–76.0 kg/hl (46–59 lb/bu) is satisfactory for swine. For sheep, similar feeding results were obtained by Thornton et al (1969). Corn of 45 kg/hl (35 lb/bu) had a lower feeding value, but even the standard discount schedule overpenalized the low-test-weight corn (Hill, 1975). Leeson and Summers (1975) reported that test weight of immature corn showed a positive correlation with metabolizable energy for poultry (broilers) but a negative correlation with protein and sugar content. Relationships between test weight and animal feeding results are not consistent, probably because of the influence of other contributing factors relating to the causes of the test weight reduction and the differences in corn composition.

E. Test Weight in Milling Applications

Laboratory trials by the author (S. A. Watson, unpublished data, 1975) have shown that starch yield and other wet-milling properties are not appreciably affected unless corn test weight drops below about 61.7 kg/hl (48 lb/bu). Wet-milling separations were conducted on corn samples from the 1974 crop that ranged in test weight from 52.1 to 73.3 kg/hl (41–57 lb/bu) (H. McMullen, A. E. Staley Manufacturing Co., unpublished data, 1975). Multiple correlation coefficients with test weight were -0.98 for starch yield, $+0.87$ for solubles yield, and -0.64 for oil in kernels. For each decrease of 1.29 kg/hl (1 lb/bu) in test weight, starch yield declined 1.1%, solubles increased 0.088%, and oil decreased 0.075%. In commercial processing, low-test-weight corn may reduce mill throughput because the steep tanks hold less dry substance than they do with normal corn. In another study, wet milling was not significantly affected below a test weight of about 61.7 kg/hl (48 lb/bu) (R. A. Anderson, Northern Reg. Res. Center, Peoria, IL, unpublished data, 1975). Oil content of corn decreased with decreasing test weight below 59.2–61.7 kg/hl (46–48 lb/bu), so one would expect lower oil yield in both the wet- and dry-milling processes. Also, solubles increased with decreasing test weight, which would imply greater yield of steepwater dry substance, a negative factor, in the wet process.

Test weight has been an important criterion for dry millers. Rutledge (1978) states that corn of lower test weight has a lower percentage of hard endosperm and therefore produces a lower yield of prime, large grits. Brekke et al (1972) found that dry milling of low-test-weight corn resulting from the 1970 corn blight epidemic gave decreased yield of prime grit products. A dry miller cannot make a satisfactory yield of grits with corn testing less than 66.9 kg/hl (52 lb/bu) (H. Anderson, Quaker Oats Company, unpublished data, 1975). For instance, a 24-hr plant run with white corn having test weights of 64.3–65.6 kg/hl (50–51 lb/bu) gave 16% grits, 27.4% meal, and 56.6% flour, compared with 40, 25, and 35%, respectively, for normal corn with tests weights of 70.7–73.33 kg/hl (55–57 lb/bu). Laboratory milling tests on nine specific corn hybrids dried at high and low temperatures over three years (Stroshine et al, 1986) showed that the highest correlation was between test weight and the desirable yield of largest size endosperm particles. Although all values varied among years, the relative ranking among hybrids remained the same. With the same procedures, very high correlations were found between test weight and dry-milling yield with samples selected for wide differences in density (A. W. Kirleis, Purdue University, unpublished data, 1986). The same relationships were found in commercial dry-milling tests using commercial corn conducted by Paulsen and Hill (1985a). The yield of coarse grits (retained on 4.76-mm, U.S. No. 4 sieve), a prime product used for cornflake manufacture, showed highly significant correlation coefficients with test weight ($r = +0.95$) and the floaters test ($r = -0.98$) described above. Correlation between coarse grit yield and true kernel density (measured in ethanol) was also positive ($r = +0.75$) but was not statistically significant. Although test weight appears to be a crude test, it appears to measure significant quality differences for some uses.

VI. HARDNESS

A. Significance

Hardness is an important intrinsic property of corn kernels because it affects grinding power requirements, nutritive properties, dust formation, kernel and bulk density, production of special foods, and yields of dry-milled and wet-milled products. Hardness is not a significant cause of differences in susceptibility to storage insect infestation in corn, especially by the maize weevil (*Sitophilus zeamais* L.), according to Schoonhoven et al (1975).

Corn hardness is an inherited characteristic that is modified by cultural conditions and postharvest handling conditions. Hardness influences the extent of damage caused by improper postharvest treatments. Intrinsic kernel hardness is decreased by artificial drying of shelled corn, which produces internal fissures (stress cracks). These fissures weaken the kernel, resulting in greater breakage susceptibility (discussed under Physical Damage). Corn dried on the ear or shelled corn dried very slowly does not develop internal fissures. Intrinsic hardness is difficult to measure because of the complexities of kernel structure, but it is closely related to the ratio of horny to floury endosperm.

Flint and popcorn types, which have a high proportion of horny endosperm, are hard. Flour corn and opaque corn, which have no horny endosperm, are soft, and the dent corn kernel is intermediate (Pomeranz et al, 1984). However, within each type are ranges of hardness related to differences in horny-floury ratios and to pericarp thickness (Szaniel et al, 1984) and cell structure. Differences in compactness of cellular components, cell sizes, and cell wall thickness within the endosperm play a part in hardness differences (Bennett, 1950b; Wolf et al, 1952). The structure of the endosperm cells is most important because, in horny cells, protein the matrix continuously surrounds each starch granule (Christensen et al, 1969). Although the protein matrix also surrounds starch granules in floury endosperm cells, it is thinner there (Wolf et al, 1967). During desiccation, the thin matrix ruptures, causing air pockets that are points of weakness (Duvick, 1961). These air pockets reduce translucency, producing an opaque appearance. Stress cracks caused by rapid drying or rewetting are also lines of weakness, which influence perceived hardness because they occur in horny endosperm.

Hamilton et al (1951) hand-dissected dent corn kernels grown under a wide range of soil fertility treatments and measured weights of horny and floury endosperms. They found average horny-floury ratios of 0.4 to 1.4 depending on the apparent nitrogen fertility level. In all cases, the horny endosperm was slightly higher in total protein and significantly higher in zein, a major component of the protein matrix (Christensen et al, 1969). Hinton (1953) found that the horny endosperm was 1.5–2.0 percentage points higher in protein than the floury endosperm. In high-lysine corn (opaque 2), the protein is mostly glutelin with very little zein, which produces only a floury, opaque-textured endosperm. Progress is being made in breeding high-lysine cultivars having some horny endosperm (CIMMYT, 1982).

B. Measurements

Estimation of the horny-floury ratio by hand dissection is very time-consuming. Better methods are being developed to estimate horny-floury ratios by visually comparing or measuring respective areas of horny and floury endosperm on cut surfaces of kernels. Kirleis et al (1984) embedded sorghum kernels in epoxy resin and then sanded away half of the embedded kernels. The area of horny (vitreous) endosperm was measured and expressed as percent of the entire endosperm area of 20–30 kernels. This method has recently been successfully applied to corn (A. W. Kirleis, Purdue University, private communication, 1985). Paulsen and McClure (1985) have experimented with computer vision methods of examining whole kernels under different types of illumination. The vitreousness range of 10–88% correlated well with other measures of hardness, including pearling in an abrasive-wheel barley pearler, density, percent floaters, and particle size index. This last measure is obtained by measuring the weight of particles retained on screens after grinding under standard conditions. For sorghum, the particle size index and pearling index methods of measuring hardness were superior to the visual measurement of horny endosperm. This visual estimation of horny-floury ratio is time-consuming and subject to a large coefficient of variability.

Methods of measuring hardness that are faster and more objective than measuring the horny-floury ratio have been sought by many investigators. Bennett (1950a) developed a specialized crushing device that measured resistance to crushing by a unique hydraulic system. The apparatus showed distinct differences between hard and soft genotypes. Szaniel et al (1984) developed the molograph, a laboratory hammer mill with a screen equipped with a means of measuring electrical energy consumed on grinding a given mass of grain in a given time. Small seeds and round seeds (at 9–10% moisture) were shown to be harder than large, medium, or flat seeds. They found a highly positive correlation of grinding resistance to total protein and oil content. Loesch et al (1977), using the Kramer shear press, found differences in hardness between normal and opaque and among opaque-2 families. The Instron Universal Testing Machine was used by Jindal et al (1978), who demonstrated that repeated freezing and thawing of ear-dried corn kernels, regardless of moisture contents of 25, 15, and 9%, did not influence hardness measurements. Jindal and Mohsenin (1978) found that, in a closed, rigid hammer mill, the damage to ear-dried kernels due to impact was a first order reaction with time and could be described as a specific breakage rate. The weight of particles retained on a 2.36-μm, or larger, screen could be used as a hardness index. Tran et al (1981) measured hardness differences using Instron compression data, breakage susceptibility (Stein breakage tester), pearling (Strong-Scott), and grinding through an attrition mill (Fisher-Quaker City mill). They found that grinding energy (power), grinding resistance (maximum torque), and grinding index in the attrition mill (weight of ground material passing through a 1.76-μm [10-mesh] sieve) showed the best differentiation of hardness differences.

Pomeranz et al (1984) used experience with wheat hardness determination as a starting point for corn hardness determination (Miller et al, 1982). They found that particle size analysis following grinding in a falling number mill is a reliable index of hardness. Average particle size (APS) was highly correlated with density determined by the air comparison pycnometer and with particle attributes measured by NIR at 1,680 λ. Commercially dried corn containing stress cracks showed poor correlation between NIR attributes, APS, and density. Breakage susceptibility as measured with the Stein breakage tester showed low correlations with other measures of hardness. The authors conclude that

> corn hardness determinations (density, NIR, and APS) can be interpreted properly only if the history of the grain (i.e., heat treatment) is considered. . . because breakage susceptibility and hardness are related and affect utilization of corn, both must be considered in its evaluation.

Instruments designed to measure breakage susceptibility (see discussion under Physical Damage) do not measure intrinsic hardness and are not correlated with hardness measurements on low temperature or field-dried corn.

More recently, Pomeranz et al (1985, 1986a, 1986b) have studied corn hardness with the Stenvert hardness test developed for wheat (Stenvert, 1974). The Stenvert test was applied using a microhammer mill (Glenn Mills, Inc., Maywood, NJ), to some of the same corn samples as used in the previous study (Pomeranz et al, 1984). Evaluations of time to grind, height of column of ground corn (as an index of fluffiness), and a ratio of coarse to fine particles from sieving showed significant correlations with NIR values. Kernel shape factors appeared to have important effects on hardness evaluation because of differences in density among the size fractions.

As with breakage susceptibility, moisture content of the kernels has a strong influence on hardness measurements (Shelef and Mohsenin, 1949; Herum and Blaisdell, 1981; Tran et al, 1981; Paulsen, 1983; Pomeranz et al, 1986a), with corn showing lower resistance to grinding as moisture content is reduced below 16%. For comparative hardness measurements, corn samples should be carefully adjusted to equal moisture levels (usually in the range of 11–14%) or a moisture correction factor applied to the results. Temperature effects are small but corn becomes slightly more brittle as temperature is lowered (Herum and Blaisdell, 1981).

VII. PHYSICAL QUALITY CHANGES

Two kinds of physical damage to corn kernels are recognized, i.e., exterior and interior. Exterior injury is generally visible and is usually referred to as "mechanical damage." The internal type of damage (fissures), termed "stress cracks," is not readily visible. Exterior damage involves a break, chip, scratch, or puncture in the kernel surface. Visible damage can range from a piece of kernel too large to pass through the sizing screen used in grading, to injuries that are just barely visible (Chowdhury and Buchele, 1976b; Pierce and Hanna, 1985). Some workers also recognize a category of injuries termed "invisible damage." These are surface injuries that are difficult to see with the naked eye and must be visualized with appropriate magnification or by staining. In all cases, exterior damage involves disruption of the integrity of the pericarp or loss of the tip cap.

A. Mechanical Damage

Pericarp damage can be rather superficial, penetrating only into the upper layers of pericarp tissue. More serious are lesions that penetrate the aleurone layer, exposing germ or endosperm tissue. Tuite et al (1985) have shown that any injuries to the pericarp make the kernel more susceptible to invasion by storage molds when conditions of temperature and moisture are conducive to mold growth. However, injuries over or around the germ are several times more serious than injuries over the endosperm.

External damage can begin in the field during development. Birds often peck at dough-stage kernels at the ear tips. Corn borer or ear worm larvae may cut into developing kernels in their feeding activities. The most important source of damage is caused by harvesting with corn combines (Chapter 4). Shelling in these machines involves the application of mechanical pressure on ears to dislodge kernels. Some breakage is inevitable in the best adjusted combine at the optimum kernel moisture content of 22–23%, but a great deal more is inflicted in poorly adjusted combines and at higher moisture levels (Byg and Hall, 1968; Ayres et al, 1972). The least amount of damage is found in corn harvested and dried before shelling. Some damage to the crown of dry kernels may occur during shucking, augering into cribs or dryers, and shelling.

Mechanical damage can be determined by examining each kernel for breaks and cuts in the kernel surface, categorizing kernels into classes such as severe damage, minor damage, invisible damage, and no damage. Visualization is aided by immersing the grain in a 0.1% solution of Fast Green FCF dye for 5–10 min (Ayres et al, 1972). After rinsing, breaks in the pericarp are stained a deep green. Chowdhury and Buchele (1976a) developed a numerical damage index with this staining procedure. Racop et al (1984) used the numerical index method to compare grain damage from three combines. Chowdhury and Buchele (1976b) adapted this method to a rapid colorimetric method by desorbing the dye from the damaged surfaces and measuring the released dye colorimetrically. Results are highly correlated with total damage determined visually and are more objective.

In a survey of performance of 80 combines in Iowa in 1971, Ayres et al (1972) found that visual damage ranged from 1.0 to 13.2%, with an average of 5.1%. Invisible damage that was made visible after staining with Fast Green FCF dye ranged from 0.8 to 66%, with an average of 26%. Even with the best combining operation, total damage amounted to a minimum of 12%. Steele et al (1969) found an average of 30%. The actual amount of corn in the damaged category measured by the U.S. grade standards averaged only 0.7% (range of 0.1–3.8%). Saul (1968) estimated that damaged kernels are two to three and a half times more vulnerable to loss of grade due to mold damage than intact kernels.

B. Stress Cracks

Drying does not inflict physical damage to kernels, but it may make them more susceptible to mechanical damage in later handling operations. This phenomenon was first studied by Thompson and Foster (1968), who showed that the formation of stress cracks or fissures in the horny endosperm was caused by rapid drying of kernels with heated air. The cracks can be seen when kernels are examined with transmitted light. Stress cracks can also be observed with low-power X-ray (Chung and Converse, 1970) and by laser technology (Gunasekaran et al, 1986). With the transmitted light method, kernels can be classified as to severity of stress crack formation, using categories such as single, multiple, and crazed (Fig. 5). Stress crack formation is caused by stresses inside the kernel, resulting primarily from differential zones of moisture content as water moves from the interior of the kernel and evaporates at the surface (Sheng and Gustafson, 1985), but temperature differentials also contribute (Gustafson et al, 1979). The cracks first become visible as grain dries down through the 18–16% moisture range—but not until the grain has been removed from the dryer to cool. The time required for stress crack formation decreases with a decrease in the moisture level of corn as it is removed from the dryer (White et al, 1982). Therefore, holding the corn at the existing temperature with no air movement (tempering) before it is cooled or during the drying cycle significantly reduces the frequency of stress cracks (Gustafson et al, 1983). Stress cracks start at the center of the kernel and extend toward the periphery underneath the pericarp. They are 35–90 μm at the widest point at the interior of the kernel (Gunasekaran et al, 1985). Of all the cereal grains, only corn and rice have serious stress crack problems. Stresses from differential temperature zones within the kernel are not of much significance because temperature equilibrium is reached much more quickly than is moisture equilibrium (Ekstrom et al, 1966).

Stress cracks can also form in corn kernels due to moisture stress from rewetting (Brekke, 1968; White et al, 1982) or due to an impact on equilibrated kernels. Moreira et al (1981), using a laboratory impacting device, showed that kernels impacted at a velocity of 18 m/sec at three moisture levels of 29, 20, and 13.4% showed internal stress cracks; most cracks were produced at the lowest moisture. Velocities this high can be achieved in the combine cylinder (Paulsen and Hummel, 1981), in discharging grain from an elevator leg (Fiscus et al, 1971), or in air-propelled transport (Magee et al, 1983). However, no cracks were formed below velocities of 10.8 m/sec, which is the terminal velocity of single kernels in free fall.

C. Breakage Susceptibility

Stress cracking, a form of hidden damage, predisposes kernels to rupture upon impact. Because visual quantitation of stress cracks is a slow and subjective procedure, Stephens and Foster (1976) demonstrated that laboratory impact devices were able to correlate corn breakage susceptibility with degree of stress damage. The Stein breakage tester ([SBT], Stein Laboratories, Inc., Atchison, KS) gave the best results of instruments then available. The original instrument (McGinty, 1970) was modified by Miller et al (1981a). A standard procedure for using the SBT for breakage susceptibility has been described (Watson et al, 1986). This instrument has been very useful in studies on the influence of drying and handling on corn breakage properties (Miller et al, 1981b; Gustafson et al, 1983).

Although the SBT has been marketed for over 20 years, it has not been adopted by the grain trade for evaluating breakage susceptibility in commerce. It is too slow for use in evaluating corn received by elevators, since the procedure requires about 5 min for each determination and needs a number of manual

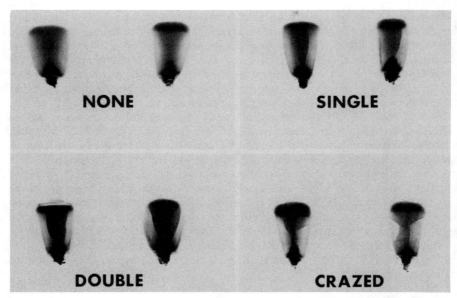

Fig. 5. Corn kernels showing different degrees of stress crack formation. (Reprinted, with permission, from Thompson and Foster, 1968)

manipulations. Furthermore, little marketing interest has been shown until recently due to lack of data on the cost of broken corn and dust in the marketing system. Since most of the stress crack damage is inflicted by adverse drying conditions on farms and country elevators, availability of a rapid, reliable test for corn breakage susceptibility could elicit commercial interest. Several research programs have produced alternative breakage devices. Seven conceptually different breakage susceptibility devices have been compared with the SBT in a collaborative study (Watson and Herum, 1986). Several of the instruments gave quite acceptable results, but one in particular, the Wisconsin breakage tester (WBT) (Finner and Singh, 1983; Singh and Finner, 1983), stood out on the basis of precision, rapid throughput, sturdy design, and unique rotor.

When a breakage susceptibility test is conducted with either the SBT or the WBT, the corn samples may all first have to be slowly equilibrated to a constant moisture content in the range of 12–13%, or a moisture breakage adjustment factor may be applied if samples differ by more than 0.5% MC, because breakage results are markedly influenced by moisture content (Fig. 6). A weighed sample is passed through the breakage tester and sieved, and the percentage of fines is calculated. The WBT discharges at a rate of 600 g/min and has the possibility of being automated. Gunasekaran and Paulsen (1985), Herum and Hamdy (1981), and Pomeranz et al (1986b) have found that the SBT gives a more reliable estimate of stress crack percentage and breakage found in commercial handling situations than does the WBT. However, Paulsen and Hill (1985a) have

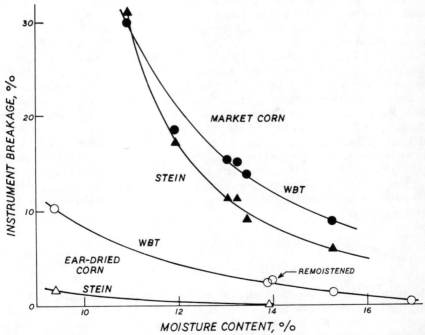

Fig. 6. Effect of moisture content on breakage susceptibility of ear-dried and market (stressed) corn, measured with the Stein breakage tester and the Wisconsin breakage tester (WBT).

demonstrated the usefulness of the WBT in selecting commercial corn samples having low breakage and superior dry-milling results. Development of a successful breakage susceptibility method that is rapid and accurate enough for market testing of corn has the potential of being useful to select lots with superior resistance to breakage or to discount lots that have been abusively dried.

Foster and Holman (1973) and Fiscus et al (1971) demonstrated that physical damage to stressed corn kernels can occur in commercial handling when kernels are impacted against hard surfaces, including other corn kernels. (Breakage from dropping onto corn is about 75% of that onto concrete.) The effect of multiple impacts is additive, so broken corn in stressed lots accumulates through the marketing system and must be removed to stay within the desired grade limits (under 3% for U.S. No. 2 corn). Corn breakage on impact is increased sharply with decreasing moisture content below about 16% (Herum and Blaisdell, 1981; Paulsen, 1983) (Fig. 6) and to some extent by lower temperatures. A rule of thumb, developed by C. R. Hurburgh (Iowa State University, Ames, personal communication) states that breakage susceptibility increases about 40% for every 1% decrease in moisture content in the range of about 12-16% MC. As a result of these findings, grain elevator managers are changing design and operation of grain-handling equipment to reduce damage by slowing transport machinery, decreasing drop height into tanks by the use of grain ladders, padding impact surfaces with resilient plastic sheeting, and other measures.

Another way breakage susceptibility can be lessened is by design and operation of grain dryers to include a tempering step at one or more places in the drying cycle to allow moisture to equilibrate (Chapter 4; Gustafson et al, 1983). This practical approach not only improves quality but reduces energy consumption. Pierce and Hanna (1985) found that reducing mechanical damage to kernels before drying reduces breakage susceptibility. Breakage susceptibility may eventually be reduced by breeding for kernels that are less susceptible to breakage than current average hybrids. Several research projects (Paulsen et al, 1983b; Russell et al, 1986; Stroshine et al, 1986) are currently examining the differences in drying rates and breakage susceptibility of selected corn inbreds and their hybrids. They have demonstrated significant differences that are consistent over several years when cultural influences are eliminated. Russell et al (1986) have concluded that "recurrent selection based on breakage in S_1 or S_2 lines, could be used to improve grain quality in a source breeding population such as a synthetic but should include yield in a selection index to prevent decrease in yield."

VIII. FUNGAL INVASION

A. "Damage" Grade Factor

As discussed in the section on grades and standards, the "damaged" category is largely the result of mold invasion of the kernel. The inspector is instructed to examine the germ for brown discoloration, "blue-eye," or other unnatural condition and to peel the pericarp away from the germ surface if necessary to make a judgment (FGIS, 1980c). The grade standards permit 3.1-5.0% damaged kernels in U.S. Grade No. 2 but only 0.2% heat-damaged kernels (Table I).

As described earlier, the official definition (FGIS, 1984a) of the "damaged kernels" category includes kernels or pieces of kernels that are heat damaged, sprouted, frosted, badly ground damaged or weather damaged, moldy, diseased, or "otherwise materially damaged." Under the "otherwise materially damaged" category is dryer damage, characterized by

> discolored, wrinkled and blistered appearance, or which are puffed or swollen and slightly discolored and which often have damaged germs; or whose seed coats are peeling off or have checked appearance.

Such kernels are considered damaged "resulting from external heat caused by artificial drying." Insect-bored kernels are considered damaged if they

> bear evidence of boring or tunneling indicating the presence within the kernels of insects and kernels in which noticeable weevil-bored holes have been eaten. . . .

However,

> kernels which have been partially eaten by insects or rodents but which are entirely free of refuse, webbing, insects, or other forms of damage shall be considered as sound kernels.

With respect to mold damage, the *Grain Inspection Handbook* (FGIS, 1980c) states that

> a germ affected by blue-eye mold, regardless of amount, shall be considered damaged. . . . a germ affected with mold caused by a fungus, regardless of amount, shall be considered damaged. . . . [Surface mold on kernels] which have mold caused by corn leaf blight on them which appears to be only on the surface, but which actually penetrates the seed coat shall be considered damaged.

This description shows that determining the damage category instrumentally may be very difficult or impossible. The FGIS relies on a set of standard color slides for training and for reference in doubtful cases. The definitions for damage in grading systems of other countries differ somewhat from the U.S. system. For example, South Africa includes pieces of kernels that will pass through a 6.35-mm (16/64-in.) sieve as defective kernels, along with kernels that have mold, insect, or mechanical damage. However, a liberal allowance of 9.0% for Grade 1 and 20% for Grade 2 are allowed (Table II).

To learn more about the characteristics of damaged kernels, Christensen et al (1971) collected 10 samples of 100% damaged corn kernels from each of 14 official USDA grain inspection stations. The kernels were evaluated for internal mold by culturing surface-disinfected kernels on an appropriate medium and observing mold growth.

The predominant fungi that grew from kernels were *Aspergillus glaucus* group, 54%; *Penicillium* species, 22%; *A. flavus-oryzae*, 9%; and *Fusarium*, 4%. An *A. candidus* group grew from 30% of the samples, but a maximum of 8% of the kernels in any one sample were infected. The authors found both *Aspergillus* and *Penicillium* growing from the same kernel in some cases. The authors state,

Many kernels, sometimes as many as 60–80% of some samples, yielded no (viable) fungi, although the embryos were visibly decayed and the cavities in the embryos filled with mycelium and spores.

In spite of such extensive decay, test weights were only slightly reduced.

Quasem and Christensen (1960) found that extensive mold growth is required to produce the brown color specified by the U.S. grade standards to classify a kernel as "damaged." The brown color is developed faster by some fungal species than others and is also affected by temperature and moisture content (Christensen and Sauer, 1982). The color is probably the result of Maillard compounds formed when reducing sugars react in germ cells with amino acids or ammonia released by the action of fungal proteolytic enzymes.

Assessment of damaged kernels is still the most subjective of the tests used for determination of U.S. grade and has defied instrumentation. Christensen and Quasem (1959) found that boiling corn and wheat kernels in a sodium hypochlorite solution made the damaged germs easier to see, due to darkening of the germ color and bleaching of the pericarp. Weak et al (1972) found that although the NaOCl treatment did make germ damage easier to see, it removed surface damage from corn kernels, making it necessary to evaluate a sample twice. The only grain in which the NaOCl treatment was of practical value was grain sorghum.

B. Role of Microorganisms in Quality Deterioration

Microorganisms are found in soil, decaying plant material, equipment dust, elevator dust, etc., and are a natural contaminant on the surface of corn kernels. However, only a few fungal species actually invade the kernels to the extent that quality is reduced. Bacteria and yeasts are of little importance unless very high moisture conditions are present. Then they contribute to heat generation in hot spots and produce the sour or fermented odors that devalue the grain to Sample Grade.

Fungi invading corn are often divided into two groups: field fungi, which invade corn under field conditions and generally die out in shelled or stored corn, and storage fungi, which invade corn in storage. An exception is *A. flavus,* which invades in both storage and field conditions and can produce aflatoxin in both situations. *Penicillium* species, normally a field fungi, also grow in storage under high moisture conditions (Table VII).

Each fungal species has specific limiting conditions for growth and a narrower set of conditions for optimum growth and competition. The minimum growth conditions shown in Table VII were determined in pure cultures; in natural situations, the more vigorous species may prevent growth of other species present in the inoculum (Horn and Wicklow, 1983). The factors that affect fungal growth on corn kernels include: the equilibrium relative humidity (ERH) of the air in the grain mass, temperature, time, condition of the grain (i.e., breaks in the pericarp, genetic makeup, presence of BCFM, how dried, etc.), mold inoculum level, oxygen content of interseed air, and prior storage history (Tuite and Foster, 1979b). Invasion of a grain lot by insects may initiate or exacerbate mold development as a source of inoculum and lead to the generation of heat, moisture, and debris resulting from their activity.

Fungal spores germinate and mycelium grows in response to relative humidity. Interseed air in stored shelled corn attains an ERH status in relation to the moisture content of the corn. However, corn condition influences this ERH. For example, corn that has been naturally dried has a lower ERH than corn dried at a high temperature (Tuite and Foster, 1979a) and, because of the hysteresis effect, corn that has been dried and then rewetted has a lower ERH than corn dried down to the same moisture content. Therefore, different lots of corn of the same moisture content may support different mold populations. Chapter 4 has a complete explanation of these principles and provides graphs that indicate "safe" and "unsafe" conditions of moisture, ERH, and temperature for storing "average" corn.

C. Field Fungi

The fungi that invade corn while it is still growing in the field have environmental requirements that do not permit much growth in stored corn. The principle genera are *Cephalosporium, Fusarium, Gibberella, Nigrospora*, and occasionally *Alternaria* and *Cladosporium*. *Gibberella* may continue to grow on ear corn stored at high moisture in a leaky crib. All field fungi require 90–100% RH for growth, which means corn moisture content greater than 22–23%. The field molds die out rapidly in moisture contents of 16–20% but may live for years in dry corn. Christensen and Kaufmann (1969) state that the proportion of wheat kernels containing *Alternaria* and *Cladosporium* to kernels containing

TABLE VII
Temperature and Relative Humidity Conditions of Growth for Storage and Field Storage Fungi on Corn[a]

Fungal Species	Growth Temperature, °C			Relative Humidity	
	Lower Limit	Optimum	Upper Limit	Lower Limit	Lower Corn Moisture Equivalent[b]
Storage species					
Aspergillus restrictus	5–10	30–35	40–45	70	13.5–14.5
A. glaucus	0–5	30–35	40–45	73	14.0–14.5
A. candidus	10–15	45–50	50–55	80	15.0–15.5
Penicillium cyclopium	−2	20–24	30–32	81	16.0–16.5
P. brevi-compactum	−2	20–24	30–32	81	16.0–16.5
P. viridicatum	−2	20–24	34–36	81	16.0–16.5
Storage and field species					
A. flavus	10–15	40–45	48–50	82	16.0–16.5
Field fungi					
P. oxalicum	8	31–33	35–37	86	17.0
P. funiculosum	8	31–33	35–37	91	19.0
Alternaria[c]	−4	20	36–40	91	19.0–20.0
Gibberella zeae[c]					
(*Fusarium graminearum*)	4	24	32	94	20.0–21.0
F. moniliforme[c]	4	28	36	91	19.0–20.0

[a] Data from Christensen and Sauer (1982) and Mislivec and Tuite (1970).
[b] Approximate equilibrium moisture content of minimum percent relative humidity at 25°C at which fungus can germinate would be below the moisture content that the fungus would ordinarily be able to grow and compete on grain.
[c] Occasionally found on stored high-moisture corn.

only storage mold species indicates the ratio of new and old corn blended in the lot. Absence of field fungi in dry, new corn may indicate adverse storage conditions before the corn was dried or drying with air at about 93° C (200° F) or higher.

The penetration by field fungi is generally limited to the pericarp, where they grow in response to high moisture conditions during grain maturation. *Nigrospora*, an exception, is favored by dry conditions (J. Tuite, Purdue University, personal communication). In seasons when harvest is delayed due to excessive rain, field fungi may cause discoloration of the grain and may affect germination and seedling vigor. Most of the quality loss is caused by *Fusarium* species. *Gibberella zeae* (the perfect stage of *F. graminearum*) produces corn ear rots that can seriously deteriorate kernels. In addition, *G. zeae* produces a number of toxins, as discussed later. Visual evidence of Gibberella ear rot is a pink or red to brown stain starting at the pedicel end and working up the sides. (This color is not to be confused with "red streak," a red coloration on the pericarp of sound corn). "Gib" infection starts at silking and may become epidemic under wet cold conditions during the harvest period (Tuite et al, 1974). In some seasons, the crop may be rendered completely useless for feeding to hogs because of the mycotoxins produced. *F. moniliforme* is believed to invade corn when warm high-moisture conditions occur during the harvest period, but the evidence is based only on animal disease symptoms as discussed in the section on toxins (Ley, 1985).

D. Storage Fungi

Storage fungi grow vigorously on shelled corn stored in bins or piles if moisture and temperature conditions are conducive to their growth. The majority of storage problems are caused by *Aspergillus* or *Penicillium* molds. The principal species are listed in Table VII, together with their temperature and moisture requirements (Tuite and Foster, 1979b). Generally, sound uninfested corn can be safely stored at slightly higher humidity and temperature than corn previously infected, because spore germination requires slightly more favorable conditions than growth of established mycelium.

Mold invasion of grain involves growth of the mold mycelium within the grain tissue. Nutrients for this growth are obtained from fungal enzymes, excreted to digest grain substance and thus provide nutrients and energy for mold growth. By-products of aerobic respiration are heat and carbon dioxide (CO_2). If mold growth is rapid, as in high-moisture corn, the heat that is generated increases the local temperature because grain is a poor conductor of heat. Such grain is said to be heating and can reach temperatures that produce heat-damaged kernels and eventually "spontaneous" combustion. Although organisms are killed in the region of intense heat, the migration of moisture and higher temperature stimulates mold growth in an expanding mass. This situation can be especially serious if the corn contains pockets of wet grain (surrounded by drier grain), which may be undetected until heating is noticed.

The amount of CO_2 generated in a lot of corn with an active mold population is directly proportional to the loss in dry matter (DM) by the grain. The relationship of storage conditions to DM loss was studied intensively by Steele et al (1969). These authors studied respiration in high-moisture shelled corn and related the carbon dioxide evolution to increase in germ damage. They calculated that loss of 14.7 g of CO_2 was equivalent to a 1.0% loss in corn DM. They showed that in corn having an average of 30% mechanically damaged kernels, the damaged germ count lowered the grade from U.S. No. 2 when the DM loss exceeded 0.5%. They derived a set of curves (Fig. 7) plotting moisture content, temperature, and germ damage to show the conditions of storage that would keep DM loss below the 0.5% value. Perez et al (1982) recently verified the observations that using up storage life by slow deterioration at low temperatures can cause rapid deterioration when the corn is moved to warmer conditions.

Christensen and Kaufmann (1969) contended that most of the respiration activity in nonsterile stored corn is due to growth of fungi, but Seitz et al (1982b) showed that in fungus-free corn that was rewetted, respiration of the grain itself contributed significantly to DM loss. Their data show that the fungus-free corn at 24% MC and 29.4°C (85°F) reached 0.5% DM loss in six days; however, bacterial growth at those moisture contents was not determined and could have contributed to CO_2 production. They also found that corn inoculated with *A. flavus* had 0.5% DM loss and that significant levels of aflatoxin had accumulated before the 3% germ damage level was reached (Seitz et al, 1982a). They suggested that the Steele et al (1969) curves should be used cautiously. Fernandez et al (1985) recently conducted experiments that confirmed Steele's findings under conditions where fungal populations were measured. They found that DM loss, measured as CO_2 production in freshly harvested corn at 22% MC with 15% physical damage, is well correlated with the number of fungal propagules and the number of infected kernels found. Corn that had been

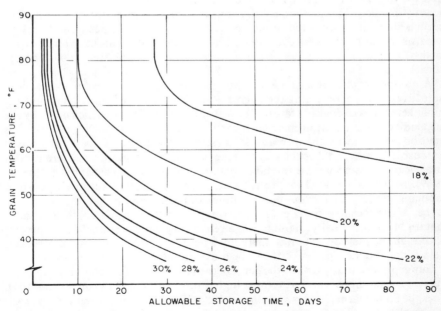

Fig. 7. Allowable storage time for loss of 0.5% dry matter in shelled corn at various temperatures and moisture contents. (Reprinted, with permission, from Shove, 1969)

prestored 70 days at 10° C (50° F) before test storage at 26° C (78.8° F) gave CO_2 production about equal to that of freshly harvested corn. Drying and rewetting to 22% MC before test storage increased CO_2 production and mold development slightly. The greatest increase in CO_2 production and mold growth during test storage was in corn prestored 70 days at 3° C (37° F) and was due to mold growth during prestorage.

Both of these recent studies agree with Steele et al (1969) that an increased amount of mechanical damage to kernels increases the rate of DM loss by promoting fungal invasion. Recent data by Tuite et al (1985) verify observations by Steele et al (1969) that cuts in the pericarp over the germ make the kernel much more vulnerable to storage mold invasion than cuts on the back or crown of the kernel. Removal of the tip cap also increased fungal invasion. Differences in resistance to mold invasion measured on intact kernels of various corn genotypes were retained in the injured seeds even though all were invaded more rapidly.

E. Detection of Fungal Invasion

Presence of fungi is an important measure of quality. The preferred procedures for evaluation are shown in Fig. 8. Visual inspection with a hand lens by a trained observer is adequate in many instances to determine the extent of mold invasions, but counting fungal propagules in a ground corn sample by a dilution technique is the most accurate. Counting the number of infected kernels after culturing kernels that have been surface-sterilized with sodium hypochlorite can be quite useful for detecting early mold growth and for fungi that do not sporulate, but it does not distinguish between samples that are heavily invaded. This method is subject to large error due to the small number of kernels that can be cultured on one petri dish, but multiple petri dishes can be used, as was done by Sauer et al (1982, 1984) to evaluate the mold populations in farm-stored and export corn. Error was reduced by plating 100 sterilized kernels.

A number of chemical and enzymatic methods have been proposed as more objective means of detecting the degree of mold invasion of corn. The ability of fungal enzymes to degrade fats to free fatty acids has long been measured by the fat acidity value titration method (now method 02-03A [AACC, 1983]). This method has been useful in many studies, but it is not as reliable as desired because different mold species hydrolyze fat in the corn germ at different rates and because they apparently metabolize fatty acids (Christensen and Kaufmann, 1969). Bottomley et al (1952) found that the conversion of sucrose in corn to reducing sugars was a more sensitive index of mold growth than the fat acidity value. More recently, Seitz et al (1982a) reported that ergosterol, a secondary metabolite of practically all fungi, was a reliable measure of fungal development. Although the method does not distinguish between individual fungi or between field fungi and storage fungi, it can be a useful index of fungal invasion.

Fungi preferentially invade the germ because it is richest in relatively available nutrients and also because it retains higher moisture levels during drying than does the endosperm (Herter and Burris, 1984). Reduced germination of the seed is correlated with advanced fungal invasion of the germ, but because seed viability can be destroyed by heat, freezing, chemicals, etc., germination is not a good primary index of fungal invasion. Measurement of glutamic acid decarboxylase activity (Bautista and Linko, 1956) is not a useful test for the same reason.

All of these methods require removal of a sample of grain for evaluation, but sampling from large storages may be an impractical way to monitor the grain for mold development. Insertion into the grain of probes bearing instrumentation to detect changes in temperature have long been a practical means of monitoring for development of mold and insects, which usually develop in localized portions

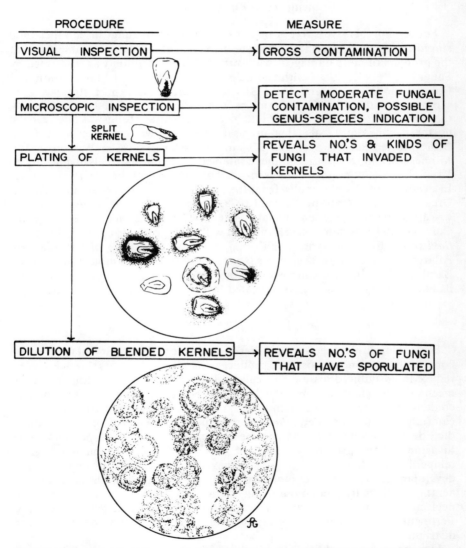

Fig. 8. Preferred procedures for examination of seed for fungal damage. (Courtesy J. Tuite, Purdue University, Lafayette, IN)

of a grain mass and produce hot-spots. Improvement in instrumentation to measure carbon dioxide permits the use of an increase in CO_2 in the air in a grain bin as an early indication of mold or insect development (Muir et al, 1985). Changes in temperature can only be detected near the hot spot because heat does not diffuse readily. Carbon dioxide does move freely in the interseed air, and any reading of CO_2 above the ambient air concentration suggests that some deterioration is occurring.

F. Incidence of Fungi in Stored Corn

Few comprehensive surveys of the general level of fungi in commercially stored corn in the United States have been made, but scattered data indicate that the problem of mold damage is widespread. This perception was confirmed by Sauer et al (1982, 1984) with the help of the Agricultural Stabilization and Conservation Service and the FGIS when farm and export storages were sampled. A total of 1,000 corn samples from farm bins in 27 states representing the 1976–1979 crop years showed that invasion of storage fungi was higher in the northern and eastern states than in the southern and western states in the Corn Belt. Infected kernels averaged 22% in Nebraska to 32% in Wisconsin and Minnesota. Only 8% of the samples contained 5% or more kernels invaded by *A. flavus*, but 84% contained kernels invaded by members of the *A. glaucus* group. In export corn, 1,299 samples from the four export regions (Pacific, Atlantic, Great Lakes, and Gulf) in 1977 and 1978, they found that the *A. glaucus* group was again the most frequent fungal invader (14% at Houston to 27% at Toledo). Corn had 10 times more kernels infected with storage molds than did wheat. *A. flavus* was found in 0.6% (California) to 2.8% (Virginia) of the samples. Aflatoxin was found in 33 samples from Virginia with 10 samples over 10 ppb. Barak and Harein (1981), surveying farm storages in Minnesota, found that 80% had severe fungal invasions as indicated by the presence of insect fungivores.

G. Loss Prevention

Because mold-damaged corn is objectionable for all uses, fungal invasion of corn results in economic loss, and producers, handlers, and users should strive to reduce mold-induced losses. This can be done by careful harvesting to reduce mechanical damage and prompt drying under gentle conditions to moisture content levels below the minimum for mold growth. Stored corn should be carefully monitored to detect increases in moisture, temperature, and carbon dioxide, important criteria that may be used to detect early development of mold invasion. Then, prompt aeration of the stored grain to maintain grain temperature below 10°C (50°F) whenever possible will prevent hot-spot development and give better moisture uniformity (Chapter 4). The proper use of aeration and the frequent monitoring of grain condition are probably the most neglected, but most important, aspects of grain storage management on farms. Pertinent to good aeration is removal of excessive BCFM or its uniform distribution.

In some instances, chemical inhibitors of fungal growth have some practical value. The only inhibitors that have been approved by the U.S. Environmental

Protection Agency for use on corn are propionic acid, sulfur dioxide, and ammonia, but formic and acetic acids and methylene-bis propionate have also been used. Propionic acid alone or in combination with the other protectants has been used for storing wet corn to be fed to animals (Hall et al, 1974), but most mixtures are not superior to propionic acid alone. Although the economics are marginal and handling is difficult (Hall et al, 1974), some producers have found worthwhile advantages. Introduction of gaseous fungicides, SO_2 or NH_3, into the air being passed through cleaned wet corn in bin drying with ambient air can permit the use of higher-moisture corn or lower airflow rates with less danger of mold growth in the undried portion of the grain (Nofsinger et al, 1977; Eckhoff et al, 1983). An application of 0.05% NH_3 in repeated doses every few days up to a total of 0.5% prevents microbial growth for the drying season. Sulfur dioxide used in a similar manner, but requiring less frequent application, has the advantage of keeping the corn color bright yellow, whereas ammonia results in the development of brown color. Both gases require good management to be successful. On U.S. farms, ammonia has more appeal because of its widespread use as a nitrogen fertilizer. Application of ammonia to bins that have begun heating has controlled the microbial growth, but SO_2 failed in this use because the gas was unable to penetrate the molded mass (Tuite et al, 1986). Grain treated with these chemicals should be fed to animals and kept out of commercial channels because it would not be suitable for most commercial uses. Treated grain entering the market probably would be graded U.S. Sample Grade because of the presence of foreign odors.

A longer-range way of reducing mold invasion of corn is to select genotypes that resist mold growth. Atlin et al (1983) and Hart et al (1984) found significant heritable differences in Gibberella ear rot among corn inbreds and hybrids. Cantone et al (1983), under controlled laboratory conditions, confirming work of Moreno-Martinez and Christensen (1971), found significant differences in resistance to invasion by *Aspergillus* and *Penicillium* species of storage molds among five popular corn inbreds and hybrids. Although important genetic differences in susceptibility to *A. flavus* and aflatoxin are known, useful resistant genotypes have not been found in spite of much research (Widstrom and Zuber, 1983).

H. Mycotoxins

Fungi produce secondary metabolites during growth on a substrate. Some of these compounds, such as citric acid and penicillin, are useful, whereas others are toxic. Young and Fulcher (1984) listed 15 fungal species that produce compounds toxic to animals and possibly to humans. Hesseltine (1979) lists 22 mycotoxic diseases known to be caused by fungal metabolites and 25 diseases suspected to be of mycotoxic origin. The number of identified known mycotoxins amounts to several dozen, and new ones are identified frequently. However, only a few are of any economic significance in moldy corn.

AFLATOXINS

The aflatoxins produced by *A. flavus* and *A. parasiticus* are of great importance. Corn is an excellent substrate for the growth of these two mold species and for the production of aflatoxin. *A. flavus* has the most widespread distribution and grows at relatively high moisture and temperature conditions (Table VII). Although *A. flavus* had been observed as an ear corn mold (Taubenhaus, 1920), it was thought to be only a storage mold until scientists with the Quaker Oats Company demonstrated that *A. flavus* grows and produces aflatoxins in growing corn ears (Anderson et al, 1975). Payne (1983) states that high temperature is the most important factor in the infection process; water stress, i.e., dry, droughty conditions during kernel maturation, increases infection apparently by increasing spore load. This explains why the southeastern United States and tropical corn-growing countries such as Thailand and Indonesia experience high incidences of *A. flavus* infection and high aflatoxin in corn. Although the cooler regions of the United States, where most of the corn is raised (the Corn Belt), experience only infrequent, localized occurrence of aflatoxin, a hot dry summer in 1983 caused widespread occurrence in Indiana and other parts of the northern Corn Belt (Cote et al, 1984; Tuite et al, 1984). The point of entry of the fungus into the corn kernel is not known, but damage by insects that carry a high spore load is a significant source (McMillian, 1983). Tsuruta et al (1981) present interesting electronmicrographs of fungal invasion into stored corn. Wicklow et al (1982) postulate that the major source of inoculum may come from specialized overwintering bodies known as sclerotia, which develop in kernels that form in standing corn but overwinter in the soil (Wicklow, 1983).

Aflatoxins are particularly important fungal metabolites (the six important isomers are known as B_1, B_2, G_1, G_2, M_1, and M_2), not only because they are acutely toxic but also because they are highly carcinogenic. Therefore, developed countries have enacted laws limiting the aflatoxin level in corn used for food or feed to a maximum of 5–50 ppb. The United States has chosen a legal limit of 20 ppb in food, based on tissue toxicological assay, although the actual exposure that will produce cancer in humans has not been established. Although the threshold doses of aflatoxin B_1 for clinical signs vary greatly, ruminant and other adult meat animals have been shown to tolerate several hundred parts per billion in feed. Muscle tissue does not contain aflatoxin residues, but livers may contain some residual aflatoxin (Hoerr and D'Andrea, 1983). During severe outbreaks of aflatoxin in corn, the U.S. Food and Drug Administration has permitted corn containing up to 200 ppb to be shipped and used in feeds for adult nonbreeding, nonlactating animals (Hamilton, 1984).

Aflatoxin in corn can be partially destroyed by roasting, but treatment with gaseous ammonia has reduced aflatoxin by 99% (Brekke et al, 1978; Anderson, 1983). The method has been conducted successfully on a large scale and fed to several species of animals over an extended time with no ill effects. However, severe browning and quality degradation occur.

Aflatoxin in corn can be detected with three levels of accuracy. The least rigorous is the presumptive test, in which whole or coarsely ground corn is examined with a high-intensity ultraviolet (UV) lamp (365 nm), which produces a bright greenish yellow fluorescence in pieces of corn (primarily floury endosperm) infected by *A. flavus* (Shotwell and Hesseltine, 1981; Calvert et al, 1983). Barabolok et al (1978) showed that whole corn could be conveniently examined in a moving stream illuminated by a UV lamp. Their data indicated that if one or more fluorescing particles were observed in 2.5–5.0 kg of commercial corn, 30–50% of the lots would contain more than 20 ppb aflatoxin B_1 and must be analyzed. Screening analysis has been by a minicolumn method calibrated to indicate which samples contain more than 10–20 ppb aflatoxin (AACC method 45-14 [AACC, 1983]). More rapid methods based on enzyme immunoassay or rocket immune assay techniques show promise for much more rapid screening assays (Hart, 1986). For more precise results, quantitative methods using thin-layer chromatography or high-performance liquid chromatography should be used. The analytical methods are reviewed by Shotwell (1983). With any method, adequate sampling must be used, because aflatoxin is not uniformly distributed among kernels (Dickens and Whitaker, 1983).

FUSARIUM TOXINS

Two other important types of toxins, zearalenone and the trichothecenes, are produced by the *Fusarium* species, mainly *G. zeae*, that produce ear rots in corn. Zearalenone is an estrogenic compound causing infertility and other reproductive problems in swine. Zearalenone can be assayed by thin-layer chromatography or by high-performance liquid chromatography (AACC, 1983). The trichothecenes are a group of about 40 toxins, one of which was first discovered in a culture of the fungus *Trichothecium roseum*. Some of the trichothecenes are extremely toxic when ingested or when in contact with the skin. The principal trichothecenes found in corn are primarily deoxynivalenol (DON), and less frequently T-2 toxin, diacetoxyscirpanol, and nirvalenol (Mirocha et al, 1980). A severe epidemic of Gibberella ear rot occurred in 1972 in the northern part of the U.S. Corn Belt. A very wet cool fall delayed harvest and resulted in heavy *Gibberella* infection of field corn, but cool wet weather at silking time initiated the fungal invasion. Tuite et al (1974) examined weather data in affected areas in Indiana and concluded that at least seven to nine days of rain during corn silking and a mean temperature below 21°C (70°F) induced Gibberella ear rot. During the 1972 epidemic, swine offered corn containing 5% or more kernels visibly damaged by *Gibberella* refused it, and ingestion of corn with 3% resulted in lower feed consumption and loss of weight. A water extract of highly infected corn (40–60% infected kernels) given by intubation into a pig's stomach or by interperitoneal injection caused immediate vomiting. A trichothecene compound known as vomitoxin, or preferably DON, was identified and found to be responsible for both the swine refusal and emesis (Vesonder et al, 1976).

F. moniliforme is not a storage fungus, but it does infect corn and is associated with a disease of horses known as equine leucoencephalomalacia (ELEM). ELEM occurs sporadically (Ley, 1985) and has been reproduced by culturing corn with certain strains of *F. moniliforme* (Haliburton et al, 1979), but a specific toxin has not been identified. *F. moniliforme* and other *Fusarium* species produce a toxin known as moniliformin that is toxic to ducklings (Rabie et al, 1982), to baby chicks (Vesonder et al, 1976), and to plants but is not the ELEM toxicant (Cole et al, 1973). The study of fungal metabolites and their effect on animal and humans has given us a better understanding of the influence of fungi on corn quality. Much more research is needed with livestock animals and poultry with respect to performance at low mycotoxin levels.

IX. QUALITY IN CORN UTILIZATION

A. Seed Corn

The seed corn industry has the most stringent quality requirements because acceptance of proprietary hybrids begins with good germination and a high percentage of vigorous seedlings (Craig, 1977). The seed producer must be sure the seed is genetically pure and true to type. Most producers prefer to harvest at full maturity but before freezing can occur. Seed corn is generally harvested on the ear, but if field shelling is done, moisture content must be below 20% to minimize mechanical damage. Ear harvesting is preferred as it permits visual sorting for off-types and elimination of weed seeds.

Drying of seed corn is begun soon after harvest; the drying temperature is held between 38 and 46°C (100–115°F). Higher temperatures may damage the embryo, producing reduced germination and lower seedling vigor. Although harvesting and drying at or after full kernel maturity has been considered safest, increasing volumes of seed corn are harvested above 35% MC. Navratil and Burris (1984) have shown a significant reduction in cold test emergence of corn dried at 32–34% MC and 50°C (121°F) and a peak emergence at 40–45% MC that is about 20% higher than for corn harvested at 25% MC. Some inbred lines are more resistant to heat damage than others. Adverse effects on seed quality can also be caused by overdrying. Obendorf (1972) showed that corn dried carefully to 6% MC showed a significant reduction in seedling vigor when compared with seed dried to 13% MC.

After drying, the seed is cleaned to remove foreign material and sized into as many as 14 size groups to fit a variety of planting machines. Most seed corn is treated with a fungicide, usually captan, to protect it from seedling diseases, and with one or more insecticides.

State laws require that every lot be labeled with a warm temperature germination percentage and a minimum level of noxious weed seeds. Reduction in warm germination of kernels can be caused by high drying temperature, mechanical damage, and mold or insect invasions. However, the warm germination test does not reflect true field conditions because usually the seed must germinate under adverse field conditions. For this reason, most corn seed is subjected to a cold germination test, which provides an index of the extent of mechanical damage and the adequacy of the fungicide treatment.

Seedling vigor can also be predicted by use of the tetrazolium test (Moore, 1958). The test which, although rapid (3–4 hr), is difficult to conduct on large scale and requires experience for adequate interpretation. The chemical 2,3,4-triphenyltetrazolium chloride produces red coloration in living tissue. Evaluation of white areas and different shades of red can be used to diagnose causes of viability loss such as freezing and mechanical injury.

Other tests that can be used to measure the health and vigor of corn seed are: growth rate measurement, seedling weight, glutamic acid decarboxylase activity (Bautista and Linko, 1962), measurement of leachates when seed is soaked in water as an index of membrane permeability, and accelerated aging (Seyedin et al, 1984). Equipment is commercially available for measuring permeability by electrical conductivity of leachate relatable to germination percentage.

B. Beverage Alcohol

Next to seed corn companies, distillers of whiskeys and other high-proof alcoholic drinks may be the most particular with respect to corn quality. Distillers are interested in obtaining the highest yield of whiskey consistent with acceptable flavor. Bourbon must be made with at least 51% corn, but the typical mash contains 70–80% corn, with the remainder made up of barley malt and other grains (Chapter 19). Distillers prefer hybrids having higher starch contents, which, therefore, have low average protein and fat contents. They prefer naturally ear-dried corn but will use low-temperature dried corn of superior quality. The nearly complete absence of moldy kernels is important, as is high test weight and a low level of stress cracks.

Corn quality evaluation is an index of possible value for beverage alcohol, but final comparisons must be made in a laboratory process, such as that described by Thomas (1966). This test duplicates the larger commercial operation on a small scale. The results may then be correlated with the best indices of quality for future corn evaluation.

C. Dry Milling and Snack Foods

Dry millers require corn that is free of detectable mycotoxins and has a minimum of kernels invaded by mold. Absence of insects and of rodent and bird excreta is also important. The presence of rat pellets has been a serious problem because, being nearly the same size and shape as corn, they are difficult to separate before milling and impossible to separate after milling. Fortunately, this problem is diminishing as the quality of corn storage structures is improved. Besides freedom from toxic weed seeds, the absence of soybeans is also important to the dry miller, because soybeans carry through the milling operation and show up in prime grits. Dry milling is described in Chapter 11.

Other attributes of corn quality such as low BCFM, low stress-crack level (Brekke 1968), high test weight, and high percentage of horny endosperm have been discussed earlier, along with appropriate test methods (Brekke 1970; Paulsen and Hill 1985a). Excessive levels of physical damage, including split and large broken pieces, are undesirable because, in the tempering process, they pick up moisture at a different rate than whole kernels. Kernel size and shape are also important. Too large a proportion of small kernels, such as those from the tip of the ear, are objectionable (R. Wichser, Quaker Oats Co., personal communication). Tough bran (pericarp plus aleurone layers) that will not release from the endosperm is undesirable; this was the case with opaque-2 corn (Wichser, 1966). Pericarp and endosperm colors are important in corn chip manufacturing because the alkali cooking process (Chapter 13) modifies the colors (Ellis et al, 1983). In some cases, the manufacturer blends yellow and white corn or selects a light yellow variety to give the desired color in the end product. Neither the dry miller nor the snack producer wants to see wide swings in moisture content as it causes variation in moisture absorption during tempering or cooking. In fact, best results are obtained in any of these processes by using corn with a quality that is uniform and constant in all respects.

Although many ways of evaluating corn quality have been described, all must be correlated with results obtained in a commercial milling operation. Since full-scale testing is not always available or practical, laboratory or pilot-plant dry milling has been used for research by USDA scientists (Brekke, 1968; Brekke et al, 1972, 1978). A simpler but effective procedure has been developed for research at Purdue University (Stroshine et al, 1986). In this procedure, corn is conditioned to 24% MC in three steps. It is then degermed in a horizontal drum degerminator, screened, and aspirated to remove hulls. The remaining material is dried to 17% MC and screened through a sieve-stack containing 3½W, 5W, 7W, 10W, and 16W sieves. Recovered fractions are aspirated and suspended in a solution of sodium nitrate (sp. gr. 1.275) to float off pieces of germ. The endosperm fractions are dried and weighed. Since the primary goal of dry milling is usually to achieve the highest yield of prime flaking grits (+7W) and high endosperm extraction, the following equation provides an expression of these goals in the form of a milling evaluation factor (MEF):

$$\text{MEF} = \frac{[(+3\tfrac{1}{2}\text{W \% endo}) + (+5\text{W \% endo}) + (+7\text{W \% endo})] (\text{total endo})}{100}$$

This procedure has shown high correlation coefficients between MEF and test weight, density, percent floaters, kernel length, horny-floury endosperm ratio, and breakage susceptibility (Stroshine et al, 1986).

D. Wet Milling

As described in Chapter 12, the corn wet-milling process involves a prolonged steep at elevated temperature in a low-pH medium containing sulfur dioxide. These conditions destroy fungi and many other microorganisms entering with the grain. After steeping, the grain is ground to recover germ and starch. Contaminating materials, such as insects and debris of all kinds are found in the feed by-products. Mycotoxins such as aflatoxin (Yahl et al, 1971) and zearalenone (Bennett et al, 1978) that would be contained in the corn initially are not present in the food (cornstarch or oil) after refining but are concentrated in feed by-products. For this reason, wet millers reject shipments that show the presence of mycotoxins.

The corn quality needs of the wet miller have been thoroughly discussed by Freeman (1973). Corn is purchased for wet milling usually through central markets, such as boards of trade, or direct from large grain companies, often with unit-train deliveries. The desired quality is selected by the buyer on the basis of U.S. grade factors with the emphasis on U.S. No. 2. Undesirable properties are high levels of BCFM, breakage susceptibility, and damage. Broken pieces must be removed from the corn by screening before processing because it interferes with the flow of the steeping medium through the corn, and because sloughed-off starch and protein enter the steepwater and cause gelling problems during steepwater evaporation. Broken corn and dust removed by aspiration become a part of the feed by-product streams, which bring a lower price than the refined products.

High levels of mold damage are undesirable because germ recovery and crude oil quality are low from corn with moldy germs. Freeman et al (1970) compared wet milling of corn (picked out of commercial lots) in which 50% of the kernels had mold-damaged germs with that of sound corn in a laboratory wet-milling process. The damaged kernels from this experiment contained 50% less fat than normal kernels and were three or four times higher in free fatty acid value. The quantity of germ recovered from the damaged lot was about one half that of the sound lot; the gluten yield was larger by a similar amount and contained higher oil content. This indicated that the majority of moldy germs disintegrated during the milling step. The oil content of germ recovered from the damaged lot was 10% below that of the sound lot and would probably produce less refined oil because of a high level of free fatty acid. Although corn oil is only a minor product of wet milling compared with starch, it has a much higher market value. Therefore, wet millers usually impose a rather high discount scale to discourage delivery of highly damaged lots. For the same reason, wet millers have encouraged the breeding of high-oil corn varieties (Watson and Freeman, 1975) and have on occasion paid small premiums to purchase hybrids with higher oil.

Aside from aflatoxin, several nongrade factors are important. Excessive drying temperatures not only cause breakage but interfere with starch recovery. Corn dried at excessive temperatures gives starch lower in yield and higher in protein content; the gluten recovered during separation of starch has a lower protein content (higher starch) (Foster, 1965; Le Bras, 1982). It is generally recommended that for best results, corn used in wet milling should not be dried over 60° C (140° F).

Freeman (1973) reviewed various quick tests of corn that have been tried as tests for determining suitability for wet milling, but none proved very satisfactory. For example, high values in a warm germination (viability) test indicate that the lot will give good wet-milling performance, but some lots with a low viability may also be good. As with all other end-use evaluations, the most reliable tests are those that simulate a commercial process. Several laboratory wet-milling procedures have been published (Watson, 1964; Saint-Lebe, 1965). Brown et al (1982) used a set of artificially dried corn samples to compare three tests (viability [germination], test weight, and stress crack intensity) with a steeping index test. For this last test, subject corn in wire baskets was placed in commercial corn steeps for the 32 hr of the commercial steeping and was evaluated for degree of kernel softening to derive a "steeping index." The steeping index was highly correlated with commercial starch recovery but was not well correlated with the results of the three quality tests made on the corn before steeping. Results of evaluation of raw corn for all three factors give indications of adverse wet-milling properties. Their data indicated that the most severe dryer damage occurred with 28% MC corn.

E. Feed Manufacturing

Although a feed manufacturer may not be as selective in quality attributes as a food manufacturer, significant concern for corn quality is shown. Changes in quality that affect animal acceptance and performance can affect customer acceptance because manufactured feeds are 50–90% corn. Since feed products are generally of relatively low value, with low profit margins, the feed manufacturer generally buys on U.S. grade with preference for U.S. No. 2. Discount schedules are used to make up for deviations from No. 2 (Nesheim,

1968). The specific quality requirements depend on the kind of feed being manufactured and whether only one type or many types are prepared in a particular facility. For example, corn with excessive BCFM is acceptable in mash or pelleted feeds because the nutrient content is close to that of corn. But in a rolled or flaked product, excessive fines give the feed an appearance that might be unacceptable to a person feeding an animal by hand (Stewart, 1978). Corn is used in feed because it is the lowest priced source of energy, supplying 65–80% of the energy in rations. It is also a source of protein, B-vitamins, B-carotene, and xanthophyl pigments for egg yolk and poultry skin pigmentation (Chapter 15).

In the modern feed plant, ingredients are blended from a linear computer program designed to select, for every nutrient, the lowest priced ingredient that is a significant source of that nutrient. For most nutrients, the program uses average established values for each ingredient. If a major ingredient like corn deviates very far from the standard value, the feed will be deficient, animal performance will suffer, and the customer will object. This is particularly true in the case of energy, a difficult entity to measure. Energy density of corn can be decreased by excessive moisture (as a diluent), low test weight, excessive damaged-kernel levels, and excessive foreign material (cobs and other corn by-products). Protein is the next most important nutrient from corn. For example, in a 15%-protein broiler ration, corn may contribute 40% of the protein. With the advent of rapid NIR measurement of protein content, many feed mills are now measuring the protein content of corn before incorporating it into a feed rather than relying on published values for corn protein content.

Moisture content of corn is important in many respects to feed manufacturers because it is a diluent to nutrients. Also, it has fundamental influence on the storage stability of the finished product. In the past, feed manufacturers specified a maximum of 15.5% MC because that was the maximum allowed in U.S. No. 2 corn. It will be interesting to see how feed manufacturers respond to the removal of moisture as a U.S. grading factor. A level of 15.5% MC in feed is too high for freedom from fungal deterioration because: 1) a corn lot that averages 15.5% MC may range from 14.5 to 16.5% MC in different parts of the lot, 2) ground corn is much more vulnerable to fungal development than whole corn, and 3) the high-protein ingredients and the vitamin and mineral supplements make feed an ideal medium for growth of many kinds of fungi.

The growth of fungi in a mixed feed may alter nutrient ratios and may produce odors and flavors that reduce its acceptance by animals. More important is the possibility of mycotoxin development during storage or in the feeding trough. The development of aflatoxin in poultry feeds has been an especially important problem in the warm southeast of the United States, where most of the broiler and egg industry is located (Edds and Bortell, 1983). To counteract this problem, antifungal agents, especially propionic, benzoic, and acetic acids, have been added to poultry mashes (Hamilton and Dixon, 1981). Many feed manufacturers now examine every lot of corn received with UV lamps for the bright greenish yellow fluorescence shown by the aflatoxin-presumptive test and reject lots that show positive levels by minicolumn tests. This is especially important for dairy rations because aflatoxin B_1 is excreted in the milk as the equally toxic aflatoxin M_1. Only 0.5 ppb of M_1 is allowed in milk. The only other major mycotoxin, DON, from *Fusarium* molds mainly affects swine performance (Cote et al, 1984), as discussed in Section VIII of this chapter. During the *Gibberella* epidemic of 1972, feed companies in the U.S. Midwest had to import corn from the South.

X. CONCLUSIONS AND TRENDS

This chapter has shown that the word "quality" applied to corn is a many-faceted property and is determined by many factors. Some factors, such as soil, climate, and some diseases, cannot be modified. However, modification is possible in practices on farms, in elevators, and during handling and transportation to provide quality that is close to the original preharvest quality. The incentive for better management must be monetary. Some users provide this incentive by offering simple premiums above the going average market price or by using contracts and "identity preserved" delivery arrangements. These purchasing agreements are too expensive for uses producing products with low profit margin, such as animal feeds, cornstarch, and sweeteners. Purchases for these uses, both U.S. domestic and export, are generally made on the basis of an official grading system. In the United States, the large grain companies, through which farmers must market their grain to volume users, claim they cannot affect quality because quality is determined at the farm level. But the purchasers do influence farm management of quality by the market signals given by purchasing strategies. As discussed earlier, purchasers, by providing discounts above U.S. No. 2 grade but no premiums below No. 2, tell the farmer that the highest price will be paid for corn exactly at 3% BCFM, 5% damage, 0.2% heat damage, 54 lb/bu test weight, and formerly at 15.5% MC. Most farmers will be unwilling to invest in equipment or to use techniques that produce better quality than that unless doing better will prevent greater losses. An example is drying corn to 14% MC or lower for long-term storage. Some farmers, however, seem unwilling to use management techniques that ensure against greater losses, such as proper use of aeration equipment, frequent monitoring of stored grain, and proper adjustment of harvesting and drying equipment.

In the recent decade, overseas buyers have made many complaints about poor quality corn purchased from U.S. grain companies. Some complaints are valid because of the high level of BCFM, which segregates during ship loading and unloading, because the U.S. grade standards seem to be too loose in some respects, and because of the practices of some foreign buyers. Some importers contract for poorer grades (at lower prices), hoping for (and often getting) better quality than was paid for. However, when quality comes up to contract, the corn may spoil during shipment. Thus, if the quality of corn in the marketplace is poorer than what farmers can produce, there is plenty of blame to go around to the various groups involved.

In the last two years, many pleas have been made to improve corn quality in the marketplace. Worldwide growing conditions have created larger crops than ever before in all corn-growing countries, resulting in intense competition for customers. Consequently, corn exports from the United States have declined, starting in 1979 with a grain embargo by the U.S. President against the Soviet Union for political reasons. An increase in the value of the U.S. dollar in relation to the value of other currencies and poorer corn quality than that from other corn-exporting countries influenced the decline in the exports of U.S. grain (corn, wheat, and soybeans).

Several conferences have been held to discuss the importance of corn quality and what can be done to improve it. Legislators have been planning to enact legislation to dictate changes in grades and trading rules that they believe would improve quality in the marketplace. Fortunately, a better method was developed, led by the North American Grain Export Association. Representatives of farm groups, grain company associations, and research people formed a task force to develop a consensus on changes needed in the U.S. grade system to provide the correct signals to farmers and importers and lead to improvement of corn quality in the marketplace. Recommendations of this task force are being implemented through the Federal Grain Inspection Service.

This kind of teamwork among researchers, extension agents, farmers, elevator operators, grain merchandisers, and buyers is needed to bring about a significant improvement in the quality of corn in the marketplace. Research information accumulated for the last 15 years by a group of Midwest researchers on ways to improve corn quality has been an important resource. This group, known as the NC-151 Committee, is a USDA North Central regional research committee made up of nearly 100 scientists and engineers from 12 state universities and several USDA laboratories. It is headquartered at Ohio State University, Wooster, and is funded by federal, state, and private grants from grain companies, associations, and farm organizations. This group is mentioned because it exemplifies the ongoing cooperative effort between academia and industry that is so necessary to provide intelligent progress in the desirable goal of improving corn quality.

In addition to the improvements in grades and marketing techniques, improvements in machinery for harvesting, drying, and handling corn are being made available commercially. Better prices for better-quality corn could stimulate use of these better machines. Likewise, corn breeders, after long years of low interest in quality, are showing that breeding corn for greater resistance to damage from harvesting, drying, insects, and molds is possible. The availability of good-yielding hybrids that will produce better-quality corn is a future possibility.

LITERATURE CITED

AACC. 1983. Approved Methods of the American Association of Cereal Chemists, 8th ed. Method 02-03A, approved Nov. 1984 (replacement of Method 02-03, approved April 1961); Method 44-15A, approved Oct. 1975, revised Oct. 1981; Method 44-19, approved April 1961, revised Oct. 1975, reviewed Oct. 1982; Method 45-14, approved Oct. 1979, revised Oct. 1981; Method 45-20, approved Oct. 1975, reviewed Oct. 1982; Method 45-21, approved Oct. 1986. The Association, St. Paul, MN.

AFUAKWA, J. J., and CROOKSTON, R. K. 1984. Using the kernel milk line to visually monitor grain maturity in maize. Crop. Sci. 24:687-691.

ANDERSON, H. W., NEHRING, E. W., and WICHSER, W. R. 1975. Aflatoxin contamination of corn in the field. J. Agric. Food Chem. 23:775-782.

ANDERSON, R. A. 1983. Detoxification of aflatoxin-contaminated corn. Pages 87-90 in: Aflatoxin and *Aspergillus flavus* in Corn. R. L. Diener, R. L. Asquith, and J. W. Dickens, eds. Southern Coop. Ser. Bull. 279. Ala. Agric. Exp. Stn., Auburn.

ANONYMOUS. 1981. Caution pledged on grain grade moves. Milling Baking News (Nov. 10):1, 56, 58, 59.

ASAE. 1983. ASAE standard S352, grains and

seeds, moisture measurement. Pages 327-328 in: Agricultural Engineers Yearbook of Standards, 31st ed. Am. Soc. Agric. Eng., St. Joseph, MI.

ATLIN, G. N., EMERSON, P. N., McGIN, L. G., and HUNTER, R. B. 1983. Gibberella ear rot development and zearalenone and vomitoxin production as affected by maize genotype and *Gibberella zeae* strain. Can. J. Plant Sci. 63:847-853.

AYRES, G. A., BABCOCK, C. E., and HULL, D. O. 1972. Corn combine field performance in Iowa. Pages 12-28 in: Grain Damage Symposium. Ohio State University, Columbus.

BARABOLAK, R., COLBURN, C. R., JUST, D. E., KURTZ, F. A., and SCHLEICHERT, E. A. 1978. Apparatus for rapid inspection of corn for aflatoxin contamination. Cereal Chem. 55:1065-1067.

BARAK, A. V., and HAREIN, P. K. 1981. Losses associated with insect infestation of farm-stored, shelled corn and wheat in Minnesota. Misc. Publ. 12. Minn. Agric. Exp. Stn., St. Paul.

BARHAM, H. N., Jr., and BARHAM, H. N. 1980. Method for adsorption of solids on seeds. U.S. patent 4,208,433. June 17.

BAUTISTA, G. M., and LINKO, P. 1962. Glutamic acid decarboxylase activity as a measure of damage in artificially dried and stored corn. Cereal Chem. 39:455-459.

BENNETT, E. H. 1950a. Kernel hardness in corn. I. A machine for the rapid determination of kernel hardness. Cereal Chem. 27:222-231.

BENNETT, E. H. 1950b. Kernel hardness in corn. II. A microscopic examination of hard and soft types of dent corn. Cereal Chem. 27:232-238.

BENNETT, G. A., VANDEGRAFT, E. E., SHOTWELL, O. L., WATSON, S. A., and BOCAN, B. J. 1978. Zearalenone: Distribution in wet-milling fractions from contaminated corn. Cereal Chem. 55:455-461.

BERMINGHAM, S. C., and HILL, L. D. 1978. A fair average quality for grain exports. AE-4459. Dept. Agric. Econ., Ill. Agric. Exp. Stn., Univ. Ill., Urbana. 6 pp.

BLOOME, P. D., BRUSEWITZ, C. H., and ABBOTT, D. C. 1982. Moisture adsorption by wheat. Trans. ASAE 25:1071-1075.

BOTTOMLEY, R. A., CHRISTENSEN, C. M., and GEDDES, W. F. 1952. Grain storage studies. X. The influence of aeration, time, and moisture content on fat acidity, nonreducing sugars, and mold flora in stored yellow corn. Cereal Chem. 29:53-64.

BREKKE, O. L. 1968. Corn dry-milling: Stress crack formation in tempering of low-moisture corn, and effect on degerminator performance.
Cereal Chem. 45:291-303.

BREKKE, O. L. 1970. Corn dry milling industry. Pages 262-291 in: Corn: Culture, Processing, Products. G. E. Inglett, ed. Avi Publ. Co., Westport, CT.

BREKKE, O. L., PEPLINSKI, A. J., GRIFFIN, E. J., Jr., and ELLIS, J. J. 1972. Dry-milling of corn attacked by southern leaf blight. Cereal Chem. 49:466-478.

BREKKE, O. L., STRINGFELLOW, A. C., and PEPLINSKI, A. J. 1978. Aflatoxin inactivation by ammonia gas: Laboratory trials. J. Agric. Food Chem. 26:1383-1389.

BROWN, R. B., FULFORD, G. N., DAYNARD, T. B., MEIERING, A. G., and OTTEN, L. 1979. Effect of drying method on grain corn quality. Cereal Chem. 56:529-532.

BUTLER, L. A. 1983. The history and background of NIR. Cereal Foods World 28:238-240.

BYG, D., and HALL, G. E. 1968. Corn losses and kernel damage in field shelling of corn. Trans. ASAE 11:164-166.

CALVERT, O. H., LILLEHOJ, E. B., KWOLEK, W. F., ZUBER, M. W., and LAUVER, E. L. 1983. Variability of bright greenish-yellow fluorescent particles and aflatoxin in ground blends of *Zea mays*. Can. J. Microbiol. 29:558-562.

CANTONE, F. A., TUITE, J., BAUMAN, L. F., and STROSHINE, R. 1983. Genotypic differences in reaction of stored corn kernels to attack by selected *Aspergillus* and *Penicillium* spp. Phytopathology 73:1250-1255.

CARTER, M. W., and PONELEIT, C. G. 1973. Black layer and filling period variation among inbred lines of corn (*Zea mays* L.). Crop Sci. 13:436-439.

CHOWDHURY, M. H., and BUCHELE, W. F. 1976a. Development of a mechanical damage index for critical evaluation of mechanical damage of corn. Trans. ASAE 19:428-432.

CHOWDHURY, M. H., and BUCHELE, W. F. 1976b. Colorimetric determination of grain damage. Trans. ASAE 19:807-808.

CHRISTENSEN, C. M., and KAUFMANN, H. H. 1969. Grain Storage: The Role of Fungi in Quality Loss. Univ. Minn. Press, Minneapolis. 153 pp.

CHRISTENSEN, C. M., and QUASEM, S. A. 1959. Note on a rapid method of detecting germ damage in wheat and corn. Cereal Chem. 36:461-464.

CHRISTENSEN, C. M., and SAUER, D. B. 1982. Microflora. Pages 219-240 in: Storage of Cereal Grains and Their Products, 3rd ed. C. M. Christensen, ed. Am. Assoc. Cereal Chem., St. Paul, MN.

CHRISTENSEN, C. M., MIROCHA, C. J., and MERONUCK, R. A. 1971. Some biological and chemical characteristics of damaged corn. J. Stored Prod. Res. 7:287-291.

CHRISTENSON, D. D., NIELSEN, H. C., KHOO, U., WOLF, M. J., and WALL, J. S. 1969. Isolation and chemical composition of protein bodies and matrix proteins in corn endosperm. Cereal Chem. 46:372-381.

CHUNG, D. S., and CONVERSE, H. H. 1970. Internal damage of wheat analyzed by radiographic determination. Trans. ASAE 13:295-297.

CIMMYT. 1982. CIMMYT Review. 1982. Centro Internacional de Mejoramiento de Maiz y Trigo, El Batán, Mexico.

COLE, R. J., KIRKSEY, J. W., CUTLER, H. G., DOUBNIK, B. L., and PECKHAM, J. C. 1973. Toxin from *Fusarium moniliforme*: Effects on plants and animals. Science 179:1324-1326.

COTE, L. M., REYNOLDS, J. D., VESONDER, R. F., and BULK, W. B. 1984. Survey of vomitoxin-contaminated feed grains in midwestern United States, and associated health problems with swine. J. Am. Vet. Med. Assoc. 184:189-192.

CRA. 1983. Standard Methods of the Member Companies of the Corn Refiners Association. Moisture methods. A-12. The Association, Washington, DC.

CRAIG, W. F. 1977. Production of hybrid corn seed. Pages 671-720 in: Corn and Corn Improvement. G. F. Sprague, ed. Am. Soc. Agron., Madison, WI.

DAYNARD, T. B., and DUNCAN, W. G. The black layer and grain maturity in corn. Crop Sci. 9:473-476.

DICKENS, J. W., and WHITAKER, T. B. 1983. Aflatoxin testing procedures for corn. Pages 35-37 in: Aflatoxins and *Aspergillus flavus* in Corn. U. L. Diener, R. L. Asquith, and J. W. Dickens, eds. Southern Coop. Ser. Bull. 279. Ala. Agric. Exp. Stn., Auburn.

DUVICK, D. N. 1961. Protein granules of maize endosperm cells. Cereal Chem. 38:374-385.

ECKHOFF, S. R., TUITE, J. F., FOSTER, G. H., KIRLEIS, A. W., and OKOS, M. R. 1983. Microbial growth inhibition by SO_2 or SO_2 plus NH_3 treatments during slow drying of corn. Cereal Chem. 60:185-188.

EDDS, G. T., and BORTELL, R. A. 1983. Biological effects of aflatoxins—Poultry. Pages 56-61 in: Aflatoxins and *Aspergillus flavus* in Corn. U. L. Diener, R. L. Asquith, and J. W. Dickens, eds. Southern Coop. Ser. Bull. 279. Ala. Agric. Exp. Stn., Auburn.

EKSTROM, G. A., LILJEDAHL, J. B., and PEART, R. M. 1966. Thermal expansion and tensile properties of corn kernels and their relationship to cracking during drying. Trans. ASAE 9:556-561.

ELAM, T. E., and HILL, L. D. 1977. Potential role of sampling variation in the measurement of corn grading factors. Ill. Agric. Econ. (Jan.):14-18.

ELLIS, E. B., FRIEDEMANN, P. D., and MEHLBERG, L. O. 1983. Grain quality for food processing. Pages 153-160 in: Proc. Annu. Corn and Sorghum Res. Conf., 38th. Seed Trade Assoc., Washington, DC.

FERNANDEZ, A., STROSHINE, R., and TUITE, J. 1985. Mold growth and carbon dioxide production during storage of high-moisture corn. Cereal Chem. 62:137-143.

FGIS. 1979. Report on adequacy of existing official U.S. standards for grain. Fed. Grain Inspection Serv., U.S. Dept. of Agric., Washington, DC. 61 pp.

FGIS. 1980a. Historical review of changes in the grain standards of the United States. Publ. FGIS-5. Fed. Grain Inspection Serv., U.S. Dept. of Agric., Washington, DC.

FGIS. 1980b. Grain Inspection Handbook. Book I. Grain Sampling. Fed. Grain Inspection Serv., U.S. Dept. of Agric., Washington, DC.

FGIS. 1980c. Grain Inspection Handbook. Book II. Grain Grading Procedures. Fed. Grain Inspection Serv., U.S. Dept. of Agric., Washington, DC.

FGIS. 1981. FGIS notice 81-41. Tentative corn moisture conversion charts for the Motomco moisture meter. Fed. Grain Inspection Serv., U.S. Dept. of Agric., Washington, DC.

FGIS. 1984a. The Official United States Standards for Grain. Fed. Grain Inspection Serv., Washington, DC.

FGIS. 1984b. Revision to the U.S. standards for corn, U.S. standards for sorghum, and U.S. standards of soybeans. Fed. Regist. 49:35743-35745. (Sept. 12)

FGIS. 1985. Annual report to Congress. Fed. Grain Inspection Serv., U.S. Dept. of Agric., Washington, DC.

FINNER, M. F., and SINGH, S. 1983. Apparatus for testing grains for resistance to damage. U.S. patent 4,422,319. Dec. 27.

FISCUS, D. F., FOSTER, G. H., and KAUFMANN, H. H. 1971. Grain-stream velocity measurements. Trans. ASAE 14:162-166.

FOSTER, G. H. 1965. Drying market corn. Pages 75-85 in: Proc. Annu. Hybrid Corn Ind. Res. Conf., 20th. Am. Seed Trade Assoc., Washington, DC.

FOSTER, G. H. 1982. Drying cereal grains. Pages 79-116 in: Storage of Cereal Grains and Their Products, 3rd ed. C. M. Christensen, ed. Am. Assoc. Cereal Chem., St. Paul, MN.

FOSTER, G. H., and HOLMAN, L. E. 1973. Grain breakage caused by commercial handling methods. Mark. Res. Rep. 96B. Agric. Res. Serv., U.S. Dept. of Agric., Washington, DC.

FREEMAN, J. E. 1973. Quality factors affecting value of corn for wet milling. Trans. ASAE 16:671-678, 682.

FREEMAN, J. E. 1980. Quality preservation during harvesting, conditioning, and storage of grains and oilseeds. Pages 187-224 in: Crop Quality, Storage, and Utilization. C. S. Hoveland, ed. Am. Soc. Agron., Madison, WI.

FREEMAN, J. E., HEATHERWICK, H. J., and WATSON, S. A. 1970. Evaluation of some grain quality tests for corn and effects of germ damage on wet milling results. Pages 1-29 in: Proc. Corn Conditioning Conf. AE-4251. Univ. Ill., Urbana.

FREY, N. M. 1981. Dry matter accumulation in kernels of maize. Crop Sci. 21:118-122.

FUGATT, A. N. 1978. Elevator explosions. Pages 47-50 in: Proc. Int. Symp. on Grain Elevator Explosions, Vol. 2. Natl. Advisory Board Publ. 352-2. Natl. Acad. Sci., Natl. Res. Counc., Washington, DC.

GRAHAM, R. J. 1981. Mission: Saving Lives and Property. Natl. Grain and Feed Assoc., Washington, DC. 16 pp.

GUNASEKARAN, S., and PAULSEN, M. R. 1985. Breakage resistance of corn as a function of drying rates. Trans. ASAE 28:2071-2075.

GUNASEKARAN, S., DESHPANDE, S. S., PAULSEN, M. R., and SHOVE, G. C. 1985. Size characterization of stress cracks in corn kernels. Trans. ASAE 28:1668-1672, 1685.

GUNASEKARAN, S., PAULSEN, M. R., and SHOVE, G. C. 1986. A laser optical method for detecting corn kernel defects. Trans. ASAE 29:294-298, 304.

GUSTAFSON, R. J., and HALL, G. E. 1970. Density and porosity of shelled corn during drying. Trans. ASAE 15:523-525.

GUSTAFSON, R. J., THOMPSON, D. R., and SOKHANSANJ, S. 1979. Temperature and stress analysis of corn kernel-finite element analysis. Trans. ASAE 22:955-960.

GUSTAFSON, R. J., MAHMOUD, A. Y., and HALL, G. E. 1983. Breakage susceptibility reduction by short term tempering of corn. Trans. ASAE 26:918-922.

GUTHEIL, R. A., KRAUSE, G. F., BROOKER, D. B., and ANDERSON, M. E. 1984. Effect of corn cultivar and sample variance on the performance of three electronic moisture meters. Cereal Chem. 61:267-269.

HALIBURTON, J. C., VESONDER, R. F., LOCK, T. F., and BUCK, W. B. 1979. Equine leucoencephalomalacia (ELEM): A study of *Fusarium moniliforme* as an etiologic agent. Vet. Hum. Toxicol. 21:348-351.

HALL, G. E. 1972. Test weight changes of shelled corn during drying. Trans. ASAE 15:320-323.

HALL, G. E., and HILL, L. D. 1973. Test weight as a grading factor for shelled corn. Publ. 124. Dept. Agric. Econ., Agric. Exp. Stn., Univ. Ill., Urbana. 18 pp.

HALL, G. E., and HILL, L. D. 1974. Test weight adjustment based on moisture content and mechanical damage of corn kernels. Trans. ASAE 17:578-579.

HALL, G. E., HILL, L. D., HATFIELD, E. E., and JENSEN, A. H. 1974. Propionic-acetic acid for high moisture corn preservation. Trans. ASAE 17:379-382, 387.

HALL, L. L., and ROSENFELD, A. 1982. Price-quality relationships for grains: An evaluation of buyers' discount behavior. Res. Bull. 82-37. Dept. Agric. Econ., N.Y. Agric. Exp. Stn., Cornell Univ., Ithaca.

HALLAUER, A. R., and RUSSELL, W. A. 1966. Effects of selected weather factors on grain moisture reduction from silking to physiological maturity in corn. Agron. J. 53:225-229.

HAMEDA, M. A., HURBURGH, C. R., and BERN, C. J. 1982. Effects of corn variety, mechanical damage, and drying temperature on electronic moisture meters. Paper 80-3549. Am. Soc. Agric. Eng., St. Joseph, MI.

HAMILTON, P. B. 1984. Determining safe levels of mycotoxins. J. Food Prot. 47:570-575.

HAMILTON, P. B., and DIXON, R. C. 1981. Evaluation of some organic acids as mold inhibitors by measuring CO_2 production from feed ingredients. Poult. Sci. 60:2182-2188.

HAMILTON, T. S., HAMILTON, B. C., JOHNSON, B. C., and MITCHELL, H. H. 1951. The dependence of the physical and chemical composition of the corn kernel on soil fertility and cropping system. Cereal Chem. 28:163-176.

HART, J. R., and NEUSTADT, M. H. 1957. Application of the Karl Fischer method to grain moisture determination. Cereal Chem. 34:27-37.

HART, L. P. 1986. Analysis of foods and feeds for aflatoxin B_1 by enzyme immunoassay. NC-151. Grain Quality Newsl. (Ohio Agric. Res. Dev. Cent., Wooster) 8(1):12.

HART, L. P., GENDLOFF, E., and ROSSMAN, E. C. 1984. Effect of corn genotypes on ear rot infection by *Gibberella zeae*. Plant Dis. 68:296-298.

HERTER, U., and BURRIS, J. S. 1984. Moisture distribution within corn seed during the drying process. Iowa Seed Sci. 6(1):1-3.

HERUM, F. L., and BLAISDELL, J. L. 1981. Effects of moisture content, temperature and test variables on results with grain breakage testers. Paper 81-3030. Am. Soc. Agric. Eng., St. Joseph, MI.

HERUM, F. L., and HAMDY, M. Y. 1981. Actual grain handling breakage compared to predictions by breakage susceptibility testers. Paper 81-3031. Am. Soc. Agric. Eng., St. Joseph, MI.

HESSELTINE, C. W. 1979. Introduction, definition, and history of mycotoxins of importance to animal production. Pages 3-18 in: Interactions of Mycotoxins in Animal Production: Proceedings of a Symposium. Natl. Acad. Sci., Washington, DC. 195 pp.

HICKS, D. R. 1975. Test weight of corn not feed value. Crops Soils (March):10-12.

HILL, L. D. 1975. Test weight of corn as a measure of quality. Pages 169-174 in: Corn Quality in World Markets. L. D. Hill, ed. Interstate Printers and Publishers, Inc., Danville, IL.

HILL, L. D. 1982a. Who pays the piper with high moisture grain? Pages 9-14 in: Evaluation of Issues in Grain Grades and Optimum Moistures. L. D. Hill, ed. AS-4548. Dept. Agric. Econ., Agric. Exp. Stn., Univ. Ill., Urbana-Champaign.

HILL, L. D. 1982b. The standardized bushel—A solution to the grain moisture debate. Pages 25-29 in: Evaluation of Issues in Grain Grades and Optimum Moistures. L. D. Hill, ed. AE-4548. Dept. Agric. Econ., Agric. Exp. Stn., Univ. Ill., Urbana-Champaign.

HILL, L. D. 1982c. Grain standards for corn exporting countries. Pages 37-41 in: Evaluation of Issues in Grain Grades and Optimum Moistures. L. D. Hill, ed. AE-4548. Dept. Agric. Econ., Agric. Exp. Stn., Univ. Ill., Urbana-Champaign.

HILL, L. D. 1985a. Removal of moisture as a determinant of numerical grade. No. 85. E-330. Dept. Agric. Econ., Agric. Exp. Stn., Univ. Ill., Urbana-Champaign. 9 pp.

HILL, L. D. 1985b. An international reference method for measuring moisture content of corn. AE-4602. Dept. Agric. Econ., Univ. Ill., Urbana-Champaign. 17 pp.

HILL, L. D., LEATH, M. N., SHOTWELL, O. L., WHITE, D. G., PAULSEN, M. R., and GARCIA, P. 1982. Alternative definitions for the grade factor of broken corn and foreign material. Bull 776. Dept. Agric. Econ., Univ. Ill., Urbana-Champaign. 36 pp.

HILL, L. D., PAULSEN, M. R., SHOVE, G. C., and KUHN, T. J. 1985. Changes in quality of corn between U.S. and Japan. AE-4609. Dept. Agric. Econ., Univ. Ill., Urbana-Champaign. 15 pp.

HILL, L. D., BEHNKE, K. C., PAULSEN, M. R., and SHOVE, G. C. 1986. Moisture variability in corn for export. AE-4610. Dept. Agric. Econ., Univ. Ill., Urbana-Champaign. 12 pp.

HINTON, J. J. C. 1953. The distribution of protein in the maize kernel in comparison with that in wheat. Cereal Chem. 30:441-445.

HLYNKA, I., and BUSHUK, W. 1959. The weight per bushel. Cereal Sci. Today 4:239-240.

HOERR, F. J., and D'ANDREA, G. H. 1983. Biological effects of aflatoxin in swine. Pages 51-55 in: Aflatoxin and *Aspergillus flavus* in Corn. U. L. Diener, R. L. Asquith, and J. W. Dickens, eds. Southern Coop. Ser. Bull. 279. Ala. Agric. Exp. Stn., Auburn.

HOFFMAN, K. D., and HILL, L. D. 1976. Historical review of the U.S. grades and standards for grain. Ill. Agric. Econ. 16(1):1-9.

HORN, B. W., and WICKLOW, D. T. 1983. Factors influencing inhibition of aflatoxin production in corn by *Aspergillus niger*. Can. J. Microbiol. 29:1087-1091.

HUNT, W. H., and PIXTON, S. W. 1974. Moisture—Its significance, behavior, and measurement. Pages 1-45 in: Storage of Cereal Grains and Their Products, 2nd ed. C. M. Christensen, ed. Am. Assoc. Cereal Chem., St. Paul, MN.

HURBURGH, C. R., Jr. 1984. On-site testing of a sampling strategy for country elevators. Paper 84-3020. Am. Soc. Agric. Eng., St. Joseph, MI.

HURBURGH, C. R., Jr., and BERN, C. J. 1983. Sampling corn and soybeans. I. Probing methods. Trans. ASAE 26:930-934.

HURBURGH, C. R., Jr., and HAZEN, T. E. 1983-1984. Performance of electronic moisture meters in corn. Drying Technol. 2:149-156.

HURBURGH, C. R., J., and MOECHNIG, B. W. 1982. Corn quality changes in commercial elevators. ASAE Paper 82-3547. Am. Soc. Agric. Eng., St. Joseph, MI.

HURBURGH, C. R., Jr., BERN, C. J., and ANDERSON, M. E. 1982. Shrinkage and corn quality changes in on-farm handling operations. ASAE Paper 82-3550. Am. Soc. Agric. Eng., St. Joseph, MI.

HURBURGH, C. R., Jr., HAZEN, T. E., and BERN, C. J. 1985. Corn moisture measurement accuracy. Trans. ASAE 28:634-640.

JINDAL, V. K., and MOHSENIN, N. N. 1978. Dynamic hardness determination of corn kernels from impact tests. J. Agric. Eng. Res. 23:77-84

JINDAL, V. K., HERUM, F. L., and MENSAH, J. K. A. 1978. Effects of repeated freezing-thawing cycles on the mechanical strength of corn kernels. Trans. ASAE 21:362-370, 374.

JOHNSON, R. M., GREENAWAY, W. T., and GOLUMBIC, C. 1969. Sampling stored corn for aflatoxin assay. Cereal Foods World 14:25-29.

JONES, F. E., and BRICKENKAMP, C. S. 1981. Automatic Karl Fischer titration of moisture in grain. J. Agric. Eng. Res. 7:185-191.

KAMEYANA, Y., LAI, F. S., SAYAMA, H., and FAN, L. T. 1982. The risk of dust explosions in grain processing facilities. J. Agric. Eng. Res. 27:253-261.

KIRLEIS, A. W., CROSBY, K. D., and HOUSLEY, T. L. 1984. A method for quantitatively measuring vitreous endosperm area in sectioning sorghum grain. Cereal Chem. 61:556-558.

LAI, F. S. 1978. Analysis of variation of fine material and broken kernels in grain. J. Agric. Eng. Res. 23:221-230.

LAI, F. S., MILLER, B. S., MARTIN, C. R., STOREY, C. L., BOLTE, L., SHOGREN, M., FINNEY, K. F., and QUINLAN, J. K. 1981. Reducing grain dust with oil additives. Trans. ASAE 24:1626-1632.

LAI, F. S., MARTIN, C. R., and MILLER, B. S. 1982. Examining the Use of Additives to Control Grain Dust. OCA-82-036. Natl. Grain and Feed Assoc., Washington, DC. 126 pp.

LEESON, S., and SUMMERS, J. D. 1975. Effect of adverse growing conditions on corn maturity and feeding value for poultry. Poultry Sci. 55:588-593.

LE BRAS, A. 1982. Maize drying and its resulting quality for wet milling industry. Pages 95-127 in: Maize: Recent Progress in Chemistry and Technology. G. E. Inglett, ed. Academic Press, New York.

LEY, W. B. 1985. Mycotoxins in stored corn linked to fatal equine disease. Feedstuffs (Jan. 28):7.

LOESCH, P. J., Jr., GRINDELAND, R. L., HAMMOND, E. G., and PAEZ, A. V. 1977. Evaluation of kernel hardness in normal and high-lysine maize (Zea mays L.). Maydica 22:197-212.

MAGEE, K. J., STROSHINE, R. L., FOSTER, G. H., and BAKER, K. D. 1983. Nature and extent of grain damage caused by pneumatic conveying systems. Paper 83-3505. Am. Soc. Agric. Eng., St Joseph, MI.

MARTIN, C. R. 1981. Characterization of grain dust properties. Trans. ASAE 24:738-742.

MARTIN, C. R. 1985. Variables affecting dust emissions from corn. ASAE Paper 85-3061. Am. Soc. Agric. Eng., St. Joseph, MI.

MARTIN, C. R., and LAI, F. S. 1978. Measurement of grain dustiness. Cereal Chem. 55:779-792.

MARTIN, C. R., CZUCHAJOWSKA, Z., and POMERANZ, Y. 1986. Aquagram standard deviations of moisture in mixtures of wet and dry corn. Cereal Chem. 63:442-445.

McGINTY, R. J. 1970. Development of a standard grain breakage test. (A progress report.) ARS 51-34. Agric. Res. Serv., U.S. Dept. Agric., Washington, DC. 13 pp.

McMILLIAN, W. W. 1983. Role of arthropods in field contamination. Pages 20-23 in: Aflatoxin and Aspergillus flavus in Corn. U. L. Diener, R. L. Asquith, and J. W. Dickens, eds. Southern Coop. Ser. Bull. 279. Ala. Agric. Exp. Stn., Auburn.

MILLER, B. S., HUGHES, J. W., ROUSSER, R., and BOOTH, G. D. 1981a. Effects of modification of a model CK2 Stein Breakage Tester on corn breakage susceptibility. Cereal Chem. 58:201-203.

MILLER, B. S., HUGHES, J. W., ROUSSER, R., and POMERANZ, Y. 1981b. Measuring the breakage susceptibility of shelled corn. Cereal Foods World 26:75-80.

MILLER, B. S., AFEWORK, S., POMERANZ, Y., BRUINSMA, B. L., and BOOTH, G. D. 1982. Measuring the hardness of wheat. Cereal Foods World 27:61-64.

MIROCHA, C. J., PATHRE, S. V., and CHRISTENSEN, C. M. 1980. Mycotoxins. Pages 159-225 in: Advances in Cereal Science and Technology, Vol. 3. Y. Pomeranz, ed. Am. Assoc. Cereal Chem., St. Paul, MN.

MISLIVEC, P. B., and TUITE, J. 1970. Temperature and relative humidity requirements of species of Penicillium isolated from yellow dent corn kernels. Mycologia 62:75-88.

MOHSENIN, N. N. 1970. Physical Properties of Plant and Animal Materials. Gordon and Breach Sci. Publ., New York.

MOORE, R. P. 1958. Tetrazolium tests for determination of injury and viability of seed corn. Pages 13-20 in: Proc. Annu. Hybrid Corn Ind. Res. Conf., 13th. Am. Seed Trade Assoc., Washington, DC.

MOREIRA, S. M. E., KURTZ, G. W., and FOSTER, G. H. 1981. Crack formation in corn kernels subject to impact. Trans. ASAE 24:889-892.

MORENO-MARTINEZ, E., and CHRISTENSEN, C. M. 1971. Differences among lines and varieties of maize in susceptibility to damage by storage fungi. Phytopathology 61:1498-1500.

MUIR, W. E., WATERER, D., and SINHA, R. N. 1985. Carbon dioxide as an early indicator of stored cereal and oilseed spoilage. Trans. ASAE 28:1673-1675.

NAVRATIL, R. J., and BURRIS, J. S. 1984. The effect of drying temperature on corn seed quality. Can. J. Plant Sci. 64:487-496.

NELSON, S. O. 1980. Moisture-dependent kernel and bulk density for wheat and corn. Trans. ASAE 23:139-143.

NELSON, S. O. 1981. Review of factors influencing the dielectric properties of cereal grains. Cereal Chem. 58:487-492.

NESHEIM, R. O. 1968. Our corn quality needs and problems: Feed manufacturers. Pages 16-17 in: Proc. Corn Quality Conf. Dept. Agric. Econ., Ill. Agric. Exp. Stn., Urbana.

NGFA. 1979. A Practical Guide to Elevator Design. Natl. Grain and Feed Assoc., Washington, DC. 518 pp.

NOFSINGER, G. W., BOTHAST, R. J., LANCASTER, E. B., and BAGLEY, E. B. 1977. Ammonia-supplemented ambient temperature drying of high moisture corn. Trans. ASAE 20:1151-1154, 1159.

OBENDORF, R. L. 1972. Factors associated with early germination in corn under cool conditions. Pages 132-139 in: Proc. Annu. Corn and Sorghum Res. Conf., 27th. Am. Seed Trade Assoc., Washington, DC.

PARKER, P. E., BAUWIN, G. R., and RYAN, H. L. 1982. Sampling, inspection and grading of grain. Pages 1-35 in: Storage of Cereal Grains and Their Products, 3rd ed. C. M. Christensen, ed. Am. Assoc. Cereal Chem., St. Paul, MN.

PAULSEN, M. R. 1983. Corn breakage susceptibility as a function of moisture content. Paper 83-3078. Am. Soc. Agric. Eng., St. Joseph, MI.

PAULSEN, M. R., and HILL, L. D. 1985a. Corn quality factors affecting dry milling performance. J. Agric. Eng. Res. 31:255-263.

PAULSEN, M. R., and HILL, L. D. 1985b. Quality attributes of Argentine corn. Appl. Eng. Agric. 1:42-46.

PAULSEN, M. R., and McCLURE, W. F. 1985. Illumination for computer vision systems. Paper 85-3546. Am. Soc. Agric. Eng., St. Joseph, MI.

PAULSEN, M. R., HILL, L. D., and DIXON, B. L. 1983a. Moisture meter-to-oven comparisons for Illinois corn. Trans. ASAE 26:576-583.

PAULSEN, M. R., and HUMMEL, J. W. 1981. Corn damage from combines. Ill. Res. (Spring):10-11.

PAULSEN, M. R., HILL, L. D., WHITE, D. G., and SPRAGUE, G. F. 1983b. Breakage susceptibility of corn belt genotypes. Trans. ASAE 26:1830-1836, 1841.

PAYNE, G. A. 1983. Nature of field infection of corn by *Aspergillus flavus*. Pages 16-19 in: Aflatoxin and *Aspergillus flavus* in Corn. U. L. Diener, R. L. Asquith, and J. W. Dickens, eds. Southern Coop. Ser. Bull. 279. Ala. Agric. Exp. Stn., Auburn.

PEREZ, R. A., TUITE, J., and BAKER, K. 1982. Effect of moisture, temperature, and storage time on subsequent storability of shelled corn. Cereal Chem. 59:205-209.

PIERCE, R. O., and HANNA, M. A. 1985. Corn kernel damage during on-farm handling. Trans. ASAE 28:239-241, 245.

POMERANZ, Y., MARTIN, C. R., TRAYLOR, D. D., and LAI, F. S. 1984. Corn hardness determination. Cereal Chem. 61:147-150.

POMERANZ, Y., CZUCHAJOWSKA, Z., MARTIN, C. R., and LAI, F. S. 1985. Determination of corn hardness by the Stenvert hardness tester. Cereal Chem. 62:108-112.

POMERANZ, Y., CZUCHAJOWSKA, Z., and LAI, F. S. 1986a. Gross composition of coarse and fine fractions of small corn samples ground in the Stenvert hardness tester. Cereal Chem. 63:22-26.

POMERANZ, Y., CZUCHAJOWSKA, Z., and LAI, F. S. 1986b. Comparison of methods for determination of hardness and breakage susceptibility of commercially dried corn. Cereal Chem. 63:39-43.

QUASEM, S. A., and CHRISTENSEN, C. M. 1960. Influence of various factors on the deterioration of stored corn by fungi. Phytopathology 50:703-709.

RABIE, C. J., MARASAS, W. F. O., THIEL, P. G., LUBBEN, A., and VLEGGAAR, R. 1982. Moniliformin production and toxicity of different *Fusarium* species from South Africa. Appl. Environ. Microbiol. 43:517-521.

RACOP, E., STROSHINE, R., LIEN, R., and NOTZ, W. 1984. A comparison of three combines with respect to harvesting losses and grain damage. Paper 84-3016. Am. Soc. Agric. Eng., St. Joseph, MI.

RUSSELL, W. A., JOHNSON, D. Q., and LEFORD, D. R. 1986. Plant breeding studies with grain quality traits in corn. Pages 125-141 in: Proc. Annu. Corn and Sorghum Res. Conf., 40th. Am. Seed Trade Assoc.,

Washington, DC.

RUTLEDGE, J. H. 1978. The value of corn quality to the dry miller. Pages 158-162 in: Proc. 1977 Corn Quality Conf. AE-4454. Dept. Agric. Econ., Univ. Ill., Urbana.

SAINT-LEBE, P. L., JOSSOUD, M., and ANDRE, C. 1965. Extraction de L'amidon de maiz. Staerke 11:341-346.

SAUER, D. B., STOREY, C. L., ECKER, O., and FULK, D. W. 1982. Fungi in U.S. export wheat and corn. Phytopathology 72:1449-1452.

SAUER, D. B., STOREY, C. L., and WALKER, D. E. 1984. Fungal populations in U.S. farm-stored grain and their relationship to moisture, storage time, regions, and insect infestation. Phytopathology 74:1050-1053.

SAUL, R. A. 1968. Effects of harvest and handling on corn storage. Pages 33-36 in: Proc. Annu. Corn Sorghum Res. Conf., 23rd. Am. Seed Trade Assoc., Washington, DC.

SCHMIDT, J. L., and HALLAUER, A. P. 1966. Estimating harvest date of corn in the field. Crop Sci. 6:227-231.

SCHNAKE, L. O. 1981. Grain dust: Problems and utilization. ESS-6. Econ. Stat. Serv., Natl. Econ. Div., U.S. Dept. Agric., Washington, DC. 17 pp.

SCHOONHOVEN, A. V., WASSOM, C. E., and MILLS, R. B. 1975. Selection for resistance to the maize weevil in kernels of maize. Euphytica 24:639-644.

SEITZ, L. M., SAUER, D. B., MOHR, H. E., and ALDIS, D. F. 1982a. Fungal growth and dry matter loss during bin storage of high-moisture corn. Cereal Chem. 59:9-14.

SEITZ, L. M., SAUER, D. B., and MOHR, H. E. 1982b. Storage of high-moisture corn: Fungal growth and dry matter loss. Cereal Chem. 59:100-105.

SEYEDIN, N., BURRIS, J. S., and FLYNN, T. E. 1984. Physiological studies on the effects of drying temperatures on corn seed quality. Can. J. Plant Sci. 64:497-504.

SHAW, R. H. 1977. Climate requirement. Pages 591-624 in: Corn and Corn Improvement. G. F. Sprague, ed. Am. Soc. Agron., Madison, WI.

SHELEF, L., and MOHSENIN, N. N. 1969. Effect of moisture content on mechanical properties of shelled corn. Cereal Chem. 46:242-253.

SHELLENBERGER, J. A. 1982. Development of an international grain quality certificate. Pages 431-435 in: Proc. World Cereal and Bread Congr., 7th. Prague.

SHENG, C. T., and GUSTAFSON, R. J. 1985. Surface mass transfer coefficients for corn kernels. Paper 85-3541. Am. Soc. Agric. Eng., St. Joseph, MI.

SHOTWELL, O. L. 1983. Aflatoxin detection and determination in corn. Pages 38-45 in: Aflatoxin and *Aspergillus flavus* in Corn. U. L. Diener, R. L. Asquith, and J. W. Dickens, eds. Southern Coop. Ser. Bull. 279. Ala. Agric. Exp. Stn., Auburn.

SHOTWELL, O. L., and HESSELTINE, C. W. 1981. Use of bright greenish yellow fluorescence as a presumptive test for aflatoxin in corn. Cereal Chem. 58:124-127.

SHOVE, G. C. 1969. Wet grain aeration for holding and drying shelled corn. Ill. Res. 11(2):12-13.

SINGH, S. S., and FINNER, M. F. 1983. A centrifugal impactor for damage susceptibility evaluation of shelled corn. Trans. ASAE 26:1858-1863.

STEELE, J. L., SAUL, R. A., and HUKILL, W. V. 1969. Deterioration of shelled corn as measured by carbon dioxide production. Trans. ASAE 12:685-689.

STENVERT, N. L. 1974. Grinding resistance, a simple measure of wheat hardness. Flour Anim. Feed Milling 12:24.

STEPHENS, L. B., and FOSTER, G. H. 1976. Breakage tester predicts handling damage in corn. ARS-NC-49. Agric. Res. Serv., U.S. Dept. Agric., Washington, DC. 6 pp.

STEWART, L. D. 1978. The significance of corn quality for the feed manufacturer. Pages 163-168 in: Proc. 1977 Corn Quality Conf. AE-4454. Dept. Agric. Econ., Ill. Agric. Exp. Stn., Urbana.

STROSHINE, R. L., KIRLEIS, A. W., TUITE, J. F., BAUMAN, L. F., and EMAM, A. 1986. Differences in corn quality among selected corn hybrids. Cereal Foods World 31:311-316.

SZANIEL, J., SAGI, F., and PALVOLGYI, I. 1984. Hardness determination and quality prediction of maize kernels by a new instrument, the molograph. Maydica 29:9-20.

TAUBENHAUS, J. J. 1920. A study of the black and yellow molds of ear corn. Bull. 270. Tex. Agric. Exp. Stn., College Station. 38 pp.

THOMAS, A. T. 1966. Evaluation of *opaque-2* maize for fermentations with *Saccharomyces cerevisiae*. Pages 128-140 in: Proc. High Lysine Corn Conf. E. T. Mertz and O. E. Nelson, eds. Corn Refiners Assoc., Washington, DC.

THOMPSON, R. A., and FOSTER, G. H. 1968. Stress cracks and breakage in artificially dried corn. Marketing Res. Bull. 631. Agric. Mark. Serv., U.S. Dept. Agric., Washington, DC.

THOMPSON, R. A., and ISAACS, G. W. 1967. Porosity determinations of grains and seeds with the air comparison pycnometer. Trans. ASAE 10:693-696.

THORNTON, J. H., GOODRICH, R. D., and MEISKE, J. C. 1969. Digestibility of nutrients and energy value of corn grain of various maturities and test weight. J. Anim. Sci. 29:983-986.

TRAN, T. L., DEMAN, J. M., and RASPER, V. F. 1981. Measurement of corn kernel hardness. Can. Inst. Food Sci. Technol. J. 14:42-48.

TSURUTA, O., GOHARA, S., and SAITO, M. 1981. Scanning electron microscopic observation of a fungal invasion of corn kernels. Trans. Mycol. Soc. Jpn. 22:121-126.

TUITE, J., and FOSTER, G. H. 1979a. Effect of artificial drying on the hygroscopic properties of corn. Cereal Chem. 40:630-637.

TUITE, J., and FOSTER, G. H. 1979b. Control of storage diseases of corn. Annu. Rev. Phytopathol. 17:343-366.

TUITE, J., SHANER, G., RAMBO, G., FOSTER, J., and CALDWELL, R. W. 1974. The Gibberella ear rot epidemics of corn in Indiana in 1965 and 1972. Cereal Sci. Today 19:238-241.

TUITE, J., SENSMEIER, R., KOH-KNOX, C., and NOEL, R. 1984. Preharvest aflatoxin contamination of dent corn in Indiana in 1983. Plant Dis. 68:893-895.

TUITE, J., KOH-KNOX, C., STROSHINE, R., CANTONE, F. A., and BAUMAN, L. F. 1985. Effect of physical damage to corn kernels on the development of *Penicillium* species and *Aspergillus glaucus* in storage. Phytopathology 75:1137-1140.

TUITE, J., FOSTER, G. H., ECKHOFF, S. R., and SHOTWELL, O. L. 1986. Sulfur dioxide treatment to extend drying time. Cereal Chem. 63:462-464.

VESONDER, R. F., CIEGLER, A., JENSEN, A. H., ROHWEDDER, W. K., and WEISLEDER, D. 1976. Co-identity of the refusal and emetic principle from *Fusarium*-infected corn. Appl. Environ. Microbiol. 31:280-285.

WATSON, C. A., HAWK, A. L., NEFFENEGGER, P., and DUNCAN, D. 1970. Performance evaluation of grain sample dividers. Spec. Rep. ARS 51. Agric. Res. Serv., U.S. Dept. Agric., Washington, DC.

WATSON, S. A. 1964. Corn starch isolation. Pages 3-5 in: Methods in Carbohydrate Chemistry, Vol. IV. R. L. Whistler, ed. Academic Press, New York.

WATSON, S. A. 1986a. A system of national moisture meter standardization. NC-151. Grain Quality Newsl. (Ohio Agric. Res. Dev. Cent., Wooster) 8(1):16-17.

WATSON, S. A. 1986b. Corn moisture methods used by wet millers. NC-151. Grain Quality Newsl. (Ohio Agric. Res. Dev. Cent., Wooster) 8(1):20.

WATSON, S. A., and FREEMAN, J. E. 1975. Breeding corn for increased oil content. Pages 261-275 in: Proc. Annu. Corn and Sorghum Res. Conf., 30th. Am. Seed Trade Assoc., Washington, DC.

WATSON, S. A., and HERUM, F. L. 1986. Comparison of eight devices for measuring breakage susceptibility of shelled corn. Cereal Chem. 63:139-142.

WATSON, S. A., DARRAH, L. L., and HERUM, F. L. 1986. Measurement of corn breakage susceptibility with the Stein breakage tester. Cereal Foods World 31:366-367, 370, 372.

WEAK, E. D., MILLER, G. D., FARRELL, E. P., and WATSON, C. A. 1972. Rapid determination of germ damage in cereal grains. Cereal Chem. 49:653-663.

WHITE, G. M., ROSS, T. J., and PONELEIT, C. G. 1982. Stress crack development in popcorn as influenced by drying and rehydration. Trans. ASAE 25:768-772.

WICHSER, W. R. 1961. The world of corn processing. Am. Miller Process. 89(3):23-24; 89(4):29-31.

WICHSER, W. R. 1966. Comparison of dry milling properties of *opaque-2* and normal dent corn. Pages 104-116 in: Proc. High Lysine Corn Conf. E. T. Mertz, and O. E. Nelson, eds. Corn Refiners Assoc., Washington, DC.

WICKLOW, D. T. 1983. Taxonomic features and ecological significance of sclerotia. Pages 6-15 in: Aflatoxin and *Aspergillus flavus* in Corn. U. L. Diener, R. L. Asquith, and J. W. Dickens, eds. Southern Coop. Ser. Bull. 279. Ala. Agric. Exp. Stn., Auburn.

WICKLOW, D. T., HORN, B. W., and COLE, R. J. 1982. Sclerotium production by *Aspergillus flavus* on corn kernels. Mycologia 74:398-403.

WIDSTROM, N. W., and ZUBER, M. S. 1983. Sources and mechanisms of genetic control in the plant. Pages 72-75 in: Aflatoxin and *Aspergillus flavus* in Corn. U. L. Diener, R. L. Asquith, and J. W. Dickens, eds. Southern Coop. Ser. Bull. 279. Ala. Agric. Exp. Stn., Auburn.

WILLIAMS, P. C., and NORRIS, K. H. 1983. Effect of mutual interactions on the estimation of protein and moisture in wheat. Cereal Chem. 60:202-207.

WOLF, M. J., BUZAN, C. L., MacMASTERS, M. M., and RIST, C. E. 1952. Structure of the mature corn kernel. III. Microscopic structure of the endosperm of dent corn. Cereal Chem. 29:349-361.

WOLF, M. J., KHOO, U., and SECKINGER, H. L. 1969. Distribution and subcellular structure of endosperm protein in varieties of ordinary and high-lysine maize. Cereal Chem. 46:253-263.

YAHL, K. R., WATSON, S. A., SMITH, R. J., and BARABOLOK, R. 1971. Laboratory wet-milling of corn containing high levels of aflatoxin and a survey of commercial wet-milling products. Cereal Chem. 48:385-391.

YOUNG, J. C., and FULCHER, R. G. 1984. Mycotoxins in grains: Causes, consequences, and cures. Cereal Foods World 29:725-728.

YOUNGS, V. L., KUNERTH, W. H., and CRAWFORD, R. D. 1984. Oil added to wheat for dust control: Effects on milling and baking quality. (Abstr.) Cereal Foods World 29:499.

CHAPTER 6

EFFECT AND CONTROL OF INSECTS AFFECTING CORN QUALITY

CHARLES L. STOREY (*retired*)
U.S. Grain Marketing Research Laboratory
U.S. Department of Agriculture
Agricultural Research Service
Manhattan, Kansas

I. INTRODUCTION

If corn is stored for an extended period, steps must be taken to preserve its quality and prevent economic loss due to insect infestation. Insects damage corn directly by feeding on the kernels and indirectly by contaminating the grain with their waste, cast skins, webbing, and body parts. They may also contribute to the conditions that cause fungal proliferation, resulting in mold and heat damage. The market value of infested corn may be substantially reduced if the number of insect-damaged kernels is sufficient to lower the grade or if the number of insects in the corn causes it to be designated "weevily" or "infested" on the grade certificate. Also, discount penalties against the price paid per bushel are often assessed by the buyer if live insects are present in the grain at the time of sale.

Insects in farm-stored corn may affect its eligibility in government storage loan programs. When a loan is approved, the producer is made responsible for any loss in quantity or quality of the corn caused by insect infestation. If not controlled, even low levels of insect infestation in farm-stored corn can develop into damaging populations before the grain reaches its final destination. Therefore, corn moving from farm storage through the market system from country elevators through terminal distribution centers must be kept free of these pests to ensure its acceptance by both domestic and foreign corn buyers.

II. IMPORTANT INSECTS AFFECTING CORN QUALITY

Grain storage insects are generally divided into two groups: those species that develop inside kernels and comprise the "hidden infestation" in a grain mass and those species that develop outside the kernel by feeding on broken kernels, germs, grain dust, or other cereal products (Cotton and Wilbur, 1982). Another group is those insects associated with high-moisture grain, which feed primarily on mold but may also damage kernels. Common representatives of each group are listed in Table I. The U.S. Department of Agriculture, in its Agriculture Handbook No. 500 (Anonymous, 1979), identifies five primary or internally developing species that cause most of the damage to grain in storage and shipment. This reference may be purchased from the Superintendent of Documents, U.S. Government Printing Office, Washington, DC 20402.

The internally developing species, the primary insect pests, are: the rice weevil, *Sitophilus oryzae*; the maize weevil, *S. zeamais*; the granary weevil, *S. granarius*; the lesser grain borer, *Rhyzopertha dominica*; and the Angoumois grain moth, *Sitotroga cerealella*. The larval stage of each of these species develops within a grain kernel, where the insect feeds unseen and usually unsuspected. These internal stages cannot be removed from grain by ordinary cleaning but must be controlled by other means, such as chemical pesticides. They also contribute most of the insect fragments found in milled cereal products.

Externally developing species of importance are: the flat grain beetle, *Cryptolestes pusillus*; the rusty grain beetle, *C. ferrugineus*; the confused flour beetle, *Tribolium confusum*; the red flour beetle, *T. castaneum*; the saw-toothed grain beetle, *Oryzaephilus surinamensis*; the Indian-meal moth, *Plodia interpunctella*; and the almond moth, *Ephestia cautella* (Table I).

These external, or secondary, insect pests can cause much damage to grain in storage conditions favoring their proliferation. Although some can attack whole kernels, population development of secondary pests is aided by grain dust or

TABLE I
Common Insect Pests of Stored Corn

Scientific Name	Common Name
Species that develop inside kernels	
Sitophilus oryzae (L.)	Rice weevil
S. zeamais Motschulsky	Maize weevil
S. granarius (L.)	Granary weevil
Rhyzopertha dominica (F.)	Lesser grain borer
Sitotroga cerealella (Olivier)	Angoumois grain moth
Species that develop outside kernels	
Cryptolestes ferrugineus (Stephens)	Rusty grain beetle
C. pusillus (Schoenherr)	Flat grain beetle
Plodia interpunctella (Hübner)	Indian-meal moth
Tribolium castaneum (Herbst)	Red flour beetle
T. confusum duVal	Confused flour beetle
Oryzaephilus surinamensis (L.)	Saw-toothed grain beetle
Ephestia cautella	Almond moth
External species associated with high-moisture storage	
Ahasverus advena (Waltl.)	Foreign grain beetle
Typhaea stercorea (L.)	Hairy fungus beetle
Cynaeus angustus (LeConte)	Larger black flour beetle
Liposcelis spp.	Psocids or booklice
Platydema ruficorne (Strum)	Red-horned grain beetle
Murmidius ovalis (Beck)	Minute beetle
Carpophilus dimidiatus (F.)	Corn sap beetle

broken kernels produced by mechanical injury during handling or by the feeding activity of the primary insect pests. Some secondary species will eat into the germ area of a kernel, where they are not easily detected or dislodged during cleaning; thus, in effect they become "internal feeders."

Recent studies by Storey et al (1983) identified the principal insect species in samples from nearly 3,000 bins of corn stored on farms across 19 states. One or more live, stored-product insect species were found in 79.7% of the corn samples. Twenty-four species or groups of species were identified in the combined samples, and individual samples commonly contained as many as five or six different species. *Cryptolestes* spp. were the most universally distributed storage pests found in corn, occurring in 57.7% of the samples at an average sample density (number) of 18 per 1,000 g. The Indian-meal moth was the second most frequent pest, occurring in 27.6% of the samples at an average density of 107 per 1,000 g. *Tribolium* spp. were found in 19.4% of the samples, *Ahasverus advena* (the foreign grain beetle) in 16.0%, *Sitophilus* spp. in 9.4%, and the saw-toothed grain beetle in 9.3%. The Angoumois grain moth was detected in only three corn samples, each of which was obtained from corn stocks stored "on the ear," and none were found in samples from bulk-stored corn. These data suggest that use of the picker-sheller, and the resulting bulk storage of corn, has downgraded the importance of this storage pest.

Data obtained in the farm study also documented the close relationship between the occurrence and successful development of specific insect populations and the moisture at which the corn was stored. Insect species that prefer high-moisture conditions or feed on molds or decaying vegetable matter were found in 28.8% of the farm-stored corn. These species included: *Murmidius ovalis* (the minute beetle), *Liposcelis* spp. (psocids), *Platydema ruficorne* (the red-horned grain beetle), *Tenebrio molitor* (the yellow mealworm), *Alphitophagus bifasciatus* (the two-banded fungus beetle), *Carpophilus dimidiatus* (the corn sap beetle), *Cynaeus angustus* (the larger black flour beetle), *Typhaea stercorea* (the hairy fungus beetle), and *A. advena* (the foreign grain beetle). The presence of significant numbers of these species in stored corn suggests that excessive moisture and the fungal development that results from this condition are a universal storage problem throughout the corn-producing area.

The frequent occurrence of insect pests in farm-stored corn is reflected in the numbers of insects found in corn exported from the United States. During a two-year period, 1977–1978, one or more live insects were found in 22.4% of over 2,300 samples obtained from corn exported from 79 port elevators (Storey et al, 1982a). *Sitophilus* spp. were the predominant insects found, occurring in 14.4% of the corn samples at an average density of 5.8 weevils per 1,000 g. *Cryptolestes* spp. were the second most frequently detected pest, occurring in 9.7% of the samples at an average of 2.4 insects per 1,000 g. Grain pests that develop outside the grain kernel and those species that feed on molds or prefer high-moisture conditions disappeared or were significantly lower in average density in corn samples obtained at export than in samples taken from farm storage. Possibly, these species are adversely affected by the handling of the corn and the resulting changes in storage environment as the grain moves through the marketing system.

III. DEVELOPMENT OF INSECTS IN STORED GRAIN

Most of the common pests of stored grain are cosmopolitan species found throughout the world where grain is harvested and stored. The major pests of stored corn are either beetles or moths that undergo changes in form, or metamorphosis, during development from egg to larval, pupal, and adult stages.

A. Developmental Stages

The insect egg stage may last from only a few days to several weeks, depending on the grain temperature. As temperatures fall below 15°C, most eggs stop hatching, but when temperatures increase to an optimum range of 30–33°C, hatching may occur in as little as three or four days.

The insect's growth occurs during the larval stage, in which it consumes many times its own weight in food and expands in size by shedding its skin (molting) to allow for the increased growth. During the pupal stage, the insect stops feeding and begins a gradual transformation to the adult—the stage most readily seen in or about stored grain. Adult stages of most beetle species are equipped with strong mouthparts suitable for feeding directly on grain and will often cause as much damage to grain as the larval stages. Adult moths, however, do not have the same type of mouthparts and are unable to feed on grain. Their primary function is reproduction. Most moth species mate within a few hours or days after emergence, and death follows shortly after the female deposits her eggs. In contrast, adult beetles mate and deposit eggs over an extended period of several weeks, depending on environmental conditions.

B. Factors Affecting Development

Most stored-grain insects have a relatively short development period, a high rate of reproduction, and a long adult lifespan. The two environmental factors that most influence these characteristics are temperature and moisture. Stored-grain insects generally require temperatures above 15°C to survive and reproduce, and many require temperatures above 21°C before damaging populations can be developed. Temperatures above 35°C become increasingly unfavorable for insect development—adults cease laying eggs and are short-lived. Although some grain insects are more hardy than others, winter temperatures common in the grain-producing areas (except for the South) are generally lethal to many stored-grain insects when the low temperature extends throughout the grain mass. Temperatures not low enough to kill insects directly may so restrict their activity that they are unable to feed, thus causing many to die of starvation. Granovsky and Mills (1982) found that granary weevils acclimated to cold temperatures (4.4°C) were unable to penetrate the pericarp of whole kernels.

Stored-grain insects depend on their food to supply the water needed for their life processes. If the moisture content of grain is low, generally under 10%, insects must obtain water by breaking down the grain components or by using their own energy reserves in the fatty tissues of the body. Under these conditions, fewer insects survive, and damaging populations are avoided. Up to a point, increasing grain moisture favors a rapid increase in the number of insects, but at levels generally above 15%, competition from mold and bacterial development adversely affects insect development. A close relationship between the occurrence and successful development of specific insect populations in farm-stored grain and the moisture at which the grain is stored was demonstrated by Storey et al (1983). They found that the incidence of high-moisture species such as the foreign grain beetle and the larger black flour beetle was more than five times higher in farm-stored corn with 13% moisture than in that with 10% moisture. Also, the incidence of *Cryptolestes* spp. and *Sitophilus* spp. increased progressively through successively higher moisture levels. Population densities of the principal internal or external species found in corn were highest in the middle moisture range (10-12%), with the lowest densities at either extreme.

IV. SAMPLING AND MEASUREMENT OF INSECT POPULATIONS IN CORN

The probability of corn becoming infested with damaging populations of insects or spoiled because of mold growth increases when the corn is stored and left undisturbed in the same locations for several months. It is important, therefore, to establish and maintain a regular monthly inspection routine to determine the corn's general condition and to detect early insect infestations and dampness or heating. Inspection equipment typically includes a grain probe and sieves for sifting insects from the grain samples. Temperature-sensing cables installed as a permanent part of the storage bin provide an excellent means to identify areas of the grain bulk in which conditions are favorable for insect development and also to locate areas in the grain that are heating. During warm weather, infestations generally begin near the corn surface, particularly in areas directly below the point of grain entry, the spoutline, where foreign material and broken grain particles have accumulated during loading. During cold weather, the grain temperatures may be used as a location guide for sampling a bin, with emphasis being placed on grain areas where the temperatures are above 18°C. Samples within the grain mass are obtained by inserting the probe vertically, at a slight angle, to its full length, and top-level samples are obtained by laying the probe horizontally on the grain and pushing it a few centimeters beneath the surface to collect the sample. To complete the inspection, one must look for insects on the exposed inside and outside bin surfaces, especially around the base, doors, and aeration ducts.

Techniques for detecting insects in grain based on the physical trapping of insects have been developed and tested in recent years (Loschiavo and Atkinson, 1967, 1973; Barak and Harein, 1982; Wright and Mills, 1983), and the use of sex-, aggregation-, and food-attractant chemicals for augmenting the trapping techniques have shown promising results (Burkholder and Dicke, 1966; Cogburn et al, 1983; Burkholder, 1984). One type of trap marketed commercially consists of a 2.54 × 33-cm perforated plastic tube fitted with a funnel-shaped inner receptacle for directing insects that enter through the perforations into a small plastic tube located at the base of the trap (Burkholder, 1984). The trap may be baited with one of the behavior-modifying chemicals or used alone. Studies by Wright and Mills (1983) found that trap locations at about two thirds of the distance toward the center of the bin and a few centimeters below the grain surface gave insect densities per trap that were close to the number of insects per kilogram of grain placed in small bins containing approximately 1.5 t of grain. Barak and Harein (1982) indicated that the use of traps increased the ability to detect insects in grain but suggested that probe samples would still be needed to provide grain for other quality determinations and to check for evidence of internal infestation.

The presence of any live stored-grain insect in probed corn samples or in traps is significant, particularly if several weeks of warm weather remain before the onset of cold temperatures. Emphasis is generally placed on insect species such as the weevils, the lesser grain borer, or the Angoumois grain moth, the development of which occurs inside grain kernels. The presence of adults of those species may signal larger populations developing later as "hidden infestations" in the corn. The frequent appearance of high-moisture species such as foreign grain beetles, fungus beetles, or corn sap beetles, which typically feed on molds or decaying organic material, suggests the presence of a wet spot in the corn.

V. PROTECTIVE MEASURES AND CONTROL OF INSECTS IN CORN STORAGE

The control of insects in stored grain is generally divided into two categories, preventive measures to limit the occurrence and development of insects and remedial measures to control existing infestations. Each of these categories consists of both chemical and nonchemical methods of control.

A. Environmental Protective Measures

The initial steps in preventing insect problems must be taken before newly harvested corn is placed in storage. The most important source of insects likely to contaminate the grain is from breeding sites in the immediate storage area. Perhaps the most important bin structure that affects insects infestation is the false floor. Grain dust and debris sifting through this generally nonremovable structure provide a continuous source of insect breeding sites for inoculation of grain placed in the bin. Ideally, clean-out access ports should be provided into the plenum area to permit vacuuming this material from the subfloor area or to allow sufficient access to apply a residual insecticide such as malathion or methoxychlor. If no access is possible, the area should be fumigated before the bin is loaded. Quinlan and McGaughey (1983) found chloropicrin, the only fumigant currently labeled for treatment of empty bins, effective at the label-directed rate, but they noted that emphasis should be placed on sealing all fan openings and bin doors before treatment. They also found that phosphine-producing tablets placed directly on the perforated floor and covered with a plastic sheet taped to the side walls were effective at a dosage of 60 tablets in a bin with a 5.49-m (18-ft) diameter.

Other important sites requiring clean-out before harvest include all grain-handling equipment such as augers, combines, and trucks or wagons. Grain and feed accumulations outside the bin near grain doors also harbor insects, as do grain residues.

The recommendation to apply residual sprays to the inside and outside surfaces of grain bins a few weeks before binning the grain has been a standard feature of pest management in stored grain for over 40 years. Many of the chlorinated hydrocarbon insecticides originally recommended for this purpose are no longer permitted. Only three compounds are currently labeled for this use: methoxychlor, pyrethrins combined with piperonyl butoxide, and malathion. The effectiveness and persistence of residual insecticides on metal, wood, and concrete surfaces have been examined by several researchers (Burkholder and Dicke, 1966; Waters and Grussendorf, 1969; Waters, 1976). In general, the insecticides were less effective and persistent on concrete surfaces than on wood or metal. The loss has been attributed to the increased porosity of concrete surfaces, which results in a lower overall deposit of insecticide per unit of surface area (Waters and Grussendorf, 1969), and to the high pH values of concrete, which contribute to a more rapid degradation of the insecticide than occurs on more neutral surfaces such as wood or metal (Okwelogu, 1968). The actual contribution of bin wall sprays in limiting invasion by storage pests is not well documented. Quinlan (1977) reported that sprays on the grain surface in combination with interior wall sprays reduced insect populations in stored corn observed over a 15-month period. Exterior wall sprays alone were not effective in limiting infestation. McGregor and Morris (1959) sprayed the exterior surface of grain bins and the ground area around the bins at five storage sites in Kansas with mixtures of DDT and malathion, chlordane, and dieldrin. Monthly examinations of grain in the bins taken from May to August indicated that the outside residual treatments were not effective in restricting infestations by the Indian-meal moth (*P. interpunctella*) or by other externally developing insect species. However, Harein (1982) suggested that exterior bin sprays applied before fumigation of the grain might be effective against insects migrating from the treated area to avoid fumigant exposure.

B. Prevention by Chemical Treatment of the Corn

One of the most important steps in preventing insects from developing into damaging populations in farm-stored corn is to apply a grain protectant insecticide to the corn as it is binned. Such treatments are often not a matter of choice. If infestation and the resulting loss in quality cannot be tolerated and if alternate control measures offer no practical or cost-relevant solution, chemical control achieved through the application of a residual insecticide may be the only realistic solution. Fortunately, many grain-protectant chemicals have proven to be effective against a broad range of storage pests, are comparatively safe to use and apply, are recognized internationally, and have established residue tolerances in raw cereal grains and in milled products made from grain.

For more than 25 years, only two chemical insecticides, synergized pyrethrins and malathion, were approved in the United States for direct application to stored corn (Anonymous, 1982). The toxicity of pyrethrins, a constituent of pyrethrin flowers, was found to be greatly increased by the addition of piperonyl butoxide (Goodwin-Bailey and Holborn, 1952), and mixtures marketed in the United States for application to stored grain typically contain a 1:10 ratio of pyrethrins to piperonyl butoxide. Malathion, an organophosphate insecticide introduced into the United States in the late 1950s, has largely replaced the use of pyrethrin formulations in stored grain. However, irrespective of malathion's approved status and long history of recommendation in grain pest control bulletins, it is not used frequently in stored corn. Storey et al (1982b, 1984) found that only 7.7% of the corn arriving at port terminals in the United States and 8.2% of the corn stored in farm bins across 19 states contained biologically active deposits of malathion. When malathion treatment was part of the pest management practice in farm-stored grain, the incidence and density of most insect species found in the treated grain was generally less. One species not materially affected by malathion treatments was the Indian-meal moth, *P. interpunctella*. This species was detected nearly twice as often in malathion-treated corn as in untreated corn (Storey et al, 1984). Beeman et al (1982) reported that nearly 90% of the Indian-meal moth strains collected in nine midwestern states were more than 17-fold resistant to malathion. The resistance was attributed to a malathion-degrading carboxylesterase enzyme that was estimated to be present in over half of the moths infesting grain in the North Central United States. Haliscak and Beeman (1983) also reported measurable tolerances to malathion among several strains of beetles infesting stored grain.

Quinlan (1982) observed that most failures with malathion occur because of excessive grain moisture or temperature or both. He recommended that grain not be treated if it is above 13% moisture and the temperature is above 32°C (90°F). Table II shows how rapidly malathion residues decline as the moisture content of the grain increases.

Corn harvested at moisture levels too high for safe storage and therefore requiring drying (immediately after storage) is generally unsuitable for the application of grain protectants at the time of binning. Insecticide treatments under these conditions are likely to be adversely affected by the excessive water content of the corn and by the high-velocity air movement and/or heated air used to dry the grain. However, high-moisture corn is especially attractive to storage pests during this initial phase of storage, and unless steps are taken to treat the corn after it is dried, insects invading the corn before the onset of cold weather can develop into damaging populations as the corn warms during the following spring. Applying a grain protectant in the conventional manner after the corn is already in place requires the added expense of rehandling the grain plus having to provide adequate turning space. An alternative to turning the grain for treatment has been proposed by Quinlan (1979, 1980), who studied

TABLE II
Malathion Residues (ppm) on Wheat at Four Moisture Levels Following Application Rate of 7.5 ppm[a]

Interval After Treatment	Moisture Content			
	10%	12%	14%	16%
24 hours	6.2	6.1	5.7	5.0
6 weeks	6.4	4.9	2.0	0.4
3 months	4.6	4.2	1.1	0.2
6 months	2.8	2.6	0.3	0.1
9 months	2.7	2.5	0.3	<0.1

[a] Source: Quinlan et al (1980).

application techniques employing thermally generated insecticide aerosols that were pulled and/or pushed into the grain with air movement created by aeration fans or with a drying system. The aerosol method was found to be effective against natural infestations and test insects placed in the grain, but Quinlan noted that malathion residues deposited with the grain mass represented only about 10% of the amount applied.

Two new grain protectant materials, chlorpyrifos-methyl (Reldan) and pirimiphos-methyl (Actellic) have been under consideration by the Environmental Protection Agency (EPA) for registration for domestic use in corn in the United States. Each of these organophosphorus compounds has an established reputation as an effective residual insecticide for stored grain pests, including such malathion-resistant species as the Indian-meal moth (Morallo-Rejesus, 1973; Cogburn, 1975; Lahue, 1975, 1976, 1977; Quinlan et al, 1979, 1980). Maximum residue limits for these compounds have been established by the joint pesticide committees of the Food and Agriculture Organization of the U.N. and the World Health Organization (FAO/WHO, 1976, 1978) for a number of raw cereal grains and milled products. Chlorpyrifos-methyl was granted registration by the EPA in June 1985 (EPA, 1985) for domestic and export use on wheat, grain sorghum, barley, oats, and rice. Approval for use on corn was not granted at that time pending development and review of additional residue data. Suppliers of this product anticipate approval for use on corn to be granted in 1987. Pirimiphos-methyl was registered in the fall of 1984 (EPA, 1984) to treat corn, rice, wheat, and grain sorghum intended for export only. Manufacturers of pirimiphos-methyl supplied additional residue data required by the EPA, and registration for domestic (U.S.) use in stored corn and sorghum was approved in July 1986 (EPA, 1986).

C. Remedial Control Measures for Existing Populations

GRAIN FUMIGATION

Fumigation has been the principal remedial procedure used for controlling developed insect infestations in grain and milled products for more than 50 years. Fumigants are chemical pesticides that are distributed through the grain as gases to kill destructive insects attacking the grain. Fumigants may be applied directly to the grain as gases (e.g., methyl bromide) that form when pressurized liquids are released into the atmosphere, as liquids (e.g., chloropicrin) that vaporize when exposed to the air, or as solids (e.g., aluminum phosphide) that produce gases on exposure to moisture in the air. The combination of high volatility and high toxicity makes fumigants uniquely suited for short-term treatment of enclosed spaces. Grain fumigations, however, are effective only when the storage structure is sufficiently tight to hold gas concentrations in the grain long enough to be lethal to storage pests (about one to five days, depending on the chemical used and on grain temperatures). Fumigants affect grain pests only during the time in which the gas is present in the insect's environment. After the fumigant diffuses out of the grain, no residual protection is left behind and the grain is susceptible to reinfestation. Fumigation becomes an essential control measure when no other pesticide treatment or other control means can reach the infestation deep within the grain mass.

Because fumigant chemicals are highly toxic and hazardous to use, they are classified as restricted-use pesticides requiring special equipment, training, and certification before they can be purchased and applied. Persons requiring certification should contact their local county agent, Cooperative Extension Service specialists, or State Board of Agriculture for further information about pesticides that are designated for "restricted use." It is often safer, less expensive, and more effective to have grain fumigated by a licensed professional fumigator (Storey et al, 1979).

Liquid fumigants composed primarily of carbon tetrachloride, carbon disulfide, or ethylene dichloride have been voluntarily withdrawn from registration in the United States by the manufactures of these chemicals. In an agreement with the EPA, marketing of these materials was stopped at the end of 1985 (Federal Register, 1985). Only three fumigants remain for treating bulk-stored grain: phosphine-producing materials, methyl bromide, and chloropicrin.

Aluminum phosphide formulations that release hydrogen phosphide (phosphine) gas when exposed to moisture and heat are the predominant fumigants used for the treatment of bulk-stored grain throughout the world. They are available in solid formulations in the form of tablets, pellets, or powder packed in paper sachets. Distribution of phosphine gas in bulk grain is generally dependent on the placement of pellets or tablets throughout the grain bulk by adding them to the grain stream during bin filling or by using a hollow tube designed to probe the material into the grain. Manufacturers of phosphine-producing fumigants indicate that a delay occurs before phosphine is evolved in large quantities from these formulations—usually 1-2 hr with pellets and 2-4 hr with tablets. With grain temperatures above 15°C, decomposition should be nearly complete in three days. With lower grain temperatures, release of the gas may be slowed significantly, and adequate concentrations may not be reached or may be delayed for several days. Banks and Annis (1984) examined phosphine gas distribution patterns in large-scale fumigations in concrete and steel bins and concluded that completely successful phosphine fumigations require the following criteria: 1) the average concentration of phosphine in the fumigated system must be one half of that expected theoretically, 2) the concentration at the end of the exposure period must be greater than the minimum effective against insects, and 3) the ratio of minimum to maximum concentration must exceed 0.25 after not more than 25% of the exposure period as a measure of the "evenness" of phosphine distribution. In actual practice, many phosphine fumigations do not achieve these stringent guidelines, primarily due to fumigant loss because of inadequate sealing of the grain storage before treatment. Other factors adversely affecting the efficiency of phosphine fumigations include low grain temperatures (manufacturers do not recommend treatment when the commodity temperature is below 5°C) and reduced exposure periods that do not allow sufficient time for the gas to fully evolve, diffuse through the commodity, and be taken into the respiratory system of the insect.

Methyl bromide is a gaseous fumigant marketed as compressed liquefied gas in cans or cylinders. The use of methyl bromide in bulk grain reached its peak in the late 1950s, but then provisions of Public Law 518, the Miller Amendment, limited the amount of inorganic bromide residue resulting from treatment with methyl bromide to 50 ppm for most grain crops. Thereafter, much of the methyl bromide used in commercial grain storages was replaced with phosphine-producing fumigants. Methyl bromide is rarely used for the treatment of farm-stored grains, except by commercial fumigators. Treatment with this fumigant can result in injury to germination of cereal grains (Whitney et al, 1958), particularly under conditions of high temperature and high moisture content, and its use for treating seed stocks containing more than 12% moisture is not recommended. Methyl bromide is generally applied to bulk grain, utilizing existing grain aeration systems to push, pull, or recirculate the gas through the grain.

Chloropicrin is a nonflammable liquid fumigant that vaporizes to a gas on exposure to air. It is added to other fumigants as a warning agent and is marketed in pressurized and nonpressurized containers as a space, grain, and soil fumigant. Chloropicrin is an effective material but difficult to use because of its irritating tear-gas properties. It may be applied by recirculation through an aeration system (Storey, 1971) or by several direct application methods, including applying a prorated portion of the total dosage directly into the grain stream as the grain is loaded and by pouring the material through probes inserted into the grain.

ALTERNATIVES TO PESTICIDES

The insecticides and fumigants used to treat stored corn have the advantage of simplicity, availability, low cost, and adaptability to a variety of storage situations. However, the use of pesticides in food supplies is coming under increased scrutiny and challenge. The reduction or elimination of dependence on chemical pesticides will bring increased emphasis on nonchemical methods of preventing insect damage to agricultural products.

Modified atmosphere treatment of bulk grain involves alteration of the proportions of the normal gaseous constituents of air (oxygen, nitrogen, carbon dioxide) to provide an insecticidal atmosphere. This represents an effective substitute for chemical fumigation of grain (Storey, 1985). Technologies for producing, handling, and applying low-oxygen combustion gases from an inert atmosphere generator or vaporized carbon dioxide have been developed (Storey, 1973; Jay, 1980; Lessard and Fuzeau, 1983). Studies have confirmed the effectiveness of these atmospheres and their compatibility with high-quality maintenance of stored commodities (Bailey and Banks, 1980; Storey, 1980). Commercial adoption of modified atmosphere technology may require 1) a substantial restructuring of existing storage facilities to improve their gas-tightness (a step also made necessary by the increased reliance on aluminum phosphide fumigants), 2) capital investment in specialized equipment to generate or to store and vaporize the modified atmospheres, and 3) the establishment of new and more sophisticated operational procedures (Storey, 1985).

Physical controls such as aeration cooling and reduction in levels of broken grain and foreign material to provide a less favorable habitat for insect development must become a more important segment of a total pest management program. They are not, however, direct replacements for chemical disinfestation of grain. Lower levels of broken grain and foreign material can reduce the amount of pesticide required for treatment by removing materials from the grain that increase sorption loss and hinder distribution of fumigant gases through the grain mass. Increased use of aeration of grain can extend the residual life of protectant chemicals (Desmarchelier, 1978) and reduce insect activity through lower grain temperatures.

Methods of heating cereal products and bulk-stored grain have been investigated, including the use of infrared, microwave, and dielectric heating and shallow fluidized-bed heating systems for continuous in-line treatment during bulk grain handling (Nelson and Stetson, 1974; Dermott and Evans, 1978). None of these systems have been developed into commercial operations in the United States; however, conventional processing-plant heating systems are being used to raise temperatures to lethal levels throughout an entire processing area for controlling insects within various types of milling equipment (Cooney, 1985; J. W. Heaps, Nabisco Brands, Minneapolis, MN, personal communication, 1986).[1]

Several forms of biological control of stored grain pests have been under investigation for more than 20 years, with little commercial success. Microbial control of insects, which involves the use of microorganisms and their by-products to regulate insect populations, has great potential, but such potential has not been realized in stored grain. The bacterium *Bacillus thuringiensis* has been prepared in commercial formulations expressly for moth control, but results obtained with this spore material have been mixed (McGaughey, 1985a, 1985b). Another form of biological control is the use of hormones and enzymes manufactured by the insect itself to control its growth and development. These materials have been synthesized into a new class of insecticides—insect growth regulators. Because vertebrates in general and humans in particular are so genetically and morphologically distant from insects, these specialized compounds have shown no adverse side effects. Their principal drawbacks appear to be high cost, slow action, and the precision necessary for their proper application to bulk grain. The insect growth regulator methoprene was registered in 1981 by the EPA for the control of stored product pests in peanuts and as a space spray in food warehouses, food processing areas, and tobacco warehouses. At present, however, no growth regulators have been approved for direct application to stored grain, and none of these materials are being used routinely in grain pest management programs in other grain-producing countries.

One of the more publicized alternatives to chemical pesticides is the irradiation of grain and grain products for insect control. High-speed electron beams, X-radiation, and gamma rays from radioactive cobalt or cesium have proved to be equally effective in controlling insects at comparable doses of absorbed ionizing radiation (Tilton and Nelson, 1984). At the level of electron energy used for disinfestation, no radioactive nuclides are formed in the grain. Irradiation has been approved by the U.S. Food and Drug Administration, for disinfestation of wheat and wheat flour since the 1960s; however, little commercial application has been developed because more economical chemical control methods have remained available. A new food irradiation rule (Federal Register, 1986) established by the Food and Drug Administration broadens the permitted uses of irradiation in feed products and also increases the maximum dosage for treatment of grain and grain products from 15 krads to 100 krads.

[1] J. W. Heaps. Super heating of a processing plant for insect elimination. Presented at a meeting of the North Central Branch of the Entomological Society of America, March 25–27, 1986, Minneapolis, MN.

VI. SUMMARY

Stored corn provides an excellent environment for the invasion and development of insect pests. Left unchecked, these pests can reproduce into damaging populations that adversely affect the market value of corn. *Cryptolestes* spp. (flat and rusty grain beetles), *Plodia interpunctella* (the Indian-meal moth), and *Tribolium* spp. (red and confused flour beetles) are the most frequent insect pests found in farm-stored corn; *Sitophilus* spp. (rice and maize weevils) and *Cryptolestes* spp. are the most frequent in corn exported from the United States. Insect species that prefer high-moisture conditions or feed on molds or decaying vegetable matter are also common in corn stored too wet. Early detection of insect pests is an important step in preventing damage. Inspection equipment, including a grain probe and sieves for sifting insects from the corn samples, may be augmented by the use of perforated insect traps placed just below the surface of the grain. The control of insects in stored corn begins with a series of preventive measures, including the cleaning out of residual grain before bin loading, application of a residual spray to the inside and outside surfaces of the bin, disinfestation of the plenum area beneath the bin floor, and treatment with a grain protectant insecticide as the corn is binned. Corn already infested requires treatment with fumigant chemicals (aluminum phosphide, methyl bromide, chloropicrin), which are classified as restricted-use pesticides that require certification training before they can be purchased and applied. Future pest management strategies place increased emphasis on nonchemical methods of control to reduce or eliminate the present dependence on chemical pesticides.

LITERATURE CITED

ANONYMOUS. 1979. Stored-grain insects. Agriculture Handbook 500. U.S. Dept. Agric., Sci. Educ. Admin., Washington, DC. 57 pp.

ANONYMOUS. 1982. Guidelines for the Control of Insects and Mite Pests of Foods, Fibers, Feeds, Ornamentals, Livestock and Households. Agric. Handbook 584. U.S. Dept. Agric., Agric. Res. Serv., Washington, DC. 734 pp.

BAILEY, S. W., and BANKS, H. J. 1980. A review of recent studies of effects of controlled atmospheres on stored-product pests. Pages 101-118 in: Controlled Atmosphere Storage of Grains. J. Shejbal, ed. Elsevier, Amsterdam.

BANKS, H. J., and ANNIS, P. C. 1984. On criteria for success of phosphine fumigations based on observation of gas distribution patterns. Pages 327-341 in: Controlled Atmosphere and Fumigation in Grain Storages. R. E. Ripp, ed. Elsevier, Amsterdam.

BARAK, A. V., and HAREIN, P. K. 1982. Trap-deletion of stored-grain insects in farm-stored, shelled corn. J. Econ. Entomol. 75:108-111.

BEEMAN, R. W., SPEIRS, W. E., and SCHMIDT, B. A. 1982. Malathion resistance in Indianmeal moths (Lepidoptera: Pyralidae) infesting stored corn and wheat in the North-Central states. J. Econ. Entomol. 75:950-954.

BURKHOLDER, W. E. 1984. Use of pheromones and food attractants for monitoring and trapping stored-product insects. Pages 69-86 in: Insect Management for Food Storage and Processing. F. Baur, ed. Am. Assoc. Cereal Chem., St. Paul, MN.

BURKHOLDER, W. E., and DICKE, R. J. 1966. Evidence of sex pheromones in females of several species of Dermestidae. J. Econ. Entomol. 59:540.

COGBURN, R. R. 1975. Stored rice insects research. Rice J. 78:78.

COGBURN, R. R., BURKHOLDER, W. E., and WILLIAMS, H. J. 1983. Efficacy and characteristics of dominicalure in field trapping lesser grain borers, *Rhyzopertha dominica* (F). Pages 629-630 in: Proc. Int. Working Conference on Stored-Product Entomology, 3rd. Kansas State University, Manhattan.

COONEY, K. 1985. If you can't stand the heat. Food Sanit. (March):24-25, 39.

COTTON, R. T., and WILBUR, D. A. 1982. Insects. Pages 281-318 in: Storage of Cereal Grains and Their Products, 3rd ed. C. Christensen, ed. Am. Assoc. Cereal Chem., St. Paul, MN.

DERMOTT, T., and EVANS, D. E. 1978. An evaluation of fluidized-bed heating as a means of disinfesting wheat. J. Stored Prod. Res. 14:1-12.

DESMARCHELIER, J. M. 1978. Loss of fenitrothion on grains in storage. Pestic. Sci. 9:33-38.

EPA. 1984. Reg. 10182-87. U.S. Environ. Protect. Agency, Washington, DC.

EPA. 1985. Reg. 7501-41. June. U.S. Environ. Protect. Agency, Washington, DC.

EPA. 1986. Reg. 10182-79. July 31. U.S. Environ. Protect. Agency, Washington, DC.

FAO/WHO. 1976. Chlorpyrifos-methyl. In: 1975 Evaluation of Some Pesticide Residues in Food. FAO/AGP: 1975/m/13. Food and Agric. Org. of the U.N., Rome.

FAO/WHO. 1978. Pirimiphos-methyl. In: 1977 Evaluations of Some Pesticide Residues in Food. FAO/AGP: 1977/m/14. Food and Agric. Org. of the U.N., Rome.

FEDERAL REGISTER. 1985. Regulatory status of grain fumigants. 50(182):38092-38095.

FEDERAL REGISTER. 1986. Irradiation in the production, processing and handling of food; Final Rule. 21 CFR. Part 179, April 19. 13376-13399.

GOODWIN-BAILEY, K. E., and HOLBORN, J. M. 1952. Laboratory and field experiments with pyrethrins/piperonyl butoxide powders for the protection of grain. Pyrethrum Post 2:7-17.

GRANOVSKY, T. A., and MILLS, R. B. 1982. Feeding and mortality of *Sitophilus granarius* (L) adults during simulated winter farm bin temperatures. Environ. Entomol. 11:324-326.

HALISCAK, J. P., and BEEMAN, R. W. 1983. Status of malathion resistance in five genera of beetles infesting farm-stored corn, wheat and oats in the United States. J. Econ. Entomol. 76:717-722.

HAREIN, P. K. 1982. Chemical control alternatives for stored-grain insects. Pages 319-363 in: Storage of Cereal Grains and Their Products, 3rd ed. C. Christensen, ed. Am. Assoc. Cereal Chem., St. Paul, MN.

JAY, E. G. 1980. Methods of applying carbon dioxide for insect control in stored grain. Advances in Agricultural Technology, Southern Series, S-13. U.S. Dept. Agric., Agric. Res. Serv., Washington, DC. 7 pp.

LAHUE, D. W. 1975. Angoumois grain moth: Chemical control of infestation of shelled corn. J. Econ. Entomol. 68:769-771.

LAHUE, D. W. 1976. Grain protectants for seed corn. J. Econ. Entomol. 69:652-654.

LAHUE, D. W. 1977. Grain protectants for seed corn: Field test. J. Econ. Entomol. 70:720-722.

LESSARD, F., and FUZEAU, B. 1983. Disinfection of wheat in an harbor silo bin with an exothermic inert gas generator. Pages 481-485 in: Proc. Int. Working Conf. on Stored Product Entomology, 3rd. Kansas State Univ., Manhattan.

LOSCHIAVO, S. R., and ATKINSON, J. M. 1967. A trap for the detection and recovery of insects in stored grain. Can. Entomol. 99:1160.

LOSCHIAVO, S. R., and ATKINSON, J. M. 1973. An improved trap to detect beetles (Coleoptera) in stored grain. Can. Entomol. 10S:437-440.

McGAUGHEY, W. H. 1985a. Evaluation of *Bacillus thuringiensis* for controlling Indian-meal moths (Lepidopteran: pyralidae) in farm grain bins and elevator silos. J. Econ. Entomol. 78:1089-1094.

McGAUGHEY, W. H. 1985b. Insect resistance to the biological insecticide *Bacillus thuringiensis*. Science 229:193-195.

McGREGOR, H. E., and MORRIS, O. W. 1959. The application of residual insecticides to CCC bins and bin sites to prevent reinfestation of calcium cyanide-treated wheat by dermestids. U. S. Dept. Agric., Agric. Mark. Serv., Mark. Res. Div. Spec. Rep. A-241. 6 pp.

MORALLO-REJESUS, B. 1973. Evaluation of five insecticides as protectants of shelled corn during storage. Annu. Rep. Corn. Project Philippines.

NELSON, S. O., and STETSON, L. E. 1974. Possibilities for controlling insects with microwaves and lower frequency RF energy. IEEE Trans. Microwave Theory and Tech. 22(12):1303-1305.

OKWELOGU, T. N. 1968. The toxicity of malathion applied to washed concrete. J. Stored Prod. Res. 4:259-260.

QUINLAN, J. K. 1977. Surface and wall sprays of malathion for controlling insect populations in stored shelled corn. J. Econ. Entomol. 70:335-336.

QUINLAN, J. K. 1979. Malathion aerosols applied in conjunction with vertically placed aeration for the control of insects in stored corn. J. Kans. Entomol. Soc. 52:648-652.

QUINLAN, J. K. 1980. A preliminary study with malathion aerosols applied with a corn drying system for the control of insects. J. Ga. Entomol. Soc. 15:252-257.

QUINLAN, J. K. 1982. Grain protectants for insect control. Marketing Bull. 72. U.S. Dept. Agric., Agric. Res. Serv., Washington, DC.

QUINLAN, J. K., and McGAUGHEY, W. 1983. Fumigation of empty grain drying bins with chloropicrin, phosphine, and liquid fumigant mixtures. J. Econ. Entomol. 76:184-187.

QUINLAN, J. K., WHITE, G. D., WILSON, J. L., DAVIDSON, L. J., and HENDRICKS, L. H. 1979. Effectiveness of chlorpyrifos-methyl and malathion as protectants for high moisture stored wheat. J. Econ. Entomol. 71:90-93.

QUINLAN, J. K., WILSON, J. L., and DAVIDSON, L. I. 1980. Pirimiphos-methyl as a protectant for high moisture stored wheat. J. Kans. Entomol. Soc. 53:825-832.

STOREY, C. L. 1971. Distribution of chloropicrin used alone or mixed with 80:20 to fumigate wheat and sorghum. U.S. Dept. Agric. ARS Marketing Res. Rep. 894. 19 pp.

STOREY, C. L. 1973. Exothermic inert-atmosphere generators for control of insects in stored wheat. J. Econ. Entomol. 65:511-514.

STOREY, C. L. 1980. Functional and end-use properties of various commodities stored in a low oxygen atmosphere. Pages 311-317 in: Controlled Atmosphere Storage of Grains. J. Shejbal, ed. Elsevier, Amsterdam.

STOREY, C. L. 1985. Modified atmosphere technology. Pages 14-21 in: Proc. Barley Insect Conf. American Malting Barley Assoc., Milwaukee, WI.

STOREY, C. L., SPEIRS, R. D., and HENDERSON, L. S. 1979. Insect control in farm stored grain. U.S. Dept. Agric. Farmer's Bull. 2269. 18 pp.

STOREY, C. L., SAUER, D. B., ECKER, O., and FULK, D. W. 1982a. Insect infestations in wheat and corn exported from the United States. J. Econ. Entomol. 75:827-832.

STOREY, C. L., SAUER, D. B., QUINLAN, J. K., and ECKER, O. 1982b. Incidence, concentrations and effectiveness of malathion residues in wheat and corn exported from the United States. J. Stored Prod. Res. 18:147-151.

STOREY, C. L., SAUER, D. B., and WALKER, D. 1983. Insect populations in wheat, corn, and oats stored on the farm. J. Econ. Entomol. 76:1323-1330.

STOREY, C. L., SAUER, D. B., and WALKER, D. 1984. Present use of pest management practices in wheat, corn, and oats stored on the farm. J. Econ. Entomol. 77:784-788.

TILTON, E. W., and NELSON, S. O. 1984. Irradiation of grain and grain products for insect control. Counc. Agric. Sci. Technol., Ames IA. 6 pp.

WATERS, F. L. 1976. Persistence and uptake in wheat of malathion and bromophos applied on granary surfaces to control the red flour beetle. J. Econ. Entomol. 69:353-356.

WATERS, F. L., and GRUSSENDORF, O. W. 1969. Toxicity and persistence of lindane and methoxychlor on building surfaces for stored-grain insect control. J. Econ. Entomol. 62:1101-1106.

WHITNEY, W. K., JANTZ, O. K., and BULGER, C. S. 1958. Effects of methyl bromide fumigation on the viability of barley, corn, grain sorghum, oats and wheat seeds. J. Econ. Entomol. 51:847-861.

WRIGHT, V. F., and MILLS, R. B. 1983. Estimation of stored-product insect populations in small bins using two sampling techniques. Pages 672-679 in: Proc. Int. Working Conf. on Stored-Product Entomology. Kansas State Univ., Manhattan.

CHAPTER 7

ECONOMICS OF PRODUCTION, MARKETING, AND UTILIZATION

MACK N. LEATH
U.S. Department of Agriculture
Economic Research Service
National Economics Division
Washington, DC

LOWELL D. HILL
Department of Agricultural Economics
University of Illinois at Urbana-Champaign
Urbana, Illinois

I. INTRODUCTION

The United States annually produces almost half of the world's corn crop; the U. S. acreage planted to corn has accounted for about 24% of the acreage planted to principal crops in recent years. On the average, about 11% of the acreage is harvested for silage and forage, and the balance is harvested for grain. In the five years preceding 1983, the grain acreage averaged almost 73 million acres (29.5×10^6 ha) annually. Harvested acreage dropped to a record low of 51 million acres (20.6×10^6 ha) in 1983, when over 32 million corn acres (13.0×10^6 ha) were taken out of production under the Payment-In-Kind (PIK) program. The acreage harvested for grain rebounded to almost 72 million acres (29.1×10^6 ha) in 1984.

In the United States, corn is used for livestock feed and seed and in the production of a variety of foods, alcoholic beverages, and industrial products. Domestic use reached a record 5.4 billion bushels (137×10^6 t) in marketing year 1982/83 (Oct. 1 to Sept. 30), and livestock feed accounted for about 83% of that total. The amount used for food and industrial purposes has expanded every year since 1968/69 due to increasing demand and now totals over one billion bushels. The products that exhibited the greatest growth were corn sweetener products and alcohol fuels (ethanol); however, future growth will be contingent upon changes in import restrictions and price-support programs for sugar, as well as tax incentives affecting ethanol production.

The most dramatic change in demand has been the volume of corn exported from the United States. Exports grew from about 300 million bushels (7.6×10^6 t) in 1960/61 to over 2.4 billion (60.96×10^6 t) in 1979/80. Many observers expected a continuation of this upward trend in the 1980s, but at a slower rate. Total world trade in corn declined during the early 1980s, and exports of U.S. corn declined by 24%. The U.S. share of world exports declined from a record 79% in 1979/80 to about 70% in 1984/85. The declining U.S. share reflects the strong dollar of recent years, higher U.S. price-support rates, high price supports for grain in the European Community (EC), and greater competition from other grain-exporting nations.

Several kinds of corn are grown in the United States. Sweet corn and popcorn are specialty crops grown primarily for food use. A limited acreage of waxy corn is grown under contract with wet-corn millers. White dent corn is a specialty corn preferred in the manufacturing of selected food products. Most field corn harvested for grain is yellow dent corn.

This chapter focuses on the production, marketing, and utilization of field corn produced for grain. Topics discussed in this chapter include production and utilization statistics, marketing systems, marketing patterns and transportation, world trade, pricing systems, and government policies affecting the industry.

II. TRENDS IN SUPPLY AND PRODUCTION COSTS

The United States is the world's leader in corn production; annual production generally approaches 50% of the world total. The annual U.S. supply available for domestic use and exports consists primarily of current production and carryover stocks. Imports have been small and do not significantly affect annual supply. Carryover stocks are inventories of corn remaining in storage on October 1, the start of the new marketing year.[1] This date traditionally signifies the start of harvest and is the transition point from old-crop to new-crop corn.

The costs associated with corn production in the United States increased substantially during the 1970s. The rise in nonland costs reflected substantial increases in the price of energy and related inputs such as fertilizer and pesticides. However, returns per acre were well above total cost during the middle of that decade, and expectations of continued profits were quickly capitalized into land values. High inflation rates, in combination with relatively low interest rates, also contributed to large increases in land values. The economics of producing corn are reviewed following the discussion of acreage, yield, and production.

A. Acreage, Yield, and Production

The corn acreage planted for all purposes (grain, silage, and forage) has exhibited sizeable variations over time. This acreage averaged 80 million acres (32.38×10^6 ha) annually during the 1950s. Acreage, yield, and production statistics for crop years 1960/61 through 1984/85 are presented in Table I. The prospects for huge carryover stocks of corn in 1961/62 (Table II) led to the

---

[1]The marketing year has been changed to September 1 to August 31, starting with the 1986/87 marketing year.

resumption of planting restrictions, and acreage planted dropped 15.5 million acres (6.27×10^6 ha) in 1961. Restrictions on planted acreage remained in effect through 1973. Acreage planted expanded rapidly during the next three years and averaged 83 million acres (33.59×10^6 ha) during 1976/77–1982/83. The PIK program reduced planted acreage in 1983 to the lowest level recorded in more than 100 years in the United States.

About 90% of the corn planted in the United States is harvested for grain. With the exception of 1983, the acreage harvested for grain has exceeded 70 million acres (28.34×10^6 ha) since 1976. It totaled 71.1 million acres (28.79×10^6 ha) in 1984, in addition to 4.2 million corn acres (1.7×10^6 ha) that were set aside under the government program for corn. Without this set-aside, the grain acreage in 1984 would probably have rebounded to the 1981 level of 74.5 million acres (30.16×10^6 ha).

The trend in grain yields has been upward over time, and yields have doubled since 1960. The increases were dramatic before 1973, when acreage restrictions

TABLE I
Corn Acreage, Yield, and Production, United States, 1960–1984[a]

Year Beginning October 1	Acreage (million acres)[b]			Average Yield per Acre (bu)[c]	Grain Production (million bu)
	Set-Aside and Diverted	Planted for All Purposes	Harvested for Grain		
1960	0	81.4	71.4	54.7	3,907
1961	19.1	65.9	57.6	62.4	3,598
1962	20.3	65.0	55.7	64.7	3,606
1963	17.2	68.8	59.2	67.9	4,019
1964	22.2	65.8	55.4	62.9	3,484
1965	24.0	65.2	55.4	74.1	4,103
1966	23.7	66.3	57.0	73.1	4,166
1967	16.2	71.2	60.7	80.1	4,860
1968	25.4	65.1	56.0	79.5	4,450
1969	27.2	64.3	54.6	85.9	4,687
1970	26.1	66.8	57.4	72.4	4,152
1971	14.1	74.1	65.1	88.1	5,646
1972	24.4	67.0	57.5	97.0	5,580
1973	6.0	71.9	62.1	81.3	5,671
1974	0	77.9	65.4	71.9	4,701
1975	0	78.7	67.6	86.4	5,841
1976	0	84.6	71.5	88.0	6,289
1977	0	84.3	71.6	90.8	6,505
1978	6.1	81.7	71.9	101.0	7,268
1979	2.9	81.4	72.4	109.5	7,928
1980	0	84.0	73.0	91.0	6,639
1981	0	84.1	74.5	108.9	8,119
1982	2.1	81.9	72.7	113.2	8,235
1983	32.2	60.2	51.5	81.1	4,175
1984	4.0	80.5	71.9	106.7	7,674

[a] Data from USDA (1972, 1983a, 1986b).
[b] To convert acres to hectares, multiply by 0.4047.
[c] To convert bu/acre to kg/hectare, multiply by 49.97.

were in effect. Average yield reached 97 bu/acre (6.09 t/ha) in 1972. That year was followed by a period of acreage expansion that curbed the trend toward higher yields because marginal, less-productive land was placed in corn production. In 1978, average yields surpassed the 100-bu mark (6.28 t/ha) for the first time in history. Average yields have been impressive since that time, except for substantial reductions in 1980 and 1983 because of summer droughts in major corn-producing areas.

The general increase in yields has been due mainly to changes in technology and improved production practices, including development of improved hybrids, increased fertilization rates, higher seeding rates, and improved weed, insect, and disease control methods (Chapters 1 and 2.) Yield variations are primarily the result of weather conditions during the growing season, as illustrated by the contrast between average yield in 1982/83 and 1983/84 (Table I).

TABLE II
Corn Supply (million bu),[a] United States, 1960–1984[b]

Year Beginning October 1	Carryover on Oct. 1				Production	Imports	Total Supply
	Farm	Interior Mill, Elevator, and Warehouse	Commodity Credit Corp. Bin Sites	Total			
1960	452	736	599	1,787	3,907	1	5,695
1961	588	813	615	2,016	3,598	1	5,615
1962	579	590	484	1,653	3,606	1	5,260
1963	534	435	396	1,365	4,019	1	5,385
1964	681	442	414	1,537	3,484	1	5,022
1965	581	258	308	1,147	4,084	1	5,232
1966	532	176	134	842	4,168	1	5,011
1967	572	156	98	826	4,860	1	5,687
1968	788	277	104	1,169	4,450	1	5,620
1969	732	243	143	1,118	4,687	1	5,806
1970	576	318	111	1,005	4,152	4	5,161
1971	427	215	25	667	5,646	1	6,314
1972	751	349	26	1,126	5,580	1	6,708
1973	405	284	20	709	5,671	1	6,380
1974	288	196	0	484	4,701	2	5,187
1975	192	169	0	361	5,841	2	6,204
1976	234	166	0	400	6,289	2	6,691
1977	448	438	0	886	6,505	3	7,394
1978	666	445	0	1,111	7,268	1	8,380
1979	795	509	0	1,304	7,928	1	9,233
1980	920	697	0	1,617	6,639	1	8,258
1981	490	544	0	1,034	8,119	1	9,154
1982	1,243	931	0	2,174	8,235	1	10,410
1983	1,519	1,610	0	3,120	4,175	2	7,297
1984	348	375	0	723	7,674	3	8,400

[a] One bushel of corn = 56 lb. To convert bushels to metric tons, multiply by 25.40×10^{-3}.
[b] Data from USDA (1972, 1983a, 1986b).

Several studies have focused on the factors that contribute to increasing corn yields in the United States (Thompson, 1969; Butell and Naive, 1978; Sundquist et al, 1982). The study by Sundquist et al (1982) evaluated three factors: average nitrogen application rate, July precipitation in the Corn Belt, and technology. The study, for 1954 through 1980, revealed that the contribution of nonnitrogen technologies has been constant over time, approximately 1 bu/acre (63 kg/ha) per year. In contrast, the contribution of nitrogen has fallen over time. Increasing the nitrogen application rate contributed approximately 2 bu/acre (125 kg/ha) per year during the 1950s and 1960s. Nitrogen application rates approached optimum levels during the 1970s, and the marginal impact of additional nitrogen was very small.

Iowa continued to be the leading corn-producing state in 1985, accounting for 18.0% of the acreage harvested for grain and 18.8% of U.S. production. Illinois was second, with 15.1% of the acreage and 16.2% of the production. Indiana, Nebraska, and Minnesota together account for about 27% of the nation's corn acreage and production each year. These five states accounted for about 60% of the nation's acreage and production in 1985. The second five states accounted for another 20% of the total (USDA, 1986a).

B. Production Costs and Returns

The management of a modern commercial farm involves decision making about the application of technology, the proper combination of crop and livestock enterprises, the procurement of inputs, and the marketing of products. Effective decision making in these areas requires accurate information.

Land-grant universities have long worked with costs-of-production (COP) data obtained through farm record-keeping projects. Under a mandate of the Agriculture and Consumer Protection Act of 1973, the Economic Research Service (ERS), an agency of the U.S. Department of Agriculture (USDA), has conducted periodic surveys of producers in major production areas to collect data needed to develop enterprise budgets for each of the major states. Because the USDA procedures for estimating the costs associated with fixed resources have been modified over time, it is difficult to use the estimates for analyzing the trend in total cost. Consequently, a homogeneous sample of Illinois cash-grain farms that have participated in the Illinois Farm-Business-Farm-Management (FBFM) record system are used here to illustrate changes that have occurred over time. This is followed by a summary of average COP estimates at the national level, which are developed by the USDA.

FARM RECORD SYSTEMS

Interest in COP estimates grew rapidly in the early 1970s, when the economics of crop production in the United States were undergoing rapid and far-reaching changes. In 1973, researchers at the University of Illinois initiated a series of annual analyses using FBFM data to determine the average cost per tillable acre to grow corn and soybeans on a homogeneous sample of farms participating in the FBFM system (Illinois Farm Business, Farm Management Service, 1974–1983).

The farms selected for these annual analyses include only farms of more than 260 acres (105.2 ha) on the more productive and nearly level soils in northern and central Illinois. The sample used for the analysis each year is restricted to farms without livestock that plant 90% or more of the tillable acres to corn and soybeans. Over the years, the farms in this sample have consistently used about 96% of their tillable land to grow corn and soybeans, with 53% of the acreage devoted to corn and 43% to soybeans. The group averaged about 300 acres (121.4 ha) of corn and 250 acres (101.2 ha) of soybeans during the years that the studies were conducted.

The annual cost summaries for the farms in the sample include some factors that farmers consider a cost of doing business, but which some other sole proprietors of businesses may not. These factors are not used as expense items on income tax returns. Examples include the labor charge for work done by the farm operator, a rent charge for all the land (both owned and rented), and an interest charge on equity in grain inventories. The average costs per tillable acre to grow corn on these farms since 1973 are illustrated in Fig. 1.

VARIABLE COSTS

Variable expenses are those that are incurred only if production takes place in a given year. The variable expenses associated with corn production include fertilizer, pesticides, seed, drying, storage, machinery repairs, machinery hire, and gas and oil. Combined, these items increased from $55 per acre ($136/ha) in 1973 to $141 per acre ($348/ha) in 1982, an increase of 156%. Increases in the early 1970s reflected substantial increases in the price of energy and related inputs such as fertilizer and pesticides. During 1979–1981, energy prices increased substantially, and variable expenses increased rapidly following four

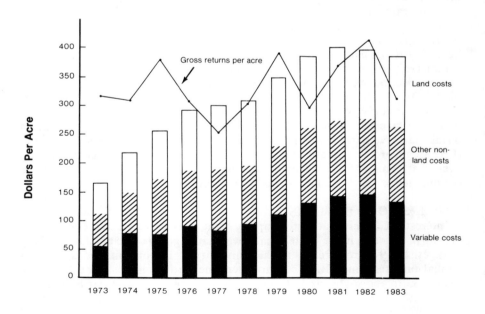

Fig. 1. Trends in corn production costs and returns, Northern and Central Illinois.

years of relative stability. Total variable cost dropped in 1983, reflecting lower prices for some input items. About half of the decline resulted from lower expenses for drying and storing because of drought-reduced yields. Energy and fertilizer prices have stabilized since 1980, and the total variable cost per acre averaged $136 ($336/ha).

OTHER NONLAND COSTS

Nonland costs include labor, building repairs, machinery depreciation, interest, and overhead. The labor charge includes hired labor plus family and operator labor. The charge per month has increased over time to reflect current wage rates. In 1983, the monthly labor charge was $1,100 per month, up from $575 in 1973. No added charge has been included for management. Interest on nonland capital covers the interest charged at the current rate on the sum of 1) one half of the average beginning and ending grain inventory value each year, 2) the average inventory of nonland capital, and 3) one half the cash-operating expense, exclusive of interest paid. The interest rate increased dramatically in 1979, 1980, and 1981, and nonland interest surged to $55 per acre ($136/ha) in 1980. Total nonland costs have averaged $268 per acre ($662/ha) during the 1980s.

LAND COSTS

Land costs include real estate taxes and an adjusted net rent that reflects the average rate earned on investment in bare land by landlords under a crop-share rental arrangement. In 1982, for example, the adjusted net rent of $97 per acre ($240/ha) represented a rate of return of 2.67% to the landlord from owning land at current land values. The lower net rents during 1982 and 1983 were due mainly to declining land values. Total land costs peaked at $125 per acre ($309/ha) in 1981, and many farmers were paying cash rents around this level.

GROSS RETURNS

Gross returns per acre were also computed for the farms in the sample by multiplying the average yield per acre reported by the sample farms by the seasonal average price received by Illinois farmers for corn each year. These results are also plotted in Fig. 1. The difference in gross returns and total costs represents residual returns to management. The impact of drought-reduced yields in 1980 and 1983 is very evident in Fig. 1. These results also clearly illustrate that the 1970s were generally very prosperous years for Illinois corn farmers.

ENTERPRISE BUDGET SYSTEMS

The national COP data are developed annually by the ERS, using an enterprise budgeting approach. Surveys of corn farmers in all major corn-producing states are conducted periodically to determine the quantity of various inputs used, field operations performed, the size and type of machinery and equipment used, fertilization and seeding rates, costs of custom operations, and costs of hired labor. The survey data are supplemented with price and quantity data available from other sources. The data collected from other sources include crop acreages, crop yields, prices received for products, prices paid for inputs, etc. A computerized budget generator combines these data with engineering relationships to develop detailed enterprise budgets for each major corn-producing state. The enterprise budget is specified on a per-acre basis and represents production in a contiguous area under homogeneous production methods.

The individual state budgets are combined by weighting according to production in each state to estimate average cost and returns at the national level (USDA, 1984a). The COP estimates for recent years are summarized in Table III.

CASH RECEIPTS

Cash receipts are estimated using average yield-per-acre values and the harvest months' average prices received by farmers for corn. The budgets exclude any additional revenues that farmers may have earned by storing the crop for later sales, as well as the costs associated with storage.

CASH EXPENSES

Cash expenses consist of both variable and fixed expenditures. Variable cash expenses include seed, fertilizer, lime, chemicals, custom operations, fuel and lubrication, repairs, and drying. The miscellaneous category includes purchased irrigation water and hired management fees. Variable cash expenses have averaged about $130 per acre ($321/ha) in recent years. Fixed cash expenses, including general farm overhead, taxes and insurance, and interest, have totaled about $80 per acre ($198/ha) in recent years.

CAPITAL REPLACEMENT

The capital replacement amount represents a charge sufficient to maintain machinery and equipment investment through time. Depreciation is a noncash expense that must be covered if the production capacity of the farm is to be maintained.

ECONOMIC COSTS

The economic costs associated with corn production include variable expense, overhead, taxes and insurance, and capital replacement. Once these costs are covered, the balance represents net returns to owned inputs. Owned inputs include operating capital other than nonland capital, land, and labor. A return to these three inputs is imputed to determine total economic costs. In the ERS accounting system, returns in excess of total economic costs are considered residual returns to management and risk. This residual has been negative in recent years (Table III).

The return to operating capital is the opportunity cost of operating capital used for all variable inputs during the six-month production period. The charge for operating capital is based on the six-month U.S. Treasury Bill rate.

Returns to other nonland capital are based on the average of the previous 20-year total return to production assets in the agricultural sector. The imputed rate of return to nonland capital has averaged 4.4% during the 1980s.

The allocated return to land is a composite cash/share rental value for corn land. The composite rental value is computed by taking the per-acre cash rental rate and the per-acre share rental rate from survey data and weighting each by its respective share of total acreage rented.

The labor charge is estimated by calculating the total hours of labor required to operate machinery and irrigation equipment valued at the average hourly rate for "all hired farm workers" in each state, plus the employer's share of social security taxes. No value is placed on available operator labor that is not required to operate machinery and equipment.

TABLE III
Corn Production Costs (dollars per planted acre), United States, 1981-1983[a]

Item	1981	1982	1983
Cash receipts			
Primary crop	260.17	245.55	258.70
Total	260.17	245.55	258.70
Cash expenses			
Seed	16.16	17.15	17.40
Fertilizer	51.93	49.29	46.33
Lime and gypsum	1.48	1.44	1.51
Chemicals	15.56	16.62	17.43
Custom operations	5.43	5.69	5.56
Fuel and lubrication	20.17	20.11	17.69
Repairs	11.98	13.24	13.26
Drying	8.30	8.38	6.32
Miscellaneous	0.21	0.23	0.27
Total, variable expenses	131.22	132.15	125.77
General farm overhead	15.43	16.11	16.39
Taxes and insurance	14.92	16.55	17.23
Interest	41.93	49.43	51.24
Total, fixed expenses	72.28	82.09	84.85
Total, cash expenses	203.50	214.24	210.62
Receipts less cash expenses	56.67	31.31	48.07
Capital replacement	28.01	30.90	31.65
Receipts less cash expenses and replacement	28.66	0.41	16.43
Economic costs			
Variable expenses	131.22	132.15	125.77
General farm overhead	15.43	16.11	16.39
Taxes and insurance	14.92	16.55	17.23
Capital replacement	28.01	30.90	31.65
Allocated returns to owned inputs			
Return to operating capital	7.42	6.00	4.64
Return to other nonland capital	11.07	12.24	12.27
Net land rent	61.32	61.43	61.43
Labor (paid and unpaid)[b]	14.87	15.33	14.50
Total, economic costs	284.26	290.71	283.88
Residual returns to management and risk	−24.09	−45.16	−25.18
Net returns to owned inputs	70.59	49.84	67.66
Harvest period price ($/bu)	2.38	2.14	3.21
Yield (bu/planted acre)	109.28	114.66	80.50

[a] Includes operator and landlord expenses or costs.
[b] Hired labor (a cash expense) and unpaid labor could not be separately identified given available survey data (USDA, 1984a).

The net returns to owned inputs have been low in the period covered by Table III, ranging from a high of $70.59 per acre ($174/ha) in 1981 to a low of $49.84 per acre ($123/ha) in 1982. The 1982 figure was about equal to average interest expense per acre, and the average net return to equity capital was essentially zero in 1982. Higher harvest-period prices in 1983 helped offset the 30% drop in average yields resulting from the summer drought. Net returns to owned inputs improved slightly in 1983; however, the residual return to management and risk remained negative.

III. TRENDS IN UTILIZATION

The quantity of U.S. corn used annually has averaged about seven billion bushels during the 1980s. Domestic use has accounted for about 70% of total annual corn disappearance, and the balance has moved into export channels. Livestock and poultry feed have accounted for 82% of domestic use during this period (USDA, 1986a). Although much smaller in terms of volume, the amount used in the production of food and industrial products has grown over time and surpassed the one-billion-bushel mark in the 1984/85 marketing year. Seed, the other major domestic use, is small compared with feed and industrial use.

A. Domestic Livestock Feed

Corn generally accounts for about 80% of the total quantity of grain fed to livestock and poultry in the United States. Since 1960, feed use has ranged from a low of about three billion bushels (76.2×10^6 t) in 1964/65 to a record high of 4.5 billion (114.3×10^6 t) in 1982/83 (Table IV). Variation in feed use over time reflects the change in the number of animals fed as well as ration adjustments made by livestock and poultry producers in response to relative prices and availability of corn and competing feed ingredients.

A sizeable proportion of the cattle and hogs produced in the United States are located on farms that also produce corn. Consequently, about 55–60% of the corn used as animal feed is fed on the farms where it is produced. The balance is purchased from elevators and grain dealers as whole corn or from feed manufacturers as prepared animal feeds. Before feeding, corn is usually ground and mixed with high-protein oilseed meals and vitamin and mineral supplements to balance the ration according to the nutritional requirements of the animals being fed. The corn by-products produced by dry-corn millers (hominy feed), wet-corn processors (corn gluten feed and meal), and distillers (distiller's dried grain) are used by manufacturers of prepared animal feeds (Chapter 15).

During the 1980s, livestock and poultry feeds have accounted for about 83% of the domestic use of corn (Table IV). Although feed use is highly correlated with the number of grain-consuming animal units (GCAU) fed annually, livestock producers make significant adjustments in the feeding rates when corn prices are high and/or livestock prices are low. When grain prices reached the then-record levels in 1973 and 1974, cattle feeders reduced the number of cattle on feed by almost 30% between 1972 and 1974 (Table V). The number of hogs slaughtered dropped substantially also. Price controls on meat, in combination with high grain prices, resulted in large losses for livestock producers. Feeding rates also dropped substantially in 1974 (to 1.69 t per GCAU). Feeding rates did not return to normal levels (about 1.95 t) until 1978. The high grain prices of 1980 following the drought had a similar impact on feeding rates.

The data in Table V suggest that GCAU numbers may not accurately measure feed requirements when the livestock industry is liquidating breeding herds. Between 1973 and 1974, cattle on feed declined 25%. In comparison, hogs slaughtered declined 14%, and these data include data for breeding stock (Table V). Since cattle and hog feeding account for over 50% of corn consumption, it is reasonable to assume that feed requirements dropped more in 1974 than the 11% drop in GCAU numbers would indicate.

B. Food and Industrial Products

Domestic use of corn for food and industrial products has been relatively small compared with the annual volume used for livestock and poultry feed. In 1984/85, domestic use for those purposes totaled 1,046 million bushels (26.6 ×

TABLE IV
Quantity of Corn (million bu) Used for Various Purposes, United States, 1960–1984[a]

Year Beginning October 1	Food and Industrial Products[b]	Seed	Feed and Residual	Total	Exports	Total Use
1960	283	11	3,092	3,387	292	3,679
1961	304	11	3,213	3,528	435	3,963
1962	312	11	3,156	3,479	416	3,895
1963	328	11	3,009	3,348	500	3,848
1964	338	11	2,956	3,305	570	3,875
1965	347	13	3,362	3,722	687	4,409
1966	350	14	3,333	3,697	487	4,184
1967	349	13	3,524	3,886	633	4,519
1968	347	12	3,607	3,966	535	4,501
1969	352	13	3,825	4,190	611	4,801
1970	368	17	3,593	3,978	517	4,495
1971	395	15	3,982	4,391	796	5,187
1972	434	16	4,292	4,742	1,258	6,000
1973	454	18	4,181	4,653	1,243	5,896
1974	478	19	3,180	3,677	1,149	4,826
1975	503	20	3,570	4,093	1,711	5,804
1976	530	20	3,572	4,122	1,684	5,806
1977	570	20	3,744	4,334	1,948	6,282
1978	600	20	4,323	4,943	2,133	7,076
1979	655	20	4,508	5,183	2,433	7,616
1980	715	20	4,133	4,868	2,355	7,223
1981	792	19	4,202	5,013	1,967	6,980
1982	883	15	4,522	5,420	1,870	7,290
1983	954	19	3,736	4,709	1,865	6,574
1984	1,046	19	4,117	5,182	1,838	7,020

[a] Data from USDA (1972, 1983a, 1986b).
[b] Includes corn and corn products processed into ethanol, alcohol, and corn sweetener products.

212 / Corn: Chemistry and Technology

TABLE V
Concentrates Fed, Grain Prices, Livestock Numbers, and Feeding Rates, United States, 1971–1983[a]

Item	Year Beginning Oct. 1												
	1971	1972	1973	1974	1975	1976	1977	1978	1979	1980	1981	1982	1983
Quantity fed (MMT)[b]													
Corn	101	109	106	81	91	91	95	110	115	105	107	115	95
Sorghum	17	16	17	11	13	10	12	14	12	8	11	13	9
Barley and oats	17	15	14	12	12	11	12	12	11	10	9	12	13
Wheat and rye	8	4	1	1	2	7	6	4	3	3	4	6	11
Oilseed meals[c]	14	13	15	13	16	15	17	18	19	17	18	19	17
Total	157	157	153	118	134	134	142	158	160	143	149	165	145
Average farm prices ($/bu)													
Corn	1.08	1.57	2.55	3.02	2.54	2.15	2.02	2.25	2.52	3.11	2.50	2.68	3.20
Sorghum	1.04	1.37	3.14	2.77	2.36	2.03	1.82	2.01	2.34	2.94	2.39	2.52	2.75
Wheat	1.34	1.76	3.95	4.09	3.55	2.73	2.33	2.97	3.78	3.91	3.65	3.55	3.54
Livestock numbers (million)													
Hogs slaughtered[d]	94.0	86.5	90.0	76.6	78.7	84.6	85.7	95.7	104.4	99.6	89.8	92.1	91.2
Cattle on feed (Jan. 1)	13.9	14.4	13.6	10.2	12.9	12.6	13.5	13.3	12.2	11.6	10.3	9.9	10.6
Animal units (GCAU)[e]	80.7	80.1	78.4	70.0	74.7	76.1	77.7	80.4	82.3	80.7	77.5	78.5	78.1
Feeding rate (t per GCAU)	1.95	1.96	1.95	1.69	1.79	1.76	1.83	1.97	1.94	1.77	1.92	2.10	1.86

[a] Data compiled from USDA (1983a, 1984b, 1984f).
[b] Million metric tons.
[c] Includes soybean, cottonseed, peanut, linseed, and sunflower seed meals.
[d] Includes commercial slaughter, farm slaughter, deaths and exports for year beginning December 1.
[e] Grain-consuming animal units, in which various classes of livestock and poultry are converted to milk-cow equivalents.

10^6 t), about 20% of total domestic use (Table IV). Most of the corn moving into food and industrial uses is processed by either the dry-milling industry (Chapter 11) or the wet-milling industry (Chapter 12). A large proportion of the primary products of these industries (meal, grits, flour, and starch) is further processed into breakfast foods, corn-sweetener products, ethanol alcohol, pet foods, malt beverages, and many other products.

The 1960s was a period of slow but rather steady growth in the quantity of corn used for food and industrial purposes; the volume totaled about 350 million bushels (8.89×10^6 t) by the end of that decade (Table IV). Food and industrial use doubled during the 1970s, mostly because of expanding markets for sweetener products. The sweetener market has continued to expand during the 1980s. Increased utilization during this decade also reflects the increased use of corn and corn products in the production of alcohol fuels. Total use for food and industrial purposes has grown at an annual rate of about 10% during the 1980s.

DRY-MILLING INDUSTRY

The volume processed by dry-corn millers was relatively stable during the 1960s, ranging from 113 million bushels (2.87×10^6 t) in 1960/61 to 130 million bushels (3.30×10^6 t) in 1966/67 (Table VI). The volume processed into breakfast foods and other dry-milling products moved upward during the 1970s and 1980s and totaled 176 million bushels (4.47×10^6 t) in 1984/85. Breakfast foods accounted for 34 million bushels (864,000 t), and other dry-milling products accounted for 145 million bushels (3.68×10^6 t). Historically, the leading product is brewer's grits, although the demand for brewer's grits has trended downward as the malt beverage industry has promoted brands that require less of adjuncts such as corn grits, broken rice, and corn syrups. In contrast, the dry-milling industry has been increasing the volume of cornmeal produced in recent years to meet a growing domestic and export demand. Cornmeal may become the leading product of dry-corn millers in the near future.

WET-MILLING INDUSTRY

The volume of corn processed annually by wet-corn millers has increased every year except one since 1960/61. The volume processed into wet-process products in 1984/85 totaled 645 million bushels (16.38×10^6 t), up from 155 million bushels (3.94×10^6 t) in 1960/61 (Table VI). Rapid growth occurred during the 1970s, following the development of a process to convert starch into high-fructose corn syrup (HFCS). This product has the sweetness characteristics that make it a viable substitute for common sugar (sucrose) in the manufacturing of soft drinks and many other processed food products (Chapter 17). This development came at the time of the sugar shortage and record high prices for sucrose in 1974. Sugar prices in the United States have been supported through government loan programs and import restrictions for a number of years. The level of sugar prices in recent years has given HFCS a pricing advantage relative to sugar.

Therefore, major food manufacturers and soft-drink bottlers have substituted increasing amounts of HFCS for sugar in their products. Major soft-drink manufacturers completely replaced sucrose with HFCS in their manufacturing operations in 1984. The growing importance of HFCS to the wet-milling industry is clearly evident in Table VII. Shipments of sweetener products totaled 14.7 billion pounds (6.66×10^6 t) in 1982, up 60% from 1977. In 1982, HFCS accounted for 56.4% of total sweetener shipments, up from 34.8% in 1977. Cornstarch was the only wet-milling product to register a decline in volume between 1977 and 1982.

ALCOHOL INDUSTRY

The amount of corn used in the production of fuel, industrial, and beverage alcohol has expanded rapidly in the 1980s. During the five years preceding 1979/80, the amount of corn used to produce alcohol averaged about 20 million bushels (508,000 t) per year (Table VI). The rapid escalation of fuel prices in 1979, 1980, and 1981 and the associated concerns about energy shortages led to the emergence of the fuel alcohol industry. Growth of this industry has been encouraged by federal and state tax incentives. Production of fuel alcohol was

TABLE VI
Volume (million bu) of Corn Processed into Various Food and Industrial Products, United States, 1960–1984[a]

Year Beginning October 1	Breakfast Foods	Other Dry-Milling Products[b]	Wet-Process Products	Alcohol and Distilled Spirits	Total	Domestic Use of Products	Export of Products
1960	17	96	155	33	300	283	17
1961	18	102	169	36	324	304	20
1962	18	105	179	28	330	312	18
1963	19	110	195	26	350	328	22
1964	19	110	201	28	358	338	20
1965	20	110	204	30	364	347	17
1966	21	109	208	33	371	350	21
1967	21	104	210	34	369	349	20
1968	22	101	297	33	363	347	16
1969	22	99	216	31	368	352	16
1970	23	106	231	24	384	368	16
1971	24	121	236	25	406	395	11
1972	24	127	271	28	450	434	16
1973	24	130	286	30	476	454	22
1974	25	134	306	27	492	478	14
1975	26	139	326	25	516	503	13
1976	26	141	351	25	543	530	13
1977	27	147	381	30	585	570	15
1978	28	137	411	35	611	600	11
1979	29	145	446	50	670	655	15
1980	31	151	476	75	733	715	18
1981	32	144	511	120	807	793	14
1982	33	150	536	180	899	883	16
1983	33	145	591	200	969	954	15
1984	34	142	645	240	1,061	1,046	15

[a] Data from USDA (1972, 1983a, 1986a).
[b] Estimated quantities used in producing cornmeal, flour, hominy grits, brewer's grits, and flakes.

begun in 1979/80, and corn use in alcohol production totaled 240 million bushels (5.08×10^6 t) in 1984/85.

C. Seed Use

Seed use is a relatively small but important component of domestic use. The annual quantity used for seed has grown from 11 million bushels (279,000 t) in the early 1960s to about 20 million bushels (508,000 t) in recent years (Table IV). The rapid increase in seed use during the 1960s and early 1970s was the result of expanding acreage (Table I) and an increase in the average seeding rate per acre. Farmers have increased the seeding rate per acre in response to the development of hybrids that tolerate higher plant populations per acre, resulting in higher average yields. In recent years, producers in the Corn Belt have seeded at a rate of about 23,000 kernels per acre. This generally results in about 19,500 plants per acre at harvest (Illinois Cooperative Extension Service, 1982). Producers growing corn under irrigation in western states may plant as many as 30,000 kernels per acre. In contrast, producers growing nonirrigated corn in the drier Northern Plains may plant only about 16,000 kernels per acre.

TABLE VII
Quantity (million pounds) of Selected Products Shipped by Wet-Corn Processors, United States, 1977 and 1982[a]

Product	1977	1982
Corn sweeteners		
Glucose syrup (corn syrup)		
Type I (20, up to 38 DE[b])	207.7	769.9
Type II (38, up to 58 DE)	2,980.5	1,799.7
Type III (58, up to 90 DE)	1,383.4	1,692.4
Type IV (90 DE or more)		601.1
Dried glucose	154.5	251.8
Dried dextrose and fructose	1,266.7	1,277.5
High-fructose corn syrup		
25, up to 50% fructose	3,202.9	4,546.2
50% fructose or more		3,737.4
Total sweeteners	9,195.7	14,676.0
Manufactured starch		
Corn starch	5,486.4	5,026.9
Other starch	251.4	452.0
Total starch	5,737.8	5,478.9
Corn oil		
Crude	N.A.	650.8
Refined	N.A.	340.8
Total oil	N.A.	991.6
By-products		
Corn gluten feed	4,199.8	5,850.9
Corn gluten meal	1,000.7	1,164.9
Other	1,699.5	1,478.1
Total by-products	6,900.0	8,493.9

[a] Data from USDC (1984).
[b] Dextrose-equivalent.

D. Exports

A major component of total U.S. disappearance of corn is export demand. Exports have had a significant impact on U.S. corn markets during the 1970s and 1980s. Exports averaged 540 million bushels (13.71×10^6 t) from 1960/61 to 1971/72, an average of about 13% of total disappearance (Table IV). Since 1972, exports have averaged approximately 28% of annual disappearance or about 1.8 billion bushels (45.71×10^6 t) per year. Exports of corn and of corn products (expressed in corn equivalents) reached a record 2,433 million bushels (61.79×10^6 t) in 1979/80. The annual volume has declined since that time; it totaled only 1,838 million bushels (46.68×10^6 t) in 1984/85.

EXPORT MARKETS FOR U.S. CORN

The volume of U.S. corn exported to various world areas has undergone a substantial change in recent years. The tonnage shipped to various world regions is shown in Table VIII. Shipments to North America (Canada) have declined over time, with shipments averaging 444,000 t during the last five years.

Shipments to Latin American destinations totaled 2.9×10^6 t in 1984/85, about 5.1×10^6 t below the record level of 1979/80. Many of the changes over time reflect variations in the volume shipped to Mexico. Almost 4×10^6 t were shipped to Mexico in 1982/83, following a serious drought that affected the 1982 crop in that country. Shipments to Mexico have declined in more recent years.

Western Europe has traditionally been an important market for U.S. corn. Shipments to that destination leveled off after peaking in 1979/80 at 15.6×10^6 t. The volume shipped to that region plunged to 6.6×10^6 t in 1984/85, a decline of 56.9% since 1981/82. The 10 nations of the EC have been the dominant importers of U.S. corn in that region for many years. The volume imported by the EC has been declining for several years, and Western European nations other than members of the EC have imported more than half of U.S. corn exports to that region since 1980/81. The decline in imports by the EC is the result of EC agricultural policies that have encouraged the production of barley and feed wheat, which have been substituted for imported feed grains (Jones and

TABLE VIII
U.S. Corn Exports (thousand metric tons) by Destination Area, for Marketing Years 1980/81 Through 1984/85[a]

Destination Area	Marketing Year				
	1980/81	1981/82	1982/83	1983/84	1984/85
North America	551.4	524.9	384.4	297.7	460.1
Latin America	7,447.8	2,453.2	6,358.8	5,253.4	2,884.9
Western Europe	14,667.6	15,337.2	9,584.2	8,353.8	6,607.9
Eastern Europe	6,741.6	3,672.8	1,496.3	703.3	715.0
USSR	4,946.7	7,646.3	3,159.2	6,499.8	14,938.5
Asia	18,403.2	17,326.1	23,617.2	20,810.1	17,132.5
Oceania	3.6	8.2	0.3	0.0	0.0
Africa	2,765.3	2,600.0	2,505.1	5,062.1	3,530.6
Unknown	3,841.0	40.1	0.0	5.0	5.7
World	59,368.2	49,608.8	47,105.5	46,985.2	46,275.2

[a] Data from USDA (1984c, 1984d, 1986c).

Thompson, 1978). Expansion of the volume of wheat exported from the EC has also had a substantial impact on export markets for U.S. wheat and feed grains (Paarlberg and Sharples, 1984).

Eastern Europe was once viewed as a region of relatively steady growth in import demand for and consumption of feed grains. Financial problems in several Eastern European nations have affected their ability to finance imports of feed grain to continue the expansion of livestock and poultry production. Those nations have apparently reassessed plans for expanded livestock production. Consequently, exports of U.S. corn to that region declined from a record 7.3×10^6 t in 1979/80 to about 715,000 t in 1984/85. The Soviet Union has been an unstable but important market for U.S. corn. Shipments to that destination totaled over 9.8×10^6 t in 1978/79. An embargo on grain sales to the Soviet Union in excess of the amount specified in existing trade agreements was announced in January 1980. The Soviet Union imported a record 15.5×10^6 t of corn in 1980/81 and only 35% came from the United States (USDA, 1985a). In more recent years, shipments of U.S. corn to the Soviet Union have varied from a high of 14.9×10^6 t in 1984/85 to a low of 3.2×10^6 t in 1982/83 (Table VIII).

Asia has replaced Western Europe as the leading market for U.S. corn. Shipments of U.S. corn to that region declined in 1981/82 as Japan purchased larger quantities from other sources. The U.S. share of the Japanese market rebounded to 91% in 1982/83 when a severe drought in South Africa virtually eliminated that country as an exporter. South Korea and Taiwan are important markets in that region; together they imported about 5.7×10^6 t in 1983/84. Japan, South Korea, and Taiwan accounted for about 93% of U.S. corn exports to Asian destinations in 1983/84. China appeared to be a market with great potential a few years ago; however, that country did not purchase U.S. corn in 1983/84.

Exports of U.S. corn to African nations during the four years preceding 1983/84 averaged 2.5×10^6 t each year. The Republic of South Africa experienced severe droughts that substantially reduced production in both 1983 and 1984. In 1983, for example, production totaled 4.1×10^6 t, down from 14.6×10^6 t in 1981 (USDA, 1985b). Corn output in 1984 totaled only 4.4×10^6 t. As a result, that nation entered the world market and purchased 141,000 t of U.S. corn in 1982/83. Those purchases soared in 1983/84, and total exports from the United States to African nations totaled 5.1×10^6 t. The Republic of South Africa accounted for 99.7% of the increase in corn shipments to Africa in 1983/84.

EXPORTS OF PRODUCTS AND BY-PRODUCTS

In addition to corn, the United States exports a variety of corn products, by-products, and corn-related specialty items. The quantities of various items exported in recent years are shown in Table IX. The commercial export market for cornmeal has exhibited a stable upward trend in recent years. Fluctuations in the annual volume of cornmeal exported mainly reflect changes in the quantity exported for relief purposes. Starch and dextrose are the leading wet-milled products exported from the United States.

Corn by-products in general and gluten feed and meal in particular have received much attention in recent years. Growth of the wet-milling industry in the United States to meet the expanding requirements for corn starch, which is used in the manufacturing of corn sweeteners and ethanol, has greatly increased the supply of corn gluten feed and meal in the United States.

TABLE IX
Exports (thousand metric tons) of Corn and Corn-Related Products from the United States for Marketing Years 1980/81 to 1984/85[a]

Item	Marketing Year				
	1980/81	1981/82	1982/83	1983/84	1984/85
Unmilled corn					
Yellow	59,170.3	49,398.3	46,967.3	46,781.6	45,894.9
Other	166.5	181.2	121.0	152.8	229.1
Relief	31.4	29.4	17.2	50.7	151.2
Subtotal	59,368.2	49,608.9	47,105.5	46,985.1	46,275.2
Dry-milled products					
Cornmeal					
Relief	70.6	51.3	75.9	84.8	117.4
Nonrelief	66.3	52.4	56.5	55.1	39.4
Other	17.2	35.6	71.0	9.8	24.7
Corn grits and hominy	28.3	19.3	11.1	14.4	13.6
Subtotal	182.4	158.6	214.5	164.1	195.1
Wet-milled products					
Starch	33.2	30.1	27.4	34.8	32.6
Dextrose	24.2	16.2	13.1	13.1	10.0
Glucose syrup	19.9	4.8	4.6	3.5	2.4
High-fructose syrup	IG[b]	1.0	1.1	1.3	0.4
Subtotal	77.3	52.1	46.2	52.7	45.4
Corn oil					
Crude, once refined	53.1	71.9	62.7	92.3	69.9
Other	28.8	19.9	39.1	48.6	41.3
Subtotal	81.9	91.8	101.8	140.9	111.2
Corn by-products					
Gluten feed and meal	2,686.5	2,796.3	3,590.4	3,555.6	867.5
Other	337.0	384.5	449.8	608.8	456.5
Subtotal	3,023.5	3,180.8	4,040.2	4,164.4	1,324.0
Specialty items					
Seed corn					
Sweet	3.2	4.5	3.5	4.6	5.0
Other	24.6	30.9	54.7	26.6	31.1
Sweet corn					
Frozen	35.2	35.4	39.9	35.8	34.0
Fresh	18.6	15.5	15.3	18.3	17.3
Popcorn	55.6	48.4	45.9	31.8	39.2
Subtotal	137.2	134.7	159.3	117.1	126.6
Grand total	62,870.5	53,226.9	51,769.3	51,624.3	48,077.5

[a] Data from USDA (1981b, 1982, 1983b, 1984h, 1985b, 1986c).
[b] Included with glucose.

Over one fourth of the volume of end products of the wet-milling process is recovered in the form of corn gluten feed and meal. The success of the industry requires that markets be developed for these by-products. Domestic use has been fairly stable, and much of the increased output has moved into export markets. The value of corn by-products exported from the United States totaled $630 million in 1982/83 (USDA, 1984e).

Over 97% of by-product exports have been imported by EC nations in recent years. Grain by-products imported by the EC have been exempt from the variable levy that is imposed on grains. Consequently, feed manufacturers in the EC have found gluten feed and meal to be economical and competitive sources of protein for use in feed manufacturing operations. The volume of corn by-products moving to the EC is expected to increase in the future; however, the EC has proposed subjecting corn by-products to a variable levy, and that would have an adverse impact on future growth. This is a subject of serious debate by agricultural officials in the United States and the EC.

Traditionally, popcorn has been the leading specialty item exported from the United States. The volume has been fairly stable at about 44,000 t in recent years. Sweet corn and seed corn are other specialty items for which the volume exported is trending upward. Seed corn shipments totaled 58,000 t in 1982/83, making seed corn the most important specialty item.

IV. THE MARKETING SYSTEM

The marketing of corn involves the execution of activities that permit corn and corn products to reach the consumer at the time, place, and in the form desired. Marketing involves physical facilities for drying and conditioning, handling and storing, merchandising, and processing as corn moves from farms to final users. Transportation and pricing are important activities that are addressed in separate sections of this chapter.

Price differentials over time, form, and space stimulate the marketing activities of storage, assembly, processing, and distribution that increase the value at each successive stage of the marketing chain from farmer to final consumer. The corn market is the sphere within which the price-making forces of supply and demand interact to generate and motivate individual firms to perform these marketing functions. The same functions are required in all countries, regardless of the simplicity or sophistication of the market and of the country in which they operate. In all major corn-producing areas, the crop is harvested in a relatively short period of time and must be stored to provide a continuous supply throughout the year. It must be transported from the areas of concentrated production to the areas of consumption or export, and it must be processed into forms of greater value as intermediate products or consumer goods. The firms that provide these marketing services vary from country to country, but the task of the market remains basically the same.

A. Overview of Marketing Flows

About one third of the corn produced annually in the United States is fed to livestock and poultry on the farm where it is grown; the balance is sold to marketing firms. The flow of corn through the marketing process is illustrated in

Fig. 2. Country elevators are the primary assemblers of corn sold from U.S. farms, annually accounting for about 80% of the volume (Leath et al, 1982). Subterminal and terminal elevators account for about 15% of farm sales of corn, and the remaining 5% is sold to feedlots and other farmers. Even though a sizeable volume of corn bypasses country elevators, these facilities continue to be the primary source of corn handled by subterminal and terminal elevators at most locations.

Country, subterminal, and terminal elevators handle, store, blend, and merchandise corn according to official grade specifications and are the primary sources of corn shipped to feed manufacturers, processors, and exporters. The rapid expansion in the volume marketed from U.S. farms during the 1960s and 1970s led to the construction of large subterminal elevators in the regions of concentrated production (Leath et al, 1982). Sales of corn from U.S. farms grew from 2.0 billion bushels in 1965/66 to an estimated 5.0 billion bushels in 1984/85. Many of the new facilities were designed for transferring grain from trucks to unit trains and barges. Other inland terminals were designed as assembly points for loading unit trains destined for export markets. These unit-train facilities rely on nearby country elevators for a continuous supply of corn during the marketing season.

Country elevators in major corn-producing areas have responded to the growing demand for marketing services by adding new drying and storage capacity, upgrading rail load-out facilities to handle multiple-car shipments, and offering additional services to their farm customers. These efforts have allowed the country elevators to handle larger volumes and maintain a fairly stable market share over time, but the number of country elevators operating in the major corn-producing areas has declined (Leath et al, 1982). This reflects the

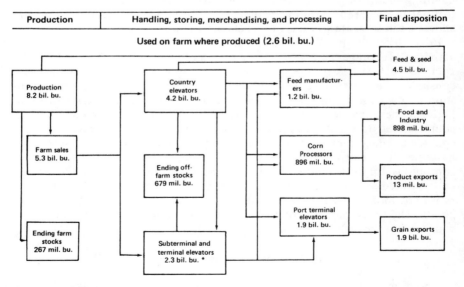

*Includes dealers and brokers.

Fig. 2. Path of corn through the marketing process, 1982/1983 marketing year.

closing of many small, obsolete facilities as well as the merger of individual plants that previously competed at the same location.

The most dramatic expansion of marketing facilities has occurred at U.S. ports. The handling capacity at U.S. ports was severely tested in 1973, when grain exports surged. Expectations of a growing export demand in the remainder of that decade led to the construction of additional handling capacity at U.S. ports, and the record grain exports of 1979/80 were handled with little difficulty. The declining export market during the 1980s has resulted in excess grain load-out capacity at many U.S. ports.

B. Farm Drying and Conditioning

Corn in the United States is harvested at moisture contents that generally exceed safe levels for storage. The higher machinery costs associated with combine harvesting and field shelling and the larger acreage per farm have encouraged early harvesting. Therefore, the first function required in the marketing process is conditioning the corn to lower the moisture content to a level that is considered safe for storage and transportation. This is accomplished either on the producing farm or in commercial dryers. The use of high-temperature dryers has had a major impact on the timing and speed of harvest in the United States, Argentina, Europe, and other production areas. Harvesting time has been shortened; susceptibility to breakage, amount of yield, and the use of fossil fuels in the grain industry have all increased. The high cost of energy and the quality effects of high-speed drying have been incentives for the introduction of new dryer designs and the use of low-temperature dryers in the major corn-producing countries (Chapters 4 and 5). Climatic conditions at harvest encourage the use of artificial dryers in most corn-producing regions of the world, with the exception of South Africa.

The cost of drying varies with the cost of labor, energy, and capital, and the cost per bushel is lower in large-volume installations such as commercial elevators. The drying cost on the farm can be illustrated with data from Illinois. Total costs for drying corn from 22 to 15.5% moisture content varied from 88.7 ¢/bu ($34.93/t), for a continuous-flow dryer with a 10,000-bu (254-t) annual volume, to 26.6 ¢/bu ($10.47/t), for a batch-in-bin stir dryer with a 100,000-bu (2,540-t) annual volume (Table X). Bin-type dryers are the most popular in Illinois because, in comparison with other dryers, the original investment per bushel dried is generally lower, flexibility in altering capacity is greater, and the bin serves for storage as well as drying. Even with bin-type dryers, the cost of drying small volumes (less than 635 t per year) is generally higher than drying charges at country elevators. Therefore, the relevant cost data are for bin-type dryers for farms where on-farm storage is part of the marketing strategy and where the annual volume dried is 25,000 bu (635 t) or more. Within these restrictions, drying costs varied from 25 to 35 ¢/bu ($9.84 to $13.78/t). Ownership costs (depreciation, interest, and maintenance) account for nearly 60% of the total cost of drying, even at annual volumes of 100,000 bu. At lower volumes, ownership costs may represent 75% of the total cost. Rising fuel prices have, therefore, had only a limited effect on farmers' decisions to dry corn. Once grain-drying equipment is purchased, commercial drying charges must be quite low before a farmer can justify using elevator-drying services. It is difficult for an elevator to dry corn for less than the farmer's out-of-pocket variable costs, despite the economies of scale that exist in the industry.

Farmers have expanded on-farm drying capacity for several reasons, in addition to a direct comparison of costs. The opportunity to maintain physical control of the grain on the farm, the need to use excess labor, previous investments in equipment, and high charges for drying at the country elevator have encouraged rapid expansion in the drying capacity on American farms. For

TABLE X
Annual Ownership and Energy Costs (cents) per Dry Bushel of Corn with Hot-Air Drying Systems and Bin Storage for Selected Quantities, 1982[a]

Costs	Bushels Annually				
	10,000	25,000	50,000	75,000	100,000
Batch-in-bin dryer					
Storage: ownership	21.1	17.3	16.0	14.6	13.3
Auxiliary					
Ownership	4.8	3.6	3.2	2.1	1.6
Electricity	0.5	0.5	0.5	0.5	0.5
Drying[b]					
Ownership	7.4	3.0	3.0	2.0	1.5
LP gas + electricity	10.5	10.5	10.5	10.5	10.5
Total	44.3	34.9	33.2	29.7	27.4
Batch-in-bin dryer					
Storage: ownership	21.1	17.3	15.1	14.0	12.3
Auxiliary					
Ownership	4.8	3.6	3.2	2.1	1.6
Electricity	0.5	0.5	0.5	0.5	0.5
Drying					
Ownership	17.8	7.1	3.6	2.4	1.8
LP gas + electricity	10.4	10.4	10.4	10.4	10.4
Total	54.6	38.9	32.8	29.4	26.6
Automatic batch dryer					
Storage: ownership	16.9	12.7	11.8	11.8	11.3
Auxiliary					
Ownership	9.4	3.8	2.1	2.1	1.6
Electricity	0.5	0.5	0.5	0.5	0.5
Drying					
Ownership	30.6	12.2	7.8	6.8	5.1
LP gas + electricity	12.9	12.9	12.9	12.9	12.9
Total	70.3	42.1	35.1	34.1	31.4
Continuous flow dryer					
Storage: ownership	16.9	12.7	11.8	11.8	11.3
Auxiliary					
Ownership	9.4	3.8	2.1	2.1	1.6
Electricity	0.5	0.5	0.5	0.5	0.5
Drying					
Ownership	51.6	20.6	13.1	8.7	6.5
LP gas + electricity	10.3	10.3	10.3	10.3	10.3
Total	88.7	47.9	37.8	33.4	30.2

[a] Data from Schwart (1982).
[b] Drying costs are for drying from 22% moisture content to 15%.

example, recent data for Illinois, Indiana, and Iowa show that 50% or more of the total corn crop in 1980 was artificially dried on the farms where it was produced (Illinois Cooperative Crop Reporting Service, 1981).

C. Farm Storage and Marketing

About 60% of the U.S. grain storage capacity is located on the farms where grain is produced (Leath et al, 1982). Producers in the major corn-producing states generally store about 65% of the corn crop on the farm at harvesttime. The balance is either sold directly from the field or placed in off-farm storage. About 80% of the amount stored on the farm is artificially dried on the farm before being stored. The economies of on-farm drying and conditioning were discussed above.

Farm storage capacity has expanded in response to expanding yields and production. The relative importance of farm storage to corn producers is revealed by the position of corn stocks on January 1 in recent years (Table XI). Producers were provided with economic incentives to build farm storage to store corn in the Farmer-Owned Reserve (FOR) program, which was introduced in 1978 (Johnson and Ericksen, 1978). As a result, expansion of farm storage capacity has mirrored the rapid buildup of corn stocks during the last 10 years, and the proportion of January 1 stocks stored on-farm has remained fairly constant, ranging from 68% in 1977 to 73% in 1979 and 1985. Farm stocks were reduced substantially in 1984 as FOR stocks were used to pay participants in the 1983 PIK program.

The marketing channel used by corn producers varies greatly from one location to another and reflects the structure of the local marketing system. The percentage of sales moving to various channels is shown in Table XII for selected corn-producing states. The percentage sold to local elevators ranged from 64 to 93% in the major states, and was 78% for all states surveyed. Terminal elevators were an important outlet in eastern Corn Belt states and also in Michigan. An interesting contrast is found in the Northern Plains, where sales directly to terminal elevators were not important. Sales to other farmers, grain dealers, and commercial feedlots account for most of the sales not moving to local elevators.

TABLE XI
Position of Corn Stocks on January 1, United States, Selected Years[a]

Year	Stocks (million bushels)			Percentage On-Farm Stocks
	On-Farm	Off-Farm	Total	
1975	2,561	1,080	3,641	70.3
1976	3,196	1,271	4,467	71.5
1977	3,346	1,544	4,890	68.4
1978	3,824	1,679	5,503	69.5
1979	4,638	1,681	6,319	73.4
1980	5,036	1,844	6,880	73.2
1981	4,139	1,718	5,857	70.7
1982	4,986	1,935	6,921	72.0
1983	5,936	2,268	8,204	72.4
1984	3,080	1,833	4,913	62.7
1985	4,304	1,560	5,864	73.4

[a] Data from USDA (1981a, 1984g, 1985a).

D. Commercial Handling and Storage

Country and inland terminal elevators perform the marketing functions of assembly and storage. These firms assemble, dry, and blend corn into the uniform lots demanded by processors and exporters. The major exporting and processing firms also own a number of country, inland terminal, and river elevators and control the flow of corn from the primary assembly point to the point of processing or export. Producers who do not own farm storage and drying facilities often rely on country elevators to store and condition corn until it is either sold or returned to the farm for livestock feeding.

Although many country elevators distribute grain back to farmers, a large proportion of the receipts from farmers are purchased by the elevator owners. Many country elevators that are owned by or affiliated with exporting firms will ship directly to ports. The independent country plants usually sell to local feed manufacturers, local processors, and inland terminals. The subterminals and terminals generally have the capability to blend and ship large, uniform lots of grain by barge and unit train directly to port locations. Processors who have precise quality requirements will buy from terminal operators. Country and inland terminals provide most of the storage services not provided on the farm. Processors, river elevators, and port elevators usually have the capacity to store working inventories only, making them dependent upon a continuous flow of corn from farms and inland elevators to maintain their working inventories.

The rated capacity of off-farm grain storage facilities totaled 8.1 billion bushels (205.74×10^6 t) on January 1, 1985, up from 7.1 on January 1, 1980. The states of Illinois, Iowa, Kansas, Nebraska, and Texas accounted for 52% of the off-farm rated capacity in 1984 (USDA, 1986e). Each of these states accounted for 680 million bushels (17.27×10^6 t) or more of capacity; Illinois was the

TABLE XII
Percentage of Corn Sales by Producers, by Marketing Channel, by State in Major Corn-Producing Regions[a]

Region and State	Channel				
	Local Elevators	Terminal Elevators	Other Farmers	Grain Dealers	Commercial Feedlots
Corn Belt					
Illinois	82	16	0	2	0
Indiana	65	25	5	5	0
Iowa	83	11	4	2	0
Missouri	93	3	2	2	0
Ohio	82	17	1	0	0
Lake states					
Michigan	70	28	1	1	0
Minnesota	89	3	7	1	0
Wisconsin	69	9	12	10	0
Northern Plains					
Kansas	75	1	1	1	22
Nebraska	81	0	16	0	3
South Dakota	64	2	14	0	20
23 states	78	13	4	3	2

[a] Data from Leath et al (1982).

leading state, with 1,027 facilities with a capacity of 974 million bushels (24.74 × 10^6 t). The average capacity per facility in the top states ranged from 881,000 bu (22,379 t) in Kansas, up to 953,000 bu (24,207 t) in Nebraska.

V. DOMESTIC TRANSPORTATION AND FLOW PATTERNS

The need to move corn from areas of concentrated production to the many locations where it is processed, fed to livestock, and exported gives rise to a large demand for transportation. In the case of corn, 10 states in the Midwest account for 80% of production and an even larger percentage of the volume entering the marketing system.

Although many ports participate in corn exports, about 60% of the annual volume exported is channeled through ports located on the Gulf of Mexico. Export facilities located along the Mississippi River in Louisiana handle most of that volume. Those ports are served by barge carriers operating on the inland waterway system. The inland waterway system provides an economical means of transporting bulk commodities such as corn from major shipping points in the Midwest to Louisiana ports.

These data illustrate how a changing export market, coupled with shifts in domestic and ocean freight rates, can dramatically change the competitive position of various U.S. export regions. The growing Asian market, along with more favorable rail rates (for unit trains) to Pacific ports, have dramatically increased the share handled by those ports. A reduction in imports by European markets has adversely affected the share handled by the Great Lakes and Atlantic regions. The competitive position of Great Lake ports is adversely affected by draft limitations on the St. Lawrence Seaway as well as by tolls imposed on vessels moving through the Seaway.

A. Modal Shares

The competitive position of U.S. corn in world markets is highly dependent upon an efficient and competitive domestic transportation system. Analysis of corn-flow patterns can reveal the geographic nature of grain markets and the type and amount of transportation services required by the grain industry. The transportation system often determines marketing opportunities and dictates where new handling, storage, and processing facilities are located. For example, the existence of the inland waterway system is the primary reason why export movements of corn are focused on Gulf ports.

Information about the origin and destination of corn movements and about the modes of conveyance used to transport corn is important for many policy and investment decisions. Recognition of the need for better information in this area led to a nationwide survey of grain shippers and receivers in 1978 (Hill et al, 1981). Statistics about the relative importance of various modes of transport in alternative types of movement are summarized in Table XIII. Trucks generally enjoy a competitive advantage in shorter intrastate movements, and in 1977, commercial trucks accounted for 75% of those movements. Farm trucks accounted for another 13%. In contrast, rail was the predominant mode in interstate movements to domestic markets, accounting for 60% of receipts at all domestic destinations receiving corn from interstate sources. Only 3.5% of interstate receipts moved by barge.

The distance of export areas from major origins and the availability of alternative modes of transportation largely determine the mode used to convey corn to export regions. Port elevators at the Gulf are the destinations for most barge movement, and that mode accounted for almost 78% of the volume moving to those ports in 1977 (Leath and Hill, 1983). Atlantic and Pacific ports are located a long distance from major production areas and, as a result, rely mainly on rail movements. Ports along the Great Lakes are near concentrated areas of production and are supplied primarily by truck shipments originating in adjacent states.

B. Movements to Domestic Markets

Grain-marketing firms reported shipping an estimated 5.1 billion bushels (129.5×10^6 t) of corn in 1977 (Table XIV). That total included movements to domestic destinations and to U.S. ports; however, shipments by port elevators to foreign destinations are excluded. A portion of the total is accounted for by the same corn being shipped in sequence by several firms. The volume sold from U.S. farms in 1976 and 1977 averaged about 4.0 billion bushels (102×10^6 t); therefore, about 22% of the total volume shipped to all destinations was shipped more than once. These data suggest that over 1 billion bushels (25.4×10^6 t) were shipped from country elevators to inland terminals for reshipment to U.S. ports by unit trains and barges. These terminals rely on country elevators for about two thirds of their volume—the balance comes directly from farms and bypasses the country elevator link in the marketing chain.

Shipments to domestic destinations within the originating states accounted for 2.37 billion bushels (60.2×10^6 t), or 46.5% of the total. Illinois, Iowa, and Minnesota together accounted for 58% of the intrastate shipments. A sizeable share of these movements are flows from country elevators to river elevators, from which the corn is reshipped to ports. Nebraska was the only other state where intrastate movements exceeded 200 million bushels (5.08×10^6 t), and most of that volume moved to terminals for reshipment to destinations in other states.

The demand for transportation equipment to move corn is reflected in the volume shipped across state boundaries to domestic destinations and export

TABLE XIII
Relative Importance (%) of Various Modes of Transportation in Moving Grain in the United States, 1977[a]

Type of Movement	Mode of Transportation			
	Rail	Truck	Barge	Farm Truck
Intrastate shipments	11.2	75.0	0.4	13.4
Interstate receipts	60.1	28.5	3.5	7.9
Lake port receipts	16.4	73.3	2.0	8.3
Atlantic port receipts	88.1	8.9	1.4	1.6
Gulf port receipts	22.0	9.4	77.6	0.0
Pacific port receipts	79.2	1.6	0.0	19.2

[a] Data from Hill et al (1981).

TABLE XIV
1977 Shipments (thousands of bushels) of Corn to Domestic Destinations and Export Regions by Marketing Firms in Each Originating State or Port Area[a]

Originating State or Port Area	Volume Shipped to Domestic Destinations		Volume Shipped to Export Regions	Total
	Within the State	In Other States		
States				
Alabama	17,425	1,281	3,442	22,148
Arizona	2,656	0	0	2,656
Arkansas	21,209	60	0	21,269
California	20,861	0	429	21,290
Colorado	7,621	9,322	0	16,943
Delaware	7,448	4,795	1,397	13,640
Florida	6,300	0	1,189	7,489
Georgia	37,354	9,210	215	46,779
Illinois	611,003	179,871	653,303	1,444,177
Indiana	178,768	175,756	212,147	566,671
Iowa	499,632	133,508	291,139	924,279
Kansas	51,467	29,465	6,015	86,947
Kentucky	11,813	34,789	46,769	93,371
Louisiana	2,846	6	180	3,032
Maryland	7,963	7,334	6,676	21,973
Michigan	13,869	21,246	49,078	84,193
Minnesota	252,839	25,185	108,895	386,919
Mississippi	4,275	1,572	216	6,063
Missouri	49,198	31,785	30,549	111,532
Nebraska	238,297	220,923	36,508	495,728
New Jersey	0	731	2,159	2,890
New Mexico	0	312	0	312
New York	1,155	2,642	1,092	4,889
North Carolina	52,908	2,434	20,286	75,628
North Dakota	0	6,653	248	6,901
Ohio	77,719	62,235	188,959	328,913
Oklahoma	0	780	97	877
Pennsylvania	34,598	2,904	2,367	39,869
South Carolina	5,960	2,627	4,354	12,941
South Dakota	0	4,833	1,789	6,622
Tennessee	13,144	4,887	2,231	20,262
Texas	129,682	11,601	3,791	145,074
Vermont	4	0	0	4
Virginia	2,425	0	1,611	4,036
Wisconsin	8,391	19,742	22,571	50,704
Wyoming	45	0	0	45
Port Areas				
Chicago-Milwaukee	540	1,708	14,904	17,152
South Atlantic	779	2,789	95	3,663
Louisiana Gulf	1,764	786	0	2,550
Total volume	2,371,958	1,013,772	1,714,701	5,100,431
Percentage of total volume	46.5	19.9	33.6	100.0

[a]Source: Leath and Hill (1983); used by permission.

regions. A majority of these shipments move long distances by either rail or barge. In 1977, marketing firms shipped more than 1 billion bushels (25.4×10^6 t) to domestic markets in other states (Table XIV). The pattern of corn flows to domestic destinations is illustrated in Fig. 3.

Almost all states shipped some corn to domestic markets in other states; in terms of volume, however, the origins of these shipments were highly concentrated. Four states (Illinois, Indiana, Iowa, and Nebraska) accounted for 710 million bushels (18×10^6 t) or 70% of the total. Nebraska, the leading state, sent over half (112 million bushels) of its out-of-state shipments to California and Colorado (Fig. 3). Arkansas and Kansas were also important markets for Nebraska shippers.

The primary domestic markets for Iowa shippers were located in Arkansas, Illinois, Missouri, and Wisconsin; shipments to each of these destinations totaled over 20 million bushels (50,800 t). The grain-deficient southeastern states are the primary markets for Illinois and Indiana firms. Processors and terminal elevators in Illinois were important buyers for Indiana corn, with shipments totaling 54 million bushels (1.37×10^6 t) in 1977. Ohio shippers participated to a lesser extent in supplying southeastern markets.

C. Movements to Export Regions

The geographic area from which corn moved to export regions was even more restricted than that for domestic movements. Five states (Illinois, Indiana, Iowa, Minnesota, and Ohio) originated 1.45 billion bushels (36.8×10^6 t), or 85% of the

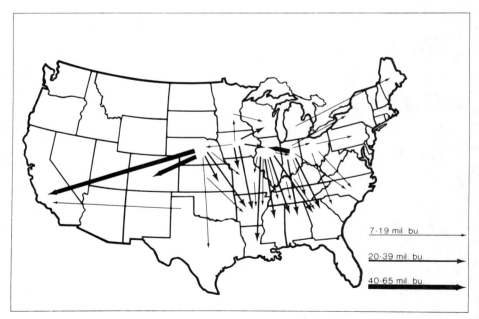

Fig. 3. Patterns of corn flows to domestic destinations in 1977. (Reprinted, with permission, from Hill et al, 1981)

total (Table XIV). Illinois firms shipped 653 million bushels (16.9×10^6 t), or 38% of the total, and 77% of shipments from Illinois moved by barge (Hill et al, 1981). Iowa and Minnesota were the other major places of origin of barge movements to ports. Indiana and Ohio were other states that originated large volumes destined for export regions. Shipments from these states to export regions moved predominantly in trainload units to Atlantic ports. Smaller quantities were trucked to Great Lakes ports from these states.

D. Developments in Recent Years

Changes in the volume exported and the share handled by the various ports have had the greatest impact on flow patterns since 1977. The most notable change is the volume exported from Pacific Coast ports. The volume inspected for export at those ports increased from 15 million bushels (381,000 t) in 1977 to 449 million bushels in 1983/84. Although California ports draw some corn from nearby origins, most of this grain is shipped from distant origins by rail. The introduction of unit train service to Pacific Coast ports is the most important reason for the growth of exports from that region. Consequently, a substantial increase in flows from Nebraska, Iowa, and South Dakota occurred after 1977. The volume exported from Pacific ports is sensitive to changes in the relative cost of transporting corn from production points to foreign destinations through alternative port locations. The foreign destinations being served by Pacific Coast ports are in Southeast Asia, a rapidly growing market that was previously supplied with shipments from Gulf ports.

Two legislative acts have been passed since 1977 that affect the grain transportation system. First, the Inland Waterways Revenue Act of 1978 contained provisions for imposing an escalating fuel tax on commercial waterway users (U.S. Congress, House of Representatives, 1978). The tax was set at four cents per gallon in 1980 and escalated to 10 cents in 1985. Other forms of taxation are also being considered to fully recover operation and maintenance expenditures on the inland waterway system. The 1985 tax level recovers 25–30% of these expenditures. The imposition of user fees to fully recover operation and maintenance expenditures would result in even higher barge rates, which could adversely affect the volume of corn moving by barge.

The volume of corn transported by barge surpassed the one-billion-bushel mark during the 1980/81 shipping season (Apr. 1 to Mar. 31) (Leath and Hill, 1983). The annual volume has averaged 998 million bushels (25.4×10^6 t) in more recent years. The apparent leveling off in the volume of corn moving by the barge mode may be due in part to the imposition of user fees. The fact that the barge mode has been able to maintain its volume during a period of declining corn exports is a reflection of its competitive advantage in moving bulk commodities such as corn. The data imply that the barge mode has increased its market share of corn movements to port locations during the 1980s. The market share has rebounded to the 50% level, approximately the same as reported for calendar year 1977 (Hill et al, 1981).

The other legislation that could impact on modal shares is the Staggers Rail Act of 1980 (U.S. Congress, Senate, 1980). The goal of this act was to improve the financial position of railroads by increasing their competitive strength relative to other modes of transport. The means to this end was increased flexibility on the part of railroads in setting rail rates. Railroads are permitted to enter into contracts with shippers that may include differentials based on seasonal or other demand changes. The effect of this legislation was somewhat less seasonal variability in the volume of corn transported by rail. Contract rates involving unit-train movements to various ports generally specify a minimum number of train loads, and the lowest rates are achieved when a shipper ships continuously throughout the year. The greater flexibility afforded the railroads in setting rates under this legislation may also be a factor in the leveling off of the volume of corn shipped via barge during the 1980s.

Other legislation proposed in recent years would allow the Federal Government to recover a portion or all of the operation and maintenance expenditures as well as capital improvement costs incurred in keeping the nation's deep-draft facilities navigable. These proposals generally contain provisions that would allow local port authorities to establish port user fees to recover designated costs. The adoption of port-specific user fees would affect the relative competitive position of various ports and could alter modal shares. Adoption of a uniform fee for all ports would potentially have less of an impact on corn flows and modal shares.

VI. TRENDS IN WORLD CORN MARKETS

The world market for grain has undergone some rather dramatic changes during the last two decades. During the 1960s, the focus of international trade in grains was on wheat, with the annual volume of wheat traded averaging about 50×10^6 t. In comparison, world trade in coarse grains (all grains except wheat and rice) averaged about 35×10^6 t annually during the 1960s (USDA, 1985b). In the 1970s, the volume of wheat traded annually increased to an average of 66×10^6 t, an increase of 32% over the previous decade, and the volume of coarse grains traded averaged 73×10^6 t, an increase of 109%. The volume of coarse grains traded surpassed that of wheat for the first time in 1973/74, when world imports of coarse grains surged 20% above the previous year and totaled 71×10^6 t. Following a decline in 1974/75, world trade in coarse grains soared to a record 109×10^6 t in 1980/81, a 68% increase in only six years. During the following five years, world trade in coarse grains declined 23%, a development of major importance for the U.S. corn industry.

A. Major Coarse-Grain Markets

The relative importance of various importers in the world coarse-grains trade since 1970 is illustrated in Fig. 4. Importing countries use coarse grains primarily for feeding livestock and poultry, and the rising trade volume during the 1970s reflected the efforts of many nations to increase the production of livestock and poultry products. Generally, rising incomes around the world enabled people to include more meat and other livestock products in their diets. Major importers in Asia (Japan, Republic of Korea, and Taiwan) have represented an important and growing market for feed grains. Imports by those nations have continued to rise during the 1980s despite an overall slump in the total volume of coarse grains traded in international markets. In 1984/85, those three nations imported 28×10^6 t of coarse grains, or almost 31% of total world imports of coarse grains (USDA, 1985a).

Perhaps the most significant development in recent years has been the decline of the EC as a major importer of feed grains. The EC represented a fairly stable market in the early 1970s, even though its share of world imports was declining. A shortfall in EC grain production in 1976 because of low yields resulted in imports of over 23×10^6 t of feed grains by the EC in 1976/77. Since that time, high internal support prices have encouraged production of feed wheat and barley, and import requirements for coarse grains have declined every year. EC imports of coarse grains dropped below 6×10^6 t in 1983/84, down 75% from the record imports of 1976/77.

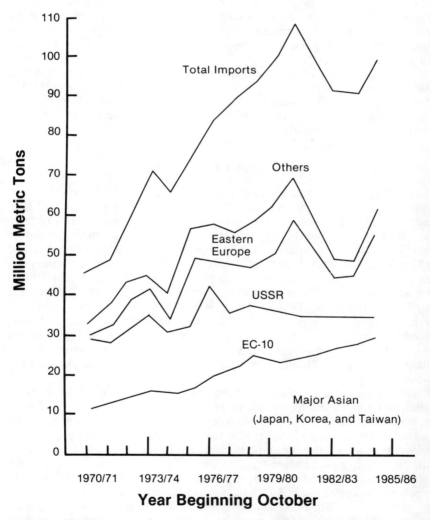

Fig. 4. Imports of coarse grains, selected regions, 1970–1984. EC-10 refers to the 10 nations of the Economic Community before the accession of Portugal and Spain in 1986.

The Soviet Union entered the world grain market in the early 1970s as it began to place greater emphasis on livestock production. Most grain-exporting nations were caught by surprise when the Soviet Union imported 15.6×10^6 t of coarse grains in 1975/76, up from 2.7×10^6 t the preceding year. The variability of Soviet purchases since that time has been the major destabilizing influence on world grain markets. During recent years, variations in the total volume of coarse grains traded in world markets have reflected annual changes in the volume imported by the Soviet Union.

Eastern Europe is the other major market for coarse grains. Coarse-grain imports by Eastern European nations increased during the 1970s and reached a record 11.5×10^6 t in 1979/80. Most observers felt that the market would continue to grow as those nations placed greater emphasis on livestock production. Financial problems during the 1980s have led many of those nations to reassess plans to expand livestock output, and coarse-grain imports have fallen as a result.

The above discussion of trade in coarse grains is important to corn-exporting nations because corn generally accounts for about 70% of the total world trade in coarse grains. Corn is the underpinning of the large international trade in feed grains, and any declines in exports of coarse grains are absorbed by the corn-exporting nations that supply the world grain market.

B. Corn Supply and Demand

The volume of corn produced around the world increased from less than 200×10^6 t in 1960/61 to over 400×10^6 t in 1979/80 (Table XV). The drought of 1980 halted a strong upward trend that began in 1973/74 in response to record price levels. The drought reduced U.S. production by 33×10^6 t, and the higher prices encouraged production outside the United States. World production dropped only 17×10^6 t in 1980/81. Normal crops in the United States during the next two years pushed total world production up to a record 437×10^6 t. The United States accounted for 47% or more of the world total both years. The acreage planted to corn in the United States was reduced substantially through government programs in 1983, and this, in combination with weather problems, reduced U.S. production to only 30% of the world total. Again, higher prices encouraged production around the world, and the 103×10^6 t drop in U.S. production was partially offset by a 15×10^6 t increase in output in other countries.

World consumption of corn has trended upward over time. In 1979/80, world consumption topped the 400×10^6 t mark and stood 56% above the level of 10 years earlier. World consumption has leveled off during the 1980s. Most of the major corn-consuming nations other than the United States have government policies that insulate domestic markets from major changes in world prices. As a result, the U.S. livestock industry makes most of the adjustment to tight world supplies and high prices. For example, in 1974/75, world consumption of corn dropped about 28×10^6 t and 89% of the decline was accounted for by lower usage in the United States. Likewise, in 1983/84, consumption in the United States declined about 20×10^6 t in response to high prices, while consumption outside the United States increased by 11×10^6 t.

C. Major Competing Exporters

Many nations around the world produce corn; however, a majority consume more corn than they produce. These deficits are filled by imports from a limited number of countries where production generally exceeds domestic requirements. The global requirements for imported feed grain have grown over time, and in 1980/81, when the volume of corn traded in international markets reached record levels, world exports of corn represented about 20% of global corn consumption.

The United States has been the world's leading corn exporter for many years. The other major exporters competing in the world market are Argentina, South Africa, and Thailand (Table XVI). Since they are in the Southern Hemisphere and their growing seasons are opposite those of the United States, the corn crop is generally planted after the U.S. harvest period. This fact allows producers in those countries to respond to strong prices that are usually associated with tight

TABLE XV
World Production, Consumption, Exports, and Ending Stocks for Corn, in Million Metric Tons,[a] 1960/61–1985/86[b]

Crop Year[c]	Production	Consumption	Exports[d]	Ending Stocks	Stocks-to-Use (%)
1960/61	197.6	190.6	13.7	58.8	30.8
1961/62	201.8	207.2	19.9	50.6	24.4
1962/63	205.0	212.3	20.0	42.9	20.2
1963/64	215.8	211.2	21.8	47.7	22.6
1964/65	213.0	219.2	23.9	38.7	17.7
1965/66	223.9	229.7	27.8	30.3	13.2
1966/67	244.9	239.7	26.3	34.4	14.4
1967/68	258.5	250.4	29.0	41.0	16.4
1968/69	248.8	253.7	26.3	37.5	14.8
1969/70	266.1	265.8	30.9	34.5	13.0
1970/71	261.4	265.7	31.4	27.1	10.2
1971/72	298.1	385.4	34.9	37.7	13.2
1972/73	293.6	302.8	44.5	26.9	8.9
1973/74	319.4	318.4	53.5	25.1	7.9
1974/75	288.6	290.6	46.3	23.3	8.0
1975/76	325.6	320.4	60.6	25.2	7.9
1976/77	353.5	336.0	60.6	39.4	11.7
1977/78	363.1	352.8	66.1	46.3	13.1
1978/79	389.3	385.7	71.1	48.2	12.5
1979/80	424.2	414.0	78.6	58.5	14.1
1980/81	406.8	415.1	83.2	50.3	12.1
1981/82	438.9	412.9	78.9	76.4	18.5
1982/83	439.9	419.2	63.2	106.8	25.5
1983/84	347.5	414.1	60.8	40.2	9.7

[a] To convert to bushels, multiply by 39.368.
[b] Data from Lin (1984) and USDA (1984c, 1986d).
[c] Based on aggregate of differing local marketing years.
[d] Includes trade within the European Community; July to June before 1979/80, thereafter, October to September.

supplies in the United States. In 1980/81, for example, the U.S. drought pushed prices up to near-record levels, and those countries responded by increasing production by 60% and exports by 37% (Table XVI).

These major competitors are aggressive sellers of corn in the international market. Consequently, they seem to be able to maintain a market share once it is established. The decline in the U.S. market share in 1980/81 (over six percentage points) was a reflection of the increasing market share claimed by the three major competitors, as well as restrictions placed on grain exports to the Soviet Union by the United States. The improved market share for the United States in 1983/84 is the result of a prolonged drought in South Africa. Argentina and Thailand have maintained their pre-1980 export levels in recent years when the import demand for corn was falling (Fig. 5).

Exports from South Africa have fluctuated greatly over time, reflecting wide swings in production. Production totaled only 4.1×10^6 t in 1982/83, down from a record 14.6×10^6 t in 1980/81. A return to a more normal production level in South Africa would likely displace some U.S. corn in world markets, since

TABLE XVI
Corn Production and Exports (million metric tons), Major Foreign Exporters and Total Foreign, 1960–1985[a]

Crop Year	Argentina		South Africa		Thailand		Total Foreign	
	Production	Exports	Production	Exports	Production	Exports	Production	Exports
1960/61	4.8	1.7	5.3	1.6	0.5	0.5	98.4	6.7
1961/62	5.2	2.9	6.0	2.6	0.6	0.6	110.4	9.4
1962/63	4.4	2.6	6.1	2.8	0.7	0.7	113.4	9.9
1963/64	5.4	3.4	4.3	1.1	0.8	0.9	113.7	9.8
1964/65	5.1	2.7	4.6	0.5	0.9	0.9	124.5	9.7
1965/66	7.0	6.4	5.1	0.5	1.0	1.1	119.7	11.0
1966/67	8.0	4.1	9.8	2.9	1.1	1.2	139.0	14.2
1967/68	6.6	3.2	5.3	2.7	1.3	1.2	135.1	13.5
1968/69	6.9	3.8	5.3	0.8	1.5	1.3	135.8	13.0
1969/70	9.4	5.6	6.1	1.1	1.7	1.5	147.0	15.4
1970/71	9.9	6.4	8.6	2.6	1.9	1.7	155.9	18.6
1971/72	5.9	2.5	9.5	3.6	2.3	2.1	154.7	15.1
1972/73	9.0	4.7	4.2	0.2	1.3	1.0	151.9	13.0
1973/74	9.9	5.7	11.1	3.2	2.4	2.1	175.4	22.3
1974/75	7.7	3.5	9.1	3.2	2.5	2.0	169.2	17.1
1975/76	5.8	3.2	7.3	1.5	2.9	2.4	177.2	17.2
1976/77	8.3	5.2	9.7	2.5	2.7	2.1	193.8	17.9
1977/78	9.7	5.9	10.2	3.0	1.7	1.2	197.9	16.7
1978/79	9.0	6.0	8.3	2.3	2.8	2.1	204.7	17.0
1979/80	6.4	6.0	10.8	3.4	3.3	2.1	220.3	16.5
1980/81	12.9	9.0	14.6	3.9	3.2	2.1	238.2	18.7
1981/82	9.6	4.9	8.4	4.7	4.3	3.3	232.7	17.9
1982/83	9.0	6.5	4.1	2.3	3.4	2.1	230.8	15.7
1983/84	9.2	5.9	4.4	0.1	4.0	3.0	241.5	13.4
1984/85	11.5	7.0	7.8	0.5	4.4	3.0	263.4	19.7
1985/86	12.3	7.4	8.0	1.4	5.2	3.7	256.4	22.6

[a] Data from Lin (1984) and USDA (1984c, 1986d).

South African white corn is preferred by many nations for a variety of food products.

Thailand, a well-known competitor in world rice markets, has emerged as a significant exporter of corn. Its location makes it a strong competitor in Asian markets. It is a reliable supplier (Table XVI); in contrast with South Africa, Thailand's production and exports have not varied greatly from year to year.

Argentina is a nation with great production potential. Corn yields in that country have been below those achieved in the United States because fertilizer has not been widely applied. Mielke (1984) observed that if existing production technology were adopted by producers in that country, output would jump considerably. High prices in the future could also lead Argentinian producers to convert existing pasture land into corn production. The country has vast amounts of pasture land that would be well suited for grain production. As a result, any efforts by the United States to support grain prices above market-clearing levels will stimulate production in other countries and reduce the export potential for U.S. corn. Argentina has signed long-term trade agreements with several countries, so its position as a major supplier of corn, sorghum, and wheat for the world would appear to be relatively secure at this time.

In addition to competition from major corn exporters, U.S. corn also faces competition from other grains, since other grains and carbohydrate sources may be substituted for corn in livestock and poultry rations. Exports of sorghum from Argentina and Australia have increased substantially in recent years. Australia, Canada, and the EC export sizeable quantities of feed barley and feed wheat that compete with corn in the international markets for feed grain. The

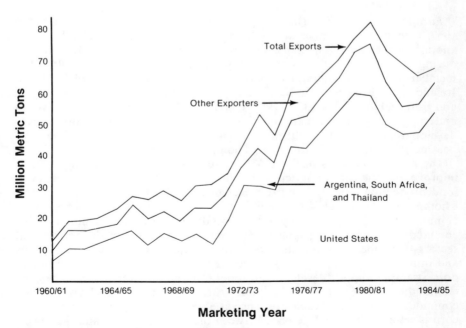

Fig. 5. Corn exports by major exporters, 1960–1983.

volume of coarse grains other than corn moving in world trade has been very stable, having averaged about 24×10^6 t in recent years. This volume was maintained in a period when the annual volume of corn traded in the world market declined by 16×10^6 t.

D. Policies of Major Trading Countries

Since the early 1950s, the United States has become increasingly dependent upon world markets for corn and other grains. This has come about primarily because U.S. agriculture has undergone a technological transformation that greatly expanded the production potential for grain. Capital-intensive inputs such as machinery, chemicals, and fertilizer have been substituted for land and labor. Labor resources have shifted out of U.S. agriculture; farm size has expanded; and resource productivity has increased rapidly.

This technological transformation of U.S. agriculture has had a tremendous impact on the evolution of U.S. farm policy. The agricultural industries of other developed countries have also undergone technological change. The policy responses of these countries have been highly individualistic. Domestic objectives dominate the policies of most major trading countries, and most of the major importers of grain have policies that insulate their agriculture from world market forces and interfere with trade in grain (Johnson, 1975). Selected policies of major importers and exporters that influence world trade in corn and, consequently, have an impact on the competitive position of U.S. agriculture are reviewed in the following section.

MAJOR IMPORTERS

European Community. The European Economic Community, established in 1957, included six member nations. Three other nations joined in 1973 and another in the 1980s. The name was shortened to the European Community during the interim and was commonly referred to as the EC-10. In 1985, members included Belgium, Denmark, France, Federal Republic of Germany, Ireland, Italy, Luxembourg, Netherlands, United Kingdom, and Greece. Spain and Portugal joined the EC in 1986, increasing the membership to 12 nations and changing its common name to EC-12. The member nations initiated steps to develop a Common Agricultural Policy (CAP) in the early 1960s. The CAP became effective in 1967 (Jones and Thompson, 1978).

The major objectives of the CAP are to 1) protect producer incomes, 2) improve the efficiency of agriculture, 3) increase the level of self-sufficiency in food production, and 4) stabilize prices of agricultural products (Jones and Thompson, 1978). The CAP stabilizes producer prices and incomes by a complex system of price supports for producers, government purchases of commodities, and direct payments to producers. To protect the agricultural sectors of member nations from outside competition, a system of variable levies and import licenses is used to keep lower-priced imports from undermining domestic price-support levels. A threshold or minimum import price for each commodity is set at Rotterdam each year, and the variable levy for each commodity is computed each day. The variable levy for a specific grain such as corn is set equal to the difference between the threshold price and the lowest import price offered on corn delivered to Rotterdam. When the threshold price is above the import price, the variable levy effectively buffers the internal corn price from fluctuations in world corn prices.

As a result of the CAP, the EC has become self-sufficient in some agricultural products that were traditionally imported, of which wheat is a leading example. The EC continues to import the higher-protein hard wheats in order to produce the wheat products desired for domestic consumption. However, high internal support prices have encouraged production of domestic varieties, and an increasing quantity is purchased by intervention agencies each year in price-support operations. Domestic wheat is subsidized to encourage domestic consumption as livestock feed, and an increasing share is subsidized as it is moved into export channels. In 1983/84, for example, the EC imported 3.6×10^6 t of wheat and exported 16.0×10^6 t (USDA, 1984c). The impact of this policy is to reduce import requirements of corn and other coarse grains by EC members. The wheat exports compete directly with U.S. soft wheats in world markets and also have an impact on the volume of feed grains moving into international markets.

The high internal support prices under the CAP have also encouraged the production of barley and corn in the EC. The EC has been a net exporter of barley in recent years, with production averaging about 40×10^6 t annually during the 1980s (USDA, 1984c). Corn production has also increased over time and is approaching 20×10^6 t on an annual basis. France and Italy are the major corn-producing members, accounting for over 85% of annual output during the 1980s (USDA, 1983b). The expanded corn output is a major factor associated with the declining volume of corn imported by nations in the EC (Fig. 4).

The CAP has an impact on corn markets in another way. Imports of soybeans and soybean meal are not subject to the variable levy and have increased dramatically. Grain by-products such as corn gluten meal and feed and feed-grain substitutes such as tapioca chips are not subject to variable levies, and imports of these products have also increased substantially. These products that enter the EC without the imposition of levies have displaced a large amount of feed grains in EC livestock rations. Rations generally contain high levels of low-protein carbohydrate sources such as tapioca chips, which are supplemented by large amounts of high-protein meals such as corn gluten meal and soybean meal. The recent accession of Spain and Portugal into the EC will result in expanded Spanish importation of nongrain feed ingredients. These will displace corn in Spain, a major market for U.S. corn in recent years.

In summary, the CAP has 1) encouraged domestic production of wheat, corn, and barley in member countries, 2) discouraged domestic consumption through higher prices, 3) reduced the import demand for feed grains, 4) encouraged the substitution of other feed ingredients for imported feed grains in livestock rations, and 5) created internal price stability by insulating internal feed-grain markets from changes in world price levels. The EC demand is not responsive to changes in world price levels, and this makes the U.S. grain industry more vulnerable to changes in world supplies and demands for corn and other grains.

Japan. Japan is the largest buyer of U.S. corn. The goals of domestic agricultural programs in Japan are to promote self-sufficiency in food production, maintain farm incomes, and stabilize domestic food prices. The country is self-sufficient in rice production, and the Japan Food Agency (JFA) controls food-grain imports to protect its rice industry. Agents of the JFA buy rice from producers at the support level, mill it, and resell the milled rice to consumers at a lower but more stable price. To protect the rice industry, wheat prices are also supported well above world market levels. The country produces a small percentage of its wheat requirements; the rest is purchased by private firms at world price levels for the JFA and resold to millers at internal support levels. Profits from the wheat tariffs help to finance the rice program. Rice is the key commodity in the food program, and the price supports for other commodities are set relative to rice (Jones and Thompson, 1978).

Price ranges are set for beef and pork. Price minimums are maintained at support levels through government purchases, and import quotas are used to balance meat supplies and keep prices within the desired range. All feed-grain requirements are imported by private firms at world market levels for the JFA. Imported feed grains are resold to feed manufacturers at the internal support level. Japan has achieved a high degree of price stability for food products, just as the EC has with the CAP. As a result, food consumption levels are maintained regardless of changing world supply-and-demand conditions. These food policies tend to cause greater price instability in the major grain-exporting countries such as the United States, where prices reflect world conditions.

Centrally Planned Economies. This group of countries includes the Soviet Union, the Eastern European nations, and the People's Republic of China. These countries have become significant importers of grains in recent years. They have placed greater emphasis on maintaining or increasing the production of livestock, poultry, and dairy products, and in years when domestic grain production has dropped below levels required to support planned livestock production, these nations—particularly the Soviet Union—have entered world markets and purchased grain to fill shortfalls in domestic production.

Grain imports by a country with a centrally planned economy are generally purchased by an official state trading agency or foreign trade organization that represents the country (Jones and Thompson, 1978). The foreign trade organization can negotiate long-term trade agreements with major grain-exporting nations, and it enjoys substantial monopolistic power in purchasing grain on world markets. These countries generally maintain control over the information about domestic crop prospects each year. This permits the trading agency to make large purchases from several suppliers before the actual import requirements are fully realized. The large year-to-year variations in exports to the Soviet Union from U.S. origins led the two countries to negotiate a five-year trade agreement in 1975. That and subsequent agreements call for a minimum and maximum tonnage that can be traded without prior approval. A larger tonnage could be approved if U.S. supplies are ample.

Agreements of this nature were intended to stabilize U.S. grain sales to the Soviet Union. They do stabilize the amount of U.S. grain imported by the Soviet Union each year; however, they have done little to stabilize world grain markets. When the United States places restrictions on additional sales to the Soviet Union, as was done in 1980, the import needs are filled with purchases from other exporters. Large year-to-year changes in grain imports by the Soviet Union and other centrally planned economies have been the major source of instability in world grain trade in recent years. Purchases by these countries are usually not influenced by the price level, and if large purchases are made when world supplies are tight, grain prices in the United States can climb rapidly.

Consequently, the U.S. livestock industry makes a large share of the adjustment to tight supplies and high prices.

MAJOR COMPETING EXPORTERS

Argentina. The trade policies of Argentina include exchange-rate controls and export taxes. Exchange-rate controls are used to help control domestic prices and maintain foreign reserves. Over-evaluation of the peso is an indirect taxation on agricultural exports that keeps domestic agricultural prices below world market levels. This reduces the incentives to farmers to produce corn and sorghum for the export market.

Export taxes have been used by Argentine authorities over time to generate revenue. Before 1977, ad valorem export duties on corn ranged from 8 to 56% of the export price of the commodity. The tax was eliminated in 1977, and floor prices were set for corn based on the world price. Corn was exported duty-free from March 1977 until April 1981, when a 12% tax was imposed. The export tax was raised to 25% in July 1982 (Mielke, 1984).

Argentina does not have adequate storage capacity for long-term storage of grain, and most exportable supplies move into export channels within a few months of harvest. Efforts are under way to improve the grain storage and transportation system so that grain exports can be used to generate foreign exchange in order to alleviate the huge foreign debt that has accumulated.

South Africa. South Africa has been a major exporter of corn for many years. The country produces a high-quality white corn that is preferred by many importers that process corn into food products. South Africa controls the marketing of corn through its Maize Board (MB). The MB sets producer prices and pays farmers the established price when production is delivered to the local elevators at harvest. The MB also sets prices on corn consumed domestically and handles all export sales. The MB maintains a stabilization fund, and all profits earned on export transactions go into the fund.

Thailand. The government of Thailand negotiates annual bilateral export agreements for corn and rice, specifying the volume, delivery schedule, and procedure for calculating the monthly export prices. The country has an open market for corn, and few restrictions or incentives are placed on its production and export. The government has had a long-term bilateral trade agreement with Taiwan and negotiates annual agreements with other southeast Asian importers. A high percentage of its annual production is exported at prices competitive with those of other major exporters.

International trading in grain is highly competitive. Major competing exporters have taken steps to protect or increase their market share, and each nation generally maintains only pipeline stocks at the end of each marketing season. Consequently, most grain-exporting nations follow pricing practices to ensure that production in excess of domestic needs is absorbed by world markets. Most of these nations have also entered bilateral trade agreements with major importers. The CAP of the EC uses export subsidies to move excess supplies of competing grains into the export markets. These policies and agreements tend to make the United States a residual supplier in the world grain market. Consequently, in periods when world production of coarse grains exceeds requirements, excess supplies build up in the United States. In 1983, the United States was the only country that spent large sums of money on government programs to control production. The higher prices that resulted from the smaller U.S. production actually triggered an increase in grain production outside the United States.

E. Implications for U.S. Exports

Corn producers in the United States are dependent upon a growing export market if they are to fully utilize the U.S. production potential. Domestic monetary and fiscal policies that have strengthened the U.S. dollar relative to the currency of other countries have made U.S. corn more expensive to foreign buyers. This has adversely affected the competitive position of U.S. feed grains in world markets, and importers have turned to alternative sources of supply whenever possible.

The EC has declined as a market for U.S. feed grains, and developing and centrally planned nations have become increasingly important. In fact, the growth that has occurred in world corn trade has reflected expanded imports by the developing and centrally planned countries. As centrally planned countries increase their share of world corn imports, the variations in their actual imports have increased potential to disrupt U.S. corn exports and prices. Most of the developed countries that export and import grains have policies that insulate domestic markets from changes in world market conditions, and the impact of variations in imports by centrally planned countries is borne almost entirely by the U.S. grain industry.

Corn exports to developing countries depend upon continued access to financing and upon the terms of credit available. The mounting debt situation facing a number of developing countries makes it difficult for them to earn enough foreign exchange to service their debts and buy grain at the same time. An increase in U.S. interest rates increases the annual debt service requirements of these countries and adversely affects their ability to import grain.

Price-support loans from Commodity Credit Corporation (CCC) have been a part of U.S. agricultural policy for many years. When the corn loan rates are above the world market-clearing prices, production is stimulated in all exporting countries and stocks accumulate primarily in the United States. The rapid buildup of stocks in the 1950s led to costly production control programs during the 1960s. The U.S. drought of 1974, in combination with a surging demand, pushed market prices above loan rates. Loan rates were raised dramatically in 1976 and 1977 in response to rising production costs that reflected inflation as well as capitalization of higher earning levels into land prices. By the fall of 1977, the U.S. loan rate set a floor under a declining grain price level, and stocks began to accumulate in the United States. The trend of stock accumulation was halted briefly in 1980 when a drought reduced U.S. production. However, about one billion bushels of corn were added to U.S. carryover stocks during each of the next two years. Consequently, the high loan rates in 1981 and 1982 encouraged U.S. production, reduced exports, and stimulated production in competing countries.

The United States must develop policies that will be more responsive to market conditions. Macroeconomic policy should be coordinated with farm policy so that the competitive position of U.S. grains in the world market will be improved. High, fixed loan rates stimulate production while restricting exports.

Loan rates should be more responsive to market conditions so that the negative effects on exports can be avoided. Many of these issues were addressed in the Food Security Act of 1985.

The United States seems to have the opportunity to build a strong market for its products by building a reputation for supplying a quality product at a competitive price. The quality of U.S. farm products moving in international markets could be improved in many areas. When prices are competitive, buyers seek out the supplier offering the highest quality.

VII. CORN PRICING SYSTEM

Price directs the use of production resources among competing farm enterprises, determines the income derived from the ownership of these resources, and directs the flow of grain moment by moment throughout the marketing system. The interaction of supply and demand determines the overall price level for corn, but price variations over time and space dictate storage decisions by many individual marketing firms and direct the geographical flow of grains among states, ports, and areas of the world. The search by thousands of independent, individual firms for the least-cost source of corn results in a marketing system in which prices respond quickly to any imbalance in supply and demand. Grain movements respond equally quickly to any imbalance in prices.

The organized grain exchanges are an important marketing institution in the discovery of prices. These organized exchanges allow interaction between buyers and sellers and, given the information network associated with the exchanges, they maintain instantaneous responses throughout the world. Actual supply and demand for corn and the resulting price are difficult to assess, since corn is produced and consumed throughout the world on different time sequences. During the month preceding harvest and as harvest progresses, traders are constantly estimating crop size. Each new appraisal is generally accompanied by a new evaluation of market prices. As new information on demand prospects or on supply becomes available, traders change their judgment of what prices are likely to be and trade accordingly. For a seasonally produced commodity like corn, the price is usually lowest at harvest and rises during the year by an amount sufficient to cover storage and other carrying charges. Divergence from this general theoretical pattern are frequent, however, as a result of changes in demand, unusual conditions, and expectation of future supplies.

In many countries, seasonal price relationships are established by government agencies. If these prices provide adequate returns to storers, supplies are released uniformly into the market. If cost of storage is not adequately covered by seasonal price differentials, the result is often a glut on the market at harvest, followed by a scarcity as the crop year progresses. The ability of market prices to reflect storage costs, as well as changes in demand over the crop year, is essential for efficient allocation of supplies over time.

A. Organized Grain Exchanges

The primary organized market for trading U.S. corn is the Chicago Board of Trade (CBT), where the bulk of trading in corn futures contracts occurs

(Hieronymus, 1971). A futures contract specifies a standard grade of the commodity that must be delivered in fulfillment of the contract at some future date. In addition, spot or cash trading in corn is done on that market and on several other organized exchanges in the country. The Bolsa de Cereals in Argentina provides a similar function, and other exchanges around the world are important in establishing corn prices. Many of these, however, are tied back to the Chicago futures market as a base. Marketing systems in almost every major producing country rely heavily upon the prices at the CBT as a basis for their pricing and marketing decisions. Even countries such as South Africa, where a marketing board establishes prices to producers, uses the CBT prices in making marketing decisions and establishing export prices.

The primary aim of an organized exchange such as the CBT is to provide a regulated marketplace so that members have facilities for trading in futures contracts. The organized exchange provides an impersonal method of price discovery that aids in the allocation of supplies and inventories over time. Futures contracts for corn at the CBT are based on a standard 5,000-bu unit—quality, method of payment, and place of delivery are all standardized. The delivery months are December, March, May, July, and September. Contracts are required to be bought and sold by an open outcry auction, with only one buyer and one seller for every contract sold. The exchanges themselves do not trade, but cover operating costs by charging a commission on each trade. Each exchange has a limited number of seats for memberships, which may be purchased only by private individuals who meet the exchange's financial and moral requirements. All trading must be done through members of the exchange.

B. Country Elevators

The rapid dissemination of price information throughout the grain industry results in a rapid response to any changes in basic supply and demand factors. The competitive nature of the grain industry throughout the world results in relatively small margins for the various services required in performing the marketing function. As a result, prices differ between geographical points by no more than the cost of transportation between those points. Any change in transportation rates results in a rapid adjustment in prices to compensate for the change. Aggregate price relationships, determined by the forces of supply and demand, are the summation of individual decisions and pricing strategies of many individual firms. Specific information on pricing practices used by country elevators may be found in a study of pricing strategies used by Illinois country elevators (Hill et al, 1983). The study describes pricing strategies typical of the alternatives used in most corn markets throughout the United States.

Because the corn market is international in scope, many of the pricing techniques and strategies are common across all major producing regions. In Argentina, for example, the base price is established at each major port elevator through daily operations of their Board of Trade. Country elevators use this price as their base price and subtract charges for services and operating expenses to generate a net payment to producers. Separate charges are assessed for cleaning, drying, storage, transportation from farm to elevator, transportation from elevator to port, and commission to cover overhead costs; in some cases, an additional percentage is withheld for capitalization fees for future building. Delayed pricing is a common technique in Brazil and Argentina, being especially attractive in periods of high rates of inflation.

The net payment to farmers must also take into account differences in quality. Most major producing countries have established quality characteristics in their grading standards that are used for adjusting price (see Chapter 5 for a description of grading standards in various countries). These grade factors are generally associated with price discounts that are based on factors in the grade standards and the limits that have been established for each factor. Grade factors and limits are generally established by a federal agency in each country.

VIII. GOVERNMENT PROGRAMS AFFECTING THE INDUSTRY

The policies of major trading nations that influence the world market for corn were discussed in section VI. The policies of those nations are generally designed to accomplish domestic objectives; however, they have had and continue to have a substantial impact on the U.S. corn industry, which is the major supplier in world markets. Domestic farm programs of the United States have also had a direct impact on the competitive position of U.S. corn in world markets. The more important government programs that have influenced the U. S. corn industry over time are reviewed in this section.

Abundance, price and income enhancement, stability, and security have been continuing aims of farm programs since the first national legislation dealing with price support for agricultural products was passed in 1933. A general objective of equality of opportunity for agriculture, coupled with the severe economic problems of the farm sector during the Depression, led to the passage of that legislation (Rasmussen and Baker, 1979).

The American farmer has adopted science, technology, and management practices that have greatly increased the production efficiency of U.S. agriculture. Output-increasing technology has led to overproduction, low prices, depressed farm income, and high government program costs for many years. Consequently, price-support efforts have generally required production adjustment to keep production in line with changing domestic needs or changing demands in the world market. The domestic agricultural policies affecting the corn industry may generally be classified as price-support programs, production adjustment programs, commodity storage programs, and commodity disposal programs.

A. Price-Support Programs

Price-support programs for corn were first introduced with the passage of the Agricultural Adjustment Act of 1933 (Rasmussen and Baker, 1979). Support prices in that legislation were specified in terms of parity prices. Parity price for corn refers to an average price that would give the commodity the same purchasing power—in terms of goods and services bought by farmers—that the commodity had in the 1910–1914 base period. The concept of parity was retained by all farm programs for corn before the passage of the Agricultural and Consumer Protection Act of 1973.

The 1933 act made price supports mandatory for designated "basic" (storable) commodities—corn, cotton, and wheat (USDA, 1967). Later that year, the CCC was incorporated under the laws of the state of Delaware. The Corporation was empowered to engage in broad operations to administer supply, price, and disposal programs for the basic commodities. The Corporation continued until 1948 when the CCC Charter Act provided a federal charter for the CCC and established it as an agency of the USDA.

Before 1963, price support for corn was available to eligible producers only in the form of nonrecourse loans. Under this method, eligible producers pledged the commodity as collateral for the loan. At maturity, the producer had the option to repay the loan with interest or forfeit the loan and deliver the commodity to the CCC. If the producer chose to forfeit, title to the commodity was transferred to the CCC with no further obligations for the borrower. Nonrecourse loans have continued to be used in CCC price-support operations in more recent years; however, a direct price-support payment was introduced in the corn program in 1963. It was paid on acreage planted in compliance with the program provisions. Accordingly, the loan rate was lowered by the amount of the price-support payment. The objective of this approach was to make U.S. corn competitive in world markets by lowering loan rates to world-market price levels and supporting producer incomes through direct payments. The basic support methods introduced in 1963 continued for 11 years. The program was successful in that CCC stock acquisitions were low and exports were expanded; however, program payments to participants were large. A detailed review of price-support operations for corn is available elsewhere (Leath et al, 1982; Lin, 1984).

The Agriculture and Consumer Protection Act of 1973 represented a new direction in American farm policy (Rasmussen and Baker, 1979). This was a market-oriented program designed to meet the growing world demand for corn and other grains. It also represented an attempt to give producers greater freedom in making production decisions and reduce the high federal expenditures associated with previous programs. The old concept of support price based on parity was replaced by a new concept of target price based on cost of production. A target price of $1.38 per bushel was established for the 1974 and 1975 crops. Target prices for the 1976 and 1977 crops were to be adjusted upward, based on an index of production costs (Johnson and Ericksen, 1977). The loan rate was to be set by the Secretary of Agriculture at a level of at least $1.10 per bushel but not more than 90% of parity.

Support payments under this legislation were equal to the difference in prices received by farmers during the first five months of the marketing year (October through February) and the target price for that crop. No payments were made in years when average market prices exceeded target levels.

The price-support methods introduced in the 1973 act were retained in farm bills passed in 1977 and 1981. The high market prices in the mid-1970s resulted in record returns for grain farmers, which were rapidly capitalized into the value of land. Inflationary pressure pushed up prices of other inputs used in producing corn, and target prices escalated rapidly in response to rising production costs. The higher target prices gave U.S. producers the incentive necessary to expand production more rapidly than demand during the 1975–1977 period, and market prices began to fall. In an effort to avoid large deficiency payments, the loan rate was increased substantially in 1976 and 1977. This philosophy was continued for the 1978–1981 crops under legislation passed in 1977 (Fig. 6). The 1977 act contained a significant departure from prior programs in that deficiency payments were based on current plantings of corn. Previous programs used an allotment based on historical planting patterns (Johnson and Ericksen, 1977).

With the exception of the drought year of 1980, U.S. production exceeded use every year, and at the end of the period covered by the 1977 legislation, stocks owned by the CCC or controlled under loan and reserve programs totaled 1.95 billion bushels (49.53 × 10^6 t), a level exceeding the record set on September 30, 1961. The Farm Bill enacted in 1981 severed the direct tie of target prices to cost of production (Table XVII). However, expectations about continued inflation and rising land values resulted in mandated loan and target-price minimums (Johnson et al, 1982). Target prices were mandated to increase at almost 6% per year for the 1982 through 1985 crops, and a minimum loan rate of \$2.55/bu (\$100.40/t) was set for corn. Program participation was minimal in 1982, and production hit a record 8.2 billion bushels (208.3 × 10^6 t). U.S. support prices were well above world levels in 1981/1982 and 1982/1983, and U.S. stocks reached a record 3.1 billion bushels (78.74 × 10^6 t) on September 30, 1983.

The prospects of that huge carryover led to the implementation of the Payment-In-Kind (PIK) program in 1983. U.S. producers removed 31.6 million acres (12.78 × 10^6 ha) from corn production under the PIK program, and participants received surplus stocks from CCC-owned and farmer-owned reserve (FOR) inventories. Production was reduced further by drought

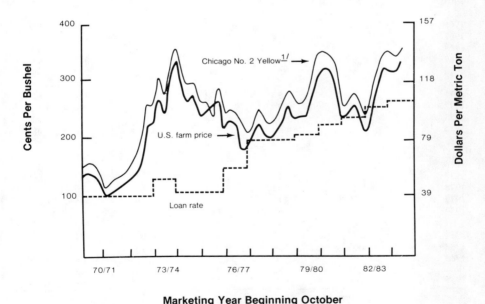

$^{1/}$St. Louis No. 2 Yellow beginning in 1981

Fig. 6. Corn prices and loan rates, by quarter, United States, 1970–1983. Corn was Chicago No. 2 yellow until 1981, then St. Louis No. 2 yellow.

conditions, which reduced average yields by 28%. Market prices increased dramatically, and domestic consumption in the form of livestock feed fell 18%.

B. Production Adjustment Programs

As noted in the previous section, corn prices in the United States have generally been above market-clearing levels because of price-support programs. As a result, production potential has usually exceeded domestic and export

TABLE XVII
Government-Owned, Government-Controlled, and Privately Owned Stocks of Corn, 1956–1985

As of Oct. 1	CCC[a]-Owned	Under Price Support[b]	Extended Loan and Reserve[c]	Total Government	Free	Total U.S.
1956	818	242	...	1,060	105	1,165
1957	932	363	...	1,295	124	1,419
1958	1,101	254	...	1,355	114	1,469
1959	1,153	247	...	1,400	124	1,524
1960	1,286	389	...	1,675	112	1,787
1961	1,327	563	...	1,890	126	2,016
1962	888	647	...	1,535	118	1,653
1963	810	465	...	1,275	90	1,365
1964	828	472	...	1,300	237	1,537
1965	540	384	...	924	233	1,157
1966	148	55	234	437	403	840
1967	139	134	101	374	449	823
1968	182	324	208	714	448	1,162
1969	295	148	293	736	377	1,113
1970	255	52	293	600	399	999
1971	105	30	203	338	329	667
1972	160	29	533	722	404	1,126
1973	84	39	48	171	538	109
1974	6	2	0	8	476	484
1975	0	2	0	2	359	361
1976	0	22	0	22	378	400
1977	0	117	0	117	769	886
1978	13	404	224	641	470	1,111
1979	100	117	592	809	495	1,304
1980	256	66	681	1,003	614	1,617
1981	238	187	185	610	424	1,034
1982	302	338	1,310	1,950	336	2,174
1983	1,166	0	1,550	2,716	404	3,120
1984[c]	201	30	430	661	62	723
1985	240	624	430	1,294	87	1,381

[a] Commodity Credit Corporation.
[b] Quantity outstanding under loan from preceding crop that has not been redeemed, delivered to CCC, or placed in the extended (reseal) loan farmer-owned reserve programs.
[c] Quantity outstanding under the extended (reseal) loan program from all previous crop years or the farmer-owned reserve. Quantities outstanding under price-support loans and old grain resealed were combined on Sept. 30 before 1966.

requirements, and programs to control production have been necessary. These programs include acreage allotments, marketing quotas, land retirement, commodity acreage diversion, and acreage set-aside programs.

The Agricultural Adjustment Act of 1938 authorized marketing quotas in conjunction with acreage allotments if at least two thirds of corn producers voting on a referendum approved the quotas (USDA, 1967). A formula to determine whether quotas were needed was specified in the law. However, production adjustment for corn was not a major concern until after the Korean conflict, when CCC-owned stocks began to accumulate. The main program to control acreage during the 1950s was the Soil Bank, which was established in the 1956 act, with the objective of reducing the amount of land planted to allotment crops such as corn (Rasmussen and Baker, 1979). The program was divided into two parts, an acreage reserve and a conservation reserve. Under the acreage reserve, corn farmers reduced planted acreage below their base acreage and received payments for diverting this acreage to conservation uses. The program attracted 21 million acres (8.5×10^6 ha) of cropland (not all from corn) in 1957; it was terminated in 1958. The conservation reserve allowed farmers to place land in conservation use up to a maximum of 10 years and receive payments. This option had attracted about 28 million acres (11.33×10^6 ha) by the summer of 1960.

Acreage diversion programs for specific crops were introduced in 1961. To be eligible for price supports, corn producers had to divert a specified percentage of base acreage to soil-conserving crops or practices. The Agriculture Act of 1965 replaced the acreage diversion program with a cropland adjustment program. The Secretary of Agriculture was authorized to enter into contracts with farmers for periods of five to 10 years to remove acreage from crops and place them into conservation uses. Payments to participants were limited to 40% of the value of the crop normally produced on the land. The acreage removed from corn production ranged from 16 to 27 million acres ($6.4-10.9 \times 10^6$ ha) during the 1960s (Table I).

The Agriculture Act of 1970 introduced the concept of set-aside acreage. Corn producers were required to set aside a certain percentage of cropland for conservation practices to qualify for price support under the program. Program participants could then grow whatever crop they chose on the cropland remaining in production. However, producers were required to plant corn or an eligible substitute crop to protect the farm's corn allotment. The corn acreage set-aside program attracted 24 million acres (9.7×10^6 ha) in 1972.

In 1973, the emphasis shifted from production control to expansion of grain production to meet a rapidly exanding world demand for grain. The Secretary of Agriculture was authorized to establish a set-aside program under farm programs passed in 1973 and 1977 if supplies were expected to be excessive. Set-aside programs were established for the 1978 and 1979 crops. The participation rate was not high, and only about 3 million acres were removed from corn production each year.

The set-aside feature of programs during the late 1970s was not effective in achieving crop-specific acreage reduction, since a producer's set-aside acreage was based on total acreage planted for harvest. Legislation passed in 1981 required crop-specific acreage reduction programs to divert a portion of crop-specific acreage base from production. In 1982, only 2 million acres were removed from corn production under the program. However, increasing yields more than offset this effort to control output, and production increased by 116 million bushels in 1982.

The rapid buildup of corn supplies in the early 1980s led to the PIK program of 1983. The PIK program, in combination with a 10% acreage reduction program and a 10% paid land diversion, removed about 32 million acres (12.96 $\times 10^6$ ha) from corn production in 1983. The program was effective in reducing output; however, program costs were high. Because of high costs, a program of this type is unpopular in a period of increasing concern about budget deficits and government spending. Only 4 million acres (1.62×10^6 ha) were diverted in 1984.

C. Commodity Storage Programs

In general, programs to control production have been ineffective in controlling corn supplies, and commodity storage programs have been used to remove excess supplies from marketing channels. In earlier years, the CCC purchased storage bins for storing corn and other grains acquired through loan forfeitures and purchases. The CCC Charter Act allowed the it to purchase bins if it determined that existing, privately owned storage facilities in an area were inadequate to store CCC-owned grain stocks. Bin-site stocks of corn reached a record 615 million bushels (15.6×10^6 t) in 1961 (Leath et al, 1982). The bin-site storage program was discontinued in the early 1970s, and the remaining grain bins were sold in 1973.

The CCC Charter Act directed the corporation to use commercial grain storage facilities to the maximum extent possible. Thus, the CCC-owned grain is usually stored in warehouses that enter into a Uniform Grain Storage Agreement with the CCC. Grain pledged under a CCC loan and stored off-farm must be stored in a warehouse operating under such an agreement.

A substantial part of the grain placed under price-support loans each year is held in storage on the farm where it is produced. Congress has directed the CCC to encourage increased farm storage capacity over time. In response, the CCC has made recourse loans available to farmers to finance new farm drying and storage facilities for corn and other grains under the Farm-Facility Loan program. These loans may finance up to 85% of the cost of the facilities, and borrowers usually receive favorable interest rates.

The 1977 act contained provisions that required the Secretary of Agriculture to administer a FOR program for wheat and, at his discretion, a similar program for feed grains. The goals of the FOR program were to 1) protect grain and livestock producers by stabilizing grain prices, 2) increase carryover stocks to meet emergencies, and 3) reduce the costs associated with storing CCC-owned inventories (Sharples, 1982). Eligible corn producers who complied with the acreage set-aside requirements of the feed-grain program were allowed to place corn into the reserve at the maturity of regular CCC loans. Producers participating in the FOR program agreed to hold their grain in storage until maturity of the contract (three to five years later), or until market prices exceeded a specified release price. In return, participants received payment for storing their grain and interest on the loans was waived after the first year.

Following the suspension of grain sales to the Soviet Union, producers who did not participate in the 1979 feed-grain program were permitted to place a limited quantity of corn into the FOR. The loan rate was raised to $2.55/bu ($100.39/t) in 1981 (the regular loan rate was $2.40) and to $2.90 ($114.17/t) in 1982 (compared to $2.55 for regular loans). The FOR became an instrument for supporting farm prices. Producers were given economic incentive to produce corn for the reserve program, and by the end of the 1982/83 marketing year, reserve stocks had reached record levels. The reserve loan rate and regular loan rate were set at the same level for the 1983 and 1984 crops (Lin, 1984).

D. Commodity Disposal Programs

The CCC has been granted authority in legislation enacted over the years to sell corn and other grains in its inventory. Minimum resale prices are normally specified in relation to the existing loan rate. When CCC stocks are used as payments under a PIK program, they are usually valued at the current support price made available through loans and purchases. The Secretary of Agriculture has the authority to make CCC-owned stocks available under emergency livestock feeding programs. These stocks are sold at not less than 75% of the current support price.

Stock disposal through CCC export operations has been a more common disposition method in many years. The CCC has engaged in operations in past years designed to promote commercial sales of agricultural commodities for dollars. In contrast with domestic sales, grain sold by the CCC from its inventories for commercial export is not subject to price restrictions. Consequently, grain may be sold to exporters at competitive world prices.

The CCC also finances the sale and exportation of grain under the Agricultural Trade Development and Assistant Act of 1954 (commonly known as Public Law 480). This legislation was passed in response to the buildup of surplus grain stocks in the early 1950s, and it became a major vehicle for sending farm products abroad. Exports under P.L. 480 have been used to assist developing nations meet their basic food needs. Importing nations may purchase P.L. 480 commodities with their currency or with dollars under long-term credit arrangements. The act also authorized donation of farm products for emergency relief. The CCC also has the authority to exchange CCC inventories for strategic and other materials. Barter transactions are enacted through regular commercial trade channels under agreements specifying the products and materials involved in the exchange.

IX. SUMMARY

The United States is the leading corn-producing nation in the world, accounting for almost half of the world's total each year. Corn has been grown on one of every four acres of U.S. cropland planted to principal crops in recent years. The adoption of new technology and improved production practices has increased output over time. Average yield per acre has trended upward over time and has become more variable in recent years. Output per acre reached a record 118 bu (7.4 t/ha) in 1985. In contrast, the drought-plagued 1983 corn crop showed a per-acre yield of only 81 bu (5.1 t/ha). Year-to-year variations of this magnitude illustrate that the successful application of existing production technology is highly dependent upon favorable weather conditions during the growing season.

Generally, rising incomes around the world have enabled people to include more meat and other livestock and poultry products in their diets. The concerted efforts of many nations to produce more of these products led to a rapidly expanding world demand for corn and other coarse grains during the 1970s. The United States had the productive capacity to meet this expanding demand, and world consumption of corn increased more than 50% during that decade. World exports, led by American corn, rose 150% during the same period. The U.S. share of the world corn exports reached 79% in 1979/80, up from 50% in 1969/70. By 1985/86, the United States's share had retreated to 58%.

Production shortfalls in the United States in 1980 and 1983 increased prices around the world and provided other nations with economic incentives to produce more corn and other coarse grains. With the production base in place, these competing exporting countries continued to increase output and exports to maintain foreign exchange earnings. The global consumption of coarse grains has stabilized during the 1980s. Support prices were maintained above market-clearing levels in the United States before 1986, and other coarse-grain exporters were able to underprice U.S. corn. The support rates have provided U.S. producers with ample incentives to expand output. Supporting U.S. farm prices at the established level required that large quantities of corn be isolated from the market in the Farmer-Owned Reserve program. The need to reduce supplies to more manageable levels led to the Payment-In-Kind program of 1983. This program, in combination with the drought that year, reduced U.S. production by 49%. Even though the surplus stocks in the United States were eliminated by the end of 1983/84, the capacity to produce in excess of market requirements still exists, and U.S. stocks have rebounded to record levels.

There is optimism that the world demand will rebound during the remainder of the decade. A continuation of a strong dollar relative to other currencies will weaken the competitive position of U.S. grains in world markets. Full utilization of U.S. capacity to produce corn will require a high level of exports for many years to come. An expanded level of exports is dependent upon economic growth around the world, which would weaken the dollar and make U.S. corn more competitive in world markets. The value of the dollar relative to the value of other currencies began to decline in 1986. The price-support loan rates for U.S. corn were lowered in 1986 in an effort to make U.S. corn more competitive in world markets.

LITERATURE CITED

BUTELL, R., and NAIVE, J. J. 1978. Factors affecting corn yields. Pages 14-17 in: Feed Situation. FdS-269, May. U.S. Dep Agric., Econ. Stat. Coop. Serv.

HIERONYMUS, T. A. 1971. Economics of Futures Trading. Commodity Research Bureau, Inc., New York.

HILL, L. D., LEATH, M. N., and FULLER, S. W. 1981. Corn Movements in the United States—Interregional Flow Patterns and Transportation Requirements in 1977. Ill. Bull. 768. Ill. Agric. Exp. Stn., Urbana-Champaign.

HILL, L., KUNDA, E., and REHTMEYER, C. 1983. Price Related Characteristics of Illinois Grain Elevators, 1982. AE-4561. Ill. Agric. Exp. Stn., Urbana-Champaign.

ILLINOIS COOPERATIVE CROP REPORTING SERVICE. 1981. Corn Harvesting, Handling, and Drying Methods. Bull. 81-2. Springfield, IL.

ILLINOIS COOPERATIVE EXTENSION SERVICE. 1982. Agronomy Handbook, 1983–84. Circ. 1208. Univ. of Ill., Urbana-

Champaign.

ILLINOIS FARM BUSINESS, FARM MANAGEMENT SERVICE. 1974-1983. Annual Summary of Illinois Farm Business Records. Ill. Coop. Ext. Serv., Univ. of Ill., Urbana-Champaign.

JOHNSON, D. G. 1975. World agriculture, commodity policy, and price variability. Am. J. Agric. Econ. 57:823-32.

JOHNSON, J., and ERICKSEN, M. 1978. Commodity Program Provisions Under the Food and Agriculture Act of 1977. Agric. Econ. Rep. 389. U.S. Dep. Agric., Econ. Res. Serv.

JOHNSON, J. D., RIZZI, R. W., SHORT, S. D., and FULTON, R. T. 1982. Provisions of the Agriculture and Food Act of 1981. Agric. Econ. Rep. 483. U.S. Dep. Agric., Econ. Res. Serv.

JONES, B. F., and THOMPSON, R. L. 1978. Interrelationships of domestic agricultural policies and trade policies. Pages 37-58 in: Speaking of Trade: Its Effects on Agriculture. Spec. Rep. 72. Agric. Ext. Serv., Univ. of Minn., St. Paul.

LEATH, M. N., and HILL, L. D. 1983. Grain Movements, Transportation Requirements, and Trends in United States Grain Marketing Patterns During the 1970's. Ill. Bull. 777. Ill. Agric. Exp. Stn., Urbana-Champaign.

LEATH, M. N., MEYERS, L. H., and HILL, L. D. 1982. U.S. Corn Industry. Agric. Econ. Rep. 479. U.S. Dep. Agric., Econ. Res. Serv.

LIN, W. 1984. Corn: Background for 1985 Farm Legislation. Agric. Info. Bull. 471. U.S. Dep. Agric., Econ. Res. Serv.

MIELKE, M. J. 1984. Argentine Agricultural Policies in the Grain and Oilseed Sectors. Foreign Agric. Econ. Rep. 296. U.S. Dep. Agric., Econ. Res. Serv.

PAARLBERG, P. L., and SHARPLES, J. A. 1984. Japanese and European Community Agricultural Trade Policies: Some U.S. Strategies. Foreign Agric. Econ. Rep. 204. U.S. Dep. Agric., Econ. Res. Serv.

RASMUSSEN, W., and BAKER, G. L. 1979. Price-Support and Adjustment Programs from 1933 through 1978: A Short History. Agric. Info. Bull. 424. U.S. Dep. Agric., Econ. Stat. Coop. Serv.

SCHWART, R. B. 1982. Investment, Ownership, and Energy Costs of New Bin Storage and of Hot Air Drying Facilities for Corn, Farm Economics Facts and Opinions. Dept. of Agric. Econ., University of Illinois, Urbana-Champaign (September).

SHARPLES, J. A. 1982. An Evaluation of U.S. Grain Reserve Policy, 1977-80. Agric. Econ. Rep. 481. U.S. Dep. Agric., Econ. Res. Serv.

SUNDQUIST, W. B., MENZ, K. M., and NEUMEYER, C. F. 1982. A Technological Assessment of Commercial Corn Production in the United States. Stn. Bull. 546. Minn. Agric. Exp. Stn., St. Paul.

THOMPSON, L. M. 1969. Weather and technology in the production of corn in the U.S. corn belt. Agron. J. 41:453-456.

U.S. CONGRESS, HOUSE OF REPRESENTATIVES. 1978. Inland Waterway Revenue Act of 1978. P.L. 95-502. 95th Congress, 2nd sess.

U.S. CONGRESS, SENATE. 1980. The Staggers Rail Act of 1980. P.L. 96-448. 96th Congress, 2nd sess.

USDA. 1967. Farm Commodity and Related Programs. Agric. Handbook 345. U.S. Dep. Agric., Agric. Stab. Cons. Serv.

USDA. 1972. Agricultural Statistics. U.S. Dep. Agric., Washington, DC.

USDA. 1981a. Stocks of Grains, Oilseeds, and Hay, Final Estimates by States 1974-79. Stat. Bull. 649. U.S. Dep. Agric., Stat. Rep. Serv. (January).

USDA. 1981b. U.S. Foreign Agricultural Trade Statistical Report, Fiscal Year 1980. U.S. Dep. Agric., Econ. Res. Serv. (March).

USDA. 1982. U.S. Foreign Agricultural Trade Statistical Report, Fiscal Year 1981. U.S. Dep. Agric., Econ. Res. Serv. (April).

USDA. 1983a. Agricultural Statistics. U.S. Dep. Agric., Washington, DC.

USDA. 1983b. U.S. Foreign Agricultural Trade Statistical Report, Fiscal Year 1982. U.S. Dep. Agric., Econ. Res. Serv. (January).

USDA. 1984a. Economic Indicators of the Farm Sector: Cost of Production, 1983. ECIFS 3-1. U.S. Dep. Agric., Econ. Res. Serv.

USDA. 1984b. Feed Outlook and Situation Report. FdS-294. U.S. Dep. Agric., Econ. Res. Serv. (November).

USDA. 1984c. Foreign Agriculture Circular, Grains. FG-14-84. U.S. Dep. Agric., Foreign Agric. Serv., (November).

USDA. 1984d. Foreign Agricultural Trade of The United States (FATUS). U.S. Dep. Agric., Econ. Res. Serv. (November/December).

USDA. 1984e. Grain and Feed Market News, Weekly Summary and Statistics. Vol. 32, No. 45. U.S. Dep. Agric., Agric. Mark. Serv. (November 9).

USDA. 1984f. Livestock and Meat Statistics, 1983. Sta. Bull. 715. U.S. Dep. Agric., Econ. Res. Serv.

USDA. 1984g. Stocks of Grains, Oilseeds, and Hay, Final Estimates by States, 1978-83. Stn. Bull. 707. U.S. Dep. Agric., Stat. Rep. Serv.

(June).

USDA. 1984h. U. S. Foreign Agricultural Trade Statistical Report, Fiscal Year 1983. U.S. Dep. Agric., Econ. Res. Serv. (March).

USDA. 1985a. Crop Production, 1984 Summary. CrPr 2-1. U.S. Dep. Agric., Stat. Rep. Serv. (January).

USDA. 1985b. Feed Outlook and Situation Report. FdS-296. U.S. Dep. Agric., Econ. Res. Serv. (May).

USDA. 1985c. Foreign Agricultural Trade of the United States, Fiscal Year 1984 Supplement. U.S. Dep. Agric., Econ. Res. Serv. (May).

USDA. 1986a. Crop Production, 1985 Summary. CrPr 2-1. U.S. Dep. Agric., Stat. Rep. Serv. (February).

USDA. 1986b. Feed Outlook and Situation Report. FdS-299. U.S. Dep. Agric., Econ. Res. Serv. (March).

USDA. 1986c. U.S. Foreign Agricultural Trade Statistical Report, Fiscal Year 1985. U.S. Dep. Agric., Econ. Res. Serv. (March).

USDA. 1986d. Foreign Agricultural Circular, Grains. FG-13-86. U.S. Dep. Agric., Foreign Agric. Serv. (November).

USDA. 1986e. Grain Stocks. GrLg 11-1. U.S. Dep. Agric., Stat. Rep. Serv. (February).

USDC. 1984. Census of Manufacturers. Wet Corn Milling. MC82-I-20D-4. U. S. Dep. Commerce, Bur. of Census (July).

CHAPTER 8

CARBOHYDRATES OF THE KERNEL

CHARLES D. BOYER
JACK C. SHANNON
Department of Horticulture
The Pennsylvania State University
University Park, Pennsylvania

I. INTRODUCTION

The major chemical constituents of the maize kernel are carbohydrates. These carbohydrates are the reason cereals are so highly valued as agricultural commodities. Not surprisingly, the metabolism and accumulation of carbohydrates in maize kernels have been popular subjects of study, and many reviews concerning maize kernel carbohydrates have appeared. Reviews have covered the process and regulation of carbohydrate biosynthesis (Preiss and Levi, 1980; Boyer, 1985; Echeverria et al, 1987), genetic modification of maize carbohydrates (Nelson, 1980; Boyer and Shannon, 1983; Shannon and Garwood, 1984), and the industrial modification and uses of maize carbohydrates (Rutenberg and Solarek, 1984). It is beyond the scope of this chapter to cover all of these aspects of maize kernel carbohydrates. However, we feel that having an integrated, developmental view of the maize kernel is fundamental to forming a complete picture of kernel carbohydrates.

The maize kernel is more than a rich source of carbohydrates for food, feed, and industry; it is a source of enzymes for the study of biosynthesis, and of genetic markers for genetic, biochemical, and genetic engineering studies. The developing maize kernel is a fascinating, highly organized structure. A better understanding of the formation of carbohydrates during kernel development and the nature of the carbohydrates in the mature kernel will allow the continued genetic improvement of maize for many purposes.

This chapter begins with a description of carbohydrate movement into developing kernels and distribution of the carbohydrates within the kernel. In the balance of the chapter, we consider the various simple carbohydrates and both structural and storage complex carbohydrates. Because the endosperm is the primary storage tissue of the kernel, endosperm carbohydrates are described in detail; carbohydrates in other kernel tissues are described briefly. Some effects of endosperm genotypes on the quantity or types of endosperm carbohydrates are also described to further demonstrate the variability in maize kernel carbohydrates.

II. GENERAL CONSIDERATIONS

Because the accumulation of carbohydrates in developing maize kernels involves many tissues and is an active process, it presents a multifaceted problem. The movement of carbohydrates into the kernels and the ultimate distribution of the carbohydrates in the kernel seem to be of primary importance in this process.

A. Transfer into the Kernel

To fully understand the movement of carbohydrates into the kernel, a brief description of the tissues involved is necessary. The base of the kernel contains the pedicel,[1] the placento-chalazal tissue, and the basal endosperm transfer cells (BETC) (Fig. 1). The closing layer between the pedicel parenchyma and the placento-chalazal tissue forms late in development. The vascular elements terminate in the pedicel (Kiesselbach and Walker, 1952; Felker and Shannon, 1980), and the unloaded assimilates must move through the placento-chalazal tissue before reaching the endosperm. At their termini, the sieve elements of the vascular strands are indistinguishable from pedicel parenchyma cells (Felker and Shannon, 1980). Numerous plasmodesmata are found among pedicel parenchyma cells and the compressed cells that will eventually form the closing

[1]The maternal tissue at the base of the kernel (the tip cap) is generally called the pedicel. However, according to Galinat (1979), this tissue is actually the upper part of the rachilla.

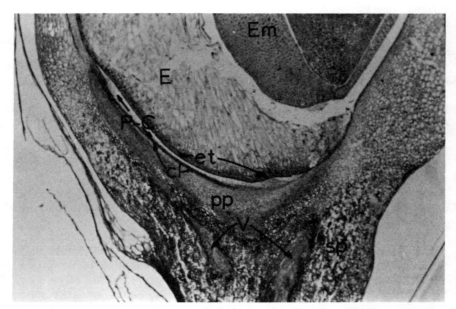

Fig. 1. Photomicrograph of the basal region of a developing maize kernel 28 days after pollination, showing the pedicel parenchyma (pp), placento-chalazal tissue (P-C), basal endosperm transfer cells (et), endosperm (E), pedicel spongy parenchyma (sp), vascular bundles (V), embryo (Em), and closing layer (cl).

layer. No plasmodesmata are observed between the placento-chalazal tissue and the endosperm. Furthermore, placento-chalazal cells die early in development, beginning eight days after pollination in the cells nearest the endosperm. By 21 days after pollination, most of the cells of the placento-chalazal tissue are devoid of recognizable cytoplasm. Therefore, the assimilates unloading from the sieve elements must pass through the pedicel cells into the free space (apoplast) of the pedicel parenchyma and the placento-chalazal tissue before entering the endosperm.

The basal endosperm cells are modified as special transfer cells, BETC. A transfer cell is characterized by extensive wall ingrowths and is considered to be a bridge between the apoplast and the symplast of a tissue (Gunning and Pate, 1969). The BETC of maize meet this description (Felker and Shannon, 1980). Wall ingrowths are most extensive in the outer basal endosperm cells that are continuous with the aleurone. Cell wall ingrowths decrease gradually in cells farther from the base of the endosperm, until essentially smooth-walled cells are observed a few cell layers into the endosperm. In addition, BETC are rich in mitochondria, rough endoplasmic reticulum, ribosomes, and dictyosomes and have prominent nuclei, indicating highly metabolically active cells (J. C. Shannon and H.-M. Cao, unpublished data). BETC begin to develop by seven days after pollination and by 20 days after pollination have developed extensive wall ingrowths, thus paralleling endosperm expansion and dry matter accumulation.

Based on anatomical and ultrastructural studies, the following pathway of assimilate movement has been proposed (Echeverria et al, 1987). Assimilates arriving through the phloem move into the pedicel parenchyma cells and through the pedicel via plasmodesmata. From the pedicel, assimilates are unloaded into the apoplast of the pedicel parenchyma and placento-chalazal tissue, from which they are absorbed by the BETC. During the movement of sugars, the major translocated sugar, sucrose, is inverted to glucose and fructose in the apoplast of the pedicel parenchyma and placento-chalazal tissue (Shannon 1968a, 1972; Shannon and Dougherty, 1972; Porter et al, 1985). Although there is good evidence that sucrose is passively unloaded into the pedical apoplast (Porter et al, 1985), we do not know yet whether sugar uptake into the BETC is by an active or passive process. However, sugar movement throughout the starchy endosperm appears to be symplastic. Finally, sucrose may be resynthesized and temporarily stored in the vacuole of endosperm cells until needed for starch synthesis (Echeverria et al, 1987).

TABLE I
Carbohydrate and Ash Composition of Maize Kernel Components[a]

Kernel Fraction	Percentage of Kernel Dry Weight	Composition (% of component dry weight)		
		Starch	Sugar	Ash
Whole kernel	100	72	2	1
Endosperm	82	86	1	1
Embyro	12	8	11	10
Bran	5	7	1	1
Tip cap	1	5	2	2

[a] Adapted from Inglett (1970).

B. Distribution Within the Kernel

The carbohydrates of maize kernels are distributed among many tissues of the kernel. The major carbohydrate of the whole kernel is starch (72% of the kernel dry weight). Of course, most of the starch is in the endosperm, but significant levels are found in the other three major fractions of the kernel, namely the embryo, bran, and tip cap (Table I; see also Chapter 3). Sugars are also found in all four kernel fractions, with the greatest amount in the embryo.

C. Endosperm Carbohydrates

The endosperm of a developing maize kernel is a population of cells of varying physiological ages. During kernel development, the cells in the central crown region of the endosperm begin starch accumulation first (about eight to 12 days after pollination); the lower endosperm cells begin starch synthesis and accumulation much later (Boyer et al, 1977). The peripheral endosperm cells are the last formed; they remain relatively small and produce small starch granules late in kernel development (Fig. 2). Thus, a major gradient of cell maturity forms from the central crown region (most mature) to the basal endosperm, and a minor gradient forms from the central crown region to the peripheral cells adjacent to the aleurone. Since all endosperm cells are not the same age, an assay of kernel homogenates for sugars, starch, or even enzyme activity represents the average content of the particular population of cells at the time of harvest. In physiologically young endosperm cells, sugar content is high and starch content is low. Thus it follows that in kernels about 12 days after pollination, sugars are relatively high and starch low. As the proportion of cells synthesizing starch

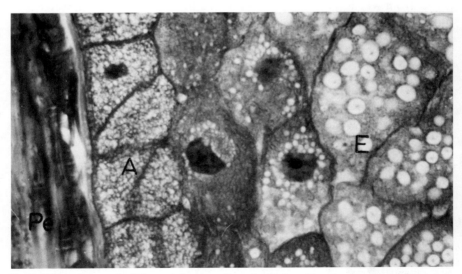

Fig. 2. Photomicrograph of the peripheral portion of a 36-day-old maize kernel near the crown region. Note the decreasing size of starch granules in cells nearest the aleurone. Pe = pericarp, A = aleurone, E = starchy endosperm.

increases, kernel sugar content declines and starch increases (Creech, 1965). In the mature endosperm, starch comprises 86% of the total dry weight and sugars about 1% (Table I).

III. SIMPLE CARBOHYDRATES

Carbohydrates play an essential role in intermediary metabolism in all tissues. Storage and structural polymers and a variety of simple carbohydrates (sugars) are synthesized in the developing maize kernel. By maturity, when synthesis is complete, sugars comprise only about 2% of the kernel dry weight. Although the simple carbohydrates generally occur in low levels, they are important in energy transfer and synthesis.

A. Monosaccharides

The primary free monosaccharides of the endosperm, D-fructose and D-glucose, occur in approximately equal proportions. Holder et al (1974) reported that 21-day-old normal kernels contained 1.8 mg of glucose and 1.6 mg of fructose per kernel. The highest concentration of reducing sugars (9.4% of kernel dry weight) was measured in 16-day-old kernels, the youngest kernels tested, and the content declined dramatically with increasing age (Creech, 1965).

Other monosaccharides generally occur as nucleotide sugars (Table II). These are continuously used as substrates for polymer biosynthesis, and thus only low levels are present in the kernels (Feingold and Avigad, 1980).

B. Disaccharides and Trisaccharides

Sucrose is the major disaccharide in maize kernels. Sucrose concentrations (in milligrams per gram of dry weight) peak 15–18 days after pollination, reaching 4–8% of kernel dry weight (Creech, 1965). Sucrose content per kernel remains relatively high, 2–3 mg per endosperm, until near maturity (Tsai et al, 1970). Endosperm genotypes have dramatic effects on sucrose levels. Standard sweet corn varieties (with the sugary [*su*] gene) have sucrose levels twice as high as those of field corn during the same period, and extra sweet varieties (*sh2* gene) can have three to four times as much sucrose (Boyer and Shannon, 1983).

Besides sucrose, developing maize kernels also contain low levels of maltose,

TABLE II
Primary Nucleotide Sugars and Their Function in Higher Plants[a]

	Biosynthesis of		
Starch	Cell Walls	Glycoproteins	Sucrose
ADP[b]-Glucose	UDP-Glucose	ADP-Mannose	UDP-Glucose
UDP[c]-Glucose	UDP-Xylose	UDP-Galacturonic acid	
	UDP-Galactose	UDP-Glucuronic acid	
	UDP-Arabinose		
	UDP-Rhamnose		

[a] Adapted from Feingold and Avigad (1980).
[b] Adenosine diphospho-.
[c] Uridine diphospho-.

which is generally found at less than 0.4% of the dry weight. However, in new sweet corn varieties combining the sugary gene with the sugary enhancer gene (*su se*), maltose levels as high as 1% have been reported (Carey et al, 1982b). The *su se* genotypes also have elevated sucrose concentrations (see Chapter 14).

Trisaccharides and higher oligosaccharides are very minor constituents of the maize kernel. Low levels of the trisaccharide raffinose have been reported (Inglett, 1970). Maltotriose and maltooligosaccharides would also be expected, although Gentinetta and Salamini (1979) were able to detect these saccharides in only one (amylose-extender [*ae*]) of several genotypes tested. Additional studies with more sensitive experimental procedures are needed to accurately determine the levels of the maltooligosaccharide series.

C. Sugar Alcohols

Until recently, phytate, the phosphate storage form of *myo*-inositol, was the only sugar alcohol reported in maize kernels. Sorbitol was recently found in *su se* sweet corn varieties (Carey et al, 1982b). More thorough examination has shown that sorbitol is present in kernels of several genotypes (Carey et al, 1982a).

D. Phytate

Phytate, the hexakis-*o*-phosphate of *myo*-inositol (Johnson and Tate, 1969), is widely distributed in higher plants. Seeds accumulate up to 90% of stored organic phosphate as phytate. Synthesis of phytate starts with the formation of *myo*-inositol-1-P from glucose-6-P, followed by complete phosphorylation by a kinase (Majumder and Biswas, 1973). Phytate is deposited in protein bodies of the aleurone and scutellum; in maize, about 90% of the kernel phytate is found in the scutellum and 10% in the aleurone (O'Dell et al, 1972). In mature maize kernels, phytate accounts for about 0.9% of the dry weight (O'Dell, 1969). In some species, the phytate aggregates as globoids in association with the protein bodies, but protein bodies of the maize aleurone do not have globoids (Pernollet, 1978). Hence, the phytate must be distributed throughout the maize protein body, although no direct evidence for this is available.

E. Metabolic Intermediates

In addition to the nucleotide sugars listed in Table II, many other carbohydrate intermediates occur at very low levels in maize kernels. These are primarily the triose phosphates and phosphorylated hexoses of the glycolytic or gluconeogenic pathways (Table III). Pools of intermediates may be compartmented at different concentrations in the cytosol and organelles.

IV. COMPLEX CARBOHYDRATES—STRUCTURAL

Many different polysaccharides play an important role in the structure of maize kernels. These can be classified as pectic substances, hemicelluloses, and cellulose. From a human dietary standpoint, the most important cell wall components are the bran and tip cap, which are important sources of dietary fiber (Kies et al, 1982). Sandstead et al (1978) found that corn bran was composed of 70% hemicellulose, 23% cellulose, and 0.1% lignin on a dry-weight basis. No detailed structural analysis of maize kernel cell wall polysaccharides has been done, but a general description can be given based on our present understanding of cell walls.

A. Cell Walls

Cell walls are classified as primary or secondary (Esau, 1960). Primary cell walls are formed during cell growth. Endosperm cells have thick primary cell walls, and embryo cells have thinner primary cell walls. In contrast, cells of the pericarp (bran) and tip cap also have secondary cell walls, which become thicker after cell expansion ceases (Esau, 1960). Cellulose fibers are the basic structural unit of the cell walls of all higher plants. A number of other cell wall polysaccharides, the matrix polysaccharides, are associated with cellulose in both primary and secondary cell walls. These wall polymers contain several sugars, including glucose, xylose, arabinose, galactose, rhamnose, and mannose. Because the bonds between the different sugars can also vary, cell wall polysaccharides have a variety of different structures.

B. Cellulose

Cellulose is a linear homopolymer of D-glucose linked β-(1→4). Its degree of polymerization is approximately 10,000 glucose units (Aspinall, 1982). Corn bran fiber has been reported to contain 23% cellulose (Sandstead et al, 1978). Burke et al (1974) determined the primary wall components of suspension-cultured monocotyledonous tissues, including ryegrass endosperm. Corn cell walls were not examined; however, in ryegrass endosperm, the amount of cellulose varied from 9 to 14% of the weight of the cell wall preparation, and corn presumably is similar.

C. Pentosans

Polymers of the five-carbon sugars, arabinose and xylose, are very important constituents of cell wall xylans (Aspinall, 1982). In general, xylans have a

TABLE III
Maize Kernel Metabolic Intermediates Associated with Nonaqueously Isolated Starch Granules[a]

Hexoses	Amount (nmol/mg of starch)	Per-centage[b]	Trioses	Amount (nmol/mg of starch)	Per-centage[b]
Glucose 1-phosphate	0.10	25	3-Phosphoglycerate	0.82	7
Glucose 6-phosphate	4.44	21	Dihydroxyacetone phosphate	0.20	27
Fructose 6-phosphate	0.93	27	Glyceraldehyde 3-phosphate	0.03	7
Fructose 1-6-bisphosphate	0.24	16	Phosphoenol pyruvate	0.34	19
			Pyruvate	0.11	14

[a] Adapted from Liu and Shannon (1981).
[b] Of total cellular constituents.

backbone of β-(1→4)-linked D-xylose. Highly branched arabinoxylans are a major component of the primary walls of suspension-cultured monocotyledonous cells (Burke et al, 1974) and cereal endosperms. Up to 30–40% of the arabinoxylan is composed of (1→3)-linked L-arabinose on the xylose backbone. Xylose and arabinose account for 90–95% of corn seed hemicellulose (Oomiya and Imazoto, 1982). Maize pericarp hemicellulose contains 54% xylose, 33% arabinose, 11% galactose, and 3% glucuronic acid (Wolf et al, 1953; Whistler and BeMiller, 1956).

D. Other Cell Wall Components

The primary cell walls of cultured monocotyledonous cells contain arabinogalactans, xyloglucans, and one or more arabinans, either as separate wall polymers or as side chains on other polymers. The presence of uronic acids and rhamnose in the walls of cultured monocotyledonous cells also suggests that the walls contain pectic polysaccharides. Although 1% or more of a hydroxyproline-rich glycoproline appears to be characteristic of the primary cell walls of dicotyledons, the primary cell walls of suspension-cultured ryegrass endosperm cells contained little hydroxyproline (Burke et al, 1974).

V. COMPLEX CARBOHYDRATES—STORAGE

Starch is the ubiquitous energy-storage polysaccharide of higher plants. Water-soluble polysaccharides, which are related to starch, also occur in high concentrations in some maize genotypes. Although great progress has been made toward understanding the structure, synthesis, and regulation of starch (and water-soluble polysaccharides), many unanswered questions remain. This section describes maize storage polysaccharides and points out unanswered questions that remain.

A. Water-Soluble Polysaccharides

Nonmutant maize endosperms contain about 2% water-soluble polysaccharides during development (Creech, 1965). In contrast, water-soluble polysaccharides account for up to 35% of the dry weight of standard sweet corn varieties (su).

Many researchers have analyzed the characteristics of the water-soluble polysaccharides from sweet corn (Morris and Morris, 1939; Sumner and Somers, 1944). These studies all conclude that the water-soluble fraction contains a highly branched polysaccharide, named phytoglycogen (Greenwood and DasGupta, 1958). Phytoglycogen is an α-(1→4)-glucan with α-(1→6) branch points, similar in structure to animal glycogens. The α-(1→4) unit chains have an average length of 10–14 glucose molecules, and outer chains range from six to 30 glucose units (Marshall and Whelan, 1974). Not all phytoglycogen is truly water-soluble. Differential centrifugation has demonstrated the presence of phytoglycogen in small particles with amylose and amylopectin (Matheson, 1975; Boyer et al, 1981). Phytoglycogen is synthesized and accumulates in the amyloplast (Boyer et al, 1977), but the precise pathway of biosynthesis remains uncertain. Erlander (1958) suggested that phytoglycogen is the precursor of the starch polymer amylopectin, which is formed by debranching phytoglycogen. Pan and Nelson (1984) reported that *su* endosperm is deficient in one of three isozymes of debranching enzyme and has reduced activity of the others, which could explain its accumulation of phytoglycogen. However, fine-structure analysis of amylopectin and phytoglycogen has shown that simple debranching of phytoglycogen cannot yield amylopectin (Marshall and Whelan, 1974). In addition, anatomical studies have shown that in *su* endosperm, starch granules are formed first and then disintegrate into small particles; finally, phytoglycogen is produced from these particles (Boyer et al, 1976). However, not all phytoglycogen is produced via the mobilization of preformed starch because both starch and phytoglycogen accumulate ^{14}C when *su* kernels are incubated in ^{14}C-glucose (Shannon, 1968b). Hence, our understanding of phytoglycogen biosynthesis is clearly not complete, and further study is needed.

B. Starch

Starch is the most abundant storage glucan in the world. Although it is made up of one sugar, glucose, and two different linkages, α-(1→4) and α-(1→6), the starch granule is composed of two glucan polymers, amylose and amylopectin.

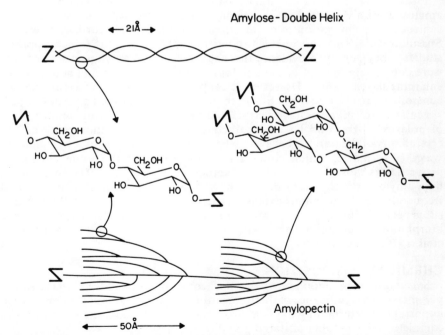

Fig. 3. Schematic representation of amylose and amylopectin molecules as they may exist in starch granules. Amylose molecules are represented as a double helix of two molecules. Single helixes probably exist as well. Amylopectin is represented as part of a growth ring of a starch granule. Individual branch chains of amylopectin molecules may also be found as double helixes. The center diagrams show the α(1→4) and α(1→6) linkages of the amylose and amylopectin molecules.

Amylose, which makes up 25–30% of the starch, is an essentially linear molecule of glucose units linked α-(1→4) (Fig. 3). Corn amylose has a degree of polymerization of 100–1,000 glucose units. In the starch granule, the conformation of amylose is unclear; however, random amylose chains and double and single helixes may all be present (French, 1984). Amylopectin, composing 70–75% of the starch, is a branched molecule with α-(1→6)-linked branch points and linear regions of α-(1→4)-linked glucose units (Fig. 3). The structure of amylopectin is complex; unit chains, linked 1→4, are of two lengths, 12–20 and 40–60 glucose units (Marshall and Whelan, 1974).

GRANULE STRUCTURE

Amylose and amylopectin in the endosperm cell are structurally arranged in an insoluble granule. The starch granule is formed inside the cellular organelle called the amyloplast (described below). Maize starch granules differ in size, and starch granules as large as 25 μm in diameter are found in mature seed. In general, maize starch granules are round (Fig. 2), but they take on polygonal shapes as the endosperm cells become packed with expanding starch granules.

Although composed of only two glucan polymers, the starch granule is a complex structure. It is organized into "growth rings," which are most clearly visualized after partial acid digestion (Fig. 4A). Within the growth rings, lamellae are found to be parallel to the ring (Fig. 4B).

Little is known of the coordinate synthesis and structural incorporation of amylose and amylopectin into the starch granule. Currently, synthesis is thought to proceed by elongation of chains already in the granule surface. However, Shannon et al (1970) were unable to confirm this using in vivo ^{14}C incorporation studies. They presented evidence that the amylose and amylopectin molecules were completely formed in the amyloplast matrix before being added to the enlarging starch granule. However, a clear picture of starch synthesis and factors controlling growth ring formation in the starch granule has not been developed.

Maize starch granules have a crystalline structure, demonstrate birefringence of polarized light, and have a characteristic X-ray diffraction pattern. The crystallinity of the starch is believed to be due to amylopectin. Starch granules of waxy (wx) maize, which contain no amylose, have the same crystalline properties as those from nonmutant kernels (Brown et al, 1971). In contrast, high-amylose starches (ae) do not have a high degree of crystallinity and frequently have amorphous extensions arising from a more crystalline "head" (Boyer et al, 1976, 1977), which results in granules with irregular shapes. Granule morphology of ae starch varies with the genetic background, however (Boyer et al, 1976; Yeh et al, 1981).

CHEMICAL COMPOSITION

Starch granules contain other chemical constituents in addition to the starch polymers. The significance of these constituents is not clear, but they may prove important in determining physical properties and may hold the key to a more complete understanding of starch granule biogenesis.

Lipids, often associated with endosperm starch, occur internally in all maize starch granules except wx (Morrison and Milligan, 1982). The work of Morrison (1981) cleared up considerable confusion about which lipids are truly starch-granule lipids. Morrison defined two types of lipids associated with starch: lipids on the starch surface and lipids within the granules. Starch surface lipids are adsorbed to the surface of the starch granule. These lipids include the amyloplast membrane lipids galactosyldiglycerides and diacylphospholipids (Fishwick and Wright, 1980). Treatments to remove the starch granule surface proteins also remove most of the adsorbed lipids. Other surface lipids, primarily monoacyl lipids, are more tightly absorbed and are retained even after all surface protein has been removed (Morrison, 1981).

Characterization of internal starch lipids requires the complete removal of surface lipids first. This is best accomplished by extraction of undamaged starch granules with water-saturated n-butanol for 10–30 min at room temperature. The internal starch lipids can then be extracted in hot (boiling water bath)

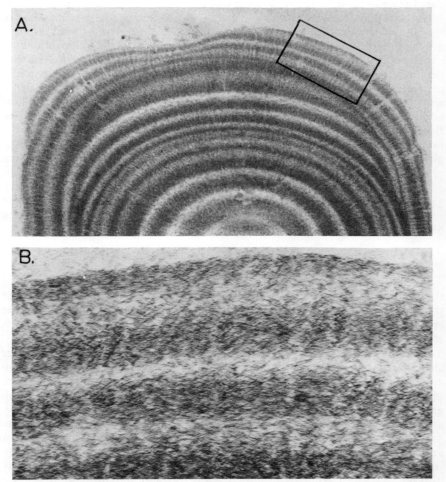

Fig. 4. A, electron micrograph of an acid-treated waxy maize starch granule, showing concentric "growth rings" (×14,000). B, enlargement of boxed area of A, showing stacks of lamellae perpendicular to growth rings (×85,000). (Courtesy K. C. Liu and C. D. Boyer)

water-saturated *n*-butanol in a nitrogen atmosphere or with hot *n*-propanol-water (Morrison and Milligan, 1982).

Morrison and Milligan (1982) concluded that maize starch granules contain free fatty acids and the three lysophospholipids: lysophosphatidylcholine, lysophosphatidylethanolamine, and lysophosphatidylglycerol. Localization of the internal starch lipids in the starch granule is not completely clear. However, most of these lipids are thought to be inclusion complexes with V-type, single-helix amylose in the native starch (Morrison and Milligan, 1982).

Proteins are also associated with the starch granule. As with the starch granule lipids, care must be taken to differentiate between starch granule-associated proteins adsorbed during granule purification and truly internal starch granule proteins. Treating purified starch granules with cold buffer containing denaturing reagents (urea or sodium dodecyl sulfate) is effective in removing surface proteins (Schwartz and Echt, 1982). Subsequent partial gelatinization of the starch granule in hot denaturing buffer yields extracts that contain polypeptides from the interior of starch granules. Because the polypeptides are denatured, no enzymatic activities can be detected. Electrophoretic analysis with denaturing conditions, however, demonstrates that four distinct polypeptides are present in the starch granule. The most abundant peptide has an approximate molecular weight of 60,000 (Schwartz and Echt, 1982). Three additional minor polypeptides are of higher molecular weights. Schwartz and Echt (1982) compared polypeptides from *wx* starch granules with those from nonmutant starch granules and demonstrated the absence of the major polypeptide in *wx* starch. Because *wx* endosperm lacks the major starch granule-bound enzyme, starch synthase (Nelson and Rines, 1962), the major polypeptide appears to be this enzyme. No roles in starch biosynthesis or starch granule structure have been suggested for the minor polypeptides. The possible role of these polypeptides, as well as of internal starch lipids, should continue to be a rewarding area of future study.

PHYSICAL PROPERTIES

Several physical properties of starch granules are important in determining the biological and economic value of the starch granule. Biologically, the dense starch granule provides a relatively unreactive form of energy storage, yet the enzymatic conversion of the starch granule to sugar during germination is efficient enough to provide the energy for subsequent seedling metabolism.

The physical properties of the starch granule are determined by the fine structure of the polysaccharides and the percentage distribution of amylose and amylopectin. Shannon and Garwood (1984) thoroughly reviewed the effects of maize genotype on starch granule properties. Important features are starch granule morphology, amylose content, crystallinity, gelatinization temperature, and digestibility. All of these characteristics are altered by different endosperm genotypes. Amylose content ranges from zero in *wx* to 70% in *ae*. The high-amylose starch granules of *ae* endosperm are generally less crystalline in structure than nonmutant and *wx* starch granules and have a weak B X-ray diffraction pattern, compared to the A X-ray diffraction pattern of nonmutant and *wx* starch granules. High-amylose starches gelatinize at higher temperatures and can be used to form gels and films. In contrast, *wx* starch granules are similar to nonmutant starch granules with respect to morphology, X-ray diffraction, and gelatinization temperature. However, wx starch granules are more rapidly digested by animal amylase (Fuwa et al, 1978) and thus have been used to improve feed digestibility. Continued genetic modification of the physical properties of starch seems certain in the future.

BIOSYNTHESIS

Although starch is relatively simple in structure, the mechanisms regulating the synthesis of starch molecules and the simultaneous incorporation of these molecules into the growing starch granule have not been adequately explained. Knowledge of certain features of the biosynthesis of starch has progressed to the point where a speculative but rational picture can be proposed. Three significant features of starch biosynthesis must be described: the enzymes in the biosynthetic pathway; the compartmentation of the enzymatic reactions in the cell and within the amyloplast; and the regulation and modification of this important process.

Pathway. Enzymes capable of synthesizing starch polysaccharides are well known and characterized. A simple description of starch polymerization includes four enzymatic reactions:

Reaction 1: ADPG pyrophosphorylase (EC 2.7.7.27)

$$ATP + G\text{-}1\text{-}P \rightleftharpoons ADPG + PPi$$

Reaction 2: Starch synthase (EC 2.4.1.21)

$$ADPG + \alpha\text{-}(1\rightarrow 4)\text{-glucan}_n \rightarrow ADP + \alpha\text{-}(1\rightarrow 4)\text{-glucan}_{n+1}$$

Reaction 3: Starch branching enzyme (EC 2.4.1.18)

$$\alpha\text{-}(1\rightarrow 4)\text{-glucan} \rightarrow \alpha\text{-}(1\rightarrow 4)\text{-glucan branched by a } (1\rightarrow 6) \text{ linkage}$$

Reaction 4: Starch phosphorylase (EC 2.4.1.1)

$$G\text{-}1\text{-}P + \alpha\text{-}(1\rightarrow 4)\text{-glucan}_n \rightarrow Pi + \alpha\text{-}(1\rightarrow 4)\text{-glucan}_{n+1}$$

where ADP = adenosine diphosphate, ADPG = adenosine diphosphoglucose, ATP = adenosine triphosphate, G-1-P = glucose-1-phosphate, PPi = inorganic pyrophosphate, and Pi = inorganic phosphate.

Reaction 1 produces the substrate, ADPG, for starch synthase, and reactions 2–4 produce the starch polymers amylose and amylopectin. Polymerization of α-(1→4)-glucan chains can be catalyzed by starch synthase (reaction 2) or phosphorylase (reaction 4). Based on metabolite compartmentation and concentration and the known enzyme kinetics, it seems unlikely that phosphorylase (reaction 4) is active in starch synthesis in maize endosperm (Liu and Shannon, 1981).

ADPG pyrophosphorylase (reaction 1) is an allosteric enzyme that provides the substrate, ADPG, for starch synthase. The activity of ADPG pyrophosphorylases from higher plants is inhibited by inorganic phosphate and is activated largely by 3-phosphoglycerate and to a lesser extent by the glycolytic intermediates phosphoenolpyruvate, fructose bisphosphate, and fructose-6-P (Preiss et al, 1967; Sanwal et al, 1968). ADPG pyrophosphorylases from nonphotosynthetic tissues have the same allosteric activators and inhibitors, but the regulation is not as sensitive as in leaf enzymes (Sanwal et al, 1968; Dickinson and Preiss, 1969; Preiss et al, 1971). The maize endosperm enzyme was activated 50% of maximum ($A_{0.5}$) by 2.2 mM 3-phosphoglycerate and showed 50% inhibition of activity by 3-mM phosphate (Dickinson and Preiss, 1969). The maize enzyme has proved difficult to purify because it is highly unstable. However, partial purification of the enzyme from maize endosperm using protamine sulfate precipitation showed the presence of two enzyme fractions (Hannah and Nelson, 1975), which differ in native molecular size and heat lability (Hannah et al, 1980). A complete understanding of the maize endosperm ADPG pyrophosphorylases will require further purification of the enzymes and continued characterization.

Starch synthase (reaction 2) transfers the glucose unit from ADPG to the nonreducing end of a growing α-(1→4)-glucan via a (1→4) bond. Starch synthase is found in two distinct physical forms, one soluble and one tightly bound to the growing starch granule. Uridine diphosphoglucose (UDPG), as well as ADPG, can serve as a substrate for the granule-bound starch synthase, but ADPG appears to be the preferred substrate (Recondo and Leloir, 1961). A K_M of 2.85 mM for ADPG was observed for the enzyme from maize endosperm (Tsai, 1974). Early attempts to free this activity from the starch granule were generally unsuccessful. However, Macdonald and Preiss (1983) recently used an α-amylase and glucoamylase treatment of damaged starch granules to release the granule-bound starch synthase from maize starch. Once starch synthase was released, these researchers were able to measure its activity in the presence of added amylase inhibitors. The released starch synthase was separated into two enzyme forms by diethylaminoethyl (DEAE)-sepharose chromatography. As noted above, the granule-bound starch synthase can transfer glucose from both ADPG and UDPG. After solubilization, however, neither solubilized fraction could use UDPG as a substrate.

Starch synthase activity also exists in a soluble form. Unlike the granule-bound activity, the soluble synthase activity has a strict requirement for ADPG (Tsai, 1974). This enzyme from maize endosperm (Ozbun et al, 1971; Boyer and Preiss, 1978) can be separated into two forms by DEAE-cellulose chromatography. Only one of the two fractions showed unprimed activity when assayed in the presence of 0.5M citrate and without added α-(1→4)-glucan primer. Thus two different forms of soluble starch synthase also exist.

Starch branching enzyme (reaction 3) catalyzes the formation of the branch points in amylopectin. The reaction occurs by the cleavage of a (1→4) bond and transfer of the resulting α-(1→4)-glucan to the same or a different α-(1→4)-glucan via the formation of a (1→6) bond (Drummond et al, 1972). Most commonly, branching enzyme activity is measured by the branching of amylose, which is followed by the decrease in iodine-amylose absorbance at 660 nm and the shift in the wavelength maximum of the iodine-amylose complex (Drummond et al, 1972). This assay, however, is insensitive, difficult to interpret, and not quantitative. A second assay, which measures the stimulation of α-(1→4)-glucan synthesis from glucose-1-P by added phosphorylase-a (Brown and Brown, 1966; Hawker et al, 1974), is more sensitive and semiquantitative. By using both assays, Boyer and Preiss (1978) identified three forms of branching enzyme from developing maize kernels.

Recently, Pan and Nelson (1984) reported the separation of debranching enzyme from nonmutant maize into three forms. The significance of debranching enzymes in starch biosynthesis is unknown at this time, but Pan and Nelson (1984) speculated that "the amylopectin component of starch is the result of an equilibrium between branching enzyme activity and debranching enzyme activity."

Clearly, the starch biosynthetic pathway is complicated by the presence of multiple enzyme forms. Although the multiple forms of the enzymes can be characterized in vitro, little is known about the specific roles these enzymes play in amylose or amylopectin biosynthesis and incorporation into the starch granule. Note, however, that certain genetic mutants are associated with the reduction or loss in activity of particular isozymes (see Shannon and Garwood [1984] for a review). The specific roles these enzymes play in determining starch granule structure remain to be elucidated.

Cellular Compartmentation. Most pathways for the conversion of sucrose (or glucose and fructose) to starch in the developing endosperm do not consider the compartmentation of starch synthesis in the amyloplast. Liu and Shannon (1981) inferred the compartmentation of the enzymes responsible for the conversion of sucrose to starch in maize endosperm, based on the metabolites associated with nonaqueously isolated starch granules. Recent studies by Macdonald and ap Rees (1983) of enzymes in amyloplasts isolated from suspension cultures of soybean support most of the conclusions drawn by Liu and Shannon (1981).

Macdonald and ap Rees (1983) suggested the following distribution of enzymes: ADPG pyrophosphorylase and starch synthase are confined to the amyloplasts; UDPG pyrophosphorylase, alkaline invertase, and sucrose synthase are exclusively cytosolic; and glyceraldehyde-3-phosphate dehydrogenase, triose-phosphate isomerase, aldolase, fructose-1-6-bisphosphatase, glucose phosphate isomerase, and phosphoglucomutase, although predominantly cytosolic, are also present in the amyloplast. Using aqueously isolated amyloplasts from developing maize endosperm (Echeverria et al, 1985, 1987), we have found some variations from the results of Macdonald and ap Rees (1983). First, the enzyme UDPG pyrophosphorylase was present in the maize amyloplast as well as in the cytosol. Furthermore, its activity in the amyloplast was four times that of ADPG pyrophosphorylase. The significance of this is not clear, but we cannot rule out the involvement of UDPG as a glucosyl donor for starch synthesis during early stages in endosperm development. Based on the compartmentation of metabolic intermediates reported by Liu and Shannon (1981) and the enzyme studies of Macdonald and ap Rees (1983) and Echeverria et al (1985), we propose the pathway shown in Fig. 5. Echeverria et al (1987) demonstrated that intact isolated amyloplasts incorporate exogenously supplied triose-P into starch. The hexoses, glucose and fructose, were not incorporated into starch in similar experiments. Therefore, triose-P appears to be the preferred substrate transported across the amyloplast membrane. However, this pathway should not be considered confirmed without a rigorous study of the transport capabilities of the amyloplast membrane. Such studies are currently being conducted in our laboratory.

Regulation and Limitations. Many stages in the life cycle of maize are critical to the ultimate production of grain carbohydrates and total yield. Likewise, many physiological processes are important in seed development, including photosynthesis, carbohydrate metabolism in leaves, translocation, assimilate partitioning between competing sinks, and carbohydrate metabolism within the kernels. It is not unreasonable to assume that any of these processes might be the first rate-limiting process and that different processes are rate-limiting in different maize genotypes (Shannon, 1982). Within the maize kernel, several possible rate-limiting steps need to be further characterized to more fully understand the limitations to carbohydrate accumulation and yield. Is the uptake of sugar by the BETC an active or a passive process, and is this rate-limiting? Are sugar movement and availability through the endosperm saturating? Is the movement of substrates into the amyloplast energy-dependent or rate-limiting? Do the allosteric properties of ADPG pyrophosphorylase regulate starch synthesis? What roles do the multiple enzyme forms of the starch-synthetic enzymes play in starch synthesis and starch granule structure?

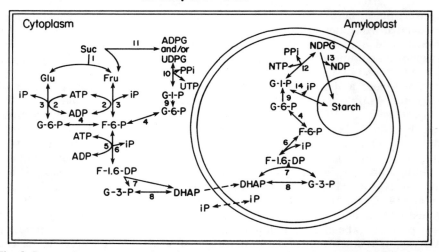

Fig. 5. Proposed compartmentation of enzymes and metabolites of starch biosynthesis in the developing maize endosperm cell. Enzymes are indicated by number: 1 = invertase, 2 = hexokinase, 3 = hexose-6-phosphatase, 4 = glucose phosphate isomerase, 5 = 6-phosphofructokinase, 6 = fructose-1-6 bisphosphatase, 7 = fructose bisphosphate aldolase, 8 = triosephosphate isomerase, 9 = phosphoglucomutase, 10 = uridine diphosphoglucose (UDPG) pyrophosphorylase, 11 = sucrose synthase, 12 = adenosine diphosphoglucose (ADPG) pyrophosphorylase, 13 = starch synthase, 14 = starch phosphorylase.

Intermediate products are: inorganic pyrophosphate (PPi), inorganic phosphate (iP), adenosine diphosphate (ADP), adenosine triphosphate (ATP), uridine triphosphate (UTP), nucleotide diphosphate (NDP), nucleotide triphosphate (NTP), nucleotide diphosphoglucose (NDPG), dihydroxyacetone phosphate (DHAP), fructose-6-phosphate (F-6-P), fructose-1,6-bisphosphate (F-1,6-P), glyceraldehyde-3-phosphate (G-3-P), glucose-6-phosphate (G-6-P).

The answers to these questions should provide further insights into the ultimate limitations on carbohydrate accumulation in maize kernels.

VI. SUMMARY

The maize kernel is a rich source of a variety of carbohydrates. The continued study of carbohydrate accumulation and metabolism in developing maize kernels should prove both rewarding in a practical sense and intellectually stimulating. The wealth of genetic mutants of the endosperm will continue to provide tools for studies of synthesis and polysaccharide structure. In addition, both nonmutant and genetically modified starches can be expected to provide an increasing array of polysaccharides or modified polysaccharides for food, feed, and industrial use.

ACKNOWLEDGMENT

This chapter is Contribution 59 of the Department of Horticulture, The Pennsylvania State University, and was authorized for publication as Publication 7082 in the journal series of the Pennsylvania Agricultural Experiment Station.

LITERATURE CITED

ASPINALL, G. O. 1982. Chemistry of cell wall polysaccharides. Page 473-500 in: The Biochemistry of Plants, Vol. 3. J. Preiss, ed. Academic Press, New York.

BOYER, C. D. 1985. Synthesis and breakdown of starch. Pages 133-153 in: The Biochemical Basis of Plant Breeding, Vol 1. C. A. Neyra, ed. CRC Press, Boca Raton, FL.

BOYER, C. D., and PREISS, J. 1978. Multiple forms of $(1\rightarrow 4)$-α-D-glucan, $(1\rightarrow 4)$-α-D-glucan-6-glycosyl transferase from developing Zea mays L. kernels. Carbohydr. Res. 61:321-334.

BOYER, C. D., and SHANNON, J. C. 1983. The use of endosperm genes for sweet corn improvement. Pages 139-161 in: Plant Breeding Reviews, Vol. 1. J. Janick, ed. Avi Publishing Co., Westport, CT.

BOYER, C. D., DANIELS, R. R., and SHANNON, J. C. 1976. Abnormal starch granule formation in Zea mays L. endosperms possessing the amylose-extender mutant. Crop Sci. 16:298-301.

BOYER, C. D., DANIELS, R. R., and SHANNON, J. C. 1977. Starch granule (amyloplast) development in endosperm of several Zea mays L. genotypes affecting kernel polysaccharides. Am. J. Bot. 64:50-56.

BOYER, C. D., DAMEWOOD, P. A., and SIMPSON, E. K. G. 1981. The possible relationship of starch and phytoglycogen in sweet corn. I. Characterization of particulate and soluble polysaccharides. Staerke 33:125-130.

BROWN, B. I., and BROWN, D. H. 1966. α-1,4-Glucan: α-1,4-glucan 6-glycosyltransferase for mammalian muscle. Pages 395-403 in: Methods of Enzymology, Vol. 8. E. F. Neufeld and V. Ginsburg, eds. Academic Press, New York.

BROWN, R. P., CREECH, R. G., and JOHNSON, L. J. 1971. Genetic control of starch granule morphology and physical structure in developing maize endosperms. Crop Sci. 11:297-302.

BURKE, D., KAUFMAN, P., McNEIL, M., and ALBERSHIEM, P. 1974. The structure of plant cell walls. VI. A survey of the walls of suspension-cultured monocots. Plant Physiol. 54:109-115.

CAREY, E. E., DICKINSON, D. B., WEI, L. Y., and RHODES, A. M. 1982a. Occurrence of sorbitol in Zea mays. Phytochemistry 21:1909-1911.

CAREY, E. E., RHODES, A. M., and DICKINSON, D. B. 1982b. Postharvest levels of sugars and sorbitol in sugary enhancer (su se) and sugary (su Se) maize. HortScience 17:241-242.

CREECH, R. G. 1965. Genetic control of carbohydrate synthesis in maize endosperm. Genetics 52:1175-1186.

DICKINSON, D. B., and PREISS, J. 1969. ADP-glucose pyrophosphorylase from maize endosperm. Arch. Biochem. Biophys. 130:119-128.

DRUMMOND, G. S., SMITH, E. E., and WHELAN, W. J. 1972. Purification and properties of potato α-1,4-glucan, α-1,4-glucan-6-glycosyltransferase (Q enzyme). Eur. J. Biochem. 26:168-176.

ECHEVERRIA, E., BOYER, C., LIU, K.-C., and SHANNON, J. 1985. Isolation of amyloplasts from developing maize endosperm. Plant Physiol. 77:513-519.

ECHEVERRIA, E., BOYER, C., and SHANNON, J. 1987. Regulatory aspects of starch biosynthesis in developing maize kernels. In: Recent Advances in Frontier Areas of Plant Biochemistry. R. Singh, ed. Indian Soc. Plant Physiol. Biochem. In press.

ERLANDER, S. R. 1958. A proposed mechanism for the synthesis of starch from glycogen. Enzymologia 19:273-283.

ESAU, K. 1960. Anatomy of Seed Plants. John Wiley & Sons, New York.

FEINGOLD, D. S., and AVIGAD, G. 1980. Sugar nucleotide transformations. Pages 101-170 in: The Biochemistry of Plants, Vol. 3. J. Preiss, ed. Academic Press, New York.

FELKER, F. C., and SHANNON, J. C. 1980. Movement of ^{14}C-labeled assimilates into kernels of Zea mays L. III. An anatomical examination and microautoradiographic study of assimilate transfer. Plant Physiol. 65:864-870.

FISHWICK, M. J., and WRIGHT, A. J. 1980. Isolation and characterization of amyloplast envelope membrane from Solanum tuberosum. Phytochemistry 19:55-59.

FRENCH, D. 1984. Organization of starch granules. Pages 183-247 in: Starch: Chemistry and Technology. R. L. Whistler, J. N. BeMiller, and E. F. Paschall, eds. Academic Press, Orlando, FL.

FUWA, H., GLOVER, D. V., NISHIMURA, R., and TANAKA, M. 1978. Comparative susceptibility to amylases of starch granules of several single endosperm mutants and their double-mutant combinations with opaque-2 in four inbred lines of maize. Staerke 30:367-371.

GALINAT, W. C. 1979. On the usage of the terms pedicel and rachilla in description of the cob, the female spikelet and the grain in maize. Maize Genet. Coop. Newsl. 53:100.

GENTINETTA, E., and SALAMINI, F. 1979. Free sugar fraction of the amylose-related mutants of maize. Biochem. Genet. 17:405-414.

GREENWOOD, C. T., and DASGUPTA, P. C. 1958. Physiochemical studies on starches. Part XI. The granular starch of sweet corn, Zea mays. J. Chem. Soc. 707-710.

GUNNING, B. E. S., and PATE, J. S. 1969. "Transfer cells." Plant cells with wall ingrowths specialized in relation to short distance transport of solutes. Their occurrence, structure, and development. Protoplasma 68:107-133.

HANNAH, L. C., and NELSON, O. E. 1975. Characterization of adenosine diphosphate glucose pyrophosphorylase from developing maize seeds. Plant Physiol. 55:297-302.

HANNAH, L. C., TUSCHALL, D. M., and MANS, R. J. 1980. Multiple forms of maize endosperm ADP-glucose pyrophosphorylase and their control by shrunken-2 and brittle-2. Genetics 95:961-970.

HAWKER, J. S., OZAKI, H., GREENBERG, E., and PREISS, J. 1974. Interaction of spinach leaf adenosine diphosphate glucose α-1,4-glucan α-4-glucosyl transferase and α-1,4-glucan, α-1,4-glucan-6-glycosyl transferase in synthesis of branched glucan. Arch. Biochem. Biophys. 160:530-551.

HOLDER, D. G., GLOVER, D. V., and SHANNON, J. C. 1974. Interaction of shrunken-2 with five other carbohydrate genes of corn endosperm. Crop Sci. 14:643-646.

INGLETT, G. E. 1970. Kernel structure, composition and quality. Pages 123-137 in: Corn: Culture, Processing, Products. G. E. Inglett, ed. Avi Publishing Co., Westport, CT.

JOHNSON, L. F., and TATE, M. E. 1969. Structure of "phytic acids." Can. J. Chem. 47:63-73.

KIES, C., BALTERS, S., KAN, S., LO, B., WESTRING, M. E., and FOX, H. M. 1982. Impact of corn bran on the nutritional status on gastrointestinal tract function of humans. Pages 33-43 in: Maize: Recent Progress in Chemistry and Technology. G. E. Inglett, ed. Academic Press, New York.

KIESSELBACH, T. A., and WALKER, E. R. 1952. Structure of certain specialized tissues in the kernel of corn. Am. J. Bot. 39:561-569.

LIU, T. T., and SHANNON, J. C. 1981. Measurement of metabolites associated with nonaqueously isolated starch granules from immature Zea mays L. endosperm. Plant Physiol. 67:525-529.

MACDONALD, F. D., and AP REES, T. 1983. Enzymic properties of amyloplasts from suspension cultures of soybean. Biochim. Biophys. Acta 755:81-89.

MACDONALD, F. D., and PREISS, J. 1983. Solubilization of the starch-granule-bound starch synthase of normal maize kernels. Plant Physiol. 73:175-178.

MAJUMDER, A. L., and BISWAS, B. B. 1973. Further characterization of phosphoinositol kinase isolated from germinating mung bean

seeds. Phytochemistry 12:315-319.
MARSHALL, J. J., and WHELAN, W. J. 1974. Multiple branching in glycogen and amylopectin. Arch. Biochem. Biophys. 161:234-238.
MATHESON, N. K. 1975. The (1-4)(1-6) glucans from sweet and normal corns. Phytochemistry 14:2017-2021.
MORRIS, D. F., and MORRIS, C. T. 1939. Glycogen in the seed of *Zea mays* (variety Golden Bantam). J. Biol. Chem. 130:535-544.
MORRISON, W. R. 1981. Starch lipids: A reappraisal. Staerke 33:408-411.
MORRISON, W. R., and MILLIGAN, T. P. 1982. Lipids in maize starch. Pages 1-18 in: Maize: Recent Progress in Chemistry and Technology. G. E. Inglett, ed. Academic Press, New York.
NELSON, O. E. 1980. Genetic control of polysaccharide and storage protein synthesis in the endosperms of barley, maize, and sorghum. Pages 41-71 in: Advances in Cereal Science and Technology, Vol. 3. Y. Pomeranz, ed. Am. Assoc. Cereal Chem., St. Paul, MN.
NELSON, O. E., and RINES, H. W. 1962. The enzymatic deficiency in the waxy mutant of maize. Biochem. Biophys. Res. Commun. 9:297-300.
O'DELL, B. L. 1969. Effect of dietary components upon zinc availability. Am. J. Clin. Nutr. 22:1315-1322.
O'DELL, B. L., BOLAND, A. R., and KOIRTYOHANN, S. R. 1972. Distribution of phytate and nutritionally important elements among the morphological components of cereal grains. J. Agric. Food Chem. 20:718-721.
OOMIYA, M., and IMAZATO, S. 1982. Corn seed hemicellulose. Pages 19-32 in: Maize: Recent Progress in Chemistry and Technology. G. E. Inglett, ed. Academic Press, New York.
OZBUN, J. L., HAWKER, J. S., and PREISS, J. 1971. Adenosine diphosphoglucose-starch glucosyltransferases from developing kernels of waxy maize. Plant Physiol. 48:765-769.
PAN, D., and NELSON, O. E. 1984. A debranching enzyme deficiency in endosperms of the *sugary-1* mutants of maize. Plant Physiol. 74:324-328.
PERNOLLET, J.-C. 1978. Protein bodies of seeds: Ultrastructure, biochemistry, biosynthesis and degradation. Phytochemistry 17:1473-1480.
PORTER, G. A., KNIEVEL, D. P., and SHANNON, J. C. 1985. Sugar efflux from maize (*Zea mays* L.) pedicel tissue. Plant Physiol. 77:524-531.
PREISS, J., and LEVI, C. 1980. Starch biosynthesis and degradation. Pages 371-423 in: The Biochemistry of Plants, Vol. 3. J. Preiss, ed. Academic Press, New York.
PREISS, J., GHOSH, H., and WITTKOP, J. 1967. Regulation of the biosynthesis of starch in spinach leaf chloroplasts. Pages 131-153 in: The Biochemistry of Chloroplasts, Vol. 2. T. W. Goodwin, ed. Academic Press, New York.
PREISS, J., LAMMEL, C., and SABRAW, A. 1971. A unique adenosine diphosphoglucose pyrophosphorylase associated with maize embryo tissue. Plant Physiol. 47:104-108.
RECONDO, E., and LELOIR, L. F. 1961. Adenosine diphosphate glucose and starch synthesis. Biochem. Biophys. Res. Commun. 6:85-88.
RUTENBERG, M. W., and SOLAREK, D. 1984. Starch derivatives: Production and uses. Pages 311-388 in: Starch: Chemistry and Technology. R. L. Whistler, J. N. BeMiller, and E. F. Paschall, eds. Academic Press, Orlando, FL.
SANDSTEAD, H. H., MUNOZ, J. M., JACOB, R. A., KLEVAY, L. M., RECK, S. J., LOGAN, G. M., DINTZIS, F. R., INGLETT, G. E., and SHUEY, W. C. 1978. Influence of dietary fiber on trace element balance. Am. J. Clin. Nutr. 31:S180.
SANWAL, G. G., GREENBERG, E., HARDIE, J., CAMERON, E. D., and PREISS, J. 1968. Regulation of starch biosynthesis in plant leaves: Activation and inhibition of ADP-glucose pyrophosphorylase. Plant Physiol. 43:417-427.
SCHWARTZ, D., and ECHT, C. S. 1982. The effect of AC dosage on the production of multiple forms of *Wx* protein by the *Wx*-m-9 controlling element mutation in maize. Mol. Gen. Genet. 187:410-413.
SHANNON, J. C. 1968a. Carbon-14 distribution in carbohydrates of immature *Zea mays* kernels following $^{14}CO_2$ treatment of intact plants. Plant Physiol. 43:1215-1220.
SHANNON, J. C. 1968b. A procedure for the extraction and fractionation of carbohydrates from immature *Zea mays* kernels. Res. Bull. 842. Purdue Univ. Agric. Exp. Stn., West Lafayette, IN.
SHANNON, J. C. 1972. Movement of ^{14}C-labeled assimilates into kernels of *Zea mays* L. I. Pattern and rate of sugar movement. Plant Physiol. 49:198-202.
SHANNON, J. C. 1982. A search for rate-limiting enzymes that control crop production. Iowa State J. Res. 56:307-322.
SHANNON, J. C., and DOUGHERTY, C. T. 1972. Movement of ^{14}C-labeled assimilates into kernels of *Zea mays* L. II. Invertase activity of the pedicel and placento-chalazal

tissues. Plant Physiol. 49:203-206.

SHANNON, J. C., and GARWOOD, D. L. 1984. Genetics and physiology of starch development. Pages 25-80 in: Starch: Chemistry and Technology. R. L. Whistler, J. N. BeMiller, and E. F. Paschall, eds. Academic Press, Orlando, FL.

SHANNON, J. C., CREECH, R. G., and LOERCH, J. D. 1970. Starch synthesis studies in *Zea mays*. II. Molecular distribution of radioactivity in starch. Plant Physiol. 45:163-168.

SUMNER, J. B., and SOMERS, G. F. 1944. The water-soluble polysaccharides of sweet corn. Arch. Biochem. 4:7-9.

TSAI, C. Y. 1974. The function of the *waxy* locus in starch synthesis in maize endosperm. Biochem. Genet. 11:83-96.

TSAI, C. Y., SALAMINI, F., and NELSON, O. E. 1970. Enzymes of carbohydrate metabolism in developing endosperm of maize. Plant Physiol. 46:299-306.

WHISTLER, R. L., and BeMILLER, J. N. 1956. Hydrolysis components from methylated corn fiber gum. J. Am. Chem. Soc. 78:1163-1165.

WOLF, M. J., MacMASTERS, M. M., CANNON, J. A., ROSEWALL, E. C., and RIST, C. E. 1953. Preparation and some properties of hemicelluloses from corn hulls. Cereal Chem. 30:451-470.

YEH, J. Y., GARWOOD, D. L., and SHANNON, J. C. 1981. Characterization of starch from maize endosperm mutants. Staerke 33:222-230.

CHAPTER 9

PROTEINS OF THE KERNEL

CURTIS M. WILSON[1]
U.S. Department of Agriculture
Agricultural Research Service
Department of Agronomy
University of Illinois
Urbana, Illinois

I. INTRODUCTION

Corn is often considered only as a major source of calories, mostly derived from its high starch content. Yet the sheer bulk of corn consumed as animal feed in the United States and Europe and as human food in many areas around the world makes it essential to consider the protein supplied with the calories (see Chapter 15). The combination of corn protein and protein from soybean or other legumes produces a diet that is more nutritionally balanced, in terms of amino acid composition, than protein from either source alone. If the protein content of the U.S. 1984/1985 corn crop of 194.9×10^6 t (Chapter 7, Table II) is calculated at 8.3% protein (15.5% moisture), the total protein production was 16.18×10^6 t. Similar calculations for the soybean crop (USDA, 1986) of 55.9 $\times 10^6$ t (35% protein at 13% moisture) give a protein yield of 17.73×10^6 t. The 1983/1984 worldwide protein production was 28.7×10^6 t for corn and 28.8×10^6 t for soybeans (USDA, 1985). Thus, corn is an important source of protein.

Twenty years ago, kernels of the mutant opaque-2 (o_2) were discovered to possess below-normal levels of zein and above-normal levels of lysine (Mertz et al, 1964). Unfortunately, agronomic qualities such as yield, disease resistance, and kernel strength were adversely affected, so that relatively small amounts of o_2 corn are grown now. Much research has been done, and hopes are still high that corn can be converted into an agronomically acceptable crop with at least moderate overall protein quality (Cimmyt-Purdue, 1975). Long-term selection for high protein (not high-quality protein) produced a line of corn with 25% protein (Dudley et al, 1977). However, the protein has low nutritional value and the yields are low, so this line of work has also been disappointing.

The past decade has seen a burgeoning of interest in the biochemistry of corn (and other seed) proteins. These proteins have begun to yield their secrets as

---

[1] Present address: USDA/ARS, Northern Regional Research Center, Peoria, IL.

newer, more precise procedures separate and give identity to previously intractable materials. Current reviews and books concerned with corn protein, as well as seed protein in general, are appearing at a rapid rate (Wall and Paulis, 1978; Larkins, 1981; Daussant et al, 1983; Wilson, 1983; Shewry and Miflin, 1985). Meanwhile, the new techniques of molecular genetics, complementary DNA (cDNA) cloning, recombinant DNA, DNA hybridization, restriction enzyme digestion, in vitro protein synthesis, etc., are producing new insights into storage protein structure and synthesis and new tools for corn improvement through breeding (Nelson, 1980; Messing, 1983; Larkins et al, 1984).

This review attempts to bring together the results of traditional biochemistry, exemplified by the Osborne (1924) classification of proteins based on solubility, with the newer techniques of protein identification based on gel electrophoresis, subcellular compartmentalization, genetics, amino acid analysis, and recombinant DNA. These new approaches to seed proteins do not fit neatly into the old concepts. It has become obvious that the Osborne fractions are composed of mixtures of proteins and that the dividing lines are not as sharp as once assumed. The Osborne fractions, however, are firmly rooted into our ideas about corn protein and do provide a framework for the examination of these proteins.

II. PROTEIN FRACTIONATION OVERVIEW

Seventy years ago, Osborne and Mendel (1914) were able to conclude that zein was the protein that limited the ability of corn to supply nutritionally essential amino acids and that substitution of other corn proteins for zein would improve the nutritional value of corn. This problem can be illustrated by the data of Table I, showing the essential amino acid composition of a recommended diet compared to those of corn germ flour, nonzein endosperm proteins, and zein.

TABLE I
Essential Amino Acids (mol %) in Corn Protein Classes as Compared with the FAO/WHO Human Nutritional Requirements

Amino Acid	1973 FAO Provisional Pattern[a]	Nonzein Endosperm Proteins[b]	Zein[c]	Germ Flour[d]
Lysine	4.3	4.7	0.1	4.6
Threonine	3.9	4.9	3.0	4.4
Valine	4.2	7.1	3.6	6.6
Cysteine + methionine	3.0	4.5	1.9	3.2
Isoleucine	3.5	4.0	3.8	3.2
Leucine	6.2	9.0	18.7	7.7
Phenylalanine + tyrosine	4.0	6.2	8.7	5.6
Tryptophan	0.5	0.6	0	1.0
Total	29.6	41.0	39.8	36.3

[a] Calculated from FAO/WHO (1973).
[b] Weighted average for reduced soluble protein (Table III), albumin, globulin, and glutelin (Table V).
[c] From Table III (A- + B-zeins).
[d] From Table V.

Only zein falls short, for it lacks tryptophan and lysine and is low in threonine, valine, and the sulfur amino acids. The high total of essential amino acids in zein does not signify high quality, for the excess of leucine may antagonize the utilization of isoleucine.

The classical separation of proteins (measured as N) into fractions based on solubility is presented in Table II, for whole kernels, endosperm, and germ. The original Osborne classes are albumins (water-soluble), globulins (salt-soluble), zein or prolamin (aqueous alcohol-soluble), and glutelins and residue (not soluble in water, saline solutions, or aqueous alcohol), with the glutelins being the proteins extracted by dilute alkali. A major improvement was introduced by Landry and Moureaux (1970), whose use of the reducing agent 2-mercaptoethanol (ME) to improve the extractability of the glutelins (and some zeins) was widely adopted. The reports of several laboratories are gathered into one set of figures (the B columns), with the five major fractions of the Landry-Moureaux (LM) system being identified by Roman numerals. The salt-soluble N fraction (LM-I) is frequently subdivided into nonprotein N, albumins, and globulins. The LM-II fraction of modern data, here called zein-I, corresponds

TABLE II
Nitrogen Distribution in Classical and Landry-Moureaux (LM) Protein Fractions from Whole Kernel, Endosperm, and Germ (% of total N within each part)

Protein Fraction	LM Fraction[a]	Whole Kernel A[b]	Whole Kernel B	Endosperm A	Endosperm B	Germ A	Germ B
Nonprotein N			6		3		20
Albumin			7		3		35
Globulin			5		3		18
Total salt-soluble N	I—0.5M NaCl	22	18	7.8	9	77.2	73
Zein-I	II—55% isopropanol	41	42	50	48	2.0	4
Zein-II	III—as II + 0.6% ME		10		12		1
Prolamin (total zein)			52		60		5
Glutelin-2	IV—pH 10 buffer + 0.6% ME		8		9		3
Glutelin-3	V—as IV + 0.5% SDS		17		17		15
Total glutelin[c]		31	25	38.2	26	0.6	18
Residue		6	5	4.0	5	20.2	4
Total protein (% of whole kernel)[d]			100		78		18
Dry weight (% of whole kernel)[d]			100		80		13

[a] One set of solvents used by Landry and Moureaux (1970, 1980, 1981). Fraction III is also known as glutelin-1 or alcohol-soluble glutelin. Differences in alcohols, salts, concentrations, volumes, temperature, etc., give some changes in N recoveries, as will be noted in various reports on the use of the "Landry-Moureaux" system. SDS = sodium dodecyl sulfate, ME = mercaptoethanol.
[b] A, From Osborne and Mendel (1914); B, from Wilson (1983), a consensus of recent literature.
[c] Osborne-Mendel glutelin would be equivalent to Landry-Moureaux fractions III + IV + V.
[d] The hull, tip cap, etc., contain about 4% of the N and 7% of the dry weight.

well to that reported by Osborne and Mendel as zein; it contains about half of the endosperm N. LM-III and LM-IV are subdivisions of the classical glutelin fraction, but contain little or no lysine. LM-III contains variable amounts of zein-I not previously extracted, giving rise to the term zein-II. Frequently, but not always, the LM fractions II and III are combined to give a total zein or prolamin fraction. Because this combined fraction is always extremely low in lysine, it is a useful subdivision. LM-IV protein is sometimes extracted with the LM-III protein and sometimes as a separate fraction extracted by a high-pH buffer plus a reducing agent, a potential cause of confusion (Landry et al, 1983). The remaining fraction (V) contains only 17% of the total endosperm N (22% if the residue is added). These results are from several types of corn, grown under different environmental conditions, but they can be considered representative of "average" corn.

Byers et al (1983) showed that the glutelin fraction in wheat varied with treatment: heat, defatting procedures, temperature of extractions, buffers, solvents, etc. They concluded that the classical division of gluten into alcohol-soluble (prolamin) and alcohol-insoluble (glutelin) proteins was no longer valid. They suggested three classifications of protein in seeds: metabolic (cytoplasmic), storage, and structural. Miflin and Shewry (1979) proposed that seed storage proteins be defined as proteins that: 1) have no metabolic function other than to provide N for the germinating seedling, 2) are formed relatively late in seed development, 3) increase preferentially if the N supply to the plant is increased, 4) may be stored in a separate package, usually known as a protein body, and 5) may be composed of a limited number of similar polypeptides. The metabolic and structural proteins, such as the albumins, globulins, and some glutelins, are also hydrolyzed during germination to release amino acids for utilization by the seedlings, but this is secondary to their metabolic functions during seed development.

The protein fractions of the germ are quite different from the endosperm. Most of the corn germ N is salt-soluble, and little is alcohol-soluble. The two major tissues of the seed are so different that they need to be considered separately. High-temperature extractions improve the extraction of some proteins but may denature other proteins and include them in the glutelin fraction. In addition, proteolytic enzymes may be active at some stages; the presence or absence of salts influences fractions other than the first; samples are stored differently; plants are raised under different levels of N fertilization; inbreds and hybrids differ, etc. Thus, the results presented here are only a broad generalization of protein fractionation in corn.

Landry and Moureaux (1980, 1981) presented a particularly thorough example of their system for fractionation of the proteins of both endosperm and germ, with amino acid analyses of all fractions. They suggested that these fractions could be best grouped into two groups: 1) basic (metabolically essential) proteins (i.e., salt-soluble proteins, G-3 glutelins, and residue proteins) and 2) and the endosperm-specific proteins (zein and the G1- and G2-glutelins). These endosperm-specific proteins are the only ones so far reported to occur in corn protein bodies (see Section V). Landry and Moureaux (1980, 1981) also discussed overlapping fractions, incomplete extractions, and other problems.

Another view of corn proteins is that provided by sodium dodecyl sulfate polyacrylamide gel electrophoresis (SDS-PAGE), which separates on the basis of size (Fig. 1). The embryo proteins in lane 2 reveal a large number of different polypeptides, with no predominant single protein or groups of proteins. This finding is consistent with the idea that the embryo consists of metabolically active proteins with many different functions to perform, and it also suggests the relative lack of specific storage proteins. In contrast, the endosperm proteins (lane 1) reveal fewer total bands, and some of them, especially the zeins, are present in large amounts. Protein bodies isolated from endosperms contain only those proteins that predominated in the total endosperm extract (Fig. 1B). Protein bodies are small membrane-bound bodies, seemingly homogeneous within the membranes, which have polyribosomes associated with the outer surface (Fig. 1C) during the developmental period (see Section V).

III. ENDOSPERM PROTEINS

A. Prolamins and Protein Body Components

FRACTIONATION, IDENTIFICATION, AND NOMENCLATURE

Several sources show that the storage proteins of the endosperm, which are the major proteins, are located exclusively (or almost so) within subcellular bodies known simply as protein bodies. These were first described by Duvick

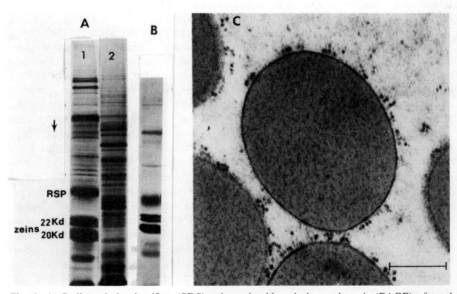

Fig. 1. A, Sodium dodecyl sulfate (SDS)-polyacrylamide gel electrophoresis (PAGE) of total proteins extracted with an SDS sample buffer containing a reducing agent. Lane 1, endosperm; lane 2, embryo. (Reprinted, with permission, from Galante et al, 1983) B, SDS-PAGE of total proteins extracted from a protein body preparation with an SDS sample buffer containing a reducing agent. (Reprinted, with permission, from Press and Vitale et al, 1982; copyright Oxford University Press) C, An electron micrograph of protein bodies in endosperm 22 days after pollination. Bar = 0.5 μm. (Reprinted, with permission, from Burr and Burr, 1976)

(1961), who concluded that they were the sites of zein deposition. It now appears that corn protein bodies contain five or six major protein fractions, which can be individually identified by SDS-PAGE (Fig. 2). These protein fractions have been identified by various workers by different names. The author recommends a nomenclature for the alcohol-soluble proteins, which is here related to the various designations previously used. This analysis depends primarily upon specific works of four authors (Tsai, 1980, 1983; Wilson et al, 1981; Vitale et al, 1982; Landry et al, 1983) and secondarily upon other papers cited in those works. Specific citations to the above references are not given, because of the interconnected nature of the research. Workers from six different laboratories, three in Europe and three in the United States, contributed to the research. Any nomenclature is arbitrary and reflects the biases of the author. It is hoped that adoption of one system, or at least its use as a cross-reference, will simplify the study of corn proteins and will assist newcomers to the field. The proposed nomenclature for endosperm storage proteins separated by SDS-PAGE is shown in Fig. 2. Each lane represents the results obtained when particular extracts are analyzed.

The distinctive property of extractability (not solubility) in aqueous alcohol solutions has allowed easy preparation of the protein fraction commonly known as zein. If no special treatments are employed, one obtains a rather homogeneous set of polypeptides with very similar properties. As Osborne (1924) stated, "it is one of the best characterized groups yet found in either plants or animals." For the proposed nomenclature, a simple extract made with alcohol (50–70%, v/v) is termed zein-I. Shewry and Miflin (1985) recommend the use of Roman numerals to avoid confusion with the nomenclature of prolamin structural loci. If the proteins in this fraction are reduced and separated by SDS-PAGE, they separate into two major size classes (lane 1). The common practice of naming the bands by apparent molecular masses can lead to confusion because different laboratories have applied overlapping estimates of molecular mass to the two classes. Therefore, the author proposes to name the larger, slower-running zein "A-zein" and the smaller, faster-running zein "B-zein." Sometimes, but not always, the two major classes can be separated into two or more subclasses by SDS-PAGE. The separations obtained so far have not been very reproducible, and the number of potential subclasses is unknown. At present, subclass identifications may need to be made on an ad hoc basis. Second-dimension separations by isoelectric focusing (IEF) and by acid-urea gel electrophoresis, which more clearly separate the A- and B-zeins into subclasses, are discussed later in this chapter. The new nomenclature avoids the use of an incorrect size designation, or the confusion that results because a zein protein with a molecular mass of 23,329 by the cDNA sequencing procedure (discussed later under Recombinant DNA Studies) apparently belongs in the 19-kDa rather than the 22.5-kDa size class (Geraghty et al, 1981). Apparent size designations by SDS-PAGE are as arbitrary as the proposed nomenclature, because they depend upon exact reproduction of experimental conditions and upon the standard proteins used for calibration.

A zein I preparation that is not reduced before electrophoresis gives a pattern known as "native" zein, which means that reducing agents are not used either in extraction from the endosperm or in preparation of the sample for SDS-PAGE.

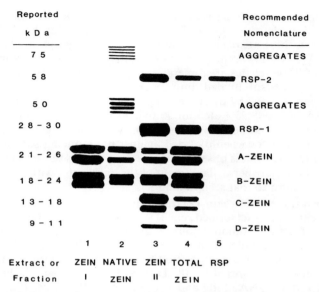

Fig. 2. Outline of recommended nomenclature for corn endosperm alcohol-soluble proteins or protein-body proteins, based on sodium dodecyl sulfate-polyacrylamide gel electrophoresis (SDS-PAGE) (compare with Fig. 3).

Across, Protein extracts or fractions (represented by lanes). 1, Zein-I, the most common fraction, representing proteins extracted by aqueous alcohol under mild conditions, and reduced before SDS-PAGE (also known as zein-1, Landry-Moureaux [LM] fraction II, or reduced zein). 2, "Native" zein, extracted without reducing agents and not reduced before SDS-PAGE. Zein-I gives a similar pattern if the reducing agent is removed before SDS-PAGE. Similar to α-zein, extracted with 90–95% ethanol. 3, Zein-II, extracted from the residue after zein-I has been removed, using alcohol plus a reducing agent (also known as LM-III, G-1 glutelin, alcohol-soluble glutelin [ASG], zein-2, zeinlike protein). 4, Total zein, extracted by alcohol plus a reducing agent and representing zein-I + zein-II. Obtained from the residue after salt-soluble nitrogen has been removed, or from isolated protein bodies. 5, Reduced-soluble proteins (RSP), a component of zein-II or of LM fraction IV. Readily separated from other alcohol-soluble proteins because it is water-soluble when reduced. (Also known as G-2 glutelin, water-soluble [WS] ASG, prolaminlike proteins, or proline-rich zein.)

Down, Individual proteins separated by SDS-PAGE (in order of decreasing size). 1. Aggregates: dimers (about 50 kDa) or trimers (about 75 kDa) formed from A-Zein and B-Zein monomers. Thought to involve disulfide bonds that are broken by reducing agents to yield monomers. 2, RSP-2, a 58-kDa protein soluble in several solvents after reduction. Has high glutamic acid and histidine, low proline. (Also known as H1 and H2 from WS-ASG.) 3, RSP-1, a 28–30-kDa protein soluble in several solvents after reduction, having high proline and high histidine. (Also known as G-2, Z, Z1-a, or H5 and H6 from WS-ASG.) 4, A-zein, one of two major alcohol-soluble proteins, which has high leucine and is most likely about 26 kDa in size. Two or three subclasses are sometimes shown by SDS-PAGE, and a larger number are found by isoelectric focusing (IEF). (Also known as Z1, α-zein [not α-zein extract], or large, heavy, or slow subunit.) 5, B-zein, the second of two major alcohol-soluble proteins, which has high leucine and is most likely about 23 kDa in size. At least three subclasses have been found by SDS-PAGE and a larger number by IEF. (Also known as Z2, β-zein, or small, light, or fast subunit.) 6, C-zein, a small zein, probable size near 18 kDa, with high methionine. (Also known as Z3, Z4, small zein, zeinlike cryoprecipitate from Zein-II, or M2 and M3 from water-insoluble [WI] ASG.) At least two subclasses have been reported. 7, D-zein, the smallest prolamin, with size near 10 kDa, which usually occurs with C-zein and has similar properties, including high methionine. (Also known as Z5, Z6, and M4 from WI-ASG.)

Major References: Tsai (1980, 1983), Wilson et al (1981), Vitale et al (1982), Landry et al (1983). Additional references are found in these papers and in Wilson (1983).

As shown in Fig. 2, lane 2, native zein displays high-molecular-mass aggregates and lower relative levels of A- and B-zeins. In spite of the low level of sulfhydryl groups in zein, these aggregates apparently have sulfhydryl bonds that can be broken by ME, dithiothreitol, or other reducing agents. The apparent sizes suggest that the aggregates are dimers, trimers, and sometimes even larger sets. Each set may be subdivided into bands apparently representing mixed aggregates of AA, AB, and BB, for example. These aggregates are also found in extracts made with 90–95% alcohol, known as α-zein. However, zein that has been separated into monomers by treatment with reducing agents will reform into similar aggregates when dialyzed against nonreducing solvents. Thus we cannot be sure that what is observed in lane 2 represents the actual state of zein within the intact endosperm. Treatment of the aggregates with reducing agents changes the pattern to that seen in lane 1.

When the residue from a zein-I extract is further extracted with alcohol plus a reducing agent, an extract termed zein-II is obtained (lane 3). Besides A-zein and B-zein, this fraction contains four additional proteins with characteristic properties. Slightly smaller than B-zein is a protein, here called C-zein, with a prolamin-type amino acid composition but different from the A- and B-zeins. Apparent molecular masses of 13–18 kDa have been reported. The highest figure is derived from a cDNA clone and is the most likely value. Polypeptides of C-zein with different mobilities on SDS-PAGE have been reported in corn hybrids or among inbreds, but results are preliminary at this time.

The smallest and least common alcohol-soluble protein, D-zein, has an apparent molecular mass near 10 kDa.

The two larger alcohol-soluble proteins (shown in Fig. 2, lane 5, as reduced soluble proteins, or RSP-1 and RSP-2) are sufficiently different from the others that they should not be named zeins (and perhaps not even prolamins). These proteins are extractable by aqueous, aqueous-alcohol, or high-pH solutions, but only after reduction (Wilson et al, 1981). The smaller of these proteins, with an apparent molecular mass of 28 kDa, is here arbitrarily termed RSP-1 because it is present in considerably larger amounts. The molecular mass deduced from a cDNA sequence is only 21.8 kDa, a change from the SDS-PAGE estimate opposite to that found for the A-, B-, and C-zeins (Prat et al, 1985). The RSPs are readily separated from the other alcohol-soluble proteins of a zein-II extract by dialysis against water, or are extracted from isolated protein bodies with water containing a reducing agent.

RSP-1 and RSP-2 can be separated on a diethylaminoethyl-Sephacel column, and RSP-1 also separates into two fractions that cannot be distinguished by amino acid analyses or by SDS-PAGE. Some charge heterogeneity is shown by IEF. RSP-1 and RSP-2 differ considerably in amino acid composition (see discussion under Amino Acid Composition). The RSPs have been located in the periphery of the protein bodies by immunocytochemical labeling (Ludevid et al, 1984). If the RSPs are removed from protein bodies by a treatment with a buffer containing a reducing agent, the protein bodies appear partly disrupted under the electron microscope. Thus the RSPs may not be storage proteins but may be structural proteins required for maintenance of the protein bodies within which the other alcohol-soluble proteins are stored. The relative amount of RSP is not as great as it seems after SDS-PAGE, because the higher content of basic amino acids should give these proteins a color yield in the gels much higher than that of the zeins.

An extract of "total" zein (zein I + zein II) may be obtained by extracting with alcohol plus a reducing agent. This extract may contain most of the alcohol-soluble proteins, which contain little or no lysine. All of the individual fractions mentioned for the separate extracts are present, except that aggregated zeins and those proteins extractable only with reducing agents are not found together.

Several problems are yet to be solved before the amounts of the different alcohol-soluble proteins can be determined with reasonable accuracy. None of the extraction procedures come close to extracting all of an individual protein without also extracting a portion of the other proteins to produce a mixture. Conversely, the extracts that contain only one class of protein (zein-I or RSP) do not contain all of that protein, for some appears in other fractions. The C- and D-zeins and the RSPs pose particular problems because they act "sticky" and thus may be found associated with other proteins, depending upon treatment, or may become insoluble during treatment. For example, C-zein was obtained by

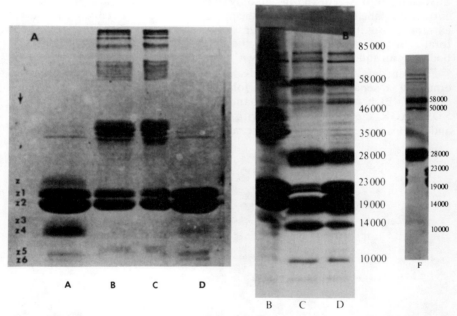

Fig. 3. Examples of sodium dodecyl sulfate-polyacrylamide gel electrophoresis (SDS-PAGE) of various corn endosperm extracts. Compare with Fig. 2.

A, 70% ethanol extracts of endosperm. Lane A, zein extracted with mercaptoethanol (ME) and dialyzed with ME (total zein); lane B, zein extracted without ME and dialyzed without ME (native zein); lane C, zein extracted with ME (total zein) and dialyzed without ME; and lane D, same as lane B, but dialyzed against ME. (Reprinted, with permission, from Tsai, 1980)

B, Extracts from protein body preparations. Lane B, protein bodies in an SDS sample buffer without ME; lane C, the residue from the extract in lane B, extracted with the same buffer plus ME; lane D, total proteins extracted with a sample buffer plus ME; and lane F, reduced soluble proteins extracted from protein bodies by dithiothreitol. (Reprinted, with permission, from Vitale et al, 1982; copyright Oxford University Press)

Melcher and Fraij (1980) by a cryoprecipitation treatment. The RSPs were not detected by IEF unless they were reduced and alkylated to prevent back-oxidation (Wilson et al, 1981). The behavior of zein preparations may be affected by the presence or absence of these proteins, which would depend upon treatment procedures.

Examples of the extraction and SDS-PAGE separation of the protein-body and alcohol-soluble proteins of the endosperm are given in Fig. 3, with some of the original designations of the authors. These illustrate most of the protein separations idealized in Fig. 2. Figure 3A shows the effects of either using or withholding the reducing agent from the extracting solvent or from the SDS sample buffer used to prepare the extracts for electrophoresis. Lanes B and C show some similarities, although the initial extraction in the presence of ME (lane C) would have dissociated the aggregates. In lane D, the extract made without ME contains small amounts of C- and D-zeins, but no RSP is seen. Direct extraction of protein bodies with an SDS buffer (not alcohol) reveals the presence of aggregates (Fig. 3B, lane B). When this extract was run in a second-dimension SDS-PAGE in the presence of ME, the aggregates dissociated into A-zein and B-zein only (not shown). The residue from the first extraction (lane C) contains all of the proteins that require reducing agents for extraction, but it has reduced amounts of A- and B-zein when compared to the total zein extract of lane D. Only the RSPs were removed from the protein bodies by an aqueous extract made with a reducing agent, dithiothreitol (lane F).

Further details of the alcohol-soluble proteins are revealed by two-dimensional separations using IEF followed by SDS-PAGE (Fig. 4). A zein-I extract, which reveals about 10 bands by the IEF alone and only two by SDS-PAGE, shows 14–15 bands with a two-dimensional separation. Both A- and B-zeins are represented along the range of pH separation (Fig. 4A). A zein-II

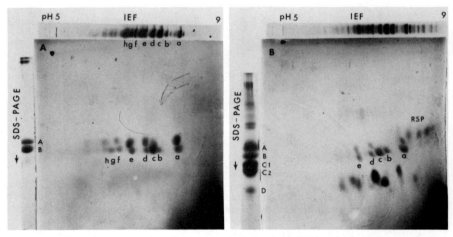

Fig. 4. Two-dimensional separation of reduced and alkylated zein-I (A) and zein-II (B) extracts, with the first dimension (isoelectric focusing) being horizontal and the second dimension (sodium dodecyl sulfate-polyacrylamide gel electrophoresis) being vertical. (Reprinted, with revised labels, from Wilson et al, 1981)

extract contains a number of A- and B-zeins and also has several C-zeins and a series of four basic RSP spots (Fig. 4B). A better two-dimensional separation is achieved when the individual polypeptides are removed from the IEF gel for transfer to separate lanes for SDS-PAGE (Fig. 5) (Vitale et al, 1980). This procedure makes it possible to track an individual polypeptide and assign values for both molecular mass and relative pI. In some cases, a single band on IEF produces two bands upon SDS-PAGE. The numbers relate to actual zein bands found in several inbreds. Isoelectric points cannot be assigned reliably because zeins do not reach equilibrium (Wilson, 1984).

Recently, the use of agarose as the gel in IEF has resulted in improved separations of total zein after a one-dimensional run, as illustrated in Fig. 6. Figure 6 shows the separations achieved with several standard corn inbreds. At least 30 bands were detected among 43 inbreds, and each inbred has from seven to 12 major bands. The patterns formed by these bands are reproducible and

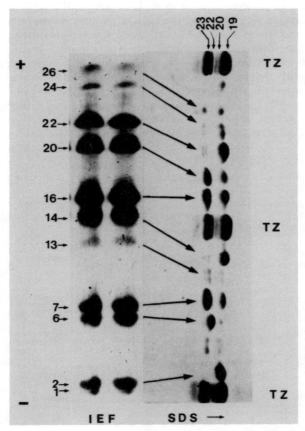

Fig. 5. Isoelectric focusing (IEF) and sodium dodecyl sulfate-polyacrylamide gel electrophoresis (SDS-PAGE) of zeins from inbred Oh43. Bands were excised after IEF in a polyacrylamide gel and run in individual lanes in SDS-PAGE, with marker total zein (TZ). (Reprinted, with permission, from Vitale et al, 1980)

characteristic for each inbred. The numbering system employed is artificial, being dependent upon the ampholyte used and experimental conditions, but it is reproducible and will allow additional bands to be added as they are discovered. Combining agarose IEF with SDS-PAGE allows the assignment of a unique IEF designation and an SDS-PAGE mobility designation to each polypeptide in the mixture of A- and B-zeins (Wilson, 1985b).

Wall et al (1984) presented a new two-dimensional separation of zein, using IEF in the first dimension and urea with aluminum lactate, pH 3.5, in the second. This technique gives a wider spread of spots than is seen when SDS-PAGE is used for the second dimension. Column chromatography can be used to separate zein extracts into fractions enriched in certain proteins. Abe et al (1981) separated zein into A-zein and B-zein fractions, and Esen et al (1985) obtained a 17-kDa methionine-rich polypeptide (a C-zein).

Bietz (1983) separated zein into 16 or more peaks by reversed-phase high-performance liquid chromatography, using acetonitrile as the major solvent. When individual fractions were run on IEF gels, additional heterogeneity was found. Separation was largely on the basis of surface hydrophobicity, providing another characteristic of proteins that can be used in classification systems.

These separations have shown that zein is a family of several individual polypeptides that differ in size, in proportion of charged amino acids, in hydrophobicity, and in relative amounts. Estimates of the number of genes for zein range up to 100, though not all would be present in any one inbred (Soave and Salamini, 1984). The suggestion has been made that different numbers of copies of each gene are present, possibly leading to the variable levels of the different polypeptides. Other types of regulation may also exist.

AMINO ACID COMPOSITION

The various alcohol-soluble proteins of corn may also be characterized by

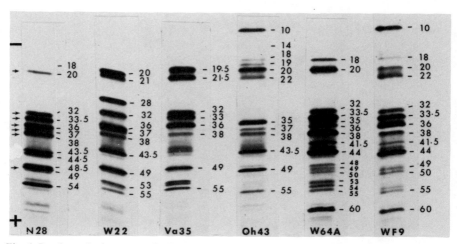

Fig. 6. Band numbering system for isoelectric focusing of total zein in agarose, using six inbreds containing a large number of different bands. The numbers give approximate millimeters from cathode of a "typical" run. Arrows at the left mark the most common bands. (Reprinted, with permission, from Wilson, 1985b)

their amino acid compositions. A number of reports presented rather similar data from well-separated protein fractions, such that averaging the amino acid compositions for presentation seemed appropriate. More recently, additional data has been obtained from the nucleotide sequences of recombinant DNA clones, which specify individual zein polypeptides. Good agreement exists between the two sets of data obtained in two quite different ways (Table III).

The amino acid compositions reported from traditional procedures are those of combined zein-I, i.e., the weighted average of the dozen or so major A- and B-zein polypeptides plus any of the other alcohol-soluble proteins that are carried through the isolation procedures. Until recently, these were the only data available. The first column of Table III presents data for the average of several zein-I type preparations, which have the well-known zein composition: low basic amino acids, moderate levels of proline and glutamic acid (glutamine), high levels (for prolamins) of aspartic acid (asparagine), and high levels of alanine

TABLE III
Amino Acid Composition (mol %) of Alcohol-Soluble Protein Body Proteins from Corn[a]

Amino Acid	Zein-I					C-Zein		Reduced Soluble Protein	
	A + B	A-Zein		B-Zein					
	Average[b]	Combined[c]	DNA[d]	Combined[c]	DNA[d]	Combined[e]	DNA[d]	1[f]	2[g]
Lysine	0.1	0.2	0	0.4	0	0.2	0	0.2	2.6
Histidine	1.0	1.1	1.5	1.1	2.0	1.0	0	7.2	11.0
Arginine	1.2	1.1	1.4	1.1	1.2	2.6	3.1	2.7	2.1
Aspartic acid	5.1	5.7	0.1	5.7	0.2	2.4	0.6	0.5	4.3
Asparagine	···	···	4.9	···	4.6	···	1.9	···	···
Threonine	3.0	3.1	2.5	3.1	2.6	3.0	2.5	4.3	2.5
Serine	6.3	6.4	6.7	7.4	7.2	5.3	5.0	4.3	7.3
Glutamic acid	21.4	21.4	0.5	20.7	0.5	20.2	1.9	16.0	33.6
Glutamine	···	···	20.1	···	19.0	···	16.2	···	···
Proline	10.7	8.7	9.0	10.0	10.3	11.6	8.8	24.9	4.7
Glycine	2.2	2.5	1.4	2.9	3.8	8.4	8.8	7.3	4.9
Alanine	13.3	13.6	14.2	12.8	13.7	12.6	13.8	5.7	3.5
Cysteine[h]	1.0	0.8	0.5	0.4	0.9	4.6	4.4	5.5	4.4
Valine	3.6	5.6	5.4	3.2	2.7	3.1	2.5	7.5	5.1
Methionine[h]	0.9	1.3	1.5	0.4	0.2	6.7	11.2	0.8	1.3
Isoleucine	3.8	3.7	4.0	3.6	4.3	1.2	0.6	2.0	2.2
Leucine	18.7	18.4	18.8	18.7	20.1	11.0	10.0	10.0	4.7
Tyrosine	3.5	2.7	3.3	3.2	3.7	6.2	8.8	2.1	3.5
Phenylalanine	5.2	3.7	4.1	5.3	6.0	1.9	0	1.3	1.9
Tryptophan	···	0	0	0	0	···	0	···	···

[a] See Fig. 2 for description of proteins.
[b] A consensus amino acid composition calculated from six recent references (Wilson, 1983).
[c] From Abe et al (1981). Zeins were separated by gel filtration.
[d] Average of A-, B-, or C-zein amino acid compositions of Table IV, from recombinant DNA nucleotide sequences.
[e] Average of data from Esen et al (1985), Gianazza et al (1977), Landry et al (1983), and Melcher and Fraij (1980).
[f] Average of data from Esen et al (1981), Landry et al (1983) (WS-ASG and G2pH3$_s$), Paulis and Wall (1977), Vitale et al (1982), and Wilson et al (1981).
[g] From Vitale et al (1982).
[h] Values for cysteine and methionine are uncertain for hydrolysate data because of possible oxidation.

and leucine. The level of cysteine is just enough to permit formation of sulfhydryl bonds between the chains. These amino acid proportions are characteristic of zein and of a family of similar prolamins from related races, species, and genera: teosinte, *Tripsacum*, sorghum, and various millets (Wilson, 1983). The prolamins of other Gramineae are considerably different.

When the A- and B-zeins are separated (Abe et al, 1981), the amino acid compositions compare favorably with the average compositions obtained from three or four recombinant DNA clones (Table III). Small but consistent differences can be seen between the two size classes of zein. A-zein contains relatively larger amounts of valine and methionine and smaller amounts of proline and phenylalanine. This information from the cDNA clones confirms the assumption that almost all of the aspartic acid and glutamic acid groups are amidated.

The amino acid composition of combined C-zein has been reported from several sources, but only one report has been made on a cDNA sequence (Pedersen et al, 1986). In addition to the smaller size of C-zein, it has an amino acid composition that differs from those of A- and B-zeins in several respects. The most striking feature is the high level of both sulfur amino acids, over 10% on average and as much as 15% in one report (Esen et al, 1985). This protein is much more likely to form intermolecular sulfhydryl bonds than are A- and B-zeins. However, C-zein does not seem to form aggregates in protein bodies (Vitale et al, 1982). C-zein also has high glycine and tyrosine. It differs from A- and B-zeins by having only about half as much leucine, although the level of alanine is almost the same. C-zein is less hydrophobic than most other prolamins (Wilson, 1983). These differences produce appreciable effects on the extractability/solubility properties of C-zein. C-zein is not extracted unless a reducing agent is present, and it is precipitated from solution upon chilling (Melcher and Fraij, 1980). Losses from solution also appear to occur during handling of nonalkylated preparations (Wilson, unpublished). Alkylation of zeins is often employed to prevent reoxidation and improve stability (Paulis, 1981; Wilson et al, 1981) and may be especially important for C-zein. Reliable estimates of the relative amount of C-zein are not yet available because of these solubility/extractability problems.

D-zein is only half the size of B-zein. Somewhat variable amino acid compositions have been reported for D-zein, but it differs most from A- and B-zeins by a methionine content of over 10% and also by high glycine and moderate leucine contents (Phillips and McClure, 1985). These authors also reported that an inbred corn line with elevated levels of C- and D-zeins had an increased content of the essential amino acid methionine.

The two RSPs (Table III) have quite different amino acid compositions from the other alcohol-soluble proteins and from each other. RSP-1 (apparent molecular mass about 28 kDa) is characterized by high histidine, proline, and cysteine, but is low in lysine. It is separated into two groups by column chromatography (Esen, 1982; Vitale et al, 1982). It also separates into several groups by IEF, but other properties appear to be similar. RSP-1 is a very interesting protein because it possesses a hexapeptide, Pro-Pro-Pro-Val-His-Leu, which is tandemly repeated six or more times (Esen et al, 1982). RSP-2 has a moderate level of lysine, a high level of histidine, low proline, and the highest glutamic acid (glutamine) content of any corn protein (Vitale et al, 1982). Cysteine is also relatively high in RSP-2.

RECOMBINANT DNA STUDIES

Recent work with recombinant DNA has produced nucleotide sequences specific for individual zein polypeptides. Membrane-bound polysomes (Fig. 1C) provide an enriched source of zein messenger RNA, which is used to generate zein cDNA in recombinant plasmids (Marks and Larkins, 1982). The cDNA can be inserted into *Escherichia coli* and clones selected that produce cDNA that will hybridize to zein messenger RNA. The strength of cross-hybridization between various zein DNAs is a measure of their relatedness or homology. Sequence homology is about 90% among different cDNAs for A-zeins and about 60% between the cDNAs for A- and B-zeins, but little sequence homology is found between these two and C-zein cDNA. The sequence of nucleotides in a cDNA can be determined, and from this the amino acid sequence is determined. A consensus amino acid sequence is shown in Fig. 7, using data from two laboratories. The amino terminal end has a signal peptide of 21 amino acids. The signal peptide may be needed for the zein to pass through the membrane into the protein body, but it is removed later. Under certain circumstances, zeins made in vitro still have this peptide and thus are larger than expected. The zein polypeptide starts with the amino terminal sequence of 36 amino acids, followed by a repeating sequence of 20 amino acids. Nine copies of this sequence occur in tandem, although the repeats are not perfect. Finally, the carboxy terminal sequence consists of either 10 or 29 amino acids, the major difference between the A- and B-zeins. Various deletions and substitutions occur along the sequence of different zeins, giving rise to variations in molecular mass. The basic overall similarity of all A- and B-zeins sequenced to date suggests that the genes originated by duplication of an ancestral gene followed by subsequent mutations. The multigene family of zein genes may total 80 (Larkins et al, 1984), although other estimates have been made.

In Table IV, several individual zein DNA sequences are reduced to a listing of the number of each amino acid occurring in each polypeptide, with data coming

Signal Peptide (21)
Met-Ala-Thr-Lys-Ile-Phe-Cys-Leu-Ile-Met-Leu-
Leu-Ala-Leu-Ser-Ala-Ser-Ala-Ala-Asn-Ala-
Amino Terminal Sequence (36)
Ser-Ile-Phe-Pro-Gln-Cys-Ser-Gln-Ala-Pro-Ile-Ala-Ser-Leu-
Leu-Pro-Pro-Tyr-Leu-Pro-Pro-Val-Met-Ser-Ser-Val-Cys-
Glu-Asn-Pro-Ala-Leu-Gln-Pro-Tyr-Arg-
Repeating Sequence (9 × 20)
Gln-Gln-Leu-Leu-Pro-Phe-Asn-Gln-Leu-Ala-Ala-
Ala-Asn-Ser-Pro-Ala-Tyr-Leu-Gln-Gln-
Carboxy Terminal Sequence (10 or 29)
Gln-Gln-Gln-Leu-Leu-Pro-Tyr-Asn-Arg-Phe-Ser-
Leu-Met-Asn-Pro-Val-Leu-Ser-Arg-Gln-Gln-Pro-
Ile-Ile-Gly-Gly-Ala-Leu-Phe

Fig. 7. Consensus sequence of the four domains of zein proteins. The amino acid sequences were derived from clones of complementary DNA for both A- and B-zeins (Data from Larkins, 1983, and Messing, 1983)

from two laboratories. As noted in Table III, these new amino acid data agree with the findings derived from bulk analyses. In addition, they give information on the high degree of amidation of aspartic acid and glutamic acid and confirm the absence of lysine. The low levels previously reported are probably the result of contamination. The true molecular mass and net charge for each polypeptide can now be determined. The zeins reported here have from two to seven more basic groups than acidic groups. These differences, plus variations in isoelectric point due to the differences among basic groups, account for the large number of zein bands seen upon IEF. Studies in this area are expanding rapidly and may make some of these comments outdated rather soon. Brief reviews have been given by Larkins et al (1984) and Messing (1983) and more detailed reviews by Rubenstein and Geraghty (1986) and Heidecker and Messing (1986).

TABLE IV
Amino Acid Composition (number of amino acids) of Zein Derived
from the Nucleotide Sequences of cDNA or Genomic Clones

Amino Acid	A-Zein			B-Zein				C-Zein[f]
	pZ22.1[a]	pZ22.3[a]	Z4[b]	A30[c]	pZ19.1[d]	A20[e]	zG31A[b]	
Lysine	0	0	0	0	0	0	0	0
Histidine	3	3	5	2	3	1	3	0
Arginine	2	4	4	2	3	3	2	5
Aspartic acid	0	0	1	0	1	1	0	1
Asparagine	12	13	11	10	9	10	10	3
Threonine	8	7	4	5	4	7	6	4
Serine	16	17	16	15	15	17	15	8
Glutamic acid	1	2	1	1	1	1	1	3
Glutamine	50	50	47	41	39	43	40	26
Proline	22	22	22	23	21	22	23	14
Glycine	4	2	4	5	3	2	5	14
Alanine	34	35	35	29	31	30	28	22
Cysteine	1	1	2	2	2	2	2	7
Valine	15	16	9	5	6	7	5	4
Methionine	5	5	1	0	1	1	0	18
Isoleucine	8	11	10	9	9	10	9	1
Leucine	44	42	52	43	45	42	43	16
Tyrosine	8	7	9	8	8	8	8	14
Phenylalanine	9	8	13	13	13	12	13	0
Tryptophan	0	0	0	0	0	0	0	0
Total amino acids	242	245	246	213	214	219	213	160
Molecular mass	26,532	26,996	27,136	23,329	23,535	24,027	23,371	17,279
Basic amino acids	5	7	9	4	6	4	5	5
Acid amino acids	1	2	2	1	2	2	1	4
Net charge	4	5	7	3	4	2	4	1
Inbred	W64A	W64A	W22	IHP	W64A	IHP	W22	

[a] Marks and Larkins (1982).
[b] Hu et al (1982).
[c] Geraghty et al (1981).
[d] Pedersen et al (1982).
[e] Geraghty et al (1982).
[f] Pedersen et al (1986).

B. Nonprolamin Proteins

Although the nonprolamin proteins traditionally have been divided into albumins, globulins, and glutelins, the dividing lines are difficult to draw and depend heavily upon the methods of separation used. Because solubility/extractability are seldom all-or-none properties, protein extracts of these fractions are heavily cross-contaminated. In particular, high temperatures or defatting solvents may affect the solubility of some albumins or globulins so that they are extracted in one of the glutelin fractions. Some glutelins may, in time, be identified as structural proteins (cell walls, membranes, etc.) or as metabolic proteins (enzymes) that are insoluble or have become so as a result of maturation reactions, changes during drying, denaturation, or association with other seed components during extraction, etc.

ALBUMINS

Albumins are defined as water-soluble proteins (Osborne, 1924), but in too many cases the actual practice has been to define them as water-extractable proteins. Two problems can be noted. Some albumins may be situated in combination with other cell constituents such that they are not readily extracted, although once extracted and separated, they are soluble in water. On the other hand, seeds contain various salts that can cause some salt-soluble proteins to be extracted. Osborne's original procedure (1924) separates albumins from globulins on the basis that, after an extraction with a salt solution, the albumins remain soluble after dialysis against water, whereas the globulins are precipitated.

When separated by SDS-PAGE, endosperm albumins reveal a large number of bands (Fig. 8A). This is not surprising, because the cell should contain

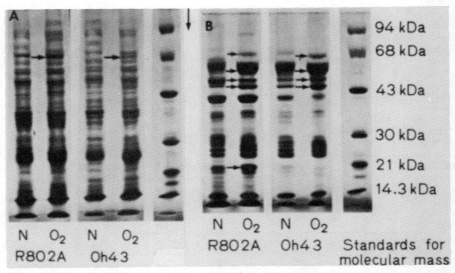

Fig. 8. Sodium dodecyl sulfate-polyacrylamide gel electrophoresis of endosperm albumins (A) and globulins (B) from normal (N) and opaque-2 (O_2) versions of R802A and Oh43. Arrows point to changes in opaque-2 proteins. (Reprinted, with permission, from Dierks-Ventling, 1981)

thousands of different enzymes for metabolism during the development and filling of the seed. In this example, one albumin band was noted (arrow) that appeared in the o_2 versions of the inbreds Oh43 and R802A. Small differences are observed on SDS-PAGE of different inbreds, but they do not appear useful for identification of inbred lines. More band differences were noted when albumins of different inbreds were separated by nondenaturing PAGE (Wilson, 1981). One major band was identified as an enzyme (sucrose synthase), but another enzyme (RNase) could not be detected by a protein stain at the levels usually found in corn endosperms. Some of the visible bands may be starch synthesis enzymes, required in large amounts because starch synthesis is the major metabolic reaction of the endosperm.

The amino acid compositions of albumin fractions reported in the recent literature are presented in Table V. No features are outstanding, which is not unexpected in view of the large number of proteins present in a single extract. The lysine levels are substantial; thus, the albumins are a good dietary source of essential amino acids.

GLOBULINS

SDS-PAGE patterns of endosperm globulins (Fig. 8B) differ from those of albumins in having fewer bands, with several bands present in relatively large

TABLE V
Amino Acid Composition (mol %) of Nonprolamin Corn Seed Proteins and Free Amino Acids

Amino Acid	Endosperm[a]					Germ
	Albumin	Globulin		Glutelin[d]	Free Amino Acids, Range	Flour[e]
		Sequential[b]	Dialysis[c]			
Lysine	5.9	5.0	5.5	5.4	2.5–10	4.6
Histidine	2.1	2.2	3.3	2.7	0.6–2.4	2.4
Arginine	5.6	5.4	9.5	4.7	1.1–7.3	6.0
Aspartic acid	9.2	9.1	7.8	8.3	24–33	8.1
Threonine	5.6	5.0	3.9	5.0	1.0–3.6	4.4
Serine	6.2	6.4	7.4	6.0	3.5–6.3	6.1
Glutamic acid	11.5	11.9	14.6	12.2	12.5–37	14.0
Proline	6.1	7.3	4.4	6.8	5.2–19.6	6.9
Glycine	11.2	10.5	9.9	8.6	1.2–8.9	10.1
Alanine	11.0	10.5	8.5	9.8	2.2–22.5	10.1
Cysteine	0–4.6	0–2.9	0–2.0	0–2.9	0–3.5	1.5
Valine	6.9	6.3	6.6	7.2	1.1–2.3	6.6
Methionine	1.3	1.4	1.3	2.2	0–0.6	1.7
Isoleucine	3.9	3.8	3.4	4.4	0.5–1.8	3.2
Leucine	7.1	7.7	6.6	9.7	0.6–1.5	7.7
Tyrosine	2.5	2.7	2.5	2.5	1.3–2.3	2.4
Phenylalanine	2.9	3.5	4.1	4.0	0.7–1.0	3.2
Asparagine	...	...	...	...	(7.2–9.1)	...
Glutamine	...	...	...	...	(4.9–15.9)	...
Tryptophan	...	...	...	...	...	1.0

[a] Wilson (1983), consensus data from recent literature.
[b] Globulins extracted after albumins had been extracted by water.
[c] Globulins separated from albumins by dialysis against water, after extraction by a saline solution.
[d] Fraction V (G_3) isolated by the Landry-Moureaux procedure.
[e] Consensus values calculated from five reports from the USDA Northern Regional Research Center (Blessin et al, 1979; Nielsen et al, 1979, and references therein) and O'Dell and de Boland (1976).

amounts. Again, small changes in pattern were induced by the o_2 mutation (arrows) and some differences are apparent between inbreds. Although, as a group, the globulins are proteins with a high lysine content, relatively little is known about the actual function of these proteins in the developing corn seed. Globulins are storage proteins in the endosperm of oats and in the cotyledons of legumes; thus, they are potential storage proteins in other species. Studies are needed of the globulins of the o_2 mutation, where total globulins are often increased, or of corn grown with high N fertilization, to determine whether any of the globulins are specifically increased under these conditions.

Globulins are sometimes characterized by a relatively high amount of arginine, especially globulins prepared by dialysis of salt extracts (Table V). In contrast, several globulins prepared by a sequential extraction procedure (salt after water) do not contain elevated arginine, possibly because of the presence of albumins not previously extracted. Differences of this sort are highly dependent upon experimental conditions.

GLUTELIN

Osborne (1924) commented that "seeds of cereals contain a large proportion of (residual) proteins which have been provisionally assigned to the group of glutelins." The "modern methods" he was waiting for now allow us to reconsider the assignment. Sodek and Wilson (1971) concluded that many glutelin extracts reported in the literature contained sizeable amounts of zein that had not been extracted into the prolamin extract. The use of reducing agents now permits complete extractions of A- and B-zeins and also can extract the C- and D-zeins and the RSPs (the LM fractions III and IV or G1 and G2). As described earlier, these proteins can now be placed in a class consisting of prolamins and other alcohol-soluble protein body proteins (Shewry and Miflin, 1985).

Other glutelins (LM-V) seem similar to proteins in the albumin and globulin fractions by amino acid composition (Table V) and by SDS-PAGE (DiFonzo et al, 1977, Wilson et al, 1981). Duvick (1961) first showed that, in mature endosperms, the starch grains and protein bodies were surrounded by "a transparent glue (clear viscous cytoplasm)," as seen in the light microscope. A series of papers from the USDA Northern Regional Laboratory extended this picture. Khoo and Wolf (1970) used the electron microscope (EM) to reveal some of the fine structure of developing endosperm cells and some relationships of protein body formation. Christianson et al (1969, 1974) isolated protein bodies and matrix proteins from developing endosperms by sucrose density gradient centrifugation. Free protein bodies contain zein. The bulk of the zein-containing protein bodies sediment with a surrounding matrix rich in protein. This protein has an amino acid composition similar to that of LM-V. EM pictures show a rather amorphous structure for this protein matrix in mature endosperms (Christianson et al, 1969; Wolf et al, 1969). Scanning EM pictures of protein bodies and matrix proteins after rehydration (simulated steeping or tempering) reveal differences between normal and o_2 endosperms, with the protein coming off in sheets from around the starch granules of mutant endosperms (Christianson, 1970). The bonds holding the matrix proteins together are loosened by treatments with alkali or reducing agents, such as ME and the sulfite of the wet-milling process (Wall and Paulis, 1978). This releases the starch grains, and the protein bodies and much of the matrix proteins are recovered in the corn gluten meal fraction. More detailed studies of the fine structure of mature endosperms and of the proteins that are involved could lead to improved methods for separating the components during milling.

MISCELLANEOUS PROTEINS

Corn seeds contain small amounts of a protein that inhibits both trypsin and activated Hageman factor (a human proteolytic enzyme). This protein has been characterized and sequenced (Mahoney et al, 1984). It has a molecular mass of 12 kDa, and the amino acid sequence is unlike that of any other plant trypsin inhibitor. This protein occurs in higher levels in o_2 seeds than in normal seed (Mitchell et al, 1976). However, from rat-feeding trials, the same authors concluded that this protein does not affect the nutritional value of opaque corn adversely. There is no good evidence for any physiological role for this protein in developing or germinating seeds. It is low in lysine but has adequate quantities of the other essential amino acids, especially the sulfur-containing amino acids (Mahoney et al, 1984).

Several unusual proteins are extracted from corn seed with $0.05M$ sulfuric acid. These are thought to be similar in chemical and physical properties to wheat purothionins but are different in structure (Jones and Cooper, 1980) and do not have the toxic properties of the wheat purothionins. Six different proteins in the extract were all high in basic amino acids and cysteine.

In the future, we will continue to isolate, identify, and characterize individual proteins or groups of proteins: enzymes, structural proteins, membrane proteins (especially from starch granules and protein bodies), mitochondrial and microsomal proteins, ribosomal proteins, etc. The number of possible proteins is quite high, and thus the amount of any random protein selected for study is probably quite low. As examples, a salt buffer is required to completely extract the major corn endosperm RNase (an albumin), which accounts for less than 0.1% of the total protein (Wilson, 1967). Sucrose synthase, presumably required in large amounts because of the large flow of carbohydrates into the endosperm, makes up an estimated 2.8% of the soluble protein at 22 days after pollination (Su and Preiss, 1978). Calculations suggest that as much as 1% of the total protein of sweet corn (at the eating stage) may be trypsin inhibitor (Hojima et al, 1980) and that the purothionins (Jones and Cooper, 1980) may make up 0.1% of the total protein of mature seeds. Will the time come when we can identify 90% of the total protein, starting with the 50% that is zein?

NONPROTEIN NITROGEN

For the present discussion, the nonprotein fraction is assumed to consist mostly of free amino acids. This fraction is usually only a minor portion of the total kernel or endosperm N, but may be an appreciable portion of the germ N (Table II). Reports in the literature on total nonprotein N are quite variable, possibly because this fraction is susceptible to environmental conditions and treatment of the sample such as drying conditions, age at harvest, etc. The relative levels of the different amino acids are also quite variable (Table V). Lysine is usually present in levels higher than it averages in total protein, but the major free amino acids are aspartic acid-asparagine, glutamic acid-glutamine, proline, and alanine. Glutamine and aspartic acid are the major amino acids in the vascular sap moving into the ear (Arruda and da Silva, 1979).

The free amino acids are increased considerably in o_2 endosperm, accounting for up to 20% of the total N in some inbreds (Misra et al, 1975; Sung and Lambert, 1983). The o_2 lines reported so far contain much higher levels of free amino acids than do the normal lines, but the actual level of free amino acids is not proportional to the extent of inhibition of zein synthesis or to the total lysine content (Sung and Lambert, 1983).

IV. EMBRYO PROTEINS

The well-known fact that corn germ protein has a higher nutritional value than does endosperm protein can be related to its much better balance of essential amino acids (Table I). However, the nutritional value of a corn germ fraction can be greatly reduced by processing, as noted by Osborne (1924). For example, Wall et al (1971) reported that various milling fractions from germ may have high total lysine values (determined after acid hydrolysis), but that nutritionally available lysine was reduced by processing. The expeller method for oil removal was particularly harmful. Removal of oil by solvents had much less effect on the available lysine, and some fractions were close to casein in their protein efficiency ratios.

The distribution of N among the various classical protein fractions of the germ is quite different from that in the endosperm (Table II). The largest fraction is the albumins, which contain about one third of the total N, although the distribution among nonprotein N, albumins, and globulins varies with the technique used. As might be expected, the amino acid compositions of these fractions are not very different and tend to be balanced in most amino acids. Landry and Moureaux (1980) present a complete picture of the germ protein fractions, with comparisons to similar fractions from the endosperm. Small amounts of zein, identified by alcohol-extractability and amino acid composition, were found. Tsai (1979) found that 10% of the germ protein was in proteins that migrated as zein on SDS-PAGE. Dierks-Ventling and Ventling (1982) could find no cross reactions between embryo tissue and antibodies against zein, although a globulin antibody reacted against both tissues.

The average amino acid composition of whole germ flour, taken from a number of reports, is presented in Table V. No unusual features are apparent. In one experiment, about three fourths of the total N was found in fractions with overall similar amino acid compositions: albumin, globulin, G-2 and G-3 glutelins, and insoluble-protein (Landry and Moureaux, 1980). Globulins had the highest arginine content. Most of the remaining N was in the nonprotein fraction. Corn germs do not contain more than minor amounts of proteins with unusual amino acid patterns, and the major fractions all consist of many proteins. Thus it is no surprise that the amino acid compositions of germs from high-lysine mutants and of various germ milling fractions are similar (Blessin et al, 1979). Contributions of germ proteins to improved nutritional value of corn (i.e., higher lysine) will be made through increased germ weight and/or increased germ protein as a percent of total seed weight and/or protein.

Cross and Adams (1983b) compared the SDS-PAGE patterns from albumins and globulins from several inbreds and a hybrid (Figs. 9A and 9B). As with endosperm albumins, germ albumins revealed a large number of bands, consistent with the idea that this fraction contains many of the enzymes needed for development. A few differences are apparent among the different inbreds, with bands from both parents appearing in the hybrid. The SDS-PAGE patterns of the globulins had fewer bands, with some being predominant. Several bands appear to be characteristic of certain inbreds and possibly could be used for inbred identification. Again, the hybrid contained bands from both parents. The globulin fraction was fractionated for molecular mass by column chromatography, producing two major peaks with estimated sizes of 127 kDa and >200 kDa (Cross and Adams, 1983a). On SDS-PAGE, the smaller fraction yielded polypeptides of 21–22, 29–30, and 50 kDa, whereas the larger fraction yielded polypeptides of 63, 65, and 71 kDa. The nature of the aggregates was not determined. The 71-kDa polypeptide was formed relatively early in development; the 50-, 63-, and 65-kDa polypeptides were formed between 18 and 40 days after pollination; and the 21–22- and 29–30-kDa polypeptides appeared later, a suggestion that they are storage proteins.

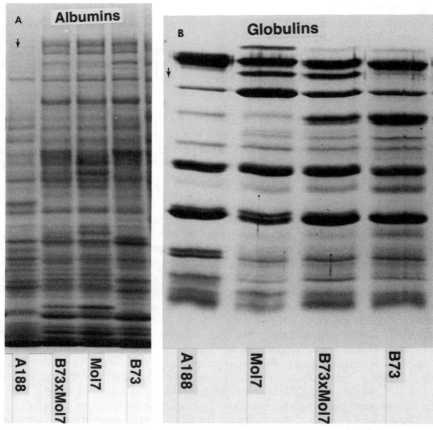

Fig. 9. Sodium dodecyl sulfate-polyacrylamide gel electrophoresis of embryo albumins (A) and globulins (B). (Reproduced [A] and adapted [B] from Cross and Adams, 1983b, by permission of the Crop Science Society of America)

Khavkin et al (1978) suggested that globulins are the major storage proteins of the embryo. During the initial stages of germination, these proteins may furnish the amino acids required for the synthesis of new enzymes that function in hydrolysis of endosperm storage materials and in the interconversion of compounds coming from the endosperm to the developing seedling. However, it is difficult to determine that these globulins are true storage proteins. Tsai et al (1978) did find an increase in germ protein as a result of N fertilization, but their techniques put most of the germ protein into the albumin fraction. The small zein fraction did not respond to N fertilization, in contrast to the result in the endosperm.

Two nonprotein components of corn germ are often thought to affect the extractability and solubility of proteins. Phytic acid may make up almost 1% of the dry weight of the total kernel, but 88% of the phytate is found in the germ (O'Dell et al, 1972). O'Dell and de Boland (1976) reported that phytate is extracted from corn germ with a water wash and that the proteins extracted with it are not complexed with phytate (in contrast to soybean proteins, which do complex with soybean phytate). Craine and Fahrenholtz (1958) reported that proteins in water extracts of whole kernels react with phytate, and that phytate affects the fractionation of albumins and globulins. It cannot be determined whether the different results from these two investigations represent differences between germ and endosperm proteins or whether they are a result of different experimental conditions. Phytate complexes sufficiently with corn alcohol dehydrogenase to affect the electrophoretic mobility (Altschuler and Schwartz, 1984).

The average corn kernel contains about 4.3% oil, with the concentration in the germ being 35% (Chapter 10). Thus, the influence of oil upon protein analysis depends upon whether or not the endosperm is analyzed separately or as part of the whole. As noted above, commercial procedures for removing oil from corn germ can cause reactions with certain amino acids, especially lysine. The assumption that lipids react with corn proteins and affect extractability and solubility is often made (Landry and Moureaux, 1981), yet no recent information demonstrates the effect of lipid on corn protein properties. The first consideration is that separation of germ from endosperm may permit analysis of endosperm protein without the tissue having to be defatted. In addition, the act of removing lipids by solvents with or without heat may denature certain proteins. Byers et al (1983) found that defatting whole-wheat grain shifted some of the salt-soluble protein into the insoluble or glutelin fraction. The literature contains examples of defatting with acetone (not a good solvent for corn lipids) or with butanol, which has also been used for the extraction of zein. Presently, personal opinions rather than experimental fact seem most involved in the choice of whether or not to defat.

When corn scutella are homogenized and centrifuged without defatting, a fat layer collects on top. This layer contains carbohydrases with extremely high specific activities relative to those of other fractions (Hanson et al, 1960). The fat layer also contains a minor nuclease free of the major RNase of the tissue (Wilson, 1968). A pair of major embryo proteins with apparent molecular masses of about 68 kDa are also complexed by the lipid layer (Schwartz, 1979). A genetic analysis of this protein was also made. Lipids obviously influence the isolation of some embryo proteins unless countermeasures are taken, but it is not known if the effect is serious when whole kernels are analyzed, or how much of the total protein is affected.

V. PROTEIN BODIES

Duvick (1961) recognized that protein granules (now called protein bodies) are the major sites of zein deposition, that the average volume of the protein bodies increases during development, and that their sizes vary in different regions of the endosperm. Protein bodies in normal endosperm and in some starch mutants average 1.5–1.8. μm in diameter (Wolf et al, 1969), although their average drops to less than 1 μm a few cells in from the aleurone (Baenziger and Glover, 1977). Protein bodies may be 0.1 μm (Wolf et al, 1969) or invisible (Robutti et al, 1974; Baenziger and Glover, 1977) in o_2 endosperm and may be distorted or absent from floury-2 (fl_2) endosperm (Christianson, 1970; Christianson et al, 1974). The protein bodies are tightly packed against the starch grains in normal horny endosperm, and indentations due to the protein bodies are visible in scanning electron micrographs (Fig. 10A). These authors all commented on the occurrence of matrix protein, which surrounds the protein bodies and the starch grains (see discussion under Glutelins). The protein matrix is also seen in o_2 endosperms (which lack visible protein bodies), where it encloses the starch grains (Fig. 10B).

The processes occurring during protein body development have been reviewed recently by Larkins (1981), Miflin et al (1981), and Miflin et al (1983). Zein is synthesized by ribosomes that are attached to the rough endoplasmic reticulum (RER) (Khoo and Wolf, 1970) and is initially produced with a signal peptide of 1–2 kDa that enables the molecule to pass through the membrane into the lumen of the RER. The signal peptide is then cleaved off, and the zein is

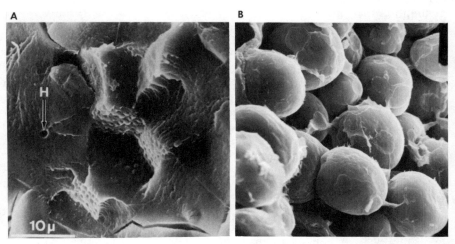

Fig. 10. Scanning electron photomicrographs of endosperm. A, Normal hard endosperm, showing damaged starch granules (H), zein bodies, and indentations made by zein bodies (×1,700). B, Opaque-2 endosperm, showing matrix protein on starch grains and no zein bodies (×1,850). (Reprinted, with permission, from Robutti et al, 1974)

deposited inside a membrane, thus producing a protein body attached to the RER and surrounded by a portion of the RER membrane, which may still have ribosomes attached (Fig. 1). Argos et al (1982) presented a hypothetical model based on the nine repeating units of 20 amino acids found in each zein amino acid sequence. Nine adjacent helices would cluster into a distorted cylinder, and the polar residues of the surface amino acids would contribute to stacking of the zein molecules in the protein body (Fig. 11). Circular dichroism spectra show up to 59% α-helix in zein, which supports the proposed model. The model suggests that mutations that would affect the formation of a compact unit, such as additional basic lysine groups in the interior of the molecule, would be disruptive. The spontaneous nature of the formation of protein bodies is reinforced by the finding that zein messenger RNA injected into *Xenopus* (South African clawed frog) oocytes causes the synthesis of zein within protein bodies that are similar to those in corn endosperm (Hurkman et al, 1981).

Among common crop species, several types of seed protein bodies contain different types of proteins. Legume protein bodies are quite different from cereal protein bodies and appear to consist of globulins within vacuoles (Miflin et al, 1981). Barley and wheat storage proteins are deposited within protein body

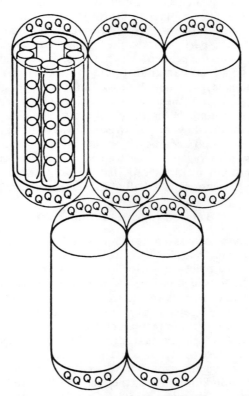

Fig. 11. A possible model for the arrangement of zein proteins. Q = glutamine. (Reprinted, with permission, from Argos et al, 1982)

membranes, but late in seed development they separate from the membranes for the most part. Corn protein bodies are found within an intact membrane (Miflin and Burgess, 1981) that protects the protein from digestion by proteinase-k (Miflin et al, 1982). Corn protein bodies appear to change in structure if the reduced soluble proteins (RSP) are removed (Ludevid et al, 1984). If RSPs are located on the protein body surface, a decrease in this fraction should take place in mutants with small or no protein bodies. However, the protein content of the extract that contains most of the RSPs is not decreased in o_2 endosperm (Landry and Moureaux, 1980). Reductions in RSP were found in fl_2 and opaque-7 (o_7) endosperms (Fornasari et al, 1982). In spite of the membrane around the protein bodies, the A- and B-zeins are extracted by nonreducing aqueous alcohol solutions, whereas little of the C- and D-zeins are extracted without a reducing agent (Miflin et al, 1981; Vitale et al, 1982). Extraction with alcohol leaves material randomly distributed throughout the protein bodies (Taylor et al, 1984). Subsequent extraction with alcohol and a reducing agent produces a zein-II fraction that contains all four classes of zein, as shown by SDS-PAGE.

VI. PROTEIN ASSAY

By their very nature, proteins present problems to the analyst that cannot be completely overcome. The use of recombinant DNA techniques may allow us to completely define single proteins. Usually we start with a complex mixture of countless proteins as a part of a complex living organism. We end with a (possibly) specific assay for protein, which works, more or less, on a partially purified mixture, or we have an almost pure protein (or proteins) after many manipulations that have resulted in the loss of most of the sought-after material. The assumptions made along the way to correct for the deficiencies in the assays are not always stated, and the useful points in a new assay are often overemphasized and the faults downplayed. We are always balancing different needs for (or availability of) speed, accuracy, precision or reproducibility, cost, sample size and number, specificity, apparatus, trained and experienced personnel, etc. The number of protein methods reported keeps growing, which is a sure indication that we have not satisfied all of the requirements.

A. Total Nitrogen and Amino Acids

The least ambiguous assay is that for total N, usually determined by Kjeldahl assay, although other methods for N are available and perhaps equally suitable. This chapter is not the place for a discussion of the ways to ensure complete determination of N, but one must be aware of the problems of relating N to protein content. The custom is to multiply N by 6.25 and call the product "protein content," which is satisfactory for routine analyses. However, for most cereals, a factor of about 5.6 is recommended (Tkachuk, 1969; Anonymous, 1979). (Corn was not tested in those studies.) My calculations from the data in Tables II and IV suggest that zein contains about 16.5% N (factor = 6.06), whereas the nonzein proteins would have an even smaller factor, depending upon the extent of amidation. Thus, the true N factor for corn may not be different from that of the other cereals. This factor, of course, makes no allowance for nonprotein amino acids or for other N-containing substances.

This problem may not be important for routine assays but is important for research purposes.

Another problem is that of expressing amino acid contents. In this chapter, mole percent has been used because it is unambiguous and can be converted into other forms without problems. Most standard solutions for amino acid analyzers are made up in molar concentrations, so the initial data appear in this form. It may bother those used to terms such as "grams of amino acid per 100 g of sample N," or "grams of amino acid per 16 g of sample N" (assuming 16% N in protein). An error of about 16% is introduced if the molecular masses of the whole amino acids are used in calculations, because protein mass comes from the amino acids minus one molecule of water per amino acid.

B. Assays for Improved Nutritional Value

The most difficult problem facing corn protein analysts is providing the breeders with satisfactory procedures for determining the nutritional value of new corn lines. Breeders must search simultaneously for improved agronomic characters, higher yields, and improved nutritional value, and thus they need rapid screening methods (Protein-Calorie Advisory Group of the U.N. System, 1976). Usually, the nutritional value can be related directly to the lysine content of the kernel. As Osborne and Mendel (1914) noted long ago, replacement of zein by other corn kernel proteins would result in an improvement in the ability of corn to support the growth of nonruminant animals. Zein provides about half of the N of the kernel but none of the lysine. On the average, the other proteins of the kernel contain satisfactory levels of lysine, and no major fraction is noted for outstanding or subaverage lysine contents (Tables I and V). Actual animal growth tests may provide the answer that is really wanted, but they are not feasible for routine screening. A recent example of correlated chemical and nutritional testing to demonstrate the superiority of o_2 hybrids is given by Eggum et al (1983). These hybrids also had good grain yields.

Unfortunately, the direct assays for lysine have various weaknesses. Acid hydrolysis followed by assay on an automatic amino acid analyzer is probably the most accurate method, but it is slow and expensive. Gas chromatographic and HPLC methods may also be satisfactory but are not yet ready for mass use. Assays after hydrolysis measure total lysine, but some of the lysine thus analyzed may not be nutritionally available because of treatments of the sample. Although various colorimetric assays claim specificity for lysine, it is difficult to prove that nothing else in a corn kernel will react.

Quick estimates of protein quality in corn can be made, provided that several simplifying correlations are accepted: 1) lysine content on a dry weight basis increases as the total N content increases (Arnold et al, 1977); 2) lysine content as percent of protein decreases when zein as percent of protein increases; 3) zein as percent of dry weight increases as the total protein content increases; and 4) the o_2 and other high-lysine mutants in corn act by decreasing the synthesis of zein (see data of Tsai et al, 1978, 1983, for examples). Lysine as percent of protein is inversely proportional to the ratio of zein to total protein; and lysine as percent of dry weight is proportional to the content of nonzein protein.

Rhodes et al (1979) evaluated two methods based on these correlations. The method of Pollmer and Fromberg (1973), which separates zein by a butanol extraction, is more sensitive than the method of Paulis et al (1974a, 1974b), which measures the turbidity of isolated zein. A third screening assay tested by Rhodes et al (1979) measured the amount of free amino acids as determined by the ninhydrin assay (Mertz et al, 1974). This procedure is based on the increase in free amino acids found in all o_2 endosperms. It has value in helping to distinguish high-lysine kernels from normal kernels, but the level of free amino acids varies independently from the levels of lysine in the mutants (Misra et al, 1975). High free amino acids may be a necessary consequence of the inability of a mutant to incorporate the available amino acids into proteins other than zein. A very desirable mutant would be one that incorporated amino acids into another storage protein, such as a globulin or "true" glutelin with high lysine content. If such an efficient mutant exists, it would not be picked out by a ninhydrin assay screen.

Zein lacks tryptophan as well as lysine, so that this essential amino acid is the second limiting amino acid in corn. Tryptophan contents are seldom reported, for acid hydrolysis of proteins destroys tryptophan. However, tryptophan assays may be as useful as lysine assays for recognizing corn lines with improved nutritional qualities (Protein-Calorie Advisory Group of the U.N. System, 1976). Tryptophan may be assayed by hydrolysis of proteins in $Ba(OH)_2$, followed by customary amino acid analysis systems (Knox et al, 1970). Tryptophan may also be released by enzymatic digestion, then determined by a colorimetric procedure involving glyoxylic acid (Opienska et al, 1963). However, the assay is affected by reagent purity (Dalby and Tsai, 1975) and by light (Nkonge and Ballance, 1982). Interlaboratory variation for the determination of tryptophan is high compared to that for other amino acids (Sarwar et al, 1985).

Another type of assay determines the dye-binding capacity (DBC) of a sample, with the assumption that an acid dye binds to the basic amino acids (Udy, 1971). The high-lysine lines of corn have increased levels of proteins with basic amino acids, and thus an increase in DBC signals the presence of a high-lysine sample. The original procedure is used to determine total N content, and for many seeds it gives results quite comparable to those of Kjeldahl N. The high-lysine mutants do not have more protein, just different protein, with the result that different standard curves are required for normal and high-lysine varieties (Fornasari et al, 1975). The actual protein content of lines with high zein levels, such as Illinois High Protein and some samples grown with high N fertility, may be underestimated by the DBC assay. The DBC method, in conjunction with a total N assay, can give an estimate of nonzein protein, because zein has as little as one fifth the basic amino acid content of nonzein proteins (Tables III and V). As a result, zein has a low color yield.

Esen (1980) used Coomassie blue R to stain zein and nonzein proteins that had been extracted separately and then precipitated onto filter paper. The adsorbed dye is a measure of each protein, although the color yield for zein is about one fourth that of most other proteins and separate standard curves are required. Zein as percent of total extractable protein (both by DBC) is correlated negatively with lysine as percent of protein for normal and o_2 samples. A very popular and simple dye-binding assay for protein is based on the color change by Coomassie blue G when bound to protein (Bradford, 1976). A modified assay (Read and Northcote, 1981) was found to work with zein, when the zein was dissolved in 55% (v/v) isopropanol (Wilson, unpublished). Again, zein had about one fourth the color yield of most other proteins. These assays can be affected by differences in dye purity among batches (Wilson, 1979). Also, Coomassie blue G and R may bind by other mechanisms to nonbasic amino acids, possibly to hydrophobic areas on the molecule. It is not safe to assume that dye-binding is always a measure of basic amino acids, let alone lysine.

The most critical problem is that of separating zein and nonzein proteins by a reproducible method that works in a similar fashion for all samples. Factors that can influence zein extractability include: genotype (normal vs. o_2), kernel hardness, age and storage conditions of the sample, grinding procedure, removal of lipids, type of alcohol and its concentration, time and temperature of extraction, shaking, and the presence of salt and/or reducing agents. Without a reducing agent, alcohols do not extract all of the zein, but with a reducing agent, they may extract other proteins with different color yields (i.e., compared to zein, RSP-1 has four times the content of basic amino acids, to which acid dyes would bind). Because of differences in physical conditions of the kernels, reproducible grinding is difficult.

Beckwith et al (1975) found a high correlation between lysine content (dry weight basis) and the ninhydrin color of cornmeal that had been extracted to remove free amino acids. Ninhydrin reacts with the ϵ-amino acid group of lysine and with the small number of terminal α-amino acid groups.

An interesting new procedure involves an enzyme-linked immunosorbent assay (ELISA) of zein (Conroy and Esen, 1984). This could bypass the need to separate zein from other proteins, a weak point in the assays mentioned above.

VII. GENETICS

Interest in the genetics of corn protein was stimulated by the discovery that the o_2 mutant has an increased level of lysine in the endosperm (Mertz et al, 1964). It was soon discovered that the o_2 endosperm contained lower-than-normal levels of zein. Because the other proteins, on the average, contain higher levels of lysine (Table I), the overall lysine proportion in the endosperm is increased. It is generally recognized that the amino acid compositions of the major protein fractions of the o_2 endosperm are not changed appreciably from those of the normal endosperm but that the proportions of the different fractions change (Sodek and Wilson, 1971; Landry and Moureaux, 1982). The changes induced by the mutation are quite variable, depending upon genotype and cultural conditions. Thus, presenting data for a typical o_2 seed is difficult. However, comparisons between normal and o_2 seed grown by the Purdue research group at two levels of nitrogen fertilizer illustrate the type of differences commonly found (Table VI). For experiment A, the normal and o_2 seeds were taken from heterozygous ears and thus received identical nutrients. The dry weights of o_2 endosperms are only slightly smaller than normal when grown with low N, but the o_2 seed does not increase in size as much as does the normal seed when adequate N is provided. Total protein per endosperm increases by 55% in the normal endosperm but by only 41% in the o_2 endosperm as a result of adequate N fertilization. However, zein increases as percent of total protein in the normal but remains about the same in the o_2 endosperm. These results suggest that most

o_2 effects are specific for the endosperm and that the vegetative tissue may not change. However, o_2 mutants have been reported to have less efficient photosynthesis (Morot-Gaudry et al, 1978).

Experiment B (Table VI) followed the changes in lysine in whole kernels with and without N fertilization. Protein increased significantly in both normal and o_2 seeds with added N, as did lysine as percent of dry weight, with the o_2 seeds maintaining their superiority. However, the increasing proportion of zein caused lysine to drop significantly as percent of protein in normal seed; it was essentially unchanged in the o_2 seed. Because of yield differences, lysine yield per hectare showed little difference for normal and o_2 seeds and was much higher with N fertilization. Tryptophan was also assayed (not shown), with results similar to those for lysine.

Tsai et al (1983) also examined reciprocal crosses of normal and o_2 hybrids. Kernel weights were greatly increased when o_2 plants were pollinated by normal pollen to give phenotypically normal seed with two doses of the o_2 genotype. Several heterozygous hybrids produced seed of the same weight as the fully normal seed. This further confirms the idea that the major limitation to yields of o_2 corn resides in the kernel, not in the vegetative portion of the plant. Tsai et al (1983) suggest that the accumulation of free amino acids and the higher osmotic potential in developing o_2 seeds inhibit the accumulation of dry matter, leading to reduced seed size and yield. They further suggest that zein synthesis, by serving as a sink for N, is necessary for maximum seed yields.

Cromwell et al (1983) found that o_2 hybrids yielded less grain than normal hybrids and that yield, protein percentage, and lysine percentage were equally increased by N fertilization. However, after fertilization, only the o_2 hybrids gave improved gains and improved feed-gain ratio when fed to chicks.

Several other mutants inhibit zein synthesis, but the ones of major interest are those that inhibit zein to a much greater extent than they do other endosperm

TABLE VI
Comparison of Normal and Opaque-2 Hybrid Seed Produced Under Conditions of Low and High Nitrogen Fertility

	No Added N		N at 201 kg/ha	
	Normal	Opaque-2	Normal	Opaque-2
Experiment A[a]				
Dry weight, mg/endosperm	214	200	268	228
Zein, mg/endosperm	6.0	4.0	10.6	5.3
Nonzein protein, mg/endosperm	8.3	8.3	11.5	12.0
Total protein, mg/endosperm	14.3	12.3	22.1	17.3
Zein, % of total	42.0	32.5	48.0	30.1
Experiment B[b]				
Protein, % dry wt	8.0	8.9	10.0[c]	9.7[c]
Lysine, % dry wt	0.31	0.42	0.34[c]	0.47[c]
Lysine, % of protein	3.9	4.7	3.4[c]	4.9
Lysine, kg/ha	13.4	12.1	27.2	29.6

[a] Data from Tsai et al (1980), used with permission. Endosperm only, taken from heterozygous plants with normal and opaque-2 kernels on the same ears.
[b] Data from Tsai et al (1983), used with permission. Whole kernels produced on normal and opaque-2 plants.
[c] Significantly different from the same genotype with no added N.

storage compounds. The fl_2 mutant appears to inhibit the synthesis of most zein polypeptides, whereas o_2 is specific for some of the A-zeins and o_7 is more specific for some of the B-zeins (Lee et al, 1976; DiFonzo et al, 1980). The double mutant with o_2 and fl_2 is not much different from o_2 alone, but the double mutant with o_2 and o_7 effectively blocks most zein synthesis.

From the beginning of studies of o_2 seed, increased amounts of nonzein protein as compared to that in normal seed have often been reported. The data of Table VI suggest that this may be only apparent. A slight increase in nonzein protein per endosperm occurred in the o_2 under high N fertility, but this difference would have been considerably exaggerated had the data been presented as percent of dry weight, because of the much lower dry weight of the o_2 endosperm. On a whole-kernel basis, Bjarnason and Pollmer (1972) found that the germ, without increasing in weight per seed, increased as a proportion of the o_2 seed. Given the higher protein content of the germ and its high content of nonzein proteins, an increase of nonzein proteins as percent of dry weight would be found. Mutations that change the proportions of protein, zein, and/or lysine as percent of either dry weight or protein are of less value if the changes are accomplished by reductions in starch (i.e., dry matter yield) or total protein (i.e., zein, the major protein fraction).

Higher lysine yields would be achieved by mutants that increase the production of globulins or glutelins, which serve as the major storage proteins of oats and rice, respectively. Higher yields of high-quality protein might also be obtained from varieties with large germs. Long-term selection for high oil increased germ size, but not protein percentage in the germ, whereas selection for high protein increased protein in both germ and endosperm (Dudley et al, 1977). The Illinois High Protein strain has about 25% protein by weight, but protein yield is not improved in proportion. The importance of the endosperm is shown by the percent of total seed protein found in the endosperm: 70% in Illinois High Oil, 75% in Illinois Low Protein, and 84% in Illinois High Protein. High endosperm protein usually means high zein, so the quality does not increase in proportion to quantity.

In summary, searches for desirable high-lysine corn varieties need to consider two steps. The first step, best demonstrated by o_2, is the reduction in zein, the low-lysine protein. The second step, not yet achieved, is that of achieving a desirable balance of amino acids, without affecting overall yields and with improved kernel hardness.

The genetics of zein gene regulation has been reviewed recently by Soave and Salamini (1984). The genes for zein have been mapped to chromosomes 4, 7, and 10, and many of the polypeptides have been identified as to size (by SDS-PAGE) and location on an IEF gel. Chromosome 4 specifies mostly A-zein polypeptides but also some B-zein polypeptides. Chromosome 7 specifies B-zein polypeptides and a D-zein. Chromosome 10 specifies an A-zein. An unexpected finding is that the o_2 gene, which regulates A-zeins, is located on chromosome 7, whereas the fl_2 gene is on chromosome 4. The o_7 gene is on chromosome 10, near the site of an A-zein, whereas this mutant affects B-zeins most drastically. These results suggest that o_2 and o_7 act by inhibiting the production of some mobile factor needed for zein synthesis. The o_2 gene appears to control the synthesis of a protein called "b-32," and in its absence zein synthesis is inhibited. Thus, the relationships of size, DNA homologies, chromosomal location, and susceptibility to the action of the o_2 or o_7 mutants will not be clear-cut.

The next few years should see a great increase in our knowledge about the identity of zein and the regulation of its synthesis. Complete identification can be made by the amino acid sequence (see references in Table IV), but this is obviously unwieldy. It should be possible to identify individual zein polypeptides by a number of techniques. The data could be presented by extending the system suggested by Soave and Salamini (1984), in which one polypeptide is identified as "Zp 20/1" for its apparent molecular mass and its IEF position (Fig. 5). Reproducible standards are needed, such as reported for IEF in agarose (Fig. 6). The following elements are available now for zein polypeptide identification: 1) apparent molecular mass class by SDS-PAGE (Fig. 2); 2) band position by IEF (Fig. 6); 3) inbred source; 4) chromosome location for structural gene (Soave and Salamini, 1984); 5) regulatory gene affecting zein, such as o_2 (Wilson, 1985b); 6) zein cDNA clone (see Table IV, also Marks and Larkins, 1982; Messing, 1983); 7) molecular mass determined from amino acid composition (Table IV); 8) peak on reversed phase HPLC (Bietz, 1983); 9) peak on phosphocellulose column (Esen, 1982); and 10) band on acid-urea PAGE (Wall et al, 1984). Wilson (1986), using only elements 1 and 2, identified over 70 different zein polypeptides in 18 corn inbreds. Bringing this type of data together will be valuable, for the whole can have more value than the sum of the disconnected parts.

Corn is the highest-yielding crop in the fertile temperate areas of the world. The ability to manipulate the protein composition in the seed should lead to an increase in the nutritional and thus the monetary value of this crop. Other chapters of this book have examples of other components of the corn seed that can be genetically controlled. We should see great advances as the molecular biologists pinpoint the means of genetic control, the biochemists identify the products of gene action, the geneticists work out the details of the genetic "manipulations," and the breeders utilize these findings in the production of new high-yielding corn lines with specific abilities to produce more of certain desired products and less of those with less value. The combined talents of these diverse researchers are needed to bring about a genetic engineering revolution in corn production.

VIII. USES OF CORN PROTEIN

A. Food and Feed

Most corn protein is consumed directly by humans or animals along with the other seed components. Yet nearly half a million metric tons of corn protein was marketed by the U.S. wet-milling industry in 1984 as corn gluten meal, with 60–70% protein. Much of this was also used directly as feed products (Chapters 12 and 15), but it can serve as a resource for various high-protein products. During the process of steeping in SO_2, about half of the cysteine and cystine are converted to S-sulfocysteine as the inter- and intramolecular disulfide bonds are cleaved (Neumann et al, 1984a). During the wet-milling process, much of the protein usually found in the albumin, globulin, and some glutelin fractions is removed in various soluble fractions. The corn gluten meal is enriched in zein, which is now almost completely soluble in 70% ethanol without the use of reducing agents. Corn gluten meal can be blended with defatted soy flour to produce food products in which the high sulfur amino acid content of corn protein improves the nutritional value of the soy flour (Neumann et al, 1984b). Corn gluten and zein have recently been approved as safe for use in food products (Federal Register, 1985).

B. Industrial Uses and Potential

Throughout the past 20 or 30 years, the industrial use of zein has fluctuated considerably, as uses for its unique film-forming properties were found (Reiners et al, 1973), then displaced by less expensive synthetic compounds (Anonymous, 1978). Zein is purified from corn gluten by solvent processes that start with hot alcohol extraction (Reiners et al, 1973). Details of the process presently being used by Freeman Industries (Tuckahoe, NY), the only company now known to be producing zein, are proprietary. Zein is used as a coating for pharmaceutical tablets and for coating nuts and candies, where it forms a moisture-resistant barrier. Preliminary electrophoretic studies on several samples of commercial zein revealed that the properties were not greatly changed from samples prepared in the laboratory and that extensive deamidation had not occurred (C. M. Wilson, unpublished). Thus, knowledge gained in the research laboratory may apply with little change to the commercial product.

ACKNOWLEDGMENT

This chapter is the product of cooperative investigations of the U. S. Department of Agriculture, Agricultural Research Service, and the Illinois Agricultural Experiment Station, Department of Agronomy, University of Illinois, Urbana.

LITERATURE CITED

ABE, M., ARAI, S., KATO, H., and FUJIMAKI, M. 1981. Electrophoretical analysis of zein and isolation of its components. Agric. Biol. Chem. 45:1467-1472.

ALTSCHULER, M. I., and SCHWARTZ, D. 1984. Effects of phytate on maize alcohol dehydrogenase isozymes. Maydica 29:77-87.

ANONYMOUS. 1978. Zein: A natural moisture barrier. Food Eng. (May) pp. 104-105.

ANONYMOUS. 1979. J. Assoc. Off. Anal. Chem. 62:370.

ARGOS, P., PEDERSEN, K., MARKS, M. D., and LARKINS, B. A. 1982. A structural model for maize zein proteins. J. Biol. Chem. 257:9984-9990.

ARNOLD, J. M., BAUMAN, L. F., and AYCOCK, H. S. 1977. Interrelations among protein, lysine, oil, certain mineral element concentrations, and physical kernel characteristics in two maize populations. Crop Sci. 17:421-425.

ARRUDA, P., and da SILVA, W. J. 1979. Amino acid composition of vascular sap of maize ear peduncle. Phytochemistry 18:409-410.

BAENZIGER, P. S., and GLOVER, D. V. 1977. Protein body size and distribution and protein matrix morphology in various endosperm mutants of Zea mays L. Crop Sci. 17:415-421.

BECKWITH, A. C., PAULIS, J. W., and WALL, J. S. 1975. Direct estimation of lysine in corn meals by the ninhydrin color reaction. J. Agric. Food Chem. 23:194-196.

BIETZ, J. A. 1983. Separation of cereal proteins by reversed-phase high-performance liquid chromatography. J. Chromatogr. 255:219-238.

BJARNASON, M., and POLLMER, W. G. 1972. The maize germ: Its role as a contributing factor to protein quantity and quality. Z. Pflanzenzuecht. 68:83-89.

BLESSIN, C. W., DEATHERAGE, W. L.,

CAVINS, J. F., GARCIA, W. J., and INGLETT, G. E. 1979. Preparation and properties of defatted flours from dry-milled yellow, white, and high-lysine corn germ. Cereal Chem. 56:105-109.

BRADFORD, M. M. 1976. A rapid and sensitive method for the quantitation of microgram quantities of protein utilizing the principle of protein-dye binding. Anal. Biochem. 72:248-254.

BURR, B., and BURR, F. A. 1976. Zein synthesis in maize endosperm by polyribosomes attached to protein bodies. Proc. Natl. Acad. Sci. U.S.A. 73:515-519.

BYERS, M., MIFLIN, B. J., and SMITH, S. J. 1983. A quantitative comparison of the extraction of protein fractions from wheat grain by different solvents, and of the polypeptide and amino acid composition of the alcohol-soluble proteins. J. Sci. Food Agric. 34:447-462.

CIMMYT-PURDUE, eds. 1975. High-Quality Protein Maize. Dowden, Hutchinson & Ross, Stroudsburg, PA. 524 pp.

CHRISTIANSON, D. D. 1970. Genetic variation in cereal grains and processing effects on cereal flours as evaluated by scanning electron microscopy. Pages 161-168 in: Proc. Ann. Scanning Electron Microscope Symp., 3rd. Ill. Inst. Technol. Research Inst., Chicago.

CHRISTIANSON, D. D., NIELSEN, H. C., KHOO, U., WOLF, M. J., and WALL, J. S. 1969. Isolation and chemical composition of protein bodies and matrix proteins in corn endosperms. Cereal Chem. 46:372-381.

CHRISTIANSON, D. D., KHOO, U., NIELSEN, H. C., and WALL, J. S. 1974. Influence of *opaque-2* and *floury-2* genes on formation of proteins in particulates of corn endosperm. Plant Physiol. 53:851-855.

CONROY, J. M., and ESEN, A. 1984. An enzyme-linked immunosorbent assay for zein and other proteins using unconventional solvents for antigen adsorption. Anal. Biochem. 137:182-187.

CRAINE, E. M., and FAHRENHOLTZ, K. E. 1958. The proteins in water extracts of corn. Cereal Chem. 35:245-259.

CROMWELL, G. L., BITZER, M. J., STAHLY, T. S., and JOHNSON, T. H. 1983. Effects of soil nitrogen fertility on the protein and lysine content and nutritional value of normal and *opaque-2* corn. J. Anim. Sci. 57:1345-1351.

CROSS, J. W., and ADAMS, W. R., Jr. 1983a Embryo-specific globulins from *Zea mays* L. and their subunit composition. J. Agric. Food Chem. 31:534-538.

CROSS, J. W., and ADAMS, W. R. 1983b. Differences in the embryo-specific globulins among maize inbred lines and their hybrids. Crop Sci. 23:1160-1162.

DALBY, A., and TSAI, C.-Y. 1975. Acetic anhydride requirement in the colorimetric determination of tryptophan. Anal. Biochem. 63:283-285.

DAUSSANT, J., MOSSE, J., and VAUGHAN, J., eds. 1983. Seed Proteins. Academic Press, New York. 335 pages.

DI FONZO, N., FORNASARI, E., SALAMINI, F., and SOAVE, C. 1977. SDS-protein subunits in normal, *opaque-2* and *floury-2* maize endosperms. Maydica 22:77-88.

DI FONZO, N., FORNASARI, E., SALAMINI, F., REGGIANI, R., and SOAVE, C. 1980. Interaction of maize mutants *floury-2* and *opaque-7* with *opaque-2* in the synthesis of endosperm proteins. J. Hered. 71:397-402.

DIERKS-VENTLING, C. 1981. Storage proteins in *Zea mays* L.: Interrelationships of albumins, globulins and zeins in the *opaque-2* mutation. Eur. J. Biochem. 120:177-182.

DIERKS-VENTLING, C., and VENTLING, D. 1982. Tissue-specific immunofluorescent localization of zein and globulin in *Zea mays* (L.) seeds. FEBS Lett. 144:167-172.

DUDLEY, J. W., LAMBERT, R. J., and DE LA ROCHE, I. A. 1977. Genetic analysis of crosses among corn strains divergently selected for percent oil and protein. Crop Sci. 17:111-117.

DUVICK, D. N. 1961. Protein granules of maize endosperm cells. Cereal Chem. 38:374-385.

EGGUM, B. O., DUMANOVIC, J., MISEVIC, D., and DENIC, M. 1983. Grain yield and nutritive value of high oil, opaque and waxy maize hybrids. J. Cereal Sci. 1:139-145.

ESEN, A. 1980. Estimation of protein quality and quantity in corn (*Zea mays* L.) by assaying protein in two solubility fractions. J. Agric. Food Chem. 28:529-532.

ESEN, A. 1982. Chromatography of zein on phosphocellulose and sulfopropyl sephadex. Cereal Chem. 59:272-276.

ESEN, A., BIETZ, J. A., PAULIS, J. W., and WALL, J. S. 1981. Fractionation of alcohol-soluble reduced corn glutelins on phosphocellulose and partial characterization of two proline-rich fractions. Cereal Chem. 58:534-537.

ESEN, A., BIETZ, J. A., PAULIS, J. W., and WALL, J. S. 1982. Tandem repeats in the N-terminal sequence of a proline-rich protein from corn endosperm. Nature 296:678-679.

ESEN, A., BIETZ, J. A., PAULIS, J. W., and WALL, J. S. 1985. Isolation and characterization of a methionine-rich protein from maize endosperm. J. Cereal Sci. 3:143-152.

FAO/WHO. 1973. Nutrition Report Series No. 52. Expert Committee on Energy and Protein Requirements, Food and Agriculture Organization of the U.N., Rome. (Also, World Health Organization Tech. Report Series No. 522, Geneva.)

FEDERAL REGISTER. 1985. Wheat gluten, corn gluten, and zein: Affirmation of GRAS status. 50:8997-8999.

FORNASARI, E., GENTINETTA, E., MAGGIORE, T., SALAMINI, F., STANCA, A. M., and LORENZONI, C. 1975. Efficacy of the DBC test in the identification of maize inbreds with high-quality proteins. Maydica 20:185-195.

FORNASARI, E., DI FONZO, N., SALAMINI, F., REGGIANI, R., and SOAVE, C. 1982. *Floury-2* and *opaque-7* interaction in the synthesis of zein polypeptides. Maydica 27:185-189.

GALANTE, E., VITALE, A., MANZOCCHI, L., SOAVE, C., and SALAMINI, F. 1983. Genetic control of a membrane component and zein deposition in maize endosperm. Molec. Gen. Genet. 192:316-321.

GERAGHTY, D., PEIFER, M. A., RUBENSTEIN, I., and MESSING, J. 1981. The primary structure of a plant storage protein: Zein. Nucl. Acids Res. 9:5163-5175.

GERAGHTY, D. E., MESSING, J., and RUBENSTEIN, I. 1982. Sequence analysis and comparison of cDNAs of the zein multigene family. EMBO J. 1:1329-1335.

GIANAZZA, E., VIGLIENGHI, V., RIGHETTI, P. G., SALAMINI, F., and SOAVE, C. 1977. Amino acid composition of zein molecular components. Phytochemistry 16:315-317.

HANSON, J. B., HAGEMAN, R. H., and FISHER, M. E. 1960. The association of carbohydrases with the mitochondria of corn scutellum. Agron. J. 52:49-52.

HEIDECKER, G., and MESSING, J. 1986. Structural analysis of plant genes. Annu. Rev. Plant Physiol. 37:439-466.

HOJIMA, Y., PIERCE, J. V., and PISANO, J. J. 1980. Hageman factor fragment inhibitor in corn seeds: Purification and characterization. Thromb. Res. 20:149-162.

HU, N.-T., PEIFER, M. A., HEIDECKER, G., MESSING, J., and RUBENSTEIN, I. 1982. Primary structure of a genomic zein sequence of maize. EMBO J. 1:1337-1342.

HURKMAN, W. J., SMITH, L. D., RICHTER, J., and LARKINS, B. A. 1981. Subcellular compartmentalization of maize storage proteins in *Xenopus* oocytes injected with zein messenger RNAs. J. Cell Biol. 89:292-299.

JONES, B. L., and COOPER, D. B. 1980. Purification and characterization of a corn (*Zea mays*) protein similar to purothionins. J. Agric. Food Chem. 28:904-908.

KHAVKIN, E. E., MISHARIN, S. I., MARKOV, Y. Y., and PESHKOVA, A. A. 1978. Identification of embryonal antigens of maize: Globulins as primary reserve proteins of the embryo. Planta 143:11-20.

KHOO, U., and WOLF, M. J. 1970. Origin and development of protein granules in maize endosperm. Am. J. Bot. 57:1042-1050.

KNOX, R., KOHLER, G. O., PALTER, R., and WALKER, H. G. 1970. Determination of tryptophan in feeds. Anal. Biochem. 36:136-143.

LANDRY, J., and MOUREAUX, T. 1970. Hétérogénéité des glutélines du grain de maïs: Extraction sélective et composition en acides aminés des trois fractions isolées. Bull. Soc. Chim. Biol. 52:1021-1037.

LANDRY, J., and MOUREAUX, T. 1980. Distribution and amino acid composition of protein groups located in different histological parts of maize grain. J. Agric. Food Chem. 28:1186-1191.

LANDRY, J., and MOUREAUX, T. 1981. Physicochemical properties of maize glutelins as influenced by their isolation conditions. J. Agric. Food Chem. 29:1205-1212.

LANDRY, J., and MOUREAUX, T. 1982. Distribution and amino acid composition of protein fractions in *opaque-2* maize grains. Phytochemistry 21:1865-1869.

LANDRY, J., PAULIS, J. W., and FEY, D. A. 1983. Relationship between alcohol-soluble proteins extracted from maize by different methods. J. Agric. Food Chem. 31:1317-1322.

LARKINS, B. A. 1981. Seed storage proteins: Characterization and biosynthesis. Pages 449-489 in: The Biochemistry of Plants, Vol. 6. A. Marcus, ed. Academic Press, New York.

LARKINS, B. A. 1983. Genetic engineering of seed storage proteins. Pages 93-118 in: Genetic Engineering of Plants. T. Kosuge, D. P. Meredith, and A. Hollaender, eds. Plenum Press, New York.

LARKINS, B. A., PEDERSEN, K., MARKS, M. D., and WILSON, D. R. 1984. The zein proteins of maize endosperm. Trends Biochem. Sci. 9:306-308.

LEE, K. H., JONES, R. A., DALBY, A., and TSAI, C. Y. 1976. Genetic regulation of storage protein content in maize endosperm. Biochem. Genet. 14:641-650.

LUDEVID, M., TORRENT, M., MARTINEZ-IZQUIRDO, J. A., PUIGDOMENECH, R., and PALAU, J. 1984. Subcellular localization of glutelin-2 in maize (*Zea mays* L.) endosperm. Plant Molec. Biol. 3:227-234.

MAHONEY, W. C., HERMODSON, M. A., JONES, B., POWERS, D. D., CORFMAN, R. S., and REECK, G. R. 1984. Amino acid sequence and secondary structural analysis of the corn inhibitor of trypsin and activated Hageman factor. J. Biol. Chem. 259:8412-8416.

MARKS, M. D., and LARKINS, B. A. 1982. Analysis of sequence microheterogeneity among zein messenger RNAs. J. Biol. Chem. 257:9976-9983.

MELCHER, U., and FRAIJ, B. 1980. Methionine-rich protein fraction prepared by cryoprecipitation from extracts of corn meal. J. Agric. Food Chem. 28:1334-1336.

MERTZ, E. T., BATES, L. S., and NELSON, O. E. 1964. Mutant gene that changes protein composition and increases lysine content of maize endosperm. Science 145:279-280.

MERTZ, E. T., MISRA, P. S., and JAMBUNATHAN, R. 1974. Rapid ninhydrin color test for screening high-lysine mutants of maize, sorghum, barley, and other cereal grains. Cereal Chem. 51:304-307.

MESSING, J. 1983. The manipulation of zein genes to improve the nutritional value of corn. Trends Biotechnol. 1(2):1-6.

MIFLIN, B. J., and BURGESS, S. R. 1982. Protein bodies from developing seeds of barley, maize, wheat and peas: The effect of protease treatment. J. Exp. Bot. 33:251-260.

MIFLIN, B. J., and SHEWRY, P. R. 1979. The biology and biochemistry of cereal seed prolamins. Pages 137-158 in: Seed Protein Improvement in Cereals and Grain Legumes, Vol. I. Int. Atomic Energy Agency, Vienna.

MIFLIN, B. J., BURGESS, S. R., and SHEWRY, P. R. 1981. The development of protein bodies in the storage tissues of seeds: Subcellular separations of homogenates of barley, maize, and wheat endosperms and of pea cotyledons. J. Exp. Bot. 32:199-219.

MIFLIN, B. J., FIELD, J. M., and SHEWRY, P. F. 1983. Cereal storage proteins and their effect on technological properties. Pages 255-319 in: Seed Proteins. J. Daussant, J. Mosse, and J. Vaughan, eds. Academic Press, New York.

MISRA, P. S., MERTZ, E. T., and GLOVER, D. V. 1975. Studies on corn proteins. VIII. Free amino acid content of *opaque*-2 double mutants. Cereal Chem. 52:844-848.

MITCHELL, H. L., PARRISH, D. B., CORMEY, M., and WASSOM, C. E. 1976. Effect of corn trypsin inhibitor on growth of rats. J. Agric. Food Chem. 24:1254-1255.

MOROT-GAUDRY, J. F., THOMAS, D. A., DEROCHE, M. E., and CHARTIER, P. 1978. Growth, leaf optical properties, chlorophyll content and net assimilation rate in maize seedlings with and without the gene *opaque-2*. Photosynthetica 12:284-289.

NELSON, O. E. 1980. Genetic control of polysaccharide and storage protein synthesis in the endosperms of barley, maize, and sorghum. Pages 41-71 in: Advances in Cereal Science and Technology, Vol. 3. Y. Pomeranz, ed. Am. Assoc. Cereal Chem., St. Paul, MN.

NEUMANN, P. E., WALL, J. S., and WALKER, C. E. 1984a. Chemical and physical properties of proteins in wet-milled corn gluten. Cereal Chem. 61:353-356.

NEUMANN, P. E., JASBERG, B. K., WALL, J. S., and WALKER, C. E. 1984b. Uniquely textured products obtained by coextrusion of corn gluten meal and soy flour. Cereal Chem. 61:439-445.

NIELSEN, H. C., WALL, J. S., and INGLETT, G. E. 1979. Flour containing protein and fiber made from wet-mill corn germ, with potential food use. Cereal Chem. 56:144-146.

NKONGE, C., and BALLANCE, G. M. 1982. Colorimetric determination of tryptophan: The effect of light on the acetic anhydride requirements. Anal. Biochem. 122:6-9.

O'DELL, B. L., and DE BOLAND, A. 1976. Complexation of phytate with proteins and cations in corn germ and oilseed meals. J. Agric. Food Chem. 24:804-808.

O'DELL, B. L., DE BOLAND, A. R., and KOIRTYOHANN, S. R. 1972. Distribution of phytate and nutritionally important elements among the morphological components of cereal grains. J. Agric. Food Chem. 20:718-720.

OPIENSKA, I., CHAREZINSKI, M., and BERBEC, H. 1963. A new, rapid method of determining tryptophan. Anal. Biochem. 6:69-76.

OSBORNE, T. B. 1924. The Vegetable Proteins, 2nd ed. Longmans, Green and Co., London. 154 pp.

OSBORNE, T. B., and MENDEL, B. 1914. Nutritive properties of proteins of the maize kernel. J. Biol. Chem. 18:1-16.

PAULIS, J. W. 1981. Disulfide structures of zein proteins from corn endosperm. Cereal Chem. 58:542-546.

PAULIS, J. W., and WALL, J. S. 1977. Fractionation and characterization of alcohol-soluble reduced corn endosperm glutelin proteins. Cereal Chem. 54:1223-1228.

PAULIS, J. W., WALL, J. S., and KWOLEK, W. F. 1974a. A rapid turbidimetric analysis for zein in corn and its correlation with lysine content. J. Agric. Food Chem. 22:313-317.

PAULIS, J. W., WALL, J. S., KWOLEK,

W. F., and DONALDSON, G. L. 1974b. Selection of high-lysine corns with varied kernel characteristics and compositions by a rapid turbidimetric assay for zein. J. Agric. Food Chem. 22:318-323.

PEDERSEN, K., DEVEREUX, J., WILSON, D. R., SHELDON, E., and LARKINS, B. A. 1982. Cloning and sequence analysis reveal structural variation among related zein genes in maize. Cell 29:1015-1026.

PEDERSEN, K., ARGOS, P., NARAVANA, S. V. L., and LARKINS, B. A. 1986. Sequence analysis and characterization of a maize gene encoding a high-sulfur zein protein of M_r 15,000. J. Biol. Chem. 261:6279-6284.

PHILLIPS, R. L., and McCLURE, B. A. 1985. Elevated protein-bound methionine in seeds of a maize line resistant to lysine plus threonine. Cereal Chem. 62:213-218.

POLLMER, W. G., and FROMBERG, H. K. 1973. Improved lysine estimation in maize (*Zea mays* L.) with alcoholic extraction at high temperature. Cereal Res. Commun. 1(2):45-53.

PRAT, S., CORTADAS, J., PUIGDOMENECH, P., and PALAU, J. 1985. Nucleic acid (cDNA) and amino acid sequences of the maize endosperm protein glutelin-2. Nucleic Acids Res. 13:1493-1504.

PROTEIN-CALORIE ADVISORY GROUP OF THE U.N. SYSTEM. 1976. Protein methods for cereal breeders as related to human nutritional requirements. (Guideline 16—Cereal Breeders' Protein Methods). Pages 378-408 in: Advances in Cereal Science and Technology, Vol. 1. Y. Pomeranz, ed. Am. Assoc. Cereal Chem., St. Paul, MN.

READ, S. M., and NORTHCOTE, D. H. 1981. Minimization of variation in the response to different proteins of the Coomassie Blue G dye-binding assay for protein. Anal. Biochem. 116:53-64.

REINERS, R. A., WALL, J. S., and INGLETT, G. E. 1973. Corn proteins: Potential for their industrial use. Pages 285-302 in: Industrial Uses of Cereals. Y. Pomeranz, ed. Am. Assoc. Cereal Chem., St. Paul, MN.

RHODES, A. P., BJARNASON, M., and POLLMER, W. G. 1979. An evaluation of three methods for the selection of high lysine genotypes of maize. J. Agric. Food Chem. 27:1266-1270.

ROBUTTI, J. L., HOSENEY, R. C., and WASSOM, C. E. 1974. Modified *opaque-2* corn endosperms. II. Structure viewed with a scanning electron microscope. Cereal Chem. 51:173-180.

RUBENSTEIN, I., and GERAGHTY, D. E. 1986. The genetic organization of zein. Pages 297-315 in: Advances in Cereal Science and Technology, Vol. 8. Y. Pomeranz, ed. Am. Assoc. Cereal Chem., St. Paul, MN.

SARWAR, G., BLAIR, R., FRIEDMAN, M., GUMBMANN, M. R., HACKLER, L., PELLETT, P. L., and SMITH, T. K. 1985. Comparison of interlaboratory variation in amino acid analysis and rat growth assays for evaluating protein quality. J. Assoc. Off. Anal. Chem. 68:52-56.

SCHWARTZ, D. 1979. Analysis of the size alleles of the *Pro* gene in maize—Evidence for a mutant protein processor. Molec. Gen. Genet. 174:233-240.

SHEWRY, P. R., and MIFLIN, B. J. 1985. Seed storage proteins of economically important cereals. Pages 1-83 in: Advances in Cereal Science and Technology, Vol. 7. Y. Pomeranz, ed. Am. Assoc. Cereal Chem., St. Paul, MN.

SOAVE, C., and SALAMINI, F. 1984. Organization and regulation of zein genes in maize endosperm. Phil. Trans. R. Soc. London, Ser. B. 304:341-347.

SODEK, L., and WILSON, C. M. 1971. Amino acid composition of proteins isolated from normal, *opaque-2*, and *floury-2* corn endosperms by a modified Osborne procedure. J. Agric. Food Chem. 19:1144-1150.

SU, J. C., and PREISS, J. 1978. Purification and properties of sucrose synthase from maize kernels. Plant Physiol. 61:389-393.

SUNG, T. M., and LAMBERT, R. J. 1983. Ninhydrin color test for screening modified endosperm *opaque-*2 maize. Cereal Chem. 60:84-85.

TAYLOR, J. R. N., SCHUSSLER, L., and LIEBENBERG, N. V. D. W. 1984. Location of zein-2 and crosslinked kafirin in maize and sorghum protein bodies. J. Cereal Sci. 2:249-255.

TKACHUK, R. 1969. Nitrogen-to-protein conversion factors for cereals and oilseed meals. Cereal Chem. 46:419-423.

TSAI, C. Y. 1979. Tissue-specific zein synthesis in maize kernel. Biochem. Genet. 17:1109-1120.

TSAI, C. Y. 1980. Note on the effect of reducing agent on zein preparation. Cereal Chem. 57:288-290.

TSAI, C. Y. 1983. Genetics of storage protein in maize. Pages 103-138 in: Plant Breeding Reviews, Vol. 1. J. Janick, ed. Avi Publ., Westport, CT.

TSAI, C. Y., HUBER, D. M., and WARREN, H. L. 1978. Relationship of the kernel sink for N to maize productivity. Crop Sci. 18:399-405.

TSAI, C. Y., HUBER, D. M., and WARREN,

H. L. 1980. A proposed role of zein and glutelin as N sinks in maize. Plant Physiol. 66:330-333.

TSAI, C. Y., WARREN, H. L., HUBER, D. M., and BRESSAN, R. A. 1983. Interaction between the kernel N sink, grain yield and protein nutritional quality of maize. J. Sci. Food Agric. 34:255-263.

UDY, D. C. 1971. Improved dye method for estimating protein. J. Am. Oil Chem. Soc. 48:29A-33A.

USDA. 1985. Agricultural Statistics 1985. U.S. Dept. Agric., Washington, DC. 351 pp.

USDA. 1986. Oil Crops Situation Yearbook. U.S. Dept. Agric., Econ. Res. Serv. OCS-11, July. 34 pp.

VITALE, A., SOAVE, C., and GALANTE, E. 1980. Peptide mapping of IEF zein components from maize. Plant Sci. Lett. 18:57-64.

VITALE, A., SMANIOTTO, E., LONGHI, R., and GALANTE, E. 1982. Reduced soluble proteins associated with maize endosperm protein bodies. J. Exp. Bot. 33:439-448.

WALL, J. S., and PAULIS, J. W. 1978. Corn and sorghum grain proteins. Pages 135-219 in: Advances in Cereal Science and Technology, Vol. 2. Y. Pomeranz, ed. Am. Assoc. Cereal Chem., St., Paul, MN.

WALL, J. S., JAMES, C., and CAVINS, J. F. 1971. Nutritive value of protein in hominy feed fractions. Cereal Chem. 48:456-465.

WALL, J. S., FEY, D. A., PAULIS, J. W., and LANDRY, J. 1984. Improved two-dimensional electrophoretic separation of zein proteins: Application to study of zein inheritance in corn genotypes. Cereal Chem. 61:141-146.

WILSON, C. M. 1967. Purification of a corn ribonuclease. J. Biol. Chem. 242:2260-2263.

WILSON, C. M. 1968. Plant nucleases. I. Separation and purification of two ribonucleases and one nuclease from corn. Plant Physiol. 43:1332-1338.

WILSON, C. M. 1979. Studies and critique of amido black 10B, Coomassie blue R, and fast green FCF as stains for proteins after polyacrylamide gel electrophoresis. Anal. Biochem. 96:263-278.

WILSON, C. M. 1981. Variations in soluble endosperm proteins of corn (*Zea mays* L.) inbreds as detected by disc gel electrophoresis. Cereal Chem. 58:401-408.

WILSON, C. M. 1983. Seed protein fractions of maize, sorghum, and related cereals. Pages 271-307 in: Seed Proteins: Biochemistry, Genetics, Nutritive Value. W. Gottschalk and H. P. Muller, eds. M. Nijhoff/Junk, The Hague.

WILSON, C. M. 1984. Isoelectric focusing of zein in agarose. Cereal Chem. 61:198-200.

WILSON, C. M. 1985a. Mapping of zein polypeptides after isoelectric focusing on agarose gels. Biochem. Genet. 23:115-124.

WILSON, C. M. 1985b. A nomenclature for zein polypeptides based on isoelectric focusing and sodium dodecyl sulfate polyacrylamide gel electrophoresis. Cereal Chem. 62:361-365.

WILSON, C. M. 1986. Serial analysis of zein by isoelectric focusing and sodium dodecyl sulfate gel electrophoresis. Plant Physiol. 82:196-202.

WILSON, C. M., SHEWRY, P. R., and MIFLIN, B. J. 1981. Maize endosperm proteins compared by sodium dodecyl sulfate gel electrophoresis and isoelectric focusing. Cereal Chem. 58:275-281.

WOLF, M. J., KHOO, U., and SECKINGER, H. L. 1969. Distribution and subcellular structure of endosperm protein in varieties of ordinary and high-lysine maize. Cereal Chem. 46:253-263.

CHAPTER 10

LIPIDS OF THE KERNEL

EVELYN J. WEBER
U.S. Department of Agriculture
Agricultural Research Service
Department of Agronomy
University of Illinois
Urbana, Illinois

I. INTRODUCTION

Although corn is not considered an oilseed, the U.S. corn crop is so large that, despite the relatively low oil content (4.4%, dry basis; Watson, 1984) of the kernel, the total amount of corn oil produced is enormous. Corn production in the United States usually exceeds the combined total production of all the other major domestic grains (wheat, oats, barley, rye, and soybeans). Soybean, the major oilseed in the United States, contains about 20% oil (Erickson, 1983), but the average yield per hectare of corn is more than three and one-half times that of soybeans. Production figures (Table II of Chapter 7) show that the 1984/85 corn crop contained 8.38×10^6 t of oil, compared to 10.13×10^6 t of oil in the soybean crop (USDA, 1985). Only a small fraction of this large supply of corn oil is available, because the major portion of the corn crop is fed to farm animals. In the 1984/85 crop year, 53.6% of the U.S. corn crop was used domestically for livestock feed, 24% was exported, and 13.9% was consumed for food, alcohol, and industrial uses (USDA, 1986), the only sources of isolated corn oil (Table IV of Chapter 7). Over the past 20 years, the average oil content of corn processed at milling plants in the U.S. Midwest has declined from about 4.8 to 4.4% (Earle, 1977). This decline may be due to genetic modifications of commercial hybrids over the years (Weber, 1983a).

Triacylglycerols (triglycerides) make up the major fraction (98.8%) of refined, commercial corn oil (Anderson and Watson, 1982) and are the predominant storage lipids in the kernel, but many other types of lipids are also present in the corn grain. Lipids are a chemically heterogeneous group of substances that have in common their solubility in organic solvents such as petroleum ether, hexane, or chloroform-methanol. Phospholipids, glycolipids, hydrocarbons, sterols, free fatty acids, carotenoids (vitamin A precursors), tocols (vitamin E), and waxes are all lipids that are found in corn grain. These lipids are present in smaller amounts than the triacylglycerols (TG), and although the exact functions of these lipids are still relatively unknown and may be diverse, many are believed to be important membrane constituents (Raison, 1980). The phospholipids, carotenoids and tocols also have antioxidant properties (Ames, 1983; Hildebrand et al, 1984).

Morrison (1978, 1983) and Weber (1973, 1978, 1983a) have previously written reviews that included corn lipids. This review attempts to integrate recent literature into an overall view of our knowledge of the lipids of the corn kernel.

II. OIL CONTENT

Among the cereal grains, only pearl millet, with 5.4% oil (Weber, 1983b), and oat groats, with 7% oil, have higher oil contents, on the average, than commercial corn hybrids (4.4%). Brown rice (2.3% oil), wheat (1.9%), barley (2.1%), and sorghum (3.4%) all have lower oil contents. Traditionally, oil content has been measured as the amount of lipid extracted from ground grain by hexane, petroleum ether, or diethyl ether. These nonpolar solvents extract mainly TG. Hexane is the solvent commonly used commercially to extract the oil from corn germs (Mounts and Anderson, 1983).

Corn oil occurs almost exclusively in the cells of the scutellum portion of the germ (Chapter 3). Specifically, the oil is deposited in microscopic droplets known as oil bodies. A thin, single-line electron-dense layer, 2.5–4.0 nm in thickness, surrounds each oil body (Trelease, 1969). This layer is believed to be composed of one half of a normal, tripartite (protein-lipid-protein) unit membrane, with the lipophilic side oriented inward toward the lipid matrix (Yatsu and Jacks, 1972). The diameter of the oil bodies as measured by Wang et al (1984) averaged 1.31 μm in the Illinois High Oil strain of corn (IHO—18% oil), whereas in the Illinois Low Oil strain (ILO—<0.5% oil), the mean diameter was 1.09 μm (statistically different at a 0.01 level of significance). No data were given, but it may be assumed that the mean oil body size for conventional, dent corn would lie between these two extremes. Although these investigators did not determine the relative number of oil bodies, IHO must certainly have a greater number, because the oil bodies of IHO and ILO had only a 1.7-fold difference in size, whereas the two strains have nearly a 40-fold difference in oil content. Oil bodies in corn endosperm are much smaller than in the germ and were measured by Ratković et al (1978) at less than 0.1 μm in diameter.

Oil content per kernel is affected by the position of the kernel on the corn ear. Lambert et al (1967) found that oil content may vary by 0.1–0.6% between kernels from the tip, middle, and basal parts of the ear. The highest percentage of oil occurred in the middle of the ear. The middle kernels are the most uniform in size, with the largest kernels at the base and the smallest kernels at the tip of the ear. When kernels are selected for oil analysis, samples should be taken from the middle of the ear.

A. Effect of Agronomic Practices on Oil Content

Earle (1977) obtained data from three commercial processors on the variation in oil content of corn by crop years from 1917 to 1972. The range was only from 4.0 to 4.9%. The data from the three companies were in good agreement, indicating that the variations in oil content were real, but no correlations between the oil contents and variations in temperature, rainfall, or fertilization rates were found.

Welch (1969) found that the addition of each of the fertilizer elements N, P, and K increased the oil content of the grain slightly, but the most important effect was the increased grain yield, which produced more oil per unit of land area. Jellum et al (1973) observed that increasing rates of N increased protein percentages but had no effect on oil percentage or fatty acid composition of the oil. In the same study, boron fertilization had no effect on protein or oil content. Genter et al (1956) found that N, P, and K all increased yield but had no appreciable effect on the percentage of oil in corn grain. The corn hybrids were a more important source of variation in oil percentage than were fertilizer, location, or rate of planting. Jellum and Marion (1966) also established that genetic factors had a greater influence on oil content than did environmental factors such as planting dates, location, and year.

Eleven different herbicides have been tested singly and in combinations for their effects on corn oil quantity and composition (Penner and Meggitt, 1974; Wilkinson and Hardcastle, 1973, 1974). None of the treatments, whether incorporated before planting or applied pre- or postemergence, affected oil percentage. When effects on fatty acid composition were noted, the changes were minor. Competition from failure to control weeds did not alter oil content or fatty acid composition.

Severe weather and disease may lower the oil content of corn grain. When corn plants were subjected during grain fill to drought that reduced kernel weight to 52% of that of the control (Jurgens et al, 1978), the drought increased the protein content of the grain from 8.3 to 11.0% but decreased the oil content from 3.8 to 3.1%. In 1970, when an epidemic of southern corn leaf blight (*Helminthosporium maydis*) occurred, no effect was noted on the starch and protein content of the grain, but oil content was reduced essentially in proportion to the amount of blight damage (Freeman, 1973).

B. Genetic Control of Oil Content

Oil content is a highly heritable trait in corn. The classic experiment in breeding corn for high and low oil contents was started at the University of Illinois in 1896 and is still being conducted (Dudley, 1974, 1977). The original Burr's White corn had 4.7% oil. In 1984 after 85 generations of mass selection among ears for high and low oil contents, the IHO strain had 20.4% oil and the ILO strain 0.3% oil (J. W. Dudley, personal communication). An important finding from this long-term experiment is that significant variability for oil still exists in the IHO strain (Dudley, 1977) and further increases in oil should occur as selection is continued. Both germ size and oil percentage in the germ have increased in IHO, but endosperm and total grain weight have decreased (Curtis et al, 1968). With selection only for oil, the yield of IHO has fallen to about 30% of that of commercial hybrids.

A significant development in breeding for higher-oil corn has been the adaptation of wide-line nuclear magnetic resonance (NMR) spectroscopy to nondestructive analysis of oil content in corn grain (Bauman et al, 1963; Alexander et al, 1967; Watson and Freeman, 1975). Large numbers of samples can be screened because the scan time may be as brief as 2 sec. Single-kernel

NMR screening and recurrent selection have facilitated the development of populations with higher oil. D. E. Alexander (Univ. of Illinois, Urbana, personal communication) has achieved an increase in oil content in Alexho Synthetic from 4.4 to 16.4% in only 23 cycles, a gain of about 0.5% per year. IHO, in the long-term experiment originally based on mass selection among ears and destructive chemical analysis for oil, has shown an increase of only 0.17% per year. Using high intensity selection by NMR within half-sib families, Miller et al (1981) were able to increase the oil content of Reid Yellow Dent corn from 4.0 to 9.1% in only seven cycles and with no reduction in yield.

Interest in breeding higher-oil corn is worldwide. Trifunović et al (1975) at the Maize Research Institute in Yugoslavia have used NMR analysis to develop lines that vary in oil content. Among 490 inbred lines from their breeding program, the oil content ranged from 2.7 to 12.5% with a mean value of 6.1%. Lico (1982) has isolated inbred lines from native populations in Albania that show promise for breeding hybrids with higher oil contents. In Iraq, Baktash et al (1982), using five cycles of modified mass selection from a synthetic corn variety Neelum, increased oil content 0.17% per cycle.

Alexander (1982) has released three high-oil inbreds to the seed trade (R802A—7% oil, R805—9%, and R806—9%). A commercial hybrid with 6.5–7.0% oil is now being marketed to U.S. farmers. Additional hybrids are being selected at the University of Illinois for 6–8.5% oil content and yields equivalent to those of commercial varieties. For example, R806 × B73 (6.7% oil) gave the same average yield (9.34 t/ha; 149 bu/acre) as the well-known hybrid Mo17 × B73 (4.3% oil) over a six-year period, 1979–1984 (Alexander, 1986). Higher-oil hybrids that yield as much grain per hectare as commercial hybrids will produce more energy (calories) per hectare. The energy content of oil is 37.7 J/g (9 kcal/g); that of protein or carbohydrate is only 16.8 J/g (4 kcal/g).

Significant positive correlations between oil content and yield components have not been found consistently. Raman et al (1983) examined correlations among oil percentage and yield components in a study with 27 inbred lines and three testers in a line-by-tester design. Oil content was positively correlated with grain yield, plant height, ear height, ear length, number of kernels per row, moisture content, and 100-kernel weight, but the correlations were significant only for 100-kernel weight and grain yield. Alexander and Seif (1963) found no correlation between 100-kernel weight and oil content.

The oil percentage of a corn kernel can be increased by a larger germ size, higher percentage of oil in the germ, or a smaller endosperm. Some endosperm mutants such as brittle-2, floury-2, and sugary-2 have higher oil percentages than does conventional corn, but the increase is due mainly to reduced endosperm size (Flora and Wiley, 1972; Arnold et al, 1974; Roundy, 1976).

Opaque-2 kernels tend to have larger ratios of embryo to endosperm than conventional corn does (Arnold et al, 1974; Valois et al, 1983). Eggum et al (1983) have developed four opaque-2 hybrids with 5.8–6.5% oil (Table I). Although the grain yields of the opaque hybrids were all lower than that of the conventional corn, the oil yields per hectare were higher for three of the four opaque types. The waxy corn with 5.6% oil also yielded more oil per hectare than the conventional corn with 5.1% oil. The highest grain yield in this study was attained by a high-oil hybrid (5.5% oil), but the authors cautioned that the grain yields were calculated from yields obtained in small field plots. All of the high-oil corns showed more utilizable protein in rat feeding trials than did the conventional corn.

C. Value of High-Oil Corn in Animal Feeding

A feeding trial of high-oil corn in growing-finishing pigs was reported by Adams and Jensen (1981). These researchers compared the utilization of high-oil corn with 7.5% oil (Diet II) to that of conventional corn with 3.5% oil (Diet I) (Table II). The experiment was also designed to test whether the oil in intact

TABLE I
Compositions and Yields of Conventional Corn and Opaque, Waxy, and High-Oil Corns[a]

Hybrid ZPSC	Type	Oil (%)	Protein[b] (%)	Yield			
				Grain (t/ha)	Oil (kg/ha)	Utilizable[c] Protein (kg/ha)	Lysine (kg/ha)
704	Conventional	5.1	9.6	14.3	733	805	42.0
071	Opaque	5.8	9.0	11.4	659	792	39.5
073	Opaque	5.9	10.9	14.0	823	1,085	60.7
074	Opaque	6.1	8.6	13.6	832	921	49.9
076	Opaque	6.5	8.9	11.7	762	788	44.0
757	Waxy	4.8	9.6	13.3	645	817	39.1
780	Waxy	5.6	8.9	13.9	774	845	38.3
781	High-oil	5.2	10.4	12.8	660	857	34.2
727	High-oil	5.5	11.9	15.5	855	1,071	49.7
717	High-oil	7.1	10.9	13.8	981	891	41.9
747	High-oil	8.2	12.5	12.0	982	836	37.3

[a] Adapted from Eggum et al (1983). All values on dry weight basis.
[b] Nitrogen × 6.25.
[c] Determined from rat feeding assays.

TABLE II
Performance of Growing-Finishing Pigs on Diets Containing Conventional and High-Oil Corns[a]

	Diets[b]		
	I Conventional Corn (3.5% oil)[c]	II High-Oil Corn (7.5% oil)[c]	III Conventional Corn + Corn Oil
Average daily gain, kg[d]	0.84	0.85	0.82
Average daily feed, kg[e]	1.67 a	1.55 b	1.56 b
Average gain/feed[e]	0.50 a	0.55 b	0.53 b

[a] From Adams and Jensen (1981).
[b] Diet I, conventional corn (73.89%) + soybean meal (23.46%) + vitamins and minerals (2.65%), 16.5 MJ/kg. Diet II, high-oil corn (73.74%) + soybean meal (23.60%) + lysine (0.01%) + vitamins and minerals (2.65%), 16.8 MJ/kg. Diet III, conventional corn (71.32%) + soybean meal (24.64%) + corn oil (1.39%) + vitamins and minerals (2.65%), 16.8 MJ/kg.
[c] Oil content expressed on "as is" basis.
[d] Each value is an average for three pens of eight pigs each.
[e] Values in the same row followed by different letters are significantly different ($P < 0.05$).

high-oil corn grain was used as efficiently as added oil. A diet (Diet III), isocaloric with Diet II, was prepared by adding corn oil to conventional corn. In all the diets, crude protein and lysine levels were above suggested requirements to ensure that neither would be nutritionally limiting. No significant differences in average daily gain were observed among the dietary treatments, but the daily feed intakes for the high-oil corn diet and for the conventional corn plus corn oil diet were significantly lower than for the conventional corn diet. The average gain/feed values for both the high-oil diets were similar and were significantly higher than for the conventional corn diet, indicating that the pigs used the intact high-oil grain efficiently. Feeding supplemental calories as higher oil in intact seeds would eliminate the handling, storing, and mixing problems that are associated with the addition of oils or fats to feedstuffs.

In an earlier experiment, Nordstrom et al (1972) fed 7% oil corn to growing-finishing pigs and found that 5–6% less high-oil corn than conventional corn was required per kilogram of gain. The 7% oil corn had little effect on pork quality, but diets containing 15% oil through addition of corn oil produced soft and oily carcasses that were unacceptable for conventional processing. The quality of the pork carcasses was affected by the additional oil in the diet and also by the greater polyunsaturation of the 15% oil diet (62% linoleic acid) compared to that of the high-oil corn diet (51.7% linoleic acid).

Han and Parsons (1984) have compared the nutritional value for poultry of a commercial, high-oil corn hybrid (5.7% oil, 9.5% crude protein, 14% moisture) relative to that of conventional corn (3.6% oil, 8.5% crude protein, 14% moisture). Laying hens fed 17% protein diets containing high-oil corn had significantly better egg-to-feed (wt/wt) ratios and higher body weight gains than hens fed conventional corn, regardless of whether the high-oil corn was substituted on an equal weight or an isonitrogenous basis. No differences between corn types were found for egg production, egg weight, feed consumption, and egg yield (grams per hen per day). Broiler chicks fed the high-oil diet from 8 to 22 days posthatching had improved gain/feed ratios compared to those fed the same diet containing conventional corn. The true metabolizable energy of high-oil corn was found to be 4.5% higher than that of conventional corn in a study involving adult roosters.

The nutritive value of high-oil corn for chickens has also been investigated in Yugoslavia (S. Savić, M. Latkovska, and B. Supić, personal communication). One-day old, male chicks were fed diets containing conventional corn (4% oil) or a high-oil corn hybrid (7.9% oil). After 56 days, the control group had gained 952 g and the high-oil group 1,006 g; the difference was statistically significant ($P<0.05$). Feed conversion per kilogram gain was 2.58 for the high-oil corn and 2.75 kg for the conventional corn.

III. FATTY ACID COMPOSITION

The structures of the fatty acids found in corn oil and other vegetable oils are shown in Fig. 1. The zigzag conformation of the hydrocarbon chains and the natural *cis* configuration of the double bonds are indicated. The pathway for the biosynthesis of the fatty acids (Stumpf, 1980) is also shown.

A major selling point for corn oil is its high level of the essential, polyunsaturated fatty acid, linoleic acid (18:2). Consumers have been made aware of the importance to health of polyunsaturated fatty acids in the diet. Corn oil is an excellent source. Although highly polyunsaturated, corn oil is a very stable oil because it contains high levels of natural antioxidants and very little (<1.0%) linolenic acid (18:3).

Among the commercial vegetable oils, only safflower oil (79.0% 18:2) and sunflower oil (69.5% 18:2) have higher percentages of polyunsaturated fatty acid than corn oil (61.9% 18:2) (Table III). The dominant vegetable oil in the world, soybean oil, has a lower level (50.8%) of linoleic acid, but a major problem is its higher content (6.8%) of linolenic acid. This triunsaturated fatty acid is susceptible to oxidation, the products of which produce off-flavors.

Corn oil has low levels of the saturated fatty acids, palmitic acid (16:0, 11%) and stearic acid (18:0, 2%), compared to the more saturated cotton seed (16:0, 25.2%) and palm (16:0, 44%) oils. Trace amounts (<1%) of lauric (12:0), myristic (14:0), palmitoleic (16:1), arachidic (20:0), behenic (22:0), erucic (22:1), and

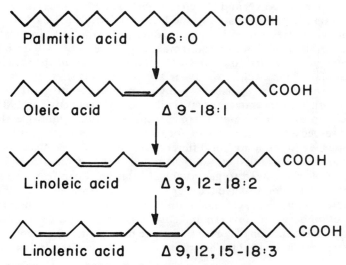

Fig. 1. Chemical structures and biosynthetic pathways of fatty acids.

TABLE III
Fatty Acid Compositions of Commercial Vegetable Oils

Oil	16:0[a]	18:0	18:1	18:2	18:3	Reference
Safflower	4.6	6.0	7.3	79.0	0.1	Fedeli (1983)
Sunflower	6.2	4.1	19.8	69.5	...[b]	Campbell (1983)
Corn	11.0	2.0	24.1	61.9	0.7	Liebovitz and Ruckenstein (1983)
Cottonseed	25.2	2.7	17.5	52.6	...	Cherry (1983)
Soybean	10.7	3.9	22.8	50.8	6.8	Pryde (1980)
Palm	44.0	4.5	39.2	10.1	0.4	MacLellan (1983)

[a] Fatty acids are identified according to number of carbon atoms and number of double bonds: palmitic acid, 16:0; stearic acid, 18:0; oleic acid, 18:1; linoleic acid, 18:2, linolenic acid, 18:3.
[b] No value given.

lignoceric (24:0) acids have been found in corn oil (Leibovitz and Ruckenstein, 1983).

Jellum (1967) investigated the influence on fatty acid composition of kernel position on the ear. Kernel position had little effect on stearic and linolenic acids. In general, the proportions of palmitic and linoleic acids increased and the proportion of oleic acid (18:1) decreased in the oil from kernels sampled from the base to the tip of the ear. Oil samples from the middle portion of the ear showed intermediate values. Sampling for fatty acid composition should be restricted to the middle portion of the ear, as suggested earlier for sampling of kernels for oil content.

A. Effect of Agronomic Practices on Fatty Acid Composition

Gallo et al (1976) examined the effect of fertilizers on yield and on the relationships between the chemical compositions of the seeds and leaves of corn in Brazil. Applications of N and P increased yield at both of the locations studied but K had a significant effect at only one site. Both fertilizer P and leaf P content were positively correlated with seed oil content. Fertilizer K and leaf K were positively correlated with seed oil content but negatively correlated with linoleic acid percentage. K tended to decrease the starch content of the seed. Both oil content and fatty acid composition were affected by location.

Corn oil from cooler regions has a higher proportion of unsaturated fatty acids than corn oil from warmer areas (Leibovitz and Ruckenstein, 1983). These differences appear to be adaptations to climatic conditions. Jahn-Deesbach et al (1975a) observed a difference of 3.7% for linoleic acid between locations for one variety, but they found a greater difference of 11% for linoleic acid among the German varieties. Casmussi et al (1980) noted that fatty acid composition in Italian varieties was linked to geographical location. Corn from the northern regions of Italy tended to have a lower content of saturated fatty acids and a higher content of linoleic acid than corn from the southern regions. However, this distinction between northern and southern types was also confounded by the historical introduction of corn into these regions. Corn was introduced in the North directly from Spain at the beginning of the 16th century. It was introduced

TABLE IV
Range of Fatty Acid Composition in Corn

No. of Strains	Source	Fatty Acid Composition, %					Reference
		16:0[a]	18:0	18:1	18:2	18:3	
788	World	6–22	1–15	14–64	19–71	0.5–2	Jellum (1970)
19	West Germany	8–12	1–2	17–31	54–69	1–3	Jahn-Deesbach et al (1975a)
490	Yugoslavia	8–22	1–4	16–43	39–68	...[b]	Trifunovič et al (1975)
102	Italy	13–18	1–4	22–42	39–54	...	Camussi et al (1980)
4	India	10–28	2–8	32–53	10–53	0.3–1	Sharma et al (1975)
Crude commercial oil	South Africa	12	2	39	44	1	Leibovitz and Ruckenstein (1983)

[a] Palmitic acid, 16:0; stearic acid, 18:0; oleic acid, 18:1; linoleic acid, 18:2; linolenic acid, 18:3.
[b] No value given.

much later in the South and, according to cytological and morphological traits, originated from a quite different source.

Jellum and Marion (1966) studied the effect of location, year, planting date, and ear position on the fatty acid compositions of nine corn hybrids. Although the year, location, and hybrid effects on fatty acid composition were all significant, the year and location effects were relatively small compared to the hybrid differences.

Several factors may be involved in location effects, but temperature is undoubtedly a major environmental factor. When corn was grown in four phytotron environments with day temperatures ranging from 18 to 30°C, the corn oil showed decreasing unsaturation with increasing temperatures (Thompson et al, 1973). However, the type and degree of response to temperature changes varied greatly among the genotypes.

B. Genetic Control of Fatty Acid Composition

Genotype has a greater influence on fatty acid composition than any environmental factor. The known variability for fatty acid composition among corn genotypes covers a wide range (Table IV). Jellum (1970) made the most extensive search for unusual fatty acid compositions. He found a range among 788 strains of from 14 to 64% for oleic acid and from 19 to 71% for linoleic acid and a strong negative correlation between these two fatty acids. Among four strains from India, Sharma et al (1975) discovered one strain with only 10% linoleic acid and 28% palmitic acid. The corn oils from Italy, India, and South Africa were more saturated than those of West Germany, Yugoslavia, and the United States.

Over the past 12 years, the polyunsaturation of U.S. commercial corn oil has increased 4% (from 57.8 to 62.0% in linoleic acid). Changes in parental lines may have caused this increase in linoleic acid (Weber, 1983a). Yield potential, disease resistance, and other desirable agronomic traits are the determining factors in the selection of parental lines. Commercial seed corn producers seldom monitor the fatty acid composition or oil content of the hybrids that they sell.

Selection for a desired fatty acid composition in corn oil should be possible. Several workers have shown single-gene or single-gene-plus-modifier effects on the synthesis of oleic and linoleic acids (Poneleit and Alexander, 1965; de la Roche et al, 1971a). Other studies (Poneleit and Bauman, 1970; Poneleit, 1972; Widstrom and Jellum, 1975; Sun et al, 1978) have indicated a more complex multigene system of inheritance for oleic, linoleic, palmitic, and stearic acids. From a practical breeding standpoint, additive genetic models are useful. Recently, Jellum (1984) has released an inbred with the highest level (18%) of stearic acid that has been reported in corn. From crosses of lines related to this inbred with standard inbred lines, Jellum and Widstrom (1983) attributed the inheritance of high stearic acid to a single major recessive gene with modifiers.

Brimhall and Sprague (1951) reported that a negative correlation existed between unsaturation of corn oil and the oil content of the kernel. C. L. Armstrong and D. E. Alexander (personal communication) found a highly significant correlation ($r = -0.51**$) between linoleic acid percentages and oil content of cycle 0 of Alexho Synthetic but no significant correlation ($r = -0.05$) between these factors in cycle XX. The original source population of Alexho

Synthetic consisted of 56 open-pollinated varieties and synthetics, which were combined and permitted to random mate for several generations before selection for higher oil content. Fatty acid composition was not monitored during the selection process. Poneleit and Bauman (1970) observed a significant, but very low, negative correlation between percentages of linoleic acid and oil. Selection for both oil content and fatty acid composition should result in progress toward desired levels.

At present, analysis for fatty acid composition by gas chromatography requires destruction of the sample. If a rapid, nondestructive method of screening individual kernels for fatty acid composition were available, it would significantly increase progress toward higher oil hybrids with acceptable combinations of oil content and linoleic acid composition. Table V shows the fatty acid compositions of some higher-oil corns. The long-term selection, IHO, has low linoleic acid (48.9%) compared to that of a current commercial hybrid, Mo17 × B73 (62% 18:2). One high-oil hybrid that is commercially available has 55.2% linoleic acid. The more saturated corn oil is advantageous for pig feeding because the soft pork problem would be alleviated.

Several studies have been made to determine the chromosomal location of the genes that control oleic and linoleic acid composition in corn germ oil. Factors located on Chromosome 2 (Plewa and Weber, 1975) and on the long arm of Chromosome 5 (Shadley and Weber, 1980; Widstrom and Jellum, 1984) were found to be involved in the determination of the oleic and linoleic acid compositions. Widstrom and Jellum (1984) also had evidence suggesting the presence on the long arm of Chromosome 4 of a recessive gene that contributed to high levels of linoleic acid and of another gene on the short arm of Chromosome 1 that controlled linoleic acid levels. Further studies will be required to determine the specific location of the genes on each chromosomal arm.

IV. LIPID CLASS COMPOSITION

Lipids are divided into two major groups, acyl lipids and nonsaponifiable lipids. The acyl lipids contain fatty acid esters that are hydrolyzed by alkali. The major classes are TG, phospholipids, and glycolipids. The nonsaponifiable lipids are those that are extracted by organic solvents after saponification. They include hydrocarbons, sterols, carotenoids, and tocols.

TABLE V
Fatty Acid Compositions of Oil from Higher-Oil Corns[a]

Corn	Oil (%)	Fatty Acid Composition (%)				
		16:0	18:0	18:1	18:2	18:3
Illinois High Oil	19.0	11.7	2.4	36.0	48.9	1.0
R802A × R805	7.7	12.8	1.9	31.4	52.4	1.0
Commercial high-oil hybrid	7.5	11.0	2.2	30.6	55.2	1.0
SK85 × Mo17	5.6	11.1	2.2	28.8	57.0	0.4
SK85A2 × Oh43[HO]	5.6	12.2	2.1	22.6	61.9	1.0
Mo17 × B73	4.3	10.5	1.6	24.7	62.0	1.3

[a]Data from Weber and Alexander (1975; and unpublished data).

A. Triacylglycerols

Weber (1969) quantified the lipid classes from three strains of corn that differed in oil content (IHO, 13.8% oil, dry wt; H51, 5.3%; K6, 2.4%). The total lipids were extracted from the grain 60 days after pollination with a mixture of chloroform-methanol-water (Bligh and Dyer, 1959). The proportion of the lipids represented by TG increased as the oil content increased (K6, 75.8% TG; H51, 79.0%; IHO, 88.1%) (Table VI). The percentages of the sterol and phospholipid-plus-glycolipid fractions declined with higher oil contents because these lipids are membrane components.

TG form complex mixtures because of the multiple combinations of fatty acids that are possible in the TG molecule (Fig. 2). Various procedures have been devised to study the molecular TG species.

Gas chromatography (GC) separates TG only according to their total number of carbon atoms (molecular weight). D'Alonzo et al (1982) were able to extend the separation to fatty acids, monoglycerides, diglycerides, and TG in a single run. The corn oil was analyzed directly by high-temperature GC on glass, capillary columns after derivatization of the lipids with (N,O)-bis(trimethyl silyl) trifluoroacetamide. This technique could be used to measure partial hydrolysis products in a vegetable oil.

High-performance liquid chromatography (HPLC) has been used increasingly in recent years to separate complex TG mixtures. The TG of corn oil have been separated on reverse-phase C18 columns with detection by refractive index. Aqueous mobile phases are generally used with reverse-phase columns. However, due to the lipophilic nature of the TG, the mobile phases for TG separations consist of mixtures of acetone and acetonitrile. This HPLC system has been described by the term nonaqueous reversed-phase chromatography. Technological advances in columns and instruments are continuing to improve HPLC. In 1977, Plattner et al separated the TG of corn oil into five peaks; in 1981, El-Hamdy and Perkins obtained eight peaks from separation both by carbon number and unsaturation. Dong and DiCesare (1983) reduced the analysis time from 30 to 8 min by using C18 columns packed with 3-μm rather than 5-μm particles. The major peak of corn oil was trilinolein. Oleoyldilinolein and palmitoyldilinolein ranked second and third.

In the TG molecule, three fatty acids are esterified to the hydroxyl groups of

TABLE VI
Lipid Class Composition of Corn[a]

Lipid Class	Strain		
	IHO[b] (13.8% oil)	H51 (5.3% oil)	K6 (2.4% oil)
Hydrocarbons + sterylesters	1.5	2.9	3.4
Triacylglycerols	88.1	79.0	75.8
Free fatty acids	0.4	1.0	1.1
Sterols	2.5	4.5	4.6
Diglycerides + monoglycerides	3.0	3.9	2.1
Phospholipids + glycolipids	4.5	8.7	13.0

[a] Adapted from Weber (1969).
[b] IHO = Illinois High Oil.

glycerol (Fig. 2). Stereospecific analysis of TG from corn inbreds indicated a nonrandom distribution of fatty acids among the three positions (Table VII) (de la Roche et al, 1971b, 1971c; Weber et al, 1971). The saturated palmitic and stearic acids were esterified predominately at the sn-1 and sn-3 positions, but a slightly higher percentage of saturated fatty acids was found in the sn-1 position than in the sn-3 position. At the sn-2 position, over 98% of the fatty acids were unsaturated and linoleic acid was the dominant fatty acid. For inbreds with different fatty acid compositions, the amount of any fatty acid at a particular position tended to be linearly related to the amount of that fatty acid in the total TG. However, although the effect is limited by the concentrations of the fatty acids available for esterification, the stereospecific distribution of the fatty acids may be under independent genetic control (de la Roche et al, 1971d; Pan and Hammond, 1983).

The placement of a specific fatty acid at each position of the TG is important. The physical properties, nutritional value, and stability of a vegetable oil are affected by its TG structure. If linoleic acid is esterified at the sn-2 position of TG, this essential fatty acid is more likely to be nutritionally available to animals. TG are not absorbed directly from the digestive tract of animals but are first partially degraded by pancreatic lipase into sn-2 monoacylglycerols (2-MG) and free fatty acids derived from the 1 and 3 positions (Raghavan and Ganguly, 1969). The integrity of the 2-MG is preserved during digestion, absorption, and resynthesis into TG or other lipids. The free fatty acids are metabolized or reesterified.

Corn oil with randomized fatty acids and corn oil methyl esters undergoes autoxidation three to four times faster than natural corn oil (Lau et al, 1982).

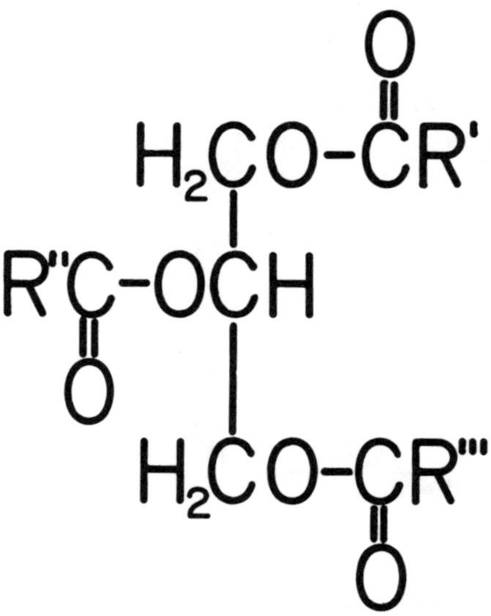

$$\begin{array}{c} \text{H}_2\text{CO-CR}'\text{=O} \\ | \\ \text{R}''\text{C(=O)-OCH} \\ | \\ \text{H}_2\text{CO-CR}'''\text{=O} \end{array}$$

Fig. 2. Triacylglycerol (triglyceride) structure. R' indicates alkyl chain of fatty acid at the sn-1 position, R" at the sn-2 position, and R''' at the sn-3 position in stereospecific numbering (sn) of the carbon atoms of glycerol.

The reason for the higher stability of the natural oil is unknown, but Raghuveer and Hammond (1967) proposed that the fatty acids on the 1 and 3 positions oxidize faster than those at the 2 position, and the preferential placement of unsaturated fatty acids on the 2 position in many natural oils increases their stability. Lau et al (1982) found three times more scission products from randomized than from natural corn oil and suggested that TG structures, through associations, may affect the rate of initiation of oxidation. Wada and Koizumi (1983) randomized mixtures of tripalmitin (PPP), tristearin (SSS), triolein (OOO), and trilinolein (LLL) and separated the TG species by HPLC. The TG having unsaturated fatty acids at the 2 position were more stable toward oxidation than the TG with unsaturated fatty acids linked at the 1 or 3 positions. After three days of oxidation at 50°C, 89.6% of SLL or LLS remained but only 72.6% of LSL; after four days, 56.5% of PLP remained but only 42.8% of LPP or PPL. These studies indicate that the stability of oils toward autoxidation is increased by the natural concentration of the polyunsaturated fatty acids at the 2 position of the TG.

Yoshida and Kajimoto (1984) have studied thermal oxidation of corn oil by heating the oil at 180°C, a temperature comparable to those attained in deep-fat frying. The TG molecular species of the thermally oxidized oils were isolated by silver nitrate-silica gel thin-layer chromatography (TLC), according to the degree of unsaturation. The oxidative decomposition of each TG species corresponded closely to its degree of unsaturation. The TG containing five or six

TABLE VII
Positional Distribution of Fatty Acids in the Lipids of Corn Inbred H51[a]

Lipid Class	Position	Fatty Acid Composition, mol %				
		16:0	18:0	18:1	18:2	18:3
Triacylglycerol		17.7	1.8	31.2	48.1	1.2
	1	26.0	3.4	30.8	38.8	1.0
	2	1.5	0.1	26.8	70.6	1.0
	3	25.4	2.0	36.1	34.9	1.6
Phosphatidylcholine		23.4	1.8	36.7	37.5	0.6
	1	42.7	3.7	25.5	27.6	0.5
	2	2.8	0.7	48.6	47.1	0.8
Phosphatidylinositol		36.9	1.7	18.8	41.4	1.2
	1	65.4	2.6	14.1	17.3	0.6
	2	7.3	1.6	24.4	65.4	1.3
Phosphatidylethanolamine		23.7	1.1	23.8	50.7	0.7
	1	45.4	2.4	17.8	34.0	0.4
	2	2.1	0.3	30.0	66.7	0.9
Phosphatidylglycerol		44.4	1.3	20.1	32.5	1.7
	1	64.8	2.2	12.5	20.0	0.5
	2	22.5	0.8	28.8	46.9	1.0

[a] Source: Weber (1983b).

double bonds were oxidized faster than those with one to four. After heating corn oil to 200° C for 48 hr, Sahasrabudhe and Farn (1964) suggested that losses in the oleoyl- and linoleoyl-glyceride fractions indicated that fatty acids in the outer positions of corn oil TG were slightly more susceptible to heat than those in the 2 position.

Hydrogenation of corn oil to produce margarine resulted in greater changes at the 2 position of TG than in the 1 and 3 positions (Strocchi, 1981). After

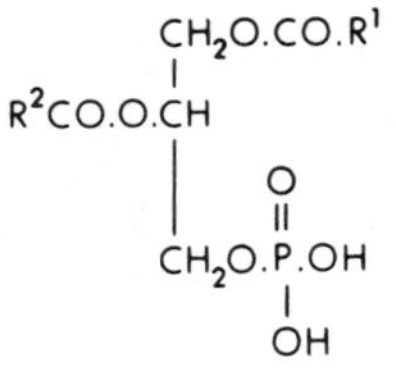

I. Phosphatidic acid (PA)

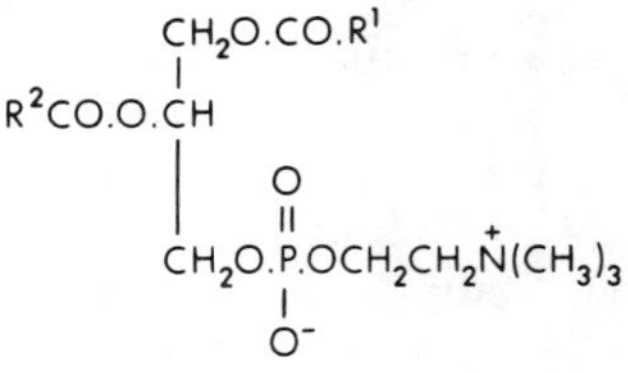

II. Phosphatidylcholine (lecithin) (PC)

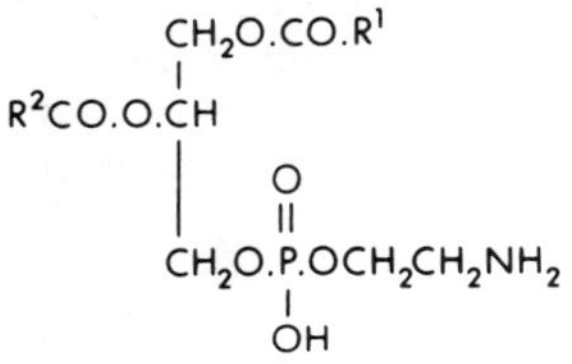

III. Phosphatidylethanolamine (PE)

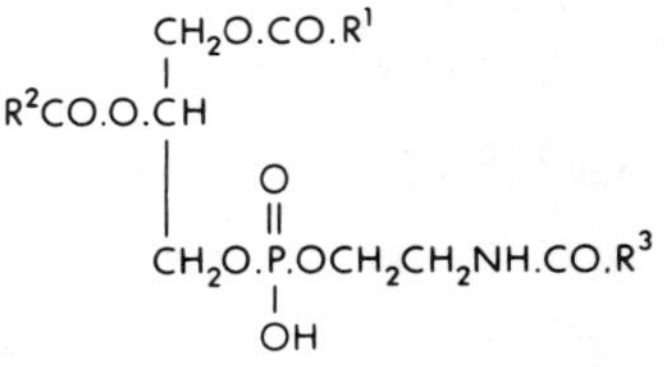

IV. N-acylphosphatidylethanolamine (APE)

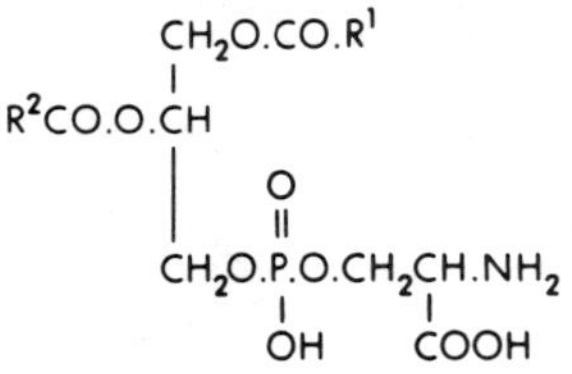

V. Phosphatidylserine (PS)

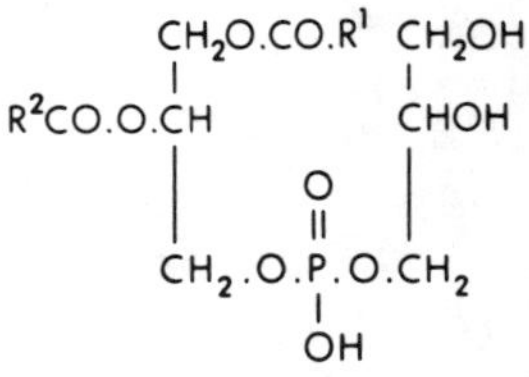

VI. Phosphatidylglycerol (PG)

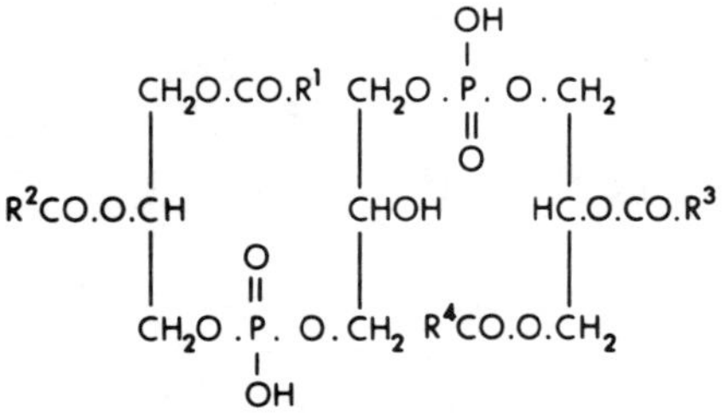

VII. Diphosphatidylglycerol (cardiolipin) (DPG)

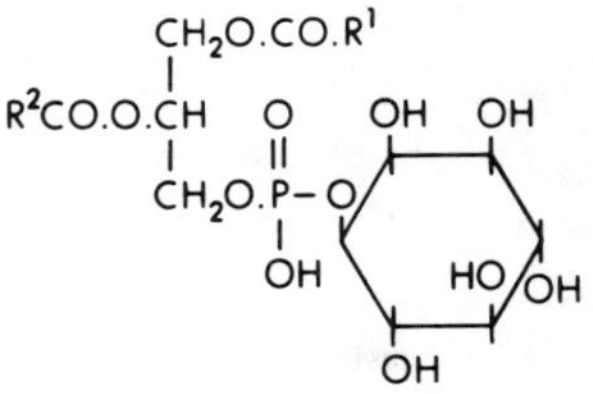

VIII. Phosphatidylinositol (PI)

Fig. 3. Phospholipid structures.

hydrogenation, the 2 position was 86% occupied by *trans*-octadecenoic acid (*trans*-18:1), whereas in the original corn oil, the 2 position was 70% occupied by *cis,cis*-linoleic acid. Hydrogenation and/or geometrical isomerization of *cis*-18:1 to *trans*-18:1 occurred preferentially in the 2 position as compared to the outer positions. In the original corn oil, 31% of the *cis*-18:1 in the TG was present in the 2 position, but in the hydrogenated corn oil, only 13% of the *cis*-18:1 remained at the middle position. The nutritional effects of *trans*-fatty acids are still under debate (Applewhite, 1983; Gottenbos, 1983; Kummerow, 1983; Ohlrogge, 1983).

B. Phosphoglycerides

In corn grain, the major phosphoglycerides (phospholipids, PhL) are phosphatidylcholine (PC), phosphatidylinositol (PI), and phosphatidylethanolamine (PE); the minor PhL are phosphatidylglycerol (PG), diphosphatidylglycerol (also called cardiolipin), phosphatidylserine, phosphatidic acid (PA), and N-acylphosphatidylethanolamine (APE). The chemical structures of these PhL are given in Fig. 3. A lysophospholipid (lysoPhL) is a monoacyl lipid formed by hydrolytic removal of only one fatty acid from a PhL. In some cases, APE, PA, and lysoPhL may be artifacts of the extraction procedures if enzymes are not inactivated completely before lipid extraction (Kates, 1972; Mudd, 1980). Extraction with boiling alcohol solvent systems is recommended because the hot alcohols usually inactivate the enzymes. Tan and Morrison (1979a) and Weber (1979) used boiling water-saturated butanol to extract corn germ and endosperm fractions and still found APE and lysoPhL (see section V-A). These PhL appear to be natural components of mature corn kernels.

Each phospholipid class has a characteristic fatty acid composition (Table VII). PC has the highest percentage of 18:1 and PE the highest percentage of 18:2. Both PI and PG have high percentages of saturated fatty acids, but PG has more 18:1 and less 18:2 than PI. All the PhL have higher percentages of 16:0 than the TG.

Table VII also shows the stereospecific distribution of the fatty acids on the glycerol molecule of the PhL. The overall fatty acid distributions for PhL were similar to those at the *sn*-1 and *sn*-2 positions of the TG, in that the saturated fatty acids were located predominately at position 1 and the highest percentage of linoleic acid was at position 2.

Genetic modification of the fatty acids of corn oil will affect not only the fatty acids of the TG but also the fatty acid compositions of the other lipid classes. This fact may have particular significance for PhL, which are essential components of all membranes. The fatty acid compositions of the TG and PhL of four corn inbreds are shown in Table VIII. Each lipid class has its own characteristic fatty acid pattern, but the genotype of the inbred has superimposed variations in fatty acid composition (Weber, 1978). When the inbreds are ranked in the same order for each class of lipid, increasing levels of linoleic acid are generally noted. The range is much larger for TG, from 42.2% for H21 to 69.6% for NY16, but differences also are apparent among the inbreds for each PhL class.

TABLE VIII
Fatty Acid Composition of Triacylglycerols and Phospholipids from Mature Kernels of Four Corn Inbreds[a]

Lipid Class	Inbred	Fatty Acid Composition, mol %				
		16:0	18:0	18:1	18:2	18:3
Triacylglycerol	H21	16.5	2.9	37.4	42.2	1.0
	IHO	12.9	2.1	35.4	48.8	0.8
	K6	11.3	1.1	22.1	64.2	1.3
	NY16	7.4	1.6	20.1	69.6	1.3
Phosphatidylcholine	H21	20.0	1.5	42.1	36.0	0.5
	IHO	19.0	1.5	37.5	41.4	0.6
	K6	23.9	1.8	28.3	45.2	0.8
	NY16	18.6	2.2	25.8	52.3	1.1
Phosphatidylinositol	H21	42.1	2.6	18.7	35.6	1.0
	IHO	39.5	2.2	20.6	37.4	0.3
	K6	40.2	1.9	12.3	44.1	1.5
	NY16	38.3	2.2	12.8	45.8	1.0
Phosphatidylethanolamine	H21	20.8	0.8	23.5	54.2	0.7
	IHO	25.1	1.0	18.9	54.3	0.7
	K6	24.0	1.0	15.3	59.0	0.7
	NY16	20.6	1.9	18.1	58.3	1.0
Phosphatidylglycerol	H21	36.3	2.0	19.7	40.3	1.7
	IHO	36.7	2.7	23.6	36.6	0.5
	K6	35.5	2.5	15.6	42.6	3.8
	NY16	34.2	3.6	16.6	44.3	1.4

[a]Source: Weber (1978).

TABLE IX
Fatty Acid Compositions and the Lipids of Corn Inbreds C103D and B73 and Their Reciprocal Crosses[a]

Lipid Class	Inbreds and Crosses	Fatty Acid Composition, mol %				
		16:0	18:0	18:1	18:2	18:3
Triacylglycerol	C103D	13.3	2.0	43.4	40.3	1.1
	C103D × B73	12.6	2.1	38.6	45.7	1.0
	B73 × C103D	12.2	2.2	33.7	50.8	1.0
	B73	11.4	1.9	29.6	55.9	1.3
Phosphatidylcholine	C103D	20.8	2.3	49.5	26.6	0.7
	C103D × B73	19.1	1.8	44.1	34.2	0.8
	B73 × C103D	20.1	1.8	37.6	39.6	0.9
	B73	20.9	1.6	30.5	45.9	1.1
Phosphatidylinositol	C103D	38.9	4.0	29.8	26.8	0.5
	C103D × B73	36.3	2.8	26.8	33.5	0.7
	B73 × C103D	37.5	2.5	20.9	38.4	0.7
	B73	33.7	2.4	19.8	42.9	1.2
Phosphatidylethanolamine	C103D	23.0	3.9	38.4	34.2	0.5
	C103D × B73	20.7	3.3	34.0	41.6	0.4
	B73 × C103D	20.6	2.2	26.7	49.8	0.7
	B73	21.5	2.7	21.9	52.8	1.0

[a]Source: Weber (1983b).

The heritability of fatty acid composition in PhL was tested by crossing two inbreds, C103D and B73, that differed widely in 18:2 (Table IX) (Weber, 1983b). The lipids were isolated from the germ only. Inheritance is more easily determined in the diploid germ, which has equal inheritance from both parents, than in the triploid endosperm. In the reciprocal crosses, the characteristic fatty acid patterns for each PhL class—high 18:1 for PC, high 16:0 for PI, and high 18.2 for PE—were present, but the genotype also had an effect on fatty acid composition. The oleic and linoleic acid percentages of each of the PhL classes were intermediate to those of the parents and tended toward the maternal parent.

Commercial mixtures of corn PhL (lecithin) have not been available since the 1950s. However, with the phenomenal growth now occurring in the demand for corn sweeteners, other products of the corn-refining industry, such as corn lecithin, may become available and competitive. Weber (1981) compared commercially prepared samples of corn and soybean lecithins. The classes of lipids were similar, but the proportions varied. The ratio of glycolipids to PhL was 0.44 for corn lecithin and 0.17 for soybean. PC was the major PhL in both lecithin samples, representing 30.4% of the polar lipids in corn and 33.0% in soybean. PI was second at 16.3% in corn and 16.8% in soybean. The percentage of PE (14.1%) nearly equaled that of PI in soybean lecithin, but in corn, the percentage of PE (3.2%) was only about one fifth the percentage of PI. The major differences in the fatty acid compositions of the commercial lecithins were a higher percentage of oleic acid and lower percentages of stearic and linolenic acids in corn than in soybean. The lower level of linolenic acid should give corn lecithin greater resistance to autoxidation and the development of off-flavors.

C. Glycosylglycerides

The glycosylglycerides (one type of glycolipid) of corn grain include monogalactosyldiglyceride (MGDG), digalactosyldiglyceride (DGDG), 6-O-acylmonogalactosyldiglyceride (AMGDG), and sulfoquinovosyldiglyceride (SQDG) (Fig. 4). DGDG, the major glycosylglyceride, occurs at two to three times the level of MGDG (Tan and Morrison, 1979a; Weber, 1979) (see section V-A). The lyso compounds, mono- and digalactosylmonoglycerides, are found in mature grain.

The fatty acid compositions of the mono- and diglycosyldiglycerides are given in Table X. The glycosylglycerides are characterized by higher levels of 18:3 than any of the other corn lipids (Weber, 1970). Tanaka et al (1984) proposed that the

TABLE X
Fatty Acid Compositions of Monoglycosyldiacylglycerol, Diglycosyldiacylglycerol, and Sterylglycoside Ester in Corn Grain[a]

	Fatty Acid Composition, %					
	16:0	16:1	18:0	18:1	18:2	18:3
Monoglycosyldiacylglycerol	9.7	2.5	2.2	37.8	30.1	17.7
Diglycosyldiacylglycerol	15.2	2.9	4.1	21.8	31.8	24.2
Sterylglycoside ester	40.9	0.1	2.1	16.3	38.0	2.0

[a]Source: Tanaka et al (1984); used by permission.

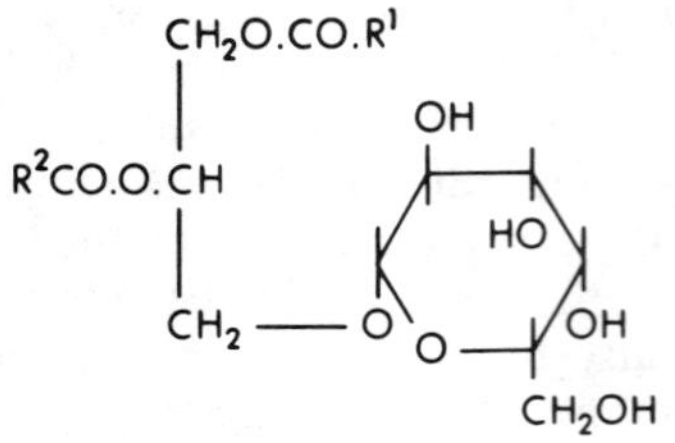

I. Monogalactosyldiglyceride (MGDG)

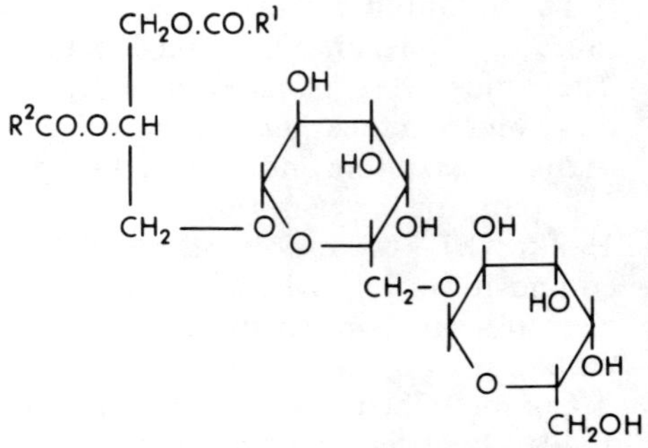

II. Digalactosyldiglyceride (DGDG)

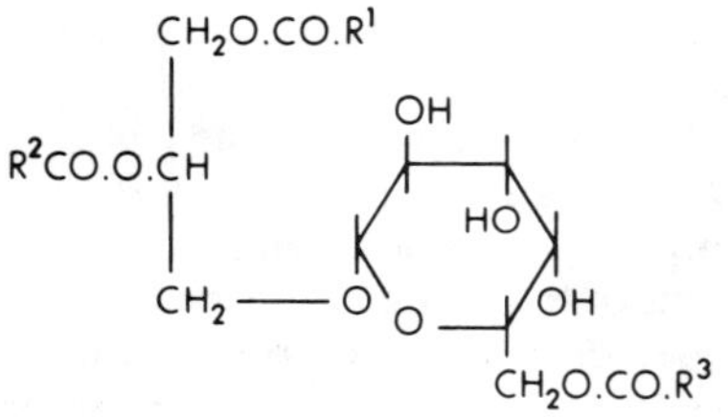

III. 6-O-Acylmonogalactosyldiglyceride (AMGDG)

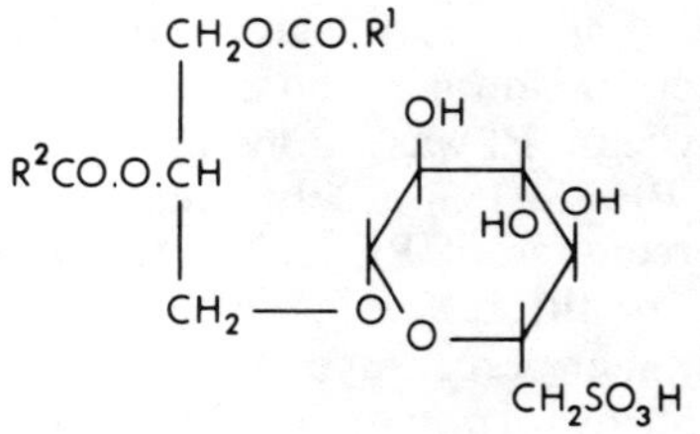

IV. Sulfoquinovosyldiglyceride (SQDG) (Sulfolipid)

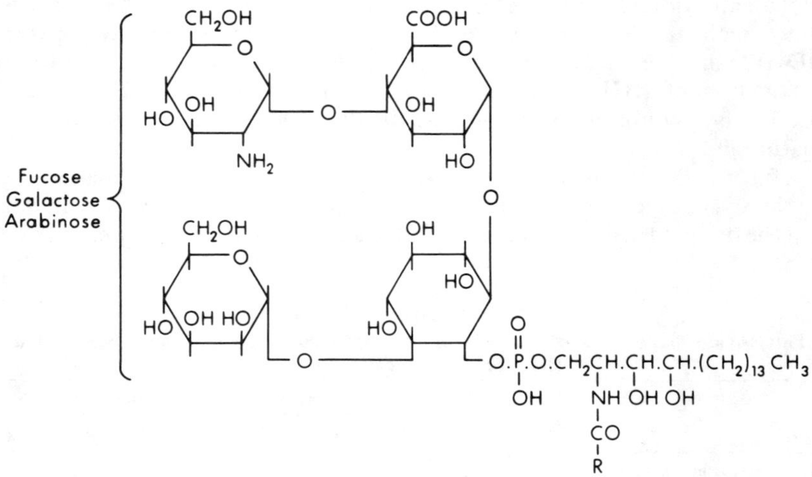

V. Structure of phytoglycolipid from *Zea mays* L.

Fig. 4. Glycolipid structures.

major molecular species of MGDG and DGDG in corn are 1-oleyl-2-linoleoyl-3-*O*-galactosyl-*sn*-glycerol and 1-linoleoyl-2-linolenoyl-3-*O*-digalactosyl-*sn*-glycerol, respectively.

AMGDG may be an artifact. It is formed readily in leaves unless the leaves are placed in boiling water to inactivate acyltransferase enzymes (Heinz, 1967). SQDG is a major lipid of chloroplasts, but it has also been identified in a nonphotosynthetic tissue, corn grain (Weber, 1979).

D. Glycosylsphingolipids

Sphingolipids contain the long-chain base sphingosine (*trans*-D-*erythro*-1,3-dihydroxy-2-amino-4-octadecene) or a related amino alcohol. The sphingolipid group includes ceramides, glycosylceramides (GCM; also called cerebrosides), and phytoglycolipids (PGL). The ceramide molecule has a fatty acid attached to the amino group of the long-chain base.

Tanaka et al (1984) identified GCM as major sphingolipids in corn grain. The predominant sugar of the GCM was glucose. Their fatty acids typically had long, saturated chains (16:0–26:0), and 88% of the total fatty acids were 2-hydroxy acids. The predominant 2-hydroxy acids were 2-hydroxyeicosanoic acid (2-OH-20:0, 34.2%), 2-hydroxytetracosanoic acid (2-OH-24:0, 31.1%), and 2-hydroxydocosanoic acid (2-OH-22:0, 12.5%). The long-chain bases of the GCM also showed considerable complexity, but *trans*-4, *cis*-8-sphingadienine (58.1%), 4-hydroxy-8-sphingenine (22.8%), and *trans*-4, *trans*-8-sphingadienine (14.7%) were the dominant forms. Tanaka et al (1984) determined that the principal molecular species of GCM in corn were 1-*O*-glucosyl-*N*-2'-hydroxyeicosanoyl-4,8-sphingadienine and 1-*O*-glucosyl-*N*-2'-hydroxytetracosanoyl-4-hydroxy-8-sphingenine.

PGL are unique among complex lipids because they have the structural features of a glycolipid and of a phospholipid. The PGL were isolated originally from commercial lecithin by Carter et al (1958) and appear to be a family of lipids in which the polysaccharide chain may vary in length and in types of sugars. The basic unit of PGL was found to be an N-acyl phytosphingosine-type long-chain base attached as a phosphate ester to inositol-glucuronic acid-glucosamine (Fig. 4). Mannose was also bound to inositol, and oligosaccharides with varying numbers of galactose, arabinose, and fucose molecules were attached to the glucosamine or mannose (Carter et al, 1969). The predominant fatty acid was 2-hydroxytetracosanoic acid. Another glycophosphoceramide that contained all the components of PGL except glucosamine was also isolated from commercial corn lecithin (Carter and Kisic, 1969).

E. Sterols

The chemical structures of the sterols discussed in this section can be found in a review of the sterols in cereal grains by Barnes (1983a). The sterols of corn oil have been separated by TLC into three groups according to their structural features—4,4-dimethylsterols (54 mg/100 g of oil), 4-monomethylsterols (62 mg/100 g of oil), and 4-demethylsterols (1,441 mg/100 g of oil) (Kornfeldt and Croon, 1981). The sterols are synthesized from squalene through the intermediates, 4,4-dimethylsterols and 4-monomethylsterols, to the more abundant 4-demethylsterols (Barnes, 1983a).

Kornfeldt and Croon (1981) determined the individual sterols by GC on capillary columns with identification by mass spectroscopy. The predominant 4,4-dimethylsterols (triterpene alcohols) in corn oil were cycloartenol (43%) and 24-methylene cycloartenol (40%). Citrostadienol (29%), gramisterol (26%), and obtusifoliol (21%) were the major 4-monomethylsterols. Gramisterol was suggested as a useful compound for characterizing corn oil. Among the common vegetable oils, only wheat germ oil had as high a percentage of gramisterol as corn oil.

Both corn and wheat germ oils were also distinguishable from other vegetable oils by their high contents of the demethylsterols. Over 90% of the corn sterols were 4-demethylsterols. The individual 4-demethylsterols are listed in Table XI. Sitosterol was the predominant sterol (60–70%). The other major sterols were campesterol (16–22%) and stigmasterol (4–10%). Although the sources varied from oil to whole kernel and also geographically, the proportions of the individual sterols were relatively constant.

A new sterol (24-methyl-E-23-dehydrolophenol) and two other minor sterols (24-methyl-E-23-dehydrocycloartanol and 24-methyl-E-23-dehydrocholesterol) were isolated from corn germ oil by Itoh et al (1981). These sterols may be intermediates in the biosynthesis of the major sterols in corn germ.

The classes of sterol lipids in corn grain are free sterols (FS), sterylesters (SE), sterylglycosides (SG), and sterylglycoside ester (SGE). Davis and Poneleit (1974, 1975) quantified the sterol classes from four strains of corn (Table XII). The FS and SE were the major fractions. SG and SGE accounted for very small amounts (2–8%) of the total sterols. The IHO strain accumulated 407 μg of total FS per kernel, which was two to four times the sterol levels of the other, lower-oil, strains.

Worthington and Hitchcock (1984) used a combination of preparative HPLC on a silica column and TLC to isolate the FS and SE without hydrolyzing the SE. The fatty acids from the SE had higher percentages of linoleic acid than the original corn oils (Mazola oil had 61.1% 18:2 and its SE had 78.1%; Kroger oil

TABLE XI
4-Demethylsterols of Corn

	4-Demethylsterols (%) in			
	Oil[a]	Oil[b]	Soap Stock	Whole Kernel
Sitosterol	60	69–70	66[c]	66–70[d]
Campesterol	17	21–22	23	16–21
Stigmasterol	6	6	6	4–10
Cholesterol	trace	...	<0.5	0.2–0.4
Δ^5-Avenasterol	10	2	4	2–9
Δ^7-Avenasterol	1	...	<0.5	...
Δ^7-Stigmasterol	trace	...	1	0.2–1
Others	6	1	<0.5	...

[a] Swedish commercial corn oil. Data from Kornfeldt and Croon (1981).
[b] Two U.S. commercial corn oils—Mazola and Kroger. Data from Worthington and Hitchcock (1984).
[c] Mexican soap stock. Data from Itoh et al (1973).
[d] Five Yugoslavian dwarf hybrid corns. Data from Mirić et al (1981).

had 58.8% and its SE, 70.1%). In contrast, other workers (Sharma et al, 1975; Mirić et al, 1981) found that the fatty acids of the SE were more saturated than those of the TG. The composition of the sterols in the SE class varied little from that of the FS class; the levels of campesterol and $\Delta 5$-avenasterol were about 5% lower and 6% higher, respectively, in SE than in FS (Worthington and Hitchcock, 1984). Billheimer et al (1983) separated the SE of corn oil by reversed-phase HPLC. The largest peak was identified as sitosteryl linoleate, but no quantitative data were given.

The sterylglycosides in corn grain are made up of SGE and steryl mono-, di-, tri-, and tetraglycosides (SG, SG_2, SG_3, and SG_4). Tanaka et al (1984) found that SGE and SG were the major classes; the principal molecular species were characterized as 6'-palmitoyl-3-O-glucosyl-sitosterol and 3-O-glucosyl-sitosterol. The fatty acids of the SGE tended to have a higher level of saturated fatty acids (43%) than other lipid classes (Table X).

F. Hydrocarbons

Worthington and Hitchcock (1984) isolated a hydrocarbon fraction along with the FS and SE fractions by HPLC from commercial corn oils. Squalene was the major hydrocarbon, with the remaining hydrocarbons consisting of a complex mixture of compounds.

G. Polyisoprenoid Alcohols

Polyisoprenoid alcohols are believed to have important roles in plants as carriers of sugars for cell wall biosynthesis and protein glycosylation (Loomis

TABLE XII
Sterol Classes of Four Strains of Corn[a]

	Strain and Kernel Weight[b] (g)			
	Oh43 174	Ky226 228	ILO[c] 345	IHO[d] 202
Free sterols				
mg/g	0.65	0.23	0.40	1.06
µg/kernel	113	52	137	214
Sterylesters				
mg/g	0.54	0.19	0.19	0.93
µg/kernel	94	43	67	187
Sterylglycosides				
mg/g	0.01	<0.01	0.02	0.02
µg/kernel	2	2	9	5
Sterylglycoside ester				
mg/g	0.02	0.03	<0.01	<0.01
µg/kernel	4	7	trace	1
Total of classes, µg/kernel	213	103	213	407

[a] Data from Davis and Poneleit (1974, 1975).
[b] Dry weight 60 days after pollination.
[c] ILO = Illinois Low Oil.
[d] IHO = Illinois High Oil.

and Croteau, 1980). When Ravi et al (1984) compared the HPLC peaks of the polyisoprenoid alcohols from mono- and dicotyledonous seeds, the monocotyledons (cereal grains) showed double peaks and the dicotyledons single peaks. The double peaks indicated that the monocotyledons contained nearly equal amounts of the α-saturated (dolichols) and α-unsaturated (polyprenols) polyisoprenoid alcohols. In corn kernels, the concentration of polyisoprenoid alcohols was 6.7 mg/100 g. The series of peaks from corn represented the range from 14 to 18 isoprene units, with the major peaks at 15 and 16.

H. Waxes

Surface wax was extracted from corn kernels by dipping mature ears in chloroform for 30 sec at room temperature (Bianchi et al, 1984). The kernel wax was composed mainly of esters (76%). The dominant chain lengths of the esters were C46 (21%), C48 (22%), C50 (9%), C52 (10%), and C54 (17%). The major components of the esters were alcohols—C22 (16%), C24 (30%), C26 (10%), C32 (15%)—and acids—C22 (48%) and C24 (37%). Corn husk wax resembled that of the kernel, but leaf and seedling waxes were more complex mixtures.

I. Cutin

Espelie et al (1979) studied the composition of cutin from the pericarp of corn kernels. Cutin was isolated as the insoluble polymeric residue remaining after thorough extraction of the pericarp with chloroform-methanol and enzymatic removal of the carbohydrates. The cutin was subjected to depolymerization with $LiAlH_4$, and the resulting monomers were identified by combined GC-mass spectroscopy. In corn, ω-hydroxylated acids constituted the major class of monomers. The major acids were ω-OH-18:1 (51%), diOH-16:0 (20%), 9,10,18-triOH-18:0 (7%), and ω-OH-16:0 (6%).

TABLE XIII
Carotenoid Composition of Corn Grain

Carotenoid	No. of OH Groups	Study 1[a]	Study 2[b]	Study 3[c]
Carotenes				
β-carotene	0	0.1– 5.4		0.4– 1.7
β-zeacarotene	0	0.1– 4.7	0.2– 5.4	0 – 0.6
ζ-carotene	0	0.1– 4.2		...
Xanthophylls				
Zeinoxanthin	1	0 – 7.8	0.1– 5.1	...
Cryptoxanthin	1	0.3– 5.2	0 – 6.9	0.9– 2.3
Lutein	2	2.0–33.1	0.1–16.1	7.0–14.9
Zeaxanthin	2	0.6–27.4	0.1–23.1	6.1–15.8
Polyoxy compounds	...	0 – 3.9	0.1– 7.5	...
Total carotenoids		0.2–57.9	0.6–57.2	...

[a] From Quackenbush et al (1963); 125 inbreds.
[b] From Grogan and Blessin (1968); 15 lines, five exotic strains, five crosses, and nine color separation series.
[c] From Cabulea (1971); five selfed lines and 10 hybrids.

J. Carotenoids

The older literature concerned with corn carotenoids has been reviewed by Watson (1962) and Morrison (1978). Two general classes of carotenoid pigments, carotenes and xanthophylls, are primarily responsible for the yellow color of corn grain. The carotenoids are important feed constituents of corn because carotenes are precursors of vitamin A and xanthophylls impart a desirable yellow color to egg yolks and the skin of poultry (Bauernfeind, 1981). The carotenoids also function as antioxidants (Ames, 1983; Burton and Ingold, 1984). The carotenes are hydrocarbons and are believed to be the biosynthetic precursors of the oxygenated derivatives, the xanthophylls (Spurgeon and Porter, 1980), but little study of carotenoid biosynthesis in seeds has been done.

The distribution of carotenoids in hand-dissected corn kernels was 74–86% in the horny endosperm, 9–23% in the floury endosperm, 2–4% in the germ, and 1% in bran (Blessin et al, 1963). The horny endosperm also made up a larger proportion (46–54%) of the kernel than the floury endosperm (28–36%).

The individual carotenoids that have been identified in corn grain are shown in Table XIII. Quackenbush (1963) also noted small and variable fractions of xanthophylls esterified with fatty acids (0.3–4.1 mg/kg). By far, the predominant xanthophylls in corn grain are lutein and zeaxanthin. β-Carotene is the principal carotene.

The chemical structures of the major corn carotenoids are shown in Fig. 5.

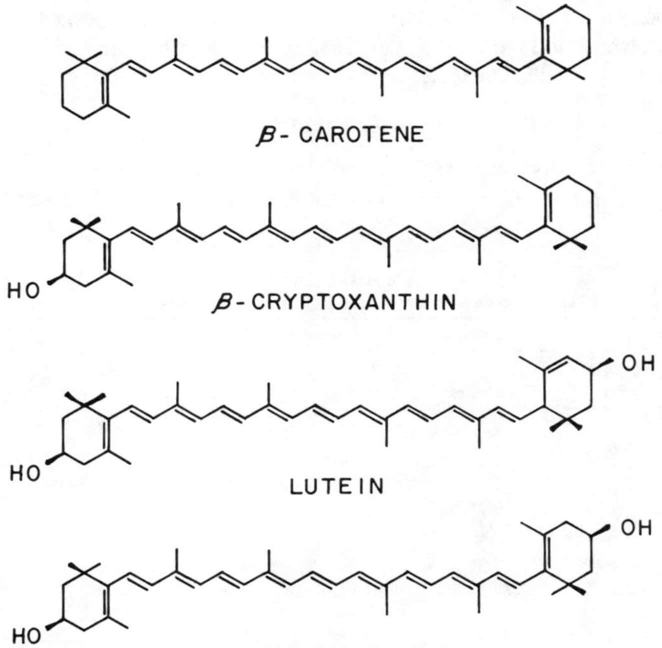

Fig. 5. Chemical structures of the major carotenoids of corn grain.

Structures of the minor carotenoids can be found in Barnes (1983a). The dominant provitamin A carotenoid is β-carotene, because vitamin A is derived from central cleavage of the β-carotene molecule. In general, any carotenoid that has vitamin A-type structure on either end of the molecule is a provitamin A, but these have only half or less of the activity of β-carotene. For example, cryptoxanthin has an OH group on only one ring; the other half of the molecule is available for conversion to vitamin A.

Because of their alternating system of double bonds, carotenoids are very sensitive to oxygen, light, heat, and acids. The loss of vitamin A activity and xanthophyll pigments during the storage of corn grain is a serious problem. Quackenbush (1963) demonstrated a logarithmic rate of loss of carotenoids from yellow corn kernels stored at 25°C. Approximately one half of the carotenes was lost in the first eight months and one half of the xanthophylls in 12 months. Ground corn lost xanthophyll pigments nearly one and a half times faster than whole grain (Dua et al, 1965).

Breeding corn with higher provitamin A and xanthophyll contents appears possible, but only a few studies of the inheritance of carotenoids have been made (Brunson and Quackenbush, 1962; Grogan et al, 1963; Cabulea, 1971; Neamtu et al, 1984). One problem has been the complex and time-consuming analysis required to quantitate the individual carotenoids. Weber (1984a) has developed an HPLC method that may be useful for genetic studies. After being extracted from the grain and saponified, the carotenoids were separated on a normal-phase silica column with hexane-isopropanol (96:4) as the mobile phase. This HPLC system separated the xanthophylls, but the carotenes were eluted as a single peak. Table XIV shows the major carotenoid fractions from 10 corn inbreds. Lutein varied from 10.9% in A632 to 64.2% in B84 and zeaxanthin from 5.8% in Oh45 to 68.3% in A632.

K. Tocols

The four naturally occurring tocopherols are designated α-, β-, γ-, and δ-tocopherol (α-, β-, γ-, and δ-T). The chemical structures of these compounds

TABLE XIV
Major Carotenoid Fractions in Grain of 10 Corn Inbreds[a]

Corn Inbred	Oil Content (%)	Carotenoid Fractions[b]				Weight (mg/kg) of Major Carotenoids
		Carotenes	Crypto-xanthin	Lutein	Zea-xanthin	
R802A	7.4	51.5	9.4	32.3	6.8	72.0
Oh45	4.6	33.0	7.8	53.4	5.8	63.2
A619	4.5	43.3	9.6	26.5	20.6	42.3
Mo17	3.8	17.0	6.5	41.2	35.3	29.5
B73	4.0	13.2	22.1	50.9	13.8	29.5
NY16	4.2	16.5	9.3	60.7	13.4	28.5
B84	3.6	9.3	4.7	64.2	21.8	26.9
A632	3.7	10.5	10.3	10.9	68.3	14.7
C105	2.3	8.6	10.2	48.4	32.8	8.5
K6	2.5	67.3	...	23.4	9.3	0.09

[a] From E. J. Weber (unpublished data).
[b] Data are percent of total carotenoids.

are shown in Fig. 6. The four closely related tocotrienols (α-, β-, γ-, and δ-T3) differ only in having three double bonds in the isoprenoid side chain. The term "tocols" includes both the tocopherols and the tocotrienols. In animals, the vitamin E biological activities of β-, γ-, and δ-T and the T3 are generally less than 15-25% of that of α-T (Scott, 1978). In plants, the functions of tocols are little known, but one role may be as antioxidants to protect the unsaturated lipids. As an antioxidant, γ-T may be superior to α-T (Cort, 1974; Wu et al, 1979; Cillard and Cillard, 1980; Hudson and Ghavami, 1984), although Burton et al (1983) proposed the same order of antioxidant activities as that of their biological activities ($\alpha > \beta \geqslant \gamma > \delta$).

In an examination of ancient corn grain from caves in the arid southwestern United States, Priestly et al (1981) found that although most of the linoleic acid was oxidized in the first 100-200 years, about 4% by weight of the linoleic acid found in modern corn was still detectable in grain more than 1,500 years old. They suggested that the tocopherols in the grain may assist in the long-term preservation of the polyunsaturated linoleic acid.

Quackenbush et al (1963) determined the total tocols of 125 corn inbred lines. Among these inbreds, the total tocols ranged from 0.03 to 0.33% of the oil. Environmental influences on total tocol contents were indicated in the analyses of Russian and German lines and hybrids by Karaiwanow et al (1982). The total tocol contents ranged from 351-859 mg/kg of oil in 1978 to 609-1,440 mg/kg of oil in 1980.

The individual tocols have been determined as their trimethylsilylethers by GC (Slover et al, 1969; Slover, 1971). In corn grain, the related saturated-unsaturated tocol pairs of the α- and γ-forms were found (α-T, 0.6 mg/100 g of grain; α-T3, 0.3; γ-T, 4.5; γ-T3, 0.5). Corn oil contained only the saturated tocopherols (α-T, 12 mg/100 g of oil; γ-T, 52).

HPLC appears to be the current, preferred method for determination of the tocols. The principal advantages of HPLC over GC are that no derivatization is required and, with lower temperatures and shorter times on the column, the tocols incur less risk of oxidative losses. HPLC is also faster and gives better separations than TLC. Cavins and Inglett (1974) were the first to demonstrate separation of the tocols of corn oil by HPLC. Separation on a silica adsorption column took 80 min, and the tocol peaks were located by an ultraviolet (UV)

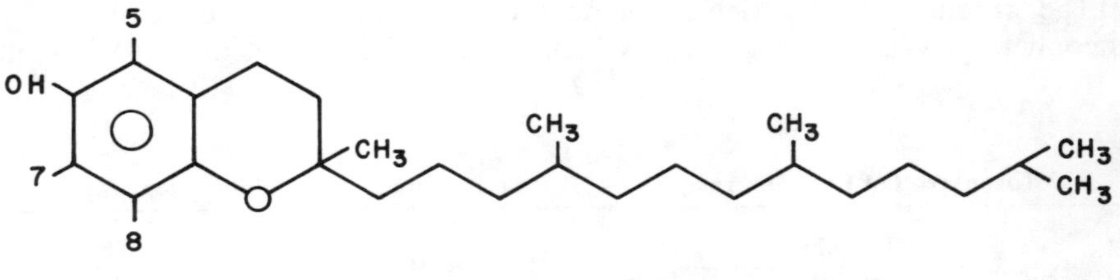

5,7,8 Trimethyl = α - tocopherol
5,8 Dimethyl = β - tocopherol
7,8 Dimethyl = γ - tocopherol
8 Methyl = δ - tocopherol

Fig. 6. Chemical structures of tocopherols.

detector set at a wavelength of 254 nm. The tocopherols were concentrated before HPLC by freezing the acetone extract and filtering off the lipids and sterols. Other workers (Carpenter, 1979; Gertz and Herrmann, 1982a; Weber, 1984b) have analyzed corn oil directly without further cleanup. In corn oil, trace amounts of β-T (Gertz and Herrmann, 1982a, 1982b) or δ-T (Carpenter, 1979; Zonta and Stancher, 1983) have been reported (Table XV).

Various improvements have been made in the HPLC analysis of corn tocols. Gertz and Herrmann (1982b) decreased the analysis time to 10 min by using a shorter column (12.5 cm × 4.9 mm i.d.) and faster flow rate (4 ml/min). Sensitivity of detection was improved 10 times by measuring the tocols with the UV detector set at 295 nm, which is closer to the adsorption maxima of the tocols than is 254 nm (Carpenter, 1979; Parrish, 1980). Because of the double spectral requirements, fluorescence detection with an excitation wavelength of 294 nm and emission cutoff at 325 nm was more specific and therefore less influenced by impurities than was UV detection (Van Niekerk, 1973; Gertz and Herrmann, 1982a). The sensitivity of fluorescence detection was further increased to 15-fold over UV at 295 nm by using a shorter, 205-nm excitation wavelength (Weber, 1984b). As little as 4 ng of a tocol was measured. The sample size could be reduced to one corn kernel or even a sample of ground corn weighing only 25 mg.

Direct saponification of ground corn grain before solvent extraction released the tocols more efficiently than saponification of a lipid extract (McMurray et al, 1980; Contreras-Guzman et al, 1982; Weber, 1984b). γ-Tocopherol has been reported to be the predominant tocol in corn grain (Barnes, 1983a, 1983b; Cort et al, 1983), but Weber (1984b) found inbreds with equal or higher levels of α-T than γ-T (Table XVI). Inbreds K6 and B37 had nearly equal percentages of α-T and γ-T, and in A632, the level of α-T (42.3%) was higher than that of γ-T (25.8%). Wide ranges were observed among the inbreds for γ-T (25.8-82.6%) and for total tocols (282-1,016 mg/kg). In general, the levels of the tocotrienols paralleled those of their respective tocopherols. No significant correlations were found between weights of the tocols and of oil or linoleic acid.

Tocotrienols were found only in the endosperm, whereas the germ contained the major proportion of the tocopherols (Grams et al, 1970; Weber, 1984b). Grams et al (1970) hand-dissected grain from four hybrids and found 93-96% of the α- and γ-T in the germ, 2-4% in the endosperm, and 2-4% in the pericarp and tip cap.

The absence of tocotrienols made the study of genetic control of the tocopherol levels in the germ less complicated. Galliher et al (1985) found

TABLE XV
Tocopherols of Corn Oil Analyzed by High-Performance Liquid Chromatography

Corn Oil	Tocopherols, % of total				Total Tocopherols (mg/kg)	Reference
	α-T	β-T	γ-T	δ-T		
Commercial	20.1	...	76.5	3.4	610	Carpenter (1979)
Commercial	27.0	...	72.0	1.0	1,040	Carpenter (1979)
Germ	37.0	1.1	61.9	...	313	Gertz and Herrmann (1982b)
Germ	49.9	...	50.1	...	1,083	Gertz and Herrmann (1982b)
Commercial	17.2	...	79.3	3.5	1,033	Zonta and Stancher (1983)

heritability estimates of 0.62 and 0.68 for α-T and γ-T, respectively, which indicated that selection would be effective. With the large variability in tocols that was observed among corn inbreds (Table XVI), breeding of hybrids with selected proportions of vitamin E forms may be possible.

An increase in the α-T content of corn used for feedstuff may be desirable because α-T is much more biologically active than γ-T. Combs and Combs (1985) analyzed 42 commercial varieties of corn and calculated the vitamin E activity of each variety by multiplying the α-T values by 1.49 and the γ-T values by 0.05. The vitamin E activities varied from 11.1 to 36.4 USP units per kilogram of air-dried corn. Feed tables used in the formulation of livestock feeds list the vitamin E content of corn as 20–26 USP units per kilogram. Of the 42 varieties analyzed, 15 had vitamin E activity that was lower than 20 USP units per kilogram.

V. DISTRIBUTION OF LIPIDS IN THE CORN KERNEL

A. Dissected Grain

The results of recent studies of the distribution of lipids in hand-dissected fractions of the corn kernel are shown in Table XVII. Previous analyses can be found in the review by Morrison (1978). When extraction procedures were tested, n-butanol gave the most complete extraction of lipids, particularly from the endosperm, where the lipids are complexed with starch and proteins (Tan and Morrison, 1979a; Weber, 1979). In addition to the kernel fractions listed in Table XVII, Tan and Morrison (1979a) also examined the tip cap of the LG-11 hybrid, high-amylose corn, and waxy corn. The tip cap made up 1.4–2.5% of the dry weight of the kernel and contained 1.8–2.0% lipid. Among the three strains, the distribution of total kernel lipids was 76–83% in the germ, 14–23% in the endosperm, 1–2% in the pericarp, and <1% in the tip cap.

Several investigators (Jahn-Deesbach et al, 1975b; Tan and Morrison, 1979a; Weber, 1979; Demir et al, 1981; Abdel-Rahman, 1983) have observed that the

TABLE XVI
Tocols in Grain of Ten Corn Inbreds[a]

Corn Inbred	Tocols, % of total				Total Tocols (mg/kg)
	α-T[b]	α-T3	γ-T	γ-T3	
C103D	25.5	14.9	39.7	19.9	282
T220	13.6	14.5	43.5	28.4	339
A632	42.3	22.4	25.8	9.4	353
Mo17	32.3	14.0	42.9	10.8	378
K6	36.2	20.6	38.0	5.1	428
W64A	2.5	6.5	82.6	8.4	570
B37	39.4	12.7	39.1	8.8	583
B73	15.0	13.1	47.0	24.9	620
R802A	19.0	4.3	61.9	14.9	914
NY16	14.8	7.5	50.7	27.1	1,016

[a] Data from Weber (1984b).
[b] T = tocopherol, T3 = tocotrienol.

fatty acid composition of the endosperm lipids is more saturated than that of the germ lipids. The endosperm lipids contain higher levels of palmitic, stearic, and linolenic acids and lower levels of oleic and linoleic acids. The fatty acid compositions reflect the distribution of the individual lipid classes among the kernel fractions.

The distribution by weight of individual lipid classes in the germ, endosperm, pericarp, and tip cap of a conventional hybrid (LG-11) corn kernel is shown in Table XVIII. In the germ, TG constituted 85% of the total lipids, PhL 3%, glycolipids 2%, and unsaponifiable lipids 4%. Therefore, the fatty acid composition of the germ lipids was governed mainly by the more highly unsaturated TG. Within the germ PhL, PC accounted for 52%, PI 17%, and PE 13%. DGDG was the major glycolipid.

The endosperm lipids were divided into nonstarch and starch lipids (Table XVIII). The nonstarch lipids were extracted with water-saturated n-butanol at room temperature for 30 min. The residual starch lipids were extracted with hot, water-saturated butanol for 6 hr. Of the lipids in the endosperm, 54% were nonstarch lipids and 46% were starch lipids. (The starch lipids are discussed in the next section.) TG made up 48% of the nonstarch lipids, free fatty acids 19%, PhL 4%, glycolipids 6%, and unsaponifiables 10%. No direct analysis of the lipids of the aleurone layer of the corn kernel has been done, but Tan and Morrison (1979a) determined the aleurone lipids by difference between the complete endosperm and the endosperm minus the aleurone and some adjacent starchy tissue. Over half of the nonstarch lipids were removed with the aleurone layer. The aleurone lipids resembled the germ lipids and contained 56% of the endosperm TG. The residual nonstarch lipids were mostly free fatty acids (48%). In a study of the lipids of developing LG-11 kernels, Tan and Morrison (1979b) observed that TG accumulated in the endosperm until 36–42 days after pollination. Maximum values for the nonstarch PhL and galactosylglycerides

TABLE XVII
Distribution of Weight and Lipids in Parts of Corn Kernel

Reference	Whole Kernel		Germ		Endosperm		Pericarp	
	Weight (%)	Lipid[a] (%)	Weight (%)	Lipid (%)	Weight (%)	Lipid (%)	Weight (%)	Lipid (%)
Tan & Morrison, (1979a)[b]	100	5.0–9.3	8.4–15.0	35.6–38.7	74.7–86.0	0.8–2.4	4.2–7.8	0.2–0.3
Weber (1979)[c]	100	4.9	10.9	35.5	82.6	1.0	...	...
Jovanic et al (1975)[d]	100	4.9–5.4	10.3–12.2	35.5–39.6	80.2–81.9	1.3	7.6–7.9	0.6
Mihajlović (1978)[e]	100	4.8–6.0	12–13	31.0–36.0	80–82	1.3–1.8	6–7	1.4–1.7
Demir et al (1981)[f]	100	4.4–6.8	8.8–15.0	18.5–25.8	...	2.1–4.4	...	...

[a] Lipids expressed as fatty acid methyl esters.
[b] LG-11 hybrid, high-amylose corn, waxy corn. Extraction with hot, water-saturated n-butanol.
[c] H51 inbred. Extraction with boiling, water-saturated n-butanol.
[d] Four Yugoslavian inbreds and three hybrids.
[e] Nine Yugoslavian hybrids. Soxhlet extraction.
[f] Ten Turkish inbred lines and 11 hybrids. Direct transmethylation of fatty acids.

were reached at 16-23 days after pollination. Extensive degradation of the nonstarch lipids occurred during the final stages of endosperm development, and lipolysis could account for the free fatty acids.

The major pericarp lipids were composed of unsaponifiables (55%), TG (20%), and SE (15%). Pericarp tissue underwent extensive senescence during the latter stages of kernel development, and most of the lipid reserves were consumed. The unsaponifiables were mainly sterols and aliphatic alcohols and were probably derived from the surface waxes. The tip cap lipids with 44% TG, 13% free fatty acids, 10% SE, and 11% unsaponifiable lipids resembled the nonstarch lipids.

TABLE XVIII
Distribution of Lipids in Kernel of LG-11 Hybrid Corn[a,b]

Lipid	Lipid (μg/kernel)					
	Germ	Endosperm Non-starch	Endosperm Starch	Pericarp	Tip Cap	Total
Sterylester	173	105	8	13	9	308
Triglyceride	7,806	739	17	17	40	8,619
Diglyceride	263	70	6	3	3	345
Free fatty acid	64	286	624	1	12	987
Monoglyceride	8[c]	15[c]	23	trace[c]	2[c]	48
Esterified sterylglycoside	15	14	14	1	2	46
Monogalactosyldiglyceride	}27	}26	}23	1	1	}79
Monogalactosylmonoglyceride				trace	1	
Digalactosyldiglyceride	92	25	19	trace	2	138
Digalactosylmonoglyceride	36	25	17	1	1	80
N-acylphosphatidylethanolamine	23	6	...	1	1	31
N-acyl lysophosphatidylethanolamine	}4	}1	...	}trace	...	}5
Diphosphatidylglycerol			...		...	
Phosphatidylglycerol	8	1	...	...	...	9
Phosphatidylethanolamine	41	2	...	...	1	44
Phosphatidylcholine	158	7	...	1	1	167
Phosphatidylinositol	52	6	...	...	1	59
Phosphatidic acid	16	2	...	trace	1	19
Lysophosphatidylglycerol	...	...	9	}trace	...	9
Lysophosphatidylethanolamine	...	2[d]	35		trace[d]	37[d]
Lysophosphatidylcholine	5	36	472	trace	1	514
Lysophosphatidylinositol[e]	...	...	17	...	1	18
Nonpolar lipids	8,314	1,215	678	34	66	10,307
Glycolipids	170	90	73	3	7	343
Phospholipids	307	63	533	2	7	912
Unsaponifiable lipids	405	155	...	48	10	618
Total lipids	9,196	1,523	1,284	87	90	12,180

[a] Source: Tan and Morrison (1979a); used by permission.
[b] Mean dry weight of LG-11 kernel = 250 mg; lipid content = 4.9%; weights of germ, endosperm, pericarp, tip cap = 21, 215, 10, 4 mg, respectively.
[c] Includes 6-O-acyl monogalactosyldiglyceride.
[d] Includes phosphatidylserine.
[e] Includes lysophosphatidylserine.

B. Starch Lipids

Only the cereal starches have been shown to contain endogenous internal starch lipids. Nonwaxy corn, barley, and rice starches contain 0.6–1.1% total lipid on a dry weight basis (Morrison et al, 1984). Starches from potato tubers, parsnip roots, and beans (*Vicia faba*, mung and red) contain less than 0.1% lipid, and these lipids are probably starch surface lipids (Morrison and Laignelet, 1983). Moreover, the major types of the starch lipids differ among the cereal grains. Corn starch has high levels of free fatty acids (51–62%) and lower levels of lysoPhL (24–46%) (Table XIX). Wheat starch contains lysoPhL almost exclusively (86–94%) with little free fatty acid (2–6%). The lipids of barley starch are also mostly lysoPhL; more equal proportions of free fatty acid and lysoPhL are found in rice, oat, millet, and sorghum starches (Morrison et al, 1984).

To alleviate contamination by starch surface lipids, Morrison et al (1984) prepared high-purity starches. The nitrogen content of the starch was used as a criterion of purity, low nitrogen (0.04–0.06%) indicating the absence of endosperm proteins and, by implication, nonstarch lipids. The fatty acid composition of the total lipids in nonwaxy corn starches is relatively constant (mean values: 37% 16:0, 3% 18:0, 11% 18:1, 46% 18:2, and 3% 18:3). The close relationship between lipid content and the proportion of amylose in the starch is illustrated by analysis of high-purity starches from corn mutants having relatively high (sugary and high-amylose), normal, and very low (waxy) amylose (Table XIX). Both the free fatty acid content and the lysoPhL content are positively correlated with the amount of amylose in the starch. The starch lipids appear to form complexes with amylose. If lipid is not removed before amylose determination, the amylose content measured by I_2-binding is lower than in lipid-free starch (Morrison and Laignelet, 1983).

The starch lipids are exclusively monoacyl lipids—free fatty acids and lysoPhL. In corn starch, lysoPC represents 89% of the lysoPhL (Table XVIII). The presence of monoacyl lipids in tissues is unusual because lysoPhL and free fatty acids are cytotoxic and cause lysis of membranes. It has been assumed that

TABLE XIX
Composition of Internal Lipids of Corn Starch from Various Endosperm Types[a]

Lipid	Commercial Starch[b] (Normal)	High-Purity Starch[b]			
		Sugary	High-Amylose	Normal	Waxy
Steryl ester, triglyceride and diglyceride	43	6	2	3	4
Free fatty acid	454	575	543	379	12
Monoglyceride	31	35	8	9	...
Glycolipid	20	50	23	13	...
Lysophospholipid	182	444	485	298	5
Total lipids	730	1,110	1,061	702	21

[a] Adapted from Morrison and Milligan (1982).
[b] In mg/100 g of dry weight.

the lipids are complexed with amylose and that the monoacyl chain lies within a helical section of the amylose 1,4-α-glucan chain (Morrison and Milligan, 1982; Galliard, 1983). The suggested evidence for the amylose-lipid complex has been the exceptional resistance of the starch acyl lipids to solvent extraction, autoxidation, and attack by chlorine. Amylose-bound lysoPC is cleaved by phospholipase D, which acts at the choline-phosphate ester group, but not by acyl hydrolase, which acts at the fatty-acyl ester bond. The polar group of the complexed lysoPhL is assumed to be partly exposed outside the helical structure and the fatty acid chain to be inside the helix. However, Morrison and Milligan (1982) suggested that the starch lipids could be protected equally well if they lay in intermolecular spaces or interstices within the dry starch granules. The inclusion complexes between amylose and lipid may be formed during any wet analytical procedure that causes the starch granule to swell.

The effects of internal and surface lipids on the physical properties of starches have not been fully evaluated. Removal of lipid from cereal starches is accompanied by significant changes in the viscosity characteristics of starch-water pastes (Galliard, 1983). Starch lipids also modify the gelatinization properties of starches and their susceptibilities to hydrolysis by enzymes or acids (Morrison and Milligan, 1982). However, more research is needed to understand these phenomena.

Galliard (1983) has questioned a direct role of internal starch lipids in the biosynthesis or degradation of starch, because lipids are absent or in low concentration in noncereal starches. He proposed that the monoacyl lipids of the cereal starches are hydrolysis products from the lipoprotein membranes involved in starch biosynthesis and that these cytotoxic products are rendered metabolically inactive by becoming bound to amylose chains. In tissues where the starch granules do not contain significant quantities of internal lipids, membranes have been degraded more rapidly and completely. For example, potato tubers contain high levels of lipolytic and lipid-oxidizing enzymes that catalyze the breakdown of mono- and diacyl lipids of membranes. The corresponding lipid-degrading enzymes are present in cereals but appear to be less active.

VI. ENZYMES AND HORMONES

Tsaftaris and Scandalios (1983a) examined the IHO and ILO lines to determine whether the amount of lipids stored in the seed affected the number of glyoxysomes and the levels of glyoxysomal enzymes present after seed germination. An active glyoxylate cycle is required to direct gluconeogenesis from the acetyl-CoA derived by β-oxidation of the fatty acids of the storage TG. Despite their 40-fold difference in lipid content, the IHO and ILO lines have similar populations of glyoxysomes and exhibit similar catalase- and malate synthase-specific activities. Only the specific activity of isocitrate lyase, the first key enzyme in the glyoxylate shunt, is higher (by two-fold) in the IHO seed than in the ILO seed. Genetic analysis indicates that the higher isocitrate lyase activity of IHO is probably not a consequence of the higher lipid content but may be due to loose linkage between genes controlling isocitrate lyase activity and genes controlling oil content (Tsaftaris and Scandalios, 1983b).

Wang et al (1984) also investigated the activities of catalase and isocitrate lyase in IHO and ILO and their F_1 generation. These workers found no differences in the activities of catalase or isocitrate lyase among the three lines. In sharp contrast, the lipase activities in the three lines are proportional to their respective lipid contents. The enzyme activity and the lipid content in the F_1 are intermediate to those of IHO and ILO. These results suggest that lipase and the glyoxysomal enzymes are under separate genetic control.

The lipase from scutella of corn seedlings has been isolated and characterized (Lin et al, 1983; Lin and Huang, 1984). The corn lipase is present in the lipid bodies of germinated but not ungerminated seed. The enzyme is active only on TG containing oleic or linoleic acids, the dominant fatty acids of corn oil. The enzyme did not cleave saturated TG, tripalmitin, or tristearin. Corn lipase differs from pancreatic lipase in that the corn lipase hydrolyzes TG to three fatty acids for β-oxidation in the glyoxysomes, whereas pancreatic lipase hydrolyzes TG to two fatty acids and a monoglyceride for intestinal absorption.

The physiological role of lipoxygenase in plants has been a mystery. The lipoxygenase-catalyzed oxidation of polyunsaturated fatty acids yields fatty acid hydroperoxides, which are highly reactive and potentially damaging to cellular components. Vick and Zimmerman (1984) have recently proposed that regulation of plant growth may be one role for lipoxygenase. First, these workers showed that oxygenated fatty acids, which are metabolites of the fatty acid hydroperoxides, are present in seedlings of corn (Vick and Zimmerman, 1979, 1982). The next discovery was that one of these oxygenated metabolites, 12-oxo-*cis,cis*-10,15-phytodienoic acid (12-oxo-PDA) is converted to jasmonic acid—3-oxo-2-(2'-pentenyl)-cyclopentaneacetic acid—by several plant species, including corn (Vick and Zimmerman, 1984). Jasmonic acid is a plant growth regulator that inhibits growth and promotes senescence. Some plant lipoxygenases catalyze oxygenation of linoleic or linolenic acids at carbon 9, others at carbon 13. The 12-oxo-PDA is formed only from the 13-hydroperoxide. Specificity for the 13-carbon varied from 17% in corn germ (Gardner and Weisleder, 1970) to 37% in 5-day-old corn leaves (Vick and Zimmerman, 1982). Another metabolite of the 13-hydroperoxide was identified as traumatin, a wound hormone (Zimmerman and Coudron, 1979). These plant metabolites resemble the potent, bioactive prostaglandins and leukotrienes synthesized from oxygenated fatty acids in mammals. Plant fatty acid metabolites may also be powerful metabolic regulators.

VII. THE FUTURE

In spite of considerable progress, much remains to be learned about the metabolism of lipids in the corn kernel. Little is known about the enzymes and hormones, particularly during development and germination. The physiological roles of the various forms of the PhL, glycolipids, tocols, and carotenoids should be investigated.

Exciting possibilities exist for genetically altering the lipids of the corn kernel. Oil content, fatty acid composition, fatty acid placement in the lipids, the proportions of the various lipid classes, tocols, and carotenoids are heritable traits. Higher-oil corn has been demonstrated to be a more efficient feed for poultry and pigs than conventional corn and should also be more profitable for corn millers. One of the problems has been segregation of the higher-oil corn from conventional corn. Now low-cost infrared analyzers are available that could be used at grain elevators to identify higher-oil corn (Hymowitz et al, 1974).

A large number of corn mutants have been identified that could serve as experimental probes to facilitate coordinated biochemical and genetic studies of lipid metabolism. Undoubtedly, a tremendous wealth of biochemical variations still remains to be discovered in corn.

LITERATURE CITED

ABDEL-RAHMAN, A. H. Y. 1983. A study on some Egyptian corn varieties. Riv. Ital. Sostanze Grasse 60:701-702.

ADAMS, K. L., and JENSEN, A. H. 1981. Evaluation of the use of high-oil corn in swine diets. Univ. Ill. Coop. Ext. Publ., Swine Res. Rep. 1981-1. 4 pp.

ALEXANDER, D. E. 1982. The use of wideline NMR in breeding high-oil corn. (Abstr.) J. Am. Oil Chem. Soc. 59:284A.

ALEXANDER, D. E. 1987. Maize. In: Oil Crops of the World. G. Robbelen, A. Ashri, and P. K. Downey, eds. Macmillan Publishing Co., New York. In press.

ALEXANDER, D. E., and SEIF, R. D. 1963. Relation of kernel oil content to some agronomic traits in maize. Crop Sci. 14:598-599.

ALEXANDER, D. E., SILVELA S., L., COLLINS, F. I., and RODGERS, R. C. 1967. Analysis of oil content of maize by wideline NMR. J. Am. Oil Chem. Soc. 44:555-558.

AMES, B. N. 1983. Dietary carcinogens and anticarcinogens. Oxygen radicals and degenerative diseases. Science 221:1256-1264.

ANDERSON, R. A., and WATSON, S. A. 1982. The corn milling industry. Pages 31-61 in: Handbook of Processing and Utilization in Agriculture. Vol. 2. I. A. Wolff, ed. CRC Press, Inc., Boca Raton, FL.

APPLEWHITE, T. H. 1983. Nutritional effects of isomeric fats: Facts and fallacies. Pages 414-424 in: Dietary Fats and Health. E. G. Perkins and W. J. Visek, eds. Am. Oil Chem. Soc., Champaign, IL.

ARNOLD, J. M., PIOVARCI, A., BAUMAN, L. F., and PONELEIT, C. G. 1974. Weight, oil, and fatty acid composition of components of normal, opaque-2, and floury-2 maize kernels. Crop Sci. 14:598-599.

BAKTASH, F. Y., YOUNIS, M. A., AL-YOUNIS, A. H., and AL-ITHAWI, B. A. 1982. Relative effectiveness of two systems of selection for protein and oil percentage of the corn grains. Mesopotamia J. Agric. 17:31-36. (Commonwealth Agric. Bur. 84:901. 1984.)

BARNES, P. J. 1983a. Non-saponifiable lipids in cereals. Pages 33-55 in: Lipids in Cereal Technology. P. J. Barnes, ed. Academic Press, London.

BARNES, P. J. 1983b. Cereal tocopherols. Pages 195-200 in: Progress in Cereal Chemistry and Technology. Vol. 1. J. Holas and J. Kratochvil, eds. Elsevier Scientific Publ. Co., Amsterdam.

BAUERNFEIND, J. C., ed. 1981. Carotenoids as Colorants and Vitamin A Precursors. Academic Press, New York. 938 pp.

BAUMAN, L. F., CONWAY, T. F., and WATSON, S. A. 1963. Heritability of variations in oil content of individual corn kernels. Science 139:498-499.

BIANCHI, G., AVATO, P., and SALAMINI, F. 1984. Surface waxes from grain, leaves, and husks of maize (Zea mays L.). Cereal Chem. 61:45-47.

BILLHEIMER, J. T., AVART, S., and MILANI, B. 1983. Separation of steryl esters by reversed-phase liquid chromatography. J. Lipid Res. 24:1646-1651.

BLESSIN, C. W., BRECHER, J. D., and DIMLER, R. J. 1963. Carotenoids of corn and sorghum. V. Distribution of xanthophylls and carotenes in hand-dissected and dry-milled fractions of yellow dent corn. Cereal Chem. 40:582-586.

BLIGH, E. G., and DYER, W. J. 1959. A rapid method of total lipid extraction and purification. Can. J. Biochem. Physiol. 37:911-917.

BRIMHALL, B., and SPRAGUE, G. F. 1951. Unsaturation of corn oil—Inheritance and maturity studies. Cereal Chem. 28:225-231.

BRUNSON, A. M., and QUACKENBUSH, F. W. 1962. Breeding corn with high provitamin A in the grain. Crop Sci. 2:344-347.

BURTON, G. W., and INGOLD, K. U. 1984. β-Carotene: An unusual type of lipid

antioxidant. Science 224:569-573.
BURTON, G. W., CHEESEMAN, K. H., DOBA, T., INGOLD, K. U., and SLATER, T. F. 1983. Vitamin E as an antioxidant in vitro and in vivo. Pages 4-18 in: Biology of Vitamin E. R. Porter and J. Whelan, eds. Pitman Press, London.
CABULEA, I. 1971. Contribution to the study of carotenoid metabolism in the maize grain. Pages 85-91 in: Eucarpia. Proc. Congr. Eur. Assoc. Res. Plant Breeding, 5th. Sept. 2-5, 1969. Akad. Kiado, Budapest.
CAMPBELL, E. J. 1983. Sunflower oil. J. Am. Oil Chem. Soc. 60:387-392.
CAMUSSI, A., JELLUM, M. D., and OTTAVIANO, E. 1980. Numerical taxonomy of Italian maize populations: Fatty acid composition and morphological traits. Maydica 25:149-165.
CARPENTER, A. P., Jr. 1979. Determination of tocopherols in vegetable oils. J. Am. Oil Chem. Soc. 56:668-671.
CARTER, H. E., and KISIC, A. 1969. Countercurrent distribution of inositol lipids of plant seeds. J. Lipid Res. 10:356-362.
CARTER, H. E., CELMER, W. D., GALANOS, D. S., GIGG, R. H., LANDS, W. E. M., LAW, J. H., MUELLER, K. L., NAKAYAMA, T., TOMIZAWA, H. H., and WEBER, E. J. 1958. Biochemistry of the sphingolipides. X. Phytoglycolipide, a complex phytosphingosine-containing lipide from plant seeds. J. Am. Oil Chem. Soc. 35:335-343.
CARTER, H. E., STROBACH, D. R., and HAWTHORNE, J. N. 1969. Biochemistry of the sphingolipids. XVIII. Complete structure of tetrasaccharide phytoglycolipid. Biochemistry 8:383-388.
CAVINS, J. F., and INGLETT, G. E. 1974. High-resolution liquid chromatography of vitamin E isomers. Cereal Chem. 51:605-609.
CHERRY, J. P. 1983. Cottonseed oil. J. Am. Oil Chem. Soc. 60:360-367.
CILLARD, J., and CILLARD, P. 1980. Behavior of alpha, gamma, and delta tocopherols with linoleic acid in aqueous media. J. Am. Oil Chem. Soc. 57:39-42.
COMBS, S. B., and COMBS, G. F., Jr. 1985. Varietal differences in the vitamin E content of corn. J. Agric. Food Chem. 33:815-817.
CONTRERAS-GUZMAN, E., STRONG, F. C., III, and da SILVA, W. J. 1982. Fatty acid and vitamin E content of Nutrimaiz, a sugary/opaque-2 corn cultivar. J. Agric. Food Chem. 30:1113-1117.
CORT, W. M. 1974. Antioxidant activity of tocopherols, ascorbyl palmitate, and ascorbic acid and their mode of action. J. Am. Oil Chem. Soc. 51:321-325.
CORT, W. M., VICENTE, T. S., WAYSEK, E. H., and WILLIAMS, B. D. 1983. Vitamin E content of feedstuffs determined by high-performance liquid chromatographic fluorescence. J. Agric. Food Chem. 31:1330-1333.
CURTIS, P. E., LENG, E. R., and HAGEMAN, R. H. 1968. Developmental changes in oil and fatty acid content of maize strains varying in oil content. Crop Sci. 8:689-693.
D'ALONZO, R. P., KOZAREK, W. J., and WADE, R. L. 1982. Glyceride composition of processed fats and oils as determined by glass capillary gas chromatography. J. Am. Oil Chem. Soc. 59:292-295.
DAVIS, D. L., and PONELEIT, C. G. 1974. Sterol accumulation and composition in developing *Zea mays* L. kernels. Plant Physiol. 54:794-796.
DAVIS, D. L., and PONELEIT, C. G. 1975. Sterols in developing seed from low and high oil *Zea mays* strains. Phytochemistry 14:1201-1203.
de la ROCHE, I. A., ALEXANDER, D. E., and WEBER, E. J. 1971a. Inheritance of oleic and linoleic acids in *Zea mays* L. Crop Sci. 11:856-859.
de la ROCHE, I. A., WEBER, E. J., and ALEXANDER, D. E. 1971b. Effects of fatty acid concentration and positional specificity on maize triglyceride structure. Lipids 6:531-536.
de la ROCHE, I. A., WEBER, E. J., and ALEXANDER, D. E. 1971c. The selective utilization of diglyceride species into maize triglycerides. Lipids 6:537-540.
de la ROCHE, I. A., WEBER, E. J., and ALEXANDER, D. E. 1971d. Genetic aspects of triglyceride structure in maize. Crop Sci. 11:871-874.
DEMIR, I., YUECE, S., KORKUT, C., and MARQUARD, R. 1981. Fat content, fatty acid composition and tocopherol content of Turkish maize genotypes with separate consideration of endosperm and embryo fractions. Getreide Mehl Brot 35:93-96. (In German)
DONG, M. W., and DiCESARE, J. L. 1983. Improved separation of natural oil triglycerides by liquid chromatography using columns packed with 3-μm particles. J. Am. Oil Chem. Soc. 60:788-791.
DUA, P. N., DAY, E. J., and GROGAN, C. O. 1965. Loss of carotenoids in stored commercial and high-carotenoid yellow corn, *Zea mays* L. Agron. J. 57:501-502.
DUDLEY, J. W., ed. 1974. Seventy Generations of Selection for Oil and Protein in Maize. Crop Sci. Soc. Am., Madison, WI. 212 pp.

DUDLEY, J. W. 1977. Seventy-six generations of selection for oil and protein percentage in maize. Pages 459-473 in: Proc. Int. Conf. on Quantitative Genetics. E. Pollak, O. Kempthorne, and T. B. Bailey, Jr., eds. Iowa State Univ. Press, Ames.

EARLE, F. R. 1977. Protein and oil in corn: Variation by crop years from 1907 to 1972. Cereal Chem. 54:70-79.

EGGUM, B. O., DUMANOVIĆ, J. D., MIŠEVIĆ, D., and DENIĆ, M. 1983. Grain yield and nutritive value of high oil, opaque and waxy maize hybrids. J. Cereal Sci. 1:139-145.

El-HAMDY, A. H., and PERKINS, E. G. 1981. High performance reversed phase chromatography of natural triglyceride mixtures: Critical pair separation. J. Am. Oil Chem. Soc. 58:867-872.

ERICKSON, D. R. 1983. Soybean oil: Update on number one. J. Am. Oil Chem. Soc. 60:351-356.

ESPELIE, K. E., DEAN, B. B., and KOLATTUKUDY, P. E. 1979. Composition of lipid-derived polymers from different anatomical regions of several plant species. Plant Physiol. 64:1089-1093.

FEDELI, E. 1983. Miscellaneous exotic oils. J. Am. Oil Chem. Soc. 60:404-406.

FLORA, L. F., and WILEY, R. C. 1972. Effect of various endosperm mutants on oil content and fatty acid composition of whole kernel corn (Zea mays L.). J. Am. Soc. Hortic. Sci. 97:604-607.

FREEMAN, J. E. 1973. Quality factors affecting value of corn for wet milling. Trans. ASAE 16:671-679.

GALLIARD, T. 1983. Starch-lipid complexes and other non-starch components of starch granules in cereal grains. Pages 111-136 in: Mobilization of Reserves in Germination. C. Nozzolillo, P. J. Lea, and F. A. Loewus, eds. Plenum Press, New York.

GALLIHER, H. L., ALEXANDER, D. E., and WEBER, E. J. 1985. Genetic variability of alpha-tocopherol and gamma-tocopherol in corn embryos. Crop Sci. 25:547-549.

GALLO, J. R., TEIXEIRA, J. P. F., SPOLADORE, D. S., IGUE, T., and de MIRANDA, L. T. 1976. Effect of fertilizer application on the relationships between the chemical composition of seeds and leaves and on yield in maize. Bragantia 35:413-432. (In Portuguese) (Chem. Abstr. 88:5330. 1978.)

GARDNER, H. W., and WEISLEDER, D. 1970. Lipoxygenase from Zea mays; 9-D-hydroperoxy-trans-10, cis-12-octadecadienoic acid from linoleic acid. Lipids 5:678-683.

GENTER, C. G., EHEART, J. F., and LINKOUS, W. N. 1956. Effect of location, hybrid, fertilizer, and rate of planting on the oil and protein content of corn grain. Agron. J. 48:63-67.

GERTZ, C., and HERRMANN, K. 1982a. High pressure liquid chromatographic separation of individual tocopherols and tocotrienols in food fats and edible oils on short columns. Mitteilungsbl. GDCh-Fachgruppe Lebensm. Gerichtl. Chem. 36:53-58. (In German)

GERTZ, C., and HERRMANN, K. 1982b. Analysis of tocopherols and tocotrienols in foods. Z. Lebensm. Unters. Forsch. 174:390-394. (In German)

GOTTENBOS, J. J. 1983. Biological effects of trans fatty acids. Pages 375-390 in: Dietary Fats and Health. E. G. Perkins and W. J. Visek, eds. Am. Oil Chem. Soc., Champaign, IL.

GRAMS, G. W., BLESSIN, C. W., and INGLETT, G. E. 1970. Distribution of tocopherols within the corn kernel. J. Am. Oil Chem. Soc. 47:337-339.

GROGAN, C. O., and BLESSIN, C. W. 1968. Characterization of major carotenoids in yellow maize lines of differing pigment concentration. Crop Sci. 8:730-732.

GROGAN, C. O., BLESSIN, C. W., DIMLER, R. J., and CAMPBELL, C. M. 1963. Parental influence on xanthophylls and carotenes in corn. Crop Sci. 3:213-214.

HAN, Y., and PARSONS, C. M. 1984. High-oil corn for poultry. Poultry Sci. 63(Suppl. 1):109-110.

HEINZ, E. 1967. Acylgalactosyl diglyceride from leaf homogenates. Biochim. Biophys. Acta 144:321-332.

HILDEBRAND, D. H., TERAO, J., and KITO, M. 1984. Phospholipids plus tocopherols increase soybean oil stability. J. Am. Oil Chem. Soc. 61:552-555.

HUDSON, B. J. F., and GHAVAMI, M. 1984. Stabilising factors in soyabean oil—Natural components with antioxidant activity. Lebensm. Wiss. Technol. 17:82-85.

HYMOWITZ, T., DUDLEY, J. W., COLLINS, F. I., and BROWN, C. M. 1974. Estimations of protein and oil concentration in corn, soybean, and oat seed by near-infrared light reflectance. Crop Sci. 14:713-715.

ITOH, T., TAMURA, T., and MATSUMOTO, T. 1973. Sterol composition of 19 vegetable oils. J. Am. Oil Chem. Soc. 50:122-125.

ITOH, T., SHIMIZU, N., TAMURA, T., and MATSUMOTO, T. 1981. 24-Methyl-E-23-dehydrolophenol, a new sterol and two other 24-methyl-E-Δ^{23}-sterols in Zea mays germ oil. Phytochemistry 20:1353-1356.

JAHN-DEESBACH, W., MARQUARD, R., and HEIL, M. 1975a. Investigations of fat-quality of maize with special consideration of linoleic acid contents. Z. Lebensm. Unters. Forsch. 159:271-278. (In German)

JAHN-DEESBACH, W., MARQUARD, R., and HEIL, M. 1975b. Investigations on fat-quality of embryo, endosperm and crush fractions of maize kernels. Z. Lebensm. Unters. Forsch. 159:279-283. (In German)

JELLUM, M. D. 1967. Fatty acid composition of corn (Zea mays L.) as influenced by kernel position on ear. Crop Sci. 7:593-595.

JELLUM, M. D. 1970. Plant introductions of maize as a source of oil with unusual fatty acid composition. J. Agric. Food Chem. 18:365-370.

JELLUM, M. D. 1984. Registration of high stearic acid maize GE180 germplasm. Crop Sci. 24:829-830.

JELLUM, M. D., and MARION, J. E. 1966. Factors affecting oil content and oil composition of corn (Zea mays L.) grain. Crop Sci. 6:41-42.

JELLUM, M. D., and WIDSTROM, N. W. 1983. Inheritance of stearic acid in germ oil of the maize kernel. J. Hered. 74:383-384.

JELLUM, M. D., BOSWELL, F. C., and YOUNG, C. T. 1973. Nitrogen and boron effects on protein and oil of corn grain. Agron. J. 65:330-331.

JOVANIC, B., ZONIC, I., and SATARIC, I. 1975. Study of the level and quality of protein and oil in the endosperm and embryo of some single cross maize kernels. Pages 454-508 in: Eucarpia. Proc. Eur. Assoc. Res. Plant Breeding, 5th. Sept. 2-5, 1969. Akad. Kiado, Budapest.

JURGENS, S. K., JOHNSON, R. R., and BOYER, J. S. 1978. Dry matter production and translocation in maize subjected to drought during grain fill. Agron. J. 70:678-682.

KARAIWANOW, G., MARQUARD, R., and PETROVSKIJ, E. W. 1982. Fat and carotene contents in corn as well as fatty acid patterns and tocopherol contents in oil of Russian maize lines and maize hybrids with very different color of endosperm. Fette Seifen Anstrichm. 84:251-256. (In German)

KATES, M. 1972. Techniques of Lipidology: Isolation, Analysis and Identification of Lipids. Am. Elsevier Publ. Co., Inc., New York. 610 pp.

KORNFELDT, A., and CROON, L.-B. 1981. 4-Demethyl-, 4-monomethyl- and 4,4-dimethylsterols in some vegetable oils. Lipids 16:306-314.

KUMMEROW, F. A. 1983. Nutritional effects of isomeric fats: Their possible influence on cell metabolism or cell structure. Pages 391-402 in: Dietary Fats and Health. E. G. Perkins and W. J. Visek, eds. Am. Oil Chem. Soc., Champaign, IL.

LAMBERT, R. J., ALEXANDER, D. E., and RODGERS, R. C. 1967. Effect of kernel position on oil content in corn (Zea mays L.). Crop Sci. 7:143-144.

LAU, F. Y., HAMMOND, E. G., and ROSS, P. F. 1982. Effect of randomization on the oxidation of corn oil. J. Am. Oil Chem. Soc. 59:407-411.

LEIBOVITZ, Z., and RUCKENSTEIN, C. 1983. Our experiences in processing maize (corn) germ oil. J. Am. Oil Chem. Soc. 60:395-399.

LICO, A. 1982. Biochemical indices in some native populations and inbred lines. Buletini I. Shkencave Bujqesore 21:31-35. (In Albanian) (Commonwealth Agric. Bur. 84:900. 1984.)

LIN, Y.-H., and HUANG, A. H. C. 1984. Purification and initial characterization of lipase from the scutella of corn seedlings. Plant Physiol. 76:719-722.

LIN, Y.-H., WIMER, L. T., and HUANG, A. H. C. 1983. Lipase in the lipid bodies of corn scutella during seedling growth. Plant Physiol. 73:460-463.

LOOMIS, W. D., and CROTEAU, R. 1980. Biochemistry of terpenoids. Pages 363-418 in: The Biochemistry of Plants. A Comprehensive Treatise. Vol. 4, Lipids: Structure and Function. P. K. Stumpf and E. E. Conn, eds. Academic Press, New York.

MacLELLAN, M. 1983. Palm oil. J. Am. Oil Chem. Soc. 60:368-373.

McMURRAY, C. H., BLANCHFLOWER, W. J., and RICE, D. A. 1980. Influence of extraction techniques on determination of α-tocopherol in animal feedstuffs. J. Assoc. Off. Anal. Chem. 63:1258-1261.

MIHAJLOVIĆ, M. 1978. The content of oil in the grains and their parts in some hybrid maize varieties. Hrana Ishrana 19:25-28. (In Croatian)

MILLER, R. I., DUDLEY, J. W., and ALEXANDER, D. E. 1981. High intensity selection for percent oil in corn. Crop Sci. 21:433-437.

MIRIĆ, M., LALIĆ, Z., and MIHAJLOVIĆ, M. 1981. The composition of certain fractions of lipids in some maize hybrids. Hrana Ishrana 22:125-128. (In Croatian)

MORRISON, W. R. 1978. Cereal lipids. Pages 221-348 in: Advances in Cereal Science and Technology, Vol. 2. Y. Pomeranz, ed. Am. Assoc. Cereal Chem., Inc., St. Paul, MN.

MORRISON, W. R. 1983. Acyl lipids in cereals.

Pages 11-32 in: Lipids in Cereal Technology. P. J. Barnes, ed. Academic Press, London.

MORRISON, W. R., and LAIGNELET, B. 1983. An improved colorimetric procedure for determining apparent and total amylose in cereal and other starches. J. Cereal Sci. 1:9-20.

MORRISON, W. R., and MILLIGAN, T. P. 1982. Lipids in maize starches. Pages 1-18 in: Maize: Recent Progress in Chemistry and Technology. G. E. Inglett, ed. Academic Press, New York.

MORRISON, W. R., MILLIGAN, T. P., and AZUDIN, M. N. 1984. A relationship between the amylose and lipid contents of starches from diploid cereals. J. Cereal Sci. 2:257-271.

MOUNTS, T. L., and ANDERSON, R. A. 1983. Corn oil production, processing and use. Pages 373-387 in: Lipids in Cereal Technology. P. J. Barnes, ed. Academic Press, London.

MUDD, J. B. 1980. Phospholipid biosynthesis. Pages 249-282 in: The Biochemistry of Plants. A Comprehensive Treatise. Vol. 4, Lipids: Structure and Function. P. K. Stumpf and E. E. Conn, eds. Academic Press, New York.

NEAMTU, G., CABULEA, I., BOTEZ, C., ILLYES, G., and IRIMIE, F. 1984. Biochemical studies of carotenoids in maize. II. Content of carotenoid pigments in normal diploid maize, in opaque-2 and in dihaploid maize. Stud. Cercet. Biochim. 27:63-69. (In Rumanian) (Chem. Abstr. 101:107540. 1984.)

NORDSTROM, J. W., BEHRENDS, B. R., MEADE, R. J., and THOMPSON, E. H. 1972. Effects of feeding high oil corns to growing-finishing swine. J. Anim. Sci. 35:357-361.

OHLROGGE, J. B. 1983. Distribution in human tissues of fatty acid isomers from hydrogenated oils. Pages 359-374 in: Dietary Fats and Health. E. G. Perkins and W. J. Visek, eds. Am. Oil Chem. Soc., Champaign, IL.

PAN, W. P., and HAMMOND, E. G. 1983. Stereospecific analysis of triglycerides of *Glycine max, Glycine soya, Avena sativa* and *Avena sterilis* strains. Lipids 18:882-888.

PARRISH, D. B. 1980. Determination of vitamin E in foods—A review. Crit. Rev. Food Sci. Nutr. 13:161-187.

PENNER, D., and MEGGITT, W. F. 1974. Herbicide effects on corn lipids. Crop Sci. 14:262-264.

PLATTNER, R. D., SPENCER, G. F., and KLEIMAN, R. 1977. Triglyceride separation by reverse phase high performance liquid chromatography. J. Am. Oil Chem. Soc. 54:511-515.

PLEWA, M. J., and WEBER, D. F. 1975. Monosomic analysis of fatty acid composition in embryo lipids of *Zea mays* L. Genetics 81:277-286.

PONELEIT, C. G. 1972. Opaque-2 effects on single-gene inheritance of maize oil fatty acid composition. Crop Sci. 12:839-842.

PONELEIT, C. G., and ALEXANDER, D. E. 1965. Inheritance of linoleic and oleic acids in maize. Science 147:1585-1586.

PONELEIT, C. G., and BAUMAN, L. F. 1970. Diallel analyses of fatty acids in corn (*Zea mays* L.) oil. Crop Sci. 10:338-341.

PRIESTLEY, D. A., GALINAT, W. C., and LEOPOLD, A. C. 1981. Preservation of polyunsaturated fatty acid in ancient Anasazi maize seed. Nature 292:146-148.

PRYDE, E. H. 1980. Composition of soybean oil. Pages 13-31 in: Handbook of Soy Oil Processing and Utilization. D. R. Erickson, E. H. Pryde, O. L. Brekke, T. L. Mounts, and R. A. Falb, eds. Am. Oil Chem. Soc., Champaign, IL.

QUACKENBUSH, F. W. 1963. Corn carotenoids: Effects of temperature and moisture on losses during storage. Cereal Chem. 40:266-269.

QUACKENBUSH, F. W., FIRCH, J. G., BRUNSON, A. M., and HOUSE, L. R. 1963. Carotenoid, oil, and tocopherol content of corn inbreds. Cereal Chem. 40:250-259.

RAGHAVAN, S. S., and GANGULY, J. 1969. Studies on the positional integrity of glyceride fatty acids during digestion and absorption in rats. Biochem. J. 113:81-87.

RAGHUVEER, K. G., and HAMMOND, E. G. 1967. The influence of glyceride structure on the rate of autoxidation. J. Am. Oil Chem. Soc. 44:239-243.

RAISON, J. K. 1980. Membrane lipids: Structure and function. Pages 57-83 in: The Biochemistry of Plants. A Comprehensive Treatise. Vol 4, Lipids: Structure and Function. P. K. Stumpf and E. E. Conn, eds. Academic Press, New York.

RAMAN, R., SARKAR, K. R., and SINGH, D. 1983. Correlations and regressions among oil content, grain yield and yield components in maize. Indian J. Agric. Sci. 53:285-288.

RATKOVIĆ, S., NIKOLIĆ, D., and FIDLER, D. 1978. Size distribution of oil bodies in maize kernel and some oil bearing seeds. I. A combined electron microscopy and nuclear magnetic relaxation study. Maydica 23:137-144.

RAVI, K., RIP, J. W., and CARROLL, K. K. 1984. Differences in polyisoprenoid alcohols of mono- and dicotyledonous seeds. Lipids

19:401-404.

ROUNDY, T. E. 1976. Effects of the sugary-2 endosperm mutant on kernel characteristics, oil, protein and lysine in Zea mays L. Ph.D. thesis, Purdue Univ., West Lafayette, IN. (Diss. Abstr. 37:1063B-1064B).

SAHASRABUDHE, M. R., and FARN, I. G. 1964. Effect of heat on triglycerides of corn oil. J. Am. Oil Chem. Soc. 41:264-267.

SCOTT, M. L. 1978. Vitamin E. Pages 133-210 in: The Fat-Soluble Vitamins. H. F. DeLuca, ed. Plenum Press, New York.

SHADLEY, J. D., and WEBER, D. F. 1980. Identification of a factor in maize that increases embryo fatty acid unsaturation by trisomic and B-A translocational analyses. Can. J. Genet. Cytol. 22:11-19.

SHARMA, B. N., GOPAL, S., PAUL, Y., and BHATIA, I. S. 1975. Fatty-acid composition of lipid classes of different varieties of Indian maize (Zea mays L.). J. Res. Punjab Agric. Univ. 12:378-381.

SLOVER, H. T. 1971. Tocopherols in foods and fats. Lipids 6:291-296.

SLOVER, H. T., LEHMANN, J., and VALIS, R. J. 1969. Vitamin E in foods: Determination of tocols and tocotrienols. J. Am. Oil Chem. Soc. 46:417-420.

SPURGEON, S. L., and PORTER, J. W. 1980. Carotenoids. Pages 419-483 in: The Biochemistry of Plants. A Comprehensive Treatise. Vol. 4, Lipids: Structure and Function. P. K. Stumpf and E. E. Conn, eds. Academic Press, New York.

STROCCHI, A. 1981. Fatty acid composition and triglyceride structure of corn oil, hydrogenated corn oil, and corn oil margarine. J. Food Sci. 47:36-39.

STUMPF, P. K. 1980. Biosynthesis of saturated and unsaturated fatty acids. Pages 177-204 in: The Biochemistry of Plants. A Comprehensive Treatise. Vol. 4, Lipids: Structure and Function. P. K. Stumpf and E. E. Conn, eds. Academic Press, New York.

SUN, D., GREGORY, P., and GROGAN, C. O. 1978. Inheritance of saturated fatty acids in maize. J. Hered. 69:341-342.

TAN, S. L., and MORRISON, W. R. 1979a. The distribution of lipids in the germ, endosperm, pericarp and tip cap of amylomaize, LG-11 hybrid maize and waxy maize. J. Am. Oil Chem. Soc. 56:531-535.

TAN, S. L., and MORRISON, W. R. 1979b. Lipids in the germ, endosperm and pericarp of the developing maize kernel. J. Am. Oil Chem. Soc. 56:759-764.

TANAKA, H., OHNISHI, M., and FUJINO, Y. 1984. On glycolipids in corn seeds. J. Agric. Chem. Soc. Jpn. 58:17-24. (In Japanese)

THOMPSON, D. L., JELLUM, M. D., and YOUNG, C. T. 1973. Effect of controlled temperature environments on oil content and on fatty acid composition of corn oil. J. Am. Oil Chem. Soc. 50:540-542.

TRELEASE, R. N. 1969. Ultrastructural characterization, composition, and utilization of lipid bodies in the maize shoot apex during postgermination development. (Abstr.) J. Cell Biol. 43:147a.

TRIFUNOVIĆ, V., RATKOVIĆ, S., MIŠOVIĆ, M., KAPOR, S., and DUMANOVIĆ, J. 1975. Variability in content and fatty acid composition of maize oil. Maydica 20:175-183.

TSAFTARIS, A. S., and SCANDALIOS, J. G. 1983a. Comparison of the glyoxysomes and the glyoxysomal enzymes in maize lines with high or low oil content. Plant Physiol. 71:447-450.

TSAFTARIS, A. S., and SCANDALIOS, J. G. 1983b. Genetic analysis of isocitrate lyase enzyme activity levels in maize lines selected for high or low oil content. J. Hered. 74:70-74.

USDA. 1985. Agricultural Statistics. U.S. Government Printing Office. Washington, DC.

USDA. 1986. Feed Outlook and Situation Report. FdS-299. U.S. Dept. Agric., Econ. Res. Serv., Washington, DC.

VALOIS, A. C. C., TOSELLO, G. A., ZONOTTO, M. D., and SCHMIDT, G. S. 1983. Analysis of grain quality in corn. Pesq. Agropec. Bras., Ser. Agron. 18:771-778. (In Portuguese)

Van NIEKERK, P. J. 1973. The direct determination of free tocopherols in plant oils by liquid-solid chromatography. Anal. Biochem. 52:533-537.

VICK, B. A., and ZIMMERMAN, D. C. 1979. Distribution of a fatty acid cyclase enzyme system in plants. Plant Physiol. 64:203-205.

VICK, B. A., and ZIMMERMAN, D. C. 1982. Levels of oxygenated fatty acids in young corn and sunflower plants. Plant Physiol. 69:1103-1108.

VICK, B. A., and ZIMMERMAN, D. C. 1984. Biosynthesis of jasmonic acid by several plant species. Plant Physiol. 75:458-461.

WADA, S., and KOIZUMI, C. 1983. Influence of the position of unsaturated fatty acid esterified glycerol on the oxidation rate of triglyceride. J. Am. Oil Chem. Soc. 60:1105-1109.

WANG, S.-M., LIN, Y.-H, and HUANG, A. H. C. 1984. Lipase activities in scutella of maize lines having diverse kernel lipid content. Plant Physiol. 76:837-839.

WATSON, S. A. 1962. The yellow carotenoid pigments of corn. Pages 92-100 in: Proc. Corn

Res. Conf., 17th. W. Hechendorn and J. I. Sutherland, eds. Am. Seed Trade Assoc., Washington, DC.

WATSON, S. A. 1984. Corn and sorghum starches: Production. Pages 417-468 in: Starch: Chemistry and Technology, 2nd ed. R. L. Whistler, J. N. BeMiller, and E.F. Paschall, eds. Academic Press, Orlando, FL.

WATSON, S. A., and FREEMAN, J. E. 1975. Breeding corn for increased oil content. Pages 251-275 in: Proc. Corn Sorghum Res. Conf., 30th. Am. Seed Trade Assoc., Washington, DC.

WEBER, E. J. 1969. Lipids of maturing grain of corn (*Zea mays* L.): I. Changes in lipid classes and fatty acid composition. J. Am. Oil Chem. Soc. 46:485-488.

WEBER, E. J. 1970. Lipids of maturing grain of corn (*Zea mays* L.): II. Changes in polar lipids. J. Am. Oil Chem. Soc. 47:340-343.

WEBER, E. J. 1973. Structure and composition of cereal components as related to their potential industrial utilization. IV. Lipids. Pages 161-206 in: Industrial Uses of Cereals. Y. Pomeranz, ed. Am. Assoc. Cereal Chem., Inc., St. Paul, MN.

WEBER, E. J. 1978. Corn lipids. Cereal Chem. 55:572-584.

WEBER, E. J. 1979. The lipids of corn germ and endosperm. J. Am. Oil Chem. Soc. 56:637-641.

WEBER, E. J. 1981. Compositions of commercial corn and soybean lecithins. J. Am. Oil Chem. Soc. 58:898-901.

WEBER, E. J. 1983a. Lipids in maize technology. Pages 353-372 in: Lipids in Cereal Technology. P. J. Barnes, ed. Academic Press, London.

WEBER, E. J. 1983b. Variation in corn (*Zea mays* L.) for fatty acid compositions of triglycerides and phospholipids. Biochem. Genet. 21:1-13.

WEBER, E. J. 1984a. High-performance liquid chromatography (HPLC) of the carotenoids in corn grain. (Abstr.) J. Am. Oil Chem. Soc. 61:672.

WEBER, E. J. 1984b. High performance liquid chromatography of the tocols in corn grain. J. Am. Oil Chem. Soc. 61:1231-1234.

WEBER, E. J., and ALEXANDER, D. E. 1975. Breeding for lipid composition in corn. J. Am. Oil Chem. Soc. 52:370-373.

WEBER, E. J., de la ROCHE, I. A., and ALEXANDER, D. E. 1971. Stereospecific analysis of maize triglycerides. Lipids 6:525-530.

WELCH, L. F. 1969. Effect of N, P, and K on the percent and yield of oil in corn. Agron. J. 61:890-891.

WIDSTROM, N. W., and JELLUM, M. D. 1975. Inheritance of kernel fatty acid composition among six maize inbreds. Crop Sci. 15:44-46.

WIDSTROM, N. W., and JELLUM, M. D. 1984. Chromosomal location of genes controlling oleic and linoleic acid composition in the germ oil of two maize inbreds. Crop Sci. 24:1113-1115.

WILKINSON, R. E., and HARDCASTLE, W. S. 1973. Commercial herbicide influence on corn oil composition. Weed Sci. 21:433-436.

WILKINSON, R. E., and HARDCASTLE, W. S. 1974. Influence of herbicide mixtures on corn oil quantity and composition. Can. J. Plant Sci. 54:471-473.

WORTHINGTON, R. E., and HITCHCOCK, H. L. 1984. A method for the separation of seed oil steryl esters and free sterols: Application to peanut and corn oils. J. Am. Oil Chem. Soc. 61:1085-1088.

WU, G.-S., STEIN, R. A., and MEAD, J. F. 1979. Autoxidation of fatty acid monolayers adsorbed on silica gel. IV. Effects of antioxidants. Lipids 14:644-650.

YATSU, L. Y., and JACKS, T. J. 1972. Spherosome membranes. Half-unit membranes. Plant Physiol. 49:937-943.

YOSHIDA, H., and KAJIMOTO, G. 1984. Changes in the molecular species of corn triacylglycerols heated in the laboratory. Nutr. Rep. Int. 29:1115-1126.

ZIMMERMAN, D. C., and COUDRON, C. A. 1979. Identification of traumatin, a wound hormone, as 12-oxo-*trans*-10-dodecenoic acid. Plant Physiol. 63:536-541.

ZONTA, B., and STANCHER, B. 1983. High-performance liquid chromatography of tocopherols in oils and fats. Riv. Ital. Sostanze Grasse 60:195-199.

CHAPTER 11

CORN DRY MILLING: PROCESSES, PRODUCTS, AND APPLICATIONS

RICHARD J. ALEXANDER
Penick & Ford, Ltd.
Cedar Rapids, Iowa

I. INTRODUCTION

A. History of Corn Dry Milling

The history of corn milling follows very closely the history of corn development, which most experts agree originated in North America; corn was probably not introduced into Europe until after Columbus discovered the New World. This history goes back several thousand years, as recorded by several investigators. Mangelsdorf and co-workers (1964) reported the discovery of 7,000-year-old plant remains, which they identified as a wild progenitor of modern corn. Over 20,000 specimens of corn, about half intact cobs, have been found in several caves in both the Tehuacan valley of Mexico and the southwestern United States. These corn samples have been dated between 4,000 and 6,500 years old (Mangelsdorf et al, 1967; Mangelsdorf, 1974).

In addition to fossil corn, artifacts used with corn have been found. Belt (1928) observed that the ancient Indians of Nicaragua buried the stones they had used for grinding corn along with their dead. Undoubtedly, these stones were considered indispensable for the person's future life. They were essentially the first rudimentary means of milling corn.

For a look at the key developments of early corn dry milling, one can examine the implements used in pioneer America. Initially, the early settlers adopted the use of the Indian metate, which was only a slight improvement over the early grinding stones. With this device, the corn was ground between a hand-held stone and a concave bedstone.

One step up from the metate was the hominy block. Early directions for making this device were quite simple, as recorded by Hardeman (1981):

> Near the cabin cut off a hardwood tree three or four feet above the ground and hollow out the top. From a springy limb of another tree extending over the stump, tie a pestle or block of wood by a strong line.

The hominy block was operated by repeatedly plunging the wood pestle into the hollow stump until the corn had been sufficiently crushed into meal.

The hominy block was eventually replaced by a single-family, stone device called a quern (pronounced kwern). This was a small, burred-stone grinding apparatus, apparently invented in ancient Rome. It was operated by pouring corn through the cone-shaped axle hole at the top. An off-set handle was used to rotate the capstone on the stationary "netherstone," and the cornmeal worked out between the stones and fell into a tub surrounding the quern.

The principle of this type of revolving stone mill was applied on a much larger scale as early as 1620 (Hardeman, 1981). It expanded to become the local grist mill, which was eventually used to process both corn and wheat. Energy to operate the mill was supplied by livestock, occasionally by humans and by water. By the mid-1800s, most of the mills in the United States were operated by water, although steam-driven mills were used in some sections of the country.

A number of grist mills are still used to grind corn today, particularly in the southern states (Larsen, 1959). As in pioneer America, the mills are relatively small, and finished product distribution is limited to a small geographic area. These milling systems have gradually given way to the more sophisticated tempering-degerming systems, which were introduced in the early 1900s.

B. Present Milling Capacity in the United States

According to Brekke (1970a), as of 1965 there were 152 dry corn mills in the United States with a daily capacity of 50 cwt (2.27 t) or more. By 1969 this number had dropped to 115. More recent statistics (Anonymous, 1984b)

TABLE I
Principal Corn Dry-Milling Companies: Plant Locations and Estimated Mill Capacities[a]

Company	Mill Location	Estimated Daily Capacity (1,000 bu)[b]
Archer Daniels Midland	Lincoln, NE (Gooch Mills)	15
Evans Milling Co.	Indianapolis, IN	30
Illinois Cereal Mills Inc.	Paris, IL	65
Krause Milling Co.	Milwaukee, WI	55
Lauhoff Grain Co.	Danville, IL	70[c]
	Crete, NE (Crete Mills)	50
Lincoln Grain, Inc.	Atchison, KS	45[d]
Martha White Foods[e]	Nashville, TN	25
Midstate Mills, Inc.	Newton, NC	15
The Quaker Oats Co.	Cedar Rapids, IA	20
	Chattanooga, TN	15
	St. Joseph, MO	10
J. R. Short Milling Co.	Kankakee, IL	30
Total		445

[a] With the exception of two companies, the figures are based on unpublished data.
[b] Divide figures in this column by 2.205 to convert to metric tons.
[c] Anonymous (1978).
[d] Anonymous (1984a).
[e] Headquarters in Nashville, TN, with five separate mills: three in Tennessee, one in Georgia, and one in West Virginia.

suggests that the decline in dry mills has tapered off, with 88 mills still operating in 1984. Of this total, 66 are smaller mills located in the southern states and California (two), with most of the remaining mills (20) in the Midwest.

Indications are that only 17 of the dry-milling plants account for most of the corn processed by the industry. A list of the principal milling companies is shown in Table I. Of the 17 mills, at least 12 are fairly large operations having daily capacities of between 10,000 and 70,000 bu (250–1,750 t) and utilizing tempering-degerming processes. Ten of the mills are located in the Midwest, and five are relatively small plants in the South.

The total of 445,000 bu (11.1×10^3 t) of corn per day (from Table I) corresponds to an annual volume of about 111 million bushels (2.77×10^6 t). This represents 92% of the corn processed by the whole industry for the 1977/1978 crop year (Anonymous, 1982) and indicates that most of the corn processed by dry millers in this country today is processed by the larger-capacity, tempering-degerming systems.

II. TEMPERING-DEGERMING SYSTEMS

A. Process with the Beall Degerminator

The Beall degerminator, introduced in 1906 (Larsen, 1959), set the stage for the development and production of refined dry-milled corn products. Although other types of degerminating equipment have been introduced since 1906, the Beall machine has remained the mainstay for most U.S. companies employing tempering-degerming systems. It has allowed the dry-milling industry to move from numerous small, locally operated grist mills with limited capacity and distribution to larger, more efficient plants processing 20,000–70,000 bu (500–1,750 t) per day with nationwide distribution. It has also allowed for the production of higher-quality, low-oil, essentially germ- and bran-free endosperm-based products with greatly extended shelf life and product stability.

Probably the major reference work cited by most of the recent authors in the field is that of Stiver (1955). This work describes the processes and equipment used by the corn dry millers in very detailed fashion, as do the more recent reviews by Brekke (1970a) and Anderson and Watson (1982). Their excellent discussions, flow charts, etc., are not reproduced here. However, the key aspects of the process are briefly reviewed to provide the reader sufficient background to the subsequent sections of this chapter.

As shown in the flow chart in Fig. 1, shelled, whole U.S. No. 2 yellow corn is received and placed in storage in standard corn silos. Today most of the corn is delivered by truck or railcar directly to the corn mills from country elevators or individual farms. Most corn mills have one to two weeks of storage capacity, which could amount to well over one million bushels (25×10^3 t), depending on the size of the mill.

The corn is first dry cleaned, which includes passage under a magnet (positioned over a belt conveyor) to remove tramp metal, aspiration to remove fines and pieces of cob, and screening to separate whole corn from broken corn. The desire is to have only whole kernels entering the corn mill. After wet cleaning to remove surface dirt, dust, rodent excreta, etc., the corn is adjusted to about

354 / Corn: Chemistry and Technology

20% moisture and placed in a tempering bin. Optimum tempering moisture and times have been reported by Brekke and co-workers (Brekke et al, 1961, 1963; Brekke and Weinecke, 1964; Brekke, 1966, 1967, 1968, 1970a, 1970b; Brekke and Kwolek, 1969).

The product is then processed in the Beall degerminator, in which the whole moist corn is essentially treated by an abrading action to strip the bran or pericarp and germ away from the endosperm while leaving the endosperm intact. The most efficient way to operate the degerminator has been reported by Brekke et al (1961, 1963), Brekke and Weinecke (1964), and Weinecke et al (1963). This is obviously an ideal picture of what the corn miller would like to have happen. In fact, some of the bran and germ remain attached to the endosperm and must be removed in subsequent aspiration and milling processes. Also, some of the endosperm remains associated with the bran and germ fractions, all of which are separated by subsequent use of aspirators and gravity tables.

Let us assume that most of the endosperm is separated from bran and germ. The Beall is set up so that the large pieces of endosperm, known as "tail hominy," proceed through the end of the degerminator. This fraction is dried, cooled, and sifted, and part of it is isolated as large flaking grits. (The conditions for obtaining maximum yield of flaking grits have been reported by Brekke [1966, 1967] and Brekke and Kwolek [1969]). The remainder is sent to the roller mills for reduction into smaller fractions, such as coarse, medium, or fine grits; meals; or flours. The bran and germ fractions (together) pass through a screen on the underside of the degerminator and become the "thru stock" stream. This stream

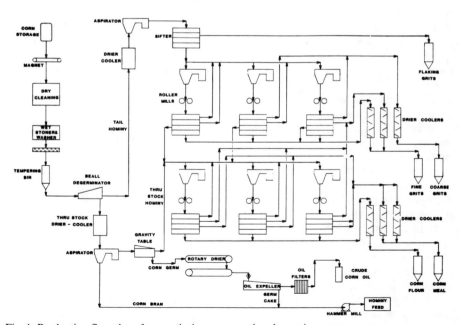

Fig. 1. Production flow chart for a typical corn tempering-degerming system.

is dried, cooled, aspirated to remove the bran, and processed on gravity tables to separate germ and endosperm.

The germ can then be expelled or hexane-extracted to remove the oil, and the spent germ or germ cake becomes one of the by-product streams. (Some of the corn dry millers do not further process the germ but sell it to other companies that do). The fines separated from the thru stock endosperm are usually high in oil, fine fiber, and tip caps; they become one of the by-product streams known as "standard meal." The bran, germ cake, standard meal, and broken corn (isolated from whole corn before entering the corn mill) are combined, dried, and ground up together to become the main by-product of the corn dry millers, which is known as "hominy feed." Since none of the dry millers refine corn oil, the crude oil obtained from either expelling or extraction is sold to one of several oil refiners in the United States. The main portion of the endosperm isolated from the thru stock is processed in the same way as the tail hominy fraction to produce prime grits, meals, and flours.

As indicated by Stiver (1955) and others, after the degerminator, the roller mills and sifters are the core of the corn milling system. The larger pieces of corn endosperm are sent to a series of roller mills, where they pass between sets of corrugated rolls; the corrugations vary in size, depending on the level of particle size reduction required. The ground endosperm stream proceeds to the sifters, which can separate the mixture into as many as 16 different fractions.

The larger-sized particles can be sent to another set of rolls for further reduction, or combined with other streams of similar particle size, aspirated, dried, cooled, and sent to a finished product bin. Similarly, the medium-sized and small-sized particles may be further milled and sifted and streams of similar size combined to give finished corn grits, meals, and flours of uniform composition.

In several of the major corn dry-milling plants, corn grits and/or flours are further processed in acid modification systems, in extrusion-cookers, or the like, to provide a wider variety of modified corn products for both food and nonfood use. These specialty products are discussed in more detail in several subsequent sections of this chapter.

B. Alternative Milling Systems

At least two alternative dry-milling systems can be employed to produce refined dry-milled corn products (Brekke, 1970a). They have been described as the Miag process, now more correctly the Buhler-Miag process (Wyss, 1974), and the Ocrim process. Both of these systems were developed in Europe and have been used to some extent throughout the world. Selected pieces of equipment from the two systems have also been incorporated into several corn mills in this country. Since the systems have been discussed in detail by Brekke (1970a), no further description is given here.

III. DRY-MILLED PRODUCTS—TYPES, VOLUMES, AND COMPOSITION

The primary products derived from the tempering-degerming process are corn grits, cornmeals, and corn flours. An infinite number of products are possible as the result of particle size reduction on the roller mills, from flaking grits as coarse as four- to six-mesh down to fine-grind corn flour with 95% passing through a 100-mesh sieve. However, most of the products can be classified into one of the six main groups shown in Table II.

At one time, brewer's grits, in particular, included numerous products covering the particle size range from 10–12 mesh grits to cornmeal. In the brewing process, the size of the particulate material remaining after cooking and enzymatic hydrolysis of the corn adjunct has a definite impact on the manner in which the hydrolyzed starch liquor, known as "wort," filters. The nature of this residual, high-protein material appears to be unique to the particular type of brewing equipment and process employed. Otherwise, essentially no differences in composition are found between coarse, regular, and fine grits.

This is also true of all products coming from the main portion of the corn endosperm, including flaking grits, coarse or fine brewer's grits, or corn cones— a very fine, uniform cornmeal. The composition of these materials is shown in Table III. Typically they contain 7–8% protein, less than 1% fat, ash, or fiber, and 77–79% starch (88–90%, dry basis).

The only material that varies somewhat in composition is corn flour, particularly that produced during normal roller milling of large grits to smaller

TABLE II
Typical Products of the Corn Tempering-Degerming System:
Granulations and Product Volumes[a]

Product	Particle Size						Annual Volume[b] (million lb)
	From			To			
	Standard U.S. Mesh	Size in.	Size µm	Standard U.S. Mesh	Size in.	Size µm	
Flaking grits	−3.5	−0.223	−5,600	+6	+0.132	+3,350	750
Coarse grits	−10	−0.0787	−2,000	+15	+0.0512	+1,290	940
Regular grits	−15	−0.0512	−1,290	+30	+0.0234	+600	1,360
Cornmeal	−30	−0.0234	−600	+60	+0.0098	+250	190
Corn cones	−40	−0.0165	−425	+80	+0.0070	+180	190
Corn flour	−60	−0.0098	−250	+325	+0.0017	+45	330

[a] Data taken in part from Alexander (1973) and Brekke (1970a).
[b] Divide figures in this column by 2,205 to convert to million metric tons.

TABLE III
Typical Composition (%, as-is basis) of Dry-Milled Corn Products[a]

Component	Flaking Grits	Coarse or Fine Grits	Corn Cones	Corn Flour
Moisture	11.7	11.5	12.0	13.0
Protein	7.0	7.5	7.9	5.2
Fat	0.6	0.7	0.6	2.0
Crude fiber	0.2	0.2	0.3	0.5
Ash	0.2	0.3	0.3	0.4
Starch	78.3	78.0	77.4	76.4
Other polysaccharides	2.0	1.8	1.5	2.5

[a] Based on unpublished data.

grits. This fine fraction (−60 mesh) is usually referred to as "break flour." It contains less protein and somewhat more fat than the other prime products. Break flour is derived from the portion of the corn kernel known as the floury endosperm. Corn flour derived from the roller milling of grits, and coming mainly from the horny endosperm, is known as "reduction flour." It has the same composition as the starting corn grits.

IV. INDUSTRIAL APPLICATIONS

A. Current Market Volumes

The principal applications of the products of the corn dry-milling industry are shown in Table IV. These figures were estimated for the calendar year 1977. Most of the figures in Table IV have been corroborated with other published data for that year, as is discussed below.

The total product volume of 6.19 billion pounds (2.81×10^6 t) should be corrected to 6.05 billion pounds (2.74×10^6 t) to account for the noncorn ingredients in fortified corn products. This corrected figure compares fairly well to the 5.87 billion pounds (2.66×10^6 t) calculated from data for the daily plant capacities of the principal dry millers in Table I, using the following equation:

TABLE IV
Estimated 1977 Product Volumes of the Corn Dry-Milling Industry[a]

Application Areas	Quantity (million lb)[b]	
Brewing, total		1,850
Food, general		
Breakfast cereals	800	
Mixes (pancake, cookie, muffin, etc.)	100	
Baking	50	
Snack foods	100	
Other foods (breadings, batters, baby foods, etc.)	75	
Total		1,125
Fortified foods (PL480), total		485[c]
Nonfood		
Gypsum board	100	
Building products (particleboard, fiberboard, plywood, etc.)	40	
Pharmaceuticals/fermentation	200	
Foundry binders	90	
Charcoal binders	75	
Other (paper, corrugating, oil well drilling fluids, etc.)	25	
Total		530
Animal feed,[d] total		2,200
Total		6,190

[a] Estimates based on unpublished data.
[b] Divide the figures in these columns by 2205 to convert to million metric tons.
[c] Corn products represent 65–70% of total shipments, or 315–340 million pounds.
[d] Data reported by Alexander (1973).

Product volume = 445 million bu/day × 55 lb/bu × 250 days × 0.96
(conversion factor[1])
= 5.874 billion lb

It also compares quite well to a figure of 6.39 billion pounds (2.90×10^6 t), which was calculated using the value of 121 million bushels of corn processed in the 1977/1978 crop year, as reported in the 1982 *Commodity Year Book* (Anonymous, 1982).

The product with the largest volume was hominy feed, the chief by-product of the corn dry millers, which is sold for animal feed. The value of 2.2 billion pounds (1.0×10^6 t) was taken from published data (Alexander, 1973), which is in good agreement with the 2.285 billion pounds (1.04×10^6 t) reported by Wells (1979). The second biggest application area is in the brewing industry, where 1.85 billion pounds (0.84×10^6 t) of corn grits and meals were used in brewing adjuncts. This agrees well with a figure of 1.77 billion pounds (0.80×10^6 t) compiled from reports in *The Brewers Digest* (Anonymous, 1974–1984).

The volume of 1.61 billion pounds (0.73×10^6 t) used in food applications (1.125 billion in general foods plus 0.485 billion in fortified foods) is reasonably close to the figure of 2.0 billion pounds (0.91×10^6 t) reported by Brockington (1970).[2] The breakdown in the general food area (Table IV) has not been published before, so no basis for comparison exists. The value of 485 million pounds (0.22×10^6 t) of fortified foods is quite similar to the figures shown in Table VI for the years 1979–1982 (Bookwalter, 1983).

The product volume of 530 million pounds (0.24×10^6 t) for nonfood uses is considerably higher than data previously published by Senti (1965) and Alexander (1973). The larger figure was the result of increases in the use of corn flours in gypsum board, foundry binders, and particularly citric acid production (in the fermentation area) during the 1975–1978 period. According to Wells (1979), this volume dropped off to 377 million pounds (0.17×10^6 t) in 1979, which is probably closer to current usage levels. Specific application areas are discussed in the following sections of this chapter.

B. Brewing

THE BREWING PROCESS

Traditionally, the brewing industry has been the largest user of prime products made by the corn dry millers. A brief description of the process will aid in understanding of the use of corn grits in making beer. When grits are used in brewing beer, the first step is cooking the grits. This is frequently accomplished in the "mashing" or "mash tun" (or tub) before treatment with barley malt. Corn grits and water are combined, heated to 90–95°C to gelatinize the starch, and cooled to 67°C. Alternatively, grits may be processed through a jet cooker or a continuous cooker. Malt is added, and starch from both malt and corn grits is hydrolyzed to fermentable sugars by the joint action of α- and β-amylases.

The solubilized carbohydrates are separated from the spent grains by filtering

---

[1] Factor converting corn with 15% moisture corn to finished products with 11% moisture.

[2] Figure obtained by adding the degermed corn meal (0.75 billion pounds) to grits for human consumption (1.28 billion pounds) (Brockington, 1970, in Table 15.1).

or lautering, and the spent grains are washed or sparged with water at 75° C. The resulting wort is combined with hops and heated to boiling, traditionally in a copper kettle. After filtration, the hopped wort is cooled and adjusted to the proper specific gravity. It is then inoculated or "pitched" with a selected strain of yeast, and the sugars are fermented to alcohol. After fermentation, the immature or "green" beer is allowed to age for seven to 10 days.

Finally, the beer is filtered, pasteurized, and packaged. More details of the brewing process may be found in such excellent reviews as that by MacLeod (1977).

INDUSTRY TRENDS

In 1977, nearly two billion pounds (0.91×10^6 t) of corn grits were employed in making beer. Unfortunately for the corn dry millers, this figure represented the peak year for brewing use. Corn grits have always had to compete with rice and corn syrups as adjuncts in the United States, and both of these cereal-derived products have increased quite dramatically during the past 10 years at the expense of corn grits. The graph in Fig. 2 shows the trends in consumption of the various products. The use of corn grits increased from 1973 to 1977. Between 1977 and 1983, it decreased on an average of 3.7% per year, for a loss of nearly 400 million pounds (0.18×10^6 t). On the other hand, both rice and corn syrups have increased, rice by 350 million pounds (0.16×10^6 t), or 61%, and corn syrups by 360 million pounds (0.16×10^6 t), or a substantial 130%, over the 10-year period. There are several reasons for these changes.

The increased use of corn syrups has been the result of several factors. Probably the most important to the brewers is economic. Corn syrups represent

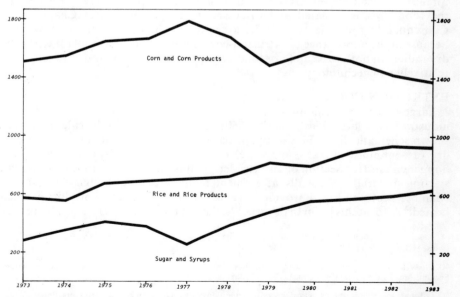

Fig. 2. Consumption of cereal adjuncts by the brewing industry for 1973–1983. Data from *The Brewers Digest* (Anonymous, 1974–1984).

a form of adjunct with a higher concentration of fermentable sugars. Therefore, increases in brewing capacity (Pfisterer et al, 1978) and production rates (Pollock and Weir, 1976; Swain, 1976) have been possible without plant expansions. Swain (1976) also cited lower production costs, whereas others (Rao and Narasimham, 1975; Pollock and Weir, 1976) have indicated that syrups result in fewer production problems plus a more uniform product with equal or higher quality. Moll and Duteurtre (1976) indicated that, when a new plant is built, lower capital costs are possible with the use of corn syrups.

The increased use of rice has not been as well documented. However, certain rice varieties can provide higher-gravity brews (Stubits and Teng, 1983) and thus increased brewing capacity similar to that for syrups. In addition, the popularity of lower-calorie beers in recent years has resulted in the increased use of enzymes, both amylolytic and debranching. One patent (Line et al, 1982) claims to use malted rice as the source of debranching enzyme as well as of starch.

C. General Food Uses

INTRODUCTION

Combining the application areas of brewing, general food uses, and fortified foods from Table IV, one obtains a total of nearly three and one half billion pounds (1.59×10^6 t). This represents 87% of the prime products sold for all uses, indicating the importance of the food-beverage industry to the dry corn milling industry. Next to brewing, the general food category is the second largest market segment, with over one billion pounds.

In recent years, several reviews have covered food applications of dry-milled corn products; these include Brockington (1970), Bailey,[3] and Wells.[4] Processes for making breakfast cereals were well documented by Matz (1959). In addition, the food area is updated and discussed more thoroughly in Chapter 13. Consequently, food uses are not described here to any great extent. However, because of the importance of extrusion-cooking as a processing tool for the corn dry millers in both food and nonfood areas, some discussion is devoted to this relatively new technology.

EXTRUSION-COOKING PROCESS

Large-volume extrusion-cookers first became available for use in the cereal industry in the late 1950s to early 1960s. Two commercial full-scale units are shown schematically in Figs. 3 and 4. Basically, the process involves feeding a dry or semimoist (15–25% moisture) corn or other cereal product into the hopper end or feed section of an extruder. At this point, the extruder screw and starting material are usually at ambient temperatures, although the mixing chamber shown in Fig. 4 provides a vehicle in which the cereal product can be heated.[5] Also, at this point in the process, the flights of the extruder screw are the

---

[3] T. Bailey. Functionality and uses of corn flour. Presented to: Central States Section, Am. Assoc. Cereal Chem., St. Louis, MO, Feb. 16–17, 1973.

[4] G. H. Wells. Cereal flours in fabricated foods. Presented to: Symposium: Fabricated Foods. Central States Section, Am. Assoc. Cereal Chem., St. Louis, MO, Jan. 30–Feb. 1, 1975.

[5] G. L. Johnston. Technical and practical processing conditions with single screw cooking extruders. Presented to: International Seminar on Cooking and Extruding Techniques, ZDS Solinger-Grofroth, West Germany, Nov. 27–29, 1978.

widest apart. As the cereal product is conveyed down the length of the extruder (compression section), the flights get closer and closer together, so that by the time the cooked product exits the end of the extruder, the material is under a pressure of 200–1,000 psi (1,300–6,900 kPa).

As the cereal product is moved through the extruder, the temperature in the extruder barrel increases, primarily as a result of 1) the internal shear forces, but also from 2) external heat that can be applied with some extruders, and 3) steam

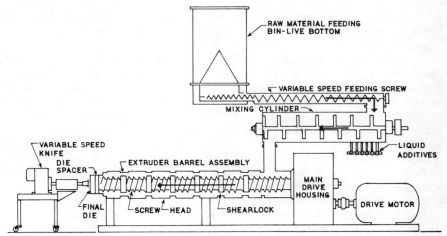

Fig. 3. Cut-away view of a single-screw extrusion-cooker. (Courtesy Anderson International, Strongsville, OH)

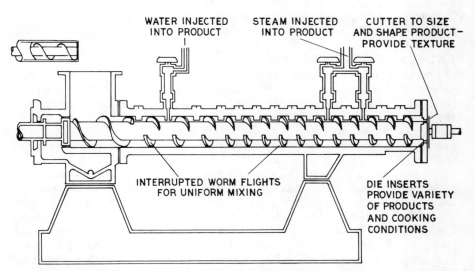

Fig. 4. Cut-away view of a complete extrusion system, including storage bin, screw-conveyor, mixing chamber, and single-screw extrusion-cooker. (Courtesy Wenger Mfg., Sabetha, KS)

injection that is possible in certain models. The heat causes the starch in the cereal product to gelatinize and swell, and the resulting extrudate becomes quite viscous. This creates additional pressure, shear, and heat at the head end of the extruder barrel.

The extrudate exits the extruder through holes in a die plate. These holes can be varied in size, shape, and number. At the end of the extruder is a variable-speed knife, which cuts the extrudate into smaller pieces called "collets," which are conveyed to a cooler/dryer. If the collets are intended for second- or third-generation snack foods, as defined by Hauck (1980) or Matson (1982), the product is packaged and made ready for shipment. If the final end use is as a precooked cornmeal or corn flour, the dried collets are ground in a hammer mill, entoleter mill, or the like, to the desired particle size before bagging and shipment.

A number of variables in screw design, extruder shape, die configuration, numbers of extruder screws, etc., are possible. For instance, in the machine in Fig. 3, water or other liquids or reagents can be added directly into the extruder barrel, and external heat can be added by steam injection into the barrel. In the machine in Fig. 4, liquid additives are added in the mixing cylinder ahead of the extruder, and part of the heat can be provided through the use of jackets suitable for circulation of water, steam, or other liquid. In addition to the Anderson and Wenger units, single-screw industrial extrusion-cookers are available from Bonnot Company and the Sprout Waldron Division of Koppers and double-screw extruders from Werner-Pleiderer Corporation and Wenger Manufacturing, just to mention a few.

Along with equipment design, cereal properties, such as degree of cooking, density, viscosity, or water absorption properties can be controlled and modified by such process variables as extrusion moisture, temperature, screw speed, and, perhaps most importantly, die configuration. This flexibility has helped make extrusion cooking the important processing tool that it is today in the food industry, and, especially for the corn dry millers, it provides a means of converting corn flour from a by-product into premium products (Roberts, 1967).

EXTRUSION APPLICATIONS

The early patent literature (Bradley and Downhour, 1970; Stickley and Griffith, 1966) describes the use of extruders in producing precooked corn and sorghum flours for use as foundry binders. Subsequent publications reported their use in breadings and croutons (Anonymous, 1983), fabricated foods,[6] fortified cereal products (Anderson et al, 1969, 1970; Bookwalter et al, 1971), and as bases for certain specialty confections and protein-fortified beverages (Hauck, 1980).

The snack food industry seems to have benefited the most from the variety of products, shapes, densities, etc., possible with modern extruders. Cornmeals or corn flours have been found to be particularly useful raw materials (Sanderude, 1969; Matson, 1982) in the production of corn chips (Sanderude, 1969; Scales, 1982; Stauffer, 1983), puffs (Williams, 1977; Toft, 1979; Scales, 1982), onion rings (Sunderude, 1969; Williams, 1977), and several second- and third-generation snacks (Toft, 1979; Hauck, 1980; Matson, 1982). Products made by

[6]See footnote 4.

direct extrusion compete with snacks made from masa flour, which involves treatment with lye before milling (Scales, 1982; Stauffer, 1983; Przybyla, 1984). Corn processed by the traditional masa process has been recently described by Bedolla and Rooney (1982) and is discussed in more detail in Chapter 13.

D. Corn-Based Fortified Foods

INTRODUCTION

During the late 1950s and early 1960s, a number of changes occurred within the government food aid programs that had a major effect on the corn dry-milling industry. In this period, the shortage of nonfat dry milk (NFDM) resulted in the search for alternate sources of low-cost protein (Senti et al, 1967). The government assisted industry with the passage of the Food for Peace Act of 1966, which broadened the range of commodities eligible for donation to underdeveloped countries. Even before passage of the new legislation, the corn dry millers had developed several corn-based prototypes (Tollefson, 1967), so that in 1966 they were ready to produce the first of several protein-fortified foods.

The first product that received worldwide distribution and acceptance was corn-soy-milk (CSM). The original formula for this material consisted of 64% partially cooked cornmeal (PCM), 24% defatted, toasted soy flour, 5% NFDM, 5% soy oil, and 2% vitamins and minerals. The product was well balanced from a carbohydrate-protein-fat standpoint, and the protein contained a particularly good amino acid profile (Cantor and Roberts, 1967). The combination of corn, soy, and milk proteins resulted in a protein efficiency ratio essentially equivalent to that of casein. It is believed that the combination of functional, flavor, and nutritive properties resulted in the success of CSM.

PURPOSE OF FOOD FOR PEACE PROGRAM

The main objective of the Food For Peace legislation (PL480) has been to provide nutrition, in the form of both total calories and high-quality protein, to the millions of people in the Third World who do not have enough food. Indications are that as many as 10,000 people die each day from malnutrition (Anonymous, 1970), while irreversible physical damage is incurred in thousands more. The emphasis has been directed, in most cases, to weaning and preschool children, in whom protein malnutrition takes its greatest toll.

To accomplish this goal, the various cereal products purchased for the program have been distributed to most of the countries in the world through the Agency for International Development, local governments, and various volunteer agencies, such as CARE, UNICEF, Catholic Relief Agency, and Church World Services. The products get to the people through various programs, including church-related schools, school-lunch programs, and mother-child care centers.

DEVELOPMENT OF CORN-BASED PRODUCTS

In the author's opinion, the development of CSM and the subsequent cereal-based products that have become part of the PL480 program was truly a cooperative effort. It involved industry and government, as represented by the

U.S. Department of Agriculture and the Agency for International Development, plus a variety of scientists, nutritionists, etc., from private industry, academia, and a variety of government and volunteer agencies. The corn industry (Tollefson, 1967) provided the process capabilities, some of the product development effort (Cantor and Roberts, 1967), and the ability to process and move the large volume of commodity ingredients involved in the manufacture of these products. The U.S. Department of Agriculture was involved in product and process development (Anderson et al, 1969, 1970; Bookwalter et al, 1971; Conway, 1971a, 1971b; Bookwalter, 1977) and in defining specifications and requirements for the products (Senti et al, 1967; Bookwalter et al, 1968, 1971; Bookwalter, 1981).

The corn-based materials developed for PL480, when they were introduced, and their compositions are shown in Table V. Quantities of the products sold to the government from 1970 through 1982 (Bookwalter, 1983) are recorded in Table VI.[7] The size of the figures points up the significant effect this program has had on the corn industry.

Before about 1967, the PCM used to make CSM was generally produced on hot rolls (Anderson, 1982), the commercial products being made on gas-fired rolls. Since 1967, most of the PCM and instant PCM (used to make instant CSM) have been made in extrusion-cookers of the type described in the

---

[7] R. J. Alexander. Creating new foods from corn and other grains. Presented to: A Workshop on Food Engineering, Texas A&M University, May 19–21, 1975.

TABLE V
Corn-Based Fortified Foods Developed for PL480 Programs

Product	Year Introduced	Composition
Cornmeal, enriched[a]	1957 (approx.)	99+% cornmeal, ¼ oz vitamin mineral premix per hundredweight.
Processed cornmeal, enriched[b]	1963	99+% PCM,[c] ¼ oz vitamin-mineral premix per hundredweight.
Ceplapro[b]	1965	58% cornmeal, 25% soy flour, 10% durum flour, 5% NFDM, 2% vitamin-mineral premix.
CSM (corn-soy-milk)	1966	59.2% PCM, 17.5% soy flour, 15% NFDM, 5.5% soy oil, 2.8% vitamin-mineral premix.[d]
Instant CSM	1971	63% instant PCM, 23.7% soy flour, 5% NFDM, 5.5% soy oil, 2.8% vitamin-mineral premix.[d]
Instant sweetened CSM	1971	53% instant PCM, 27.5% soy flour, 7.35% sucrose, 5% NFDM, 5% soy oil, 2% vitamin-mineral premix, 0.15% vanilla flavor.
Soy-fortified cornmeal	1972	85% cornmeal, 15% soy grits.
CSB (corn-soy blend)	1973	67% PCM, 25% soy flour, 5% soy oil, 3% vitamin-mineral premix.

[a] According to Tollefson (1967), between 1957 and 1967 approximately 140 million tons was sold for U.S. aid programs.
[b] Product was tested in more than thirty countries, but never purchased under PL480.
[c] PCM = processed cornmeal, NFDM = nonfat dry milk.
[d] The original formulas for these products called for 64% PCM, 24% soy flour, 5% NFDM, 5% soy oil, and 2% vitamin-mineral premix. Present requirements were described by Bookwalter (1981).

preceding section and by Conway (1971a, 1971b). Other forms of cooking have been described by Anderson et al (1969, 1970), although the extruder appears to be the most versatile and practical cooking equipment for dry-milled corn products.

Although soy flour and soy grits have been the main protein supplement used in PL480 products (alone or in combination with NFDM), other protein sources have been investigated. Cantor and Roberts (1967) described the use of fish protein concentrate in producing materials equivalent to CSM in protein efficiency ratio. They also discussed the possible use of amino acid fortification. Hayes et al (1978, 1983) examined the replacement of soy flour in both corn- and wheat-based products with both defatted cottonseed and peanut flours. Whey protein concentrate was also tested and eventually approved as a replacement for NFDM and for use in a whey-soy drink mix (Bookwalter, 1981).

RESULTS OF PL480 PROGRAM

The main thrust of the PL480 program has never been to continue indefinitely the donation of food products to less developed countries, but to provide a stop-gap measure until adequate agricultural and processed food products could be produced by local governments. However, this stop-gap program has been going on for over 30 years and will probably continue into the foreseeable future.

Nevertheless, the efforts expended in this and other programs have been quite fruitful. The author is aware of at least 60 different products developed by various companies, governments, and foundations, many of which are being sold directly to people or local governments by private industry (Anonymous, 1970). Products with such names are Areparina, BalAhar, Duryea, Incaparina, Modern Bread, Pronutro, Saridele, Superamine, Golden Elbow Macaroni, and Yoo Hoo are but a few of the cereal-based foods that have helped to relieve part of the hunger problem. Add to these the 15–20 fortified cereal products, plus the many other commodity-based foods sold to and distributed by the U.S. government, and one begins to get a picture of the total energy and resources spent in this area in the last 30 years.

TABLE VI
Quantities (million lb)[a] of Corn-Based Fortified Foods Sold to Government Agencies under PL480

Product	1970[b]	1974[b]	1977[c]	1979[d]	1980[d]	1981[d]	1982[d]
Cornmeal, enriched	...	...	...	0.3	8.1	57.4	54.3
Soy-fortified cornmeal	...	25.5	...	105.6	108.2	133.9	83.1
Corn-soy blend	...	191.6	...	7.3	...	2.4	2.0
Corn-soy-milk (CSM)	383.5	...	...	226.5	247.3	282.9	215.4
Instant CSM	...	40.7	...	65.3	82.4	70.7	45.6
Instant sweetened CSM	...	31.4	...	...	...	...	...
Totals	383.5	289.2	485.0	405.0	446.0	547.0	400.4

[a] Divide figures in this table by 2205 to convert to million metric tons.
[b] Data from R. J. Alexander, Creating new foods from corn and other grains, paper presented to A Workshop on Food Engineering, Texas A&M University, May 19–21, 1975.
[c] Unpublished data; no information available on specific products.
[d] Data from Bookwalter (1983); values converted from metric tons to pounds.

E. Nonfood Uses

The importance of nonfood or industrial uses of dry-milled corn products has grown rather dramatically in recent years. Data for 1977 (Table IV) suggested a total volume of 530 million pounds (0.24×10^6 t). This is much higher than the figures reported by Senti (1965), Senti and Schaefer (1972), or Alexander (1973), indicating a substantial growth in the pharmaceutical-fermentation area, products for the construction industry, and foundry and charcoal briquetting applications.

This growth is undoubtedly the result of research and product development activities conducted in the 1950–1975 period, which led to an increased demand for corn flour and modified corn flours made by a variety of methods, as suggested by Roberts (1967). It is also a result of the fact that corn flour is an inexpensive starch source and can be used interchangeably with the more expensive corn starch in a number of end uses. This trend has been accentuated by the cost of cereal products, which have doubled, and in some cases tripled, in the last 20 years.

Most of the specific nonfood applications have been covered to some extent in recent years by Brockington (1970), Alexander (1973), and Rankin (1982). Nevertheless, the main nonfood uses are briefly reviewed here with special emphasis on the more recent developments.

GYPSUM BOARD BINDERS

Acid-modified, dry-milled corn and sorghum flours have been the most common starch-based binders employed in the production of dry wall or gypsum board for a number of years. Traditional products have been described by Slotter (1952) and Wimmer and Meindl (1959). Chemically modified, acid-modified flours with improved properties have been described by Rankin et al (1963) and more recently by Ferrara (1976).

The acid-modified flours seem to be preferred in many gypsum board plants over related acid-modified starches because of both price advantages and differences in functional properties. Because of process differences, dry-milled flours have more residual soluble carbohydrates, which aid in the manufacture of gypsum board.

The starch-containing corn flour is gelatinized in situ during the manufacture of gypsum board; it functions by controlling the rate of water loss during drying of the board (Alexander, 1973). The soluble carbohydrates migrate to the surface and control the rate of crystallization of the gypsum, providing a strong bond between gypsum and liner.

OTHER BUILDING MATERIALS

Corn flours have also been used in a variety of applications associated with the building products or construction industry. These include insulation or fiber board (Naffziger et al, 1963; Alexander, 1973), plywood and related laminating adhesives (Senti, 1965), and compression-molded particleboard (Alexander and Krueger, 1976) and wafer board. They have also been used in ceiling tile, as edge pastes in gypsum board production, and as binders in gypsum-based taping compounds (Cummisford, 1973).

PHARMACEUTICAL FERMENTATION

As noted in Table IV, the largest volume of dry-milled corn products in the nonfood area was in the pharmaceutical-fermentation area. This volume was principally that of corn flour used in the production of citric acid and other pharmaceuticals by fermentation procedures. The large increase in volume between 1973 (Alexander, 1973) and 1977 was related to the price structure of competing starch-based ingredients during that time, as evidenced by the peak use of corn grits in brewing during the same period (Fig. 2). Based on the general decline of nonfood uses of dry-milled corn products in 1979 (Wells, 1979), use in this category today is probably less than 200 million pounds.

Fermentation of corn products is also employed in processes for making industrial alcohol, while grain neutral spirits and whiskey production generally use whole corn. The potential use of dry-milled corn products for the production of various corn-based sweeteners using enzymatic processes has never been economically feasible in this country because of the dominance of the corn wet-milling industry. In Europe, however, commercial processes have been developed and are summarized by Kroyer (1966).

FOUNDRY BINDERS

Although the market for cereal-based binders has been eroded over the past 30 years by a number of different products, particularly petroleum-based binder systems, precooked corn flours still command a sizeable percentage of foundry binders. These precooked flours are used particularly as core binders in a sand-cereal-linseed oil system, in which the flour serves as the primary binder before the core is baked (Alexander, 1973). It provides a so-called "green strength" to the core while it is being conveyed from the core room to the baking ovens. During baking, the linseed oil takes over as the primary binder.

Typical precooked corn flours described by Burgess and Johnson (1961), Farrel,[8] and Smith[9] compete with similarly processed sorghum flours (Stickley and Griffith, 1966), both manufactured with extrusion-cookers. Certain wet-milled, hot-rolled starches have also been used in this application (Kowall and Hadyn, 1960), although the extruded flours appear to be preferred (Wenninger, 1966; Caine and Toepke, 1969) in many core systems. Chemically modified products (Penny and Harrington, 1951; Sietsma, 1961; Fortney and Hunt, 1966) have also been prepared, with reportedly improved properties as foundry binders.

A totally new corn-flour-based core binder was recently described by Cummisford et al (1977), in which an acid-modified, extruded flour was combined with glyoxal, or related polyaldehydes, under certain conditions to provide a thermosetting resin. In this sytem, no core oil was needed, and the mixture of cereal flour and dialdehyde functioned as the binder for the sand particles both before and after drying in the core ovens.

---

[8] R. Farrel. Extrusion equipment—Types, functions, and application. Presented to symposium: Extrusion: Process and Product Development. Central States Section, Am. Assoc. Cereal Chem., St. Louis, MO, Feb. 12–13, 1971.

[9] O. Smith. Why use extrusion? Presented to symposium: Extrusion: Process and Product Development. Central States Section, Am. Assoc. Cereal Chem., St. Louis, MO, Feb. 12–13, 1971.

CHARCOAL BRIQUETS

Corn flour has been reported (Senti, 1965; Alexander, 1973) as the primary adhesive or binder in the production of charcoal briquets. In 1977 (Table IV), the volume was still reasonably high at 75 million pounds. Very few, if any, significant technical advances or product improvements have been generated in this area in recent years, as evidenced by the lack of technical publications and patents. A brief discussion of the process used to make briquets is reported by Alexander (1973).

OTHER USES

Paper. Judging from the literature, the largest area of new product development among the various "other" uses of dry-milled corn products is in paper applications. This is a fairly logical extension of the research conducted in the wet corn milling industry, since the paper industry is the largest nonfood user of starch and modified starches. Several specialized applications within the paper area have been investigated.

The first area is that of wet-end additives used in the improvement of either dry strength (Russell et al, 1962; Jones et al, 1966) or wet strength through the use of cereal flour xanthates (Russell et al, 1962, 1964). Dry-milled flours have also been used in fiber board produced from waste paper (Alexander, 1973) and as a surface size in Kraft liner board.

Another application area is that of cationic corn flours useful in improving both dry strength and pigment retention. Modified corn flours made with tertiary amino- or quaternary ammonium-etherifying agents were claimed by Alexander and Cummisford (1971). Several authors have described the preparation and properties of cereal flours treated with ethylenimine (Rankin and Russell, 1970; McClendon and Berry, 1973; McClendon, 1974), a reagent first reported by Kerr and Neukom (1952) for modifying corn starch.

The last area is that of surface sizing of paper, which has usually required an acid modification of the starch-based material. Processes for the dry acid modification of cereal flours have been reported by Rankin et al (1963, 1964) and Lancaster and Griffin (1965). Further improvements in surface sizing properties can be accomplished by treating the acid-modified flour with ethylene oxide, as described by Rankin et al (1973).

Because of paper industry requirements, significant quantities of corn flour are probably not being used in paper applications today. The demand for high-brightness papers as well as starch sizes and adhesives with low levels of sludge (usually after enzyme conversion) preclude the use of yellow corn flours containing as much as 10% nonstarch components.

Corrugating Adhesives. A number of special processes for making corrugating adhesives from corn flours have been reported. Wimmer (1959a) used corn flour plus sodium chloroacetate in the carrier portion of the adhesives to make an improved product. Horner (1961) described the use of waxy starch or flour in the carrier in combination with a cereal flour, preferably from sorghum, to produce a nonthixotropic adhesive. A more recent process (Fortney and Hunt, 1966) employs fine-grind corn flours with a special granulation profile to produce a superior adhesive. The author is aware of at least one corrugating plant that was using corn flour made according to this latter patent, although the quantities of flour employed today are not known.

Adhesives (General). The potential use of corn flour and modified corn flours as general starch-based adhesives has been described by several investigators. Rankin and Mehltretter (1959), Rankin et al (1959), Smith et al (1962, 1969), and Alexander (1974) have prepared chemically modified flours with excellent viscosity and adhesive characteristics. Waggle (1971) claims a process for producing a modified cereal flour useful in pelletizing animal feed. Bailey[10] suggests the use of corn flour as an adhesive for bag bottoms, and Senti (1965) indicates its utility in laminating adhesives. Other adhesive uses were discussed earlier in this chapter.

Ore Refining. Although Senti (1965), Senti and Schaefer (1972), and Wells (1979) have suggested that as much as 60 million pounds of corn flours have been used in ore-refining operations, Alexander (1973) has indicated that this quantity of flour is probably derived from sorghum rather than corn. The author is aware of three companies that sold unrefined sorghum flours for the beneficiation of aluminum ore (bauxite) during the 1960s and 1970s. Such products and processes have been patented by Jones (1960). Also, extruded sorghum flours have been patented as binders in pelletizing iron ore (Stickley et al, 1964).

Oil Well Drilling. This is one application area for corn flours that has definitely increased in volume since the market statistics reported in 1965 (Senti) and 1973 (Alexander) and since the 1977 estimates in Table IV. Although domestic use of crude oil was on a steady incline until the formation of OPEC in 1973–1974, the drilling of new wells in this country was on steady decline until 1979. In June of 1979, President Carter deregulated the price of domestic crude oil and natural gas. This created the setting for the biggest drilling boom in history and the need for increased quantities of precooked starch-based materials.

The total market for precooked starch for oil drilling uses increased to nearly 120 million pounds in 1981. About 40%, or 50 million pounds, was corn-flour-based materials. In 1982–1983, the number of new wells dropped considerably as a result of the recession, combined with a surplus of petroleum-based products. The volume today is thought to be around 30 million pounds.

In the oil-well-drilling application, the precooked starch or flour functions to minimize water loss in the drilling mud. The mud cools and lubricates the drill bit, suspends and removes the cuttings from the bottom of the hole, and coats the hole with an impermeable layer. It also controls subsurface pressures in the hole, supports part of the weight of drill pipe and casing, and minimizes adverse effects on the formation adjacent to the hole.

The precooked flours used in this application have been produced using either drum dryers or hot rolls (Roemer and Downhour, 1970a, 1970b) or extrusion-cookers (Bradley and Downhour, 1970; Roemer and Downhour, 1970a, 1970b). Chemically modified flours, such as described by Smith et al (1962, 1969), have been suggested to be improved agents for decreasing water loss because of their increased water-holding capacities. In extremely deep holes, where mud temperatures in the 200–300° C range are reached, corn flours (Wimmer, 1959b) or starches (Hullinger, 1967) that have been crosslinked are preferred.

Thermoplastics and Polyurethanes. A number of different corn-flour-based

[10] See footnote 3.

products have been developed for use with petrochemical-based synthetic polymers. Russell[11] discussed the use of flours as extenders in polyvinyl alcohol and polyvinyl chloride films used as agriculture mulches. Cereal flours have also been used as polyol-extenders in rigid polyurethane foams (Bennett et al, 1967), in flexible polyurethane foams (Hostettler, 1978), and in polyurethane resins (Otey et al, 1968), which can be molded and machined into furniture parts and the like. In certain polyurethane foams (Anonymous, 1967; Bennett et al, 1967), the flour appears to function as a fire-retarding agent as well as an extender.

Miscellaneous. A number of other minor uses for dry-milled corn products have been reported, including explosives, carriers for vitamins in animal feeds and for certain pesticides, and abrasive agents in industrial hand soaps (Alexander, 1973); as textile sizes (Rankin et al, 1963); and as a dry carrier in solvent-based dry cleaners.[12]

F. By-Products and Animal Feed

As indicated in Section II of this chapter, the various by-product streams from the corn dry-milling process are most frequently combined to produce a single by-product known as hominy feed. The volume of this product, the largest for the products sold by the dry corn millers, is annually about 2.2 billion pounds (Brekke, 1970a; Alexander, 1973). This material competes with similar corn by-products, such as corn gluten feed and spent brewer's grains (Shroder and Heiman, 1970), as ingredients in animal feed. This area has been the subject of fairly comprehensive reviews (Morrison, 1959; Smith, 1959; Shroder and Heiman, 1970; Ensminger and Olentine, 1978) and is not reported here in any detail.

Hominy feed provides the U.S. feed industry, as well as numerous countries throughout the world, with an inexpensive, high-fiber, high-calorie ingredient. The material is high in carotenoids (the yellow pigments in corn) and in vitamins A and D. The high carotenoid content is particularly desirable in chicken feed for providing eggs with bright yellow yolks.

In recent years, a number of investigators have taken a close look at alternative, potentially more profitable uses for some of the by-product streams, in particular, corn bran and oil-free corn germ. Corn bran has been of particular interest because of its potential as a source of dietary fiber (Wells, 1979). One researcher (DuVall, 1982) incorporated dry-milled corn bran of a certain specific particle size into an extruded, high fiber, corn-based breakfast cereal. At least one dry-milled product (Anonymous, 1977; Tabor Milling Company, 1977) was introduced as a high-fiber ingredient for food applications. Others (Alexander and Krueger, 1978) have taken advantage of corn bran's superior water absorption properties in providing an extender-viscosifier for use in urea-formaldehyde plywood adhesives.

Oil-free, dry-milled corn germ, especially that produced by hexane extraction, is a good source of high-quality protein (Blessin et al, 1972, 1973; Garcia et al,

---

[11] C. R. Russell. Cereal starches and flour products as substitutes and extenders for petroleum-based polymers and plastics. Presented to: American Corn Milling Federation meeting, Northern Regional Research Laboratory, Peoria, IL, June 1975.

[12] See footnote 3.

1972), plus a fairly good source of dietary fiber. Such material has recently received considerable attention as an ingredient in protein-fortified cookies (Blessin et al, 1972, 1973; Tsen 1976), bread (Tsen et al, 1974), and other food products, such as corn muffins and meat patties. At least one dry corn milling company reportedly produced pilot plant quantities of a corn germ flour for evaluation during the mid-1970s. The current status of this venture is not known.

By the use of alkaline extraction and dialysis techniques, Nielsen and co-workers (1973, 1977) prepared protein isolates from oil-free corn germ. Products with nutritionally valuable amino acid profiles for potential fortification of food products were described.

Typical analyses of the various by-product streams are shown in Table VII. Composition of hominy feed can vary somewhat depending on whether the corn miller incorporates germ cake into the feed, sells corn germ to another company that processes germ, or finds other ways of upgrading the various streams.

V. FUTURE OF CORN DRY MILLING

Considering the current situation with corn dry milling, it is somewhat difficult to predict what will happen to the industry in the next 10–15 years. On the negative side, the industry has not really grown in recent years. The amount of corn processed by dry millers in 1977–1982 (Anonymous, 1982) is about the same as that reported by Brekke (1970a) for the 1965–1969 period, or about 120–140 million bushels (3.0–3.5×10^6 t). Increases in product volumes achieved in certain food and nonfood applications have been offset by decreases in brewing.

On the positive side, some new developments, especially in the food area, could lead to increases in the near future. One area of particular importance has been that of ethnic foods, specifically Mexican and Latin American food products. The number of alkali-processed corn products in both the supermarket and the fast-food and Mexican food restaurants has definitely shown an increase. New corn mills devoted exclusively to the production of masa flour (Rice, 1983) have been built since 1980, and masa flours produced by

TABLE VII
Typical Composition (%, as-is basis)
of Dry-Milled Corn By-Product Streams[a]

Component	Standard Meal	Corn Germ[b]	Corn Bran	Hominy Feed
Moisture	14.0	9.6	10.0	13.5
Protein	11.0	15.8	8.0	8.0
Fat	4.5	23.8	4.5	3.4
Crude fiber	2.5	5.7	12.0	4.7
Ash	2.0	6.7	2.5	2.0
Starch	60.0	18.4	35.0	61.0
Other polysaccharides	6.0	20.0	28.0	7.4

[a] Except for hominy feed, the product streams normally do not exist as separate finished products but are combined with broken corn and subsequently dried and hammer milled together to produce hominy feed.
[b] Represents composition before oil expelling or extraction.

traditional dry millers have been described in recent years (Wimmer and Sussex, 1968; Anonymous, 1984a).

Although the use of corn products in more traditional food applications has probably declined, the production of processed corn flours has increased. As discussed earlier, this is in large part the result of the use of extrusion-cookers, which has resulted in increases in the snack food, fortified food, and specialty breakfast food areas. Nonfood areas, such as foundry binders and oil well drilling, have also benefited from the use of extrusion.

The author's prediction is that the industry will remain fairly static during the next 10-15 years, with increases in food uses being counterbalanced by decreases in brewing. The industry will probably grow at about the same rate as the population.

LITERATURE CITED

ALEXANDER, R. J. 1973. Industrial uses of dry-milled corn products. Pages 303-315 in: Industrial Uses of Cereals. Y. Pomeranz, ed. Am. Assoc. Cereal Chem., St. Paul, MN.

ALEXANDER, R. J. 1974. Art of manufacturing modified amylaceous materials with condensed phosphates and urea. U.S. patent 3,843,377.

ALEXANDER, R. J., and CUMMISFORD, R. G. 1971. Cationic cereal flours and a method for their manufacture. U.S. patent 3,578,475.

ALEXANDER, R. J., and KRUEGER, R. K. 1976. Art of manufacturing compression molded particle board with nitrogenous modified amylaceous binder. U.S. patent 3,983,084.

ALEXANDER, R. J., and KRUEGER, R. K. 1978. Plywood adhesives using amylaceous extenders comprising finely ground cereal-derived high fiber by-product. U.S. patent 4,070,314.

ANDERSON, R. A. 1982. Water absorption and solubility and amylograph characteristics of roll-cooked small grain products. Cereal Chem. 59:265-269.

ANDERSON, R. A., and WATSON, S. A. 1982. The corn milling industry. Pages 31-61 in: CRC Handbook of Processing and Utilization in Agriculture. Vol. II: Part 1. Plant Products. I. A. Wolff, ed. CRC Press, Inc., Boca Raton, FL.

ANDERSON, R. A., CONWAY, H. F., PFEIFER, V. F., and GRIFFIN, E. L., Jr. 1969. Gelatinization of corn grits by roll- and extrusion-cooking. Cereal Sci. Today 14:4-7, 11-12.

ANDERSON, R. A., CONWAY, H. F., and PEPLINSKI, A. J. 1970. Gelatinization of corn grits by roll-cooking, extrusion-cooking and steaming. Staerke 22:130-135.

ANONYMOUS. 1967. Use of flour in urethane. Southwest. Miller. 47(April 18):25.

ANONYMOUS. 1970. Fortified foods: The next revolution. Chem. Eng. News 48:35, 37, 39, 41, 43.

ANONYMOUS. 1974-1984. Materials used at breweries. Brew. Dig. (May issues)

ANONYMOUS. 1977. Corn bran flour fiber introduced by Tabor Milling Company. Brew. Bull. Feb. 14.

ANONYMOUS. 1978. Bunge's business grows, changes. Milling Baking News 57(Nov. 28):1, 27-34.

ANONYMOUS. 1982. Corn. In: The 1982 Commodity Yearbook. Commodity Research Bureau, Inc., New York.

ANONYMOUS. 1983. Extruded corn-based ingredients provide limitless new product potentials. Food Process. 44(1):39-40.

ANONYMOUS. 1984a. Lincoln grain develops masa; Eligible for relief. Milling Baking News 63(Aug. 7):13, 16.

ANONYMOUS. 1984b. Milling directory. Milling Baking News 63(Nov.):62-65.

BEDOLLA, S., and ROONEY, L. W. 1982. Cooking maize for masa production. Cereal Foods World 27:218-221.

BELT, T. 1928. The Naturalist in Nicaragua. E. P. Dutton & Co., New York.

BENNETT, F. L., OTEY, F. H., and MEHLTRETTER, C. L. 1967. Rigid urethane foam extended with starch. J. Cell. Plast. 3(Aug.):1-5.

BLESSIN, C. W., INGLETT, G. E., GARCIA, W. J., and DEATHERAGE, W. L. 1972. Defatted germ flour—Food ingredient from corn. Food Prod. Dev. 33(May):34-35.

BLESSIN, C. W., GARCIA, W. J., DEATHERAGE, W. L., CALVINS, J. F., and INGLETT, G. E. 1973. Composition of three food products containing defatted corn

germ flour. J. Food Sci. 38:602-606.
BOOKWALTER, G. N. 1977. Corn-based foods used in food aid programs; Stability characteristics—A review. J. Food Sci. 42:1421-1424.
BOOKWALTER, G. N. 1981. Requirements for foods containing soy protein in the food for peace program. J. Am. Oil Chem. Soc. 58:455-459.
BOOKWALTER, G. N. 1983. World feeding strategies utilizing cereals and other commodities. Cereal Foods World 28:507-511.
BOOKWALTER, G. N., PEPLINSKI, A. J., and PFEIFER, V. F. 1968. Using a Bostwick Consistometer to measure consistencies of processed corn meals and their CSM blends. Cereal Sci. Today 13:407-410.
BOOKWALTER, G. N., CONWAY, G. F., and GRIFFIN, E. L. 1971. Extruder-cooked cereal endosperm particles and instant beverage mixes comprising the same. U.S. patent 3,579,352.
BRADLEY, P. P., and DOWNHOUR, R., Jr. 1970. Drilling mud additives. U.S. patent 3,518,185.
BREKKE, O. L. 1966. Corn dry-milling: A comparison of several procedures for tempering low-moisture corn. Cereal Chem. 43:303-312.
BREKKE, O. L. 1967. Corn dry-milling: Pretempering low-moisture corn. Cereal Chem. 44:521-531.
BREKKE, O. L. 1968. Corn dry-milling: Stress crack formation in tempering of low-moisture corn, and effect on degerminator performance. Cereal Chem. 45:291-303.
BREKKE, O. L. 1970a. Corn dry milling industry. Pages 262-291 in: Corn Culture, Processing, Products. G. E. Inglett, ed. Avi Publishing Co., Inc., Westport, CT.
BREKKE, O. L. 1970b. Dry-milling artificially dried corn; Roller-milling of degerminator stock at various moistures. Cereal Sci. Today 15:37-42.
BREKKE, O. L., and KWOLEK, W. F. 1969. Corn dry-milling: Cold-tempering and degermination of corn of various initial moisture contents. Cereal Chem. 46:545-559.
BREKKE, O. L., and WEINECKE, L. A. 1964. Corn dry-milling: A comparative evaluation of commercial degerminator samples. Cereal Chem. 41:321-328.
BREKKE, O. L., WEINECKE, L. A., WOHLRABE, F. C., and GRIFFIN, E. L., Jr. 1961. Tempering and degermination for corn dry milling; A research project for industry. Am. Miller 89:14-17.
BREKKE, O. L., WEINECKE, L. A., BOYD, J. N., and GRIFFIN, E. L., Jr. 1963. Corn dry-milling: Effects of first-temper moisture, screen perforation, and rotor speed on Beall degerminator throughput and products. Cereal Chem. 40:423-429.
BROCKINGTON, S. F. 1970. Corn dry milled products. Pages 292-306 in: Corn: Culture, Processing, Products. G. E. Inglett, ed. Avi Publishing Co., Westport, CT.
BURGESS, H. M., and JOHNSON, J. B. 1961. Binder product and process. U.S. patent 3,159,505.
CAINE, J. B., and TOEPKE, R. E. 1969. The case for baked oil-bonded cores. Foundry (10):140-146.
CANTOR, S. M., and ROBERTS, H. J. 1967. Improvements in protein quality in corn based foods. Cereal Sci. Today 12:443-445, 460-462.
CONWAY, H. F. 1971a. Extrusion cooking of cereals and soybeans. Part I. Food Prod. Dev. (4):76-80.
CONWAY, H. F. 1971b. Extrusion cooking of cereals and soybeans. Part II. Food Prod. Dev. (5):14-22.
CUMMISFORD, R. G. 1973. Art of manufacturing dry wall taping and finishing compounds with a nitrogenous-modified amylaceous binder. U.S. patent 3,725,324.
CUMMISFORD, R. G., WASZELEWSKI, R. J., and KRUEGER, R. K. 1977. Art of catalyzing the reaction between a polyol and a polyaldehyde. U.S. patent 4,013,629.
DUVALL, L. F. 1982. Corn bran expanded cereal. U.S. patent 4,350,714.
ENSMINGER, M. E., and OLENTINE, C. G., Jr. 1978. Feeds and Nutrition. Ensminger Publishing Co., Clovis, CA.
FERRARA, P. J. 1976. Alkylene oxide modified cereal flours and process of preparing the same. U.S. patent 3,943,000.
FORTNEY, C. G., and HUNT, K. R. 1966. Cereal flour adhesive product and process. U.S. patent 3,251,703.
GARCIA, W. J., GARDNER, H. W., CAVINS, J. F., STRINGFELLOW, A. C., BLESSIN, C. W., and INGLETT, G. E. 1972. Composition and air-classification of corn and wheat-germ flours. Cereal Chem. 49:499-507.
HARDEMAN, N. P. 1981. Schucks, Shocks, and Hominy Blocks: Corn as a Way of Life in Pioneer America. L.S.V. Press, Baton Rouge.
HAUCK, B. W. 1980. Marketing opportunities for extrusion cooked products. Cereal Foods World 25:594-595.
HAYES, R. E., WADSWORTH, J. I., and SPADARO, J. J. 1978. Corn- and wheat-based blended food formulations with cottonseed or peanut flour. Cereal Foods

World 23:548-553, 556.
HAYES, R. E., WADSWORTH, J. I., SPADARO, J. J., and FREEMAN, D. W. 1983. An experimental evaluation of computer-formulated corn blends. Cereal Foods World 28:670-675.
HORNER, J. W., Jr. 1961. Non-thixotropic flour adhesives and methods therefore. U.S. patent 2,999,028.
HOSTETTLER, F. 1978. Polyurethane foams containing stabilized amylaceous materials. U.S. patent 4,156,759.
HULLINGER, C. H. 1967. Production and uses of crosslinked starches. Pages 445-450 in: Starch: Chemistry and Industry, Vol. II. R. L. Whistler and E. F. Paschall, eds. Academic Press, New York.
JONES, E. J., NAGEL, S. C., and SWANSON, J. W. 1966. Use of dry-milled corn flours in papermaking. Cereal Sci. Today 11:54-56.
JONES, R. L. 1960. Starch-borax settling aid and process of using. U.S. patent 2,935,377.
KERR, R. W., and NEUKOM, H. 1952. The reaction of starch with ethylenimine. Staerke 4:255-257.
KOWALL, N., and HADYN, S. 1960. Team of corn-based binders improves foundry cores. Iron Age. (Jan 28):94.
KROYER, K. K. 1966. Is it necessary to produce starch in order to obtain syrup, dextrose and total sugar? Staerke 18:311-316.
LANCASTER, E. B., and GRIFFIN, E. L., Jr. 1965. Anhydrous HCL modification of flours. U.S. patent 3,175,928.
LARSEN, R. A. 1959. Milling. Pages 214-216 in: Cereals as Food and Feed. S. A. Matz, ed. Avi Publishing Co., Westport, CT.
LINE, W. F., CHAUDHARY, V. K., CHICOYE, E., and MIZERAK, R. J. 1982. Method of preparing a low calorie beer. U.S. patent 4,355,047.
MacLEOD, A. M. 1977. Beer. Pages 43-137 in: Alcoholic Beverages. A. H. Rose, ed. Academic Press, London.
MANGELSDORF, P. C. 1974. Corn: Its Origin, Evolution and Improvement. Harvard University Press, Cambridge.
MANGELSDORF, P. C., MacNEISH, R. S., and GALINANT, W. C. 1964. Domestication of corn. Science 143:538-545.
MANGELSDORF, P. C., MacNEISH, R. S., and GALINANT, W. C. 1967. The Prehistory of the Tehuacan Valley, Vol. I. Univ. Texas Press, Austin.
MATSON, K. 1982. What goes on in the extruder barrel. Cereal Foods World 27:207-210.
MATZ, S. A. 1959. Manufacture of breakfast cereals. Pages 547-566 in: Cereals as Food and Feed. S. A. Matz, ed. Avi Publishing Co., Westport, CT.
McCLENDON, J. C. 1974. Cationic flours and starches. U.S. patent 3,846,405.
McCLENDON, J. C., and BERRY, E. L. 1973. Aminoethylation of flour and starch with ethylenimine. U.S. patent 3,725,387.
MOLL, M., and DUTEURTRE, B. 1976. The adjunct fermentation process. Part II: Trials in 10-hl pilot brewery. Tech. Q. Master Brew. Assoc. Am. 13:26-30.
MORRISON, S. H. 1959. Supplementation of cereal-based animal feeds. Pages 662-694 in: Cereals as Food and Feed. S. A. Matz, ed. Avi Publishing Co., Westport, CT.
NAFFZIGER, T. R., SWANSON, C. L., HOFRIETER, B. T., RUSSELL, C. R., and RIST, C. E. 1963. Crosslinked flour xanthate-wood pulp for insulating board. Tappi 46:428-431.
NIELSEN, H. C., INGLETT, G. E., WALL, J. S., and DONALDSON, G. L. 1973. Corn germ protein isolate—Preliminary studies on preparation and properties. Cereal Chem. 50:435-443.
NIELSEN, H. C., WALL, J. S., MUELLER, J. K., WARNER, K. and INGLETT, G. E. 1977. Effect of bound lipid on flavor of protein isolate from corn germ. Cereal Chem. 54:503-510.
OTEY, F. H., BENNETT, F. L., and MEHLTRETTER, C. L. 1968. Process for preparing polyether-polyurethane-starch resins. U.S. patent 3,405,080.
PENNY, N. M., and HARRINGTON, B. J. 1951. Corn meal milling in Georgia. Ga. Agric. Exp. Stn. Res. Bull. 272.
PFISTERER, E. A., GARRISON, I. F., and McKEE, R. A. 1978. Brewing with syrups. Tech. Q. Master Brew. Assoc. Am. 15:59-63.
POLLOCK, J. R. A., and WEIR, M. J. 1976. The adjunct fermentation process. Part I. Tech. Q. Master Brew. Assoc. Am. 13:22-25.
PRZYKYBLA, A. 1984. Diversity seen in new cracker introductions. Prepared Foods 153 (June):120-121.
RANKIN, J. C. 1982. The nonfood uses of corn. Pages 63-78 in: CRC Handbook of Processing and Utilization in Agriculture. Vol. II, Part 1. Plant Products. I. A. Wolff, ed. CRC Press, Inc., Boca Raton, FL.
RANKIN, J. C., and MEHLTRETTER, C. L. 1959. Process for preparing hydroxyalkyl cereal flours. U.S. patent 2,900,268.
RANKIN, J. C., and RUSSELL, C. R. 1970. Acidified ethylenimine modified cereal flours. U.S. patent 3,522,238.
RANKIN, J. C., MEHLTRETTER, C. L., and SENTI, F. R. 1959. Hydroxyethylated cereal

flours. Cereal Chem. 36:215-227.
RANKIN, J. C., RUSSELL, C. R., and SAMALIK, J. H. 1963. Process for preparing improved sizing agents from cereal flours. U.S. patent 3,073,724.
RANKIN, J. C., SAMALIK, J. H., HOLZAPFEL, M. M., RUSSELL, C. R., and RIST, C. R. 1964. Preparation and properties of acid-modified cereal flours. Cereal Chem. 41:386-399.
RANKIN, J. C., SAMALIK, J. H., RUSSELL, C. R., and RIST, C. E. 1973. Acid-modified wheat flours. Cereal Sci. Today 18:74-76, 81.
RAO, B. A. S., and NARASIMHAM, V. V. L. 1975. Adjuncts in brewing. J. Food Sci. Technol. 12:217-220.
RICE, J. 1983. Mexican-based firm in full swing with its first U. S. corn flour plant. Food Process. 44(June):58-59.
ROBERTS, H. J. 1967. Corn flour: From surplus commodity to premium product. Cereal Sci. Today 12:505-508, 532.
ROEMER, P., and DOWNHOUR, R., Jr. 1970a. Starch of good water retention. U.S. patent 3,508,964.
ROEMER, P., and DOWNHOUR, R., Jr. 1970b. Drilling mud additives. U.S. patent 3,518,185.
RUSSELL, C. R., BUCHANAN, R. A., RIST, C. E., HOFREITER, B. T., and ERNST, A. J. 1962. Cereal Pulps. I. Preparation and application of crosslinked cereal xanthates in paper products. Tappi 45:557-566.
RUSSELL, C. R., BUCHANAN, R. A., and RIST, C. E. 1964. Cellulosic pulps comprising crosslinked xanthate cereal pulps and products made therewith. U.S. patent 3,160,552.
SANDERUDE, K. G. 1969. Continuous cooking extrusion: Benefits to the snack food industry. Cereal Sci. Today 14:209-210.
SCALES, H. 1982. The U. S. snack food market. Cereal Foods World 27:203-205.
SENTI, F. R. 1965. The industrial utilization of cereal grains. Cereal Sci. Today 10:320-327, 361-362.
SENTI, F. R., and SCHAEFER, W. C. 1972. Corn: Its importance in food, feed, and industrial uses. Cereal Sci. Today 17:352-356.
SENTI, F. R., COPLEY, M. J., and PENCE, J. W. 1967. Protein-fortified grain products for world uses. Cereal Sci. Today 12:426-430, 441.
SHRODER, J. D., and HEIMAN, V. 1970. Feed products from corn processing. Pages 220-240 in: Corn: Culture, Processing, Products. G. E. Inglett, ed. Avi Publishing Co., Westport, CT.
SIETSMA, J. W. 1961. Core binder. U.S. patent 2,974,048.
SLOTTER, R. L. 1952. Dextrinization of sorghum flours. U.S. patent 2,601,335.
SMITH, B. W. 1959. Feed manufacture. Pages 404-426 in: Cereals as Food and Feed. S. A. Matz, ed. Avi Publishing Co., Westport, CT.
SMITH, H. E., RUSSELL, C. R., and RIST, C. E. 1962. Preparation and properties of sulfated wheat flours. Cereal Chem. 39:273-281.
SMITH, H. E., RUSSELL, C. R., HOLZAPFEL, M. M., and RIST, C. E. 1969. Sulfated wheat flours of low sulfur content. Northwest. Miller. (April):13-15.
STAUFFER, C. E. 1983. Corn-based snacks. Cereal Foods World 28:301-302.
STICKLEY, E. S., and GRIFFITH, E. 1966. Cold water dispersible cereal products and process for their manufacture. U.S. patent 3,251,702.
STICKLEY, E. S., GRIFFITH, E., and WILLIAMS, E. W. 1964. Process for pelletizing ores. U.S. patent 3,154,403.
STIVER, T. E., Jr. 1955. American corn milling systems for de-germed products. A.O.M. Bull. pp. 2168-2179.
STUBITS, M. C., and TENG, J. 1983. High gravity brewing using low gel point rice adjuncts. U.S. patent 4,397,872.
SWAIN, E. F. 1976. The manufacture and use of liquid adjuncts. Tech. Q. Master Brew. Assoc. Am. 13:108-113.
TABOR MILLING COMPANY. 1977. Corn bran fiber flour. Product Bull. 800. Kansas City, MO.
TOFT, G. 1979. Snack foods: Continuous processing techniques. Cereal Foods World 24:142-143.
TOLLEFSON, B., Jr. 1967. New milled corn products, including CSM. Cereal Sci. Today 12:438-441.
TSEN, C. C. 1976. Regular and protein fortified cookies from composite flours. Cereal Foods World 21:633-640.
TSEN, C. C., MOJIBAN, C. N., and INGLETT, G. E. 1974. Defatted corn-germ flour as a nutrient fortifier for bread. Cereal Chem. 51:262-271.
WAGGLE, D. H. 1971. Adhesive process. U.S. patent 3,565,651.
WEINECKE, L. A., BREKKE, O. L., and GRIFFIN, E. L., Jr. 1963. Corn dry-milling: Effect of Beall degerminator tail-gate configuration on product streams. Cereal Chem. 40:575-581.
WELLS, G. H. 1979. The dry side of corn milling. Cereal Foods World 24:333, 340-341.
WENNINGER, C. D. 1966. High efficiency mulling of clay-cereal mixtures. Modern

Castings 49:139-146.

WILLIAMS, M. A. 1977. Direct extrusion of convenience foods. Cereal Foods World 22:152-154.

WIMMER, E. L. 1959a. Adhesives and method of manufacturing the same. U.S. patent 2,881,086.

WIMMER, E. L. 1959b. Starch crosslinked with hexahydro-1,3,5-triacyrlol-S-triazine. U.S. patent 2,910,467.

WIMMER, E. L., and MEINDL, F. 1959. Art of manufacturing cold water dispersible adhesives. U.S. patent 2,894,859.

WIMMER, E. L., and SUSSEX, J. L. 1968. Process for manufacturing corn flour. U.S. patent 3,404,986.

WYSS, E. 1974. Development of corn processing. Diagram (Buhler Corp.), No. 58. Buhler Brothers, Ltd., Uzwil, Switzerland.

CHAPTER 12

WET MILLING: PROCESS AND PRODUCTS

JAMES B. MAY (*retired*)
A. E. Staley Manufacturing Co.
Decatur, Illinois

I. INTRODUCTION

Corn is abundant and relatively low in price; it has a high starch content and protein of acceptable quantity and quality. Hence its primary use is for animal feed. It is also useful for processing into valuable food and industrial products, such as ethyl alcohol by fermentation, cornmeal by dry milling, and highly refined starch by the wet-milling process. The greatest volume is processed by wet milling to produce starch products (Chapter 16) and sweetener products for foods (Chapter 17). Nonfood products such as industrial starches, corn gluten feed, and corn gluten meal are also manufactured.

The wet-milling process involves an initial water soak under carefully controlled conditions to soften the kernels. The corn is then milled and its components separated by screening, centrifuging, and washing (Fig. 1), to produce starch, oil, feed by-products, and sweeteners (by starch hydrolysis). Applications for these products have shown steady growth, which has necessitated major investment to expand production facilities in recent years.

II. THE PROCESS

A. Steeping

The first critical step in the wet milling of corn is steeping—the soaking of the corn in water under controlled processing conditions of temperature, time, sulfur dioxide (SO_2) concentration, lactic acid content, etc. These conditions have been found necessary to promote diffusion of the water through the tip cap of the kernel into the germ, endosperm, and their cellular components. Steeping softens the kernels, facilitating separation of the components.

Corn is shipped in bulk to the wet-milling plants by truck, hopper car, and barge. It is then cleaned on vibrating screens to remove coarse material (retained on 12.7-mm [1/2-in.] openings) and fine material (through 3.18-mm [1/8-in.] openings). These screenings are diverted to animal feed. If they are allowed to remain with the corn, they cause processing problems such as restricted water flow through steeps and screens, increased steep liquor viscosity, and quality problems with the finished starch.

Steeping is accomplished by putting corn into tanks (steeps) that have a capacity of 50–330 t (2,000–13,000 bu) each. The corn is then covered with steepwater, heated to 52°C (125°F) and held for 22–50 hr. Steeps have cone bottoms with screens so that the water can be separated from the corn and pumped elsewhere or recirculated back into the top of the steep. To maintain steeping temperature, the recirculated flow is heated with steam directly by

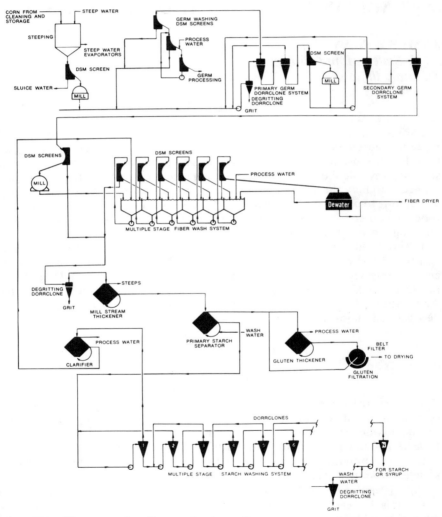

Fig. 1. Wet-milling process flow diagram, showing equipment arrangement for the separation of the major components—steepwater, germ, fiber, gluten, and cornstarch. (Courtesy Dorr-Oliver, Inc., Stamford, CT)

sparge injection or indirectly by using a heat exchanger. The water should not exceed 55°C (131°F) to avoid destroying the bacteria needed to produce lactic acid.

Steeping is a countercurrent system, utilizing a battery of six to twelve or more steep tanks. Steeps are filled one at a time as they become empty. The corn does not move—just the water, which is transferred from one steep to the next. However, steeping is accomplished in one plant by continuously adding dry corn at the top of the steep while continuously withdrawing steeped corn from the bottom (Randall et al, 1978).

Water for the steeps originates in the wet-milling process, where it accumulates corn solubles. It is treated with SO_2 to a concentration of 0.12–0.20%. The SO_2 is purchased as a liquid or manufactured on-site by burning elemental sulfur. The SO_2 increases the rate of water diffusion into the kernel and assists in breaking down the protein-starch matrix, which is necessary for high starch yield and quality.

The SO_2-treated water is added to the steep containing the oldest corn. As the water is advanced from steep to steep, the SO_2 content decreases and bacterial action increases, resulting in the growth of lactic acid bacteria. The desired lactic acid concentration is 16–20% (dry basis) after the water has advanced through the system and been withdrawn as light steepwater. Meanwhile, the SO_2 content drops to 0.01% or less.

The volume of water available for steeping is normally 1.2–1.4 m^3/t of corn (8–9 gal/bu). About 0.5 m^3/t (3.5 gal/bu) is absorbed by the corn to increase its moisture from 16 to 45% during the steeping. The remaining 0.7–0.8 m^3/t

Fig. 2. Degerminating mill opened up to show the knobby disks. Feed enters the center portion. One disk rotates while the other remains stationary. (Courtesy Dorr-Oliver, Inc., Stamford, CT)

(4.5–5.5 gal/bu) is the quantity withdrawn from the steeping system. This water contains the solubles soaked out of the corn, which is 0.05–0.06 t of solids per tonne of corn processed (2.8–3.6 lb of solids per bushel). It is evaporated to 40–50% solids, mixed with corn fiber, dried, and sold as corn gluten feed. Watson (1984) gives an in-depth presentation of the chemistry associated with the steeping process.

B. Separation of Kernel Components

GERM

A water sluice system is generally used to transport steeped corn from the discharge of a steep tank to a surge bin. The water used for transportation is screened out of the corn before the corn is put into the bin and is returned to

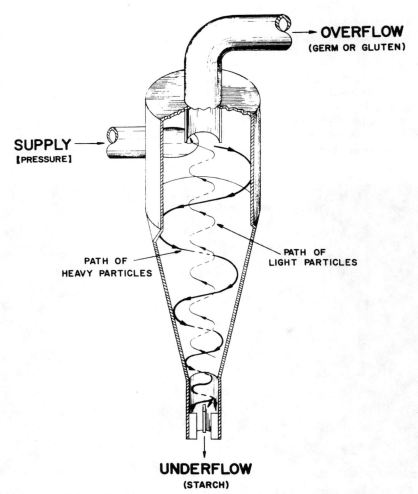

Fig. 3. Cutaway diagram of hydroclone separator. (Reprinted, with permission, from Watson, 1984)

sluice more corn. The required volume is 4.5–6.0 m³/t of corn grind (30–40 gal/bu). Dewatered corn is metered from the wet corn bin into coarse grinding mills having one stationary and one rotating disk (Fig. 2). The disks have knobs that break up the corn kernel. Clearance between the disks is adjusted so that a few whole kernels but few broken germs are in the mill discharge. Some dilution water is added to the mill feed, as well as to the mill discharge. The diluted slurry from the mills is pumped to flotation tanks or hydroclones (Fig. 3), where the oil-bearing germs are floated off the top. These are routed to a series of 50° wedge-wire screens (Fig. 4) that are used to wash the germs, with the addition of 1.2–1.3 m³ of wash water per tonne of corn (8–9 gal/bu) (Anderson, 1970; Watson, 1984). The recovered germs are then dewatered in screw presses, dried, and further processed to recover the corn oil.

The water-starch slurry from the germ wash is used to dilute the milled corn slurry being fed to the germ separation step. Other dilution water may also have to be added because the carrying medium must be carefully controlled within an optimum range of 15–17% solids for optimum germ recovery. The carrying medium is the corn slurry, with the fiber screened out for density measurement. Banks of hydroclones (Fig. 5) that are 150–200 mm (6–8 in.) in diameter are now preferred over flotation tanks because they are easier to control and more sanitary than the open-vat flotation system. This operation is controlled to obtain germ purity of 45–50% oil in washed, dried germs and optimum germ recovery by adjusting the overflow rate from the separation equipment. The lower the flow, the higher the oil content, but fewer germs are recovered and oil yield is reduced.

The underflow material from the flotation tanks or hydroclone separators is usually passed through a second disk mill with closer adjustment to release more germs with a second separation, which usually duplicates the first separation. The freed germs are routed back to the first germ separation for final recovery. The germ recovery flow system is diagrammed in Fig. 1.

FIBER SEPARATION

After germ separation with flotation or hydroclones, the corn slurry is screened, using 50 μm, 120° wedge-wire (Fig. 4) to separate fiber (mostly

Fig. 4. Simulation of unique slicing action of wedge-bar screen surface. (Courtesy Dorr-Oliver, Inc., Stamford, CT)

pericarps) from the starch and gluten. About 30–40% of all the starch is separated out and routed to mill starch.

The remaining stream includes fiber with some attached starch. Entoleter mills that sling the material against pins at high speed are used to free the starch with minimum fiber breakup (Fig. 6). Counterrotating disk mills are also used, in which the disks turn at 1,500–1,800 rpm in opposite directions (Fig. 7). Recently developed mills, 1.3 m (50 in.) in diameter, with only one disk rotating at 1,800 rpm and with 800–1,000 connected horsepower are being used for capacities as high as 1,269 t (50,000 bu) of corn milling per day. All the disk mills have knoblike protrusions or grooves that are capable of reducing the percent of starch bound to the fiber from 35% to 5–15% with minimum attrition of the fiber.

Fig. 5. Battery of hydroclones used for separation of germs from milled steeped corn. (Courtesy Dorr-Oliver, Inc., Stamford, CT)

The milled slurry is then washed and screened to separate the starch from the fiber. The most common washing equipment consists of 120° wedge-wire screens with five to seven separate stages operating in series (Figs. 1 and 4). Wash water, about 1.8–2.1 m^3/t of corn (12–14 gal/bu), is introduced ahead of the last stage, where it flows countercurrent to the fiber, finally emerging from the first stage with its accompanying starch and then being routed to mill starch. Wedge-wire screens (50 μm) have been found satisfactory for the first washing stage, and 70-μm screens can be used effectively on other stages. Screens must be thoroughly washed frequently to maintain good performance. Each stage has a separate pump and tank. Large compartmented tanks are also used. The design is such that water can cascade from one level to another as required to reduce the need for automatic controls.

Washed fiber from the last wash stage is only 10% solids, 15% at maximum. Further dewatering is necessary and is accomplished by mechanical means. Solid bowl centrifuges have been used to attain 40% solids. A disadvantage is poor starch recovery because all the free starch goes with the fiber to the feed. All other types of equipment have screens or bar spacings, allowing starch to escape with the water that is being forced out. This starch is recovered by recycling the flow upstream.

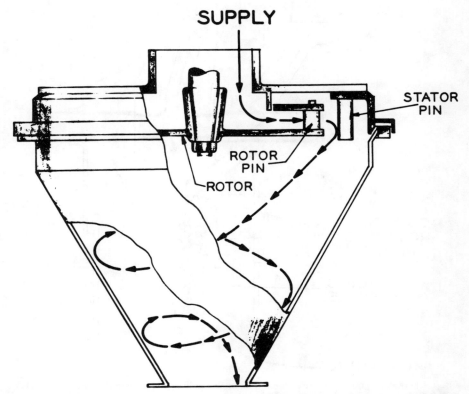

Fig. 6. Cutaway diagram of Entoleter mill used for fine milling of degerminated residue. (Reprinted, with permission, from Watson, 1984)

Dewatering is commonly done in two steps. The first is with screen centrifuges or screened reels to increase the solids to 20–30%. The advantage is greater capacity and somewhat drier cake from the second step, which utilizes screw presses (or recently introduced horizontal belt presses). Final dryness achieved is in the 42–48% range. The dewatered fiber is mixed with evaporated steepwater, and then normally dried and sold as corn gluten feed. Efforts are being made to expand sales for the wet feed.

The performance of the fiber system is frequently affected by the percentage of fine fiber present. Fine fiber is pulpy and washes (screens) poorly, resulting in wet fiber with a high starch content. The fine fiber is generally created by intensive milling, which is done to dislodge as much starch as possible. Frequently, milling is reduced somewhat to reach a compromise between starch recovery and dewatering performance. This problem is more commonly associated with plants that use disk rather than pin mills.

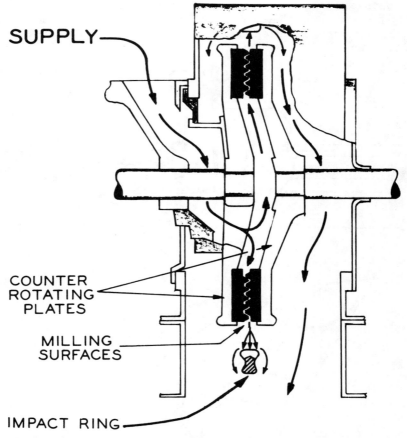

Fig. 7. Cutaway diagram of Bauer attrition mill for fine milling of degerminated residue. It is also used for degermination (first break) milling, with one rotor fixed, reduced rpm, no impact ring, and wider spacing of interlocking teeth. (Reprinted, with permission, from Watson, 1984)

PRIMARY STARCH SEPARATION

After the germs and fiber have been screened from the ground corn, only high-protein substances, gluten and corn soluble impurities, remain with the starch slurry. The gluten is separated from the starch by taking advantage of its lower density (1.06, in contrast to 1.6 sp. gr. for starch). Disk-nozzle types of centrifuges designed to separate the insoluble gluten particles by classification are used to obtain high-protein gluten (>68% total protein, solids basis).[1] This high protein level is necessary to meet finished product specifications. Some protein and impurities stay with the starch stream, which is further processed in the subsequent starch washing step.

The gluten from the primary centrifuge is dewatered with other disk-nozzle centrifuges to increase solids from 3 to 12%, then vacuum filtered to 42% solids, and finally dried to 88% solids. This corn gluten meal is sold for animal feed formulations (Chapter 15). The water separated from the dewatering operations is used as wash water in the wet-milling process.

Equipment used to dewater gluten, besides rotary, cloth-belt, vacuum filters, are plate-type pressure filters (used where labor costs are low) and solid bowl centrifuges. The dewatering ability of the centrifuge is increased by such means as heating the feed and raising its pH from the normal level of 4.5 to as high as 7.0 pH.

STARCH PURIFICATION

Starch from the primary centrifuges contains 3–5% protein. It is washed with water in countercurrent fashion, using hydroclones 10 mm in diameter that are grouped into clusters and enclosed in housings capable of holding as many as 720 cyclones. These units are then arranged into 10–14 separate stages operating in series (Fig. 8). The installations are compact and sanitary, have no moving

[1]Throughout this chapter, percent protein = percent nitrogen × 6.25.

Fig. 8. Hydroclone starch washing-separating units. Each clamshell contains as many as 480 individual hydroclone tubes 10 mm in diameter. (Courtesy Dorr-Oliver, Inc., Stamford, CT)

parts (except pumps), and are easy to automate. They replace the tables, centrifuges, and vacuum filters previously used for this operation.

About 2.1–2.5 kg of fresh water per kilogram of dry starch is used to wash out the soluble impurities in the starch; the hydroclone action separates out the remaining insoluble gluten. The washed starch can be expected to contain 0.3–0.35% total protein and 0.01% soluble protein. These results are attainable with normal dent corn and conventional steeping, milling, screening, centrifuging, and washing.

The water used for starch washing is generally deionized. It can be supplemented with a suitably pure condensate; condensates such as that from the steepwater evaporator, which can impart off-flavor to finished sweeteners, are not used. The water is heated for optimum soluble removal. The temperature is limited to 38–43°C (100–110°F). The lower temperature is now more prevalent because more stages of hydroclones are being used and more heat is generated from the additional pumps. Temperature sensors in the hydroclone systems keep temperatures well below the starch slurry pasting temperature of 63°C (145°F).

The wash water enters the washing system at the last stage, where the washed starch is exiting. The water and impurities leave the system at the first stage, along with about 25% of the starch being fed in. This stream, termed "middlings," contains solubles and gluten but is predominantly very small starch granules, some of which may have some protein attached. The middlings are generally concentrated in centrifuges, and the underflow is recycled back to the primary separation. The overflow water removed by the centrifuges is low in solubles and ideally suited for use as wash water for the primary separation.

C. Corn Varieties Processed Commercially

Normal dent corn is the predominant corn processed. Its starch has 27% amylose, a linear glucose polymer, and 73% amylopectin, a branch-chain glucose polymer.

The proportion of these fractions is constant in the starch recovered from any single variety of normal dent corn. Flint corn, a variety of normal corn grown in South America, is processed successfully even though it does not soften much, even with 50–60 hr of steeping. Its starch yield is slightly less than is regular, but its quality is good and the amylose-amylopectin ratio is the same as that of normal dent corn. No flint corn is processed in the United States.

Waxy corn, a genetic mutant, contains starch composed entirely of amylopectin. It can be processed in the same way as regular corn, with minor adjustments. The pasting temperature of the starch is lower, so the process must be cooler by about 3°C (5°F). Separation of the starch and gluten is easier, but the starch yield is only 90% of that of regular corn. About 1.5% of the corn processed by wet milling in the United States is the waxy type. The starch may be somewhat more difficult to filter because of the presence of a small amount of phytoglycogen (Freeman et al, 1975).

Another genetic mutant variety that is processed commercially is high-amylose corn, with starch that is 60–70% amylose. Like waxy corn, it commands a premium price, because the farmer must grow it in fields isolated from other varieties to prevent cross-pollination and then handle it separately after harvest to avoid contamination (Chapter 2). Unlike the lower temperature for waxy corn, the processing temperature should be 3°C higher than that for regular corn. More steep time is required, starch-gluten separation and starch filtration are more difficult, and the starch yield is only 80–90% of that of regular corn. Careful attention is necessary to prevent high-amylose starch from contaminating the regular variety because it affects syrup and starch quality (for instance, it is difficult to "paste"). Only about 0.2% of the corn wet milled in the United States is the high-amylose variety. Unique properties of the cornstarch from these genetic varieties are discussed in Chapter 16.

D. Feed Production Process

STEEPWATER EVAPORATION

Solubles extracted from the corn during steeping are routed to evaporators, where 0.6–0.7 m^3 of water per tonne of corn (4.0–5.0 gal/bu) is removed to increase the solids from 5–10 to 40–50%. The solids are then mixed with corn fiber and processed into corn gluten feed (see Chapter 15).

Falling-film recirculating evaporator systems are used to increase the light steepwater solids to 30%, using steam at a pressure of 1.2 kg/cm^2 (17 psia) or less (Fig. 9). Multiple effect design, in which 1 kg of steam can evaporate several

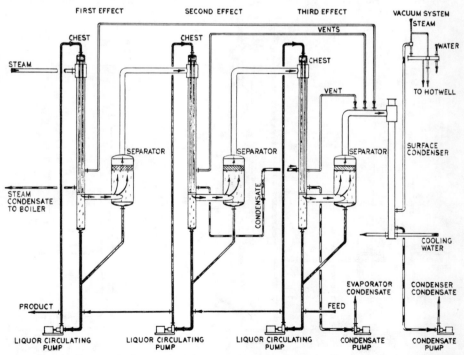

Fig. 9. Falling-film evaporator with continuous recirculation for each of its three effects. (Courtesy Dedert Corp., Olympia Fields, IL)

kilograms of water, is necessary to minimize energy costs. Energy can be further reduced with mechanical recompression, which completely recycles the vapors, compresses them, and discharges them into the evaporator steam chest (Fig. 10). Efficiencies of 167 kcal/kg (300 btu/lb) of evaporation for quadruple-effect steam evaporators can be reduced to 17–28 kcal/kg (30–50 btu/lb) at the expense of high capital investment. Compressors are radial-flow centrifugals with mechanical efficiencies in the 75% range. Electric motor drives are conventional, but steam-driven turbines are also used.

Forced-circulation multiple-effect steam evaporators are used to increase solids to 50%. These units differ from the falling-film design in that 7–11 m^3/hr (30–48 gal/min) of the steepwater is recirculated per tube instead of 0.7 m^3/hr (3 gal/min). (Tube sizes generally used are 38 mm [1.5 in.] in diameter and 3.7–9.1 m [12–30 ft] in length.) Evaporators must be boiled out with alkali to remove a proteinaceous coating that increases in thickness with time, primarily on the heating surfaces. Biweekly boilouts are common for falling-film, monthly for forced-circulation evaporators.

Condensate from the evaporators is very corrosive, about 3.0 pH, and contains condensed organic volatiles and alcohols that require extensive waste treatment before disposal.

FEED DRYING

Corn gluten feed is produced by mixing wet corn fiber with evaporated steepwater and drying to 8–11% moisture to give a finished product of 21%

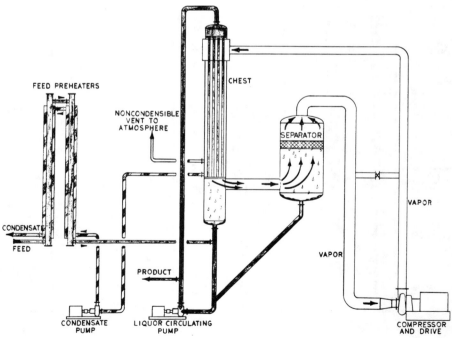

Fig. 10. Mechanical vapor recompression evaporator with forced recirculation. (Courtesy Dedert Corp., Olympia Fields, IL)

protein content. Corn cleanings and defatted corn germ meal are frequently added. Various types of dryers are used, such as direct-fired rotaries with inlet temperatures of 300–700°C (572–1,292°F) and outlet temperatures of 65–145°C (149–293°F). Rotary steam tube dryers are also effective, provided enough product recycling is used to keep the material from adhering to the tubular heating surfaces.

The dried feed is cooled to 43°C (110°F) to make it friable for milling. Pneumatic cooling-conveying systems are used in small facilities. Rotary coolers are used in the larger plants. Air, with either concurrent or countercurrent flow, is effective with 30–50% of the cooling, due to the evaporation of 1–2% moisture from the feed. Rotary water tube coolers are also utilized.

Standard swinging hammer mills with screens that have 3–8 mm (1/8–5/16 in.) openings are used to reduce the particle size of the feed. The ground feed can be sifted to assure that it meets the specifications of being finer than 12 U.S. mesh but no coarser than 10% through 100 U.S. mesh. Feed is sometimes cooled further to avoid caking and self-heating in storage and in transit, especially in hot, humid areas.

Most gluten feed manufactured in the United States is pelleted to increase its density and improve its handling characteristics for shipment to the European Economic Community. Requirements for making good pellets are proper feed moisture, small particle size, steam conditioning, pellet dies of 6–10 mm (1/4–3/8 in.) diameter, energy input (20 connected horsepower per ton of feed per hour), and cooling to 38°C (100°F).

GLUTEN DRYING

After the gluten has been dewatered as much as possible by mechanical means (to 55–60% moisture), it is dried in direct-fired, flash or rotary dryers to 12% moisture. Product recycling is necessary to overcome the sticky nature of the wet gluten. Drying temperatures should be limited to 400°C (752°F) to avoid a dark-colored product, burnt particles, and offensive odor and haze in the dryer exhaust. The rotary dryer has the advantage of making a product that is not considered dusty, whereas the flash dryer product is less dense and has the advantage of brighter color due to less dryer retention time. The natural golden pigmentation, xanthophyll, is important to poultry feeders. The pigment can be destroyed by excess dryer heat but may be partially protected by the introduction of heavy steepwater during the drying cycle. A minimum of 60% protein in the product is sometimes difficult to achieve in the wet-milling process. The causes are frequently improper steeping, excess fine fiber in the mill starch, and poor primary centrifuge operation. Dried gluten meal should be finer than 12 U.S. mesh but no more than 20% through 100 U.S. mesh.

Very small plants find it more economical to avoid investing in separate gluten drying facilities but to mix the gluten cake with wet fiber and evaporated steepwater to produce a combined feed with a 30% protein level.

E. Germ Processing

Germs separated from the corn slurry by the wet-milling process are dewatered to 50% solids in a screw press. They are then dried in rotary steam tube dryers to 3% solids and cooled in rotary water tube coolers or pneumatic cooler-conveyors to 38° C (100° F) to avoid heat degradation and self-ignition. Smaller plants frequently sell their dry germs instead of processing them on-site for oil recovery.

Pressure and heat are required to rupture the oil cells in the germ to obtain the oil. The germs should be softened by heating to 120° C (248° F) and steamed before being mechanically squeezed in conventional oil-expeller equipment. The oil content can be reduced from 45 to 6% for "full pressing," where no further processing is done to recover the remaining oil. In larger installations, the germs are squeezed only enough to break the oil cells, which reduces the oil content to 13–20% (Bredeson, 1983). They are then flaked with rolls and solvent-extracted with hexane to lower the oil content to 1.5%.

Extruders are showing promise as a lower-cost means for germ preparation before extraction. Another commercially operating process eliminates expellers by preconditioning, solvent extraction, flaking in the presence of the solvent, and then further solvent extraction.

The solvent extraction process involves soaking the oil out of the oil-bearing material. The oil-free material that is saturated with solvent must be desolventized, i.e., heat is used to drive off the solvent with direct and indirect steam. The oil-bearing solvent, called miscella, is evaporated, and the solvent is driven off the oil by the action of heat, direct steam, and vacuum. The process is relatively simple, but safety and control are critical for satisfactory operation with the flammable solvent (see Chapter 18).

The germ oil meal, containing 23–25% protein on a dry basis, is a corn gluten feed component. The crude corn oil is shipped to refiners to be degummed, stripped of free fatty acids by alkali refining, bleached with clay, deodorized, and chilled to remove waxes.

III. YIELDS, PRODUCTION, AND MARKETING OF PRODUCTS

Product yields from corn wet-milling operations (Table I) have remained constant for many years at 66% starch and 30% animal feeds. The bulk of the starch is further processed into sweeteners and ethyl alcohol (Table II). Twenty-eight wet-milling plants operate across the United States, with the bulk of them

TABLE I
Yield Distribution of Products Obtained by the Wet-Milling Process as a Percentage of the Raw Corn[a]

Product	As Sold[b]	Moisture-Free Basis
Starch	74	66.0
Oil[c]	3.9	3.9
Gluten feed, 21%[b] protein	27	24.0[d]
Gluten meal, 60%[b] protein	6.4	5.7
Unaccounted-for loss	...	0.4

[a] Adapted from Long (1982).
[b] Calculated at 11% moisture content.
[c] Unrefined.
[d] Includes fiber (pericarp), steepwater (solubles, ~7% db), and germ residue (4% db).

in the Corn Belt. There are probably more wet-milling plants outside the United States, but their total output about equals the capacity of all the wet-milling plants in the United States because many of them are quite small. Many are owned and operated by U.S. wet-milling companies. One company, CPC International, Inc., operates 30 corn wet-milling plants outside the United States. Japan has become a major importer of corn for processing into starch and sweeteners (Jones, 1983).

Production of conventional starch, glucose, and dextrose products increased a modest 39% between 1972 and 1985 in the United States (Table III), but the industry capacity increased by 335% due to alcohol and to the explosive growth of high-fructose corn syrup (HFCS), first at 42% fructose and then at 55% (Chapter 17). The latter allowed near one-to-one substitution for sucrose. The HFCS market in the United States increased by as much as 50% per year in the early 1970s, dropping into the 13–22% range in the early 1980s. This growth rate is now decreasing as HFCS nears 100% replacement of sucrose in soft drinks. The remarkable growth since 1972 can be attributed to the following: 1) technical breakthroughs in the HFCS process, 2) a sugar shortage in the early 1970s, 3) high U.S. sugar price supports in recent years, 4) the availability of abundant low-cost corn, and 5) the willingness of the wet-milling industry to capitalize on the above opportunities.

It is anticipated that HFCS market growth in the future will parallel the modest growth pattern experienced by corn starch and sweeteners over the years due to a growing market and new applications developed by wet-milling industry research. Growth outside the United States can be expected to vary,

TABLE II
Shipment of Products (thousand pounds) of the Corn Refining Industry in the United States[a,b]

	1983	1984	1985
Starch products			
(Includes corn starch, modified starch, and dextrin)	4,018,905	4,182,866	4,225,171
Refinery products			
(Includes glucose syrup, high-fructose corn syrup, dextrose, corn syrup solids, and maltodextrins)	16,005,529	17,921,126	20,341,535
High-fructose corn syrup	9,707,041	11,502,324	13,920,406
Other products			
Corn oil crude	72,612	116,142	164,382
Corn oil refined	399,919	407,456	382,234
Corn gluten feed	7,391,069	8,739,730	8,811,476
Corn gluten meal			
41% protein	19,115	20,272	18,503
60% protein	1,383,129	1,635,228	1,609,112
Corn oil meal	28,728	29,465	48,585
Steepwater	211,937	300,770	282,333
Hydrol	208,807	216,558	228,742
Ethanol (thousand gallons, 100%)[c]	325,000	375,000	425,000

[a] Source: Anonymous (1986b); used by permission.
[b] To convert to tonnes, multiply by 4.535×10^{-4}.
[c] Ethanol values calculated from values for wet-milled alcohol in Table II, multipled by ethanol yield of 2.5 gal (100%) per bushel of corn (Gill and Allen, 1985).

depending upon local economic conditions.

Industry products are mostly shipped in bulk. Liquids are handled in railroad tank cars and tank trucks. Dry products are shipped in special hopper railroad cars and trucks, although much is still handled in 100-lb paper bags. In recent years, most of the corn gluten feed manufactured in the United States has been shipped to the European Economic Community. The increasingly use of larger shipping containers for products necessitates ever-increasing sanitary and product quality standards to avoid excessive economic losses associated with off-grade shipments.

IV. FINISHING OF WET-MILLING PRODUCTS

A. Starch

A relatively pure starch slurry from the wet-milling operation contains 40% solids. It may be dried directly or treated with an array of chemicals such as bleaches and acids to modify the starch properties to meet customers' requirements. The chemical residuals are washed from the starch in nozzle-type centrifuges or vacuum filters (with sprays), using 1–2 m^3 of water per tonne (0.12–0.24 gal/lb) of dry starch. Cake from the vacuum filters at 40–50% solids is fed into belt-type dryers with 65–150° C (150–302° F) steam-heated air rising up through the dryer belt perforations. The product is agglomerated somewhat but is free flowing and easy to handle. Flash dryers are also used, generally fed from large basket-type centrifuges that dewater to 55–67% solids. The dryer heating

TABLE III
Annual Use of Corn (million bushels) in the Production of Starch and Starch-Derived Products from Wet Milling[a,b]

Year Beginning October 1	HFCS[c]	Glucose and Dextrose	Starch	Wet-Milled Alcohol	Total Wet-Milled[d]
1971/1972	10	125	100	10	245
1972/1973	15	145	110	10	280
1973/1974	20	155	110	10	295
1974/1975	30	160	115	10	315
1975/1976	45	165	115	10	335
1976/1977	65	165	120	10	360
1977/1978	80	170	130	15	395
1978/1979	105	170	135	15	425
1979/1980	140	175	130	30	475
1980/1981	165	185	125	40	515
1981/1982	190	185	135	85	595
1982/1983	215	185	135	130	665
1983/1984	255	190	145	150	740
1984/1985	310	190	150	170	795
1985/1986[e]	320	190	150	170	830

[a] Reprinted from Livesay (1985).
[b] To convert to tonnes, multiply by 2.54 × 10^{-3}.
[c] High-fructose corn syrup.
[d] Use yield data in Table I to calculate actual volumes of starch and by-products.
[e] Preliminary (Anonymous, 1986a).

medium is steam from coils or direct-fired gas furnaces furnishing 220°C (428°F) air. The products are generally ground in air-swept mills, blended, and shipped in bags, hopper trucks, and railroad hopper cars. They are used in paper, textiles, adhesives, and food. The dry starch is sometimes processed further by dry roasting, cooling, blending, and packaging for use as adhesives. Starch is cooked on hot, steam-heated rolls, cooled, and ground for use in food and adhesives. A less severe treatment using extruders results in instant laundry products (see Chapter 16).

B. Ethanol

Ethyl alcohol (ethanol) produced by the fermentation of starch from the wet-milling process has a unique economic advantage because the yeast can be recycled (there are no other solids to interfere with the separation after fermentation). Another advantage is that the light steepwater from wet milling can be used as dilution water for the fermenters to save overall evaporation (energy) costs. Although some alcohol is used in the beverage industry, about 85% of the ethanol produced from corn is blended with gasoline. Besides serving as a fuel extender, it also is an octane enhancer for unleaded fuel. In the past, wet millers have produced 65–75% of the grain ethanol, but in the 1984/85 crop year, the amount dropped to 60%. This downward trend is expected to continue in future years because most new plants use the dry (whole-corn) milling process (Anonymous, 1985). The ethanol fermentation process is covered in more detail in Chapter 19.

C. Feed By-Products and Miscellaneous

Corn steepwater (condensed fermented corn extractives) is the soluble portion of the corn kernel that is evaporated to as high as 50% solids and sold in liquid form for cattle and dairy feeding. At this concentration and its normal pH of 4.2–4.5, it is biologically stable. It is also used as a nutrient supplement in antibiotic and other fermentations. Most of the steepwater, however, is utilized in the manufacture of corn gluten feed.

Corn germ meal, the fraction remaining after the oil is removed from the germ, is frequently used as carrier of nutrient supplements such as vitamins, minerals, and medicants in animal feeds. The bulk of the meal, however, goes to corn gluten feed, where its ability to absorb oils and water is utilized.

Hydrol, or corn sugar molasses, is a by-product in the production of crystalline dextrose (D-glucose). It is used as an animal feed supplement and as a raw material for the manufacture of food ingredients such as caramel color.

The production and finishing of corn sweeteners is discussed in Chapter 17 and the processing of crude corn oil into food grade oil in Chapter 18.

V. POLLUTION CONTROL

A. Waste

The larger wet-milling and sweetener plants can generate as much waste as a medium-large city. Economics dictate the need for waste treatment facilities on the plant site to reduce the high strength to the concentration of municipal wastes. These pretreated wastes plus low-strength streams such as boiler blowdown and sanitary flows can then be routed to municipal facilities. Municipal systems commonly levy surcharges based on excess waste volume and strength.

The combined high-strength wastes are in the range of 1,000–2,000 mg/L of BOD_5 and are reduced to 200–300 mg/L of BOD_5 after pretreatment.[2] The raw waste quantity is approximately 6.3 kg of BOD_5 per tonne (0.35 lb of BOD_5 per bushel) of corn grind, which is reduced to 0.9 kg of BOD_5 per tonne (0.05 lb of BOD_5 per bushel) after pretreatment. This volume amounts to about 3.7 m^3/t (25 gal/bu), of which a third is derived from the wet-milling process and the remainder from product-finishing operations. The on-site treatment facility generally consists of equalization tanks having volumes to handle the residue of 8–24 hr of production. This capacity reduces shock loads due to variations in strength, temperature, flow rate, pH, and nutrient availability. Each of the above factors has a range that must be met by adjusting the pH, cooling or heating, and adding nutrients so that the microorganisms in the aerated, mixed, activated sludge system that follows can remain viable. Effluent mixed liquor from an activated sludge tank is clarified to remove the sludge. This can raise the efficiency of the system from 70 to 80–90%. The dilute clarifier sludge (about 1% solids) may be concentrated to 5–6% solids by the use of solid-bowl or basket-type centrifuges and then can be mixed with light steepwater and further processed as part of the animal feed ingredients.

B. Air Quality

The wet-milling process uses about 1.8 kg of SO_2 per tonne of corn (0.1 lb/bu). It is dissolved in process waters but its pungent odor is present in the slurries, necessitating the enclosing and venting of the process equipment. Vents can be wet-scrubbed with an alkaline solution to recover the SO_2 before the exhaust gas is discharged to the atmosphere. Process building interiors should be well ventilated to dissipate not only fugitive fumes, but also the heat generated by motors and process equipment (the process operates in the 50° C [122° F] range).

The most critical environmental problem associated with air pollution is the exhaust from the drying processes. Cyclones are standard for particulate containment but do not meet stringent air quality standards in some areas, necessitating the use of secondary collectors. Bag collectors are used, but they also become dirty and are fire problems. Wet scrubbers create a liquid waste disposal problem and a wet plume. Incinerators have high energy costs and high capital expenditures.

The greatest problem with dryer exhausts is odor and blue haze (opacity). Germ dryers emit a toasted smell that is not considered objectionable in most areas. Gluten dryer exhausts are satisfactory as long as the drying temperature does not exceed 400° C (752° F). Higher temperatures promote hot smoldering areas in the drying equipment, creating a burnt odor and a blue-brown haze. The drying of feeds where steepwater is present results in environmentally unacceptable odor if the drying temperature exceeds 400° C. Of concern is the

[2]BOD_5 is the strength of waste as analyzed by the five-day biochemical oxygen demand procedure.

formation of a blue haze at elevated temperatures. These exhausts contain volatile organic compounds with acrid odors such as acetic acid and acetaldehyde. Rancid odors can come from butyric and valeric acids, and fruity smells emanate from many of the aldehydes present.

The objectionable odors have been reduced to acceptable levels commercially with ionizing wet-collectors, in which particles are loaded electrostatically with up to 30,000 V. An alkaline wash is necessary before and after the ionizing sections. However, the most conventional approach is incineration at approximately 750° C (1,382° F) for 0.5 sec followed by some form of heat recovery. This hot exhaust can be utilized as the heat source for other dryers or for generating steam in a boiler specifically designed for this type of operation. The incineration can be accomplished in conventional boilers by routing the dryer exhaust gases to the primary air intake. The limitations are potential fouling of the boiler air intake system with particulates, etc., and derating the boiler capacity due to low oxygen content. Another type of system being utilized is recooperative incineration, in which dampers divert the gases across ceramic fill so that exhaust heat is used to preheat the fumes to be incinerated. The size of the incinerator can be reduced 20–40% by recycling some of the dryer exhaust back into the dryer furnace. Recycling of 60–80% of the dryer exhaust is being done by chilling it to condense the water before recycling. The volume of dryer exhaust can also be minimized by reducing the evaporative load. The steepwater can be evaporated to 50% solids. The corn fiber can be predried in a separate dryer, where the exhaust is not a problem as long as temperatures are limited to 450–500° C (842–932° F).

VI. TRENDS

A. Automation

The wet-milling process and associated equipment have matured sufficiently to permit consistent, reliable separation and product quality on a 24-hr basis with minimum operating labor. The most notable achievements are attained in the new large plants where television-like displays and control systems distributed by cathode ray tubes are utilized to start process equipment sequentially on demand and then to monitor and control the system (Martin, 1979) (Fig. 11). Further, the technician can be alerted, by rate-of-change or trend-in-measurement functions, to variables that are about to get out of control. This allows time to make the necessary adjustments to avoid spills and off-quality. These computer systems are the latest development in the search for reduced costs and better product quality. Even better computer applications are probable as improved on-line measuring devices for protein, starch, fiber, and soluble content are perfected.

B. Utilities

Fresh water usage for a typical large sweetener plant is about 4.5–6.0 m^3/t of corn processed (30–40 gal/bu), of which one fourth is consumed in wet milling. Reduction in the requirements for the wet-milling portion is limited because the usage about equals the need for properly steeping the corn. Greater water consumption for the finishing processes is likely because the trend is toward more sophisticated products that require more water.

Wet-milling operations are high energy users at 1.48 million kilocalories per tonne of corn (150,000 btu/bu), even for the efficient large sweetener plants. About 20% of the total is for electricity, which indicates that most of the energy goes into fuel to make steam. As products continue to become more sophisticated, the use of energy will increase unless new technology can reverse the trend.

The use of mechanical vapor recompression evaporators in new facilities and in the replacement of old equipment is reducing energy usage because it is five to six times as efficient as quadruple-effect steam evaporation. More imaginative use of heat exchangers can recover heat now being lost to the atmosphere via cooling towers. Steepwater evaporators now in service can utilize feed dryer exhausts as their source of energy.

Superheated steam is showing promise for use in drying feed materials (Covington, 1983). Most (60–80%) of the energy for the primary heating of steam is recovered in low-pressure steam for use elsewhere in the process. Not only is a 20–40% energy saving possible, but there is no discharge to the atmosphere, eliminating odor and particulate problems associated with feed drying. Reverse osmosis is improving due to breakthroughs in membrane technology. When operating with a pressure differential of 35–70 kg/cm^2

Fig. 11. Television-like displays used to sequentially start and stop process equipment. The processes are also monitored and operated using computer-controlled automation. (Courtesy Foxboro Co., Foxboro, MA)

(500–1,000 psi), they may replace steepwater and sweetener evaporators, achieving a major reduction in energy requirements (Cicuttini et al, 1983).

New steeping technology may be in the offing that improves the possibility of success in concentrating the solids content of steepwater; it works by reducing the low-molecular-weight components such as alcohol that reverse osmosis, up to now, has had difficulty separating. Another benefit is reduced steeping time. A technology being introduced to the industry to treat wastes relies on anaerobic digestion, which has the advantage of using less energy-intensive processes and giving energy credits via recovery of methane gas, which is utilized as a fuel source. The cost of energy is being reduced by installing cogeneration, which was popular in the industry 50–60 years ago. New coal-fired boilers operating at 40–80 atm (600–1,200 psi) drive electric generators. Steam for processing is then extracted in the 10 kg/cm^2 (150 psi) range, and steam for low-temperature heating (of water, for evaporation, etc.) is extracted at exactly 2 kg/cm^2 (30 psi) absolute.

LITERATURE CITED

ANDERSON, R. A. 1970. Corn wet milling industry. Pages 151-170 in: Corn: Culture, Processing, Products. E. E. Inglett, ed. Avi Publ. Co., Westport, CT.

ANONYMOUS. 1986a. Food and Industrial Demand. Feed Outlook and Situation Rep. U.S. Dept. Agric., Econ. Res. Serv., Washington, DC. FdS-299. March.

ANONYMOUS. 1986b. Industry Statistics 1986. Corn Refiners Assoc., Washington, DC. 4 pp.

BREDESON, D. K. 1983. Mechanical oil extraction. J. Am. Oil Chem. Soc. 60:163A-165A.

CICUTTINI, A., KOLLACKS, W. A., and RAKERS, C. J. N. 1983. Reverse osmosis saves energy and water in corn wet milling. Staerke 35:149-154.

COVINGTON, R. O. 1983. Steam dryer aimed at by-products. Food Eng. 55(2):102-103.

FREEMAN, J. E., ABDULLAH, M., and BOCAN, B. J. 1975. An improved process for wet milling. U.S. patent 3,928,631.

GILL, M., and ALLEN, E. 1985. Status of the U.S. ethanol market. Pages 14-22 in: Feed Outlook and Situation Rep. U.S. Dept. Agric., Econ. Res. Serv., Washington, DC. FdS-297. Aug.

JONES, S. F. 1983. World Market for Starch and Starch Products with Particular Attention to Cassave (Tapioca) Starch. Trop. Dev. Res. Inst., London.

LIVESAY, J., 1985. Estimates of corn use for major food and industrial products. Pages 8-10 in: Situation Rep. U.S. Dept. Agric., Econ. Res. Serv., Washington, DC. FdS-296 March.

LONG, J. E. 1982. Food-sweeteners from the maize wet milling industry. Pages 282-299 in: Processing, Utilization and Marketing of Maize. M. R. Swaminathan, E. W. Sprague, and J. Singh, eds. Indian Council of Agric. Technol., New Delhi.

MARTIN, R. G. 1979. Implementation of a highly automated corn processing plant. Annual Conference. Instrument Soc. Am., Research Triangle Park, NC.

RANDALL, J. R., LANGHURST, A. K., and SCHOPMEYER, H. H. 1978. Continuous steeping of corn for wet processing to starches, syrups and feeds. U.S. patent 4,106,487.

WATSON, S. A. 1984. Corn and sorghum starches: Production. Pages 418-468 in: Starch: Chemistry and Technology, 2nd ed. R. L. Whistler, J. N. Bemiller, and E. F. Paschall, eds. Academic Press, Orlando, FL.

CHAPTER 13

FOOD USES OF WHOLE CORN AND DRY-MILLED FRACTIONS

LLOYD W. ROONEY
Cereal Quality Laboratory
Soil and Crop Sciences Department
Texas A&M University
College Station, Texas

SERGIO O. SERNA-SALDIVAR
Centro de Investigaciones en Alimentos
University of Sonora
Hermosillo, Sonora, Mexico

I. INTRODUCTION

Maize or corn (*Zea mays* L.) is the second largest cereal grain crop produced in the world and the leading cereal crop in the United States (Anonymous, 1984). This cereal grain is a staple food for large groups of people in Latin America, Asia, and Africa. In the United States, it is generally used as animal feed and for production of cornmeal, flour, grits, starches, tortillas, snacks, and breakfast cereals. The starch is converted into syrups, sweeteners, and industrial products. The increase in popularity of Mexican foods throughout the United States has increased the utilization of corn in snack foods and tortillas. Recent references of general interest on food use of corn are available (Inglett, 1970, 1982; Gutcho, 1973; Daniels, 1974; Duffy, 1981; Pomeranz and Munck, 1981; Matz, 1984). This chapter summarizes the use of corn in foods, with special emphasis on alkaline-cooked corn products, i.e., tortillas and snacks. The terms "maize" and "corn" are used interchangeably.

II. USES OF CORN IN THE UNITED STATES

World corn production in 1984/85 was estimated at 458.3 million tonnes (Chapter 7, Table XV). Forty-eight percent of the world production was harvested in the United States. Corn has been an integral part of U.S. culture and diet since the first settlements at Jamestown in 1607 (Hardeman, 1981). More than one billion bushels (25.4×10^6 t) are used annually in the United

States for wet and dry milling, alcohol, tortillas, and snack foods. Corn food products include: sweeteners, syrups, cooking oil, starch, flours, meals, grits, puddings, and numerous convenience foods (Fig. 1). The U.S. snack food market was estimated at 20.8 billion dollars in 1984. Corn chips, extruded snacks, and popcorn had 6.4, 1.8, and 0.8% of the total snack food market, respectively, whereas potato chips and candy had 12.7 and 30.7%, respectively (Anonymous, 1985). Several food companies have built facilities to produce dry corn masa flours to meet the demands of expanding Mexican food markets.

A. Sweet Corn

A recessive gene (*sugary*) causes an alteration in the endosperm of corn that results in higher levels of soluble sugars and reduced levels of starch in the kernel. Sweet corn hybrids have been developed specifically to produce corn with desirable color, sweetness, and tenderness (Chapter 14). Its ability to retain sweetness longer than regular dent corn when picked at the kernel milk stage has made sweet corn a highly desirable food. It is a popular vegetable in the United States, where it is consumed fresh, canned, dehydrated, or frozen. Whole-kernel and cream style are major canned corn products. Corn-on-the-cob and whole kernels are popular frozen products.

B. Degerminated Corn Products

FLAKING GRITS

Flaking grits are large (U.S. mesh sieves 3.5–6) endosperm chunks that are used to produce corn flakes. Generally, yellow corn with uniform kernel size and

Fig. 1. Some corn foods.

unfissured hard endosperm is preferred. The grits are free of germ and pericarp. Each grit becomes an individual flake, so the largest possible grit is required. Selection of desirable lots of corn that will produce high yields of large grits has become difficult because most corn is artificially dried, which causes fissuring and reduces dry-milling quality (Chapter 5).

CORN GRITS

Corn grits are particles of endosperm that pass through a 1.19-mm sieve (U.S. No. 14) and ride on a 0.59-mm (U.S. No. 28) sieve. They are free of pericarp and have less than 1% oil content. Dry milling of corn is described in Chapter 11. Corn grits are consumed as a side dish for breakfast in the southern United States, where they are cooked in boiling water for 10–25 min and served with butter or margarine. The bland flavor and light color of white corn grits are generally preferred. "Instant" corn grits, which require only warm water or boiling for 5 min, are popular. Preprocessed grits are made by fissuring the grits to increase water absorption and by steaming, cooking, and drying. Cooked grits should retain individual integrity with a pleasing balance of soft grittiness and smoothness. If too much fine material is present or the grits are overcooked, they have a lumpy, sticky texture.

CORNMEAL

Meal has a greater particle size range (1.19–0.193 mm; U.S. sieves No. 14 to 75) than that of grits. It is a popular dry corn product because of its long shelf life, freedom from black specks, and bright color. Cornmeal is often enriched with thiamine, riboflavin, niacin, and iron and is used to produce an assortment of chemically leavened baked and fried products like corn bread, muffins, fritters, hush puppies, and spoon bread (Chapter 11). Corn proteins do not form gluten upon hydration and mixing, so corn bread formulas sometimes include wheat flour to produce a lighter, more aerated product. However, too much wheat flour reduces the "corn" taste. A strong corn flavor is desirable, especially in the southern United States.

Self-rising mixes containing cornmeal, sodium bicarbonate, monocalcium phosphate, salt (optional), and seasonings (optional) have been widely accepted. Many different breading mixes containing cornmeal are available. Corn bread is used to produce stuffing and other foods.

FLOUR

Corn flour consists of fine endosperm particles, of less than 0.193 mm (U.S. sieve No. 75) obtained during dry milling. It is used as an ingredient in many dry mixes such as pancakes, muffins, doughnuts, breadings, and batters. It is used extensively in ready-to-eat breakfast cereals and for snacks. Corn flour in small amounts adds important functional characteristics to many baked products and is used as a binder in some processed meats. The modification and use of corn flour and meals to produce breadings for frozen products that retain crisp texture after microwaving is currently being investigated. Corn flour is a major ingredient in blends of corn and soy milk that are used in U.S. food aid programs. Corn flour and meal are extruded to produce many different snack foods.

C. Fermentation

Alcohol, beer, and distilled beverages are produced from corn or corn products. Lager beers brewed in the United States often use corn grits or corn syrups as a source of fermentable sugars (Chapter 11). Various distilled beverages, e.g., bourbon and gin, are made from corn in the United States (Chapter 19).

III. CORN-BASED READY-TO-EAT BREAKFAST CEREALS

Between 1970 and 1980, annual consumption in the United States of ready-to-eat cereals increased from 2.72 to 3.63 kg (6 to 8.5 lb) per capita. Estimates of 1982 sales indicated that sales of all breakfast cereals would account for 4.7 billion dollars (Hayden, 1980). Despite a 14% decline in the number of children under 14, per capita consumption of ready-to-eat cereals increased more than 40% during the past decade. This occurred due to the increased number of working mothers, the convenience of breakfast foods, and the increased interest in high fiber, low cholesterol, and enriched/fortified foods. Corn products are major ingredients in breakfast cereals that are prepared by extrusion, flaking, shredding, and puffing.

Ready-to-eat breakfast foods are made by cooking the cereals to gelatinize the starch and then shaping and forming the dough or cooked particles into flakes, shreds, granules, or collets. Desirable flavor, aroma, and texture are obtained by controlled toasting. The texture becomes crisp due to dehydration, and caramelization and Maillard browning reactions develop desirable flavor and color. Conventional processes are still used to produce many of the cereals that continue to be popular, such as corn flakes.

A. Corn Flakes

Corn flakes are the most popular ready-to-eat breakfast cereals in the United States and probably in the world. The fundamental processes for corn flake production have remained relatively unchanged over the past several decades. A mixture of corn grits, syrup, sugar, malt, salt, and water is pressure-cooked at 0.044–0.067 t/cm^2 (15–23 psi) for 1–2 hr (Matz, 1970) (Fig. 2). Nondiastatic malt is added for flavoring. The grits reach 33% moisture and appear translucent when properly cooked. The starch is gelatinized during pressure-cooking, but starch granule swelling is restricted due to limited water. Thus, each grit retains its individual integrity. The proper cooking time varies with each lot of corn and is influenced by the variety and the conditions under which it was grown, harvested, dried, and stored. The cooked grits are transported by a moving belt to a countercurrent dryer set at 66° C (150° F). When the moisture content of the corn drops to 19–20%, the grits are equilibrated or tempered for 6–24 hr in a bin. Tempering permits more equal distribution of the moisture and produces uniform flakes that do not break during handling. Then, the hard, dark brown grits are flaked through a pair of counterrotating rollers. The rollers apply a pressure of 234 t/cm^2 (40 tons/in^2) and are cooled by continuous circulation of water. Flakes are toasted in a gas-fired oven for 50 sec at 302° C (575° F) or 2–3 min at 288° C (550° F). Toasting dehydrates the flakes and develops a crisp texture, brown color, and desirable flavor. The flakes are cooled, sprayed with nutrients, and equilibrated. They are packaged immediately after processing to prevent moisture absorption so that they retain crispness. Sugar-coated corn flakes are made by the same process; sweetener is sprayed on the flakes after toasting.

Modern extrusion processes can be used to make corn flakes. Vollink (1962) described an alternative process to produce corn flakes from fine particles of corn endosperm. The formula consists of the following ingredients: 90% corn flour, 8% sugar, 1% salt, and 1% malt flavoring. Water is added to form a mixture containing approximately 25% moisture, which is fed into an extruder (Chapter 11) and cooked at 177° C (350° F) for about 3 min. Sixty percent of the heat produced during the process is due to mechanical heat and 40% to steam injection. The cooked mass is cooled and transferred into a second forming

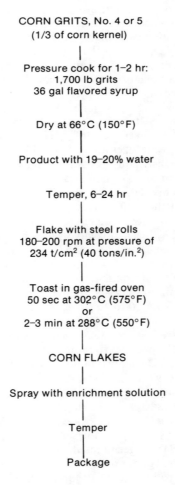

Fig. 2. A general scheme for corn flake production.

extruder, in which the screw is water-jacketed to maintain a temperature of 82° C (180° F). The material is extruded through a die to form pellets 82 cm (1/8 in.) long. The pellets are cooled to below 38° C (100° F) to eliminate stickiness and are rolled into 0.025-in. thick flakes in the conventional flaking rolls. They are then toasted at 204–260° C (400–500° F) with radiant heat or forced hot air. Larger, more uniform flakes can be obtained consistently by this and other modern processes, but the flavor and texture of extruded corn flakes differ from those of traditional corn flakes.

B. Other Corn-Based Breakfast Cereals

Extruders are often used in breakfast and snack food production (Harper, 1981). Several types of extruders exist, including: 1) forming extruders, which size and shape a precooked dough; 2) extruders that mix and cook cereal-based ingredients at ordinary pressure and then form and size the dough; and 3) short-time, high-temperature, high-pressure continuous cooking and puffing extruders. This puffing process has been described by Alexander (Chapter 11).

Corn products are major ingredients in many puffed composite breakfast foods that have been in existence for some time. Conventional processes involve cooking the dry ingredients with water and other additives to gelatinize the starch and make a dough, which is made into collets of specific size and shape by cold forming extrusion. Alternatively, the dry ingredients are cooked in an extruder at temperatures lower than 100° C (212° F) to form a soft dough that is extruded in the appropriate sizes and shapes. The collets are dried to maintain individuality and are then equilibrated. They require further processing by puffing, toasting, or a combination of methods, but they can be dried and stored for final processing as needed.

One way to expand the collets is by gun puffing (Matz, 1970). A gun is a pressure vessel approximately 15.2 cm (6 in.) in diameter and 152 cm (30 in.) long, with a steam inlet, bleed-off valves, and heaters. It is charged with the equilibrated collets, sealed, and heated. Temperatures may reach 260–427° C (500–800° F), with pressures of 7.03–14.06 t/m^2 (100–200 psi). Pressure is suddenly released by an automatic device, and the material explodes from the "gun." Five to seven minutes may be required to complete the puffing process. The expansion varies from 10 to 16 times. Continuous steam pressure puffing processes are used also.

Products that require less expansion are generally oven-puffed. Puffing is achieved by exposing the product to radiant heat on a belt or by tumbling it in a heated rotating cylinder. Oven puffing produces a three- to fourfold expansion. According to Daniels (1974), gun puffing is more expensive than oven puffing because the equipment must be built to resist high pressures and the maintenance costs are high.

Breakfast cereals, gruels, and snacks can be produced by short-time, high-temperature, high-pressure extrusion cooking and puffing (Harper, 1981; Smith, 1982; Matz, 1984). Degerminated corn flour, meal, grits, and starches are used to prepare a wide range of products. Extrusion puffing offers the advantage of producing many products with different characteristics. The selection of ingredients, granulation, additives, working conditions, and characteristics of the extruder (such as temperature, dwell time, length, pH, moisture content, rpm, type of screw, and die characteristics) determines the type of product produced. The interactions of these variables dictate the degree of starch gelatinization, protein denaturation, and expansion of the product, i.e., its texture and bulk density. The dough is in a plastic condition and is therefore capable of being shaped and puffed when it reaches the die. The degree of puffing depends on the pressure difference between the dough inside the extruder and the atmosphere outside it. The shapes and sizes of products are varied by changing the die and the cutting blade spread. The puffed extrudate must be toasted, dried, and browned to produce desirable texture and flavor.

IV. Maize Processing and Food Use Around the World

Maize is widely used directly for human food in Asia, Africa, Latin America, and parts of the Soviet Union. White maize is generally preferred for food use, although yellow maize is preferred for some products, e.g., polenta, a food popular in Argentina and Italy. The major classes of traditional foods and some of their "dialect" names are presented in Table I. These traditional foods are prepared with maize, sorghum, or millet (Rooney et al, 1986), depending upon availability. In general, maize flavor is liked by consumers. Maize porridges and alcoholic and nonalcoholic beverages are very important foods consumed extensively in Africa. Steinkraus (1983) discussed many traditional processes using fermentation techniques. Some traditional food processes (e.g., for tortillas and *arepa*) have been adapted to modern industrial technology and are discussed further below. The quality of maize for producing the various kinds of products is not well documented. Generally, hard white maize appears preferable, but, in practice, maizes that are white, yellow, and other colors are used, depending upon the country or region. The poor acceptance of high-lysine

TABLE I
Summary of the Major Types of Traditional Foods from Corn

Type	Examples
Bread	
Flat, unleavened, unfermented	Tortilla, *arepa*
Fermented and/or leavened	Pancakes, corn bread, hoe cake, blintzes
Porridges, thick/thin, fermented/unfermented	*Atole, ogi,* kenkey, *ugali, ugi, edo pap, maizena, posho, asidah*
Steamed products	Tamales, couscous, Chinese breads, ricelike products, dumplings, *chengu*
Beverages	
Alcoholic	*Koda,* chicha, Kaffir beer, maize beer
Nonalcoholic	*Mahewu,* magou, chica dulce
Snacks	*Empanada*, chips, tostadas, popped corn, corn fritters

maize (opaque-2), which has a soft, floury endosperm and improved nutritional value, demonstrated that hard, flinty maize was preferred in many areas of the world. Yet, in a few cases, soft maize was acceptable. The soft maizes have poor storage properties, break easily, and cannot be efficiently dry milled into grits, flour, or meal that contain low oil and pericarp levels. Considerable efforts to develop maize varieties with hard endosperm and higher lysine content have led to the development of quality protein maize lines (Sproule, 1985; Ortega et al, 1986) with hard endosperm texture and significantly improved nutritional and agronomic properties.

A. Traditional Milling and Grinding

The use of wooden mortars and pestles, stone metates, and *manos* (stones) to grind maize is still a common practice in many areas of the world. The pericarp and germ may be partially removed by decortication with the mortar and pestle. After the bran is removed, the endosperm and germ are reduced into flour, grits, or meal by additional pounding or stone grinding. The broken particles are hand-sieved to produce appropriate sizes for preparation of various products. The products prepared often contain the germ and become rancid rapidly, so milling is done frequently.

B. Alcoholic Beverages

Traditional beer is produced in many African and South American countries. These beers usually have a high solids content and are consumed during fermentation. They are usually produced by women for their families in relatively small quantities. However, large, modern breweries produce sorghum beer in southern Africa (Novellie, 1981), using sorghum malt and maize or sorghum grits as adjuncts. This sorghum beer (Bantu or Kaffir beer) is a thick, opaque, sour gruel containing 4–6% solids and 1–3% alcohol. During the process, the maize and sorghum extracts are soured (with *Acetobacter, Lactobacillus*, etc.) before inoculation with yeast. The beer is fermented for one to two days and consumed while actively fermenting. Pearl millet and sorghum are preferred for brewing, but maize is used frequently in Africa. Chicha and *soru* are traditional alcoholic beverages made from maize in South America (Steinkraus, 1983). Maize grits are used as common adjuncts for brewing European types of beer in many areas of the world.

C. *Arepa*

Arepas are widely consumed in Colombia and Venezuela, where they are considered the national bread (Cuevas et al, 1985). *Arepas* are traditionally produced from partially degermed maize grits or meal that is soaked, cooked in water, ground to a moist dough (masa), and shaped into flat disks approximately 7.5 cm in diameter and 1 cm thick. The disks are browned on each side and baked (Fig. 3). Sometimes, gas-fired grills with open flames are used to impart characteristic toast marks. The *arepa* is cut in half and stuffed with meat, cheese, butter, jellies, and other fillings. Sometimes it is stuffed and fried in fat.

Hallaquitas, hallacas, and *empanadas* are prepared from the same basic masa but have different shapes and ingredients and may be fried or baked. *Arepas* contain 58–64% moisture, 4% protein, 0.7% fat, 38% carbohydrates, 0.2% crude fiber, and 1% ash (Cuevas et al, 1985). Hard, white maize is preferred.

Commercially prepared, "instant" arepa flours that require only hot water to produce a dough are made by cooking the maize grits to gelatinize the starch, putting the cooked grits through flaking rolls, and grinding the dried flakes into flour or meal of acceptable particle size distribution (Fig. 4). Using instant dry flour, one can make *arepas* in 30 min instead of the 12–24 hr required by the traditional process. The shelf life and texture of *arepas* made from instant flours may not be as good as those of freshly prepared dough, but the convenience is appreciated by busy consumers. Currently, 700,000 t of maize are processed annually into *arepa* flour in nine plants in Venezuela (Cuevas et al, 1985). Smith et al (1979) proposed an alternative method for production of instant *arepa* flour by extrusion cooking moist maize grits followed by drying, grinding, sieving, and packaging.

D. Porridges

Porridges are traditional foods in Latin American and in African and Asian countries. In Latin America, thin porridges have the common name *atoles* and

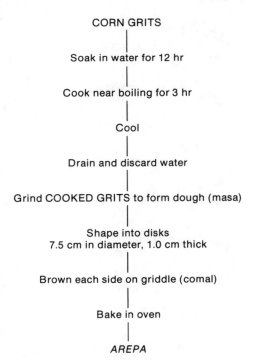

Fig. 3. The traditional method of *arepa* production.

are usually prepared from maize. *Atole* is a smooth, creamy, free-flowing product prepared from wet-milled pastes, dry-milled flours, or degermed flours. On the other hand, a roasted porridge called pinole has characteristics similar to those of *atoles*, but the maize is roasted to promote flavor changes and increase palatability. Flavor development depends upon roasting time, temperature, and the initial moisture content of the grain. Methods to prepare porridges vary among regions. A general scheme (Fig. 5) for preparation of *atole* and pinole in Mexico involves cooking and/or steeping, wet or dry milling, and addition of other ingredients such as milk, sugar, orange leaves, and cinnamon (Vivas, 1985). The same processes are applied to the production of *atoles* from degermed maize flour. The processes summarized in Fig. 5 are similar to those used in some African countries, where *ogi* (thin fermented porridge) and *tô* (thick porridge) are prepared from maize, sorghum, millet, cassava, and rice (Rooney and Murty, 1981). In Africa, the porridges are often fermented and are consumed as the major food three times a day by dipping a portion of the cooled thick porridge into a sauce composed of tomatoes, okra, onions, chiles, and other ingredients. In many areas of Africa, maize is preferred over sorghum or millet because of its flavor and relative ease of processing. For instance, maize grits and meal are available in urban centers because shelf-stable products can be prepared from degermed maize. Sorghum and millet are not as easily degermed; therefore, they do not make shelf-stable products. Thus, maize is consumed in many African countries, which sometimes exacerbates food shortages because maize is not as well adapted to hot, dry conditions as sorghum.

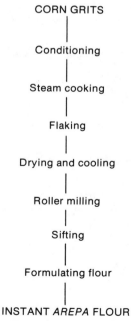

Fig. 4. Commercial production of instant *arepa* flour.

Food Uses of Corn / 409

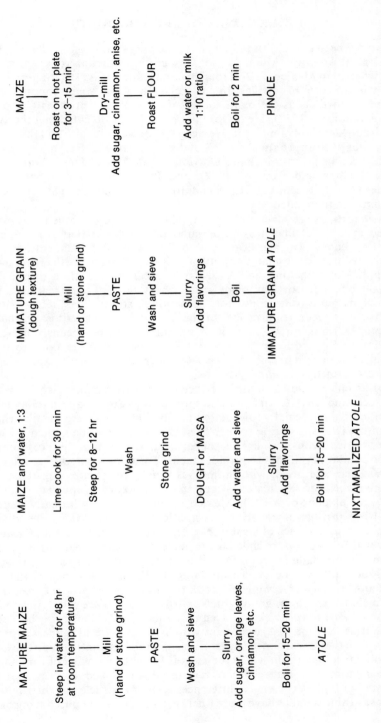

Fig. 5. Typical methods for production of thin porridges from maize in Mexico.

V. ALKALINE-COOKED PRODUCTS

Corn has been the traditional cereal for the preparation of tortillas in Mexico and Central America. About 10% of U.S. corn production is used for human food, whereas in Mexico, 72% of total maize production is used for human food, mainly in the form of tortillas. In particular, the lower socioeconomic groups depend on tortillas as the main source of calories and proteins (Cortez and Wild-Altamirano, 1972). Trejo-Gonzalez et al (1982) indicated that the average annual consumption of maize in Mexico was 186 kg per capita and that, in rural areas, it supplied approximately 70% of the caloric intake. Recent reviews of the tortilla-making process include: Cravioto et al (1952), Trejo-Gonzalez et al (1982), Bedolla and Rooney (1982), and Paredes-Lopez and Saharopulos-Paredes (1983). Recently, tortillas and other Mexican foods have increased in popularity in the United States.

Maize tortillas are made from masa, which is made by stone grinding lime-cooked maize. Tortillas are flat, circular dough pieces that are baked on a griddle, or comal. The masa can be treated with spices, condiments, and other ingredients to produce a large variety of products. A table tortilla from maize is thin, puffs during baking, and is used as bread. The term "flour tortillas" (soft tortillas) refers to tortillas made from wheat flour. Lime treatment is not involved in flour tortilla production. Sometimes, "soft tortillas" refers to unfried table tortillas made from maize. In the United States, burritos are made by wrapping a flour tortilla around a mixture of refried beans, cheese, and sometimes chili and sauce. The wheat flour used to make flour tortillas is dry-milled from hard or soft wheats. Deep-fat-fried flour tortillas sprinkled with honey are termed *sopapillas*.

Tortilla chips were originally produced from leftover tortillas that were cut into pieces and fried. Corn chips and tortilla chips are produced today in large quantities, using specially formulated masas. Taco shells are tortillas that are bent into a U-shape during frying. The inside of the tortilla is then filled with cheese, refried beans, ground seasoned meat, sour cream, lettuce, tomatoes, peppers, and other fillings to produce tacos. Tostados (*chalupas*) are flat, fried tortillas that are used with numerous kinds of meats, vegetables, and condiments. Nachos are tostados cut into small pieces and covered with melted cheese and jalapeno peppers. In Mexico, enchiladas are traditionally made with special red tortillas made from masa colored with paprika. The tortilla is wrapped around a piece of meat or cheese, covered with a sauce and heated. In the United States, most enchiladas are made with noncolored tortillas.

Tamales are made from masa that is placed around a special pork-based, spiced meat filling. The masa containing the filling is wrapped inside a corn shuck and steamed to completely cook the masa. Other products based upon masa include thin beverages or gruels called *atoles* that are often consumed for breakfast. *Pazole* (*pasole*) is a common food in Mexico made by cooking nixtamal (see Section V-A) or hominy with meat, spices, and peppers to produce a thick soup that is consumed with tortillas. *Menudo* is another typical Mexican food similar to *pazole*, in which the meat is from the stomach of ruminants.

The relationships between the various alkali-cooked corn products are presented in Fig. 6. The major differences relate to particle size distribution of the masa. Table tortillas have more finely ground masa and puff during baking, whereas the masa for tacos or tostados is more coarse. The larger particle size promotes pores in the masa that enable the steam to escape without puffing the taco or chip. Frying table tortillas produces tacos with air bubbles or pillows, which are extremely undesirable. Corn chips are made by deep fat frying the masa without baking. Thus, corn chips contain more oil, and the masa must be coarse to enable the moisture to escape readily.

Traditionally, in Mexico, white maize has been preferred for food use. However, inadequate supplies of white maize have forced the use of yellow maize. In some areas of Mexico City, white tortillas are difficult to find, but in other areas, excellent white tortillas are available. Blue corn tortillas, made from a soft, floury maize with deep blue pigmentation, are consumed in some parts of Mexico. In the United States, white or yellow tortillas are used. White maize is most often preferred in the southwestern United States. Sorghum is substituted for part or all of the maize in tortillas prepared in parts of Honduras, Guatemala, El Salvador, and Nicaragua. Good tortillas can be made from sorghum when the cooking conditions are optimized.

A. Traditional Process for Tortillas

The traditional method to process maize into tortillas (nixtamalization) was developed by Latin American Indians (Cravioto et al, 1952). In this process (Fig. 7), maize is cooked in boiling lime (ash) solution for a relatively short time (5–50 min) and steeped overnight. The steep liquor (called *nejayote*) is discarded. The cooked-steeped maize (called nixtamal) is washed to remove excess alkali and loose pericarp tissue. Then, the nixtamal is ground with a pestle and stone into dough (called masa), and flattened into thin disks that are baked on a hot griddle (comal) for 30–60 sec on each side to form tortillas (Bedolla and Rooney, 1982; Khan et al, 1982). During baking, the tortilla puffs. Today, even though most commercial plants use relatively sophisticated technology, the basic principles developed by the Indians are used. Water (containing lime, CaO) is used to cook the maize, although the leachate of wood ashes may still be used in some traditional village processes in Mexico and Central America (Rooney et al, 1986).

Alkali-cooking improves flavor, starch gelatinization, and water uptake and

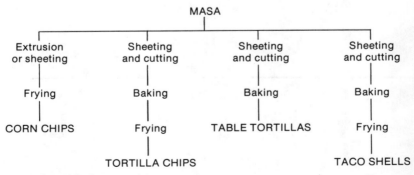

Fig. 6. Alkaline-cooked corn products.

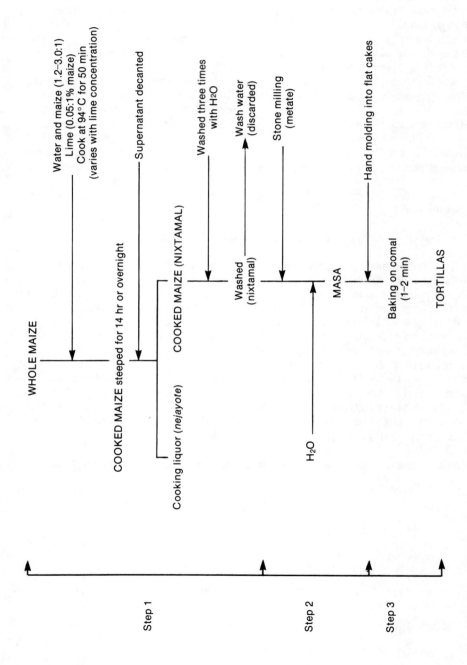

Fig. 7. Traditional processing of maize into tortillas.

partially removes the germ and pericarp of the corn kernels. The maize kernel is only partially cooked. Steeping distributes moisture and lime throughout the cooked grain. Vigorous washing of the nixtamal removes excess lime, loose pericarp, and part of the germ. The washed nixtamal is generally ground between lava stones by hand or using a small diesel-powered stone grinder. Stone grinding disrupts swollen gelatinized starch granules and distributes the hydrated starch and protein around the ungelatinized portions of the corn endosperm, forming masa. Undercooked nixtamal is difficult to grind; overcooked nixtamal forms sticky masa with poor handling properties.

Different maize varieties and even the same variety grown, harvested, handled, and stored under different conditions, have significantly different cooking times (Bedolla, 1980; Goldstein, 1983). The maize is properly cooked when the pericarp is easily removed between the thumb and forefinger. Cooking time, lime concentration, temperature, and steeping times are interrelated. Tortilla production workers who are experienced in cooking maize can subjectively determine the proper length of time to cook maize for tortilla production. Incomplete cooking of maize can be overcome by adding water and making other adjustments during grinding.

Dry matter losses during traditional tortilla processing are estimated at 8–17% of the original clean corn (Bressani et al, 1958; Bedolla et al, 1983; Choto et al, 1985). The amount lost depends on the type of maize, the kind and concentration of lime, the cooking and steeping times and temperatures, the extent of rubbing during washing, and the particular cooking vats and methods used. These conditions vary considerably in the small tortillerias and lead to considerable variation in tortilla quality. The pericarp is removed completely in some localities and only partially removed in others. Usually, most of the germ is retained, although solubles are leached from it. Product yield varies, but 120–130 lb of table tortillas are expected from 100 lb of corn, since the tortillas contain about 40% moisture.

Losses of thiamine, riboflavin, niacin, fat, and fiber during alkaline cooking averaged 60, 52, 32, 44, and 46%, respectively, of levels present in the original corn (Bressani et al, 1958). In spite of chemical losses, feeding trials have shown that rats fed tortillas without supplemental niacin had better performance than animals fed raw corn. The gain in niacin availability due to lime treatment is partially responsible for the improved animal performance (Cravioto et al, 1952; Koetz and Neukom, 1977). On the other hand, recent studies with rats and swine demonstrated that the protein quality of tortillas was slightly inferior to that of raw corn (Serna-Saldivar, 1984). Crosslinking, racemization, degradation, and formation of Maillard browning complexes with sugars lead to reduced protein digestibility and nitrogen retention and lower animal performance (Chu et al, 1976; Sanderson et al, 1978) and increased levels of dietary fiber (Reinhold and Garcia, 1979).

B. Modern Methods of Preparing Alkaline-Cooked Corn Products

In the United States, corn is cooked in industrial processes using several different cooking and steeping processes (Bedolla and Rooney, 1982). A general flow sheet summarizing typical processes for corn and tortilla chips is presented in Fig. 8. Important process variables are cooking time, temperature, kind and concentration of lime, type and frequency of agitation to keep the lime suspended, and the nixtamal washing procedures.

A common procedure for cooking corn uses Hamilton steam kettles, in which the dry corn is added to the water along with dry, powdered lime. Steam is injected until a temperature near boiling is reached. The cooking water is circulated with a pump, and the corn is stirred to suspend the lime. Alternatively, compressed air and live steam injection are used to maintain temperature and suspend the lime. The corn is held at near boiling for 5–10 min for tortilla chips and 20–30 min for corn chips. For table tortillas, longer cooking times are required. The time to reach boiling (rise time) is critical. Afterwards, the cooked corn is quenched by addition of cold water and steeped for 8–12 hr. Quenching reduces the temperature below 65° C (150° F) to stop the cooking process. For table tortillas, cooked corn is not quenched. After soaking, corn is pumped to the washers using a fluidized bed system or it is dropped into the washers by gravity. Most washers are horizontal rotating barrels or drums that spray the nixtamal with pressurized water that removes the pericarp, lime, and solubles from the cooked corn. This cooking process produces uniformly cooked corn because the slow agitation and quenching reduce hot spots and permit better temperature control. However, pumping of the nixtamal causes increased dry matter losses during washing, especially when poor quality, stress-cracked corn is used.

The washed nixtamal is ground, using lava or aluminum oxide stones that cut, knead, and mash the nixtamal to form masa. Attrition milling is essential. The masa characteristics are affected by the diameter of the stones, the clearance between them, the pressure on the stones, the depth and configuration of the grooves, the number of revolutions per minute, and the horsepower of the mill. A typical stone is 4 in. thick and 16 in. in diameter. A mill contains a stationary stone and another that is turned by a 30-hp motor that can process up to 1,360 kg (3,000 lb) of corn per hour. The stones impart considerable energy to the corn, which develops the cohesive characteristics of the masa. Additional water is often added to the nixtamal during grinding to increase the moisture level in the masa and to cool the stones. The stones rapidly overheat and "burn" when insufficient nixtamal and water go through the mill. The lava stones require constant dressing and regrooving by experienced personnel to produce optimum masa. The aluminum oxide stones reduce the need for redressing and provide a uniform grinding surface with longer service time. The trend is toward the use of synthetic stones because of their consistency and longer grinding time; however, they are expensive and may have some other disadvantages.

The masa is formed into tortillas using a sheeter that automatically presses the masa into a thin sheet of dough and cuts circular dough pieces for tortillas, or other shapes for chips. Then, the dough pieces are conveyed through a gas-fired triple-pass oven, which bakes the tortillas or tortilla chips for 15–30 sec at a temperature of 302–316° C (575–600° F). The baked tortillas are cooled and packaged if they are for table use. For tacos and tortilla chips, they are equilibrated for a few minutes and fried. The equilibration produces uniform distribution of water in the tortillas, which reduces blisters and oil absorption during frying. The fryers are designed to maintain uniform temperature (190° C, 375° F) and produce products with acceptable color and low moisture content

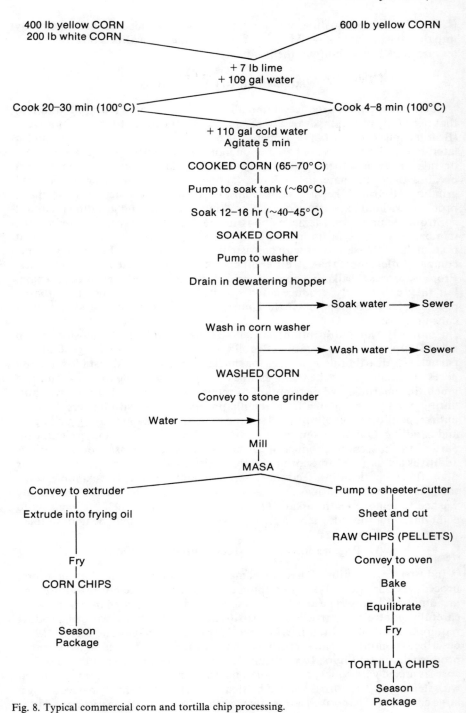

Fig. 8. Typical commercial corn and tortilla chip processing.

(<3%). Corn chips are different in that the masa is extruded or sheeted directly into the oil without baking. Moisture content of the masa affects oil absorption, texture, and acceptability of the products.

C. Nixtamalized Dry Corn Flours

Several companies produce nixtamalized dry corn flours (corn masa flours) that need only to be rehydrated to produce masa for use in tortillas and snacks (Bedolla and Rooney, 1984). The process of producing these flours is discussed later, but the best flours are based upon lime-cooked corn that is ground before drying. In Mexico, these flours have gained popularity with urban consumers because they eliminate the tedious, labor-intensive cooking, washing, and grinding of corn. Likewise, small chip and tortilla manufacturers can make products without expensive cooking, soaking, washing, and grinding facilities.

Strong demand for Mexican foods in the United States has increased the number of corn masa flours available. Several companies produce excellent nixtamalized (masa) flours with consistent, uniform quality. For example, one company offers more than 20 different masa flours formulated for specific uses. Properly processed alkaline-cooked corn (masa) flours are expensive to prepare, but they consistently produce good quality tortillas and other masa-based products. Several companies in the United States sell dry masa flours directly for consumer use.

Color, pH, water absorption, particle size distribution, and viscosity of corn masa flours vary considerably (Bedolla and Rooney, 1984). In general, the particle size distribution is coarser for snacks, taco shells, and tostados because pores are needed to let the steam escape during frying. Table tortillas require a much finer particle size distribution to retain the steam so that they will puff during baking. Frying of a table tortilla produces chips with blisters, high oil uptake, poor organoleptic properties, and excessive breakage during packaging and handling. Obtaining optimum particle size distribution is a very significant part of the successful production of high-quality, dry, masa flours. Various additives are used in some corn masa flours designed for certain foods. The pH of the flour significantly affects the taste, color, and shelf life. For example, some tortillas are made with masa prepared from nixtamal that is not thoroughly washed to remove all the alkali. The tortillas have longer shelf life, e.g., six to eight days, but they also have a strong alkaline taste that may be undesirable.

D. Preparation of Dry Nixtamalized Corn Flours

Industrial production of precooked masa flours is accomplished using several procedures based on alkaline cooking and washing of the corn, followed by grinding and drying. The dried masa is sieved and reformulated into flours with carefully controlled particle size distributions to meet the various product requirements (Sollano and Berriozabal, 1955). Montemayor and Rubio (1983) described continuous and batch cooking procedures for cooking corn to produce dry masa flours. In a continuous process, the lime (0.6–1.0% based on corn) is mixed with equal parts of corn and water in a large screw conveyor fitted with steam jets. The corn is cooked by steam injection as it slowly passes along a large conveyor. The cooked corn is then washed to remove the pericarp and lime and is ground with a hammer mill. The particles are flash-dried and fed into a sifter, and the appropriate fractions are recombined to provide flours with desired particle size distribution and other properties. In batch processes, the corn is mixed with lime, cooked, steeped, washed, ground with a hammer mill or stone mill, dried, and formulated into dry masa.

The cooking and steeping times are critically important. In general, the cooking times and temperatures are less for corn used to prepare dry masa flours than for that for wet masa because the drying process causes additional gelatinization of the ground, wet corn particles. Browning of the masa flour often occurs during drying, which can be controlled by reducing drying times and temperatures. Incorrect particle size and improper drying impart a sandy texture to the tortillas. Good process control is critical for production of good dry masa flour.

E. Experimental Preparation of Dry Masa Flours

Alternate methods to produce dry, nixtamalized corn flours have been proposed (Molina et al, 1977, Johnson et al, 1980). Molina et al (1977) cooked raw, whole ground corn flour with water (1:3) and lime (0.3% of raw corn weight). The mixture was cooked and dried in a double drum drier with a gap of 0.007 mm (0.0003 in.) and an internal pressure of 110–183 kg/m^2 (15–25 psi) at 2–4 rpm (Fig. 9). Supposedly, the tortillas produced by drum drying had physicochemical and organoleptic characteristics similar to those of tortillas prepared traditionally. Drum drying proved to be more energy efficient and had lower dry matter losses than the traditional method. Micronization of corn grits tempered in dilute alkali may have excellent potential (Bedolla, 1983), but more refinements are required to produce good products. Hart (1985) has a patent for masa production using micronization.

Extrusion puffing has not proven feasible for preparation of dry masa flours; however, the use of continuous cooking without puffing may have potential

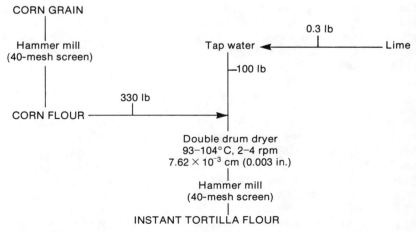

Fig. 9. Production of instant dry corn masa flours.

(Bazua et al, 1979; Bedolla, 1983; Almeida-Dominguez, 1984). In general, the experimental processes do not produce dry flours that are comparable to the best commercial dry masa flours. Subtle differences are important and must be overcome if extrusion, micronization, or other processes are to be used in the production of dry corn masa. Currently, major research efforts to develop nixtamalized corn flours are under way.

F. Hominy

The early settlers in America cooked corn in leachate from wood ashes to produce hominy (nixtamal). The process was obtained from the Indians and was a popular, practical method of corn preparation. The alkali cooking removed the pericarp and greatly improved the convenience and palatability of corn in pioneer America (Hardeman, 1981).

The commercial preparation of hominy uses lye to soften and remove the pericarp (Marden and Montgomery, 1915). Hot water containing 454 g (1 lb) of caustic soda to 39 L (10 gal) of water is added to cover the corn, which is heated and stirred. The corn remains in the lye until the pericarp is freed from the endosperm, which requires 25–40 min depending on the corn. The boiled corn is washed thoroughly with water to remove all traces of alkali and the pericarp. Then, the hominy is salted and canned. Hominy is made from both yellow and white corn varieties; U.S. Southerners prefer white corn hominy. Hominy made with lye has better flavor and acceptability than that made with other sources of alkali (Lopez, 1981). Lime-cooked corn is unacceptable as hominy.

The variety of corn and the environment during kernel maturation, harvesting, and handling, and the storage conditions greatly affect the quality of hominy. The corn should not have any fissured or cracked kernels because those kernels disintegrate during lye cooking, washing, and canning.

VI. SNACKS FROM CORN

Three major categories of corn snacks are: 1) alkaline-cooked corn, 2) extruded products made mainly from cornmeal, grits, or starches, and 3) products made from popped corn (Matz, 1984). The flavor and functional properties of corn and corn products contribute to the expanding demand for corn-based snacks.

A. Corn and Tortilla Chips

The cooking of corn for chips and tortilla chips in commercial operations has been described in the section on tortillas. In general, steam cooking procedures are used. However, cooking the corn in a pressure cooker has been used as an alternative for faster production of nixtamal (Brown and Anderson, 1966). Pressure-cooking requires less cooking and steeping time, but more lime is used. Thus, the pressure cooking methodology can produce nixtamal semicontinuously if alternating cookers are used. After washing (if required), nixtamal is stone-ground to yield a coarser dough than the one preferred for tortilla production. For corn chips, the masa is conveyed to a low-temperature extruder.

The extruded product is shaped or cut and fried in vegetable oil at 199° C (390° F) for 40–75 sec. In the case of tortilla chip production, the shaped pieces of masa are conveyed to a gas-fired oven before frying (Lachmann, 1969).

Automatic equipment for dough handling, shaping, cutting (wavy configurations, disk shapes), baking, and frying has been developed to facilitate increased production rates (Matz, 1984). Anderson et al (1964) developed batch and continuous methods to impart a bent shape to a flat piece of corn dough during frying. Alternatively, Amadon and Boren (1972) patented a forming oven for taco shells in which the pieces of masa are positioned on a conveyor so that they drape across the sides of the conveyor chain. The dough is heated while on the chain to reduce moisture content and fix the taco shape; then it is deposited in the cooking oil.

Production of an expanded corn chip snack food is described by Berg (1968). The masa and properties of the system are modified by inclusion of cornstarch, salt, and carboxymethyl cellulose. The gelatinized masa is dried to about 10% moisture before it is fried in cooking oil. The final products have fourfold expansion, absorb less oil, and develop a crispy, bubbly, cellular structure. Addition of cornstarch (94 parts corn to six parts corn starch) improves the expansion by 20%. Salt and carboxymethyl cellulose improve expansion and corn flavor and reduce oil uptake. Selective toasting (5–15 sec at 176–260° C [350–500° F]) of the dough before frying or baking produces a corn/tortilla product with a toasted flavor.

Quality control criteria for corn and tortilla chip manufacture vary considerably (SFA, 1986). The optimum oil content is 22–25% and 32–35% for tortilla and corn chips, respectively. The correct moisture content in the masa ensures that sufficient oil will be absorbed by the chips during frying. Low oil content imparts poor mouthfeel and harsh texture, whereas too much oil produces greasy chips. The chips should not have more than 15% defects, including 6% puffing or pillowing. Pillowing refers to chips with air bubbles. Other defects include burnt, folded, and broken chips and those with improper size. The moisture level of the chips must be less than 2% to ensure crisp texture. The cooking oil and frying temperatures significantly affect the taste, color, flavor, and mouthfeel of the chips. Frying temperatures below the recommended 188–210° C (370–410° F) are associated with higher oil absorption and lower chip quality. For corn chips, blends of white and yellow corn are often preferred, which produce chips with lighter color, slightly sweeter taste, and improved "corn" flavor.

B. Extruded Snacks

Extruded snacks are a growing segment of the corn-based snack market. Estimated sales in 1983 were 334 million dollars. These snacks are made mainly from cornmeal, sometimes with small amounts of cereals, oilseeds, or other ingredients. Cornmeal is processed through extrusion cooking and puffing to produce corn curls, puffs, and balls. The shape of the puffed extrudate is determined by the die, cut-off knife speed, and other factors. Extrudates are baked or fried, flavored, and packaged to produce the final ready-to-eat product. Mouthfeel is affected by the oil content of the products.

Collets or "half products" can be produced by conventional methods, i.e., macaroni presses, or by using a combination of two extruders. The first extruder cooks and puffs the raw ingredients; the extrudate is cooled, processed, formed, and sized into dense collets in a second extruder. The collets can be dried and stored or equilibrated, then immediately baked or fried. These "third generation" snacks are comprised of many different flours and starches and are formed into numerous shapes and sizes. Baked collets are crisp with a light crunchy texture. Deep-fat fried collets are also crisp, but due to the uptake of oil, usually have smoother texture. Preextrusion moisture content of the meal is critical in determining the characteristics and texture of the product. As the moisture content is increased, extrusion temperatures generally drop, and a dense, less expanded product is obtained. High-moisture meals produce hard, dense extrudates (with less starch gelatinization) that are generally fried. Meals with low moisture content produce highly expanded extrudates that are baked to form a light-textured, puffed snack. The amylose-amylopectin ratio affects the degree of expansion or puffing. High-amylopectin (waxy) corn tends to produce lighter (lower density), more fragile products. For the production of highest-quality baked, puffed corn snacks, a meal or flour with at least 80% amylopectin is recommended (Matz, 1984). The final moisture level of the processed snack food for optimum keeping quality characteristics should be less than 2%. The processed products are generally flavored with powdered cheese, barbecue, sour cream, onion, and chili flavors (Matz, 1984).

C. Popcorn

Popcorn has been a favorite traditional snack food in the United States. Production and technology of this important product is described by Matz (1984), Gutcho (1973), Duffy (1981), and Jugenheimer (1976). In 1982, the United States harvested 101,111 ha (250,000 acres) of popcorn, which yielded 3,521 kg/ha (3,142 lb/acre) with a value of more than 100 million dollars (USDA, 1983). Sales of popcorn (regular and flavored) in 1983 accounted for 163 million dollars. According to Matz (1984), about 70% of the popcorn consumed in the United States is popped at home. The remainder is sold through retail outlets, amusement parks, and theaters and by the confection industry.

Popcorn is a special kind of flint corn that was selected by Indians in early Western civilizations. Its popularity is increasing gradually in other areas of the world. Special varieties of sorghum and rice are parched, puffed, and popped in India. The popped grain is used as an ingredient for infant and other foods. Popcorn may have been the first cereal snack food prepared by humans.

Expansion volume is the most critical quality factor for popcorn. The popped volume is important because the commercial buyer purchases the popcorn by weight and sells it by volume. In addition, popcorn texture (tenderness and crispness) is positively correlated with popping volume. Most commercial popcorn has a 30- to 40-fold expansion.

Popping occurs at about 177° C (350° F), which is equivalent to a steam pressure of 2.5 t/cm^2 (135 psi) inside the kernel. The water in the kernel is superheated and at the moment of popping converts to steam, which provides the driving force for expanding the thermoplastic endosperm after the kernel ruptures (Hoseney et al, 1983). The pericarp and outer layers of the kernel participate directly in the popping action by serving as a pressure vessel enclosing the endosperm. Hybrids or varieties with high proportions of translucent, flinty, or vitreous endosperm give higher expansion rates than corn with floury endosperm.

Popped corn with a spherical shape is called mushroom, ball, or flake type. Due to its configuration, this type of popcorn is preferred in the confection industry because it is less susceptible to breakage, more resistant to handling, and more efficiently coated with flavors and confectionery syrups. The butterfly type has a higher expansion, lower apparent bulk density, and better mouthfeel and is preferred for on-premises popping, where it is sold by volume. Popcorn is usually popped in oil (wet popping) in concession stands. Large-scale operations usually pop their corn with the use of radiant heat (dry popping) at 210–221°C (410–430°F). Dry popping requires about 25% less oil than wet popping. Oil is added to the dry popped corn and is often the vehicle for addition of other flavors and colors. Coconut oil is preferred for popping and coating popcorn. Popped corn rapidly absorbs moisture and becomes tough at moisture levels above 3%.

Large quantities of popcorn are sold as formulated snacks. Regular ball popcorn is treated with coatings free of water. Butter or buttery flavored vegetable oil is generally used as the base for flavors and colors. Flavored coatings are usually applied by spraying the popcorn as it moves on a conveyor or pouring a free-flowing flavored solution on the popcorn held on a revolving coating pan. Coatings are generally composed of sugar-syrup, vegetable oil/butter, and other additives (colored, flavored compounds) that are heated at 144–149°C (290–300°F) to produce a free-flowing solution for application to the popcorn. The product is popularly known as caramel corn.

Unpopped corn can be colored with the use of edible liquid dyes. Before packaging, the colored kernels are subjected to countercurrent heated air (27–32°C, 80–90°F) to remove excess moisture (Schwarzkopf, 1970). Popped corn must be packaged in an airtight container or warmed before use to retain product freshness. Young et al (1978) produced flavored popcorn by combining the popcorn kernels in a cooking medium together with a flavoring substance. A feature of the process is the provision of a premix containing the cooking oil and the flavoring substances encapsulated in acacia, starch, or methylcellulose.

A package consisting of unpopped corn in a cake of solid fat (melting point 49°C, 120°F), containing colored agents has been described (Fujiwara, 1971). The cake is formed in the cooking pan by placing a layer of popcorn in heated colored fat. Different colors are placed in various sections of the pan. The pan is then cooled to solidify the popcorn in the colored fat to form the cake.

Packaging popcorn kernels under vacuum (in airtight, flexible, cellophane material) offers the advantage of increased product volume, tenderness, and overall acceptability. The container is designed for microwave or dielectric heating (at 232–260°C, 450–500°F). The reduced internal pressure that is maintained during popping produces a product with more volume and better attributes than conventional popped popcorn (Jones, 1971). Special packages for microwave popping have been patented (Brandberg and Andreas, 1976). The package is a flexible, expandable bag designed so the steam produced during popping will expand the bag. Eight parts of corn are mixed with one to five parts of shortening and salt (popcorn, 66.6%; coconut oil, 25%; salt, 8.4%).

D. Parching

Corn is heated on hot stones, sand, or ashes until the kernel is partially expanded and develops brown color and roasted flavor. Then the corn is ground and used in gruels, porridges, and other products. A special snack food called "CornNuts" is prepared from large corn kernels heated in alkali and washed to remove the pericarp. The washed kernels are dried, equilibrated, and fried to develop desirable texture, flavor, and color. The corn for CornNuts was developed from Corioco lines of maize from Peru. The kernels are large and white with soft texture and bland flavor. Many products are made from ground parched corn. For example, in Nigeria, parched, ground corn is mixed with red peppers and peanuts and deep-fried in palm oil.

VII. FACTORS AFFECTING FOOD CORN QUALITY

A. Corn Properties for Alkaline Processing

Both environment and genetics affect the properties of corn for alkaline cooking (Bedolla, 1980; Goldstein, 1983). The gross morphology of the corn kernel has been given in Chapter 3, and quality attributes have been discussed in Chapter 5. Corn that can be easily processed into high yields of masa probably has excellent dry-milling properties as well. In general, the properties desired in corn for alkaline cooking are: uniformly sized kernels of high density and high test weight, a high proportion of hard or flinty endosperm, intact kernels free of fissures or stress cracks, kernels without prominent dents in the crown, easily removed pericarp, clean yellow or white color, and white cobs instead of red. These properties are affected by genetics and by environmental conditions. In general, harder corns are usually obtained when hybrids with longer maturities are grown under warmer conditions. However, corn produced in northern areas may be used to produce food products by altering processing conditions. More information to document differences is required, but softer corns generally produce significantly more problems during processing than hard corns.

Rate of cooking is affected by the relative rate of water and alkali uptake by the corn kernels. Anything that reduces the variability among corn kernels helps to achieve more uniform cooking. Improper drying and handling of corn causes fissuring and breakage, which causes overcooking. Soft corn kernels, broken kernels, or kernels with fissures take up water and alkali more quickly and cook faster. Thus, some kernels are overcooked and may dissolve during handling, which increases dry matter losses and produces masa with poor properties. The harder, denser corns tend to cook more uniformly and retain their integrity during subsequent handling and processing. In general, corn samples with high test weight have a higher proportion of hard endosperm. The kernel should not have a prominent dent in the crown. Varieties with deeper dents are generally softer, and the dent provides a pathway along which water and alkali penetrate more quickly into the corn kernel. Those corns are more easily overcooked, which leads to excessive dry matter losses. Pericarp removal is related to hardness of the kernel and the infrastructure of the pericarp tissue. Some corn hybrids have pericarps that dissolve easily; others do not dissolve readily, which affects subsequent sheeting and forming operations.

The preferred color of corn tortillas, chips, and other products is based on regional and local preferences. However, the cleanest, brightest-colored products are obtained when the corn has white cobs instead of red or pink cobs (Montemayor and Rubio, 1983; Hahn et al, 1984). Corn contains phenolic compounds that affect color, especially during alkaline processing. The phenols or other as-yet-undetermined compounds produce dirty, off-white or yellow products when the corn has red cobs. Another defect called "brown banding" occurs in some hybrids when the corn is attacked by mites during kernel maturation. The mites cause red bands in the pericarp, which turn brown upon alkali cooking, producing off-colored products. Cooking conditions affect color significantly and may mask differences among corn cob colors.

Corn hybrids with outstanding alkaline cooking properties include Asgrow hybrids 404 and 405W. Conlee hybrid 202 is another yellow corn with good processing properties. In general, the grain yields and agronomic properties of the food corn hybrids produced to date are inferior to those of feed corn hybrids, which means that farmers must be paid for the reduced yield potentials of the food corns. Work on the breeding of white and yellow food corns with improved grain yields is starting. Sufficient genetic variability exists to develop food corn hybrids with outstanding yields, provided sufficient effort is expended.

B. Cooking Properties of Corn

Hardness, density, test weight, moisture uptake, and other properties of corn have been related to optimum cooking time of corn (Martinez-Herrera and LaChance, 1979; Khan et al, 1982). The utilization of the Instron Universal Tester and a specially designed shear cell and die proved to be superior to the individual kernel hardness method (Bedolla and Rooney 1982; Khan et al, 1982). The force required to shear an average of 60 cooked kernels through a shear cell was related to differences in cooking requirements of corn varieties with similar endosperm textures. Bedolla (1983) classified grain of 31 U.S. commercial white corn varieties, using the nixtamal shear force test, which classified the corn varieties into three groups. These are: 1) the floury endosperm group, for which 211–280 N (1.53–2.03 psi) of force is required to extrude the nixtamal through the cone and die shear cell, 2) the intermediate endosperm group (292–357 N, 2.11–2.63 psi), and 3) the corneous endosperm group (367–435 N, 2.66–3.15 psi). Determination of moisture in the cooked corn may prove to be one of the most practical ways of determining cooking properties, provided an accurate, quick moisture determination can be done routinely.

The extent of starch gelatinization has been used as a measurement of masa properties. Degree of gelatinization affects masa texture and dough handling properties. Maximum amylograph peak viscosity indicates the extent of cooking and masa cohesion. Increased cooking time, temperature, lime concentration, and steeping produce lower viscoamylograph peak viscosity at both 95 and 50°C (204 and 122°F). A faster and more practical way to measure degree of starch damage or gelatinization was proposed by Bedolla (1983), who reported a direct relationship between amount of starch susceptible to glucoamylase hydrolysis and optimum cooking time for different corn hybrids.

None of the methods described here have replaced the subjective determination of cooking times practiced by an experienced person.

C. Changes in Corn Kernel Structure During Cooking

Paredes-Lopez and Saharopulos (1982) evaluated cooked corn structure using scanning electron microscopy. The endosperm of the corn kernel is modified during alkaline cooking because some starch granules swell and gelatinize and some of the protein matrix becomes hydrated. However, within a kernel, some areas are not thoroughly cooked. In general, cooking occurs first in the pericarp, germ, and floury endosperm areas and then gradually moves into the flinty endosperm. The alkali partially solubilizes and weakens the pericarp, which is completely or partially removed during washing of the corn kernel. The pericarp usually breaks at the cross and tube cell area due to solubilization of hemicellulose from cell walls (Chapter 3). The aleurone cells remain intact in most kernels and effectively slow moisture and lime movement into the starchy endosperm. Thus, a properly cooked corn kernel consists of enough gelatinized, swollen starch granules and hydrated protein matrix to produce a dough when it is stone-ground. The attrition of the stone disrupts the swollen starch granules and hydrated protein and causes dough formation. The amylose, amylopectin, and protein form a continuous system, i.e., "glue," that holds the ungelatinized starch and intact endosperm cells together in a cohesive dough. Thus, undercooked corn can be salvaged by addition of water and more pressure between the stones during grinding. Overcooked corn often forms masa with a sticky consistency because too much glue is formed. The complex interaction between amylose, amylopectin, proteins, ungelatinized starch granules, and endosperm particles is not understood. More information on this complex could lead to many practical applications.

D. Nutritional Value and Protein Fortification of Corn Products

Significant changes in chemical composition occur when raw corn is cooked into tortillas. Bressani et al (1958) found that the chemical constituents of corn during tortilla preparation are lost in two ways: 1) by the physical loss of some components (i.e., germ, pericarp) and 2) by their chemical loss, destruction, or transformation. Physical loss of solids from raw corn to tortilla varies from 5 to 14%. The combined physical and chemical losses of important compounds such as thiamine, riboflavin, niacin, fat, and crude fiber average 60, 52, 32, 44, and 46%, respectively. Calcium content is increased significantly during lime processing.

Components of some corn foods are compared with bread and wheat flour tortillas in Table II. Corn chips, popped popcorn, and fried extruded (puffed) snacks have about 35% oil content, and tortilla chips have about 25% oil content. Corn tortillas and ready-to-eat breakfast foods have low oil content. Corn tortillas have a high moisture content, which affects their storage properties and nutrient density significantly. Considerable variability in composition occurs for corn tortillas from different sources (Ranhotra et al,

TABLE II
Nutritional Value of Typical Corn Snack Products, Potato Chips, Wheat Flour, Tortillas, and Bread[a]

Product	Moisture (g/100 g)	Protein (N×6.25) (g/100 g)	Fat (g/100 g)	Carbohydrates (g/100 g)	Ash (g/100 g)	Crude Fiber (g/100 g)	Calcium (mg/100 g)	Phosphorus (mg/100 g)
Potato chips, salted	2.5	6.4	35.3	50.5	3.8	1.4	23.6	152.8
Corn chips								
Salted	0.9	6.3	36.6	53.0	1.9	1.0	135.4	192.3
Barbeque	0.1	6.6	33.9	54.4	2.7	0.8	135.5	209.8
Tortilla chips								
Salted	2.0	7.1	25.2	62.3	0.6	0.9	166.0	235.5
Nacho	2.1	7.7	24.0	62.4	2.7	1.0	174.8	259.5
Taco	1.6	7.8	24.4	62.3	2.4	1.1	171.6	246.5
Corn curls								
Baked	2.3	6.3	33.2	54.4	0.3	0.2	72.2	155.7
Fried	2.1	5.8	37.0	52.6	2.2	0.2	47.8	116.0
Popcorn								
Prepopped	4.1	9.4	25.0	57.6	2.5	1.5	10.4	281.2
Prepopped with cheese	2.5	8.7	33.8	50.4	3.6	1.3	89.9	340.8
Tortillas								
Corn[b]	40.5	5.8	0.9	51.2	1.1	1.4	163.6	57.5
Corn[c]	45.2	5.2	3.1	41.1	1.4	0.7	198.0	...
Flour[b]	26.9	7.3	7.7	54.6	2.1	1.3	60.6	77.0
Bread	35.8	8.7	3.2	50.4	1.9	0.2	70.0	87.0
Corn Flakes	3.0	7.1	...	88.0	...	...	...	...
Corn Chex	3.0	7.1	...	88.0	...	...	...	...

[a] Adapted from values of Deutsch (1986) for composite samples of national brands of each type of product.
[b] Adapted from Saldana and Brown (1984).
[c] Data from Ranhotra (1985).

1984, 1985; Saldana and Brown, 1984; Ranhotra, 1985) because of processing variables. The increased calcium content of corn tortillas and lime-cooked corn snacks is an important consideration in diets. Extruded snacks from corn endosperm cooked without lime do not have that advantage.

Alkali-cooked corn snacks and tortillas may have high dietary fiber levels. Ranhotra et al (1985) found 4.09 and 1.05% dietary fiber in corn tortillas and bread, respectively. Reinhold and Garcia (1979) found an average of 6.60 ± 0.98% neutral detergent fiber and 3.75 ± 0.83% acid detergent fiber in 20 different samples of corn tortillas produced in Mexico. The higher dietary fiber values for alkaline-cooked corn products may be the result of changes in proteins, which produce complex polymers during cooking that show up as dietary fiber. More information is needed on the nutritional properties of tortillas and snack foods.

Feeding trials have shown that animals fed tortillas without supplemental niacin had better performance than animals fed raw corn (Cravioto et al, 1952). The development of pellagra among tortilla eaters in Mexico is virtually unknown because the bound nicotinic acid in corn is liberated during alkali cooking. The bound forms have been recognized as niacinogen and niacytin (Koetz and Neukom, 1977). When vitamins are not limiting, swine and rat feeding trials have demonstrated that the protein quality of tortillas is slightly inferior to that of raw corn. Serna-Saldivar (1984) found a reduction in lysine digestibility at the terminal ileum of pigs and slightly reduced apparent crude protein digestibility in studies conducted with rats. Tortillas have a lower protein quality due to crosslinking, racemization, degradation, and formation of complexes with sugars (DeGroot and Slump, 1969; Chu et al, 1976; Sanderson et al, 1978).

Corn proteins are known to be deficient in the essential amino acids lysine and tryptophan. Quality protein maize and opaque-2 corn have a higher lysine and tryptophan content and therefore better nutritional quality (Ortega et al, 1986). A recent experiment demonstrated that rats gained significantly more weight and had better feed conversion and protein efficiency when fed quality protein maize (raw and as tortillas and tortilla chips) than rats fed food-grade corn products. The rats fed tortillas and tortilla chips had slightly lower performance than rats fed raw corn.

Many efforts have been made to improve the nutritional quality of corn tortillas via protein fortification (Bressani and Marenco, 1963; Bressani et al, 1972, 1974, 1979; Del Valle and Perez-Villasenor, 1974). The addition of 5–8% soybean significantly improved performance of rats (Del Valle and Villasenor, 1974; Bressani et al, 1979) and children (Bressani et al, 1972, 1974) without changing the overall quality or characteristics of the tortillas. However, the authors did not conduct extensive consumer trials to determine flavor, acceptability, and extra cost to tortilla consumers.

VIII. FUTURE OF CORN FOODS

The high productivity of modern corn hybrids, the good flavor of corn, and the ease of processing corn into multitudes of foods and industrial products will cause more corn to be utilized directly for human consumption. Plant hormones and new enzymes to modify kernel components will provide an array of new product opportunities. Special kinds of corn with built-in properties that have increased value for specific processes will be developed through plant breeding using molecular genetics and other tools. Thus, corn will increase in value over time and will likely displace other less-efficient plants.

ACKNOWLEDGMENTS

We recognize the tremendous contribution of C. Clement, Cereal Quality Lab, in typing and preparing the manuscript. We thank O. Paredes-Lopez, IRN, Irapuato, Mexico; S. Bedolla, University of Sonora, Hermosillo, Mexico; L. Johnson, Iowa State University; and, E. F. Caldwell, University of Minnesota, for critically reviewing the manuscript. Graduate students R. Pflugfelder, C. McDonough, D. Jackson, C. Choto, M. Gomez, and N. Vivas provided constructive criticisms and considerable assistance.

LITERATURE CITED

ALMEIDA-DOMINGUEZ, H. D. 1984. Development of maize-soy sesame and sorghum soy tortilla flour using extrusion and nixtamalization. M.S. thesis, Texas A&M University, College Station.

AMADON, R. M., and BOREN, M. G. 1972. Process for preparing folded food chips. U.S. patent 3,690,895. September 12.

ANDERSON, E. E., TICHENOR, J. H., and RAYMOND, S. A. 1964. Process for cooking corn dough in the form of chips. U.S. patent 3,149,978. September 22.

ANONYMOUS. 1984. Report 2: Analysis of Changes in Third World Food and Feed Uses of Maize. CIMMYT Maize Facts and Trends. Centro Internacional de Mejoramiento de Maiz y Trigo, Londres, Mexico City, Mexico.

ANONYMOUS. 1985. 17th Annual State of the Snack Food Industries Report. Snack Food. (June):M1-M27.

BAZUA, C. D., GUERRA, R., and STERNER, H. 1979. Extruded corn flour as an alternative to lime-heated corn flour for tortilla preparation. J. Food Sci. 44:940-941.

BEDOLLA, S. 1980. Effect of genotype on cooking and texture of corn for tortilla production. M.S. thesis, Texas A&M University, College Station.

BEDOLLA, S. 1983. Development and characterization of instant tortilla flours from sorghum and corn by infrared cooking (micronizing) and extrusion cooking. Diss. Abstr. Int. B. 45(Aug 1984):502.

BEDOLLA, S., and ROONEY, L. W. 1982. Cooking maize for masa production. Cereal Foods World 27:219-221.

BEDOLLA, S., and ROONEY, L. W. 1984. Characteristics of U.S. and Mexican instant maize flours for tortilla and snack preparation. Cereal Foods World 29:732-735.

BEDOLLA, S., GONZALEZ de PALACIOS, M., ROONEY, L. W., and KHAN, M. N. 1983. Cooking characteristics of sorghum and corn for tortilla preparation by several cooking methods. Cereal Chem. 60:263-268.

BERG, I. A. 1968. Corn chip. U.S. patent 3,368,902. February 13.

BRANDBERG, L. C., and ANDREAS, D. W. 1976. Popcorn package for microwave popping. U.S. patent 3,973,045. August 3.

BRESSANI, R., and MARENCO, E. 1963. The enrichment of lime treated corn flour with protein, lysine and tryptophan, and vitamins. J. Agric. Food Chem. 11:517-522.

BRESSANI, R., PAZ y PAZ, R., and SCRIMSHAW, N. S. 1958. Corn nutrient losses: Chemical changes in corn during preparation of tortillas. J. Agric. Food Chem. 6:770-774.

BRESSANI, R., BRAHAM, J. E., and BEHAR, M., eds. 1972. Nutritional Improvement of Maize. Proc. Int. Conf. held at the Institute of Nutrition of Central America and Panama (INCAP). INCAP, Guatemala City, Guatemala. 200 pp.

BRESSANI, R., MURILLO, B., and BEHAR, M. 1974. Whole soybeans as a means of increasing protein and calories in maize based diets. J. Food Sci. 39:577-580.

BRESSANI, R., BRAHAM, J. E., ELIAS, L. G., and RUBIO, M. 1979. Further studies on the enrichment of lime treated corn with whole soybeans. J. Food Sci. 44:1707-1710.

BROWN, J. D., and ANDERSON, E. E. 1966. Method of manufacturing corn dough and corn chips. U.S. patent 3,278,311. October 11.

CHOTO, C. E., MORAD, M. M., and ROONEY, L. W. 1985. The quality of tortillas containing whole sorghum and pearled sorghum alone and with yellow maize. Cereal Chem. 62:51-55.

CHU, N. T., PELLET, P. L., and NAWAR,

W. N. 1976. Effect of alkali treatment on the formation of lysinoalanine in corn. J. Agric. Food Chem. 24:1084-1085.

CORTEZ, A., and WILD-ALTAMIRANO, C. 1972. Contribution to the technology of maize flour. Pages 161-167 in: Nutritional Improvement of Maize. R. Bressani, J. E. Braham, and M. Behar, eds. Inst. Nutr. Cent. Am. and Panama, Guatemala City, Guatemala.

CRAVIOTO, R. O., MASSIEU, G. H., CRAVIOTO, O. Y., and FIGUEROA, F. de M. 1952. Effect of untreated corn and Mexican tortilla upon growth of rats fed on a niacin-tryptophan deficient diet. J. Nutr. 48:453-459.

CUEVAS, R., FIGUEIRA, E., and RACCA, E. 1985. The technology for industrial production of precooked corn flour in Venezuela. Cereal Foods World 30:707-708, 710-712.

DANIELS, R. 1974. Breakfast Cereal Technology. Noyes Data Corp., Park Ridge, NJ.

DeGROOT, A. P., and SLUMP, P. 1969. Effect of severe alkali treatment of proteins on amino acid composition and nutritive value. J. Nutr. 106:1527-1538.

DEL VALLE, F. R., and PEREZ-VILLASENOR, J. 1974. Enrichment of tortilla with soy protein by lime cooking of whole raw corn-soybean mixtures. J. Food Sci. 39:244-247.

DEUTSCH, R. M. 1986. Seasoned snacks: A taste of good nutrition. Chipper/Snacker 4(April):24-26.

DUFFY, J. I. 1981. Snack food technology: Recent development. Food Technol. Rev. 55. Noyes Data Corp., Park Ridge, NJ.

FUJIWARA, H. K. 1971. Colored popcorn package. U.S. patent 3,556,815. January 19.

GOLDSTEIN, T. M. 1983. Effect of environment and genotype on hardness and alkaline cooking properties of maize. M.S. thesis, Texas A&M University, College Station.

GUTCHO, M. 1973. Prepared Snack Foods. Food Technol. Rev. 2. Noyes Data Corp., Park Ridge, NJ.

HAHN, D. H., ROONEY, L. W., and BOCKHOLT, A. J. 1984. The effect of phenols on the color of maize and sorghum tortillas. (Abstr.) Cereal Foods World 29:494.

HARDEMAN, N. P. 1981. Shucks, Schocks and Hominy Blocks. La. State Univ. Press, Baton Rouge. 271 pp.

HARPER, J. M. 1981. Extrusion of Foods. Vols. I and II. CRC Press Inc., Boca Raton, FL.

HART, E. R. 1985. Cereal processing. U.S. patent 4,555,409. November 26.

HAYDEN, E. B. 1980. Breakfast cereals—Trend Foods for the 1980's. Cereal Foods World 25:141-143.

HOSENEY, R. C., ZELEZNAK, K., and ABDELRAHMAN, A. 1983. Mechanism of popcorn popping. J. Cereal Sci. 1:43-52.

INGLETT, G. E., ed. 1970. Corn: Culture, Processing, Products. Avi Publ. Co. Inc., Westport, CT.

INGLETT, G. E., ed. 1982. Maize: Recent Progress in Chemistry and Technology. Academic Press, Inc., New York.

JOHNSON, B. A., ROONEY, L. W., and KHAN, M. N. 1980. Tortilla making characteristics of micronized sorghum and corn flours. J. Food Sci. 45:671-674.

JONES, S. A. 1971. Container with popcorn and process of packaging and popping. U.S. patent 3,582,363. June 1.

JUGENHEIMER, R. W. 1976. Corn: Improvement, Seed Production and Uses. John Wiley & Sons. New York.

KHAN, M. N., Des ROSIERS, M. C., ROONEY, L. W., MORGAN, R. G., and SWEAT, V. E. 1982. Corn tortillas; Evaluation of corn cooking procedures. Cereal Chem. 59:279-284.

KOETZ, R., and NEUKOM, H. 1977. Nature of bound nicotinic acid in cereals and its release by thermal and chemical treatment. Pages 305-316 in: Physical, Chemical and Biological Changes in Food Caused by Thermal Processing. T. Hoyden and O. Kvale, eds. Applied Sci. Publ., London.

LACHMANN, A. 1969. Snacks and Fried Products. Food Process. Rev. 4. Noyes Data Corp., Park Ridge, NJ.

LOPEZ, A. 1981. A Complete Course in Canning, Vol. II, 11th ed. The Canning Trade, Baltimore, MD. 473 pp.

MARDEN, J. W., and MONTGOMERY, J. A. 1915. The lye hulling of corn for hominy. J. Ind. Eng. Chem. 7:850-852.

MARTINEZ-HERRERA, M. L., and LACHANCE, P. A. 1979. Corn (*Zea mays*) hardness as an index of the alkaline cooking time for tortilla preparation. J. Food Sci. 44:377-380.

MATZ, S. A. 1970. Cereal Technology. Avi Publ. Co., Westport, CT. 388 pp.

MATZ, S. A. 1984. Snack Food Technology, 2nd ed. Avi Publ. Co., Westport, CT.

MOLINA, M. R., LETONA, M., and BRESSANI, R. 1977. Drum drying technology for the improved production of instant tortilla flour. J. Food Sci. 42:1432-1434.

MONTEMAYOR, E., and RUBIO, M. 1983. Alkaline cooked corn flour: Technology and uses in tortilla and snack products. (Abstr.) Cereal Foods World 28:577.

NOVELLIE, L. 1981. Alcoholic beverages from

sorghum. Pages 113-120 in: Proc. Sorghum Grain Quality Symposium. L. W. Rooney and D. S. Murty, eds. Int. Crop Res. Inst. for the Semi-Arid Tropics, Hyderabad, India.

ORTEGA, E. T., VILLEGAS, E., and VASEL, S. K. 1986. A comparative study of protein changes in normal and quality protein maize during tortilla making. Cereal Chem. 63:446-451.

PAREDES-LOPEZ, O., and SAHAROPULOS, M. E. 1982. Scanning electron microscopy studies of limed corn kernels for tortilla making. J. Food Technol. 17:687-693.

PAREDES-LOPEZ, O., and SAHAROPULOS-PAREDES, M. E. 1983. Maize: A review of tortilla production technology. Bakers Dig. 57:16-25.

POMERANZ, Y., and MUNCK, L., eds. 1981. Cereals: A Renewable Resource. Theory and Practice. Am. Assoc. Cereal Chem., St. Paul, MN.

RANHOTRA, G. S. 1985. Nutritional profile of corn and flour tortillas. Cereal Foods World 30:703-704.

RANHOTRA, G. S., GELROTH, J. A., NOVAK, F. A., BOHANNAN, F., and MATTHEWS, R. H. 1984. Proximate components in selected variety breads commercially produced in major U.S. cities. J. Food Sci. 49:642-644.

RANHOTRA, G. S., GELROTH, J. A., NOVAK, F. A., and MATTHEWS, R. H. 1985. Minerals in selected variety breads commercially produced in four major U.S. cities. J. Food Sci. 50:365-368.

REINHOLD, J. G., and GARCIA, J. S. 1979. Fiber of the maize tortilla. Am. J. Clin. Nutr. 32:1326-1328.

ROONEY, L. W., and MURTY, D. S., eds. 1981. Proceedings: Sorghum Grain Quality Symposium. Int. Crop Res. Inst. for the Semi-Arid Tropics, Hyderabad, India.

ROONEY, L. W., KIRLEIS, A. W., and MURTY, D. S. 1986. Traditional foods from sorghum: Their production, evaluation, and nutritive value. Pages 317-353 in: Advances in Cereal Science and Technology, Vol. VIII. Y. Pomeranz, ed. Am. Assoc. Cereal Chem., St. Paul, MN.

SALDANA, G., and BROWN, H. E. 1984. Nutritional composition of corn and flour tortillas. J. Food Sci. 49:1202-1203.

SANDERSON, J., WALL, J. S., DONALDSON, G. L., and CAVINS, J. F. 1978. Effect of alkaline processing of corn on its amino acids. Cereal Chem. 55:204-213.

SCHWARZKOPF, B. J. 1970. Process for dyeing popcorn. U.S. patent 3,537,861. November 3.

SERNA-SALDIVAR, S. O. 1984. Nutritional evaluation of sorghum and maize tortillas. Ph.D. dissertation, Texas A&M University, College Station.

SMITH, O. B. 1982. Extrusion cooking of corn flours and starches as snacks, breadings, croutons, breakfast cereals, pastas, food thickeners, and adhesives. Pages 193-219 in: Maize: Recent Progress in Chemistry and Technology. G. E. Inglett, ed. Academic Press, Inc., New York.

SMITH, O., DE BUCKLE, T. S., SANDOVAL, A. M., and GONZALEZ, G. E. 1979. Production of pre-cooked corn flours for arepa making using an extrusion cooker. J. Food Sci. 44:816-819.

SFA. 1986. Corn Manual. Snack Foods Association, Alexandria, VA.

SOLLANO, C., and BERRIOZABAL, C. 1955. Method for producing corn tortilla flour. U.S. patent 2,704,257.

SPROULE, A. M. 1985. Nutritional evaluation of tortillas and chips from quality protein maize and food grade maize. M.S. thesis, Texas A&M University, College Station.

STEINKRAUS, K. H., ed. 1983. Handbook of Indigenous Fermented Foods, Vol. IX. Marcel-Dekker, Inc., New York. 671 pp.

TREJO-GONZALEZ, A., FERIA MORALES, A., and WILD-ALTAMIRANO, C. 1982. The role of lime in the alkaline treatment of corn for tortilla preparation. Pages 245-263 in: Modification of Proteins. Advances in Chemistry Ser. No. 198. R. E. Feeney and J. R. Whitaker, eds. Am. Chem. Soc., Washington, DC.

USDA. 1983. Agricultural Statistics. U.S. Govt. Printing Office, Washington, DC.

VIVAS, N. E. 1985. Thin porridges (atoles) prepared from maize and sorghum. M.S. thesis, Texas A&M University, College Station.

VOLLINK, W. L. 1962. Process for making breakfast cereal flakes. U.S. patent 3,062,657. November 6.

YOUNG, R. W., PRUSSIN, S. B., CACCAVALE, J. L., and PIERCE, V. J. 1978. Method and composition for producing flavored popcorn. U.S. patent 4,096,281. June 20.

CHAPTER 14

SWEET CORN

STEPHEN W. MARSHALL
Crookham Company
Caldwell, Idaho

I. INTRODUCTION

For many years, maize (*Zea mays* L.) has been a very important crop in the Western Hemisphere. Its development in Central and South America has been closely related to the development of civilizations there, where maize is used both as a vegetable and a flour product.

Sweet corn is primarily grown in North America, where, relatively speaking, it recently originated. However, in the last decade, sweet corn consumption has greatly increased in Southeast Asia. Recently, European seed companies and processors have been expressing much greater interest in sweet corn products.

Sweet corn is one of the more popular vegetables grown in the United States (Boyer and Shannon, 1982). It currently ranks second in farm value for processing and fourth for fresh-market among vegetable crops.

Sweet corn differs from field corn in terms of its genetic makeup rather than in its systematic or taxonomic characterization. In 1954, Huelsen discussed the problems of drawing conclusions about the relationship between climatic requirements for sweet corn and for field corn based on the information garnered from field corn research. Sweet corn differs from field corn by a mutation at the sugary (*Su*) locus on chromosome 4; sweet corn has the gene *su* at this locus. Thus, the literature pertaining to the origin, development, and breeding of field corn is very applicable to sweet corn. However, because the sweet corn mutant type differs in seed quality and use, the evaluation procedures and breeding objectives differ greatly from those of field corn.

The pericarp and endosperm tissues of sweet corn are the primary products consumed. Recently, sweet corn breeders, seed producers, and specialist in the marketing of seed and products of sweet corn have directed much attention toward the use of various new mutants to improve sweet corn quality. This chapter discusses both the standard sweet corn and the new elevated-sugar mutants.

II. ORIGIN AND HISTORY

Galinat (1971) gave two alternative explanations of the origin of corn: 1) the present day teosinte is the wild ancestor of corn or 2) an extinct pod corn was the ancestor of corn, and teosinte mutated from the pod corn. Beadle (1981) concluded that the stronger evidence is that teosinte is the wild ancestor of corn.

One cannot pinpoint a definite time for the origin of sweet corn (Huelsen, 1954). Standard sweet corn is a mutant. The mutation at the Su locus probably occurred at different times and in different races of corn. The su mutation affects the endosperm composition by causing it to accumulate twice as much sugar and eight to 10 times more water-soluble polysaccharides (WSP) than normal field corn has at the immature state when it is normally consumed (Creech, 1968). It is generally thought that sweet corn originated from a mutation in the Peruvian race, Chullpi (Mangelsdorf, 1974). This type of mutant maize was grown by the American Indians. The first variety, Papoon, was acquired from the Iroquois in 1779 (Galinat, 1971).

Sweet corn was first listed by seed growers in 1828 as "a very sweet vegetable known as sweet corn" (Huelsen, 1954). According to Galinat (1971), Dantings Early was the first named sweet corn cultivar.

The first varieties were, of course, open-pollinated. A 1853 report of the U.S. patent office called attention to two varieties of sweet corn, Mammoth Sweet and Stowell Late Evergreen (Huelsen, 1954). Galinat (1971) described in detail the open-pollinated sweet corn varieties. The most important of these, from which many later varieties originated, was Golden Bantam, released in 1902.

Work in the late 1920s and 1930s at the Connecticut Agricultural Experiment Station by W. R. Singleton and D. F. Jones was very significant. They developed the first early fresh-market hybrid sweet corn varieties. Hybrid sweet corn was established before hybrid field corn (Singleton, 1948). Later, in the 1940s and 1950s, G. Smith at Indiana Agricultural Experiment Station, W. Huelson at the Illinois Agricultural Experiment Station, and E. Haber at the Iowa Agricultural Experiment Station were paramount in developing hybrid sweet corn varieties. They placed great importance on the characters needed for and important to processing. The most important variety to come from this later group, a release out of Iowa named Iochief, was the first release specifically bred for processing with an emphasis on a deep kernel and on cut-corn yield. Industry breeders have become increasingly more important in sweet corn breeding since the 1950s.

Sweet corn production in the United States mainly lies in the northern limits of the Corn Belt. However, sweet corn is an important crop in home and market gardens in almost all states. States of importance for sweet corn processing are Minnesota, Wisconsin, New York, Illinois, Washington, and Oregon. Florida has the largest fresh-market acreage and specializes in fall, winter, and spring markets. Other states with large acreages grown for the fresh market are New York, Pennsylvania, Ohio, Michigan, and California. Ninety-five percent of the nation's sweet corn seed is produced in Idaho.

III. ELEVATED-SUGAR ENDOSPERM MUTANT TYPES

Sweet corn (which is homozygous recessive for the endosperm gene su) has one serious limitation, however: the relatively short period the kernels remain at peak eating quality. Not only is this quality lost very quickly after harvest, but it can be lost in the field if the corn is not harvested at precisely the right time. Often the acceptable harvest window is as short as one day.

The product of consumption in sweet corn is primarily endosperm tissue. Recently, sweet corn breeders have directed much attention toward the use of various new mutants to improve the eating quality of the endosperm. Many genetic mutations affecting endosperm composition have been described in maize. Some of these mutant types that have already been incorporated and are being used for human consumption are generally classified under sweet corn.

In 1982, concern was voiced in the National Sweet Corn Breeders Association about the classification of vegetable corns and how to describe the various genetic mutants as they are commercially available in vegetable corns. Courter and Rhodes (1982) described a genetic model for vegetable corns and presented examples of commercially available vegetable corns using the terminology and genotype examples. Their genetic model is divided into three categories: sugary mutants, other mutants, and starchy corns. The category of sugary mutants is again separated into standard sweet corn and augmented-sugary sweet corn.

A. Sugary Mutants

STANDARD SWEET CORN

Standard sweet corns are those listed in seed catalogs today as "normal," "traditional," "standard," "conventional," and "sugar." The kernels of the standard sweet corns (su) accumulate more sugar than the kernels of the starchy normal (Su) maize. The predominant sugar in the standard sweet corn is sucrose, with lesser amounts of maltose, glucose, and fructose (Cobb and Hannah, 1981). The main component extracted in the WSP fraction in sweet corn is phytoglycogen, a polysaccharide consisting of glucose molecules linked by α-D-(1,4) bonds with α-D-(1,6) branch points (Shannon and Garwood, 1984). The su gene causes the accumulation of phytoglycogen to 25% or more of the kernel dry weight. Essentially no phytoglycogen is present in starchy corn. Starch, composed primarily of two glucose polysaccharides, amylose and amylopectin, is much lower in sweet corn than in starchy corn. According to Shannon and Garwood (1984), the starch concentration in sweet corn, expressed as a percentage of dry weight, increases until approximately 20 days after pollination and then remains constant. Because of continued increases in phytoglycogen concentration, the total polysaccharide concentration increases approximately 30–40 days after pollination, so that sweet corn kernels have approximately the same total carbohydrate concentration as normal kernels (Shannon and Garwood, 1984). (See Chapter 8 for more information on carbohydrates.)

The accumulation of WSP gives a creamy texture to prepared sweet corn products. However, the rapid conversion of sugars to phytoglycogen after harvest and the resultant loss of sweetness is a major problem affecting quality. Until recently, sweet corn breeders increased the kernel sugar level by selecting within the standard homozygous su genotype. Common examples of sweet corn hybrids selected for increased sugar are Spartan, Earlivee, Main Liner, Tender Treat and Silver Queen. These have higher sugar content than other standard sweet corns such as Iochief (Garwood et al, 1976). Common examples of standard sweet corn cultivars in the sugary mutant category are Golden Cross Bantam, Gold Cup, and Jubilee.

AUGMENTED-SUGARY SWEET CORN

The augmented-sugary types have higher sugar levels and are sweeter than the standard sweet corns. The seed of the augmented types is phenotypically indistinguishable from standard sweet corn seed. The classification of augmented-sugary sweet corn is further divided into partial modification and 100% modification types.

Partial Modification. The strategy for cultivars that fall into the partial modification class is to augment the *su* genotype with a second mutant, resulting in an F_1 hybrid in the heterozygous condition.

During the mid-1970s, in commercial breeding programs, it was popular to develop hybrids in the partial modification class in which one or both of two genes segregated on the ear of the hybrid. The aim of this approach was to overcome the major problem with the homozygous mutant types that had been released up until that time. Those first homozygous mutant types, to be discussed later, drastically influenced the production of high sugar and low starch content in the seed, resulting in reduced seed energy reserves. The reserves are needed for germination, emergence, and vigorous seedling growth. Thus, the quality of partially modified augmented-sugary seed would be improved because the seed would be phenotypically *su* or normal. The F_2 kernels for consumption would be at least 25% modified for higher sugar on the ear because of the 3:1 segregation of the single gene. These partially modified augmented-sugary sweet corns are sweeter to the taste than the standard sweet corns and have little detectable difference in texture.

The partial modification class of the augmented sugary category is further broken down into heterozygous shrunken, heterozygous sugary enhancer, and heterozygous shrunken and sugary enhancer.

The first group to be used commercially were the heterozygous shrunken, in which the F_1 seed is made by crossing the sugary seed parent (homozygous *su*) and a sugary shrunken pollen parent (homozygous *su sh₂*). Thus, the seed produced in the F_1 hybrid configuration is homozygous *su* and heterozygous *sh₂* and has seed quality equal to that of standard sweet corn (*su su Sh₂ Sh₂*). The F_2 kernels have 25% higher sugar (*su sh₂*). Examples of heterozygous shrunken cultivars are Synergistic Intrepid, Honeycomb, and Sugar Loaf. Hybrids in this classification have been described by different commercial breeders as "synergistic," "bi-sweet," "sugary supersweets," and "sweet genes."

The second group, the heterozygous sugary enhancer, uses the sugary enhancer (*se*) gene. The homozygous *se* gene is discussed under the 100% modification class. This mutant (*se*) differs from the ones previously discussed in that it appears to be a recessive modifier to the *su* gene. It is only detected in the homozygous recessive *su* background. The heterozygous sugary enhancer made by a *su Se* seed parent pollinated by a *su se* pollinator results in increased sugar, as does the heterozygous shrunken 3:1 ratio. However, the precise nature of the modifying factor(s) is uncertain at this time. Examples of heterozygous sugary enhancer cultivars are Terrific, Platinum Lady, White Lightning, Kandy Korn EH, and Tender Treat EH.

The third group is the heterozygous shrunken and sugary enhancer. The sh_2 gene combined with the *se* gene modifying the *su* background is a 44% modification in a 9:7 ratio. Courter and Rhodes (1982) cite the bicolor Symphony as a heterozygous shrunken and sugary enhancer.

100% Modification. The second class under the augmented-sugary category of the sugary mutants is the 100% modification. Only one mutant in this class has been described and used commercially. That one is the homozygous sugary enhancer (*se*).

The *se* gene has the advantage of increasing sugar while maintaining many of the desirable characteristics of the standard *su* sweet corn. Thus, when eaten, the homozygous *se* hybrids are of very high quality. They also maintain superior eating quality long after harvest. They start with more sucrose, but the rate of conversion of sugar to phytoglycogen is the same as for the standard sweet corn (*su Se*). Also, this hybrid maintains good quality in the field for four to five days longer than the standard *su Se* does. Qualities that are not lost by utilizing this gene in a *su* background are similar eating texture (due to phytoglycogen), germination, seedling vigor, and most other horticultural characteristics. Since *su se* is homozygous, cross pollination with standard sweet corn produces normal sugary kernels. However, the beneficial effect of the added genes (*su se*) is not maximized by cross pollination with *su Se*.

Sweet corn breeders are seldom experienced biochemists and thus find it difficult to run chemical tests for the presence or absence of the *se* gene. The simplest way to detect it is by using the xenia effect of a test cross for *se*. The 677a background is used as the female parent in the test cross because it is very sensitive to pollen from the heterozygous *se* plant. The F_1 kernels of the homozygous sugary enhancer types have the appearance of normal sugary kernels, with the exception of the finely wrinkled pericarp and lighter color. Therefore, the *se* kernels show up very well in contrast to the nonwrinkled surface and face and the faster drying of *su Se* kernels.

Homozygous sugary enhancer hybrids released to date have excellent germination, emergence, and seedling vigor. The flavor of the hybrids is unique, and the kernels have exceptionally tender pericarps. The *se su* hybrids do not require isolation from *su* varieties. Pollination by *su* varieties produces normal *su* kernels. Maximum expression of the *se su* qualities is brought out when pollination is made from pollen of *se su* types. Examples of homozygous sugary enhancer hybrids are Sugar Buns, D'Artagnan, Bodacious, Incredible, Miracle, Double Delicious, and Double Delight.

B. Other Mutants

The second division in Courter and Rhodes's (1982) classification of vegetable corns consists of two classes of other mutants, single gene and multiple gene.

SINGLE GENE CLASS—SHRUNKEN

Two types—shrunken and brittle—exist under the single gene mutant class. Shannon and Garwood (1984) classed shrunken-2 (sh_2), brittle-2 (bt_2), and brittle-1 (bt) in one group because they have a rather dramatic effect singly on endosperm composition. They accumulate sugars at the expense of starch and phytoglycogen and are also epistatic to the mutants amylose-extender (ae), dull (du), su, and waxy (wx). Laughnan (1953) first described the cultivar using the sh_2 endosperm mutant "shrunken-2" gene, which is found in the "supersweet" hybrids. The sh_2 gene, like the su in the homozygous condition, has a drastic effect on the composition of the endosperm carbohydrates. The phenotypic effect of the homozygous sh_2 is a shrunken kernel, from which it gets its name. The sh_2 gene raises the sucrose content to more than 35% of the dry weight of the kernel, which is more than twice that of the su kernels at harvest. It also lacks phytoglycogen and is lower in total carbohydrates and in dry weight than sugary types.

The main advantage of this type of sweet corn is its increased sweetness and storage life. It differs from the su type in that the sugar content of the sh_2 kernels remains high. However, because the sh_2 kernels do not have the high levels of WSP, the creamy texture often associated with standard sweet corn is absent in most of the "supersweet" hybrids. The sh_2 types seem to have a much more watery texture when eaten fresh. The texture is not necessarily objectionable, however, just different. Its inherent advantage over the su types results not only because it starts with the increased sucrose level but because the conversion of sucrose to starch is nonexistent. The sh_2 type therefore has a longer acceptable harvest period. It also maintains its sweet flavor at high levels for a relatively longer time after harvest, even without refrigeration. For most consumers, sweetness in addition to tenderness is the main component of quality that is detected. A study by Wolf and Showalter (1974) in Florida in which they substituted a sh_2 type for a su type in supermarkets resulted in a significant preference for the sh_2 type.

A major problem with the first sh_2 types was low germination and poor vigor, which resulted in poor stand establishment. Because of the mutant's drastic influence on starch synthesis, the seeds were not well suited for survival. Also, the increased sugar level may well have rendered them vulnerable to attack by microorganisms. As Boyer and Shannon (1982) pointed out, evidence suggests that the problems of seed quality associated with these endosperm mutants are not inherent to the genes themselves. Significant interactions of inbred × genotype have been observed for germination, shoot length, and seedling weight. The most recent sh_2 types, which were developed with increased selection effort for seedling vigor, germinate satisfactorily in warm soils and have good vigor.

Examples of shrunken types are Northern Xtrasweet, Crisp 'N Sweet 690, Early Xtrasweet, Crisp 'N Sweet 700, Summer Sweet 7200, Crisp 'N Sweet 710, Crisp 'N Sweet 718, Illini Xtrasweet, Summer Sweet 7600, Crisp 'N Sweet 720, Summer Sweet 7800, Sugar Sweet, Florida Staysweet, Hawaiian Supersweet No. 1, and Hawaiian Supersweet No. 5.

SINGLE GENE CLASS—BRITTLE

The second type in the single gene class is the brittle type, including the brittle-1 (bt) and the brittle-2 (bt_2) genes. These two genes behave similarly to sh_2, both in horticultural characteristics and in biochemical mechanisms. As of today, only two cultivars have been released containing one of these mutants. Hawaiian Super-Sweet No. 6 and Hawaiian Super-Sweet No. 9 are bt_2 and bt, respectively (Brewbaker, 1977). These two hybrids are not conventional F_1 hybrids but composite cultivars grown only in the tropics.

MULTIPLE GENE CLASS

The second class, with multiple genes, has not met with as much success as the single mutants; only one hybrid has been commercially released. That hybrid, Pennfresh ADX, contains the mutants *ae*, *du*, and *wx* (Garwood and Creech, 1979). The mutants seem to result from lesions in the insomatic reactions forming starch. Singly, these mutants do not produce large increases in sugars, but certain combinations can have a substantial effect. Pennfresh ADX is intermediate in sucrose and WSP content between *su* and sh_2. Yet the WSP content is still quite low. This type does not inhibit conversion of sucrose after harvest, as the single mutants do, but does maintain a storage advantage over standard *su* types because it starts storage with more sucrose. The phenotypic expression of *ae du wx* is not as extreme as that of sh_2; the appearance of the seed is more like that of *su*. Also, germination and seedling vigor have reportedly been acceptable without the selection for improved germination, as in the sh_2 types.

A generic classification is needed for the new mutant hybrids that have been released. This is important for seed retailers in advising growers of sweet corn, because the xenia effect on the edible product of sh_2, *bt*, bt_2, and *ae du wx* pollinated by pollen of standard sweet corn (*su*) will make the product starchy. The reciprocal is also true. According to Courter and Rhodes (1982), more than 40 hybrids involving one or more mutant endosperm genes other than or in addition to *su* are on the market. These various mutant hybrids of vegetable corn are all classified generally under sweet corn and are advertised under various trade names and descriptions. Therefore, sweet corn should be classified so that growers can be advised of the isolation requirements to prevent the xenia effect of foreign pollen.

Great genetic advances have been made by plant breeders, who have greatly altered the phenotype of sweet corn and improved seed production, handling processes, and seed treatments.

IV. SWEET CORN BREEDING—PAST, PRESENT, AND FUTURE

After the introduction of the hybrid Red Green in 1924, as described by Jones and Singleton (1934), and the later introduction in 1931 of Golden Cross Bantam, hybrids made up more than 75% of the total crop grown by 1947 (Singleton, 1948). Presently, very little open-pollinated sweet corn is being used. Since the release of Golden Cross Bantam in the early 1930s, only small improvements in standard *su* sweet corn hybrids have been made. Those improvements were in uniformity, adaptation to mechanization, yield, and quality.

Even though Illini Xtrasweet (sh_2) was released 25 years ago, only since the mid-1970s has economically significant achievement been made in the mutant endosperm hybrids. Only now are the high-sugar mutant endosperms coming into prominence.

Because sweet corn is consumed both as a fresh and a processed vegetable, most sweet corn seed producers have breeding programs directed to both markets. Of course, commercial breeding programs of processors breed solely for the needs of the processing industry.

Regional adaptation of hybrid sweet corn has become a very important priority for sweet corn breeders. In the eastern United States fresh market, the straight-rowed white and bicolor types prevail, whereas the yellow hybrids are dominant in the Florida winter market and the home garden.

A breeding program has by necessity a long-term commitment to envisioning future needs, so that improved products will be ready for marketing when they are needed. The sweet corn breeder must understand the present day market and technology and the changes of both to anticipate the future. Recently, the apparent success of the endosperm mutants sh_2 and se has demonstrated this balance between technology and marketing.

In the future, the sweet corn breeder will need to be increasingly aware that the standard (su) sweet corn germ plasm currently available has only a low level of resistance to many of the diseases and insects that affect maize (Sim and Garwood, 1978). This will be increasingly important because commercial sweet corn seed produced in the United States is becoming even more important in Asia and Europe. Also, the breeder must be cognizant of the ever-increasing use of the endosperm mutants, which generally have a greater problem with stand establishment and soilborne pathogens than standard sweet corn does. Sweet corn, in turn, has more difficulty with stand establishment than does field corn.

A. Insect Resistance

Because the sweet corn product is consumed directly, resistance to insect attack of the primary ear must be addressed. The factors to be considered are the number and tightness of husk, as well as the length of husk beyond the ear. In addition, the breeders' preference for long silk tissues should be reassessed in light of the silk's suitability as a substrate for bacterial and fungal infections secondary to infestation from various insects. The sweet corn breeder must realize that selecting for longer husk and silk tissue may directly affect seed set, harvestability, and husking as well as other marketing factors so important in the sweet corn industry.

Absence of resistance to the corn earworm (*Heliothis zea* Boddie) and European corn borer (*Ostrinia nubilalis* Huebner) probably causes more loss than lack of resistance to any other insects in the main sweet corn production areas. The corn earworm probably causes more loss than any other insect because it is found in essentially all sweet corn producing areas.

Wann and Hills (1975) found they were able to maintain resistance to northern and southern corn leaf blights while conducting tandem recurrent mass selection for corn earworm resistance. They selected for both husk length and tightness.

The European corn borer has become economically costly in the past 10 years. Because the kernel is the product of commercial sweet corn production, damage to the ear by either kernel feeding or cob tunneling is more damaging than lodging or ear dropping in commercial sweet corn production fields. However, in sweet corn seed production, stalk lodging is of as great a concern as it is in field corn production.

B. Disease Resistance

Seed rots and seedling blights, stalk rots, leaf diseases, leaf rusts, Stewart's wilt, smuts, and virus diseases have all become very important in the vocabulary of sweet corn breeders. Greenhouse cold tests and early nursery plantings to screen for tolerance to seedling diseases have been effective and have resulted in genotypes tolerant to seed rots and seedling blights. The most effective control method continues to be seed treatment.

Fusarium, Pythium, Diplodia, and *Gibberella* cause the stalk rots most pathogenic to sweet corn. Stalk rots affect both seed production and commercial sweet corn production.

The most common ear rots in sweet corn are caused by *Fusarium* and *Diplodia* spp. and *Gibberella zeae*.

Northern leaf blight (caused by *Helminthosporium turcicum* Pass) has determined the varieties in use in Florida and is becoming more important in the northeastern United States. The best control has been obtained from hybrids containing both monogenic (*Ht*) and multigenic resistance.

Southern corn leaf blight (caused by *Helminthosporium maydis* Nisik. and Miyake) was very important in 1970 because of the virulent race "t" attack on the Texas male sterile cytoplasm (Tatum, 1971). The disease was essentially controlled by the discontinued use of Texas cytoplasm.

Leaf rust (caused by *Puccinia sorghi* Schw.) has recently been cause for concern in the Midwest sweet corn production regions. The best resistance is from a single dominant gene, *Rp*, on chromosome 10. In addition, various sweet corn hybrids have shown partial resistance (Groth et al, 1983).

Stewart's wilt, caused by *Erwinia stewartii* (E. F. Smith) Dye, has long been a problem in Maryland, Pennsylvania, Delaware, and New Jersey. Golden Cross Bantam was one of the first resistant hybrids to be released. Many commercial hybrids now have varying degrees of resistance and tolerance to bacterial wilt.

Pathogens for both common smut (*Ustilago maydis* (DC.) Cda.) and head smut (*Sphacelotheca reiliana* (Kuehn) Clint.) infect sweet corn. Both smuts are easy to recognize and can be selected against in a breeding program. Head smut can effectively be controlled by seed treatment.

The major virus problem in the sweet corn production regions in the United States is maize dwarf mosaic virus (MDMV) strains A and B. Resistance for sweet corn has been obtained from such field corn inbreds as Pa405 and B68. Little is known about the inheritance of resistance to MDMV at this time.

C. General Hybrid Sweet Corn Improvement

Sweet corn breeding since the 1930s has changed the edible product considerably. The most drastic changes have occurred relatively recently, since 1970. Improvement in eating quality, pericarp tenderness, and sweet flavor has most recently been noted in fresh-market releases. Just in the past few years, the processing industry, canning and freezing, has become aware of the drastic changes in sweet corn eating quality. Not only do these changes produce much more tender pericarp and greatly elevated sugars, but they are available in plant types with good levels of resistance to ear worm, stalk rots, northern leaf blight, leaf rust, and Stewart's wilt. What proves to be good background for one mutant may not necessarily be good for another. As new combinations are made and selection continued, important improvements in seed quality, eating quality, and storage quality will undoubtedly be made.

Sweet corn breeding, like all plant breeding, must encompass both science and art. The sweet corn breeder must also have visions about future needs as they relate to economic use.

V. PRODUCTION AND PROCESSING OF SWEET CORN SEED AND SWEET CORN

Production of elevated-sugar sweet corn for seed and for the fresh market and processing is different in all respects from the production of standard sweet corn.

A. Seed

According to private estimates, about 95% of the total sweet corn seed used in the United States is produced in Treasure Valley, Idaho, a valley 60 miles long and 30 miles wide, through which the Snake River flows. Treasure Valley has a truly desert climate but, with controlled irrigation, makes an optimum environment for sweet corn seed production. Its hot days, cool nights, and controlled moisture provide a relatively disease-free environment. Six sweet corn seed producers are located in Treasure Valley. Approximately 11,000 acres of hybrid sweet corn seed was produced by contract growers for these producers in 1985.

The production and processing of sweet corn seed has changed dramatically from the production of normal standard sugary sweet corns, as described by Huelsen (1954), to production of the elevated-sugar (sh_2 gene and se modifier gene) sweet corn hybrids. Use of the elevated-sugar hybrids has increased at a high rate compared to that of the standard hybrids because of consumer demand. This rapid change in use is referred to in the vegetable seed industry as the "Sweet Corn Revolution."

Since the consumer has been exposed to the new elevated-sugar hybrid sweet corns and the sweet corn revolution has begun, tremendous pressure has been put on the seed producer to provide quality seed. The seed produced is the same product that is consumed as the fresh or processed vegetable.

The new elevated-sugar hybrids have extremely tender pericarp, endosperms that are predominantly concentrated sugar shrunken away from the pericarp, and relatively little or no starch content. Thus, the growing, husking, drying, processing, treating, and handling of these hybrids has all been necessarily changed to accommodate the new sweet corn, and they have become a part of the sweet corn revolution.

GROWING ELEVATED-SUGAR SWEET CORN HYBRID SEED

Standard sweet corn is still grown in much the same way as was described by Huelsen (1954) except that much of the machinery has been improved. Only the differences occasioned by the production of the new elevated-sugar hybrids (with the sh_2 gene and the *se* modifier gene) are discussed here.

The seed of the elevated-sugar types is much smaller than that of the standard sugary types. The standard seed average 2,500 kernels per pound, whereas the shrunken seed average 3,800 kernels per pound and the sugary enhancer, 3,500 kernels per pound. The much smaller seed of the elevated-sugar types has made it necessary to use new precision air or finger planters rather than the older plate-type planters used to plant standard sweet corn.

The elevated-sugar seed needs nearly twice as much moisture during the imbibition process as the standard sweet corn seed does. The extremely high concentration of sugar, the high moisture needed to effect imbibition, and the thin and brittle pericarp of the elevated-sugar seed together create an optimum medium for soil pathogens.

The pathogens of most concern at the seedling stage to the grower of sweet corn seed are *Pythium* spp., *Fusarium moniliforme* Sheld., and *Rhizoctonia zeae* Voorhees. Only recently has a good combination of seed protectant fungicides and seed coating been effective in establishing good, evenly emerging stands of the elevated-sugar sweet corn inbreds in hybrid production fields. Standard sweet corn seed is customarily treated with captan and/or thiram and Vitavax. The protectants now used on the elevated-sugar seed are captan, thiram, Difolatan, and Apron, as well as a new graphite seed coat containing potassium nitrate and phosphorus. The seed coat helps to keep the treatment on the seed as well as keeping the seed in contact with the potassium nitrate and phosphorus and giving a uniform seed size for precision planting. The Apron is specific and systemic for *Pythium* and *Rhizoctonia*, and the Difolatan in combination with captan and thiram is very effective in controlling *Fusarium*.

As mentioned earlier, ample moisture is necessary in the seed bed for the elevated-sugar sweet corn seed. It is also essential that the seed be planted no deeper than two inches, as compared to three to four inches for standard sweet corn seed. The elevated-sugar seed generally needs warmer soil temperature (55–60° F) to establish a good, even stand. Most standard sweet corns require a minimum temperature of 50° F. After the elevated-sugar sweet corn plant has passed the 6-in. stage and the stand is established, agronomic practices are the same as for standard sweet corn, as outlined by Huelsen (1954). The differences in producing the elevated-sugar sweet corn seed and standard sweet corn seed arise again at harvesttime.

HARVESTING ELEVATED-SUGAR SWEET CORN SEED

Elevated-sugar sweet corn seed is commonly harvested when the kernel moisture is between 50 and 55% moisture. This is a much higher moisture content than for standard sweet corn seed, which is commonly harvested at a moisture of 10–55%, usually 30%. The consistent high moisture content and the much more tender pericarp of the elevated-sugar kernel at harvesttime has necessitated the design and manufacture of modified pickers to harvest the ears. These new pickers have been designed and built by the growers and seed producers who grow and produce the new seed. The new pickers have modified ear stripper bars and rolls as well as low drop hoppers to lessen the mechanical damage occasioned by the more moist, tender pericarp and tougher, wetter shanks of the elevated-sugar inbreds.

PROCESSING THE ELEVATED-SUGAR SWEET CORN SEED

Because of the higher moisture, thinner pericarp, and absence of starch, the processing of elevated-sugar sweet corn seed differs from that of standard sweet corn seed as much as the processing of standard sweet corn seed differs from that of field corn seed.

Harvested ears of the elevated-sugar types are no longer husked in the field, as are ears of standard sweet corn or field corn. Ears with husks are brought into the processing plant, where the husks are removed by specialized husking beds built specifically to handle the elevated-sugar ears. The modified husking beds have shorter alternating rubber and steel rollers placed at a 24° angle, steeper than the 18° angle of standard husking beds. This gives a less aggressive system, which moves the ears off the husking beds very rapidly after the husk has been cleanly removed; therefore, the ear is not rotated over and over, damaging the pericarp. After husking, much care is taken not to damage the kernels on the ears in handling. Husked ears are placed in large bulk drying bins.

Because elevated-sugar sweet corn seed loses moisture much more slowly than standard sweet corn seed, drying the ears of the elevated-sugar sweet corn is much more time-consuming and costly. This is especially true when the artificial drying of the elevated-sugar seed is begun at 50–55% moisture instead of the 30–40% moisture of the standard seed. Also, the temperature and rate of drying are very critical to good seed quality. Temperatures and rates are some of the secrets of success of those commercial producers who produce elevated-sugar sweet corn seed.

During the drying process, the endosperm of elevated-sugar seed (especially the sh_2 types) tends to dry at a different rate than the pericarp and shrinks more. Thus, the pericarp usually blisters away from the shrunken endosperm, leaving the pericarp vulnerable to mechanical damage. Also, the embryo is more exposed and more vulnerable than that of standard sweet corn seed is.

As a result, the shelling, milling, sizing, kernel sorting, transporting, treating, and packaging processes have been modified to handle the elevated-sugar seed. The sheller has been modified by lowering the cylinder speed and reconstructing the cage in which the cylinder is housed to allow by-passing of the preshelled seed. The milling and sizing machinery has been modified to cope with seed half the weight and size of standard sweet corn seed. New electric eye instruments have been installed to cope with the color differences expressed on the dry seed, to further separate quality seed. Transporting the seed from process to process in the seed processing plant has been modified to eliminate abrasion, falls, and pressure that cause mechanical damage to the elevated-sugar seed not previously experienced with processing of standard seed. After being processed, the elevated-sugar sweet corn hybrid seed is treated with seed protectant and seed coating, as described earlier in this section.

Packaging the coated seed and transporting it have been drastically changed from the packaging and transporting of standard sweet corn seed. The treated and coated seed is placed in heavy sealed plastic pouches, put into corrugated cardboard boxes, placed on pallets, and shrink-wrapped. This eliminates the mechanical damage to the coated seed that occurs when paper bags are stacked on top of one another during shipping. In paper bags, the weight of shifting seed breaks the fragile pericarp. The breakage is nearly eliminated when the coated seed is transported in boxes.

B. Sweet Corn for Processing

Huelsen (1954) wrote in detail about the history, contracting, growing, handling of the raw product, canning, freezing, dehydrating, and nutritional value of processed sweet corn. Until the very recent commercial release of adapted and accepted processing varieties of the elevated-sugar sweet corn hybrids, little had changed from Huelsen's writing.

The growing of the elevated-sugar sweet corn hybrids for processing is essentially the same as described for the production of the seed crop, i.e., it requires a fine seed bed, high-moisture soil, precision planting, effective seed protectants, and a maximum planting depth of 1/2 to 2 in. The harvest moisture is different for shrunken hybrids for processing than for standard sweet corn. Also, the method used to determine moisture has necessarily been changed.

The optimum moisture for harvesting shrunken sweet corn for freezing and canning is no less than 76% and no more than 79%. This compares to a range for standard sweet corn of 70–72%. Because the shrunken sweet corn loses only about 1/4 percent of moisture per 24-hr period at the 76% level, compared to 1% per 24-hr period for standard sweet corn, the harvest window for shrunken corn is approximately four times longer. This is a real advantage for programming the harvest for a processing line and results in fewer bypassed fields due to planting, mistiming, or weather delay.

The most effective, accurate, and rapid method to determine moisture at the necessary moisture levels of the shrunken sweet corn hybrids is with the microwave moisture tester. The established refractometer used for years to determine harvest moisture for standard sweet corn does not work on the shrunken hybrids. This is apparently due to the very low levels of phytoglycogen and absence of starch in the shrunken sweet corn hybrids.

Consumer preference tests on the canned product show a significant preference for shrunken sweet corn processed without salt or sugar over standard sweet corn processed with added salt and sugar. According to the sweet corn processors, this accounts for substantial monetary savings.

Processing procedures vary from processor to processor, and details are closely held secrets, although it is common knowledge that cooking times have been shortened and cooking temperatures lowered for the processing of the new shrunken hybrid sweet corns. When these hybrids were processed at the higher temperatures and longer times of the standard procedure, darker color and off-flavor due to caramelization were apparent.

Consumer preference for the frozen cob as well as cut-kernel shrunken sweet corn compared to that for the standard sweet corn was well established with the first large commercial pack from the 1985 crop.

The first commercial acreage of sweet corn using adapted and commercially accepted shrunken hybrids for processing was grown in crop year 1984. In 1985, approximately 10,000 acres of shrunken sweet corn was processed. An estimated 30,000 acres of shrunken sweet corn will be planted in 1986 for processing; in crop year 1987, an estimated 100,000 acres of this corn will be processed. The U.S. total acreage in 1984 was 450,300 acres (USDA, 1984). Wisconsin and Minnesota sweet corn growers lead the nation in production, with 57% of the total acreage in 1984.

VI. MARKETING SWEET CORN

Very little has been written about marketing sweet corn, either its seed or its products. Since the 1970s, many hybrids have been developed for specific regions and uses. This has created a need for expertise in disseminating information concerning the new developments as well as in marketing the seed and the product.

In the early 1970s, the first "new" endosperm mutant (sh_2) hybrids released were not readily accepted by U.S. seed retailers nor farmers. This was because the earlier sh_2 hybrids had poorer seed quality, which resulted in poorer stand establishment, and tended to have tougher pericarps, even though they were much sweeter than normal sweet corn. During that period, a Japanese company created a whole new market in Japan by using a novel marketing approach. This approach claimed that the corn was a new vegetable, like sweet corn but with an extremely elevated, natural, high-sugar flavor. The important part of the marketing strategy was that it was aimed at the consumer rather than at the seed producer, grower, or broker of the vegetable crop. This strategy differed from the marketing of vegetable seeds as much as the new product differed from the old. Previously, a new release was marketed by the seed producer to the grower.

This new approach to marketing was not attempted in the United States until 1983. With the release of the newer generation endosperm mutants in combination with the new marketing strategy to the consumer, consumption of the new high-sugar cultivars has increased dramatically. In 1982, hybrids based on the sh_2 gene accounted for less than 2% of the fresh-market corn grown in Florida. In the 1983/1984 winter season, these cultivars accounted for 30% of the fresh-market crop. In the 1984/1985 winter season, 80% of the winter fresh-market sweet corn crop was of the shrunken varieties. The 1985/86 winter-season fresh-market sweet corn was 90% of the shrunken varieties. It was limited by seed availability, due to an untimely freeze in Idaho that caused a shortage of seed of the newer types of shrunken sweet corn hybrids. It is predicted that 95% of the 1986/87 winter fresh-market sweet corn crop will be made up of new sweet corn cultivars based on the sh_2 gene (H. Branch, personal communication). By any measure, this is an important change.

The same marketing strategy now taking place in fresh-market sweet corn is happening in the processing (canning and freezing) of sweet corn. The canning industry will be promoting the new high-sugar cultivars by advertising the product as naturally sweet, with no sugar added.

LITERATURE CITED

BEADLE, G. W. 1981. Origin of corn: Pollen evidence. Science 213(4510):890-892.

BOYER, C. D., and SHANNON, J. C. 1982. The use of endosperm genes in sweet corn improvement. In: Plant Breeding Reviews 1. Avi Publishing Co., Westport, CT.

BREWBAKER, J. L. 1977. 'Hawaiian Supersweet #9' corn. HortScience 12:355-356.

COBB, B. G., and HANNAH, L. C. 1981. The metabolism of sugars in maize endosperms. (Abstr.) Plant Physiol. 67:107.

COURTER, J. W., and RHODES, A. M. 1982. A classification of vegetable corns and new cultivars for 1983. Ill. Agric. Exp. Stn., Urbana. (Mimeographed)

CREECH, R. G. 1968. Carbohydrate synthesis in maize. Adv. Agron. 20:275-322.

GALINAT, W. C. 1971. The evolution of sweet corn. Mass. Agric. Exp. Stn. Res. Bull. 591.

GARWOOD, D. L., and CREECH, R. G. 1979.

'Pennfresh ADX' hybrid sweet corn. HortScience 14:645.

GARWOOD, O. L., MACARDLE, F. J., VANDERSLICE, S. F., and SHANNON, J. C. 1976. Postharvest carbohydrate transformations and processed quality of high sugar maize genotypes. J. Am. Soc. Hort. Sci. 101:400-404.

GROTH, J. V., DAVIS, O. W., ZEYEN, R. J., and MOGEN, B. D. 1983. Ranking of partial resistance to common rust in 30 sweet corn hybrids. Crop Prot. 2:219-223.

HUELSEN, W. A. 1954. Sweet Corn. Interscience Publishers, Inc., New York. 409 pp.

JONES, D. F., and SINGLETON, W. R. 1934. Crossed sweet corn. Conn. Agric. Exp. Stn. Bull. 361.

LAUGHNAN, J. R. 1953. The effect of the $sh2$ factor on carbohydrate reserves in the mature endosperm of maize. Genetics 38:485-499.

MANGELSDORF, P. C. 1974. Corn, its origin, evolution and improvement. Belknap Press, Harvard Univ. Press, Cambridge, MA.

SHANNON, J. C., and GARWOOD, D. L. 1984. Genetics and physiology of starch development. Pages 26-86 in: Starch: Chemistry and Industry, 2nd ed. R. L. Whistler, E. F. Paschall, and J. N. BeMiller, eds. Academic Press, Orlando, FL.

SIM, L. E., and GARWOOD, D. L. 1978. Sweet corn disease evaluation summary. Annu. Rep. of CSRS Regional Project NE-66. Penn. Agric. Exp. Stn., University Park. (Mimeographed)

SINGLETON, W. R. 1948. Hybrid sweet corn. Conn. Agric. Exp. Stn. Bull. 518.

TATUM, L. A. 1971. The southern corn leaf blight epidemic. Science 171:1113-1116.

USDA. 1984. Vegetable Outlook and Situation Report. U.S. Dept. Agric., Econ. Res. Serv., Washington, DC.

WANN, E. V., and HILLS, W. A. 1975. Tandem mass selection in a sweet corn composite for earworm resistance and agronomic characters. HortScience 10:168-170.

WOLF, E. A., and SHOWALTER, R. K. 1974. Florida-Sweet. A high quality $sh2$ sweet corn hybrid for fresh market. Fla. Agric. Exp. Stn. Circ. S-226. 13 pp.

CHAPTER 15

NUTRITIONAL PROPERTIES AND FEEDING VALUE OF CORN AND ITS BY-PRODUCTS

K. N. WRIGHT
Wright Nutrition Service
Decatur, Illinois

I. INTRODUCTION

Of all the grains commonly used in livestock and poultry rations in the United States, corn is by far the most important because it is produced in a quantity substantially over that needed for human food. Corn is palatable, is readily digested by humans and by monogastric and ruminant animals, and is one of the best sources of metabolizable energy (ME) among the grains. The availability of corn and soybean meal, as economical sources of energy and protein, has played a most important role in the rapid growth and development of the livestock and poultry industries. Initially these industries used primarily the by-product materials from the grain milling and beverage industries. As the nutritional needs of farm animals became better known and the demand for balanced rations increased, ground corn became the major ingredient.

Corn contains about 72% starch on a dry basis and is low in fiber. The starch is found in the endosperm as granules in a protein matrix (see Chapter 3). Although corn is low in protein content, the volume utilized makes it a major source of protein for the livestock industry. In 1985, the United States harvested approximately 75.1 million acres (30.4×10^6 ha) of corn having an estimated yield of 118.0 bu per acre (7.87 t/ha) (USDA, 1986). The protein in this much corn is equivalent to 46.7×10^6 tons (42.4×10^6 t) of 44% soybean meal. For comparison, the soybeans harvested during this period (2,098 million bushels [53.3×10^6 t]) would produce about 55.2 million tons (50.1×10^6 t) of 44% soybean oil meal.

Average U.S. consumption of corn by farm animals over the last five crop years (1980/81 to 1985/86) was 107.12×10^6 t ($4,217 \times 10^9$ bu). Distribution by class of animals was: swine, 34%; beef, 22.3%; dairy, 18.2%; poultry, 21.3%; other classes, 5.1%. Swine and poultry consumed an average of 55.3% of the corn, and dairy and beef cattle consumed 40.5%. Although the protein content of corn is low, corn still provides as much as 20–50% of the total protein present in many livestock and poultry rations because of the high level of corn used in these diets (USDA, 1985b).

When corn is fed to livestock, it is usually first processed in some manner to improve acceptability or nutritional value. Although whole grain may be fed to swine or cattle, grinding or roll-flaking either dry or wet corn is preferred. Dry grinding of corn permits easier blending with other ingredients but also improves conversion efficiency for swine (Beeson, 1972). Pelleting the mixed feed produces about 10% additional improvement in feed efficiency for swine (Jensen and Becker, 1965). The bulk of poultry feed is pelleted. Aside from the advantages of ease of handling and prevention of segregation of ingredients, pelleted feeds give improved growth rate and feed improvement in poultry. Of all the cereals, corn shows the most improvement from pelleting (Allred et al, 1957).

Feeding corn to feedlot beef cattle provides them a high-energy feed, especially for finishing. Dry roll-flaking the corn may produce an improvement of 2–3% in feed efficiency, but wet processing produces the greatest improvement. High-moisture corn obtained by harvesting at 25–30% moisture content or by rewetting dry corn to that same moisture content (called reconstitution) gives 2–6% improvement in feed efficiency (Hale, 1984). If the high-moisture corn is to be stored more than a few days, it must be treated with a preservative, preferably propionic acid, to prevent mold development (Hall et al, 1974). Steam flaking, which involves a steam cooking step followed by roll-flaking to a bulk density of 22–24 lb/bu (28–31 kg/hl), produces a feed efficiency improvement of 6–10%. However, energy costs must be factored into the benefit calculation (Hale, 1984; Schnake, 1984).

The composition of corn is given in Tables I–III.

TABLE I
Guaranteed and Proximate Analyses of Corn Wet-Milled By-Products[a]

Item	Corn[b] As is	Corn[b] DSB[c]	Corn Gluten Feed As is	Corn Gluten Feed DSB	Corn Meal As is	Corn Meal DSB	Germ Meal As is	Germ Meal DSB	Steep Liquor As is	Steep Liquor DSB
Guaranteed, %										
Protein (min.)	...	...	21.0	...	60.0	...	20.0	...	23.0	...
Fat (min.)	...	...	1.0	...	1.0	...	1.0	...	...	...
Fiber (max.)	...	...	10.0	...	3.0	...	12.0	...	...	...
Proximate, %										
Moisture	15.5	...	9.0	...	10.0	...	10.0	...	50.0	...
Protein (N × 6.25)	8.0	9.5	22.6	25.1	62.0	68.9	22.6	25.1	23.0	46.0
Fat	3.6	4.3	2.3	2.7	2.5	2.8	1.9	2.1	0.0	0.0
Fiber										
Crude	2.5	2.9	7.9	8.9	1.2	1.3	9.5	10.6	0.0	0.0
NDF[d]	8.0	9.5	25.4	30.0	4.1	4.8	41.6	48.0	0.0	0.0
Ash	1.2	1.4	7.8	8.6	1.8	2.0	3.8	4.2	7.3	15.6
NFE[e]	69.2	81.9	50.1	55.7	22.5	25.0	52.2	58.0	19.2	38.4
Starch	60.6	71.7	low	low	low	low	low	low	low	low
TDN[f] (ruminants)	75.5		75		86		70		40	

[a] Source: Anonymous (1982a); used with permission.
[b] Anonymous (1982b).
[c] Dry substance basis.
[d] NDF = neutral detergent fiber (Watson, 1986, and Chapter 3).
[e] Nitrogen-free extract.
[f] Total digestible nutrients.

II. NUTRITIONAL VALUE OF CORN PROTEINS

A. Normal and High-Protein Corn

The proteins in corn have a relatively high percentage of the sulfur-bearing amino acids, methionine and cystine, but are very deficient in the essential amino acids lysine and tryptophan (Table III). Soybean protein is a good source of lysine and tryptophan but a relatively poor source of the sulfur-bearing amino

TABLE II
Typical Nutrient Content of Corn and Corn Wet-Milling By-Products[a]

Item	Corn[b]	Corn Gluten Feed	Gluten Meal	Germ Meal	Steep Liquor
ME,[c] kcal/kg (DSB[d])					
Chicks	3,818	2,007	4,131	1,822	3,110
Poults	...	1,813	4,131[e]	1,711[e]	3,110[e]
Hens	3,818	2,007	4,131[e]	1,956[e]	3,110[e]
Turkeys	...	2,321	4,131[e]	2,078[e]	3,110[e]
Swine	3,762	2,635	3,907	3,296	NA[f]
Ruminants	3,420	3,249	3,510	2,850	NA
Minerals (DSB)					
Potassium, %	0.37	1.4	0.50	0.38	4.8
Phosphorus, %	0.29	1.0	0.78	0.56	3.6
Magnesium, %	0.14	0.46	0.17	0.18	1.42
Chloride, %	0.05	0.25	0.11	0.04	0.86
Calcium, %	0.03	0.2	0.02	0.04	0.28
Sulfur, %	0.12	0.18	0.92	0.36	1.18
Sodium, %	0.03	0.13	0.03	0.04	0.22
Iron, mg/kg	30	334	186	367	220
Zinc, mg/kg	14.0	97	47	118	132
Manganese, mg/kg	5.0	24	trace	4.1	58
Copper, mg/kg	4.0	10.9	24	4.9	31.2
Chromium, mg/kg	...	<1.5	<1.5	<1.5	<2.0
Molybdenum, mg/kg	...	0.9	0.67	0.56	2.0
Selenium, mg/kg	0.08	0.24	0.73	0.37	0.7
Cobalt, mg/kg	0.05	0.1	0.0	0.0	0.28
Vitamins, mg/kg (DSB)					
β-Carotene	3.0	0.0	49–73	0.0	0.0
Choline	567	2,659	2,444	1,564	6,996
Niacin	28.0	82	90	46	167
Pantothenic acid	6.6	19	3.2	4.9	30
Pyridoxine	5.3	16	6.8	6.6	18
Riboflavin	1.4	2.7	2.4	4.2	12
Thiamine	3.8	2.2	0.24	6.8	6
Biotin	0.07	0.2	0.24	0.24	0.66
Inositol	NA	5,923	2,102	NA	12,012
Xanthophyll, mg/kg (DSB)	19.0	24[e]	244–550	0.0	0.0
Linoleic acid, % (DSB)	2.05	2.4[e]	3.6	0.6[e]	0.0

[a] Source: Anonymous (1982a); used by permission.
[b] Anonymous (1982c).
[c] Metabolizable energy.
[d] Dry substance basis.
[e] Estimate.
[f] Data not available.

acids. Hence, a mixture of corn and soy proteins complement each other quite well in poultry and swine diets.

Corn protein content, and its amino acid ratios, may vary widely due to genetic manipulation by plant breeders (Chapters 2 and 9) and to a lesser degree

TABLE III
Typical Amino Acid Content of Corn and Corn Wet-Milling By-Products[a]

Item	Corn[b]	Corn Gluten Feed	Gluten Meal	Germ Meal	Steep Liquor
Protein,[c] % (DSB[d])	9.5	24.5	68.9	25.1	46.0
Amino acids,[e] % (DSB)					
Lysine	0.22	0.65	1.1	1.0	1.6
Methionine	0.15	0.55	2.1	0.7	1.0
Cystine	0.19	0.55	1.2	0.44	1.6
Tryptophan	0.07	0.11	0.33	0.22	0.1
Threonine	0.31	1.0	2.2	1.2	1.8
Isoleucine	0.34	0.7	2.6	0.8	1.4
Leucine	1.05	2.1	11.1	2.0	4.0
Phenylalanine	0.42	0.9	4.2	1.0	1.6
Tyrosine	0.33	0.7	3.2	0.8	1.0
Valine	0.38	1.1	3.0	1.3	2.4
Histidine	0.25	0.8	1.3	0.8	1.4
Arginine	0.42	1.1	2.1	1.4	2.2
Glycine	0.37	1.1	1.8	1.22	2.2
Serine	0.44	1.1	3.4	1.1	2.0
Alanine	0.78	1.7	5.8	1.6	3.6
Aspartic acid	0.68	1.3	4.0	1.6	2.8
Glutamic acid	1.77	3.7	15.3	3.6	7.0
Proline	0.84	1.9	6.1	1.4	4.0

[a] Source: Anonymous (1982a); used by permission.
[b] Anonymous (1982c).
[c] Protein = N × 6.25.
[d] Dry substance basis.
[e] Listed in approximate decreasing order of importance in feed formulation. Lysine through serine are essential amino acids or have sparing effects. Alanine through proline are nonessential.

TABLE IV
Biological Value of Protein in High- and Low-Protein Corn[a]

Corn Source	Protein (%, DSB[b])	Zein[c] (%)	Tryptophan (%)	Lysine (%)	Biological Value (BV)[d] (%)
U.S. hybrid 13					
Continuous planting	7.32	23.2	0.87	2.92	68.6
Corn, oats, clover rotation	10.73	32.9	0.75	2.72	63.1
Illinois high-protein corn					
Nitrogen deficient	13.47	46.4	0.71	2.19	46.9
Nitrogen fertilized	20.04	57.7	0.55	1.76	44.7

[a] Source: Mitchell et al (1952); used by permission.
[b] Dry substance basis.
[c] Protein extracted by 71% ethanol.
[d] $BV = \dfrac{[\text{Food N} - (\text{fecal N} - \text{metabolic N}) - (\text{urinary N} - \text{endogenous N})]100}{\text{Food N} - (\text{fecal N} - \text{metabolic N})}$

by crop year (Earle, 1977), soil fertility, crop management (especially nitrogen fertilization), and climatic conditions (Hamilton et al, 1951; Bird and Olson, 1972; Pierre et al, 1977; Asghari and Hanson, 1984). Most change in protein content is a change in the amount of endosperm proteins relative to the total protein present in the kernel.

The endosperm contains the gluten protein, which is primarily a mixture of the proteins glutelin (dilute alkali-soluble) and zein (alcohol-soluble; see Chapter 9). Zein is nearly devoid of the essential amino acids lysine and tryptophan. The zein content of whole corn and the endosperm increases linearly with total protein content (Hansen et al, 1946; Hamilton et al, 1951); hence, as total protein in whole corn increases, the level of lysine and tryptophan and the biological value of the total protein decline (Mitchell et al, 1952). That zein has a poor biological value (Table IV) has been known and documented for many years (Osborne and Mendel, 1914). Fortunately, the protein content and amino acid ratios are fairly constant in most commercial corn hybrids and in corn purchased in market channels because this corn is a blend of many hybrids from many farms.

It is important to know the essential amino acid content as well as the level of total protein present in the corn source to be used. When corn of higher protein level is used in a ration for monogastric animals already marginal in lysine and/or tryptophan, the need for more supplemental protein from soybean or other quality protein sources is increased if the diet is to remain isonitrogenous. A higher protein level in corn tends to dilute the levels of lysine and tryptophan in the diet because of the increase in zein. If a higher-protein corn is substituted for a normal corn in a ration on a unit basis, the levels of the essential amino acids present and required relevant to energy and total feed intake, not the level of total protein, should be the first consideration. Addition of synthetic essential amino acid supplements, or a natural source high in the deficient amino acids lysine and tryptophan, may be made.

Rations based on corn-soybean meal often need supplementation with methionine and/or lysine. Lysine is usually the first limiting amino acid in swine rations. Usually the most economical source of lysine is soybean meal, and the protein level in these rations is set at about two percentage points above that required to meet the second limiting amino acid—usually tryptophan or threonine. Herein lies the economic basis for supplementing swine rations with lysine when the price of an equivalent amount of corn plus lysine becomes less costly than soybean meal. Soybean meal is usually the economical major high-protein source used to supplement the protein in corn with lysine and to increase total protein level in a ration above that possible with corn alone. Naturally, a corn with a greater amount of lysine would be beneficial for certain nutritional applications.

B. High-Lysine Corn and Synthetic Lysine

High-lysine corn has been made available through the efforts of plant geneticists and breeders (see Chapter 2). The glutelin fraction is increased at the expense of zein by inserting the opaque-2 and/or floury-2 genes into the genetic pattern of the plant to increase the level of lysine in the corn grain (Mertz and Bates, 1964; Jacques and Moureaux, 1984; see also Chapter 9). The need for lysine supplementation in diets containing a high level of high-lysine corn can be significantly reduced.

However, the net cost of raising a high-lysine variety must be carefully compared with that of raising normal hybrid dent corn because grain yield is less. Soft kernel texture, poor germination, and higher grain moisture have been reported to be other problems (Alexander and Creech, 1977). Higher costs, lower quality, and availability of good lysine sources have precluded the use of high-lysine corn in food products. However, for animal feeding, some swine producers have found it profitable to grow high-lysine corn for their own feeding because it reduces purchase of soybean meal or synthetic lysine. At present, no commercial high-lysine hybrids of acceptable grain yield or kernel properties are available.[1]

The commercial availability of synthetic L-lysine at reasonable cost now makes it feasible to supplement corn-soybean oil meal rations for swine. With lysine supplementation, the protein level may be lowered by approximately two percentage points with equal or better performance, especially in grower-finisher rations. When the amino acid tryptophan becomes commercially available, the protein level probably can be dropped even further.

The use of high-lysine corn and/or lysine and tryptophan supplementation has the potential to substantially reduce the need for soybean meal in livestock rations, but their use is strictly an economic decision.

To realize the fullest potential when using high-lysine corn, the nutritionist must have precise knowledge of the amino acid content of the particular high-lysine corn source to be used. Different mutants vary widely in lysine content and may vary in the lysine-tryptophan ratio (Alexander and Creech, 1977). The minimum total protein and the essential amino acid requirement of the animal species must be known for the purpose intended. The relative bioavailability of the amino acids in the diet must be known to help establish the limiting amino acid.

III. NUTRITIONAL VALUE OF CORN LIPIDS

A. Corn Oil

The oil in corn is highly polyunsaturated and rich in linoleic acid (2.9% of the whole corn, dry basis). Corn is therefore a good source of this essential fatty acid and energy for humans and for poultry and swine. The composition of oil in the germ is very similar to that of endosperm oil (Weber, 1979; see also Chapter 10), but the germ contains 83% of the kernel oil (Earle et al, 1946). Fats contribute approximately 2.25 times more ME than starch or protein on an equal-weight basis. The oil in corn contributes about 10–12% of the total ME provided by corn.

The percent absorption of a fat largely determines the ME content of the fat (Scott et al, 1969). Among the important factors influencing absorbability are: 1) chain length, 2) number of double bonds in the fatty acids, 3) the presence of ester linkages and the position of the fatty acid on the glycerol moiety of the fat

---

[1] A. F. Troyer. Breeding for improved amino acid content in corn. Presented at the Natl. Feed Ingredient Assoc. meeting, Chicago, IL, May 2, 1984. (Copies available from the author, DeKalb Pfizer Seed Co., Dekalb, IL.)

molecule, 4) the ratio of saturated to unsaturated fatty acids, and 5) the amount and types of fat in the fat mixture consumed. Because corn oil is highly polyunsaturated, its presence in the diet not only helps to supply the essential linoleic acid requirement but aids in the absorbability of such saturated fats as beef tallow and other fat-soluble nutrients.

B. Carotenoids

CAROTENE

Yellow corn is the only grain that contains useful quantities of the carotenoid pigments (Watson, 1962; see also Chapter 10). These pigments are composed of carotenes and xanthophylls. Carotenoid pigments are mostly concentrated in the horny and floury endosperm of the kernel of yellow corn varieties but are nearly absent in white corn. Yellow corn (12% moisture) contains approximately 22 mg/kg, corn silage (60% moisture) contains 17.3 mg/kg, and stalklage (60% moisture) contains 6.5 mg/kg of carotene (Anonymous, 1982c). Quality corn silage is a good source of provitamin A for ruminants.

The molecular structure of vitamin A is identical to one half of the molecular structure of β-carotene, a provitamin A that is metabolized in the gut and tissues of the animal to vitamin A (Sebrell and Harris, 1967). In general, any carotenoid pigment that has the vitamin A carbon structure on either end is a provitamin A. The vitamin A activity of β-carotene varies with the species of animal and other important factors. The rat and chick obtain 1,667 IU of vitamin A from 1 mg of β-carotene, the cow 400 IU, the sheep 681 IU, and swine 200–500 IU (NRC, 1978, 1985). These values represent a conversion efficiency of 100% (rats and chicks), 41% (sheep), and 12–30% (swine), based on the utilization of all-*trans* β-carotene (Sebrell and Harris, 1967).

β-Carotene may possibly play a role in reproduction independent of its role as a provitamin A source (Hemken and Bremel, 1982). The carotenoids are subject to destruction by oxidation, light, minerals, heat, moisture, length of storage, etc. The rates of loss for carotene and xanthophyll in corn and in corn gluten meal (CGM) as a result of storage are illustrated in Table V. The CGM from wet milling progressively older sources of corn shows similar characteristics in this respect (Table VI).

Feeding high-concentrate diets, mature and bleached pasture or hay grown under drought conditions, and feeds stored for long periods of time may result in

TABLE V
Yellow Carotenoid Pigments in Corn Stored at 25°C[a]

Storage Time (months)	Carotenes (mg/kg)	Xanthophylls (mg/kg)
0	4.8	40.1
4	3.6	36.1
8	2.5	28.6
12	1.8	19.4
24	1.7	20.8
36	1.0	14.1

[a] Source: Watson (1962); used by permission.

a dietary vitamin A deficiency. Winter feeding, using poor quality roughages such as weathered hay or stalklage, and/or feeds high in nitrates is most likely to cause problems. With only a few exceptions, most nutritionists supplement diets with a source of preformed vitamin A for animals fed in dry lot under the conditions described above. Feeds for swine and poultry are supplemented regardless of the level of corn or corn by-products in the diet. The carotene in the diet helps provide the excess needed to offset the carotene losses for various reasons and meet vitamin A requirements over time before the corn is used.

XANTHOPHYLLS

Most xanthophylls have no provitamin A activity. These pigments primarily maintain a uniform color in and give an aesthetic appearance to egg yolks and the skin of broilers. Generally, the xanthophylls do not contribute to any nutritional need of known importance.

The most common sources of xanthophylls are: yellow corn, CGM, dehydrated alfalfa meal and grasses, marigold meal, and synthetic xanthophylls (Marusich and Wilgus, 1968; Kuzmicky et al, 1968; Halloran, 1970; Papa and Fletcher, 1985). The principal oxycarotenoids in yellow corn are lutein, zeaxanthin, and cryptoxanthin (Chapter 10). The amounts and kinds of xanthophylls found in other sources vary from those in corn. Different investigators have found differing results regarding the coloring efficiencies and effects on the visual color of egg yolks and skin by xanthophyll sources (Fletcher et al, 1985). Other factors affecting degree of coloration, such as the genetic capability to absorb and deposit xanthophylls, the presence of antioxidants in the feed, the level of xanthophylls and amount consumed, the presence of disease, etc., are important and must be considered for optimum results. Since the pigments are subject to oxidation and destruction during storage, their levels in feedstuffs should be monitored regularly if they are needed for vitamin A and pigmentation.

IV. VITAMINS

In addition to Vitamin A, corn contains some of all the important vitamins with the exception of vitamin B_{12} (Table II).

TABLE VI
Seasonal Variation in Xanthophyll of 60% Corn Gluten Meal

Month	Xanthophyll (mg/kg)
October	364 ± 34
November	471 ± 22
December	482 ± 37
January	455 ± 30
February	456 ± 27
March	427 ± 19
April	414 ± 26
May	408 ± 11
June	399 ± 16
July	368 ± 37
August	340 ± 39
September	322 ± 30

The fact that niacin, present in corn and many other cereals, is unavailable to monogastric animals including humans has been known for a long time (Christianson et al, 1968). Pellagra has been recognized to be a result of a niacin deficiency in animals and humans consuming diets containing a high level of corn.

Corn processed using a lime-cooking step, as in the preparation of tortillas in Mexico and other Latin American countries, has been shown to improve the growth of rats having a niacin-deficient diet (Laguna and Carpenter, 1951; see also Chapter 13). The niacin in reconstituted corn from wet-milled corn fractions was no more available to the rat than that from raw corn. The niacin in corn is not available to swine (Becker, 1964).

Niacin can be made available if hydrolyzed with dilute alkali or boiled (Pearson et al, 1957). The vitamin can be isolated in a bound complex state from corn (Christianson et al, 1968). The complex is extractable with a 50% ethanol-water mixture from commercial corn gluten, a source high in niacin. The essential amino acid tryptophan is a precursor for niacin, provided an adequate source of pyridoxine is available (Scott et al, 1969). Excess amounts of tryptophan reduce the need for supplemental niacin.

The available pyridoxine content of corn was estimated to be 2.69 mg/kg for the chick (Yen et al, 1976), out of a total of 4.7 mg/kg required (Anonymous, 1982c). The availability of pyridoxine in corn bran has been reported to be zero for humans (Kies et al, 1984). Corn contains about half as much choline, and somewhat less folic acid, pantothenic acid, pyridoxine, riboflavin, and thiamine than do the small grains barley, oats, and wheat (Anonymous, 1982c).

Corn is a good source of vitamin E (Ball and Ratcliff, 1978; Cort et al, 1983). Vitamin E is subject to oxidation, as are the other fat-soluble vitamins. The level of vitamin E in corn can be changed by using different genetic sources (Contreras-Guzman and Strong, 1982; Weber, 1984). The nutritional requirement for vitamin E increases as the level of polyunsaturated fatty acids increases in the diet (Sebrell and Harris, 1972).

V. MINERALS

Corn, like other cereal grains, is very low in calcium (Table II). The phosphorus, potassium, and magnesium contents of corn are also low but are about equal to those in the other common cereals.

More than 80% of the phosphorus in corn is in the form of phytate (O'Dell et al, 1972). The germ of corn contains nearly 90% of the phytate present in whole corn. A good rule of thumb for estimating availability of cereal phosphorus for monogastric animals is to assume that only 30% of the total phosphorus present is utilizable by this class of animals. The presence of phytic acid interferes with the availability of certain minerals, especially calcium, magnesium, zinc, and iron (Underwood, 1962; Momcilovic and Shahl, 1976).

The trace mineral content of corn is low compared to that in small grains. In addition to supplying supplemental sources of the macrominerals in a corn-soymeal diet, it is advisable to include supplemental sources of the trace minerals, including zinc, selenium, iron, manganese, copper, and iodine, to prevent deficiencies.

Selenium is the most recent trace mineral recognized to be essential in the diet, although it is highly toxic at a level not much higher than the requirement level. Many soils in the U.S. are deficient in this element, but some soils have an excess, causing "alkali disease" (Kubota and Alloway, 1971). Vitamin E and selenium share in an antioxidant biological function (Oldfield, 1985).

VI. ANTINUTRIENTS

Mycotoxins are metabolites produced by fungi that grow on corn kernels produced or stored under adverse conditions (Shotwell, 1977; see also Chapter 5). They are not a natural component of sound corn. When they are present at toxic levels in corn fed to animals, various disease symptoms may develop, some which are very severe. Some of the symptoms of mycotoxicosis in animals are reduced growth, feed refusal, lowered resistance to certain infections, reproductive failure, teratogenesis, and carcinogenesis. The most important mycotoxins causing economic losses in corn are aflatoxins, zearalenone, trichothecenes, and ochratoxins (Hesseltine, 1979).

Of the mycotoxins, aflatoxins are the most serious threat to animal and human nutrition because they possess acute and subclinical toxicity and carcinogenicity. Poultry, especially ducklings and turkey poults, young swine, pregnant sows, calves, and dogs are highly susceptible. The toxin is excreted in the milk. Stunting or "poor doing," liver damage, anorexia, and depression are common symptoms of aflatoxin toxicosis, and the young of most animals are more susceptible than adults to a given dosage. Aflatoxin is highly stable to the level of heat generally encountered in the processing of grain. In a laboratory wet-milling procedure, all of the toxin was found to be concentrated in the steep liquor, fiber, gluten, and germ, in that order (Yahl et al, 1971).

The U.S. Food and Drug Administration established a 20-ppb action level for aflatoxin in human food in about 1978. A periodic exemption allows corn containing 20–100 ppb of aflatoxin to be shipped interstate, provided it is to be fed solely to mature, nonlactating livestock.

Corn contains only low levels of the natural antinutrients trypsin and chymotrypsin inhibitors.

VII. CORN FEED BY-PRODUCTS FROM FOOD PROCESSES

Three major processes are used to manufacture food products from corn. These are the wet-milling and dry-milling processes and the corn distilling process for beverage alcohol. Fuel alcohol is produced by wet milling and by a dry milling process. Approximately 1,065 million bushels (27×10^6 t) of corn was used in the United States by these three industries during the 1984/85 crop year (USDA, 1986). The wet-milling industry was estimated to have used 74.6% of this corn, including 14.1% for conversion to ethanol. Direct fermentation of corn amounted to 8.45% for ethanol. About 14.1% was used for alkaline-cooked food products and the dry-milling process for food and industrial uses. These uses accounted for about 15.2% of the 1984/85 disappearance of 7,019.8 million bushels (178.2×10^6 t) out of a total supply of 8,400.5 million bushels (213.3×10^6 t). Analysts at the U.S. Department of Agriculture estimated the total U.S. corn supply for 1985/86 to be 10,274.5 million bushels (260.9×10^6 t). They estimated an increase in 1985/86 for regular wet-milled products of 25 million bushels and an increase of 20 million each for ethanol from wet milling and direct fermentation (USDA, 1986).

In each of these processes, only 65-70% of the corn is converted to primary end products. An exception is alkaline-cooked corn food products (Chapter 13), which utilize whole corn. The remainder of the corn is by-products, most of which are utilized as ingredients in animal feeds and are estimated at 9-10 million tons ($8.2-9.1 \times 10^6$ t). Most of the feed products from the fermentation and dry-milling industries are used by the U.S. feed manufacturing industry. In the case of the major feed by-product from the wet-milling process, corn gluten feed (CGF), approximately 80% of the 5 million tons (4.54×10^6 t) produced annually is exported, mainly to Europe (Wookey and Melvin, 1981; Watson, 1986). All by-product feed materials are given official definitions and international code numbers by the American Association of Feed Control Officials (AAFCO, 1986). Feed formulators must use these definitions and code numbers in dealings with the U.S. Food and Drug Administration or other official bodies and on sales contracts.

The discussion that follows deals extensively with the wet-milling industry for several reasons. One is that wet-milling products dominate the market due to their large volume. Another is that the corn wet-milling industry has been pressed to find new markets for a rapidly growing volume of by-products. This industry has experienced an 8-10% annual growth rate from 1979 to 1985, resulting in a 3.5-fold grind increase since 1972 (USDA, 1985a). The industry has conducted considerable research on product properties both in-house and at major universities. This has resulted in a significant body of literature that has not been adequately summarized heretofore. Although domestic utilization was increased, the major relief was found by developing markets overseas for CGF densified by pelleting (Watson, 1986). Most CGF sold domestically is now in the pelleted form except for that marketed in the wet state.

The Distillers Feed Research Council (Anonymous, 1982b) has also sponsored research for many years and has published most of the results in its annual proceedings of the distiller's feed research conferences and in other publications. Very little literature is available on hominy feed, the single feed by-product of the corn dry-milling industry, for reasons discussed later. Furthermore, results from feeding trials of wet-milled products are similar to those of dry-milled products and are somewhat cross-applicable due to the fact that all feeds are comprised of the fibrous and proteinaceous components of the corn kernel.

VIII. CORN WET-MILLING FEED PRODUCTS

The primary products from the wet-milling process are food and industrial starches and sweeteners. By-products include corn oil and the feed products CGF, CGM, corn germ meal, and condensed fermented corn extractives (steep liquor). (See Chapter 12 for a complete description of the process.) The derivation of the by-products is shown graphically in Fig. 1. They constitute about one third of the weight of the original corn. The germ is solvent-extracted to recover oil, and the extracted germ meal is used in feed products. The gluten is separated from starch by centrifuges, giving a stream containing 69–72% (dry substance basis) total protein, which is dried to become 60% protein CGM. The solubles removed from the corn during steeping are concentrated by evaporation and are called steep liquor or condensed fermented corn extractives. The steep liquor, corn germ meal, and bran are separate components at this point in the process and may be processed and sold as separate products. Condensed fermented corn extractives are usually sold on a 50% solids basis. The corn germ meal is also sold at times at a premium price over CGF. Its absorptive properties and its good amino acid content and balance render it a valued material for use as a carrier for micro-ingredients in formulated feed, but it is primarily a component in CGF. The corn bran (fiber), corn germ meal, and condensed fermented corn extractives are combined in the proportions obtained from the process to become CGF, which is marketed in a wet or dried form. In the dry form, it has a protein content of 21.0%. These components may also be combined in special proportions to produce two other products: condensed fermented corn extractives with germ meal and bran, dehydrated (which is 25–30% protein) and 10% protein corn gluten feed (the entire bran stream with nothing added). The bran can be sold in the wet or dry state.

Wet CGF, containing a minimum 40% dry substance, is being marketed by some corn wet-milling processors. The primary reason is the high cost of energy

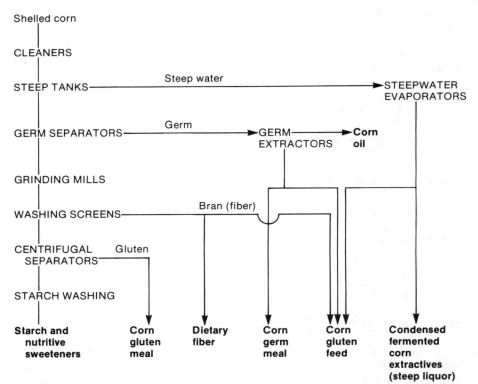

Fig. 1. The corn wet-milling process.

and equipment to dry the product. The nutritive content of the wet feed is very similar to that of the dry feed expressed on a dry basis. The maximum storage time for open wet CGF outside under some conditions approximates seven days or less. Molds will grow on wet CGF stored under certain conditions, but storage conditions should be such as to prevent mold development, which usually is deleterious to feed value.

Domestic corn utilization by the wet-milling industry increased sharply during the period 1981/82 to 1985/86 from 581 million bushels (14×10^6 t) to approximately 840 million bushels (21.3×10^6 t) annually (USDA, 1986). Most of the increase supplied the rapidly developing market for high-fructose corn sweeteners and fuel alcohol. By 1985/86, the growth in corn usage had also made available additional coproducts approximating 2.1 million tons (1.9×10^6 t) of CGF, 450,000 tons (408×10^3 t) of 60% CGM, and 480 million pounds (108×10^3 t) of corn oil per year. A new crystallizing process for pure fructose has recently been developed; the USDA (1986) has predicted that fructose will eventually compete directly with table sugar, which will increase the supplies of the coproducts still further. The nutrient contents of corn wet-milled feed products are given in Tables I–III.

A. CORN GLUTEN FEED

CGF is a feed ingredient with a medium protein level and is palatable to all classes of livestock and poultry. In spite of its name, CGF does not contain any gluten. It is comprised of the fiber fraction (bran), steep liquor, and, where available, germ meal. It commonly contains a minimum of 21% crude protein and approximately 15% starch (Reiners and Howland, 1976). The proximate analysis (Table I) of CGF reflects the effects of starch, gluten, and oil removal on the concentration of the remaining protein, ash, crude fiber, vitamins, and minerals in the corn. The amino acid profile (Table III) is similar to that of whole corn unless the germ meal is not included, as is the case in certain of the wet-milling plants. The energy content of CGF (Table II) varies with the species and application. The ME content is about 92, 71, and 52% that of corn for cattle, swine, and poultry, respectively. The high energy value of CGF for cattle reflects the ability of ruminants to readily digest the cellulose and hemicelluloses present. Unless significant quantities of gluten protein are included in the CGF, the xanthophyll content is only slightly higher than that of corn.

The high solubles content of CGF may cause shipping and storage problems. The product is prone to cake and heat in rail cars, barges, and silos if not dried and cooled properly before being loaded. It is more apt to cake and heat, even to the point of combustion, during the summer months when the temperatures and relative humidity are higher. During this period, its tendency to cake and heat can be reduced if the product is dried to less than 9–10% moisture and cooled to less than 40–43° C before being loaded.

BEEF CATTLE

Corn CGF has been fed to cattle for many years, dating back to as early as 1888 (Turk, 1951). It is a palatable feed ingredient for ruminants. Both dry and wet CGF have rumen protein escape values similar to those of soybean meal (Firkins et al, 1984). Rumen protein escape values are based on relative performance resulting from dietary protein that escapes digestion in the rumen, passes to the abomasum, and is then available for digestion in the lower gut.

Production of rumen total volatile fatty acids was found to be similar for soybean meal and CGF in fistulated steers fed a diet of cottonseed hulls and equivalent amounts of digestible nitrogen (Davis and Stallcup, 1967). The CGF is low in calcium (0.2%) and thiamine (2.2 mg/kg) but high in phosphorus (0.9%). The phytate phosphorus is readily available to the ruminant (Nelson, 1976). Calcium supplementation is required to help maintain a satisfactory Ca-P ratio of at least 1:1 for beef and 1.5:1 for diary cows with high levels of CGF dry matter intake.

When high levels of CGF dry matter, above 30–40%, are used in the total diet, it is desirable to supplement with thiamine to help prevent polioencephalomalacia (polio), which is caused by a thiamine deficiency (Siegmund, 1973). The use of SO_2 in the steeping process reduces the thiamine level in the corn during steeping and therefore in the feed produced. It may also interfere with adequate thiamine synthesis in the rumen.

CGF can be used in large quantities in beef cattle growing-fattening rations because of its high-energy dietary fiber content, high rate of digestibility, and semibulkiness.

Several feeding trials have been conducted on the use of CGF in growing and fattening beef cattle rations. The University of Illinois reported on feeding trials, evaluating both wet and dry CGF in balanced rations for growing and fattening

TABLE VII
Use of Corn Gluten Feed in Growing and Fattening Steer Diets[a,b]

	Control	Dry Corn Gluten Feed		Wet Corn Gluten Feed				Urea
		35%	50%	35%	50%	70%	90%	0.7%
Trial 1[c]								
Daily gain, kg	1.24	1.52	...	1.46	...	...	...	...
Daily feed, kg	9.61	10.42	...	9.52	...	...	...	...
Feed/gain	7.73	6.86	...	6.52	...	...	...	...
Trial 2[d]								
Daily gain, kg	1.33	...	1.35	...	1.38	...	...	1.27
Daily feed, kg	8.13	...	9.46	...	8.80	...	...	7.76
Feed/gain	6.13	...	7.01	...	6.37	...	...	6.13
Trial 3[e]								
With 10% corn silage								
Daily gain, kg	1.24	...	...	...	1.34	...	1.26	...
Daily feed, kg	6.40	...	...	...	8.80	...	8.85	...
Feed/gain	11.80	...	...	...	6.57	...	7.04	...
Condemned liver, %	11.8	...	...	...	14.7	11.8	...	...
Without corn silage								
Daily gain	...	...	...	...	1.32	1.32	1.22	...
Daily feed	...	...	...	...	8.53	8.57	8.07	...
Feed/gain	...	...	...	...	6.48	6.47	6.64	...
Condemned liver, %	...	...	...	...	35.30	23.5	32.40	...

[a] Adapted from Firkins et al (1985).
[b] Percentages of additions to feed are on dry substance basis.
[c] Trial 1: 98-day, growing, 11.5% crude protein, 275-kg steers.
[d] Trial 2: 113-day, finishing, 12.0% crude protein, 328-kg steers.
[e] Trial 3: 150-day, finishing, 12.0% crude protein, 328-kg steers.

steers (Firkens et al, 1985). Results from some of their reported trials are summarized in Table VII. These data suggest that either dry or wet CGF may be included at 50% or more of the diet dry matter without depressing feed lot performance. The relative cost of nutrients from CGF and the feeding situation should determine the level of usage.

Wet CGF reduces the versatility of use, and requires special handling, equipment, storage facilities, and management. The nutrients are diluted and may decrease to as little as 40%, wet basis, of their level in dry feed (91%). The transportation cost on a dry basis is increased accordingly.

Although corn bran is not now a product readily available to the feed industry, the feeding value of the product helps to determine the value of CGF. Corn bran (fiber) composes approximately 50–55% of the dry matter in CGF.

Two feed lot trials were conducted by the University of Illinois to determine the relative feeding value of corn, corn bran, and soyhulls as energy source in diets based on corn silage or fescue silage (Barclay et al, 1985). The results are summarized in Table VIII. No interactions were observed between ingredient level and energy source. Gain and dry matter intake were not different between the control diet containing corn and the other energy sources. The feed-gain ratio was poorer for soy hulls than for corn or corn bran. Comparison of the 25 and 50% levels of corn bran and soy hulls in the fescue silage-based diet showed that gain, dry matter intake, and feed-gain ratio were improved by increasing the levels of the energy sources. However, in corn silage-based diets, gain and feed-gain ratio were not affected by ingredient levels. Apparently, corn bran and soy hulls may be substituted for corn as energy sources in high-fiber diets with little or no depression in performance.

A recent study has been conducted to determine whether corn bran reduces the negative associative effects of fiber digestion often seen with corn supplementation (Klopfenstein et al, 1985a). Calves ate more feed, gained faster, and were more efficient when 25 or 50% of a corn cob and alfalfa haylage ration was replaced by either a corn grain or corn bran supplement. The calves ate more dry matter when fed corn but were more efficient when fed corn bran. These data

TABLE VIII
Feedlot Performance of Steers Fed Corn Bran and Soyhulls[a]

Item	Effect of Energy Source[b]			Effect of Energy Level[c]	
	Corn	Corn Bran	Soy Hulls	25%	50%
Trial 1 (fescue silage)[d]					
Daily gain, kg	1.15	1.15	1.11	0.96 a	1.32 b
Intake, kg/day	6.36	6.14	6.55	5.91 a	6.82 b
Feed/gain	5.56 a	5.43 a	6.00 b	6.17 a	5.16 b
Trial 2 (corn silage)[e]					
Daily gain, kg	1.39	1.35	1.30	1.33	1.37
Intake, kg/day	10.27 a	9.95 b	10.43 c	10.10 a	10.60 b
Feed/gain	7.39 a	7.37 a	8.25 b	7.59	7.6

[a] Courtesy R. A. Barclay.
[b] Values in a row that are followed by different letters differ at a level of $P < 0.01$.
[c] Values in a row that are followed by different letters differ at a level of $P < 0.05$.
[d] Ninety-six crossbred steers (203 kg).
[e] Seventy-two crossbred steers (360 kg).

suggest that a 25% corn bran addition does not reduce fiber digestion and that the 50% level of bran does to some extent but not as much as corn does.

DAIRY CATTLE

CGF has been fed to dairy cows probably longer than to any other species. Before the 1900s, the milk-producing properties of CGF were recognized. Turk (1951) has summarized early feeding trials on the ability of CGF to replace such protein supplements as linseed meal, cottonseed meal, peanut meal, and soybean meal in dairy rations composed of corn, oats, and wheat bran along with the roughages soybean and/or alfalfa hay and corn silage. A 1936 Connecticut experiment showed that a grain mixture containing 70% CGF had no effect on the titratable acidity of milk. According to Turk (1951), the Cornell University Agricultural Experiment Station in 1943 studied the relative biological value of CGF and other protein sources in a simple grain mixture. In a total of 86 lactations, averages of the results of five experiments showed no differences in milk production, palatability of the rations, or maintenance of body weight. In other Cornell experiments (1944), 25 and 50% levels of CGF in the concentrate mixture were compared with the Cornell test cow mixture. CGF fed at a level of as much as 50% of the concentrate did not decrease palatability; the mixtures were as efficient in producing milk as a more common mixture of farm-grown grains and by-product feeds and gave satisfactory results.

CGF was found to be a substitute for copra meal and ground yellow corn in rations for lactating dairy and Murrah cows; it can be a substitute also for copra meal in rations for growing dairy heifers but cannot substitute for ground yellow corn in rations for growing Murrah heifers (Clamohoy et al, 1968).

The feeding value of wet CGF for dairy heifers was studied by Jaster et al (1984). The wet CGF was ensiled in a plastic silo bag and showed good preservation and keeping qualities, as measured by changes in pH, temperature, and organic acid concentration. Heifers were fed the wet CGF to determine ad libitum intake and were observed to consume 2.4% of their body weight in dry matter. The apparent digestibility of wet CGF dry matter was found to be 76.6%, compared to those of alfalfa haylage (60.7%), oatlage (53.3%), and sorghum-soybean silage (54.9%). Heifers consuming wet CGF also showed higher digestibility values for neutral detergent fiber (NDF), acid detergent fiber (ADF), lignin, hemicelluloses, and crude protein than when fed the other feeds. In an 83-day performance feeding trial involving 64 dairy heifers (275 kg), weight gain and body growth measurements were distinctly superior for dairy heifers fed wet CGF than for those fed alfalfa haylage (Table IX) and other forages. However, because of excessive body weight gain and mild diarrhea, it was recommended that wet CGF not be fed free choice as the sole feed to replacement dairy heifers. Other feeding studies with lactating cows (Staples et al, 1984) showed that when wet CGF was ensiled and fed with corn silage, a 25–30% (dry matter basis) replacement of corn grain with wet CGF gave best results.

The effect of the roughage-concentrate ratio on milk production and changes in body weight is discussed by McCullough (1973). Mertens (1985) suggested that NDF and ADF contents and particle size of ingredients are related to energy content, filling effect, and chewing activity. The levels of wet CGF, on a dry matter basis, used in the above experimental rations significantly increased the NDF and ADF content of the diet above that in the control ration. This could explain the linear reduction in milk production and loss in body weight at the highest levels of wet CGF. The data suggest that wet CGF may have a roughage-sparing property.

Improvement in milk butter fat content has been observed by the author when either CGF or corn bran (wet-milled) was used at relatively high levels as a major source of energy in a concentrate fed to lactating cows. Similar results have been reported by Hutjens et al (1985). The ability of feeds containing CGF or bran to maintain or improve butter fat percentage probably is due to the low amount of starch and the higher level of digestible NDF in the diet, compared to the level in rations containing corn or hominy feed. These high-energy, high-dietary-fiber ingredients can be used very advantageously to formulate high nutrient-density dairy concentrates that maintain or increase butter fat production without decreasing milk production.

SWINE

The ME content of CGF for swine has been reported as 2,770 kcal (Yen et al, 1974) and 2,730 kcal/kg of dry matter (Young et al, 1977). These data compare favorably with the data in Table II. The estimated ME values were not significantly affected by pelleting. The amino acid pattern for CGF is similar to that of corn—a good source of the sulfur-bearing amino acids but deficient in the essential amino acids lysine and tryptophan (Table III). CGF is a much better source of vitamins than is whole corn (Table II). As in corn, the niacin present should be considered unavailable (Laguna and Carpenter, 1951).

The use of CGF in growing-fattening swine rations has been studied (Yen et al, 1971; Hollis et al, 1985). CGF may be used in 12%-protein corn-soy finishing rations, replacing corn as an energy source for up to 30% of the ration dry matter without significantly affecting performance (Table X). The protein content was allowed to increase as the level of CGF increased (Yen, 1971). A nonsignificant decrease in daily gain and gain-feed ratio was observed when CGF replaced up to 30% of the corn in a 16%-protein diet fed in meal form to growing pigs. Pelleting the diets resulted in similar gains and gain-feed ratios at all levels. However, when CGF replaced corn and soybean meal in a 12%-protein isonitrogenous diet, the daily gain, daily feed, and gain-feed ratio were significantly depressed at the 20 and 30% levels of dry CGF. It was later demonstrated (Yen, 1971) that the first limiting amino acid was tryptophan and

TABLE IX
Growth and Dry Matter Intake of Dairy Heifers Fed Alfalfa Haylage or Wet Corn Gluten Feed for 83 Days[a]

Item	Alfalfa Haylage	Wet Corn Gluten Feed
Dry matter intake, kg/day	8.5	8.4
Average daily gain, kg/day	0.45	1.1
Increased heart girth, cm	8.3	19.1
Increased height at withers, cm	4.2	6.6
Increased body length, cm	6.6	9.1

[a] Source: Jaster et al (1984); used by permission.

the second was lysine at the 30% level of CGF in the diet. The data suggest that when CGF is used to replace corn and soybean meal on an isonitrogenous basis, a maximum 10% level of CGF should be used in finishing rations without tryptophan and lysine supplementation.

The results of these trials indicate that the previously observed inefficient use of CGF by swine is not due to bulkiness and/or unpalatability but primarily to amino acid deficiencies, especially to low tryptophan and lysine availability.

Wet CGF is best fed to gestating sows because of their large intestinal capacity and relatively low daily nutrient requirements (Hollis et al, 1985). However, because of its low dry matter and nutrient levels, the sow cannot consume enough nutrients from wet CGF to meet her requirements, especially those for energy, calcium, available phosphorus, trace minerals, salt, vitamins, lysine, and tryptophan. Daily intake should be monitored, and diet supplementation with additional sources of these nutrients is required even at the maximum intake of wet CGF by the sow.

Feeding trials were also conducted on the feeding value of dried condensed fermented corn solubles with germ meal and bran (dried steep liquor concentrate [DSLC]) in swine rations (Harmon et al, 1975a, 1975b). DSLC is a blend of steep liquor solids, corn germ meal, and some corn bran. This product contains ME of 3,788 kcal/kg, dry basis, which is greater than that of corn and soybean meal (Cornelius et al, 1973). Lysine and tryptophan in DSLC were found to be at too low a level for swine, as in CGF. When DSLC was the only amino acid supplement for corn in finishing pig diets, much lower gains and efficiency resulted. A corn-DSLC diet designed to meet the lysine and tryptophan requirements could not be improved by addition of lysine or tryptophan alone but was significantly improved when both amino acids were added (Harmon et al, 1975a, 1975b).

When DSLC was used to provide up to 30% of the total lysine in the diet, performance of young pigs was equal to that of pigs receiving a corn-soybean diet. For finishing pigs, DSLC could replace up to 36% of the total dietary lysine.

POULTRY

The use of CGF is limited in poultry feeds because of its low ME, lysine, and tryptophan content, but it is a good source of methionine and cystine. The "fiber

TABLE X
Performance of Finishing Swine When Dry Corn Gluten Feed Replaces Corn in Ration (61-day trial)[a]

Item	Percent of Dry Corn Gluten Feed[b]			
	0[c]	10	20	30
Protein (estimated), %[d]	...	13.35	14.7	16.05
Daily gain, kg	0.58	0.58	0.61	0.57
Average daily feed, kg	2.14	2.17	2.31	2.19
Gain/feed	0.27	0.27	0.27	0.25

[a] Source: Yen et al (1971); used by permission.
[b] Each value is an average for 10 pigs individually fed. Initial weight was 47 kg.
[c] The corn-soy control diet contained 12% total protein.
[d] Total protein content estimated by author (N × 6.25).

fraction" is reported to contain ME at a rate of 1,266 kcal/kg and CGF to contain ME at 1,967 kcal/kg in the dry matter for chicks (Bayley et al, 1971). Corn steep liquor contains ME at 3,110 kcal/kg on a dry matter basis for chicks (Anonymous, 1982a). The ME content of CGF appears to be somewhat variable, and the data suggest that the ME value may vary with the type of bird and the amount of soluble solids. CGF normally contains approximately 25–35% corn soluble solids. Pelleting does not seem to affect the ME content of CGF for chickens or turkeys. CGF has a significant potential for use in commercial layer and breeder rations and grower rations, where energy content is not as critical.

In a Canadian experiment, CGF was used to replace meat meal in rations for growing chicks and laying and breeding birds containing no soybean meal (Slinger et al, 1944). In rations for growing chicks, CGF was satisfactory up to 18% of the diet, with an optimum at around 10%. With layers and breeders, CGF could be used in a mash to replace part of the meat meal on a protein equivalent basis up to 16%. The results reported are not surprising, since meat meal is also marginal in tryptophan content. Other experiments with laying hens have shown that CGF can be included at levels of 10–15% of the diet without decreasing egg production or feed efficiency (Heiman, 1961).

In the author's experience, when CGF is used in balanced poultry diets at levels of up to 10–15% of the ration (assuming that the deficiency of energy, lysine, tryptophan, and calcium and the poor availability of phytin phosphorus and niacin are recognized and corrected for), excellent results can be obtained for chick starters, growers, and layer-breeders (Wright, 1957). However, its use in diets for broiler chickens is rare. The intermediate ME level of CGF for poultry generally gives it poor economic value as an ingredient in a high-nutrient-density ration such as that for the broiler chicken or turkey.

B. CORN GLUTEN MEAL

CGM is the dehydrated protein stream resulting from starch separation (Fig. 1). It has a high nutrient density and usually is sold containing a minimum of 60% total protein (Tables I–III). It is highly digestible, contains ME of 4,131 kcal/kg of dry matter for the chick (slightly higher than corn ME), and is a rich source of available carotenes (49–73 mg/kg) and xanthophylls (244–550 mg/kg, dry matter basis). Its crude protein is highly digestible, a good source of methionine and cystine, but very low in lysine and tryptophan. The amino acid pattern of soybean meal complements that of CGM very well, soybean protein being deficient in methionine and cystine but rich in lysine and tryptophan.

CATTLE

The protein in CGM is insoluble in water and has high rumen bypass properties (Burroughs and Trenkle, 1978; Loerch et al, 1983; Stern et al, 1983). It is superior to most other plant protein sources as measured by using a duodenal collection technique, which gives a relative bypass value of 2 compared to 1 for soybean meal (Klopfenstein et al, 1985b). Tests indicate that approximately 57–60% of the protein will bypass the rumen, compared to about 25% of the protein in soybean meal. Methionine and lysine have been shown to be limiting in ruminant rations under various conditions because the protein synthesized by rumen-active microorganisms is deficient in these two amino acids for cattle. This can be corrected by using CGM as a source of bypass protein with another bypass protein that is high in lysine, such as soybean meal treated with additional heat or an aldehyde source, alfalfa meal, blood meal, meat meal, etc. The economics of using bypass protein sources plus nonprotein nitrogen sources, such as urea, is often quite favorable in comparison to using all native protein sources.

Rumen-degradable protein is needed because the carbon chains from carbohydrates and the nitrogen from ammonia sources serve mostly as the foundation for amino acid synthesis by rumen-active microorganisms. However, certain amino acids or their branched-chain volatile fatty acid precursors may not be produced in sufficient quantity to provide maximum protein synthesis when bypass protein is fed (Dehority et al, 1958). Corn steep liquor solids are an excellent high-energy source of the soluble branched-chain amino acids and other amino acids that are readily available to the rumen organisms for protein synthesis (Wright, 1981).

POULTRY

The high nutrient density, ME, and content of sulfur-bearing amino acids and xanthophyll of CGM makes this product a highly desirable poultry feed ingredient. These factors often make CGM more valuable than soybean meal on a protein-cost basis. The high nutrient density of CGM is taken advantage of in many well-balanced broiler feeds with high energy content, and it can be used at a 10% level or more.

Many studies have been conducted on the bioavailability of xanthophylls in CGM for pigmentation of poultry skin and egg yolks (Marusich and Wilgus, 1968; Halloran, 1970). CGM is often used in layer diets as a source of nutrients and for its available xanthophyll content for egg yolk coloration.

The source of corn, and the length of time it has been stored at the time of milling, effect the amount of xanthophyll in CGM (Tables V and VI). The peak level is reached during the period from November to January. Later in the year, when a substantial amount of corn comes out of storage, a marked reduction in the average carotene and xanthophyll contents of CGM made from this corn can be expected. Because xanthophyll is unstable, the xanthophyll content of CGM and other sources being used must be monitored on a shipment basis to help ensure a uniform degree of pigmentation in the skin or eggs.

The high digestibility of CGM makes this ingredient desirable for use in pet foods, especially where low-residue diets are desired or required.

C. Condensed Fermented Corn Extractives (Steep Liquor)

Corn steep liquor is known officially (AAFCO, 1986) as "condensed fermented corn extractives." It is the concentrated solubles obtained from the corn steeping process (Fig. 1). Its solids are rich in organic nitrogen (44–46% protein on a dry matter basis). About half the nitrogen is present as free amino acids; the balance exists as small peptides with very little intact protein (Christianson et al, 1965). It contains relatively high levels of several important vitamins, trace elements, and lactic acid. The lactic acid (10–30%, dry basis) is synthesized by desirable lactic-acid-producing organisms (Liggett and Koffler, 1948). Fermentation by these organisms takes place in the steeps and elsewhere in the process. The degree of fermentation depends on the conditions set and maintained by the processor.

Because many fermentation products are known to contain "unidentified growth factor" activity, it is not at all surprising that one finds unidentified growth factor activity in steep liquor solids. Corn steep liquor solids have been recognized for many years as a fairly consistent source of unidentified growth factor activity for poultry when fed at a 2.5–5.0% level (Russo and Heiman, 1959; Russo et al, 1960; Simon et al, 1960, Marrett et al, 1968; Potter and Shelton, 1977, 1978). Some evidence indicates that the origin of a chick growth stimulating component in steep liquor is the corn itself (Russo et al, 1960).

Dried steep liquor products are difficult to make and store if the steep liquor solids are much above 45%. The hygroscopic nature of the organic acids and salts in these solids, including a high level of potassium salt of lactic acid, is the primary cause. Several products having considerably more steep liquor solids than CGF (25–30%) have been marketed in the past for use in poultry feed where lower fiber content was desired. Feeding high levels of dry steep liquor products to laying hens has demonstrated an improvement in Haugh units, a measure of egg white integrity, measured on commercial eggs (Hazen and Waldroup, 1972; Lilburn and Jensen, 1984). The effect is not due to lactic acid. At the present time, no such products are being sold, but as the supply of CGF increases, the production of dry steep liquor products for specialized applications may prove to be economically attractive.

Dried distiller's grains with solubles (DDGS) and/or extra amounts of certain trace minerals were found to improve Haugh units (Jensen et al, 1978). Brewer's dried grains also have been demonstrated to improve interior egg quality (Jensen et al, 1976). Therefore, CGF, corn distiller's grains with solubles, and dried brewer's grains should be valuable ingredients in layer feeds, serving to help maintain interior egg quality in commercial egg quality production and storage.

IX. THE DISTILLING INDUSTRY

Fermentation of cereal grains for beverage and industrial alcohol is a very old industry. The distilling industry has long been engaged in the recovery of nutrients from the spent mash from yeast fermentation and their evaluation in animal-feeding regimes. A schematic drawing of a typical process flow, showing the derivation of distiller's feed products, is presented in Fig. 2. This process is thoroughly described in Chapter 19. Briefly, the grains are ground, slurried with water, cooked to gelatinize the starch, cooled, and saccharified with malt and fermented with yeast. The fermented mash then is distilled, and the whole spent stillage is further processed to make distiller's feeds.

The corn wet-milling industry has recently began to produce "fuel alcohol" by recovering the starch in the usual manner, saccharifying it with enzymes, and fermenting it. This process uses steep liquor solids as a major source of supplemental nutrients, and returns the still bottoms back to the CGF. The composition of the resulting feed is similar to that of the original CGF, except that it may contain more solubles, the yeast, and other by-products such as

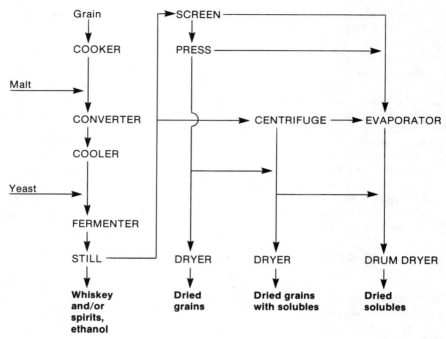

Fig. 2. Distilling process.

TABLE XI
Proximate Analysis of Corn Distiller's Feed By-Products[a]

Proximate Analysis, %	Distiller's Dried Grains		Distiller's Dried Solubles		Distiller's Dried Grains with Solubles		Condensed Distiller's Solubles			
							Product No. 1		Product No. 2	
	As is	DSB[b]	As is	DSB	As is	DSB	As is	DSB	As is	DSB
Moisture	7.5	...	4.5	...	9.0	...	25.0	...	40.0	...
Protein	27.0	29.2	28.5	29.8	27.0	29.7	6.5	8.7	11.6	19.3
Fat	7.6	8.2	9.0	9.4	8.0	8.8	2.4	3.2	2.9	4.8
Fiber										
Crude	12.8	13.8	4.0	4.2	8.5	9.3	0.9	1.2	1.3	2.2
Neutral detergent	40	43.2	21	22	40	44	NA[c]		NA	
Acid detergent	NA	NA	8	8.4	17	18.7	NA		NA	
Ash	2.0	2.2	7.0	7.3	4.5	4.9	1.7	2.3	4.4	7.3
Nitrogen-free extract	43.1	46.6	45.2	47.3	43.3	47.3	63.5	84.6	39.8	66.4
Total digestible nutrients (ruminants)	83.0	89.7	80.0	83.8	82.0	90.1	NA		NA	
Linoleic acid	3.6	3.89	4.4	4.6	3.9	4.3	NA		NA	

[a] Source: Anonymous (1982b); used by permission.
[b] Dry substance basis.
[c] Not available.

glycerol from the alcohol fermentation process. The higher soluble level in the feed, and the presence of glycerol, modify the drying and hygroscopic properties of the feed, increasing its tendency to cake and heat if not dried and cooled sufficiently.

The major distiller's feeds produced are dried distiller's grains (DDG), distiller's dried solubles (DDS), and DDGS. The major source of the cereal, such as corn, must be stated in the name of each product. The proximate analysis and the nutrient and amino acid contents of corn-derived distiller's feeds are given in Tables XI–XIII.

Much research work has been conducted and published on the feeding value of distiller's feed products in cattle, poultry, and swine diets. A major portion of

TABLE XII
Typical Analysis of Corn Distiller's Feed By-Products[a]

Item	Distiller's Dried Grains	Distiller's Dried Solubles	Distiller's Dried Grains with Solubles
Moisture, %	7.5	4.5	9.0
Energy,[b] kcal/kg (DSB[c])			
NE_{milk}	2,368	2,429	2,527
NE_m	2,000	2,105	2,198
NE_p	1,351	11,382	1,439
Metabolizable energy, kcal/kg (DSB)			
Poultry	2,162	2,880	2,879
Swine	1,989	3,120	3,725
Minerals (DSB)			
Potassium, %	0.16	2.2	1.1
Phosphorus, %	0.4	1.68	1.04
Magnesium, %	0.08	0.63	0.38
Calcium, %	0.05	0.31	0.38
Sulfur, %	0.6	0.61	0.55
Sodium, %	0.05	0.16	0.05
Iron, mg/kg	113.5	209.0	219.8
Zinc, mg/kg	54.0	104.7	87.9
Manganese, mg/kg	10.8	62.8	33.0
Copper, mg/kg	16.2	57.6	54.9
Selenium, mg/kg	0.32	0.42	0.33
Cobalt, mg/kg	<0.05	0.083	0.11
Vitamins (DSB)			
Choline, mg/kg	806.4	4,530.9	2,787.4
Niacin, mg/kg	45.4	125.6	84.6
Pantothenic acid, mg/kg	7.1	23.0	12.1
Pyridoxine, mg/kg	4.3	13.6	7.1
Riboflavin, mg/kg	3.6	23.0	9.9
Thiamine, mg/kg	2.2	7.3	3.8
Biotin, mg/kg	0.22	0.5	0.33
Inositol, mg/kg	1,027.0	9,214.7	3.5
Vitamin B-12, mg/kg	0.00027	0.0073	0.0016
α-Tocopherol, mg/kg	...	58.4	33.5
Folic acid, mg/kg	1.2	1.8	1.1

[a] Source: Anonymous (1982b); used by permission.
[b] For cattle. NE = net energy; m = maintenance; p = production.
[c] Dry substance basis.

this work has been sponsored by the Distillers Feed Research Council (Anonymous, 1982b). A summary of results and literature can readily be obtained by writing to the council.

A. Corn Distiller's Dried Grains

DDG contain most of the water-insoluble nutrients in the original corn except for starch, which has been removed during the fermentation. The content of crude protein, fat, and fiber are higher than that in CGF. For ruminants, the ME content of DDG is slightly higher than that of corn, but it is lower for poultry and swine. No starch or soluble solids are present, but as in CGF, the hemicelluloses, cellulose, and insoluble proteins still remain. The composition of the fat remains essentially that of corn because corn oil is not recovered. The amino acid pattern generally reflects the corn source. DDG is a good source of B-vitamins, some of which have been generated by the yeast during fermentation.

The mineral content of DDG is low because it does not contain the solubles. The phosphorus in DDG probably is present primarily in the phytin form and should be treated as in any other plant source for monogastric species. However, it has been reported that considerable amounts of the phosphorus in condensed

TABLE XIII
Typical Amino Acid Content of Some Corn Distiller's Feed By-Products[a]

Item	Distiller's Dried Grains	Distiller's Dried Solubles	Distiller's Dried Grains with Solubles
Moisture, %	7.5	4.5	9.0
Protein, % (DSB[b])	29.2	29.8	29.7
Amino acids,[c] % (DSB)			
Lysine	0.65	0.95	0.60
Methionine	0.54	0.50	0.60
Cystine	0.22	0.40	0.40
Tryptophan	0.22	0.30	0.20
Threonine	0.97	0.98	0.95
Isoleucine	1.08	1.25	1.00
Leucine	3.24	2.60	2.70
Phenylalanine	1.30	1.30	1.20
Tyrosine	0.86	0.95	0.80
Valine	1.41	1.39	1.30
Histidine	0.65	0.63	0.60
Arginine	1.18	1.15	1.00
Glycine	1.08	1.20	1.00
Serine	1.08	1.25	1.00
Alanine	2.16	1.75	1.90
Aspartic acid	1.82	1.90	1.70
Glutamic acid	4.32	6.00	4.20
Proline	2.81	2.90	2.80

[a] Source: Anonymous (1982b); used by permission.
[b] Dry substance basis.
[c] Listed in approximate decreasing order of importance in feed formulation. Lysine through serine are essential amino acids or have sparing effects. Alanine through proline are nonessential.

distiller's solubles is highly available (Nelson et al, 1968). The availability of phosphorus in DDGS is increased substantially because the solids from distiller's solubles are added back and dried on the insoluble DDG. The phosphorus was found to be 43% available in DDGS and 93% available in condensed distiller's solubles. During the fermentation process, the yeast apparently provides the enzyme phytase in sufficient quantities to hydrolyze the soluble phytin to an inorganic, available form of phosphorus.

Wet and dry distiller's grains were compared with wet and dry CGF at the University of Illinois (Firkins et al, 1985). Data from an in situ trial using rumen-cannulated steers indicated that the dry matter of wet and dry CGF disappeared at a faster rate than that of wet or dry distiller's grains. Lambs fed wet or dry distiller's grains had similar digestibilities of nitrogen, dry matter, and NDF. In a trial using rumen-cannulated sheep, wet CGF had higher ruminal pH and acetate-propionate ratios 3 hr after feeding than did those fed dry CGF. Steers fed DDG had dramatically improved performance compared with steers fed soybean meal-corn silage growing diets. Steers fed wet CGF supported similar daily gains but consumed less feed and had better feed-gain ratios than did those fed dry CGF (Table XIV).

Distiller's grains, with or without solubles, are used widely in rations for lactating dairy cows. These distiller's products are palatable and an excellent source of nutrients in concentrates for milk production. DDG are considered a superior source of bypass protein, having a relative value of 2; DDGS are valued at 1.6 (Klopfenstein et al, 1985b). The soluble solids have little if any protein bypass properties.

The high ME and fat levels and the low amount of starch in DDG and DDGS are quite useful in formulating high-nutrient-density concentrates relatively low in starch. The absence of starch helps to limit the level of rapidly fermentable carbohydrates for possible improvement in roughage digestion and utilization and in the acetate-propionate ratio. The latter aids in the maintenance and/or improvement of the level of milk fat content.

Results from a recent growing-fattening study with swine conducted by the University of Kentucky indicate that DDGS containing approximately 0.07% lysine could be used at up to a 10% level in a corn-soy diet on an isolysine basis (Cromwell et al, 1984). These results are similar to those reported by the Illinois workers with CGF.

TABLE XIV
Comparison of Wet and Dry Corn Gluten Feed and Dry Distiller's Grains in a Growing Trial for Steers[a,b]

Item	Dry Corn Gluten Feed	Wet Corn Gluten Feed	Soybean Meal	Dry Distiller's Grains
Initial wt, kg	275	276	274	273
Final wt, kg	422	418	394	426
Daily gain,[c] kg	1.52 cd	1.46 d	1.24 e	1.57 c
Daily feed, kg	10.42 c	9.52 d	9.61 d	8.99 d
Feed/gain	6.86 d	6.52 d	7.73 c	5.71 e

[a] Source: Firkins et al (1985); used by permission.
[b] Steer initial weight, 274 kg; 98-day trial.
[c] Means in the same row that are followed by a different letter differ at the level of $P < 0.05$.

B. Dried Distiller's Solubles

DDS contain 28.5% protein, 9% fat, 4% fiber, and 7% ash (Table XII). The dried solubles are a good source of protein, vitamins, minerals, including phosphorus and potassium, and unidentified growth factor. DDS is used extensively as a source of unidentified growth factor activity in poultry rations at a level of 2.5–5.0% of the ration. DDGS, because they contain soluble solids, are a source of unidentified growth factor for poultry. Also, the dried solubles have been used in milk replacers for calves and pigs and in starter rations. Much work has been conducted and published on the distiller's feed products, largely sponsored by the Distillers Feed Research Council (Anonymous, 1981).

X. CORN DRY MILLING

The corn dry-milling industry is the smallest of the three major corn milling industries producing products for food and industrial purposes. A detailed discussion of the process, process flow, and product composition may be found in Chapter 11.

A. Process

Briefly, the process includes cleaning, conditioning, degermination, cooling, grading, aspiration, grinding, classifying, purifying, drying, and oil recovery from the germ. The process flow diagram given in Chapter 11 shows that the only feed product is hominy feed. All large mills expel the germ or extract it to recover oil. The germcake (expelled) or germ meal (extracted) is blended with bran, corn cleanings, and "through stock," also called "standard meal," which is a stream of fine materials high in oil and fiber content. Some small mills sell the germ to others for oil extraction, and other mills include the whole germ in the

TABLE XV
Hominy Feed: Guaranteed and Typical Analyses[a]

Guaranteed Analysis		Proximate Analysis		
Item	%	Item	%, as is	% DSB[b]
Protein, min.	9.0	Moisture	10.0	..
Fat, min.	4.0	Protein	10.3	11.5
Fiber, max.	6.0	Fat	6.9	7.7
Ash, max.	4.0	Fiber		
		Crude	6.0	6.7
		Neutral detergent	20.5	23.0
		Ash	2.8	3.1
		NFE[c]	64.0	71.1
		TDN[d]	81.9	91.0
		Linoleic acid	3.3	3.7
		Xanthophyll, mg/kg	3.6	4.0

[a] Source: Anonymous (1982b); used by permission.
[b] Dry substance basis.
[c] Nitrogen-free extract.
[d] Total digestible nutrients.

hominy feed. Because of these practices and because the prime product mix may differ from day to day and from one mill to another, hominy feed may be more variable than feed from the other processes. Because of this variability, the historically large number of small mills, and their relatively small volume and slow growth, very little research has been conducted and published on the feeding values of hominy feed. In spite of this, however, hominy feed is well established as a feed ingredient of value and is noted for its high energy value.

TABLE XVI
Hominy Feed: Energy, Vitamins, Minerals, and Amino Acids[a]

Item	Amount
Metabolizable energy, kcal/kg (DSB[b])	
Poultry	3,208
Swine	3,748
Ruminants	3,740
Vitamins, mg/kg (DSB)	
β-Carotene	10.0
Choline	1,280
Niacin	52.0
Pantothenic acid	9.1
Riboflavin	2.3
Thiamine	8.9
Biotin	0.15
Folic acid	0.3
Minerals, DSB	
Potassium, %	0.65
Phosphorus, %	0.57
Magnesium, %	0.26
Chloride, %	0.06
Calcium, %	0.05
Sulfur, %	0.03
Sodium, %	0.09
Iron, mg/kg	75.0
Zinc, mg/kg	3.0
Manganese, mg/kg	16.0
Copper, mg/kg	15.0
Selenium, mg/kg	0.11
Cobalt, mg/kg	0.06
Amino acid content, % (DSB)	
Arginine	0.52
Cystine	0.16
Glycine	0.38
Histidine	0.22
Isoleucine	0.43
Leucine	0.94
Lysine	0.42
Methionine	0.18
Phenylalanine	0.36
Threonine	0.44
Tryptophan	0.12
Tyrosine	0.55
Valine	0.55

[a] Source: Anonymous (1982b); used by permission.
[b] Dry substance basis.

B. Hominy Feed

A typical yield of hominy feed is 35% of corn input, including germ expellor cake (Brekke, 1970). Composition of a typical hominy feed is given in Tables XV and XVI. Since oil yield by expelling is only 1% and by extraction only about 1.25%, much oil remains with hominy feed in any processing configuration. A low-fat product called solvent-extracted hominy feed is also produced.

Hominy feed normally contains about 10.4% protein, 6.9% fat, 6.0% fiber, 20% total (dietary) fiber, and 2,896 kcal of ME per kilogram for poultry on an "as fed" basis (Tables XV and XVI). Many of the uses of hominy feed are similar to those of CGF. The product is generally lower in protein, higher in fat, and much higher in starch than CGF. All the water-soluble substances are present because no water extraction is involved in the process. The removal of most of the fat by expelling or solvent extraction slightly increases the protein, starch, and fiber contents but decreases the fat and ME contents of hominy feed.

Nutritionally, hominy feed resembles whole corn in some respects and can replace some corn in rations. It has a lower bulk density and less starch but more protein, fat, and fiber than whole corn. The ME content for ruminants is 3,740 compared to 3,420 kcal/kg for corn on a dry basis. As with CGF and the distiller's feeds, the starch and the cellulose and hemicellulose in the bran are highly digestible by ruminants.

Poultry utilize the product less efficiently than corn, primarily because of its lower starch and higher celluose and hemicellulose contents. These are partially offset by a higher oil content. Poultry derive approximately 3,208 kcal of ME per kilogram on a dry matter basis from hominy feed compared to 3,818 kcal/kg from corn.

Hominy feed is used widely in dairy feeds because of its high level of fat and high available energy for ruminants. This ingredient is a good substitute for corn in rations for beef cattle and swine. In fattening rations for swine, if a product with a high fat content is used, a possible concern may be its tendency to produce soft pork when used at high levels, because of a higher level of polyunsaturated fat intake (Morrison, 1956). In properly balanced rations, hominy feed can be used satisfactorily to replace a portion of the corn or milo in various poultry rations.

LITERATURE CITED

AAFCO. 1986. Official Publication, 1986. Assoc. Am. Feed Control Officials, Inc., 135 pp.

ALEXANDER, D. E., and CREECH, R. G. 1977. Breeding special industrial and nutritional types. Pages 370-373 in: Corn and Corn Improvement. G. F. Sprague, ed. Am. Soc. Agron., Madison, WI.

ALLRED, J. B., FREY R. E., JENSEN, L. S., and McGINNIS, J. 1957. Studies with chicks on improvement in nutritive value of feed ingredients. Poult. Sci. 36:517-523.

ANONYMOUS. 1981. Feed Formulation. Distillers Feed Res. Council, Cincinnati, OH.

ANONYMOUS. 1982a. Corn Wet-Milled Feed Products, 2nd ed. Corn Refiners Assoc., Washington, DC.

ANONYMOUS. 1982b. Distillers Feeds Research. Distillers Feed Res. Council, Cincinnati, OH.

ANONYMOUS. 1982c. United States-Canadian Tables of Feed Composition. Nutritional Data for United States and Canadian Feeds, 3rd ed. Natl. Acad. Press, Washington, DC.

ASGHARI, M., and HANSON, R. G. 1984. Climate, management, and N effect on corn leaf N, yield, and grain N. Agron. J. 76:911-916.

BALL, G. F. M., and RATCLIFF, P. W. 1978. The analysis of tocopherols in corn oil and bacon fat by thin-layer chromatography and spot density measurement. J. Food Technol. 13:433-443.

BARCLAY, R. A., FAULKNER, D. B., BERGER, L. L., and CMARIK, G. F. 1985. Feedlot performance of steers fed corn bran and soyhulls. (Abstr.) J. Anim. Sci. 61(Suppl. 1):337.

BAYLEY, H. S., SUMMERS, J. D., and SLINGER, S. J. 1971. A nutritional evaluation of corn wet-milling by-products with growing chicks and turkey poults, adult roosters, and turkeys, rats, and swine. Cereal Chem. 48:27-33.

BECKER, D. E. 1964. Niacin in certain cereals is unavailable to the pig. Ill. Res. Winter, pp. 18-19.

BEESON, W. M. 1972. Effect of steam flaking, roasting, popping, and extrusion of grains on their nutritional value for beef cattle. Pages 326-337 in: Effect of Processing on the Nutritional Value of Feeds. Natl. Acad. Sci., Washington, DC.

BIRD, H. R., and OLSON, D. W. 1972. Effect of fertilizer on the protein and amino acid content of yellow corn and implications for feed formulation. Poult. Sci. 51:1353-1358.

BREKKE, O. L. 1970. Corn dry milling industry. Pages 265-291 in: Corn: Culture, Processing, and Products. G. E. Inglett, ed. Avi Publ. Co., Westport, CT.

BURROUGHS, W., and TRENKLE, A. 1978. Naturally protected protein (corn gluten meal) vs unprotected protein (soybean meal) in supplements for calves up to 650 pounds weight. A. S. Leaflet R269. Iowa State Univ. Coop. Ext. Serv., Ames, IA.

CHRISTIANSON, D. D., CAVINS, J. F., and WALL, J. S. 1965. Identification and determination of non-protein nitrogenous substances in corn steep liquor. J. Agric. Food Chem. 13:277-280.

CHRISTIANSON, D. D., WALL, J. S., DIMLER, R. J., and BOOTH, A. N. 1968. Nutritionally unavailable niacin in corn. Isolation and biological activity. J. Agric. Food Chem. 16:100-104.

CLAMOHOY, L. L., PALAD, O. A., NAZARENO, L. E., CASTILLO, L. S., BONTUYAN, J. S., ORDINARIA, P. L., and TURK, K. L. 1968. Feeding value of corn gluten feed in rations for lactating dairy cows and growing dairy heifers. Anim. Husb. Agric. J. Oct. pp. 29-34.

CONTRERAS-GUZMAN, E, and STRONG, F. C., III. 1982. Determination of total tocopherols in grains, grain products, and commercial oils, with only slight saponification, and by a new reaction with cupric ion. J. Agric. Food Chem. 30:1109-1112.

CORNELIUS, S. G., TOTSCH, J. P., and HARMON, B. G. 1973. Metabolizable energy of DSLC for swine. (Abstr.) J. Anim. Sci. 37:277.

CORT, W. M., VICENTE, T. S., WAUYSEK, E. H., and WILLIAMS, B. D. 1983. Vitamin E content of feedstuffs determined by high-performance liquid chromatographic fluorescence. J. Agric. Food Chem. 31:1330-1333.

CROMWELL, G. L., STAHLY, T. S., and MONEGUE, H. J. 1984. Distillers dried grains with solubles for growing-finishing swine. 1984 Swine Res. Rep. Univ. Ky., College of Agric., Agric. Exp. Stn., Dept. Anim. Sci., Lexington.

DAVIS, G. V., and STALLCUP, O. T. 1967. Effect of soybean meal, raw soybeans, corn gluten feed, and urea on the concentration of rumen fluid components at intervals after feeding. J. Dairy Sci. 50:1638-1644.

DEHORITY, B. A., JOHNSON, R. R., BENTLEY, O. G., and MOXON, A. L. 1958. Studies on the metabolism of valine, proline, leucine and isoleucine by rumen microorganisms in vitro. Arch. Biochem. Biophys. 78:15-27.

EARLE, F. R. 1977. Protein and oil in corn: Variation by crop years from 1907 to 1972. Cereal Chem. 54:70-79.

FIRKINS, J. L., BERGER, L. L., FAHEY, G. C., Jr., and MERCHEN, N. R. 1984. Ruminal nitrogen degradability and escape of wet and dry distillers grains and wet and dry corn gluten feeds. J. Dairy Sci. 67:1936-1944.

FIRKINS, J. L., BERGER, L. L., and FAHEY, G. C., Jr. 1985. Evaluation of wet and dry distillers grains and wet and dry corn gluten feeds for ruminants. J. Anim. Sci. 60:847-860.

FLETCHER, D. L., PAPA, C. M., and HALLORAN, H. R. 1985. Utilization and yolk coloring capability of dietary xanthophylls from yellow corn, corn gluten meal, alfalfa, and coastal bermudagrass. Poult. Sci. 64:1458-1463.

HALE, W. H. 1984. Comparison of "wet" processing methods for finishing cattle. Pages 90-98 in: Proc. Feed Grain Utilization Symp. (for Feedlot Cattle). Texas Tech. Univ., Lubbock.

HALL, G. E., HILL, L. D., HATFIELD, E. E., and JENSEN, A. H. 1974. Propionic-acetic acid for high moisture corn preservation. Trans. ASAE 17:379-382, 387.

HALLORAN, H. R. 1970. Review of the most important xanthophyll products in the U.S. Feedstuffs Oct. 10. p. 31.

HAMILTON, T. S., HAMILTON, B. C., CONNOR JOHNSON, R., and MITCHELL, H. H. 1951. The dependence of the physical and chemical composition of the corn kernel on soil fertility and cropping system. Cereal Chem. 28:163-176.

HANSEN, D. W., BRIMHALL, B., and SPRAGUE, G. F. 1946. Relationship of zein to the total protein in corn. Cereal Chem. 23:329-335.

HARMON, B. G., GALO, A., CORNELIUS, S. G., BAKER, D. H., and JENSEN, A. H. 1975a. Condensed fermented corn solubles with germ meal and bran (DSLC) as a nutrient source for swine. I. Amino acid limitations. J. Anim. Sci. 40:242-246.

HARMON, B. G., GALO, A., PETTIGREW, J. E., CORNELIUS, S. G., BAKER, D. H., and JENSEN, A. H. 1975b. Condensed fermented corn solubles with germ meal and bran (DSLC) as a nutrient source for swine. II. Amino acid substitution. J. Anim. Sci. 40:247-250.

HAZEN, K. R., and WALDROUP, P. W. 1972. Improvement of interior egg quality by corn dried steep liquor concentrate. (Abstr.) Poult. Sci. 51:1816-1817.

HEIMAN, V. 1961. Corn gluten feed in layer and breeder rations. Feed Profile 71968. CPC International, Englewood Cliffs, NJ.

HEMKEN, R. W., and BREMEL, D. H. 1982. Possible role of beta-carotene in improving fertility in dairy cattle. J. Dairy Sci. 65:1069-1073.

HESSELTINE, C. W. 1979. Introduction, Definition, and History of Mycotoxins of Importance to Animal Production. Natl. Acad. Sci., Washington, DC.

HOLLIS, G. R., EASTER, R. A., WEIGEL, J. D., and BIDNER, S. G. 1985. Swine. Pages 10-13 in: Corn Gluten Feed, The Future of Feeding. Ill. Corn Growers Assoc., Bloomington.

HUTJENS, M. F., WEIGEL, J. C., and BIDNER, S. G. 1985. Dairy cattle. Pages 6-7 in: Corn Gluten Feed, The Future of Feeding. Ill. Corn Growers Assoc., Bloomington.

JACQUES L., and MOUREAUX, T. 1984. A quantitative analysis of amino acid accumulation in developing grain of normal and opaque-2 maizes. J. Sci. Food Agric. 35:1051-1062.

JASTER, E. H., STAPLES, C. R., McCOY, G. C., and DAVIS, C. L. 1984. Evaluation of wet corn gluten feed, oatlage, sorghum-soybean silage, and alfalfa haylage for dairy heifers. J. Dairy Sci. 67:1976-1978.

JENSEN, A. H., and BECKER, D. E. 1965. Effect of pelleting diets and dietary components in the performance of young pigs. J. Anim. Sci. 24:392-397.

JENSEN, L. S., CHANG, C. H., and MAURICE, D. V. 1976. Improvement in interior egg quality and reduction in liver fat in hens fed brewers dried grains. Poult. Sci. 55:1841-1847.

JENSEN, L. S., CHANG C. H., and WILSON, S. P. 1978. Interior egg quality: Improvement by distillers feeds and trace elements. Poult. Sci. 57:648-654.

KIES, C., KAN, S., and FOX, H. M. 1984. Vitamin B-6 availability from wheat, rice, and corn brans for humans. Nutr. Rep. Int. 30:483-491.

KLOPFENSTEIN, T. K., GOEDEKEN, F. K., BRANDT, R. T., BRITTON, R., and NELSON, M. L. 1985a. Corn Bran as High Fiber Energy Supplement. Neb. Beef Cattle Rep. MP 58. Univ. Neb., Lincoln. 49 pp.

KLOPFENSTEIN, T., STOCK, R., and BRITTON, R. 1985b. Relevance of bypass protein to cattle feeding. Prof. Anim. Sci. 1:27-31.

KUBOTA, J., and ALLOWAY, W. H. 1971. Selenium in Micronutrients in Agriculture. Proc. Symp. sponsored by Tennessee Valley Authority. Soil Sci. Soc. Am., Madison, WI.

KUZMICKY, D. D., KOHLER, G. O., LIVINGSTON, A. L., KNOWLES R. E., and NELSON, J. W. 1968. Pigmentation potency of xanthophyll sources. Poult. Sci. 47:389-397.

LAGUNA, J., and CARPENTER, K. J. 1951. Raw versus processed corn in niacin-deficient diets. J. Nutr. 45:21-28.

LIGGETT, K. W., and KOFFLER, H. 1948. Corn steep liquor in microbiology. Bacteriol. Rev. 12:297-311.

LILBURN, M. S., and JENSEN, L. S. 1984. Evaluation of corn fermentation solubles as a feed ingredient for laying hens. Poult. Sci. 63:542-547.

LOERCH, S. C., BERGER, L. L., GIANOLA, D., and FAHEY, G. C., Jr. 1983. Effects of dietary protein source and energy level on in situ nitrogen disappearance of various protein sources. J. Anim. Sci. 56:206-216.

MARRETT, L. E., BIRD, H. R., and SUNDE, M. L. 1968. Growth promoting effect of corn fermentation condensed solubles. (Abstr.) Poult. Sci. 47:1691-1692.

MARUSICH, W. L., and WILGUS, H. S. 1968. Evaluating the pigmentation value of feedstuffs for poultry rations. Pages 9-22 in: Proc. 1986 Arkansas Formula Feed Conf., Fayetteville, Sept. 26-27. Univ. Ark., Fayetteville.

McCULLOUGH, M. E. 1973. Optimum Feeding of Dairy Animals. Univ. Georgia

Press, Athens.

MERTENS, D. R. 1985. Recent concepts useful in optimizing nutrition of dairy cows. Pages 99-123 in: Proc. Monsanto Tech. Symp., Sept. 16, 1985, Bloomington, MN. Monsanto Chem. Co., St. Louis.

MERTZ, E. T., and BATES, L. S. 1964. Mutant gene that changes protein composition and increases lysine content of maize endosperm. Science 145:279-280.

MITCHELL, H. H., HAMILTON, T. S., and BEADLES, J. R. 1952. The relationship between the protein content of corn and the nutritional value of the protein. J. Nutr. 48:461-476.

MOMCILOVIC, B., and SHAHL, B. C. 1976. Femur zinc, magnesium, and calcium in rats fed tower rapeseed (*Brassica napus*) protein concentrate. Nutr. Rep. Int. 13:135-142.

MORRISON, F. B. 1956. Feeds and Feeding, 22nd ed. Morrison Publ. Co., Ithaca, NY. 423 pp.

NELSON, T. S., FERRARA, L. W., and STORER, N. L. 1968. Phytate phosphorus content of feed ingredients derived from plants. Poult. Sci. 47:1372-1374.

NELSON, T. S. et al. 1976. Hydrolysis of natural phytate phosphorus in the digestive tract of calves. J. Anim. Sci. 42:1509-1512.

NRC. 1978. Nutrient Requirements of Dairy Cattle, 5th ed. Natl. Res. Counc., Natl. Acad. Sci., Washington, DC.

NRC. 1985. Nutrient Requirements of Sheep. Natl. Res. Counc., Natl. Acad. Press, Washington, DC.

O'DELL, B. L., de BOLAND, A. R., and KOIRTYOHANN, S. R. 1972. Distribution of phytate and nutritionally important elements among the morphological components of cereal grains. J. Agric. Food Chem. 20:718-721.

OLDFIELD, J. E. 1985. Nutritional Implications of Selenium. Monsanto Nutr. Update, Vol. 3. March. Monsanto Chem. Co., St. Louis.

OSBORNE, T. B., and MENDEL, L. B. 1914. The nutritive value of the proteins of maize kernel. J. Biol. Chem. 18:1-16.

PAPA, C. M., and FLETCHER, D. L. 1985. Utilization and yolk coloring capability of xanthophylls from synthetic and high xanthophyll concentrates. Poult. Sci. 64:1464-1469.

PEARSON, W. N., STEMPFEL, S. J., VALENZUELA, J. S., UTLEY, M. H., and DARBY, W. J. 1957. The influence of cooked vs raw maize on the growth of rats receiving a 9% casein ration. J. Nutr. 62:445-463.

PIERRE, W. H., DUMENIL, L., JOLLEY, V. D., WEBB, J. R., and SHRADER, W. D. 1977. Relationship between corn yield, expressed as a percentage of maximum, and the N percentage in the grain. I. Various N-rate experiments. Agron. J. 69:215-220.

POTTER, L. M, and SHELTON, J. R. 1977. Evaluation of corn fermentation solubles, menhaden fish meal, methionine, and hydrolyzed feather meal in diets of young turkeys. Poult. Sci. 57:1586-1593.

POTTER, L. M., and SHELTON, J. R. 1978. Evidence of an unidentified growth factor in corn fermentation solubles for young turkeys. Nutr. Rep. Int. 17:509-517.

REINERS, R. A., and HOWLAND, D. W. 1976. Note on vitamins and trace elements in feeds derived from the wet-milling of corn. Cereal Chem. 53:964-968.

RUSSO, J. M., and HEIMAN, V. 1959. Ability of corn fermentation condensed solubles to replace unidentified growth factor sources for chickens. Poult. Sci. 38:1325-1328.

RUSSO, J. M., WATSON, S. A., and HEIMAN, V. 1960. Source of chick growth stimulus in corn fermentation condensed solubles. Poult. Sci. 39:1408-1412.

SCHNAKE, L. M. 1984. Factors to consider in selecting the grain and processing method to use. Pages 113-122 in: Proc. Feed Grains Utilization Symp. (for Feedlot Cattle). Texas Tech. Univ., Lubbock.

SCOTT, N. L., NESHEIM, N. C., and YOUNG, R. J. 1969. Nutrition of the Chicken. M. L. Scott & Assoc., Ithaca, NY. 194 pp.

SEBRELL, W. H., Jr., and HARRIS, R. S. 1972. The Vitamins, 2nd ed., Vol. V. Academic Press, New York.

SHOTWELL, O. L. 1977. Mycotoxins—Corn-related problems. Cereal Foods World 22:524-527.

SIEGMUND, O. H., ed. 1973. The Merck Veterinary Manual, 45th ed. Merck & Co., Inc., Rahway, NJ. 644 pp.

SIMON, T., TSANG, L., and SCHAIBLE, P. J. 1960. The value of corn fermentation solubles in poultry nutrition. Poult. Sci. 39:251-257.

SLINGER, S. J., PETTIT, J. H., and EVANS, E. V. 1944. The use of corn gluten feed to replace meat meal in poultry rations. Sci. Agric. 24:234-239.

STAPLES, C. R., DAVIS, C. L., McCOY, G. C., and CLARK, J. H. 1984. Feeding value of wet corn gluten feed for lactating dairy cows. J. Dairy Sci. 67:1214-1220.

STERN, M. D., RODE, L. M., PRANGE, R. W., STAUFFACHER, R. H., and SATTER, L. D. 1983. Ruminal protein degradation of corn gluten meal in lactating

dairy cattle fitted with duodenal T-type cannulae. J. Anim. Sci. 56:194-205.
TURK, K. L. 1951. Corn gluten feed. Flour Feed. August. pp. 12-15.
UNDERWOOD, E. J. 1962. Trace Elements in Human and Animal Nutrition, 2nd ed. Academic Press, Inc., New York.
USDA. 1985a. Feed Situation Outlook Report. FdS-296. May. U.S. Dept. Agric., Econ. Res. Serv., Washington, DC.
USDA. 1985b. Feed Outlook and Situation Yearbook. FdS-298. U.S. Dept. Agric., Econ. Res. Serv., Washington, DC.
USDA. 1986. Feed Situation Outlook Report. FdS 300, August. U.S. Dept. Agric., Econ. Res. Serv., Washington, DC.
WATSON, S. A. 1962. The yellow carotenoid pigments of corn. Pages 92-100 in: Proc. Corn Res. Conf., 17th. Am. Seed Trade Assoc., Washington, DC.
WATSON, S. A. 1986. Coproducts overview. Pages 104-132 in: 1986 Scientific Conference Proc. Corn Refiners Assoc., Washington, DC.
WEBER, E. J. 1979. The lipids of corn germ and endosperm. J. Am. Oil Chem. Soc. 56:637-641.
WEBER, E. J. 1984. High performance liquid chromatography of the tocols in corn grain. J. Am. Oil Chem. Soc. 61:1231-1234.
WOOKEY, N., and MELVIN, M. A. 1981. The relative economics of wheat and maize as raw materials for starch manufacture. Pages 55-68 in: Cereals: A Renewable Resource. Y. Pomeranz and L. Munck, eds. Am. Assoc. Cereal Chem., St. Paul, MN.
WRIGHT, K. N. 1957. What's ahead for corn gluten feed? Feed Merchant. May, pp. 64-76.
WRIGHT, K. N. 1981. Corn wet-milling byproducts—Corn steep liquor. Pages 29-40 in: Annu. Liquid Feed Symp., 11th, Omaha, NE, Sept 15-16. Am. Feed Industry Assoc., Arlington, VA.
YAHL, K. R., WATSON, S. A., SMITH, R. J., and BARABOLOK, R. 1971. Laboratory wet-milling of corn containing high levels of aflatoxin and a survey of commercial wet-milling products. Cereal Chem. 48:385-391.
YEN, J. T., BAKER, D. H., HARMON, B. G., and JENSEN, A. H. 1971. Corn gluten feed in swine diets and effect of pelleting on tryptophan availability to pigs and rats. J. Anim. Sci. 33:987-991.
YEN, J. T., BROOKS, J. D., and JENSEN, A. H. 1974. Metabolizable energy value of corn gluten feed. J. Anim. Sci. 39:335-337.
YEN, J. R., JENSEN, A. H., and BAKER, D. H. 1976. Assessment of the concentration of biologically available vitamin B-6 in corn and soybean meal. J. Anim. Sci. 42:866-870.
YOUNG, L. G., ASHTON, G. C., and SMITH, G. C. 1977. Estimating the energy value of some feeds for pigs using regression equations. J. Anim. Sci. 44:765-771.

CHAPTER 16

CORN STARCH MODIFICATION AND USES

F. T. ORTHOEFER[1]
A. E. Staley Manufacturing Co.
Decatur, Illinois

I. INTRODUCTION

Starch is the world's most abundant worldwide commodity. Its use has shown steady, rapid growth over the past several years. Currently, approximately 4.2 billion pounds of starch are produced by the corn wet millers in the United States, who also produce about 20 billion pounds for conversion to sweeteners (Chapter 12, Table III). Future expanded uses for starch may result from the emerging ability to modify starch structure and the relatively inexpensive character of the resulting products.

The utilization and development of corn starches began in about 1844. Commercial cornstarch today is derived by the wet-milling process (Chapter 12), using about 9% of the harvested corn for both starch and sweetener production. About 65% of isolated corn starches is used in a variety of industrial products, primarily in paper, adhesives, and coatings. The remainder goes to foods for thickening and gelation.

Starch usage and demand generally follow the national income and output (e.g., the Gross National Product). A wide range of products are produced, both in native and modified forms. Their various characteristics depend upon their intended applications.

II. Properties Involved in Modification

A. Molecular Characteristics

Starch is a polymer of anhydrodextrose units (Mahler and Cordes, 1971). The basic repeating unit involves the linkage of successive D-glucose molecules by α-D-$(1\rightarrow 4)$ glycosidic bonds. Two distinct structural classes exist: linear chains named amylose and branched chains named amylopectin (Fig. 1). (See Chapter 9 for a detailed description.) The abundant hydroxyl groups on the starch molecules impart the characteristic hydrophilic properties. The polymer

---

[1]Present address: Louanna Foods, Inc., Opelousas, LA 70570.

attracts water and is self-attractive through hydrogen bonding. The self-attraction and crystallization tendencies are most readily apparent for the amylose or straight-chain component (Wurzburg, 1978).

The association between the polymer chains results in the formation of an intermolecular network that traps water. At sufficient starch concentration (>3%), gels are produced, whereas in dilute solutions, the associated forms may precipitate. Precipitation is particularly evident for amylose. Amylopectin association is interrupted because of its highly branched character. However, at low temperature, even amylopectin will associate, resulting in decreased water binding and gel formation.

Starch from common corn contains about 27% amylose and 72% amylopectin (Swinkels, 1985). The genetic mutant varieties of importance are waxy maize, which contains essentially 100% amylopectin, and high-amylose corn, having 50–70% amylose.

The characteristic firm, opaque gel produced by common corn is attributed to the amylose fraction. Properties of waxy maize starch are a result of the amylopectin sols produced having a characteristic soft translucent paste form.

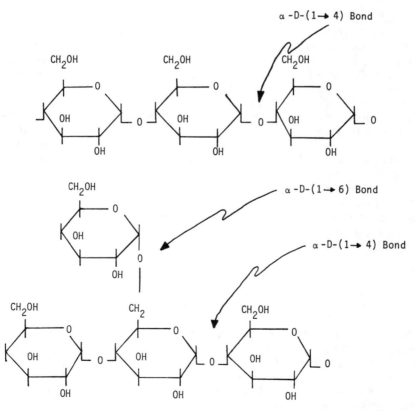

Fig. 1. Chemical structure of amylose (top) and amylopectin (bottom) fractions of the corn starch molecule.

B. Granule Structure and Character

Corn starch is recovered as minute, discrete granules that are an organized aggregation of amylose and amylopectin molecules (Fig. 2) (French, 1984). The granules vary somewhat in shape and range from 5 to 25 μm in diameter (Wurzburg, 1978). They possess a quasicrystalline structure and are insoluble in cold water. When a water slurry of the granules is heated, water is imbibed. With continued heating, the hydrogen bonds holding the granule intact are weakened, and irreversible swelling occurs. The granules eventually burst with continued heating, and solution clarity increases. The viscosity of the solution rises through a maximum and then drops as the granules rupture and disintegrate.

C. Viscosity

The viscosity profile during heating of starch paste is shown in Fig. 3 using a Brabender Visco/amylo/Graph. The pasting of the starch occurs at the major increase in viscosity (Wurzburg, 1978). At the peak viscosity, the swollen granules represent a major factor governing the rheology and viscosity of the cook. Beyond the peak viscosity, the role of the granule disappears, and the characteristics of the dispersion are governed by the size of the molecules and aggregates present (Swinkels, 1985).

Swollen granules exhibit a short, salvelike texture suited primarily for food use, where starch serves as a thickener. Control of granule integrity is dependent on cooking time, temperature, concentration, shear, and pH. As a molecular dispersion, the polymeric molecules develop a cohesive, rubbery texture and have decreased thickening power compared to that of the granular form. Starches for industrial, nonfood applications are generally utilized as molecular dispersions.

III. OBJECTIVES OF MODIFICATION

The objective of starch modification is to alter the physical and chemical characteristics of the native starch to improve functional characteristics

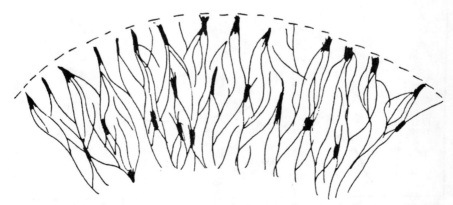

Fig. 2. Concept of the arrangement of starch molecules in a starch granule according to Meyer (1942). Dark areas represent crystalline regions.

(Orthoefer, 1984; Rutenberg and Solarek, 1984). The intent of the modification reaction is to 1) change the character of the dry granule, 2) modify granule integrity, and 3) alter the chemical characteristics. Starch derivatives include modifications that change some of the D-glucopyranosyl units in the molecules. The modifications involve oxidation, esterification, etherification, hydrolysis, and dextrinization (Wurzburg, 1978). The general methods for modification are: acid thinning, bleaching or oxidation, cross-linking or inhibition, substitution or derivatization, and instantizing. A process flow through a corn starch facility is diagrammed in Fig. 4.

Modified starches are defined by source, prior treatment, amylose-amylopectin ratio or content, measure of molecular weight or degree of polymerization, type of derivative or substituent, degree of substitution, physical form, and associated components (Radley, 1968).

Derivatives are prepared to modify the gelation and cooling characteristics of the granular starch, to decrease the tendency toward the retrogradation-crystallization and gelling of the amylose fraction, to increase the water-holding capacity of the starch dispersion, to improve the hydrophilic character, to impart hydrophobic properties, or to introduce ionic groups (Wurzburg, 1978). Modification is important for the continued and increased use of starch to provide thickening, gelling, binding, adhesiveness, and film-forming characteristics.

Multiple treatments may be used to obtain the desired combination of properties. Acid-converted starches, for example, may also be derivatized, and hypochlorite-oxidized starches also may be further converted. Derivatives may

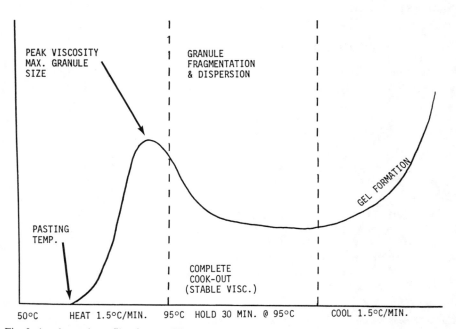

Fig. 3. Amylograph profile of unmodified corn starch.

be dextrinized, acid-converted, or oxidized to obtain the desired viscosity. Cross-linking is often used in combination with derivatization to obtain dispersion viscosity upon processing by high-temperature cooking, high shear, or acid treatment (Orthoefer, 1984).

IV. CHARACTERISTICS OF MODIFIED STARCHES

Modified starches are characterized by a number of factors, which are most often dependent on the type of derivative being produced. Some of these factors are source, molecular weight, type of derivative, nature of the substituent group, and degree of substitution (DS). DS, a characteristic of all derivatized starches, is a measure of the average number, expressed on a molar basis, of hydroxyl groups on each D-glucopyranosyl unit derivatized:

$$DS = 162W/[100 M - (M - 1) W],$$

where W = weight percent of the substituent and M = molecular weight of the substituent. The maximum number possible for starch is 3 DS since three hydroxyl units are available. Most commercially modified starches have a DS of less than 0.2 (Rutenberg and Solarek, 1984).

During derivatization, the formation of the chemical derivative may be followed chemically. Various viscosity techniques may also be employed that indirectly measure the extent of modification (Fleche, 1985). Viscosity measurements employed include the Brabender Visco/amylo/Graph, fluidity techniques, and Brookfield viscometry. Many of the viscosity methods have been modified to meet the needs of the individual starch processor. The Brabender instrument is probably the one most widely used to obtain complete cooking and cooling curves.

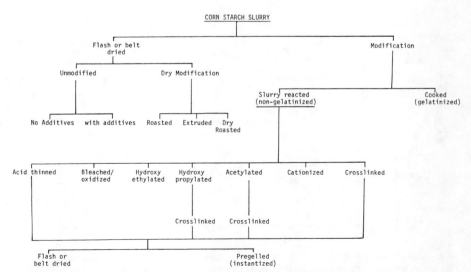

Fig. 4. Process flow chart for preparation of unmodified and modified corn starches.

The curves demonstrate gelatinization temperature, peak viscosities, ease of cooking, stability of breakdown of pastes, setback or gelling, and final viscosities (Fig. 5). Overall, for quality control, the common specifications are: moisture, foreign matter, color, ash, pH, flavor and odor for food, and particular characterization measures for functional properties (Orthoefer, 1984). The common viscosity values utilized are peak, breakdown after peak viscosity, and viscosity after cooling.

V. MANUFACTURE OF DERIVATIVES

The low-DS derivatives are processed in slurry form at 35–45% solids and at pH values of 8–12 (Orthoefer, 1984; Rutenberg and Solarek, 1984; Fleche, 1985). Sodium hydroxide and calcium hydroxide are commonly used to control alkaline pH. Reactions are performed in agitated reaction tanks, called "tubs," having a capacity of 40,000–100,000 lb or higher at temperatures up to about 60°C. Conditions are employed to prevent gelatinization, thus allowing recovery of the modified starch in the granular form. Conversions generally take 4–24 hr. Derivatives are washed, centrifuged, and dried either on a belt or in flash dryers. To prevent swelling under strongly alkaline conditions, salts such as sodium sulfate or sodium chloride are added at a 10–30% concentration. The practical upper limit to the DS is the lowering of the gelatinization temperature by the derivative being prepared.

Higher-DS derivatives have been produced using a nonswelling solvent.

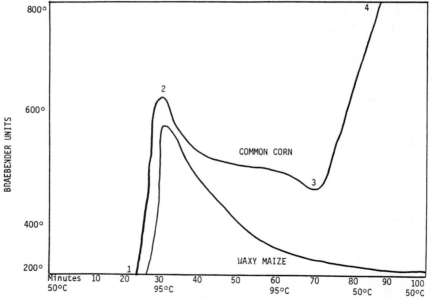

Fig. 5. Amylograph profiles of unmodified common corn starch and waxy maize starch. 1, Gelatinization temperature; 2, peak viscosity temperature; 3, breakdown viscosity; 4, final setback viscosity.

Isopropanol, ethanol, or acetone may be used to prepare derivatized starches that gelatinize when dispersed in water at room temperature (Rutenberg and Solarek, 1984).

Dry reacting of starch is also practiced. The dry starch is heated with the reactants to about 150°C in dry or nonaqueous form, yielding granular products that are readily dispersible in cold water. These derivatives generally contain unreacted reagents plus salts. Nonaqueous solvent washing has been applied to remove the excess salts.

Instant or pregelled corn starches are sometimes included on lists of modified starches (Moore et al, 1984). These dry powders hydrate in cold water and develop a thick viscosity. Production of the pregels is a continuation of the processing of starch. The process may be as simple as drum drying of the slurry after modification and washing. Other techniques used include spray drying, extrusion, and more sophisticated swelling of the starch in nonaqueous solvent-water mixtures (Eastman, 1984). Overall, the pregelled starches represent a compromise between ease of use and functional characteristics.

VI. MODIFICATION

A. Acid Thinning

Acid thinning or conversion is used to reduce the hot paste viscosity of raw or native starch (Radley, 1968; Wurzburg, 1978; Fleche, 1985). Glucosidic linkages joining one anhydroglucose unit to another are cleaved, with water being consumed in the process (Fig. 6).

Both wet and dry processes are used for production of acid-thinned starches. Controlled acid hydrolysis in aqueous suspension leads to the thin-boiling types. The thinning reduces the viscosity to permit these starches to be cooked at higher concentrations than the native starches. Textural properties and cold water solubilities are also affected. Dry acid conversions lead to dextrins, which are most often used in adhesives.

In the wet process, hydrochloric or sulfuric acids are used at 1–3% of the starch weight. The temperature is maintained to prevent gelatinization, generally near 50°C. When the desired reduction in viscosity is obtained, the acid is neutralized and the starch recovered by filtration. The acid depolymerization occurs primarily in the amorphous or noncrystalline regions of the granule, with an apparent increase in the linearity of the starch. The degree of hydrolysis by acid is limited, due to the necessity of recovering the modified granules by filtration. A series of products with viscosities suitable for various applications are prepared. Nonfood uses include paper sizing, calendering, and coating applications (Sanford and Baird, 1983; Bramel, 1986).

Overall, as a result of acid thinning, the gel strength of the cooled paste increases and gel clarity improves. In foods, acid-thinned starches are used in confectionery products, particularly starch jelly candies (Moore et al, 1984).

Dextrins are prepared primarily for nonfood uses. To produce dextrins, dry starch powders are degraded by being roasted in the presence of limited moisture and trace levels of acid, generally hydrochloric. Hydrolysis of the glucosidic bond occurs, as well as molecular rearrangement.

Dextrins are usually more soluble and possess lower viscosity than acid-modified starches. The products are classed according to color as well as water solubility. These include white dextrins, yellow or canary dextrins, and British gums. The type produced depends on the amount of acid and heat, type of equipment, and conditions used for conversion. Dextrins are used primarily in paste formulations, including box pastes, envelope adhesives, carton sealing, and gums for tapes (Fleche, 1985). In foods, dextrins are used as an adhesive for pan coatings, such as those used in jelly bean production, and as a means to develop a clinging mouthfeel in some foods (Moore et al, 1984).

B. Bleaching or Oxidation

Bleaching, or oxidation, represents a single modification producing two types of products (Rutenberg and Solarek, 1984; Wurzburg, 1978). Those treated with low levels of reagents are referred to as bleached starches. They are prepared with hydrogen peroxide, peracetic acid, ammonium persulfate, chlorine as sodium hypochlorite, potassium permanganate, or sodium chlorite. These reagents may be used to whiten corn starch by bleaching xanthophyll and related pigments. Whiteness is desired when the bleached starch is used as a fluidizing agent for dry powders such as confectioners' sugar. The reagents also allow control of microbial counts, which is particularly necessary for food applications. Low thermophile bacterial counts as well as reduction in yeast and mold counts occur.

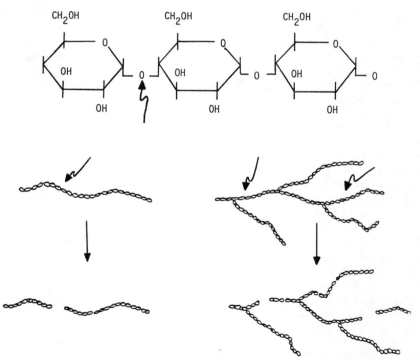

Fig. 6. Mechanism of acid thinning of starch. Arrows indicate points of attack of hydrogen ion.

Bleaching is performed on aqueous slurries. Sodium bisulfite may be used for neutralization. The color bodies solubilized by bleaching are removed during filtering and washing of the starch before drying.

Oxidized starches also are prepared by treatment of aqueous starch slurries. Chlorine as sodium hypochlorite is used at levels of up to 5.5% of the dry starch. The process results in oxidation of hydroxyls, producing a limited number of carboxyl and carbonyl groups along with scission of the glucosidic linkages (Fig. 7). Within the starch granule, oxidation occurs primarily in the amorphous regions. The bulkiness of the carboxyls and carbonyls reduces the tendency of the starch to retrograde. Oxidized starches have less tendency to form a firm gel than do acid-modified starches of comparable hot paste viscosity. Usage is primarily in applications where intermediate to low viscosities are desired and where the acid-converted starches do not show adequate stability.

In foods, bleached corn starch is used for improved adhesion of batter and breading mixes in fried foods (Moore et al, 1984). The carbonyl group probably interacts with the surface of the food, promoting adhesion. Industrially, oxidized starches have been widely used for paper sizing and coating because of their excellent film-forming and binding properties (Bramel, 1986).

C. Cross-Linking

The salvelike texture produced by swollen starch granules is desired particularly in food applications (Orthoefer, 1984). This state is difficult to

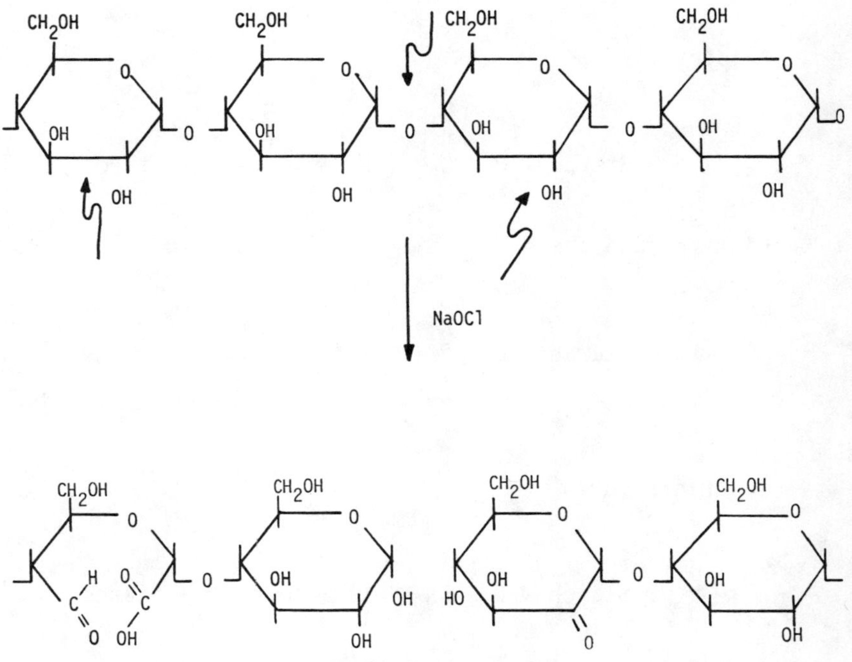

Fig. 7. Hypochlorite oxidation of corn starch, showing carbonyl and carboxyl formation.

maintain in cooking of native starches. Often the starch is undercooked, or it is overcooked and becomes elastic. Control of cooking time, temperature, concentration, shear, and pH are required.

Cross-linking is used to overcome the sensitivity of the starch granules to disruption (Fleche, 1985). It improves the strength of the swollen granules, providing resistance to rupture. No reduction in digestibility or caloric value is seen.

Cross-linking is accomplished by treating the starch in the granular state with difunctional reagents capable of reacting with hydroxyl groups on two different molecules within the granule (Fig. 8). The cross-linked starch, when cooked in water at temperatures that weaken or destroy hydrogen bonds, exhibits granule integrity by virtue of the chemical cross-links. Brabender viscosity profiles of cross-linked starches are shown in Fig. 9.

Two major types of cross-linking reagents are commercially utilized (Rutenberg and Solarek, 19840). Distarch adipate is produced using a mixture of adipic acid and acetic anhydride. Distarch phosphates are produced with phosphorus oxychloride or by roasting with sodium trimetaphosphate. Epichlorohydrin, previously widely used for cross-linking, now is used to a much lesser extent. The reactions are normally run on granular starch suspended in water in the presence of low to moderately alkaline conditions. After completion of the reaction, the slurry is neutralized and the starch is filtered, washed, and dried.

When adipic acid is the reagent, 0.12% or less is used on a commercial basis. The mixed anhydride reacts almost instantly with the starch or is hydrolyzed,

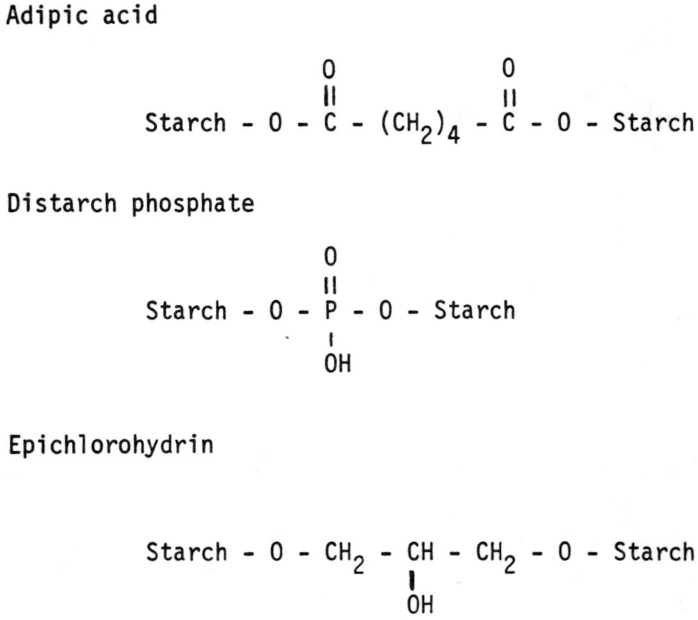

Adipic acid

$$\text{Starch} - O - \overset{\overset{O}{\|}}{C} - (CH_2)_4 - \overset{\overset{O}{\|}}{C} - O - \text{Starch}$$

Distarch phosphate

$$\text{Starch} - O - \underset{\underset{OH}{|}}{\overset{\overset{O}{\|}}{P}} - O - \text{Starch}$$

Epichlorohydrin

$$\text{Starch} - O - CH_2 - \underset{\underset{OH}{|}}{CH} - CH_2 - O - \text{Starch}$$

Fig. 8. Chemical structure of cross-linked starches.

forming the salts of adipic and acetic acids. The reaction with POCl₃ is also extremely rapid, forming the cross-linked starch or sodium phosphate and sodium chloride. After the reaction, the slurry is neutralized, washed, and dried.

Epichlorohydrin reacts with aqueous starch suspensions in the presence of caustic. The level required is very low, with 0.05% or less previously having been used for commercial starch. The starch industry voluntarily stopped using epichlorohydrin in the processing of modified food starches when its hazard to the work environment was identified.

The extent of cross-linking utilized varies, depending on the system in which the starch is used, such as pH, time, temperature, and cooking technique. Generally, the level of cross-linking giving the desired thickening is employed. Cross-linked starches designed for a given use have higher viscosities than unmodified starches under the same conditions. The actual amount of cross-linking corresponds to approximately one cross-link per 1,000 anhydroglucose residues (Wurzburg, 1978).

D. Derivatization or Substitution

Derivatization, or substitution, refers to the introduction of substituent groups on the starch by reacting some of the starch hydroxyls with monofunctional reagents (Rutenberg and Solarek, 1984; Fleche, 1985). For foods, acetate, succinate, octenyl succinate, phosphate, or hydroxypropyl groups are introduced. Industrially important starches are produced utilizing hydroxyethyl and cationic substitutes, as well as others.

Acetylated corn starches are produced with acetic anhydride in the presence of an alkaline starch slurry (Wurzburg, 1978). The reaction is terminated by neutralization. Hydroxypropylation utilizes propylene oxide also at alkaline pH.

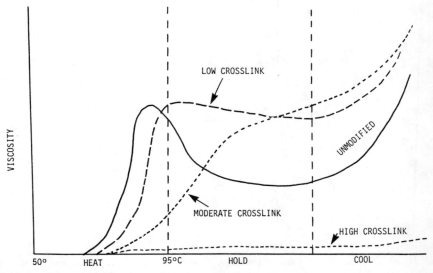

Fig. 9. Amylograph profiles of cross-linked corn starches.

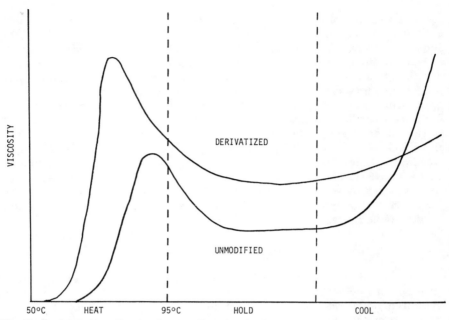

Fig. 10. Amylograph profile, showing the effect of derivatization of corn starch.

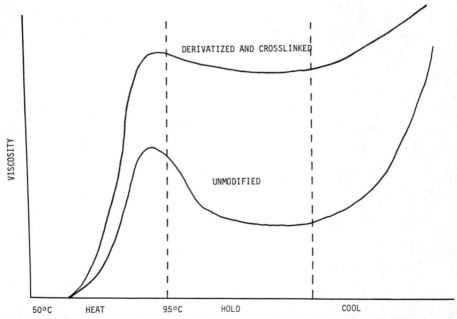

Fig. 11. Amylograph profile, showing the effect of derivatization and cross-linking on viscosity during cooking of corn starch.

Derivatization retards or inhibits the association of the gelatinized amylose chains in the starches. Improved clarity, reduced gelling, and improved water-holding capacity result (Fig. 10). When substituents such as acetate or hydroxypropyl groups are introduced into the molecule, the tendency toward interbranch associations is reduced or restricted, depending on the level of substitution. Acetylated or propylated cooked corn starch can be kept from gelling. Substituted and cooked waxy corn starch can be kept from losing hydration ability and clarity on storage at low temperatures. Succinylation is used to increase the hydrophilic character through the free carboxy moiety, resulting in improved thickening properties.

In foods, combined cross-linking and substitution are used to give thickeners that have the desired processing characteristics and texture and the desired rheological properties during storage and shipment (Fig. 11) (Moore el al, 1984).

The derivatization of starch with acetate, succinate, octenyl succinate, phosphate, or hydroxypropyl groups is cleared for use by the U.S. Food and Drug Administration under 21 CFR 172.892 (d), (e), and (f).

STARCH ESTERS

The principal starch esters are acetates, phosphates, and octenyl succinates (Fleche, 1985). The low-DS acetylated starches are made by treatment of an aqueous starch dispersion with acetic anhydride at pH 7–12 (Fig. 12). Sodium

1) With acetic anhydride

$$\text{Starch - OH} + CH_3 \overset{O}{\underset{\|}{C}} - O - \overset{O}{\underset{\|}{C}} - CH_3 \xrightarrow{\text{NaOH}}$$

$$\text{Starch - O} - \overset{O}{\underset{\|}{C}} - CH_3 + CH_3 - \overset{O}{\underset{\|}{C}} - OH$$

2) With vinyl acetate

$$\text{Starch - OH} + CH_2 = CH - \overset{O}{\underset{\|}{C}} - CH_3 \xrightarrow{\text{NaOH}}$$

$$\text{Starch - O} - \overset{O}{\underset{\|}{C}} - CH_3 + CH_3 - \overset{O}{\underset{\|}{C}} - OH$$

Fig. 12. Chemical reactions of acetylation of starch with acetic anhydride and vinyl acetate.

hydroxide is the preferred alkali, with the optimum pH from 8.0–8.4, but it is temperature dependent. Reactions of 70% efficiencies can be obtained. Alkaline-catalyzed transesterification with vinyl acetate in aqueous medium is also used to acetylate starch.

Acetylated starch having about 1.8% acetyl has a gelatinization temperature 5–7° C lower and a higher hot-peak viscosity at about 10° C lower than those of untreated corn starch. The acetylated starches are more readily dispersed on cooking. On cooling, the corn starch acetates do not reach as high a cold-paste viscosity. This is probably due to acetate interference within the amylose portion. Acetylation has been applied commercially to regular corn, waxy corn, and high-amylose variants. Levels of 0.5–2.5% acetyl groups are utilized.

Acetylation has been applied to impart the thickening needed in food starch applications (Wurzburg, 1978). Generally combined with cross-linking, food starches are prepared that withstand retort processing, high shear, and refrigeration, as well as freeze-thaw conditions, and may be used in low-pH foods. Viscosity stability and clarity are obtained with notable low-temperature storage stability. Cross-linking provides textural properties and resistance to breakdown.

Common food uses of acetylated starches are in fruit pies, gravies, salad dressings, and filled cakes. They can also be used as a filling aid in canning, where viscosity breakdown is desired after heat treatment (Moore et al, 1984; Orthoefer, 1984). Nonfood uses of acetylated starches include warp sizing for textiles, surface sizing for paper, and gummed tape adhesives (Sanford and Baird, 1983; Bramel, 1986). These applications take advantage of the convenient preparation and application of acetylated starches, as well as the film-forming, adhesion, and flexibility properties.

The phosphate monoesters of starch are prepared by roasting blends of starch and orthophosphates at pH 5–6.5 for 0.5–6 hr, at 120–160° C.

$$\text{Starch-OH} + \text{NaH}_2\text{PO}_4 \longrightarrow \text{Starch-O-}\overset{\overset{\displaystyle O}{\|}}{\underset{\underset{\displaystyle O^-}{|}}{P}}\text{-O}^-\text{Na}^+$$

$$\text{Na}^+$$

Tetrasodium pyrophosphate and tripolyphosphates have also been used. The production of cross-linked starch during phosphating is prevented by washing at pH less than 5.0.

Phosphated starches possess good clarity, high viscosity, long cohesive textures, and stability to retrogradation. The primary use for the phosphated derivatives is in the paper industry as wet-end additives (Mentzer, 1984). When combined with cationic derivatives, the phosphated starches improve drainage, increase retention of paper furnish, and improve the strength of the finished sheet. When combined with clays, they have outstanding dispersive properties. Other uses of starch phosphates include textile sizing, thickeners for textile printing, adhesives in corrugating, and flocculants. In foods, the starch phosphates are useful as emulsifiers. Only monosodium orthophosphate and sodium tripolyphosphate are permitted for food use. Residual phosphate may not exceed 0.4% as phosphorus.

The corn starches substituted with octenyl succinate are produced as emulsifying starches, most commonly on a waxy starch base (Moore et al, 1984). Principal uses are for flavor encapsulation, as cloud agents for beverages, and as salad dressing stabilizers.

HYDROXYALKYL STARCHES

The hydroxyalkyl starches possess dispersion stability similar to that of the acetylated starches (Wurzburg, 1978). Substitution with alkylene oxide is usually performed on a 35–45% starch suspension in water under strongly alkaline conditions at temperatures up to 50° C (Fig. 13). The starch reaction tanks are blanketed with inert gas due to the explosivity of the oxide-air mixtures.

Because of the alkalinity required to obtain reaction efficiency and the lowering of the gelatinization temperature with the introduction of hydroxyalkyl groups, salts such as sodium sulfate or sodium chloride are added to minimize the swelling tendency of the starch (Rutenberg and Solarek, 1984). After reaction, the starch is neutralized, filtered, washed, and dried.

Highly substituted starches are produced by "dry reacting" the starch with gaseous alkylene oxide. The alkali and salt are added to the dry powder and the gas is introduced. Ungelatinized hydroxypropylated corn starches that swell in cold water can also be prepared using phosphate salts or carboxylic acid salts as catalysts in place of the alkali.

Common degrees of substitution of 0.1–0.3 are prepared. During substitution, the 2-hydroxyl position of the glucose is favored. At low substitution levels with ethylene oxide, monosubstitution dominates. At high substitution levels, polysubstitution on hydroxyethyl substituents occurs.

The properties of the low-DS, hydroxyethyl, and hydoxypropyl starches are similar to those of the low-DS starch acetates (Wurzburg, 1978). Granule swelling on cooking increases; clarity and cohesiveness increase; and tendency to "gel" on cooling and aging is reduced.

As with the acetates, low-temperature stability, solubility, clarity, and freeze-thaw stability occur. Starch ethers, unlike the esters, are pH-stable and can be used under strongly alkaline conditions. The ether-substituted starches also may be further modified by acid conversion, oxidation, or dextrinization.

The higher the degree of substitution, the lower the gelatinization temperature

$$\text{Starch} - \text{OH} + \text{H}_2\text{C} \underset{\text{O}}{\overset{}{\diagdown\!\!\diagup}} \text{CH} - \text{R} \xrightarrow{\text{NaOH}}$$

$$\text{Starch} - \text{O} - \text{CH}_2 - \underset{\text{OH}}{\text{CH}} - \text{R}$$

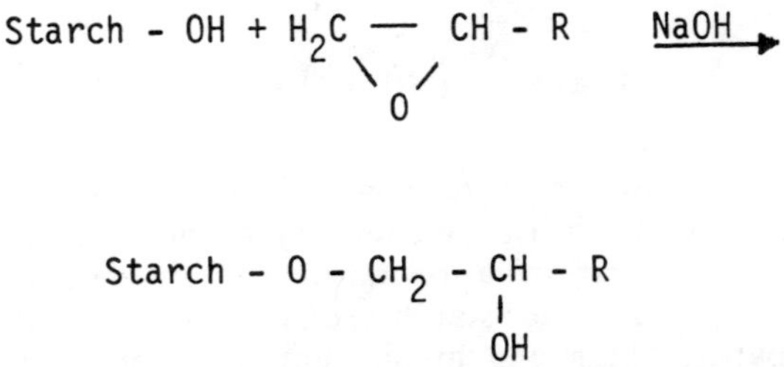

Fig. 13. Reaction of alkylene oxide with starch, forming starch hydroxyalkyl ether.

(Fleche, 1985). Lowering of the gelatinization temperature is used to monitor the progress of the reaction.

The hydroxyethyl starches are prepared for nonfood uses primarily for surface sizing and paper coatings (Bramel, 1986). In surface sizing, they give strength and stiffness to the paper, whereas in coatings, they may be used alone or in combination with other pigment binders. Viscosity stability is a major attribute of the ethylated starches, along with film-forming. The water-holding capacity is useful in preparing adhesives for bag pastes, case sealing, and label and envelope adhesives. A variety of viscosity grades of ethylated corn starches are commercially available.

The hydroxypropyl substituted corn starches are sold primarily to the food industry. Normally cross-linked, they provide the desired viscosity, texture, and stability for processing and storage. Common uses are similar to those of the acetylated starches, including use in gravies, sauces, pie fillings, puddings, and salad dressings. Industrial uses include adhesives and wall board binders.

CATIONIC STARCHES

Cationic starches are used extensively in the paper industry, primarily as wet-end additives during formation of the paper (Fleche, 1985; Bramel, 1986). Retention and drainage of the pulp and strength of the finished sheet are benefits seen. These starches are also used in paper sizing, textile sizing, and coatings, as binders in laundry starches, and as flocculants. Total usage is estimated at near 90,000 t.

The most important cationic derivatives are those containing tertiary amino or quarternary ammonium groups (Rutenberg and Solarek, 1984). Derivatives of 0.1–0.4 DS are prepared. Examples of each are:

Quarternary:

$$\text{Starch-OH} + \overset{O}{\overset{\diagup \diagdown}{CH_2-CH}}-CH_2-{}^+N-R_3 \; Cl^- \longrightarrow \text{Starch-OCH}_2\overset{OH}{\underset{|}{C}H}-CH_2-N^+-R_3 \; Cl^-$$

Tertiary:

$$\text{Starch-OH} + Cl-CH_2-CH_2-N-R_2$$
$$\downarrow$$
$$\text{Starch-O-CH}_2-CH_2-N-R_2$$
$$\downarrow$$
$$[\text{Starch-O-CH}_2-CH_2-N-R_2]^+ \; Cl^-$$

Under alkaline conditions, the tertiary derivative is a free base. Neutralization converts the free base to the cationic tertiary ammonium salt.

As noted previously, cationization may be combined with phosphating of the starch. These dually derivatized starches, or amphoteric starches, are reported to be useful in papermaking and in pigment retention. Cationic starches are prepared on base starches from common and waxy corn. The starches may be acid-thinned or bleached during derivatization to meet a particular viscosity or functional requirement.

E. Other Derivatives

Starch has important possibilities as a base for other derivatives (Sanford and Baird, 1983). Some of these are manufactured on a limited scale, whereas others show future potential. The starch xanthates are prepared under strongly alkaline conditions with carbon disulfide. Recent interest has centered on their use as encapsulants for pesticides. Dialdehyde starches have been proposed for use in paper as a wet-strength additive, in water-resistant adhesives, and in leather tanning. These are prepared by periodic acid oxidation of the 2,3-glycol structure specifically to aldehydes. The high process cost has limited their development.

F. Starch as a Source of Chemicals

Starch is a prime candidate for use as a raw material because it is readily available at low cost and can be converted into a variety of useful products by chemical and biochemical means (Rutenberg and Solarek, 1984; Swinkels, 1985). Fermentation of starch to ethanol theoretically produces 567.9 g of ethanol per kilogram of starch. Gasoline blends (90:10) have received continued commercial interest (Chapter 19).

Conversion of starch to glucose and then to cyclic and acyclic polyols, aldehydes, ketones, acids, esters, and ethers has been investigated (Fleche, 1985). Sorbitol, the hexahydroxyl derivative of glucose, is the most widely produced and utilized derivative. Methyl glucoside (methyl α-D-glucopyranoside), prepared by reacting glucose with methanol, has shown promise for making polyesters for rigid polyurethane foam production. Higher-alkyl glucosides are sold commercially as nonionic surfactants. Oil-modified urethane coatings and alkyds also have been investigated.

Starch has been proposed for use in plastics because it is a renewable resource and is biodegradable. Starch-polyvinyl chloride, starch-polyethylene, starch-polyvinyl alcohol, and others have received attention.

Graft polymerization of starch has also been investigated. The graft polymers produced have utilized the generation of a free radical on the starch backbone, using cerium catalysis and then allowing the free radical to react with a polymerizable vinyl or acrylic monomer. Styrene, isoprene, acrylonitrile, and alkyl acrylates and methacrylates have been studied (Weaver et al, 1976).

VII. CORN STARCH UTILIZATION

Corn starch utilization is split, with about 35% going to foods and the remainder to nonfood applications (Whistler, 1984). The brewing industry accounts for about 20% of the edible use, with similar quantities in chemicals, drugs, and pharmaceuticals. The next largest uses are in canning (7.5%) and confectionery (6.5%). The baking industry uses approximately 6%. Condiments, prepared mixes, and miscellaneous uses account for the remainder.

Nonfood use is primarily in the paper industries, which account for 86% of the total. Adhesives, textiles, and other miscellaneous uses such as laundry starch account for the remaining 14%. Laminating and corrugating uses make up the single largest use, 26% of the total. Consumption in other paper applications such as wet-end addition, size press, calendering, and coatings make up 68%. Building products account for the remaining 6%.

A. In Foods

In foods, corn starches are used in the granular form or as dry powders, as swollen granules, and in molecular dispersions (Table I). As dry powders, the starches are treated to provide anticaking, dusting, or molding properties. Various treatments, such as redrying or addition of flow agents and oils, impart the desired compaction or flow characteristics (Orthoefer, 1984).

When using starches as swollen granules, one must pay attention to the processing conditions of time, temperature, pH, and shear. In the selection of a particular starch, one must consider the method of preparation, the desired function of the starch, the method of preparation of the fininshed food, the storage condition, the eating quality, and the desired appearance.

Overall, the type of modification as well as the base starch must be balanced with the product and the processing equipment. Adjustment of modification to obtain fully swollen granules is generally desired. In some instances, a highly delayed gelatinization through cross-linking may be desired for efficient heat penetration during retort processing of foods. Substitution to obtain instability such as for filling aids for canning is also utilized.

Many of the differences in native starch gel characteristics carry through the starch modification process. Some examples are: 1) high clarity from substituted, cross-linked waxy starch, 2) a slightly gelled texture from a substituted cross-linked starch from common corn, and 3) a film-forming starch from an acid-thinned high-amylose corn starch. Interfacing the base starch with the use, type and degree of substitution or modification, amount of cross-linking, and use of cook-up or pregelled starch is a complex process.

B. In Paper

Approximately 90% of the starch used in paper is derived from corn. Starch addition results in improvements in the physical properties of paper and reduced

TABLE I
Food Uses of Corn Starch

Physical Form of Starch	Application
Granules	Anticaking agent
	Dusting
	Diluent
	Molding
Swollen granules	Viscosity
	Texture
	Solids suspension
	Processing aid
	Binder
Molecular dispersion	Film-forming
	Colloid protection
	Encapsulation

production costs; however, native starches produce thick and unstable cooked pastes that have only limited use in papermaking applications (Mentzer, 1984). Therefore, starch is modified to make it a more functional and versatile papermaking additive (Casey, 1983). For food packaging, "Industrial Starch Modified" is classified by the U.S. Food and Drug Administration under 21 CFR 178.3520.

Starches may be modified in the paper mill as well as by the starch producer. Modifications performed in the mill may consist of cationization, enzyme thinning, high-temperature conversion, and chemical-thermal conversion techniques. Conversion with enzymes requires special equipment capable of handling the extremely high initial viscosity of the starch paste before thinning. The uniformity of the conversion is dependent on close control of pH, time, temperature, and concentration of enzyme promoters such as calcium salts.

Modification of starches at high temperatures utilizes jet cookers. Limited thinning occurs due to shear and heat when the starch is cooked at about 160°C. Chemical modifiers are added before cooking. Ammonium persulfate at a concentration of 0.05–0.2% based on starch weight is the preferred additive, with hold times of 4–5 min at about 150°C. Other additives used have been hydrogen peroxide, sodium hypochlorite, alum, and citric acid.

Control of pH and storage temperatures is required to maintain the viscosity of the cooked starch. Avoiding the 67–87°C temperature range prevents amylose crystallization, and storage below 54°C prevents irreversible retrogradation.

Differences exist between starches converted at the paper mill and premodified starches. Low-molecular-weight starch components and sugars in premodified starches are removed by washing before drying (Bramel, 1986). These low-molecular-weight products have poor film-forming and binding properties compared to those of starch. The high-temperature conversion used in the mill gives more thoroughly dispersed starches, which are more likely to fractionate and retrograde upon cooling. With ammonium persulfate or other oxidizing chemicals used in the conversion, some carboxyl formation occurs during mill conversion. Enzyme-modified starches are less harshly treated.

Special converting corn starches are marketed by corn starch producers for conversion at the paper mill. Unmodified common corn starch with additives may be buffered to adjust pH; salts may be added for enzyme activity and antifoams added to control foaming. Other starches may be supplied that are slightly bleached and lightly acid-thinned. These provide both convenience and improved uniformity during paper mill conversion. Some starches are lightly modified to have lower peak paste viscosities, contain less protein, and have a reduced tendency to fractionate.

The commercially modified starches consist of acid-thinned, oxidized, acetylated, and hydroxyethylated types. The acid-thinned starches are used for paper sizing, calendering, and coating applications (Bramel, 1986). Acid-thinned products, although the least costly of the premodified starches, are not widely used because of their low film strength and greater tendency to retrograde.

Oxidized starches with thin to thick viscosities and carboxyl contents up to 1% are produced for sizing and coating applications. The oxidized starches exhibit excellent film-forming and binding properties plus resistance to retrogradation and congealing. The use of oxidized products in paper is declining because of their dispersing effect on fillers and fines, which also have an anionic character. This dispersing effect is seen in the wet end of the paper machine, where repulped paper containing oxidized starch lowers retention and efficiency.

Acetylated starches, sometimes referred to as gums, are used in coatings and surface sizes because of their excellent film-forming and reduced gelling tendencies. The instability of the acetate substitution somewhat limits their use, since free acid is generated during cooking and during storage, resulting in equipment corrosion and, of course, loss of acetate functionality.

Hydroxyethylated starches are produced at different levels of substitution and varying viscosities. The hydroxyethyl substitution is pH-stable as opposed to the less stable ester linkage. The films are strong, flexible, and clear. The cooked pastes are noncongealing and viscosity-stable, with desirable water-holding and rheological characteristics. Good compatibility with latex coatings is observed.

The diverse functional characteristics of starches produced for the paper industry ensure that starch will maintain its position relative to competitive products such as petroleum-based substitutes.

C. Future for Starch Utilization

The utilization of corn starch in foods and industrial markets has been challenged by substitute as well as competitive products. The use of more desirable native starches in foods has been met with processing and modification of either waxy or common corn starches. However, in nonfood utilization, the challenge is more severe. Hot-melt and pressure-sensitive adhesives have taken much of the former dextrin market. Latexes of various types have displaced starches in paper coatings and other applications.

New starch derivatives continue to be explored. Attempts are being made to penetrate markets that require functional properties outside those required by the traditional starch markets. Starch will likely maintain its position with continued technical improvement.

The possibilities of starch as an agricultural raw material are attractive. Much research attention, both academic and industrial, has focused on corn starch as a source of chemicals. The abundance, availability, and low price of corn starch make it an attractive commodity.

LITERATURE CITED

BRAMEL, G. F. 1986. Modified starches for surface coatings or paper. Tappi 69:54-56.

CASEY, J. P. 1983. Pulp and Paper Chemistry and Technology, 3rd ed. Vol. 4. John Wiley & Sons, Inc., New York.

EASTMAN, G., and MOORE, C. O. 1984. Cold-water-soluble granular starch. U. S. patent 4,456,702.

EASTMAN, J. 1984. Granular starch ethers. U.S. patent 4,452,978.

FLECHE, G. 1985. Chemical modification and degradation of starch. Pages 73-100 in: Starch Conversion Technology. G. Van Beynum and J. A. Roels, eds. Marcel Dekker, Inc., New York.

FRENCH, D. 1984. Organization of starch granules. Pages 184-248 in: Starch: Chemistry and Technology, 2nd ed. R. L. Whistler, J. N. BeMiller, and E. F. Paschall, eds. Academic Press, Orlando, FL.

MAHLER, H. R., and CORDES, E. H. 1971. Biological Chemistry. Harper and Row, New York.

MENTZER, M. J. 1984. Starch in the paper industry. Pages 543-575 in: Starch Chemistry and Technology. R. L. Whistler, J. N.

BeMiller, and E. F. Paschall, eds. Academic Press, Orlando, FL.
MEYER, K. H. 1942. Recent developments in starch chemistry. Pages 143-182 in: Advances in Colloid Science. E. O. Kramer, ed. Interscience, New York.
MOORE, C. O. TUSCHHOFF, J. V., HASTINGS, C. W., and SCHANEFELT, R. V. 1984. Applications of starches in foods. Pages 579-592 in: Starch: Chemistry and Technology, 2nd ed. R. L. Whistler, J. N. BeMiller, and E. F. Paschall, eds. Academic Press, Orlando, FL.
ORTHOEFER, F. T. 1984. Industrial modification of starch. (Abstr.) Cereal Foods World 29:507.
RADLEY J. A. 1968. Starch and Its Derivatives, 4th ed. Chapman and Hall, London.
RUTENBERG, M. W., and SOLAREK, D. 1984. Starch derivatives: Production and uses. Pages 312-388 in: Starch: Chemistry and Technology, 2nd ed. R. L. Whistler, J. N. BeMiller, and E. F. Paschall, eds. Academic Press, Orlando, FL.
SANFORD, P. A., and BAIRD, J. 1983. Industrial utilization of polysaccharides. Pages 411-490 in: The Polysaccharides, Vol. 2. G. O. Aspinall, ed. Academic Press, New York.
SWINKELS, J. 1985. Source of starch, its chemistry and physics. Pages 15-46 in: Starch Conversion Technology. G. Van Beynum and J. A. Roels, eds. Marcel Dekker, Inc., New York.
WEAVER, M. O., BAGLEY, E. B., FANTA, G. F., and DOANE, W. M. 1976. Highly absorbent starch containing polymeric compositions. U. S. patent 3,997,484.
WHISTLER, R. L. 1984. History and future expectations of starch use. Pages 1-10 in: Starch: Chemistry and Technology, 2nd ed. R. L. Whistler, J. N. BeMiller, and E. F. Paschall, eds. Academic Press, Orlando, FL.
WURZBURG, O. B. 1978. Starch, modified starch and dextrin. Pages 23-32 in: Products of the Corn Refining Industry: Seminar Proceedings. Corn Refiners Assoc., Inc., Washington DC.

CHAPTER 17

CORN SWEETENERS

RONALD E. HEBEDA
Enzyme Bio-Systems Ltd.
Bedford Park, Illinois

I. INTRODUCTION

Corn sweeteners are defined as those nutritive sweeteners manufactured from corn starch. Included in this category are crystalline monohydrate and anhydrous dextrose (D-glucose, corn sugar), high-fructose corn syrup (HFCS) (isosyrup, isoglucose), regular corn syrups (glucose syrup, starch syrup), and maltodextrins (hydrolyzed cereal solids). A variety of enzymatic and acid-catalyzed processes are used for the manufacture of corn sweeteners. For instance, dextrose is prepared by enzymatic hydrolysis of starch and purified by crystallization. HFCS is produced by partial enzymatic isomerization of dextrose hydrolysate, followed, in some cases, by enrichment to a higher fructose level by chromatographic separation. Regular corn syrups and maltodextrins are manufactured by partial hydrolysis of corn starch, using acid, acid-enzyme, or enzymatic techniques. Corn sweeteners are refined and sold in dry or syrup form for a wide range of uses depending on individual functional properties. U.S. manufacturers of corn sweeteners are listed in Table I, along with the estimated capacity of each as a percentage of the total industry.

Sweeteners are produced commercially throughout the world from other starch sources such as wheat, rice, potato, sweet potato, and tapioca. However, within the United States, corn starch is used exclusively as the raw material for production of starch-based sweeteners. This is due to the favorable cost, availability, and storage properties of corn, as well as the benefit of revenues generated from coproducts such as starch, oil, and animal feed. Typically, about 11–13% of the U.S. corn utilization has been used in wet-milling operations (Table II; see also Chapter 7). The one exception was in 1983. Of the total purified starch recovered from corn, about 75% is used for sweetener production. Consequently, corn sweeteners are major commodity products throughout the world and are used extensively in various food applications. Utilization has increased rapidly during the last 25 years, reaching 51% of the per capita consumption of corn sweeteners and sucrose in 1985 (Table III).

II. HISTORY

The study of starch-based sweeteners began in the early 19th century when a search was initiated for a sucrose replacement. During a French-English war, a British blockade of European ports caused a cane sugar shortage that led to studies on alternate sweeteners. J. L. Proust was awarded a prize by Napoleon for his work on preparing dextrose from grapes (Wichelhaus, 1913); however, Kirchoff's work in 1811 (Kirchoff, 1811) provided the basis for an eventual corn sweetener industry. While searching for a gum arabic substitute as a binder for clay, Kirchoff discovered that a sweet material was formed when a suspension of potato starch was heated in the presence of sulfuric acid. Samples were made of dry and evaporated syrups, as well as dextrose crystals separated from the syrup. For this discovery, Kirchoff was decorated by the Russian emperor and given a lifetime pension. In 1815, de Saussere reported that acid conversion of starch to sugar proceeded by hydrolysis rather than dehydration and that starch sugar was the same as grape sugar.

U.S. manufacture of dextrose by hydrolysis of potato starch began commercially in 1842, and corn sugar was first manufactured in a Buffalo, New York, plant in 1866. By 1876, 47 U.S. factories were in production and the corn sweetener industry was firmly established.

Research efforts were soon directed toward producing new and more sophisticated products. A major breakthrough occurred with the development of a process for manufacturing pure crystalline dextrose. Early efforts between 1880 and 1920 were of limited success and required repeated crystallization or crystallization from nonaqueous solvents. Large commercial quantities were not manufactured, but sufficient material was prepared to establish the therapeutic value of crystalline dextrose (Porst, 1921). In 1921, however, a commercially viable process was developed for crystallizing dextrose using carefully controlled conditions of concentration, agitation, temperature, and seed crystals (Newkirk, 1923).

Until the 1930s, acid was used exclusively for converting starch to syrups and sugar. Products and applications were often limited, however, due to the formation of off-flavors and color as a result of the nonspecific nature of acid hydrolysis. In the late 1930s, research in the area of fungal-derived enzymes led

TABLE I
Estimated 1984 U.S. Corn Sweetener Manufacturing Capacity

Company	Location	Estimated Capacity (% of total industry)[a]
ADM Foods	Cedar Rapids, IA	31
A.E. Staley Manufacturing Co.	Decatur, IL	24
Cargill, Inc.	Minneapolis, MN	17
CPC International, Inc.	Englewood Cliffs, NJ	12
American Maize Products Co.	Stamford, CT	7
Coors	Johnstown, CA	<5
Grain Processing Corporation	Muscatine, IA	<5
Hubinger	Keokuk, IA	<5
Minnesota Corn Processors	Marshall, MN	<5
Penick and Ford	Cedar Rapids, IA	<5

[a] Data from Greditor (1983).

to an acid-enzyme process (Dale and Langlois, 1940) for the production of a sweeter, less viscous syrup that found application in many areas. This improvement, along with reduced regulatory constraints on corn syrup use, resulted in an increase in corn syrup and corn sugar consumption by 86 and 68%, respectively, between 1940 and 1942 (Jones and Thomason, 1951).

The success of the acid-enzyme process inspired further research efforts to improve dextrose production using enzymes. By the 1960s, commercial processes were developed that used a fungal enzyme (glucoamylase) to convert partially acid-hydrolyzed starch slurries to higher dextrose levels than those attained with acid hydrolysis alone. By the 1970s, bacterial enzymes (α-amylases) were commercialized that were effective for replacing the acid hydrolysis step. Thus, a total enzyme process for dextrose production was achieved that provided an increase in dextrose yield.

The introduction of enzymatic processes had taken the corn sweetener industry a long way from Kirchoff's discovery in 1811; however, one goal not attained was the commercial production of a corn sweetener at least as sweet as sucrose. Such a sugar, fructose, occurs naturally in many fruits and vegetables and is about 1.8 times as sweet as sucrose. Fructose was first isolated in 1847 (Doty, 1980) and was prepared by alkaline isomerization of dextrose in 1895 (Lobry DeBruyn and Van Eckenstein, 1895). Many attempts were made to

TABLE II
Utilization of Corn in Wet Milling[a]

Crop Year[b]	Corn Disappearance[c] (bu × 10^6)	Corn Used in Wet Milling		
		Sweeteners (bu × 10^6)	Total (bu × 10^6)	Total Usage as Percent of Disappearance
1980/81	7,223.4	350	515	7.1
1981/82	6,980.2	375	595	8.5
1982/83	7,290.1	400	665	9.1
1983/84	6,573.9	445	740	11.2
1984/85	7,003.4	500	795	11.4
1985/86[d]	6,845.0	520	840	12.3

[a] Source: USDA (1986).
[b] Oct. 1 to Sept. 30.
[c] To convert to metric tons multiply by 0.0254. Values include exports.
[d] Projected.

TABLE III
Per Capita Consumption of Corn Sweeteners and Sucrose

Year	Sucrose (kg)	Corn Sweeteners (kg)	Total (kg)	Corn Sweeteners (% of total)
1900[a]	29.5	...	29.5	...
1925[a]	46.4	...	46.4	...
1960[a]	44.4	5.2	49.6	10.5
1975[b]	40.5	12.5	53.0	23.6
1980[b]	37.9	18.2	56.1	32.4
1985[b]	28.7	29.5	58.2	50.7

[a] Data from Kean (1978).
[b] Data from Anonymous (1986).

develop a commercial process for producing a fructose syrup by alkaline isomerization. In 1943, a syrup containing 20% fructose was produced commercially (Cantor and Hobbs, 1944) for use as a moisturizing agent in tobacco; however, a food product was not manufactured due to problems of color, off-flavor, degradation products, and low fructose yield.

Studies in the early 1950s indicated that enzymes previously thought only to catalyze the isomerization of xylose could also convert dextrose to fructose. This observation was reported in 1957 (Marshall and Kooi, 1957) and patented in 1960 (Marshall, 1960), although the patent was later invalidated due to misidentification of the enzyme source. Interest in a commercial enzymatic process for isomerization of dextrose decreased in the United States, but research continued in Japan and eventually led to the discovery of enzymes that produced high yields of fructose. Commercial production of HFCS began in Japan in 1966 (Takasaki, 1966), and a U.S. patent was issued in 1971 (Takasaki and Tanabe, 1971) that consolidated several Japanese applications. Japanese technology was licensed for use in the United States, and a limited amount of 15% fructose syrup was produced in 1967 (Mermelstein, 1975), followed by a 42% fructose product the next year. Initially, HFCS products were produced in a batchwise system using soluble enzyme. However, concentrated effort within the corn refining industry resulted in a continuous system using bound enzyme technology in 1972 (Lloyd et al, 1972; Thompson et al, 1974); other industrial processes were developed shortly thereafter.

In the years immediately following the development of HFCS, sucrose prices increased dramatically due to increased demand and reduced inventory. As a consequence, the less costly HFCS became increasingly popular as a sugar substitute, and sales increased significantly. U.S. shipments of HFCS in metric tons (71 wt % basis) amounted to 21 in 1967, 102 in 1970, 679 in 1975, and 1,553 in 1978 (Peckham, 1979). The development of commercial processes for HFCS production occurred at the right time to take advantage of high sucrose prices, and, therefore, HFCS has continued as a viable alternative sweetener to sucrose in many applications.

III. CHEMISTRY

A. Starch Hydrolysis

Starch used to produce corn sweeteners is generally prepared from U.S. grade No. 2 corn by the wet-milling process. The starch is about 99% pure and contains, on a dry basis (db), 0.25–0.35% protein, 0.5–0.6% lipid, and less than 0.1% minerals.

Corn starch granules are composed of two distinct fractions, i.e., amylose, a linear polymer of 1,4-linked α-D-glucopyranosyl units, and amylopectin, a branched polymer containing short 1,4-linked glucose chains connected by 1,6 branch points. Typically, the amylose fraction represents 25–30% of the starch granule and exhibits a molecular weight of about 250,000, whereas the much larger amylopectin fraction represents 70–75% of the granule and has a molecular weight of 50–500 million (Zobel, 1984).

Corn sweeteners are produced by the hydrolytic action of enzymes or acid on the α-1,4, and α-1,6 glucosidic linkages. Saccharides of various molecular weights are liberated and, depending on the extent and type of hydrolysis, sweeteners of different functional properties are produced. The number of dextrose units in a saccharide is referred to as degree of polymerization (DP), where DP-1 is dextrose, DP-2 is a disaccharide such as maltose or isomaltose, DP-3 is a trisaccharide such as maltotriose or panose, and DP-4$^+$ is a tetra- or higher saccharide. Degree of hydrolysis is expressed in terms of dextrose equivalent (DE), a measurement of the reducing content of a starch hydrolysate calculated as dextrose on a dry basis. Since starch contains only one reducing group per molecule, it exhibits essentially zero DE. Complete hydrolysis to dextrose yields a 100-DE product, and partial hydrolysis produces a product of intermediate DE. DE is determined by reacting reducing sugars with an alkaline solution of a copper salt. The amount of copper reduced is proportional to the amount of reducing sugars present (AACC, 1983). Alternatively, given the composition of a starch hydrolysate by a procedure such as high performance liquid chromatography (Brobst and Scobell, 1981; Bernetti, 1982), DE is calculated by summing the DE contribution of each saccharide, where DP-1 = 100 DE, DP-2 = 62 DE, DP-3 = 40 DE, and DP-4$^+$ = about 18 DE. Since the composition of the DP-4$^+$ fraction varies depending on the hydrolysis technique, the DE contributed by this fraction must be determined for each type of product. An approximate relation between DP and DE can be expressed as follows:

$$DP = [(20,000/DE) - 18]/162.$$

Each hydrolytic scission is accompanied by the addition of one molecule of water. The weight gain due to water addition is referred to as chemical gain; therefore, complete hydrolysis of starch to dextrose yields a dry substance increase by a factor of $(162 + 18)/162$ or 1.11. Partial hydrolysis yields a lower chemical gain factor. The relationship between DE and chemical gain factor can be described as follows:

$$\text{Chemical gain factor} = (0.00111)(DE) + 0.9973.$$

B. Acid Hydrolysis

Acid hydrolysis of starch proceeds randomly, cleaving both α-1,4 and α-1,6 linkages and releasing increasing amounts of low-DP material with time. At a given DE level, the same saccharide composition is obtained regardless of the hydrolysis conditions used. Typical compositions at varied DE levels are shown in Table IV.

Since acid is not a specific catalyst for hydrolysis only, other reactions occur as hydrolysis proceeds. For instance, dehydration of dextrose yields hydroxymethylfurfural. In turn, this compound may decompose to levulinic and formic acids or polymerize to compounds believed to be intermediates to color formation (Singh et al, 1948). In addition, acid-catalyzed recombination of saccharides forms gentiobiose and branched polymers of higher molecular weight. Other reactions occur that produce off-flavor and color components that are often difficult to remove from the finished product. Consequently, syrups produced by acid hydrolysis are limited to about 42 DE to avoid problems associated with side reactions.

C. Enzyme Hydrolysis

Enzymatic hydrolysis of starch is much more specific than acid hydrolysis. This specificity minimizes off-flavors, color, and degradation product formation, resulting in higher product yields and reduced refining costs. Enzymes used in corn sweetener production include bacterial and fungal α-amylases, glucoamylase, β-amylase, and pullulanase.

α-AMYLASES

Bacterial α-amylases are produced from *Bacillus* organisms in aerated, submerged-culture fermentation media specifically formulated to maximize enzyme production. α-Amylases are classified as endoenzymes, since hydrolysis proceeds randomly within the molecule, cleaving α-1,4 linkages only and releasing shorter fragments or saccharides retaining the α configuration. The action pattern of bacterial α-amylases varies, depending on the source of the enzyme. For example, *B. subtilis* amylases produce significant quantities of maltose, maltotriose, and maltohexose (Allen and Spradlin, 1974), whereas *B. licheniformis* amylase releases a significant amount of maltopentaose (Saito, 1973). α-Amylases from fungal sources such as *Aspergillus oryzae* are also endoenzymes and exhibit specificity for α-1,4 linkages. Unlike bacterial α-amylases, however, fungal α-amylases are considerably less thermostable and produce maltose as the main product. Consequently, these enzymes are used to make syrups with a high maltose content.

Typical use conditions for α-amylases depend on the particular enzyme and process. Enzymes from *B. licheniformis* or *B. stearothermophilus* are the most thermostable of the bacteria-derived α-amylases and can be used at 105–110°C for a short time. *B. subtilis* α-amylases are less stable and are generally used at temperatures of 85–90°C. Fungal α-amylases are much less stable than bacterial enzymes and are used in reactions conducted at 55–60°C. The best pH range for starch hydrolysis is 5.5–7.0 for bacterial α-amylases and 5.0–5.5 for fungal α-amylases.

TABLE IV
Composition of Acid-Converted Corn Starch

DE[a]	Saccharide (%, db)							
	DP[b]-1	DP-2	DP-3	DP-4	DP-5	DP-6	DP-7	DP-8+
20	6	6	5	5	5	4	4	65
25	8	8	7	7	6	5	4	55
30	11	9	9	8	7	6	5	45
35	13	11	10	9	8	6	5	38
40	18	13	11	10	8	6	6	28
45	22	15	12	10	8	6	5	22
50	25	17	13	10	8	6	5	16
55	32	18	13	9	7	5	4	12
60	36	20	13	9	6	4	4	8
65	42	21	13	8	5	3	3	5

[a] Dextrose equivalent.
[b] Degree of polymerization.

GLUCOAMYLASE

Glucoamylase (also called amyloglucosidase or γ-amylase) is produced from strains of *A. niger* in submerged fermentation (Armbruster, 1961). The fermentation broth contains two distinct glucoamylase isozymes (Svensson et al, 1982), α-amylase, possibly transglucosidase, and other enzymes such as protease and cellulase. The presence of the α-amylase is important, since it assists saccharification by hydrolyzing large starch fragments to provide additional substrate for the action of glucoamylase (Kooi and Armbruster, 1967). Transglucosidase, however, reduces dextrose yield by catalyzing the formation of isomaltose. Consequently, the removal of this enzyme from the fermentation broth is accomplished by adsorption on a clay mineral or by other techniques (Kerr, 1961, 1962; Hurst and Turner, 1962; Kooi et al, 1962; Kathrein, 1963). Complete elimination of transglucosidase is also possible, since transglucosidase-free mutants have been isolated (Norman, 1979).

Glucoamylase exhibits exoamylytic activity and releases dextrose stepwise from the nonreducing end of the substrate by cleaving both α-1,4 and α-1,6 linkages. Rates of hydrolysis are dependent on the type of linkages present and substrate size (Abdullah et al, 1963). Typically, in commercial processes, glucoamylase produces a dextrose hydrolysate containing about 96% dextrose (db). The enzyme also catalyzes the reverse reaction (reversion) of dextrose to disaccharides. At a normal commercial solids level of 30 wt %, a hydrolysate containing 87% dextrose, 11% isomaltose, and 2% maltose (db) would be produced (Subramanian, 1980) if the reaction reached equilibrium. Equilibrium dextrose level, a function of total dry solids, increases from 87% at 30 wt % solids to 92 and 96% at 20 and 10 wt % solids, respectively (Subramanian, 1980). The kinetics of hydrolysis and reversion reactions catalyzed by glucoamylase have been investigated by Shiraishi et al (1985) and Beschkov et al (1984).

Glucoamylase exhibits good stability over a wide pH range; however, in commercial processes, a pH of 4.0–4.5 is generally used. Recommended temperatures are about 58–62° C, since a higher temperature reduces enzyme stability and a lower temperature increases the possibility of microbial contamination.

β-AMYLASE

Barley β-amylase is a maltogenic enzyme used for syrup production. The enzyme, an exoamylase, releases maltose in the β configuration sequentially from the nonreducing end of a molecule. Unlike that of the α-amylases, the action of β-amylase is terminated near an α-1,6 linkage and, in the absence of endoamylase activity, yields a highly branched β-limit dextrin. The presence of α-amylase endoamylytic activity in malt, however, provides additional substrate for β-amylase activity. Reaction conditions are generally 55–60° C, pH 5.0–5.5.

PULLULANASE

Pullulanase, a debranching enzyme, is specific for hydrolyzing α-1,6 bonds in starch. When used in combination with glucoamylase, an increased rate of dextrose production is achieved, resulting in higher dextrose yields. The combined action of pullulanase and a maltogenic enzyme also yields higher maltose levels.

D. Isomerization

Glucose isomerase (D-xylose ketol-isomerase, xylose isomerase) catalyzes the isomerization of dextrose to D-fructose and of D-xylose to D-xylulose and is important in the corn sweetener industry because of its use in producing HFCS. The enzyme is produced in aerated, submerged fermentation from a wide variety of bacterial organisms (Chen, 1980a, 1980b). Xylose is frequently used as an inducer; however, constitutive mutants have been developed (Armbruster et al, 1974; Outtrup, 1976). Since the enzyme is produced intracellularly, mechanical disruption or treatment of the cells with surfactants or lysozyme is required to produce a soluble enzyme. The enzyme exhibits activity at a temperature as high as 90°C, although stability is low and commercial operations are conducted at about 60°C. Typically, optimum pH is in the range of 7.5–8.5, although 70% of the activity is retained at pH 7 (Zittan et al, 1975). Magnesium is added as a cofactor to maintain isomerase stability and prevent inhibition by trace levels of calcium.

IV. MANUFACTURING PROCESSES

A. Dextrose

Dextrose (D-glucose) has been produced commercially from starch by a wide variety of acid, acid-enzyme, and enzyme-catalyzed processes. Presently, most production schemes use a thermostable α-amylase for thinning (liquefying) starch to 10–15 DE, followed by saccharification with glucoamylase to about 95–96% dextrose (db). The hydrolysate is clarified and refined and then is further processed to crystalline dextrose, liquid dextrose, high-dextrose corn syrup, or to HFCS feed (Fig. 1).

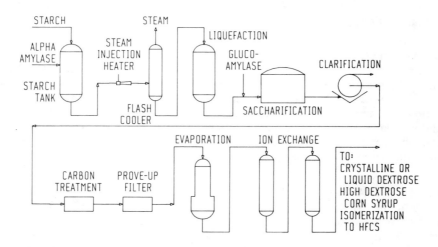

Fig. 1. Dextrose production.

STARCH HYDROLYSIS

Until about 1960, commercial dextrose processes were conducted by straight acid hydrolysis. The process consisted of acidifying a 15–25 wt % starch suspension with HCl to 0.03–0.04N and heating at 140–160°C for 5–10 min or until a maximum dextrose level was reached. However, due to acid-catalyzed formation of degradation products, a dextrose level of only about 86% (db) was achievable (Kooi and Armbruster, 1967).

The development of glucoamylase for saccharification provided a major improvement in dextrose production technology by allowing a reduction in the extent of acid hydrolysis. Liquefaction was conducted batchwise in corrosion-free equipment by adjusting a 30–35 wt % starch suspension to pH 1.8–2.0 with HCl or H_2SO_4 and thinning at 120–140°C until a 10–20 DE hydrolysate was obtained. The hydrolysate was then saccharified with glucoamylase. Control of DE is important in this type of process, since the level of acid-reversion products increases with increasing DE, resulting in reduced dextrose yield during subsequent saccharification. Conversely, if DE is too low, i.e., less than 10, starch retrogradation occurs and filtration problems may develop. Typically, an acid-enzyme process yields a dextrose level of 92–94% (db) (Woolhouse, 1976).

Total enzyme processes were developed in the 1960s, when thermostable α-amylases derived from *B. subtilis* became commercially available. The lower temperature and neutral pH used for liquefaction were effective in minimizing side reactions and increasing the dextrose yield to 95–97% (db) (Kooi and Armbruster, 1967; MacAllister, 1979). The *B. subtilis* enzyme is sufficiently stable to be used at 85–95°C for a short time (Reichelt, 1983), and processes were designed to take advantage of this property. A typical process is conducted by thinning a 30–40 wt % starch suspension with α-amylase at pH 6.0–6.5, 85–90°C, for 1 hr in the presence of 200–400 ppm calcium (db) added for enzyme stabilization. The reaction is continued by using a 5–10-min heat treatment at 120–140°C to dissociate insoluble complexes of fatty acid and amylose that are present in starch or formed during processing and are resistant to the action of enzyme (Hebeda and Leach, 1974). A heat treatment is needed, since a temperature of 100°C or higher is required to dissociate the complex, as evidenced by differential scanning calorimetry studies (Kugimiya et al, 1980; Bulpin et al, 1982). Since the high reaction temperature completely inactivates *B. subtilis* α-amylase, a second enzyme addition at a reduced temperature is required to continue the hydrolysis before the complex reforms. The second stage of the reaction is conducted at 85°C for about 1 hr to produce a 10–15-DE hydrolysate for saccharification. This process is effective in preventing the occurrence of insoluble starch particles in the final hydrolysate and, therefore, maximizing filtration rates during processing.

The next improvement in enzymatic dextrose process technology was achieved when enzymes became available that exhibited increased thermostability and retained activity above 100°C. α-Amylases derived from *B. licheniformis* (Chiang et al, 1979; Rosendal et al, 1979) or *B. stearothermophilus* (Anonymous, 1984a) are effective at a sufficiently high temperature to eliminate the need for a heat treatment step. A typical process (Slott and Madsen, 1975) is conducted by thinning a 30–40 wt % starch slurry containing 100–200 ppm calcium (db) with α-amylase through a steam injection heater at 103–107°C, pH 6.0–6.5. The instantaneous heating gelatinizes the starch rapidly and completely and dissociates the amylose-lipid complex. The temperature is maintained for 5–10 min in a holding tube; batch or continuous reaction then follows for 1–2 hr at about 95°C to yield a 10–15-DE hydrolysate. This type of liquefaction process, currently in widespread use throughout the wet-milling industry, produces an efficiently thinned low-DE hydrolysate for subsequent saccharification.

α-Amylase stability during liquefaction is a function of the combined effects of temperature, pH, dry substance, time, and calcium. Knowledge of these variables allows calculation of residual activity (Rosendal et al, 1979). For instance, at 105°C and otherwise typical liquefaction conditions, only about 20% of the α-amylase activity is lost. Since operation above 100°C is effective in achieving complete starch gelatinization and rapid dissociation of the amylose-lipid complex, the residual activity is sufficient to hydrolyze amylose rapidly and prevent reassociation of the complex; therefore, a second enzyme addition or heat treatment is not required. An additional advantage of the highly thermostable α-amylase is a low level of contaminating protease. As a result, protein solubilization is minimized during processing, allowing the use of reduced-purity starch containing protein levels that are higher than normal (Reichelt, 1983).

Other specialized processes that have been suggested as alternatives to conventional starch liquefaction techniques include the Kroyer process for hydrolyzing impure starches such as corn grits (Holt et al, 1975), low-temperature granular starch processes (Hebeda et al, 1975); and a dual-addition thermostable enzyme process for thinning wheat starch (Reichelt, 1983).

Regardless of the manner in which starch is thinned, saccharification is conducted with glucoamylase to produce a dextrose hydrolysate. Saccharification is conducted by batch or continuous operation in reactors that are often as large as several hundred thousand liters. Thinned starch is diluted to 28–30 wt % solids and is cooled to about 60°C; the pH is adjusted to 4.0–4.5, and sufficient glucoamylase is added to achieve a maximum dextrose level in 24–96 hr. Temperature control is important, since a higher temperature reduces enzyme stability and a lower temperature increases the risk of microbial contamination. At a typical final solids level of 30 wt %, a dextrose level of 96% (db) is achieved. Decreasing the solids level to 10–12 wt % is effective in increasing maximum dextrose level to 98–99% (McMullen and Andino, 1977). However, the increased dextrose level is attained at the expense of increased evaporation cost and risk of microbial contamination. Conversely, at a higher solids of 40 wt %, dextrose is reduced to 94% (Norman, 1981). Typical dextrose hydrolysate compositions attained at varied solids levels are shown in Table V. At maximum dextrose, DP-2 consists of roughly equal amounts of maltose and isomaltose with a trace of maltulose. At this point, the maltose level is at equilibrium. Isomaltose, however, is still substantially below the equilibrium level, and extending the reaction time reduces the dextrose level due to continual accumulation of isomaltose via the reversion reaction. Consequently, accurate control of glucoamylase dosage and reaction time is required to avoid loss in dextrose yield during saccharification. Typical saccharification curves (Fig. 2) show the effects of dosage and time.

An additional source of dextrose yield loss is the formation of maltulose (4-O-α-D-glucanopyranosyl D-fructose). Precursors to this disaccharide are formed when end dextrose units of oligosaccharides are isomerized to fructose by alkaline isomerization during the liquefaction step (Norman, 1979). During saccharification, the last α-1,4 linkage of the oligosaccharide is resistant to hydrolysis because of the terminal fructose and, therefore, maltulose accumulates in the hydrolysate, with an equivalent loss in dextrose. This reaction is minimized by controlling thinning pH below 6.3.

Maximum dextrose level can be increased by using a specialized glucoamylase preparation (Tamura et al, 1981b) or by using a debranching enzyme in combination with glucoamylase to increase rate of dextrose formation relative to isomaltose formation. Increased dextrose yields of 2% or more have been obtained with pullulanase (Hurst, 1975) or isoamylase (Norman, 1982). However, in both cases, enzyme instability under normal saccharification conditions limits commercial utilization. A thermostable, acidophilic debranching enzyme from *B. acidopullulyticus* (Nielsen et al, 1982) has been used at pH 4.5, 60° C in the presence of glucoamylase to increase dextrose levels. Other reported advantages include reduced glucoamylase requirement, higher solids operation, and reduced reaction time.

Commercial dextrose processes use soluble glucoamylase, although the use of immobilized glucoamylase has been studied extensively. Many different supports have been evaluated for binding glucoamylase, and the immobilized enzyme has been studied in continuous operation by many investigators, including Bachler et al (1970), Smiley (1971), Gruesbeck and Rase (1972), Park and Lima (1973), and Lee et al (1975). Its commercial utilization, however, has been prevented by several factors, including 1) the need for clarified feed, 2) a relatively short enzyme life, and 3) 1.0–2.5% lower dextrose yield (Daniels, 1980; Rugh et al, 1979) due to diffusion problems and increased reversion. Specific problem areas have been addressed and potential solutions suggested. For instance, dextrose yield can be increased by using a partially saccharified feed (Hebeda et al, 1979). Use of a nonporous carrier reduces reversion by minimizing diffusion (Wasserman et al, 1982), and use of a more stable glucoamylase (Tamura et al, 1981a) increases enzyme half-life. In addition, purification of glucoamylase is effective in providing a bound enzyme preparation of increased potency, resulting in reduced residence time requirement (Lobarzewski and Paszczynski, 1983).

REFINING

Refining of dextrose hydrolysate is required to remove insolubles contributed by the starch, as well as ash, color, and protein solubilized during processing.

TABLE V
Dextrose Hydrolysate Composition as a Function of Solids

Final Hydrolysate Solids (wt %)	Saccharide (%, db)			
	DP-1	DP-2	DP-3	DP-4+
10	98.8	0.6	0.2	0.4
15	98.2	1.1	0.2	0.5
20	97.5	1.6	0.3	0.6
25	96.9	2.1	0.3	0.7
30	96.1	2.7	0.3	0.9
34	95.5	3.1	0.4	1.0

Clarification to remove traces of insoluble fat, protein, and starch is accomplished by centrifugation or precoat filtration using a diatomite filter aid (Basso, 1982). The hydrolysate is then treated with powdered or granular carbon (van Asbeck et al, 1981) and ion-exchange resins to remove residual trace impurities such as color, color precursors, proteinaceous material, and inorganics. The decolorized liquor is evaporated to 50–55 wt % solids and may

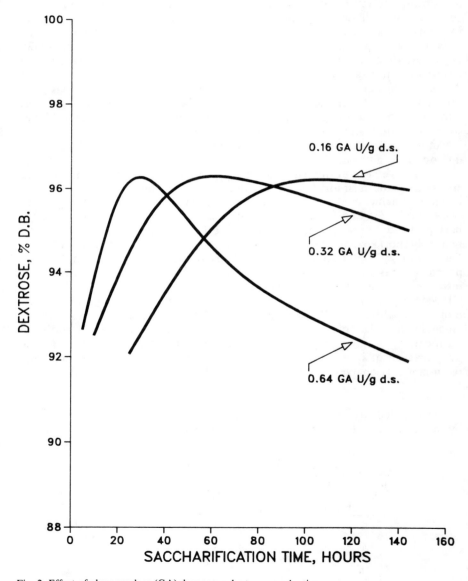

Fig. 2. Effect of glucoamylase (GA) dosage on dextrose production.

be refined additionally before being processed to crystalline monohydrate or anhydrous dextrose, liquid dextrose, or high-dextrose corn syrup.

CRYSTALLINE DEXTROSE

Crystalline dextrose monohydrate is produced by evaporating the refined dextrose hydrolysate to 70–78 wt % solids, cooling to about 46°C, and mixing with seed crystals from the prior batch in a crystallizer. The mass is agitated and cooled slowly to 20–30°C over a period of three to five days, during which about 60% of the dextrose is crystallized as the monohydrate. The magma is centrifuged to remove the mother liquor (greens); the crystals are washed with a spray of cold water and dried in a rotary dryer to about 8.5% moisture, a level slightly below the theoretical value of 9.1% for one molecule of water of crystallization per dextrose molecule. By reducing moisture to this level, the caking tendency of dextrose crystals is minimized. The mother liquor may be partially recycled to the initial crystallization step or concentrated and crystallized to obtain a second crop of crystals and corresponding second greens. The second greens contain about 60% dextrose (db), and a total dextrose crystal yield of 87.5% is obtained based on a hydrolysate dextrose level of 95% (MacAllister, 1979). USP grade dextrose, which is used for special therapeutic purposes, is obtained by dissolving the centrifuge cake and recrystallizing to achieve a very high degree of purity.

Anhydrous α-dextrose is manufactured by crystallizing dissolved dextrose monohydrate at a temperature of 60–65°C in a vacuum pan. Evaporative crystallization is required to prevent color formation at high temperature and hydrate formation at low temperature. Crystals are separated, washed, and dried.

LIQUID DEXTROSE

Liquid dextrose is prepared by dissolving crystalline dextrose to 71 wt % solids. A high-dextrose corn syrup is made by refining a dextrose hydrolysate containing 95–96% dextrose and concentrating it to 70–75 wt % solids. This particular product is technically a corn syrup, since dextrose is defined as a material that exhibits a DE of 99.5 or higher.

SHIPPING

Crystalline dextrose is shipped in 45.4-kg bags or in bulk. Liquid products are transported in insulated tank cars or trailers at approximately 50°C to prevent crystallization.

PRICE

The price of crystalline dextrose monohydrate since 1970 is shown in Table VI. Dextrose price is affected by a variety of factors, including sucrose pricing, corn and production costs, process capacity, and demand. However, the price of dextrose is generally controlled to the largest extent by the price of sucrose. In 1985, the average crystalline dextrose price on a dry weight basis was about the same as the price of refined sucrose, and 126, 114, and 214% of the prices of 42% HFCS, 55% HFCS, and corn syrup, respectively. The high price of dextrose relative to that of HFCS is due in part to limited production facilities and to the high cost of crystallizers. Mid-1984 prices for crystalline anhydrous dextrose and USP anhydrous dextrose were $0.91/kg ($0.41/lb) and $1.03/kg ($0.47/lb), respectively.

B. High-Fructose Corn Syrup

Corn syrups containing 42% (db) fructose (levulose) are manufactured from refined dextrose hydrolysate by converting a portion of the dextrose to fructose with immobilized glucose isomerase. Chromatographic separation techniques are used to produce a 90% fructose product that is blended with 42% HFCS, to yield a 55% HFCS product (Fig. 3).

IMMOBILIZED ISOMERASE

In the early stages of HFCS development, it became evident that batchwise isomerization with soluble enzyme was economically prohibitive due to high enzyme cost. In addition, the long reaction time required to minimize enzyme consumption was responsible for the production of undesirable by-products, such as mannose, psicose, color, and off-flavors, resulting in high refining costs. Consequently, to reduce reaction time, and at the same time reduce enzyme cost, a process was needed that would allow enzyme reuse. Many different types of enzymes and immobilization systems were studied, and several reached the point of commercial utilization.

Glucose isomerases used in commercial operation have been developed from a variety of bacterial sources including *Actinoplanes missouriensis* (Shieh et al, 1974), *Aerobacter levanicum* (Shieh and Donnelly, 1974), *Arthrobacter* (Lee et al, 1972), *B. coagulans* (Outtrup, 1976), *Flavobacterium arborescens* (Anon., 1983a), *Streptomyces griseofuscus* (Anonymous, 1984b), *S. olivaceus* (Brownewell, 1971), *S. olivochromogenes* (Armbruster et al, 1976), and *S. rubiginosus* (Anonymous, 1984c). Techniques for immobilizing enzymes include treating whole cells to maintain enzyme activity and structural integrity or separating the enzyme from the cells and attaching the soluble enzyme to a solid support. Examples of processes for producing commercial whole-cell

TABLE VI
Price ($/kg) of Corn Sweeteners in the United States[a]

Year	Dextrose[b]	HFCS-42[c]	HFCS-55[d]	Corn Syrup[e]
1970	0.21	0.36	...	0.15
1975	0.43	0.37	...	0.31
1980	0.59	0.38	...	0.26
1981	0.60	0.34	0.40	0.28
1982	0.55	0.26	0.32	0.25
1983	0.53	0.29	0.37	0.23
1984	0.54	0.32	0.40	0.23
1985	0.49	0.30	0.36	0.20
1986[f]	0.48	0.32	0.36	0.18

[a] Based on 1970–1975 data (Anonymous, 1976) and 1980–1986 data (Anonymous, 1986).
[b] Monohydrate basis, in 45.4-kg bags.
[c] 42% fructose, 71 wt % solids, in tank cars.
[d] 55% fructose, 77 wt % solids, in tank cars.
[e] 80.3 wt % solids, in tank cars.
[f] July price.

products include cross-linking cells with gluteraldehyde (Zienty, 1973); flocculating whole-cell aggregates with a polyelectrolyte, polyacid, or mineral hydrocolloid (Lee and Long, 1974); or rupturing cells by homogenization and reacting with gluteraldehyde (Amotz et al, 1976). Cell-free products are produced by adsorption of soluble enzyme onto ion-exchange cellulose (Sutthoff et al, 1978) or adsorption within pores of alumina or other inorganic materials (Messing, 1974; Eaton and Messing, 1976). Soluble enzyme may also be adsorbed onto an inorganic carrier followed by cross-linking with a bifunctional reagent (Levy and Fusee, 1979; Rohrbach, 1981).

HFCS-42

Regardless of the type of enzyme or immobilization technique used, process conditions for HFCS production are generally very similar. A high-quality HFCS feed stream is prepared from a 93–96% dextrose hydrolysate by clarification, carbon and ion exchange refining, and evaporation to 40–50 wt % solids. Clarification is generally conducted using traditional rotary-drum precoat filtration. Flocculation and continuous removal of insolubles has also been reported (Anonymous, 1983b). A high-quality feed is important to prevent accumulation of insolubles within the enzyme bed, resulting in a reduced flow rate.

Magnesium as $MgSO_4$ is added to the feed to activate and stabilize isomerase and also counteract the inhibitory effect of residual calcium. A magnesium level of $0.0004M$ is sufficient to overcome a maximum calcium level of 1 ppm; however, up to 15 ppm calcium can be tolerated if magnesium is added at a magnesium-calcium molar ratio of at least 20 (Anonymous, 1981). Addition of sulfite or bisulfite salts to the feed is also effective in increasing enzyme stability and reducing color formation (Cotter et al, 1976). Cobalt, another known activator of isomerase, was used in initial batch processes in which pH was maintained at a low level to minimize side reactions (Carasik and Carroll, 1983).

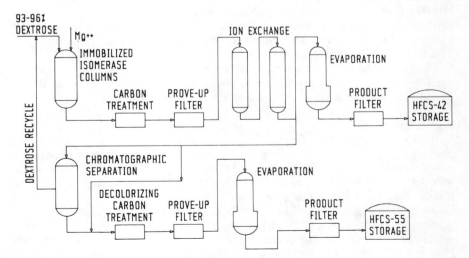

Fig. 3. High-fructose corn syrup production.

At the higher pH levels used in continuous systems, however, cobalt is not needed. An additional potential problem is the presence of dissolved oxygen in the dextrose feed. Therefore, deaeration may be necessary to prevent formation of by-products and inactivation of isomerase.

Reaction temperature and pH depend on the specific enzyme preparation used; however, conditions recommended by the enzyme suppliers are within the ranges of 55–61° C and 7.5–8.2 pH. Isomerase activity decreases below pH 7, and the enzyme is inactivated below pH 5 and above pH 9 (Anonymous, 1981). In general, the pH level should be controlled as low as possible to minimize side reactions while still maintaining optimal activity.

The isomerization reaction is conducted by passing feed through a fixed bed of immobilized isomerase at a controlled rate sufficient to yield 42–45% (db) fructose. An equilibrium fructose level of 51% (db) can be achieved at 60° C by increasing the reaction time significantly; however, this approach is not commercially viable. Minimizing nondextrose saccharides and, therefore, maximizing dextrose in feed is advantageous, since the reaction time required to attain the target fructose level is reduced. Isomerase performance is not affected by the presence of maltose or isomaltose (Hultin, 1983) or other nondextrose saccharides normally present. An exception may be reduced isomerase efficiency due to the presence of maltulose or other components formed under conditions that produce maltulose (Hurst and Lloyd, 1983).

During the isomerization process, enzyme activity decreases due to a combination of thermal inactivation and other factors such as the presence of feed impurities. Therefore, residence time is increased by reducing the flow rate through the column to maintain the desired fructose level. Enzyme half-life is normally several weeks. The enzyme is generally used for two to three months through several half-lives until its residual activity is reduced to 10% of the initial level (Carasik and Carroll, 1983).

Refining. The isomerized product is refined to an acceptable final quality by adjusting the pH to 4–5 and treating it with carbon to remove undesirable color and flavor. Additional refining by ion-exchange is conducted with a strong-acid cation resin in the hydrogen form followed by a weak-base anion resin in the free base form to remove salts and residual color. The product is then evaporated to 71% solids at a low temperature.

Shipping. The product is transported in tank cars or by rail. Storage at 30–32° C is required to prevent crystallization. If crystallization does occur, heating to 38° C is necessary to dissolve the crystals before unloading.

HFCS 55 and 90

HFCS containing 42% fructose was the first corn sweetener produced that provided a sweetness approaching that of sucrose. In certain applications such as soft drinks, however, a sweetener exhibiting sweetness equivalent to sucrose was desired. Several techniques have been studied for increasing the fructose level. These include the use of a chemically stabilized isomerase at a higher isomerization temperature (Lloyd, 1983), complexation of fructose with borate compounds, or removal of dextrose by crystallization. Current commercial processes use a chromatographic separation technique to produce products enriched in fructose. Products of this type were first introduced commercially in limited quantities in 1976.

Enrichment. Enrichment is accomplished by passing 42% HFCS through a column of adsorbent containing calcium or other cations (Long, 1978) or a nonresin, nonswelling mineral adsorbent (Broughton et al, 1977). Fructose is bound to the adsorbent and is retained to a greater extent than dextrose or oligosaccharides. In batch operations, elution with water produces a high-dextrose fraction, followed by a mixture of dextrose and fructose and, finally, by a relatively pure fructose fraction. Batch operations are not efficient; therefore, to increase production rate and reduce water consumption, continuous systems were developed (Keller et al, 1981; Teague and Arnold, 1983). Continuous operations utilize, in principle, a simulated moving-bed process in which feed and desorbent enter the column at different points, while fructose (extract) and the nonfructose fraction (raffinate) are withdrawn. Points of entry and withdrawal change in periodic fashion in correspondence with respective material flows through the column. In this way, separation of fructose and dextrose is maximized and costs are minimized. In a typical operation, a 42% fructose feed at 50 wt % solids is passed through a column to yield a product stream containing, on a dry basis, 94% fructose and 5% dextrose, and a raffinate containing 86% dextrose and 6% fructose. Fructose extraction efficiency is 91.5% (Teague and Arnold, 1983). The enriched HFCS is blended with 42% HFCS to provide a product containing 55% fructose. Raffinate at about 20 wt % solids is recycled to isomerization or blended into saccharification tanks to convert a portion of the oligosaccharides to dextrose before isomerization. Alternatively, raffinate can be saccharified using immobilized glucoamylase (Rugh et al, 1979). Reportedly, this technique is being used in commercial operations (Anonymous, 1984d).

Shipping. The 55% HFCS product is evaporated to 77 wt % solids for shipment. Crystallization is not a problem because of the high fructose level and reduced dextrose content. The 90% HFCS product can be evaporated to 80 wt % solids for shipment as a noncrystallizable syrup.

CRYSTALLINE FRUCTOSE

Pure crystalline fructose has been produced commercially by a variety of processes (Hamilton et al, 1974; MacAllister, 1980). For instance, pure fructose has been made by oxidation of glucose with glucose oxidase, followed by precipitation of sodium gluconate and crystallization of fructose from methanol (Holstein and Holsing, 1962). A more direct approach for producing commercial quantities of crystalline fructose involves the use of a sulfonated-polystyrene resin in the calcium form to separate fructose from invert followed by crystallization (Bollenback, 1983). Because of the difficulty involved in crystallizing fructose, alcohol was used as a solvent in many processes to decrease fructose solubility (Lauer et al, 1971; Nitsch, 1974). Fructose, however, can be crystallized from aqueous solution by using a high level of seed crystals (Kusch et al, 1970). Most recently, HFCS has been used as raw material for commercial production of crystalline fructose. This process involves production of 42% HFCS, chromatographic separation to obtain a fraction containing 97% fructose, evaporation to 70%, and crystallization of about 50% of the fructose in 80–100 hr (Morris, 1981).

A different approach to production of pure fructose has been suggested that involves enzymatic conversion of dextrose to D-glucosone followed by hydrogenation to fructose and recovery by crystallization (Neidleman et al, 1981). Alternates to crystallization include spray drying (Lundquist et al, 1976) or preparation of a noncrystallized dry product by adding anhydrous fructose to a highly concentrated fructose solution, kneading, and drying (Yamauchi, 1975).

Crystalline fructose is packed in multiwalled bags containing a foil liner. The product can be stored for at least 12 months without significant moisture pickup if storage conditions are at 60% rh and 25° C or less (Osberger, 1979).

PRICE

The price of HFCS (Table VI) is controlled by many of the same factors affecting dextrose pricing; however, sucrose price is probably the most important factor. Traditionally, HFCS is priced 10–20% below sucrose. In 1985, the price discounts of 42% and 55% HFCS to sugar were 23.4 and 13.6%, respectively (Anonymous, 1986). The price differential between 42% and 55% HFCS has increased from 8.5% in 1981 to 20.0% in 1985, representing the increased demand for 55% HFCS as a replacement for sucrose in soft drinks.

In mid-1984, the prices of 90% HFCS and crystalline fructose were \$1.47/kg (\$0.67/lb) and \$2.27/kg (\$1.03/lb), respectively, at commercial solids. A new product that contains a blend of crystalline fructose and sugar and that reportedly is 10–80% sweeter than sugar alone is expected to be in production in 1987 at a price of \$0.77–\$1.32/kg (\$0.35–\$0.60/lb) (Anonymous, 1986).

C. Corn Syrups

Regular corn syrups are, by definition, those products that range from 20 to 99.4 DE. Products with lower DE are classified as maltodextrins and those with higher DE are categorized as dextrose. HFCS is considered a special type of corn syrup and is not defined on a DE basis.

Corn syrups are produced by straight acid or acid-enzyme hydrolysis, although enzyme-enzyme processes can be used for some syrups. Typical commercial corn syrups are straight acid products of 26–42 DE, high-conversion acid-enzyme syrups of 64–70 DE, and high-maltose syrups. These syrups are often referred to as "glucose syrups," even though the actual D-glucose level may be very low.

ACID SYRUPS

Acid hydrolysis of starch is conducted by batch or continuous processes using a 35–40 wt % starch slurry adjusted to 0.015–$0.2N$ with hydrochloric acid (Fig. 4). The slurry is held at 140–160° C for 15–20 min or until the desired DE is reached. Batch processes are conducted in manganese bronze convertors as large as 10,000 L, using direct steam injection to heat the water in the reactor to boiling. Starch slurry containing acid is added; the reactor is pressurized; and the reaction is allowed to proceed until the desired DE is attained. At the end of the reaction, pressure is used to discharge the hydrolysate to a neutralizer tank, where the pH is adjusted to 4.5–5.0 with sodium carbonate. Continuous convertors use indirect heating and, because of improved process control, yield a more uniform product than batch operation does. The composition of acid corn syrups is shown in Table IV. Commercial acid corn syrups do not generally exceed 42 DE since, at higher DE levels, color and flavor components formed during extended acid hydrolysis are difficult to remove by traditional refining techniques.

HIGH-CONVERSION SYRUPS

High-conversion syrups are prepared commercially from acid substrates of 38–42 DE, although enzymatically liquefied starch can also be used. Saccharification is conducted with a combination of glucoamylase and fungal α-amylase at 55–60°C, pH 4.8–5.2 for 24–48 hr. A typical syrup at 63 DE contains, on a dry basis, 36% dextrose, 30% maltose, and 13% DP-3. The ratio of dextrose to maltose and levels of each can be altered by changing the ratio and amount of enzymes used for saccharification. A heat treatment step to inactivate enzymes and stop the reaction at the proper time is recommended; however, the heat treatment step can be eliminated by careful control of dosage and time (Reichelt, 1983).

HIGH-MALTOSE SYRUPS

High-maltose syrups are prepared from acid- or enzyme-thinned substrates by saccharifying with a maltose-producing enzyme such as a fungal α-amylase derived from *A. oryzae* or barley β-amylase extracted from germinated barley. Reaction conditions are about 55°C, pH 5, and 35–45 wt % solids. A hydrolysate containing 50–55% (db) maltose is obtained in a typical operation. A higher maltose level, i.e., 60–80% (db), is achieved by the addition of pullulanase to the saccharification step. The debranching enzyme hydrolyzes α-1,6 linkages and provides additional substrate for the maltogenic enzyme. Typical maltose syrup compositions are given in Table VII.

A maltogenic exo-α-amylase from *B. stearothermophilus* has been reported to be effective in producing a high-maltose syrup (Outtrup and Norman, 1984). Improved yields have been obtained by recombinant DNA techniques, and the enzyme can produce 60–70% maltose alone or 80% maltose in combination with

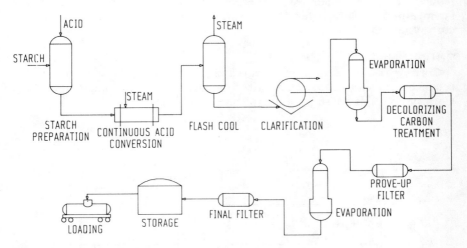

Fig. 4. Corn syrup production.

a debranching enzyme. Other bacterial enzymes from sources such as *B. cereus* (Takasaki, 1978) and *B. megaterium* (Armbruster and Jacaway, 1970) have been shown to be maltose producers. These enzymes, however, are not used commercially.

SPECIALTY SYRUPS

Many enzymes produce specialty corn syrups exhibiting unique saccharide distributions. For instance, a maltotetraose syrup is produced with a *Pseudomonas stutzeri* enzyme (Abdullah, 1972), and a maltohexose syrup is made using an enzyme from *Ae. aerogenes* (Kainuma et al, 1975). In addition, *B. licheniformis* α-amylase produces a maltopentose syrup (Saito, 1973), and a maltose-maltotriose syrup is made with a porcine α-amylase (Allen and Spradlin, 1974). None of these syrups have been produced commercially.

SYRUP PRODUCTION WITH IMMOBILIZED ENZYMES

The use of immobilized enzyme systems for corn syrup production does not have the disadvantages associated with its use for dextrose production. Therefore, in certain applications, bound enzymes provide a viable alternative to soluble enzymes. Systems have been studied for producing high-conversion syrups using immobilized glucoamylase alone (Rugh et al, 1979) or a combination of immobilized fungal α-amylase and glucoamylase (Bohnenkamp and Reilly, 1980; Hausser et al, 1983). In addition, high-maltose syrups can be produced using immobilized β-amylase (Maeda et al, 1978) or a combination of immobilized β-amylase and pullulanase (Ohba and Ueda, 1980).

REFINING

Corn syrup hydrolysate is clarified by centrifugation and/or filtration, evaporated to 60 wt % solids, carbon treated to remove color and acid degradation products (Heros and Bailey, 1977), and concentrated to 75–86 wt % solids. Sulfur dioxide is added during evaporation to some grades of syrup to reduce color development. As an alternative to evaporation, some types of corn syrup, especially those of low DE, are spray dried or roll dried and referred to as corn syrup solids.

SHIPPING

Corn syrup is transported in rail cars and tank trucks equipped with coils that can be heated with steam or hot water to reduce viscosity and aid in unloading. Corn syrup solids are shipped in moisture-proof bags.

TABLE VII
Maltose Syrup Composition[a]

Enzyme	Saccharide (%, db)			
	Dextrose	Maltose	DP-3	DP-4+
Fungal α-amylase	4	52	23	21
Malt β-amylase	2	54	19	25
Malt β + pullulanase	1	76	16	7

[a] Data from Heady and Armbruster (1971).

PRICE

Typical corn syrup prices are shown in Table VI. The price has been relatively constant since 1972, when competitive factors forced prices to a very low level of $0.124/kg ($0.056/lb) (Keim, 1979).

D. Maltodextrins

Maltodextrins are starch hydrolysis products of less than 20 DE. Commercial products are manufactured by hydrolysis of regular corn starch or waxy maize starch by straight acid, acid-enzyme, or enzyme-enzyme techniques. Acid processes are similar to those described for corn syrup. Acid-enzyme processes are conducted by thinning starch to a low DE with acid (Armbruster and Harjes, 1971; Morehouse et al, 1972), followed by neutralization to pH 6–7 and hydrolysis with bacterial α-amylase at 80–85°C for 1–3 hr to the desired DE. A typical total enzyme process uses bacterial α-amylase hydrolysis at 85–92°C, pH 6–8, 30 wt % solids for about 1 hr, followed by additional conversion at 80°C to the desired DE (Armbruster and Kooi, 1974). A heat treatment at 120°C is used to inactivate the enzyme before final processing. Another type of enzymatic approach uses a two-stage process (Armbruster, 1974). Starch is hydrolyzed at pH 7.5–8, 90–92°C for 1 hr with α-amylase to 2–15 DE, followed by a high-temperature heat treatment at 120–150°C for several minutes. Additional enzyme is added and the reaction continued at 80–85°C to the desired DE. The heat treatment is effective in improving the filtration rate and reducing the level of insoluble starch. Typical maltodextrin compositions attained by this procedure are shown in Table VIII. Higher-temperature processes using thermostable α-amylases above 93°C have also been proposed for producing maltodextrins of 10–13 DE (Coker and Venkatasubramanian, 1984).

Maltodextrins prepared from cornstarch are susceptible to haze formation during storage due to association and precipitation of linear amylose fragments. However, when waxy starch is used for maltodextrin production, increased stability is achieved, since the branched polysaccharides do not retrograde as readily as the linear saccharides derived from regular starch. However, even maltodextrins prepared with waxy starch at less than 15 DE develop haze when stored for more than about three days at 70 wt % solids (Harjes et al, 1976). Several techniques have been developed to achieve increased stability. For instance, using as substrate a starch dextrin having a degree of branching of at

TABLE VIII
Maltodextrin Composition (%, db)[a]

DP	Product			
	5 DE	10 DE	15 DE	20 DE
1	<1	<1	<1	1
2	1	3	6	8
3	2	4	7	9
4	2	4	5	7
5	2	4	5	8
6	3	7	11	14
7+	90	78	66	53

[a] Data from Armbruster (1974).

least 7% improves product stability (Harjes et al, 1976). In addition, maltodextrins stable at 70–80 wt % solids can be made from oxidized starch by thinning to 7 DE with acid or enzyme and then converting with α-amylase to the desired DE (Harjes and Wermers, 1976). Stable products are also made by simultaneously liquefying and oxidizing starch to about 4 DE with an oxidizing agent at elevated temperature and then hydrolyzing it with α-amylase (Horn and Kimball, 1976). The hazing tendency may also be reduced through derivatization by introducing nonionic, cationic, or anionic groups (Hull, 1972). Another method for production of nonhazing syrups is to hydrolyze starch to 20–40 DE and then remove low-molecular-weight material by reverse osmosis (Meyer, 1973) or molecular exclusion (Deaton, 1973) to provide stable products of 5–18 DE.

Maltodextrins are refined in the same manner as other corn sweeteners, using clarification, carbon treatment, and ion exchange. The final products are spray dried to a moisture level of 3–5% or evaporated to 75 wt % solids.

V. PROPERTIES

A. Dextrose

Dextrose crystallizes as α-D-glucose, α-D-glucose hydrate, or β-D-glucose, depending on temperature. The physical properties of each form are given in Table IX. In solution, dextrose mutarotates to an equilibrium level of about 62% β- and 38% α-dextrose. At equilibrium, the solubility of dextrose is 51.2 g/100 ml. However, initial solubility characteristics vary depending on the particular crystalline form of dextrose used. For instance, the hydrated form of dextrose dissolves rapidly at 25°C to its solubility level of 30.2%. As mutarotation to the more soluble β-form occurs, solubility increases to a level of 51.2%. Anhydrous dextrose dissolves past the limit of dextrose hydrate solubility, at which point dextrose hydrate then crystallizes from solution and solubility follows the pattern for dextrose hydrate. In solution, both α- and β-dextrose exist primarily in the pyranose form and in equilibrium with a small amount of the open-chain aldehyde form. This open-chain form is responsible for the reducing properties of dextrose.

TABLE IX
Physical Properties of D-Glucose[a]

Property	α-D-Glucose	α-D-Glucose Hydrate	β-D-Glucose
Molecular formula	$C_6H_{12}O_6$	$C_6H_{12}O_6 \cdot H_2O$	$C_6H_{12}O_6$
Melting point, °C	146	83	150
Solubility (at 25°C), g/100 g of solution	62→30.2 − 51.2[b]	30.2→51.2[b,c]	72→51.2[b]
$[\alpha]_D^{20}$[d]	112.2→52.7[b]	112.2→52.7[b,c]	18.7→52.7[b]
Heat of solution (at 25°C), J/g[e]	−59.4	−105.4	−25.9

[a] Source: Hebeda (1983); used by permission of John Wiley & Sons, Inc.
[b] Equilibrium value.
[c] Anhydrous basis.
[d] Specific rotation.
[e] To convert J to cal, divide by 4.184.

In acidic solution, dextrose yields condensation products such as isomaltose and gentiobiose. In addition, dextrose may undergo a dehydration reaction to 5-hydroxymethylfurfural, a water-soluble, high-boiling, unstable compound that can decompose to levulenic and formic acids or polymerize to dark-colored compounds. In alkaline solution, dextrose is isomerized to fructose and mannose. Other decomposition products such as saccharic acids are also formed.

Oxidation reactions yield gluconic acid with dilute alkali and saccharic, tartaric, and oxalic acids with nitric acid. Heating dextrose with methanol in the presence of anhydrous hydrogen chloride yields α-methyl-D-glucoside and a small amount of β-D-glucoside. Similarly, reactions also occur with higher alcohols, although reaction products are more difficult to crystallize. Reaction of dextrose with acid anhydrides in the presence of basic catalysts gives a mixture of esters. Catalytic hydrogenation of dextrose yields sorbitol, and hydrogenolysis produces mixtures of glycols and other degradation products. Reaction of dextrose with a reactive nitrogen, as found in amino acids or proteins, yields a variety of pigments of varied molecular weight via the Maillard reaction. The type of pigments produced is dependent on reaction conditions such as pH, temperature, type of nitrogen, and concentration of reactants.

Important functional properties of dextrose include sweetness, fermentability, and osmotic pressure. In dry and liquid forms, dextrose is about 76 and 65–70%, respectively, as sweet as sucrose (Hanover, 1982). However, perceived sweetness is controlled by many factors such as solids concentration, temperature, and the presence of other ingredients. For instance, synergistic effects in some formulations of dextrose-sucrose blends yield perceived sweetness levels that are equal to or greater than that of sucrose alone. In applications where fermentability is desired, dextrose is an excellent carbohydrate source and is fermented rapidly by yeast and other organisms. Dextrose also produces a greater osmotic effect than sucrose on an equal weight basis and offers advantages where this property is important.

B. High-Fructose Corn Syrup

Pure fructose is a ketohexose monosaccharide that has a melting point of 102–104°C and crystallizes as β-D-fructopyranose. In solution at 36°C, fructose mutarotates to an equilibrium mixture containing 57% β-fructopyranose, 31% β-fructofuranose, 9% α-fructofuranose, and 3% α-fructopyranose (Doddrell and Allerhand, 1971). The solubility of fructose at equilibrium is 80 wt % at 25°C. Fructose reacts in the same manner as other carbohydrates to form esters, ethers, and acetals. Condensation products such as difructose dianhydride are formed in aqueous solution (Binkley et al, 1971). Fructose in solution is most stable at pH 3–4; however, degradative reactions at high and low pH occur more readily than with dextrose (MacAllister and Wardrip, 1978).

Crystalline fructose exhibits a sweetness of 1.8 times that of sucrose and 2.4 times that of crystalline dextrose (Hanover, 1982); however, in solution, the less-sweet forms of fructose reduce apparent overall sweetness. The sweetness of fructose is greatest at cold temperatures, increasing by a factor of 1.8 when temperature is reduced from 60 to 5°C (Freed, 1970). This effect is apparently due to a reduced rate of mutarotation to the less-sweet forms at the lower temperature. Sweetness perception is greatest at neutral or slightly acid pH or when the fructose is in dilute solution (Olefsky and Crapo, 1980) or in the presence of organic acids (Hanover, 1982).

The most important property of HFCS is sweetness. HFCS products are less sweet than pure fructose, due to the presence of dextrose and higher saccharides. However, even though oligosaccharides reduce overall sweetness, taste is not affected (MacAllister and Wardrip, 1978). HFCS is considerably sweeter than regular corn syrups, but as is the case with pure fructose or dextrose, the intensity of sweetness is due to several factors such as temperature, pH, and concentration. HFCS containing 55% fructose exhibits about the same sweetness as sucrose, whereas, 42 and 90% HFCS are 9% less sweet and 6% sweeter than sucrose, respectively (Hanover, 1982).

Other important functional properties of HFCS include high solubility, which prevents crystallization during shipment; humectancy (ability to retain moisture), which gives increased shelf life of bakery products; decomposition during baking, which supplies color and flavor; and high osmotic pressure. Due to the low molecular weight of HFCS, viscosity is relatively low even at high concentration. At a normal commercial solids level of 71 wt %, viscosity ranges from 52 cp at 43°C to 360 cp at 16°C (MacAllister and Wardrip, 1978).

C. Corn Syrup

Important functional properties of corn syrups include fermentability, viscosity, humectancy-hygroscopicity, sweetness, colligative properties, and participation in the browning reaction.

The fermentability of corn syrups is dependent on the specific process and organism used, but in general, dextrose, maltose, and maltotriose are considered fermentable by yeast in brewing and baking. Since the levels of mono-, di-, and trisaccharides increase as syrup DE is increased, fermentability also increases with increasing DE.

The viscosity of corn syrups increases as DE decreases and is an important functional property in many food applications, affecting both organoleptic properties and product stability. Typical viscosity data for acid and acid-enzyme syrups are given in Table X.

The hygroscopicity of corn syrup is related to dextrose level, and rate of moisture absorption increases with increasing DE. Humectancy is also important; moisture retention increases with increasing DE, causing difficulties in drying high-DE syrups.

TABLE X
Viscosity (1,000 cp) of Corn Syrups

Temperature (°C)	Dextrose Equivalent and Corresponding Solids[a]			
	43	43	64	63
	80.3	82.9	81.7	83.8
27	56	285	22	79
38	15	56	6	18
49	5	15	2	5

[a] % wt basis.

The sweetness of corn syrups depends on the level of simple sugars, i.e., dextrose and maltose, and therefore increases with increasing DE. Syrups of 63 and 42 DE are 45 and 30% as sweet as sucrose and 67 and 45% as sweet as dextrose, respectively (Hanover, 1982).

Other important properties of corn syrup include contribution to browning via the Maillard reaction, flavor enhancement, freezing point depression, and osmotic pressure, all of which increase with increasing DE. Properties such as body contribution, cohesiveness, foam stabilization, and prevention of sugar crystallization increase with decreasing DE.

D. Maltodextrin

Maltodextrin is the least hygroscopic of corn sweeteners due to a low DE, i.e., a low monosaccharide content. At the same time, maltodextrins exhibit high viscosity and contribute mouthfeel and body due to the presence of higher-molecular-weight saccharides. Sweetness is not a consideration; maltodextrins are essentially tasteless. In general, maltodextrin properties are similar to those of very low-DE corn syrups.

VI. APPLICATIONS

Corn sweeteners are used in a variety of food and nonfood applications to provide a wide range of nutritive, physical, and chemical properties. Sales of dextrose, HFCS, and corn syrup to major food industries are given in Table XI.

A. Dextrose

Dextrose is used alone or in combination with other sweeteners in the beverage, confectionery, baking, and other industries. For instance, in many applications where sweetness is desired, dextrose is used in conjunction with sucrose (Godzicki, 1975), since the combination may be as sweet as pure sucrose at an equivalent concentration (Nieman, 1960). Dextrose can also be used in combination with aspartame to yield a synergistic effect relative to sweetness (Homler, 1984).

In the beverage industry, dextrose is used to supply sweetness, body, and osmotic pressure. In beverage powders, dextrose enhances flavor and reduces excessive sweetness. Dextrose is used in light beer production as a completely fermentable adjunct to reduce the residual carbohydrate level and therefore the caloric content (Hebeda and Styrlund, 1986). Dextrose is also used in wine manufacture as an additive to grape and other juices low in fermentable sugars. In the confectionery industry, dextrose supplies sweetness, softness, and crystallization control. In confections such as chewing gum and candies, it is used for coating, strength, hardness, color, and gloss. In baking applications, it reacts with protein to provide color and flavor in crusts and strength in bread for improved slicing and handling. In cookies, dextrose gives a tenderizing effect and crust coloration. The use of dextrose in cakes and cake mixes results in improved physical characteristics such as texture and color. Dextrose is also used in prepared mixes for biscuits, pancakes, waffles, doughnuts, and icings. In canning, dextrose is used in sauces, soups, gravies, fruits, and juices to provide flavor, body, and sweetness, as well as to improve texture and aesthetic quality. In dairy applications, it is used in ice cream and frozen desserts to provide sweetness and control crystallization for a smooth, creamy texture. Other food applications include use in such diversified products as peanut butter, meat, pickles, and condiments.

In pharmaceutical applications, dextrose is used for intravenous and subcutaneous injections and for tableting.

Dextrose or high-dextrose corn syrups are used in the fermentation industry as a raw material for biochemical synthesis of products such as citric acid, antibiotics, vitamins, amino acids, and enzymes. Ethanol produced by yeast fermentation of unrefined dextrose hydrolysate is used in gasoline as an octane enhancer or additive (Keim, 1983). Production of other chemicals by fermentation of dextrose has been suggested as a potential alternative to petrochemicals (Ng et al, 1983).

In chemical processes, dextrose is used for production of sorbitol (Morris, 1983) and methyl glucoside (Tokay, 1983), as well as gluconic acid, gluconates, and heptonates. Acid-catalyzed polymerization of dextrose in the presence of a polyol is used to manufacture polydextrose, a cross-linked polymer used as a water-soluble, low-calorie, bulking agent (Torres and Thomas, 1981). Polydextrose has been proposed as a replacement for sugar in various

TABLE XI
Distribution (1,000 t) by Industry of Dextrose, High-Fructose Corn Syrup (HFCS), and Corn Syrup in the United States[a,b]

Product and Year	Baking	Beverage	Canning	Confectionery	Dairy	Total
Dextrose						
1965	178	8	21	37	6	468
1970	174	8	23	52	7	547
1975	157	18	15	62	7	561
1980	51	66	4	55	2	513
1982	50	77	6	51	2	481
1983	56	82	5	63	2	488
1984	54	86	4	66	1	465
HFCS						
1970	18	39	9	0.5	5	99
1975	129	279	64	4	36	715
1980	365	1,093	235	15	140	2,659
1982	390	2,086	277	27	189	3,968
1983	385	2,621	272	31	201	4,603
1984	392	3,698	264	34	229	5,479
Corn syrup						
1965	209	36	104	410	159	1,211
1970	222	93	98	439	213	1,449
1975	315	205	174	416	278	2,278
1980	196	384	126	446	241	2,201
1982	161	359	143	458	268	2,295
1983	148	370	124	467	267	2,299
1984	110	376	139	464	276	2,305

[a] Source: Hebeda (1983); used by permission of John Wiley & Sons, Inc.
[b] Commercial basis.

confections, baked goods, and other products to provide low-calorie foods containing reduced sugar and fat levels.

Dextrose may also be used in adhesives to provide flow control, in library paste to increase open time, in wallboard as a humectant to prevent brittle edges, in concrete as a setting retardant, in resin manufacture as a modifier and plasticizer, as a reducing agent in metal treatment, and as a diluent for standardizing dyes.

B. High-Fructose Corn Syrup

HFCS sales in the United States have increased dramatically in recent years. For instance, between 1970 and 1984, HFCS sales increased from 5 to 69% of total corn sweetener sales. Much of this increase was due to the increased use of 55% HFCS. Between 1980 and 1985, sales of 55% HFCS increased from 33 to 65% of total HFCS sales (Vuilleumier, 1985). The rapid increase of HFCS use has been primarily due to the acceptance of HFCS as a replacement for sucrose in soft drinks. In 1985, about 96% of the sweetener used by the beverage industry was HFCS (Vuilleumier, 1985).

The first substitution of HFCS for sucrose in soft drinks took place in 1974, when high sucrose prices led to the approval of a 25% replacement of sucrose with 42% HFCS in several different products. With the development of a 55% HFCS product, increased substitution levels of as much as 100% were approved for some products in 1978 and 1979. The next major increase in HFCS usage occurred in 1983, due to the improved product consistency requested by soft drink producers. Quality improvements were initiated in 1982 through a cooperative effort among HFCS producers (Morris, 1984), and improved standards were achieved through better control of processing and refining to maintain consistency in ash, color, pH, solids, and saccharide composition. Currently, the approved HFCS substitution level is 100% in most major soft drinks.

The second largest use for HFCS is in the baking industry, where 42% HFCS is used to replace all or a portion of sucrose and still achieve the same product characteristics. In canning, HFCS is used as the predominant sweetener in combination with sucrose and high-conversion syrup. In dairy applications, blends of HFCS and corn syrup function as a bodying agent and also improve texture and mouthfeel. In confectionery, HFCS supplies sweetness, grain control, and humectancy.

HFCS containing 90% fructose can be used in reduced-calorie products to lower the caloric level by 30–50% in a wide range of foods, including dairy products, frozen and canned fruit, soft drinks, confections, salad dressings, baked goods, dry cereals, and table top sweeteners (Young and Long, 1982).

Crystalline fructose is essentially pure and is therefore used in low-calorie and specialty foods. Examples are frozen desserts, yogurt, puddings, dry beverage powders, ice cream, candy, baked goods, and table top sweeteners.

C. Corn Syrup

Corn syrup is often used in combination with sucrose, dextrose, or HFCS (Godzicki, 1975). Its application is based on specific functional properties, and the type of syrup used depends on the properties desired in the final product.

The primary use of corn syrup is in the confectionery industry in virtually all types of products from hard candies to marshmallows to provide viscosity, mouthfeel, sweetness, texture, grain inhibition, hygroscopicity, and resistance to discoloration.

In the beverage industry, corn syrups are used as a source of fermentables in beer and malt liquor production (Hebeda and Styrlund, 1986) and also to enhance flavor and provide body. High-conversion or high-maltose syrups are used as replacements for dry cereal adjuncts to achieve a fermentable level that approximates that produced during mashing. Corn syrups are generally not used in manufacture of carbonated beverages, although Maeda and Tsao (1979) report that the mild sweetness of high-maltose syrup has found application in soft drinks in Japan.

In the dairy industry, corn syrups provide texture, smoothness, and grain control; modify melt-down and shrinkage; and function as bodying agents. Generally 36–42 DE acid-converted syrups are used in ice cream, sherbet, iced milk, and other frozen desserts.

In baking, high-conversion or high-maltose syrups are used as a source of fermentables. Syrups are used in yeast-raised products and in cakes for viscosity and rheological properties and for maintaining moisture balance and improved shelf life. In cookies, syrups are used as a tenderizing ingredient and provide crust color and moisture retention.

The major application for syrups in canning is in canned fruit to prevent crystallization of sucrose and provide body and accentuate fruit flavor while improving color and texture. Generally, high-conversion syrups are used.

Other food applications are in icings and fillings to improve sheen and appearance; in chewing gum for sweetness and for softening the gum base; in meat products to supply flavor, humectancy, body, and viscosity; and in pickles for viscosity and mouthfeel.

Hydrogenation of corn syrup is used to produce a nonreducing sweetener that can replace regular corn syrups in various confections (Rockstrom, 1980).

In nonfood applications, corn syrup may be used in adhesives to improve stability, as a setting retardant in concrete, as a humectant in air fresheners, for evaporation control in colognes and perfumes, as a carrier and sweetener in medicinal syrups and lozenges, and as a humectant in tobacco.

D. Maltodextrins

Maltodextrins are used in applications where nonsweet, nonhygroscopic, water-holding properties are desired. For instance, maltodextrin is used as a bodying agent or bulking agent in puddings, soups, frozen desserts, and dry mixes. It is also used as a spray-drying adjunct for coffee and tea extracts, a dispersing agent in synthetic coffee whitener, and as a moisture-holding agent in breads, pastries, and meats. Maltodextrin can be used as a partial or total replacement for sweeteners in gum confection (Godzicki and Kimball, 1971; Horn and Kimball, 1971) or to replace a portion of the protein whipping agent in aerated confections (Horn et al, 1971). In addition, maltodextrin can improve hygroscopicity characteristics in hard candy (Horn and Godzicki, 1974). Nonhygroscopic, water-soluble fondants are prepared using a combination of a dry sugar and maltodextrin (Sands and Marino, 1975a, 1975b). Maltodextrin used in combination with dextrose is effective in preparing a direct-compression vehicle for tableting for use in pharmaceutical applications (Kanig, 1975; Nelson et al, 1977).

LITERATURE CITED

ABDULLAH, M. 1972. Production of high maltotetraose syrup. U.S. patent 3,654,082.
ABDULLAH, M., FLEMING, I. D., TAYLOR, P. M., and WHELAN, W. J. 1963. Substrate specificity of the amyloglucosidase of *Aspergillus niger*. Biochem. J. 89(1):35-36.
AACC. 1983. Approved Methods of the American Association of Cereal Chemists, 8th ed. Method 80-68, approved, April 1961; reviewed, Oct. 1982. Am. Assoc. Cereal Chem., St. Paul, MN.
ALLEN, W. G., and SPRADLIN, J. E. 1974. Amylases and their properties. Brew. Dig. 49(7):48-53, 65.
AMOTZ, S., NIELSEN, T. K., and THIESEN, N. O. 1976. Immobilization of glucose isomerase. U.S. patent 3,980,521.
ANONYMOUS. 1976. Sugar and Sweetener Report. U.S. Dept. Agric., Commodity Econ. Div., Washington, DC.
ANONYMOUS. 1981. Continuous Production of Fructose Syrup with Novo's Immobilized Glucose Isomerase, Sweetzyme Type Q. Brochure 175c-GB, August. Novo Industri A/S, Bagsvaerd, Denmark.
ANONYMOUS. 1983a. Taka-Sweet, immobilized glucose isomerase for high fructose syrup production. Data Sheet R883 P883 L-1160. Biotech Products Division, Miles Laboratories, Inc., Elkhart, IN.
ANONYMOUS. 1983b. Corn syrup clarifiers now operating on three continents. Sugar Azucar 78(8):10.
ANONYMOUS. 1984a. Enzyme bio-systems will focus on starch conversions. Biomass Dig. (1):7.
ANONYMOUS. 1984b. HFCS-production enzyme. Food Eng. 56(8):56.
ANONYMOUS. 1984c. Spezyme IGI450, immobilized glucose isomerase. Data Sheet IGI-101, April. Fermco Biochemics Inc., Elk Grove Village, IL.
ANONYMOUS. 1984d. Immobilized glucoamylase: Breakthrough in HFCS process. Food Eng. 56(9):160.
ANONYMOUS. 1986. Sugar and sweetener Report. U.S. Dept. Agric. Natl. Econ. Div. Washington, DC. Sept.
ARMBRUSTER, F. C. 1961. Enzyme preparation. U.S. patent 3,012,944.
ARMBRUSTER, F. C. 1974. Process for producing non-waxy starch hydrolyzates. U.S. patent 3,853,706.
ARMBRUSTER, F. C., and HARJES, C. F. 1971. Low DE starch conversion products. U.S. patent 3,560,343.
ARMBRUSTER, F. C., and JACAWAY, W. A. 1970. Process for making high maltose syrup. U.S. patent 3,549,496.
ARMBRUSTER, F. C., and KOOI, E. R. 1974. Low DE starch conversion products. U.S. patent 3,849,194.
ARMBRUSTER, F. C., HEADY, R. E., and CORY, R. P. 1974. Production of xylose (dextrose) isomerase enzyme preparations. U.S. patent 3,813,318.
ARMBRUSTER, F. C., HEADY, R. E., and CORY, R. P. 1976. Production of xylose (dextrose) isomerase enzyme preparations. U.S. patent 3,957,587.
BACHLER, M. J., STRANDBERG, G. W., and SMILEY, K. L. 1970. Starch conversion by immobilized glucoamylase. Biotechnol. Bioeng. 12:85-92.
BASSO, A. J. 1982. Vacuum filtration using filteraids. Chem. Eng. 89(8):159-162.
BERNETTI, R. 1982. Modern methods of analysis of corn derived sweeteners. Pages 1-11 in: Food Carbohydrates. D. R. Lineback and G. E. Inglett, eds. Avi Publ. Co., Inc., Westport, CT.
BESCHKOV, V., MARC, A., and ENGASSER, J. M. 1984. A kinetic model for the hydrolysis and synthesis of maltose, isomaltose, and maltotriose by glucoamylase. Biotechnol. Bioeng. 26:22-26.
BINKLEY, W. W., BINKLEY, R. W., and DIEHL, D. R. 1971. Identification of the anhydrides of D-fructose from the "fingerprint" region of their infrared spectra. Int. Sugar J. 73:259-261.
BOHNENKAMP, C. G., and REILLY, P. J. 1980. Use of immobilized glucoamylase-β-amylase and glucoamylase-fungal amylase mixtures to produce high-maltose syrups. Biotechnol. Bioeng. 22:1753-1758.
BOLLENBACK, G. N. 1983. Special sugars. Pages 944-948 in: Kirk-Othmer Encyclopedia of Chemical Technology, Vol. 21, 3rd ed. John Wiley & Sons, Inc., New York.
BROBST, K. M., and SCOBELL, H. D. 1981. Chromatographic analysis of sugars in cereals

and cereal products. Cereal Foods World 26:224-227.
BROUGHTON, D. B., BIESER, H. J., BERG, R. C., CONNELL, E. D., KOROUS, D. J., and NEUZIL, R. W. 1977. High purity fructose via continuous adsorptive separation. Sucr. Belge. 96:155-162.
BROWNEWELL, C. E. 1971. Process for production of glucose isomerase. U.S. patent 3,625,828.
BULPIN, P. V., WELSH, E. J., and MORRIS, E. R. 1982. Physical characterization of amylose-fatty acid complexes in starch granules and in solution. Staerke 34:335-339.
CANTOR, S. M., and HOBBS, K. C. 1944. Conversion of dextrose to levulose. U.S. patent 2,354,664.
CARASIK, W., and CARROLL, J. O. 1983. Development of immobilized enzymes for production of high fructose corn syrup. Food Technol. Chicago 37(10):85-91.
CHEN, W. 1980a. Glucose isomerase (a review). Process Biochem. 15(5):30-35.
CHEN, W. 1980b. Glucose isomerase (a review). Process Biochem. 15(6):36-41.
CHIANG, J. P., ALTER, J. E., and STERNBERG, M. 1979. Purification and characterization of thermostable alpha-amylase from Bacillus licheniformis. Staerke 31:86-92.
COKER, L. E., and VENKATA-SUBRAMANIAN, K. 1984. Process for the manufacture of low DE maltodextrins. U.S. patent 4,447,532.
COTTER, W. P., LLOYD, N. E., and HINMAN, C. W. 1976. Method for isomerizing glucose syrups. U.S. patent Re. 28,885. Reissue of U.S. patent 3,623,953, 1971.
DALE, J. K., and LANGLOIS, D. P. 1940. Sirup and method of making the same. U.S. patent 2,201,609.
DANIELS, M. J. 1980. Commercial starch processing using immobilized enzymes. Pages 103-109 in: Food Process Engineering, Vol. 2. P. Linko and J. Larinkari, eds. Applied Science Publ., Ltd., London.
DEATON, I. F. 1973. Process for the production of non-hazing starch conversion syrups. U.S. patent 3,756,919.
de SAUSSERE, T. 1815. Ann. Phys. 49:129.
DODDRELL, D., and ALLERHAND, A. 1971. Study of anomeric equilibrium of ketoses in water by natural-abundance carbon-13 Fourier transform nuclear magnetic resonance. D-Fructose and D-turanose. J. Am. Chem. Soc. 93:2779-2781.
DOTY, T. 1980. Fructose: The rationale for traditional and modern uses. Pages 259-268 in: Carbohydrate Sweetness in Foods and Nutrition. P. Koivistoinen and L. Hyvonen, eds. Academic Press, London.
EATON, D. L., and MESSING, R. A. 1976. Support of alumina-magnesia for the adsorption of glucose isomerase enzyme. U.S. patent 3,992,329.
FREED, M. 1970. Fructose—The extraordinary natural sweetener. Food Prod. Dev. 4(1):38-39.
GODZICKI, M. M. 1975. Engineering "sugar." Food Eng. 47(10):14-17.
GODZICKI, M. M., and KIMBALL, B. A. 1971. Gum confections containing dextrose and 5-15 DE starch hydrolysate. U.S. patent 3,589,909.
GREDITOR, A. S. 1983. Nutritive Sweetener Update, Recent Trends and Outlook. Drexel Burnham Lambert, Inc., New York.
GRUESBECK, C., and RASE, H. F. 1972. Insolubilized glucoamylase enzyme system for continuous production of dextrose. Ind. Eng. Chem. Prod. Res. Dev. 11(1):74-83.
HAMILTON, B. K., COLTON, C. K. and COONEY, C. L. 1974. Glucose isomerase: A case study of enzyme catalyzed process technology. Pages 85-131 in: Immobilized Enzymes in Food and Microbial Processes. A. C. Olson and C. L. Cooney, eds. Plenum Press, New York.
HANOVER, L. M. 1982. Functionality of corn-derived sweeteners in formulated foods. Pages 211-233 in: Chemistry of Foods and Beverages: Recent Developments. G. Charalambous and G. Inglett, eds. Academic Press, New York.
HARJES, C. F., and WERMERS, V. L. 1976. Malto-dextrins of improved stability prepared by enzymatic hydrolysis of oxidized starch. U.S. patent 3,974,033.
HARJES, C. F., LEACH, H. W., and TRP, J. M. 1976. Low DE starch hydrolyzates of improved stability prepared by enzymatic hydrolysis of dextrins. U.S. patent 3,974,032.
HAUSSER, A. G., GOLDBERG, B. S., and MERTENS, J. L. 1983. An immobilized two-enzyme system (fungal α-amylase/gluco-amylase) and its use in the continuous production of high conversion maltose-containing corn syrups. Biotechnol. Bioeng. 25:525-539.
HEADY, R. E., and ARMBRUSTER, F. C. 1971. Preparation of high maltose conversion products. U.S. patent 3,565,765.
HEBEDA, R. E. 1983. Syrups. Pages 499-522 in: Kirk-Othmer Encyclopedia of Chemical Technology, Vol. 22, 3rd ed. John Wiley & Sons, Inc., New York.
HEBEDA, R. E., and LEACH, H. W. 1974. The nature of insoluble starch particles in liquefied corn-starch hydrolysates. Cereal

Chem. 51:272-281.
HEBEDA, R. E., and STYRLUND, C. R. 1986. Starch hydrolysis products as brewing agents. Cereal Foods World 31(9):685-687.
HEBEDA, R. E., HOLIK, D. J., and LEACH, H. W. 1975. Enzymatic hydrolysis of granular starch. U.S. patent 3,922,199.
HEBEDA, R. E., HOLIK, D. J., and LEACH, H. W. 1979. Dextrose production with immobilized glucoamylase. U.S. patent 4,132,595.
HEROS, D. V., and BAILEY, C. 1977. New ways of granular activated carbon application in the corn sweetener purification. Staerke 29:422-425.
HOLSTEIN, A. G., and HOLSING, G. C. 1962. Method for the production of levulose. U.S. patent 3,050,444.
HOLT, N. C., BOS, C., and RACHLITZ, K. 1975. Method of making starch hydrolyzates by enzymatic hydrolysis. British patent 1,401,791.
HOMLER, B. E. 1984. Properties and stability of aspartame. Food Technol. 38(7):50-55.
HORN, H. E., and GODZICKI, M. M. 1974. Hard candy. U.S. patent 3,826,857.
HORN, H. E., and KIMBALL, B. A. 1971. Gum confections containing 5-15 DE starch hydrolyzate. U.S. patent 3,582,359.
HORN, H. E., and KIMBALL, B. A. 1976. Malto-dextrins of improved stability prepared by enzymatic hydrolysis of oxidized starch. U.S. patent 3,974,034.
HORN, H. E., JENSEN, E. R., and KIMBALL, B. A. 1971. Aerated confection containing 5-15 DE starch hydrolyzate. U.S. patent 3,586,513.
HULL, G. A. 1972. Low DE starch derivatives. U.S. patent 3,639,389.
HULTIN, H. O. 1983. Current and potential uses of immobilized enzymes. Food Technol. 37(10):66-82, 176.
HURST, L. S., and LLOYD, N. E. 1983. Process for producing glucose/fructose syrups from unrefined starch hydrolyzates. U.S. patent 4,376,824.
HURST, T. L. 1975. Process for producing dextrose. U.S. patent 3,897,305.
HURST, T. L., and TURNER, A. W. 1962. Method of refining amyloglucosidase. U.S. patent 3,067,108.
JONES, P. E., and THOMASON, F. G. 1951. Competitive relationships between sugar and corn sweeteners. Bull. 48. U.S. Dept. Agric, Production and Marketing Admin., Washington, DC.
KAINUMA, K., WAKO, K., KOBAYASHI, S., NOGAMI, A., and SUZUKI, S. 1975. Purification and some properties of a novel maltohexose-producing exo-amylase from *Aerobacter aerogenes*. Biochim. Biophys. Acta 410:333-346.
KANIG, J. L. 1975. Direct compression tabletting composition and pharmaceutical tablets produced therefrom. U.S. patent 3,873,694.
KATHREIN, H. R. 1963. Treatment and use of enzymes for the hydrolysis of starch. U.S. patent 3,108,928.
KEAN, C. E. 1978. Changing sources and industrial uses of sugar. Food Prod. Dev. 12(April):43, 46, 48.
KEIM, C. R. 1979. Competitive sweeteners. Sugar Azucar Yearbook 47:101-120.
KEIM, C. R. 1983. Technology and economics of fermentation alcohol—An update. Enzyme Microb. Technol. 5(March):103-114.
KELLER, H. W., REENTS, A. C., and LARAWAY, J. W. 1981. Process for fructose enrichment from fructose bearing solutions. Staerke 33:55-57.
KERR, R. W. 1961. Method of making dextrose using purified amyloglucosidase. U.S. patent 2,967,804.
KERR, R. W. 1962. Method of making dextrose using purified amyloglucosidase. U.S. patent 3,017,330.
KIRCHOFF, C. G. S. 1811. Mem. Acad. Imp. Sci. St. Petersbourg 4:27.
KOOI, E. R., HARJES, C. F., and GILKINSON, J. S. 1962. Treatment and use of enzymes for the hydrolysis of starch. U.S. patent 3,042,584.
KOOI, E. R., and ARMBRUSTER, F. C. 1967. Production and use of dextrose. Pages 553-568 in: Starch: Chemistry and Technology, Vol. 2. R. L. Whistler and E. F. Paschall, eds. Academic Press, Inc., New York.
KUGIMIYA, M., DONOVAN, J. W., and WONG, R. Y. 1980. Phase transitions of amylose-lipid complexes in starches. A colorimetric study. Staerke 32:265-270.
KUSCH, T., GOSEWINKEL, W., and STOECK, G. 1970. Process for the production of crystalline fructose. U.S. patent 3,513,023.
LAUER, K., STEPHAN, P., and STOECK, G. 1971. Process and apparatus for the recovery of crystalline fructose from methanolic solution. U.S. patent 3,607,392.
LEE, C. K., and LONG, M. E. 1974. Enzymatic process using immobilized microbial cells. U.S. patent 3,821,086.
LEE, C. K., HAYES, L. E., and LONG, M. E. 1972. Process of preparing glucose isomerase. U.S. patent 3,645,848.
LEE, D. D., LEE, Y. Y., and TSAO, G. T. 1975. Continuous production of glucose from dextrin by glucoamylase immobilized on

porous silica. Staerke 27:384-387.
LEVY, J., and FUSEE, M. C. 1979. Support matrices for immobilized enzymes. U.S. patent 4,141,857.
LLOYD, N. E. 1983. Process for isomerizing glucose. U.S. patent 4,411,996.
LLOYD, N. E., LEWIS, L. T., LOGAN, R. M., and PATEL, D. N. 1972. Process for isomerizing glucose to fructose. U.S. patent 3,694,314.
LOBARZEWSKI, J., and PASZCZYNSKI, A. 1983. Catalytic properties of immobilized crude and pure glucoamylase from *Aspergillus niger* C. Biotechnol. Bioeng. 25:3207-3212.
LOBRY DEBRUYN, C. A., and VAN ECKENSTEIN, W. A. 1895. Action of alkalis on sugars. Reciprocal transformation of glucose and fructose. Rec. Trav. Chim. 14:201-216.
LONG, J. E. 1978. The second generation high fructose corn syrups. Pages 16-119 in: Corn Annual. Corn Refiners Assoc., Inc., Washington, DC.
LUNDQUIST, J. T., Jr., VELTMAN, P. L., and WOODRUFF, E. T. 1976. Method for drying fructose solutions. U.S. patent 3,956,009.
MacALLISTER, R. V. 1979. Nutritive sweeteners made from starch. Pages 15-56 in: Advances in Carbohydrate Chemistry and Biochemistry, Vol. 36. R. S. Tipson and D. Horton, eds. Academic Press, New York.
MacALLISTER, R. V. 1980. Manufacture of high fructose corn syrup using immobilized glucose isomerase. Pages 81-111 in: Immobilized Enzymes for Food Processing. W. H. Pitcher, Jr., ed. CRC Press, Inc., Boca Raton, FL.
MacALLISTER, R. V., and WARDRIP, E. K. 1978. Fructose and high fructose corn syrups. Pages 329-332 in: Encyclopedia of Food Science. M. S. Paterson and A. H. Johnson, eds. Avi Publ. Co., Inc., Westport, CT.
MAEDA, H., and TSAO, G. T. 1979. Maltose production. Process Biochem. 14(7):2, 4-5, 27.
MAEDA, H., TSAO, G. T., and CHEN, L. F. 1978. Preparation of immobilized soybean β-amylase on porous cellulose beads and continuous maltose production. Biotechnol. Bioeng. 25:383-402.
MARSHALL, R. O. 1960. Enzymatic process. U.S. patent 2,950,228.
MARSHALL, R. O., and KOOI, E. R. 1957. Enzymatic conversion of D-glucose to D-fructose. Science 125:648-649.
McMULLEN, W. H., and ANDINO, R. 1977. Production of high purity glucose syrups. U.S. patent 4,017,363.
MERMELSTEIN, N. H. 1975. Immobilized enzymes produce high-fructose corn syrup. Food Technol. 29(6):20-26.
MESSING, R. A. 1974. Enzymes immobilized on porous inorganic support materials. U.S. patent 3,850,751.
MEYER, G. R. 1973. Process for the production of non-hazing starch conversion syrups. U.S. patent 3,756,853.
MOREHOUSE, A. L., MALZAHN, R. C., and DAY, J. T. 1972. Hydrolysis of starch. U.S. patent 3,663,369.
MORRIS, C. E. 1981. First crystalline-fructose plant in U.S. Food Eng. 53(11):70-71.
MORRIS, C. E. 1983. Computer-controlled sorbitol plant operates 24 hours per day. Food Eng. 55(7):108-109.
MORRIS, C. E. 1984. Industry effort upgrades HFCS. Food Eng. 56(3):55-56.
NEIDLEMAN, S. L., AMON, W. F., Jr., and GEIGERT, J. 1981. Process for production of fructose. U.S. patent 4,246,347.
NELSON, A. L., SKRABACZ, D. J., and YOUNG, B. 1977. Process for preparing a sugar tablet. U.S. patent 4,013,775.
NEWKIRK, W. B. 1923. Method of making grape sugar. U.S. patent 1,471,347.
NG, T. K., BUSCHE, R. M., McDONALD, C. C., and HARDY, R. W. F. 1983. Production of feedstock chemicals. Science 219:733-740.
NIELSEN, G. C., DIERS, I. V., OUTTRUP, H., and NORMAN, B. E. 1982. Debranching enzyme product preparation and use thereof. British patent 2,097,405.
NIEMAN, C. 1960. Sweetness of glucose, dextrose, and sucrose. Manuf. Confect. 40(8):19-24, 43-46.
NITSCH, E. 1974. Method of producing fructose and glucose from sucrose. U.S. patent 3,812,010.
NORMAN, B. E. 1979. The application of polysaccharide degrading enzymes in the starch industry. Pages 339-376 in: Microbial Polysaccharides and Polysaccharases. R. C. W. Berkeley, G. W. Gooday, and D. C. Ellwood, eds. Academic Press, Inc., New York.
NORMAN, B. E. 1981. New developments in starch syrup technology. Pages 15-50 in: Enzymes and Food Processing. G. G. Birch, N. Blakebrough, and K. J. Parker, eds. Applied Science Publ., Ltd., London.
NORMAN, B. E. 1982. Saccharification of starch hydrolysates. U.S. patent 4,335,208.
OHBA, R., and UEDA, S. 1980. Production of maltose and maltotriose from starch and pullulan by an immobilized multienzyme of pullulanase and β-amylase. Biotechnol. Bioeng. 22:2137-2154.

OLEFSKY, J. M., and CRAPO, P. 1980. Fructose, xylitol, and sorbitol. Diabetes Care 3(2):390-393.

OSBERGER, T. F. 1979. Tableting characteristics of pure crystalline fructose. Pharm. Technol. 3(6):81-86.

OUTTRUP, H. 1976. Production of glucose isomerase by *Bacillus coagulans*. U.S. patent 3,979,261.

OUTTRUP, H., and NORMAN, B. E. 1984. Properties and application of a thermostable maltogenic amylase produced by a strain of *Bacillus* modified by recombinant-DNA techniques. Staerke 36:405-411.

PARK, Y. K., and LIMA, D. C. 1973. Continuous conversion of starch to glucose by an amyloglucosidase-resin complex. J. Food Technol. 38:358-359.

PECKHAM, B. W. 1979. Economics and invention: A technological history of the corn refining industry of the United States. Ph.D. dissertation, Univ. of Wisconsin, Madison.

PORST, C. E. G. 1921. Chem. Age 29:213.

REICHELT, J. R. 1983. Starch. Pages 375-396 in: Industrial Enzymology, The Applications of Enzymes in Industry. T. Godfrey and J. R. Reichelt, eds. The Nature Press, New York.

ROCKSTROM, E. 1980. Lycasin hydrogenated hydrolyzates. Pages 225-232 in: Carbohydrate Sweeteners in Foods and Nutrition. P. Koivistoinen and L. Hyvonen, eds. Academic Press, London.

ROHRBACH, R. P. 1981. Support matrices for immobilized enzymes. U.S. patent 4,268,419.

ROSENDAL, P., NIELSEN, B. H., and LANGE, N. K. 1979. Stability of bacterial alpha-amylase in the starch liquefaction process. Staerke 31:368-372.

RUGH, S., NIELSEN, T., and POULSEN, P. B. 1979. Application possibilities of a novel immobilized glucoamylase. Staerke 31:333-337.

SAITO, N. 1973. A thermophilic extracellular-amylase from *Bacillus licheniformis*. Arch. Biochem. Biophys. 155:290-298.

SANDS, M. A., and MARINO, S. P. 1975a. Non-hygroscopic, water-soluble sugar products and process for preparing the same. U.S. patent 3,874,924.

SANDS, M. A., and MARINO, S. P. 1975b. Non-hygroscopic, water-soluble fondant and glaze composition and process for preparing the same. U.S. patent 3,917,874.

SHIEH, K. K., and DONNELLY, B. J. 1974. Methods of making glucose isomerase and converting glucose to fructose. U.S. patent 3,813,320.

SHIEH, K. K., LEE, H. A., and DONNELLY, B. J. 1974. Method of making glucose isomerase and using same to convert glucose to fructose. U.S. patent 3,834,988.

SHIRAISHI, F., KAWAKAMI, K., and KUSUNOKI, K. 1985. Kinetics of condensation of glucose into maltose and isomaltose in hydrolysis of starch by glucoamylase. Biotechnol. Bioeng. 27:498-502.

SINGH, B., DEAN, G. R., and CANTOR, S. M. 1948. The role of 5-(hydroxymethyl)-furfural in the discoloration of sugar solutions. J. Am. Chem. Soc. 70:517-522.

SLOTT, S., and MADSEN, G. B. 1975. Procedure for liquefying starch. U.S. patent 3,912,590.

SMILEY, K. L. 1971. Continuous conversion of starch to glucose with immobilized glucoamylase. Biotechnol. Bioeng. 13:309-317.

SUBRAMANIAN, T. V. 1980. Equilibrium relationships in the degradation of starch by an amyloglucosidase. Biotechnol. Bioeng. 22:643-649.

SUTTHOFF, R. F., MacALLISTER, R. V., and KHALEELUDDIN, K. 1978. Agglomerated fibrous cellulose. U.S. patent 4,110,164.

SVENSSON, B., PEDERSEN, T. G., SVENDSEN, I., SAKAI, T., and OTTESEN, M. 1982. Characterization of two forms of glucoamylase from Aspergillus niger. Carlsberg Res. Commun. 47:55-69.

TAKASAKI, Y. 1966. Studies on sugar-isomerizing enzyme production and utilization of glucose isomerase from Streptomyces sp. Agric. Biol. Chem. 30:1247-1253.

TAKASAKI, Y. 1976. Studies on amylases from Bacillus effective for production of maltose. Agric. Biol. Chem. 40:1515-1522.

TAKASAKI, Y., and TANABE, O. 1971. Enzyme method for converting glucose in glucose syrups to fructose. U.S. patent 3,616,221.

TAMURA, M., SHIMIZU, M., and TAGO, M. 1981a. Highly thermostable glucoamylase and process for its production. U.S. patent 4,247,637.

TAMURA, M., SHIMIZU, M., and TAGO, M. 1981b. Novel neutral glucoamylase and method for its production. U.S. patent 4,254,225.

TEAGUE, J. R., and ARNOLD, E. C. 1983. UOP technology for the production of fructose sweeteners. Sugar Azucar 78(8):18-23.

THOMPSON, K. N., JOHNSON, R. A., and LLOYD, N. E. 1974. Process for isomerizing glucose to fructose. U.S. patent 3,788,945.

TOKAY, B. A. 1983. New perspectives on biomass chemicals. Chem. Mark. Rep. (Aug. 22):8-14.

TORRES, A., and THOMAS, D. 1981.

Polydextrose and its application in foods. Food Technol. 35(7):44-49.

van ASBECK, T. M. W., GOUWEROK, M., and POLMAN, E. 1981. The evaluation of activated carbon in the purification of starch-based sweeteners. Staerke 33:378-383.

VUILLEUMIER, S. 1985. An update on high fructose corn syrup in the United States. Sugar Azucar 80(10):13-20.

WASSERMAN, B. P., BURKE, D., and JACOBSON, B. S. 1982. Immobilization of glucoamylase from *Aspergillus niger* on poly(ethylenimine)-coated non-porous glass beads. Enzyme Microb. Technol. 4(March):107-109.

WICHELHAUS, H. 1913. Der Starkezucker. Akad. Verlagsgesellschaft Leipzig.

WOOLHOUSE, A. D. 1976. Starch as a Source of Carbohydrate Sweeteners. Rep. CD 2237. Dept. of Scientific and Industrial Research, New Zealand.

YAMAUCHI, T. 1975. Production of free-flowing particles of glucose, fructose, or the mixture thereof. U.S. patent 3,929,503.

YOUNG, L. S., and LONG, J. E. 1982. Manufacture, use, and nutritional aspects of 90% high fructose corn sweeteners. Pages 195-210 in: Chemistry of Foods and Beverages: Recent Developments. G. Charalambous and G. Inglett, eds. Academic Press, New York.

ZIENTY, M. F. 1973. Enzyme stabilization. U.S. patent 3,779,869.

ZITTAN, L., POULSEN, P. B., and HEMMINGSEN, S. H. 1975. Sweetzyme—A new immobilized glucose isomerase. Staerke 27:236-241.

ZOBEL, H. F. 1984. Gelatinization of starch and mechanical properties of starch pastes. Pages 285-309 in: Starch: Chemistry and Technology, 2nd ed. R. L. Whistler, J. N. BeMiller, and E. F. Paschall, eds. Academic Press, Inc., Orlando, FL.

CHAPTER 18

CORN OIL: COMPOSITION, PROCESSING, AND UTILIZATION

F. T. ORTHOEFER[1]
R. D. SINRAM
A. E. Staley Manufacturing Company
Decatur, Illinois

I. INTRODUCTION

Among the edible vegetable oils in the marketplace, corn oil is a minor oil, since it constituted only about 9.0% of the 1984/85 U.S. vegetable oil production. The U.S. oil market is dominated by soybean oil, which amounted to 78.4% of the 1984/85 production. The other edible vegetable oils produced in the United States are cottonseed, sunflower, and peanut, with 8.0, 3.3, and 1.3% of the market, respectively. Worldwide, the major oils produced include rape, olive, cocoa, palm and palm kernel, peanut, and sunflower. Nevertheless, corn oil is an important food oil because it has a positive image with both the user and the consumer (Erickson and Falb, 1979). Its high polyunsaturated fatty acid content, which has been widely publicized, has important nutritional and health benefits. The oxidative stability of corn oil during use and its lack of precipitation under refrigeration have contributed to its market demand (Reiners, 1978). All but a minor fraction ($<5\%$) is used in foods; the largest single use is bottled oil, followed by margarine and industrial snack-frying operations.

II. PRODUCTION AND MARKETS

Corn oil production in the United States doubled from 1976 to 1986, based on the USDA (1986) projection of 1.349 billion pounds (612×10^3 t) produced in 1986. This has been due largely to expansion of the corn wet-milling industry, which has increased its capacity by 8–10% a year for the last 10 years. Of the 1.195 billion pounds (542×10^3 t) produced in 1984/85, over 90% was produced as a by-product of corn wet milling, with the remainder from dry milling (CRA, 1986; USDA, 1986). Twenty-eight major corn refining plants in the United States produce isolated corn germ, and all either process germ into oil or meal, or sell germ on an oil basis for processing. In the wet-milling industry, 14 of 16 plants produce crude corn oil and eight produce refined corn oil. Twelve

[1] Present address: Louanna Foods, Inc., Opelousas, LA 70570.

degerminating-type corn dry mills operate in the United States, and all produce corn oil (Chapter 11). The largest worldwide producers of corn oil after the United States include Brazil, China, Rumania, the USSR, Yugoslavia, and South Africa (Anonymous, 1985). The United States is the major exporter of corn oil. U.S. exports reached 300 million pounds (136 × 10^3 t) in 1984/85 (USDA, 1986), with the principal growth markets being in Asia and the Middle East. The European Economic Community is the largest importer of corn oil.

The limited production of corn oil, compared to that of soybean and other major oils, and the high market demand lead to the ready sale of all corn oil that is produced. Corn oil production has increased markedly in recent years due to increased volumes of corn being used in sweetener and starch production (USDC, 1985b). The supply of corn oil is largely dependent on the demand for the major products of corn wet milling: starches and sweeteners (CRA, 1986). The U.S. starch industry markets have been shown to grow with population and increases in the gross national product (USDC, 1985a). The market for sweeteners is nearly mature, with significant increases in capacity projected only for the next few years. Much of the increase in capacity is now in place or under construction. In addition to the starch and sweetener push on corn processing, fuel alcohol has gained attention (CRA, 1986).

III. COMPOSITION AND CHARACTERISTICS OF CORN OIL

Refined corn oil is composed of triglycerides (99%) (Sonntag, 1979a). Its chemical structure is discussed in Chapter 10. It has an average molecular weight of 326 and is a liquid at room temperature. The fatty acids are composed of hydrocarbon chains ranging from 16 to 20 carbons each, with 18 carbons being the most common. Each chain may have up to three double bonds. The typical fatty acid composition of refined corn oil is given in Table I. Because of its relatively low linolenic acid content, corn oil is inherently more stable than several other vegetable oils (Sonntag, 1979b; Erickson and List, 1985). The component lipids and their chemical characteristics have been thoroughly described in Chapter 10. This chapter emphasizes chemical characteristics of most importance to refining and utilization. Figure 1 shows four of the chemical reactions that occur in the refining process.

One property of unsaturated fatty acid components of corn and other vegetable oils is the addition of hydrogen to the double bonds. Hydrogenation is

TABLE I
Fatty Acid Composition of Corn Oil[a]

Common Acid Name	R Chain Length:No. of Double Bonds	Percent of Total Fatty Acids
Palmitic	$C_{16:0}$	11.0 ± 0.5
Palmitoleic	$C_{16:1}$	0.1 ± 0.1
Stearic	$C_{18:0}$	1.8 ± 0.3
Oleic	$C_{18:1}$	25.3 ± 0.6
Linoleic	$C_{18:2}$	60.1 ± 1.0
Linolenic	$C_{18:3}$	1.1 ± 0.3
Arachidic	$C_{20:0}$	0.2 ± 0.2
Essential fatty acids		61.2 ± 0.9

[a] Average for oil from U.S. Midwest, 1977–1985. (Personal communication, J. M. Hasman, Best Foods, Union, NJ)

performed to give fats the desired physical properties, including melting point, plasticity, and oxidative stability (Allen, 1982). Complete hydrogenation yields products that are extremely hard, with melting points above 60°C. In partial hydrogenation, saturation is selective, with the most unsaturated fatty acids being hydrogenated. Positional and geometric isomerization also occurs. Hydrogenation is manipulated by control of hydrogen gas pressure, rate of agitation, temperature, and catalyst type (Allen, 1982). Corn oil presents no unusual problems in hydrogenation, and various products, particularly frying and margarine oils, are readily made.

Linoleic acid is selectively hydrogenated to form oleic acid, which is hydrogenated in turn to form higher levels of stearic acid. The greater amounts of stearic acid ($C_{18:0}$) and oleic acid ($C_{18:1}$) add to the oil's stability and increase the melting point, giving it higher solids at a given temperature or plasticity.

1. Caustic Refining:
$$RCOOH + NaOH \longrightarrow RC\text{-}ONa + H_2O$$
(with C=O)

2. Acidulation of Soapstock:
$$RC\text{-}ONa + H_2SO_4 \longrightarrow RCOOH + NaHSO_4$$
(with C=O)

3. Complete Hydrogenation:
$$-RCH = CHR' + H_2 \longrightarrow RCH_2CH_2R'$$
Catalysts may include Pt, Pd, Ni, or Zr

4. Partial Hydrogenation:

$-CH_2CH = CHCH_2-$

↓ Adsorption on surface of catalyst

$-CH_2CH\text{-}CH\text{-}CH_2 + H*$

→ $-CH_2\text{-}CH_2\text{-}CH\text{-}CH_2-$

or

$-CH_2\text{-}C\text{-}H\text{-}CH_2\text{-}CH_2$

+H*

↑ Desorption from surface of catalyst

$-CH_2CH_2CH_2CH_2-$

C/T: $-CH_2CH_2CH = CH$

or

C/T: $-CH_2CH = CHCH_2-$

or

C/T: $-CH = CH\text{-}CH_2CH_2-$

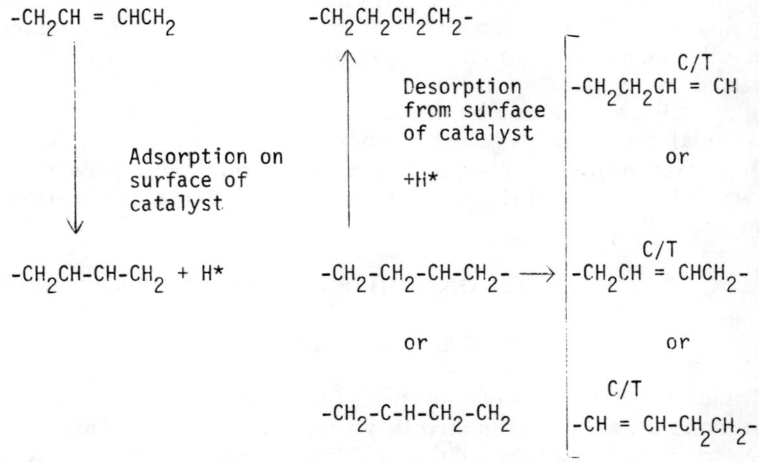

Fig. 1. Chemical reactions of corn oil refining. (Adapted from Albright, 1970)

Autoxidation of the unsaturated fatty acids leads to oxidative rancidity, which produces off-flavors and off-odors in vegetable oils. Autoxidation, or free-radical oxidation, can occur at cool temperatures, even 0°C, but is accelerated at elevated temperatures. It is accelerated in the presence of added or naturally occurring free radical promoters, such as iron, copper, benzoyl peroxide, and also light. The degradative reactions are inhibited by the presence of compounds that react with free radicals. Common synthetic antioxidants added to corn oil are *tert*-butylhydroquinone (TBHQ), butylated hydroxytoluene (BHT), and butylated hydroxyanisole (BHA). Antioxidants generally act as proton donors or free radical accepters, reacting principally with the free hydroperoxy radical ROO* (Perkins, 1984).

Thermal oxidation reactions occur at temperatures above 150°C, as in pan frying or deep fat frying. Six-membered ring structures and large polymers (30,000 mol wt) can be formed (Perkins, 1984). Thermal polymerization occurs at high temperatures (250°C) in the absence of oxygen, also forming polymerized compounds, but at this temperature the oil would probably become inedible.

IV. RECOVERY OF CRUDE OIL

Corn itself contains approximately 4.5% oil, about 85% of which is located in the germ. Corn germ isolated by the dry-milling process contains 25–30% oil, and germ recovered by the wet-milling process contains 45–50% oil.

In wet milling, after separation and washing of the germ to remove free starch, the germ is dried. Oil is recovered from the dried germ, usually by a combination of mechanical expression and solvent extraction. In mechanical expression, continuous screw expellers, under high pressure and moderate heat, press the oil from the germ. As much as 94% of the oil can be removed by expelling. Usually only about 80% is recovered with screw expellers. After the expellers, the germ cake may be flaked and the remaining oil extracted using hexane as the solvent. The solvent-oil solution is filtered and the solvent stripped in an evaporator or stripping column. Solvent in the germ flake is removed by heating and steam stripping. The solvent from both the oil and germ is condensed for recycling. The corn oil recovered by mechanical pressing is combined with the solvent-extracted crude oil.

In dry milling, germ is a by-product of the primary products, cornmeal and grits. Small mills include germ as a component of hominy feed, but the larger mills recover the oil by screw pressing, alone or followed by solvent extraction. Dry-milled crude corn oil has less color and lower refining losses than oil derived from wet-milling.

V. CORN OIL PROCESSING

A. Crude Corn Oil

Crude corn oil consists of a mixture of triglycerides, free fatty acids (FFAs), phospholipids, sterols, tocopherols, waxes, and pigments (Table II, see also Chapter 10). Nearly all the oil is used in foods, which requires removal of most extraneous components (refining) to obtain the quality necessary for food applications. During refining of the oil, substances that detract from the quality are removed or reduced in concentration. These include FFAs, phospholipids, color bodies, odors, flavors, pesticides, aflatoxin, metals, oxidative by-products, and milling residues.

The steps involved in refining the oil are: 1) degumming—removal of most of the phospholipids; 2) alkali wash—removal of FFAs, phospholipids, and color; 3) bleaching—pigment and phospholipid removal; 4) winterization—wax removal; and 5) deodorization—improvement of flavor and odor and reduction in FFA content.

Hydrogenation of the oil also may be a part of the corn oil refining process. The process flow for refining of crude oil to refined corn oil is detailed in Fig. 2.

B. Crude Oil Filtration and Degumming

The process of crude oil filtration and degumming removes solids and gums (hydrated phospholipids) from crude corn oil in preparation for caustic or physical refining. Lecithin, a coproduct of corn oil degumming, may also be recovered. Crude oil is transferred from a crude storage tank to a scale for metering the correct amount of ingredients. The oil is heated to 71–82°C and slurried with diatomaceous earth (filter aid) before the filtration. Residual fiber, dirt, metal fragments, etc., are removed. The filtered oil is mixed with approximately 1–3% soft water (depending upon the phosphatide level of the crude oil) and is allowed to hydrate in an agitated tank. The wet oil is then passed through a degumming centrifuge that separates the dense wet gums from the oil. The degummed oil is vacuum-dried and either cooled to about 38°C for storage or sent directly to refining.

While in storage, the degummed oil is kept agitated to prevent separation of residual phosphatides (Norris, 1982). Phosphoric acid may be added to degummed oil before storage to improve refining yield. This acid aids in the hydration of nonhydratable gums and improves the efficiency of primary separation (Taylor, 1975). The phosphoric reaction requires approximately 12 hr. Some refiners eliminate the degumming step and rely entirely on caustic refining for phosphorus removal.

TABLE II
Average Composition of Crude and Refined Corn Oil Components[a]

	Crude Oil Components (%)	Refined Oil Components (%)
Triglycerides	95.6	98.8
Free fatty acids	1.7	0.03
Waxes	0.05	0.0
Phospholipids	1.5	0.0
Cholesterol	0.0	0.0
Phytosterols	1.2	1.1
Tocopherols	0.06	0.05

[a] Data from Reiners (1978).

C. Caustic Refining

Caustic refining is the treatment of degummed oil with a dilute sodium hydroxide solution to remove residual gums and FFAs, color, etc. (Carr, 1976). The caustic reacts preferentially with FFAs, forming water-soluble soaps. Residual phosphatides remaining after degumming also react with caustic to form oil-insoluble hydrates. Total removal of impurities is impractical. Removal of impurities must be balanced to minimize the loss of triglycerides and ensure adequate oil quality. The major triglyceride loss occurs during alkali refining.

Before being injected into the degummed oil, caustic is diluted with soft water to approximately 17–18° Be' NaOH. The amount of caustic solution added to the oil is termed the "treat." The correct treat will produce an adequately refined oil with the highest refining efficiency (minimum loss of neutral oil). The treat is based on the percent of FFAs in the degummed oil. A 0.13% excess treat (above the quantity of NaOH required to neutralize the FFAs) is commonly used for unreacted phosphatides (Mounts and Anderson, 1983). The excess treat is determined by experience and is adjusted according to laboratory results.

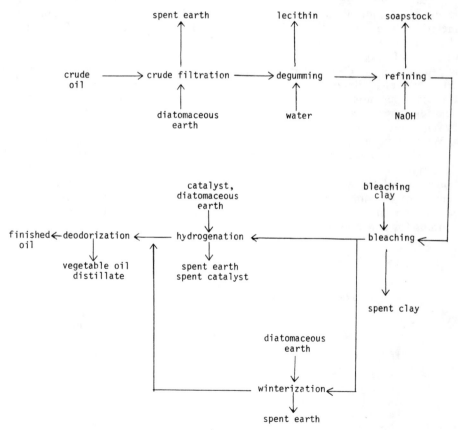

Fig. 2. Processing diagram for corn oil caustic refining. (Adapted from CRA, 1986)

After the caustic is injected, it is mixed for 3–5 min to ensure thorough contact with the fatty acids and phosphatides. Thorough mixing without formation of a stable, inseparable emulsion is required. The caustic-treated oil is then heated to create a thermal "shock," which assists in breaking the emulsion of the oil-soapstock-water mixture. The mixture is then fed to a refining centrifuge, where the lower-density oil phase and higher-density soapstock phase are separated into two layers. The main factors influencing the ease of separation include: 1) density difference between the two phases, 2) temperature of the feed stock (higher temperatures give improved separations), 3) viscosities of the two phases, 4) amount of centrifugal force, and 5) residence time in the centrifuge (Norris, 1982).

Most of the soap remaining in the oil following primary centrifugation (typically 200–600 ppm) is removed by a water-washing step. The once-refined corn oil at 82° C is mixed with 93° C soft water at a rate of approximately 15% of the oil flow. The majority of the soap distributes itself with the water. To aid separation, phosphoric acid may be added to the wash water before it is mixed with oil for adequate pH control. Thermal shock created by the greater water temperature than oil temperature also aids separation in water-wash centrifugation. The low-soap, water-washed oil is then either vacuum-dried to remove residual moisture before bleaching and sent to storage or is sent wet from the water-wash separator directly to bleaching.

D. Physical Refining by Steam

An alternative refining method is "physical" or "steam" refining. The phosphatide level must be low before physical refining, to prevent darkening of the oil and development of off-flavor. In addition, nonhydratable phosphatides are not removed by physical refining (Hvolby, 1971). A degumming step and phosphoric acid treatment are required for adequate phosphatide removal before physical refining (Taylor, 1975). The degummed oil is steam-sparged under vacuum, to remove volatile components, primarily the FFAs. Because the oil is very high in FFAs, corrosion-resistant 316 stainless steel is used in equipment fabrication. The FFA level is reduced from approximately 1.0–3.0% in degummed oil to less than 0.05% in finished oil. Additional trays or sections in deodorizers are required to give the oil longer retention time for maximum FFA removal during the steam stripping. Carotenoids are removed in the steam refining step. A typical flow diagram for physical refining of corn oil is shown in Fig 3.

E. Bleaching

The purpose of bleaching is to remove pigments and residual soap in once-refined or caustic-treated oil before hydrogenation and deodorization. Once-refined corn oil (either from storage or directly from the water-wash centrifuge) is slurried with bleaching clay, a natural hydrated aluminum silicate. The hot slurry (105° C) enters a vacuum bleacher, where excess moisture is removed and absorption of the color bodies and soap onto the clay surface occurs. After about 20 min, the dried slurry is pumped from the bleacher to leaf filters precoated with diatomaceous earth to separate the oil from the clay. The spent clay is steamed

542 / Corn: Chemistry and Technology

for oil removal, discharged to a hopper, and transported to bins for landfill disposal (Watson and Meierhoefer, 1976). Residual oil levels in spent clay may range from 30 to 70%, depending upon the extent of steaming (Patterson, 1976).

F. Winterization

Corn oil contains trace quantities of waxes that crystallize at home refrigerator temperatures (<4°C), producing a cloudy appearance. This is prevented by removal of waxes in the "winterization" process. Normally, ammonia or Freon is used in the refrigeration system. After cooling to approximately 4°C, the oil is passed through a filter coated with diatomaceous earth. Older units winterize the corn oil in batches, requiring up to three days per batch, but newer units winterize semicontinuously. Winterization is not necessary for corn oil that is to be hydrogenated.

G. Hydrogenation

In the hydrogenation process, hydrogen is chemically combined with the unsaturated double bonds of the triglycerides. The purpose of hydrogenation is to increase the oil's oxidative stability and to impart physical properties similar to those of butter or margarine. The hydrogenation reaction is done in a pressurized vessel, or converter, in the presence of a catalyst, usually nickel. The

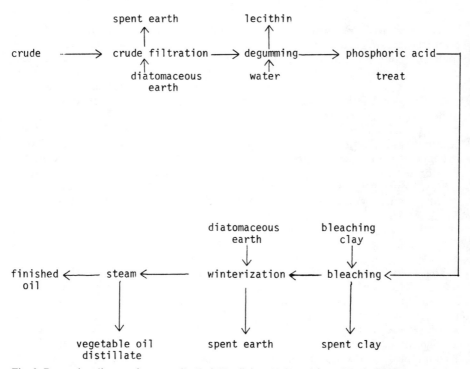

Fig. 3. Processing diagram for corn oil physical refining. (Adapted from Norris, 1982)

reaction with hydrogen gas is exothermic, making it necessary to cool the vessel during the reaction. Reaction pressures of 0.703–4.218 kg/cm^2 (10–60 psig) at 121–128°C are used. When the end point is reached, the oil is passed through a filter precoated with diatomaceous earth for removal of most of the catalyst. A chelating agent such as citric acid is added to the filtered oil to complex the remaining catalyst. Diatomaceous earth is also added. Following mixing, the chelated nickel and earth are filtered from the oil. The oil is then transferred to storage tanks before being blended.

H. Blending of Hydrogenated Oil

Certain products, such as margarine oil, require hydrogenated corn oil with very specific properties, including solid fat content, commonly called a solid fat index (SFI) profile, melting point, and fatty acid composition. Since some of these specifications are not easily met by a single set of hydrogenation conditions, blends of two or more hydrogenated stocks with different properties are necessary. In the blending process, the proper quantities of each stock are transferred to agitated mixing tanks. Following thorough agitation, the oil is analyzed for the expected specifications and is finally deodorized.

I. Deodorization

Deodorization removes volatile materials in the oil: tocopherols, sterols, FFAs, herbicides, pesticides, dissolved gases, and moisture, as well as flavor components. Deodorized oil has lighter color, improved oxidative stability, and lower FFA and peroxide values (Reiners and Gooding, 1970). The process is similar for either unhydrogenated or hydrogenated corn oils. The bleached and/or hydrogenated oil is first partially heated and deaerated. Citric acid is added to chelate metals in the oil. The oil is further heated to approximately 232°C as it enters the vacuum vessel ($\approx$2–10 mm of Hg). Steam is sparged through the oil during deodorization.

Corn oil may be deodorized continuously, semicontinuously, or in batches (Gavin, 1977; Norris, 1985). In one type of semicontinuous deodorizer, the oil enters into a large tray at the top of the vacuum vessel. At the end of about 8 min, the oil overflows into a tray below; this is repeated until the oil has passed through four trays. The oil is sparged with steam to provide agitation and to aid distillation of the FFAs. The overall retention time is about 32 min. Another type of deodorizer processes a batch of oil in a series of separate compartments. As the oil exits the deodorizer, it is cooled to approximately 38°C for corn salad oil, or 5°C above the melting point for each hydrogenated oil. Chelating agents such as citric and phosphoric acids and antioxidants such as TBHQ, BHT, and BHA may be added to the oil for improved stability following deodorization.

J. Quality Analyses for Corn Oil

Numerous tests have been developed by the American Oil Chemists Society and others to measure chemical and physical characteristics of corn and other vegetable oils (Sonntag, 1982; AOCS, 1985). Table III compares typical quality control specifications for refined corn oil and crude corn oil. These principal analyses give significant information about the oils and the efficiency of the total refining process. The tests are summarized below.

The FFA test indicates the efficiency of the caustic treatment and the deodorization. The presence of FFAs lowers the smoke point of the oil.

Lovibond color is determined by finding the closest red and yellow color matches between the oil sample and the standards for the test. Typically, the yellow component of finished oil is about 10 times the magnitude of the red component. Factors that contribute to finished oil color include degree of refining, conditions of oil extraction from the germ, whether the germ was dry- or wet-milled, level of hydrogenation, and color components (chlorophyll, carotenoids, etc.) present in the crude oil.

Peroxide value is a measure of the degree of oil oxidation. It is determined by reacting the oil with iodine, followed by titration with sodium thiosulfate in the presence of a starch paste indicator.

Flavor in partially oxidized corn oil is composed primarily of volatile aldehydes, including such compounds as pentanal, hexanal, nonanal, and 2-decenal. Corn oil flavor is evaluated on a scale of 1 to 10, with high scores corresponding to mild or bland flavors. Common flavor descriptors of acceptable corn oil include *bland, nutty, corny,* and *buttery*. Marginal descriptors include *metallic, hydrogenated,* and *grassy*. Descriptors of unacceptable oil include *rancid, burnt, painty,* and *fishy* (Jackson, 1985).

Iodine value is defined as the number of grams of iodine required to react with 100 g of oil. The higher the iodine value, the less saturated the oil is.

The melting point of triglycerides is dependent upon their chain length, degree of saturation, composition, and crystalline structure. All vegetable oils are mixtures of various triglycerides having a melting range. The melting point of a corn oil sample is seldom very sharp and generally has a range over several degrees.

The SFI measures the solid fat content of the oil at 10, 21.1, 26.7, 33.3, and 40°C. SFIs are usually measured dilatometrically. Such information is

TABLE III
Typical Specifications for Crude and Refined Oil[a]

Analytical Constants	Refined Corn Oil Specifications	Refined Corn Oil Typical Values	Crude Corn Oil Specifications
Free fatty acids, %	0.05 max.	0.02–0.03	1.0–1.5
Color			
Lovibond	3.0 red max.	2.0–2.5 red	10–12 red[b]
Gardner	6 max.	3.5	...[c]
Peroxide value, meq/kg	0.5 max.	...	<0.5
Flavor	Bland	Bland	n.a.[d]
Moisture, %	n.a.	n.a.	<0.5
Iodine value[e]	...	122–128	...
Density, g/cm^3	...	0.922–0.928	...

[a] Data from CRA (1986).
[b] From 1-in. tube reading.
[c] No value given.
[d] Not applicable.
[e] Grams of iodine absorbed per 100 g of fat.

especially important in applications where plasticity is critical, as in the manufacture of shortenings and margarines.

Smoke, flash, and fire points are reached sequentially when oil is exposed to extremely hot temperatures (204–371°C). For each determination, oil is heated in an open cup with a flame, and the specific temperatures are noted when the sample smokes, flashes, and burns. This information is important in establishing thermal oxidation stability, as for deep-fat frying applications.

VI. PROCESSING OF CORN OIL COPRODUCTS

A. Corn Lecithin

Corn oil lecithin, a potentially valuable product, is produced by removing water from the wet gums that result from the degumming process. Thin-film vacuum dryers are used. The lecithin may then be adjusted by addition of peroxide for bleaching and fatty acids or divalent salts for viscosity control to meet customer specifications. Lecithin is used as an emulsifier, antioxidant, nutrient, dispersant, etc. However, many oil refiners add the wet gums to refining soapstock rather than processing them into lecithin.

B. Soapstock

Soapstock is processed to be used in poultry and livestock feeds and in fatty acid distillation. It is usually acidulated to a neutral or slightly acidic pH or converted back to fatty acids with sulfuric acid. After sulfuric acid treatment (acidulation), the FFAs are again insoluble in water and separate into a two-phase oil-water system.

C. Vegetable Oil Distillate

Vegetable oil distillate is the condensed volatiles removed from the oil during deodorization. The distillate, composed mostly of FFAs, also contains tocopherols, carotenoids, herbicides, pesticides, and flavor and odor components. This material is used as a source of tocopherols for Vitamin E.

VII. HANDLING

A. Storage

Corn oil is held in bulk storage tanks between refining steps, during blending, and before loadout. Correct storage conditions include the absence of air, light, moisture, and heavy metals (Leo, 1985). For corn salad oil, the storage temperature should not exceed 38°C. For hydrogenated oil products, the storage temperature is kept about 5°C above the melting point. To avoid stratification, the tanks are usually equipped with agitators or recirculation facilities. The presence of air is avoided by use of nitrogen in the headspace of airtight tanks. Light exposure is not of concern if covered metal tanks are used. Moisture buildup may be a problem in cold weather, due to condensation accumulating on the top interior surface. Heavy metals, including iron and especially copper, cause the oil to quickly deteriorate, particularly in the presence of heat, moisture, and air. Ideally, stainless steel or plastic should be used to minimize the oil's exposure to active metals.

B. Loadout and Shipping

Finished corn oil, unfinished intermediate products, and coproducts are normally loaded out in tank trucks and rail cars (Erickson and List, 1985). Some products are packed in drums, pails, and glass and plastic containers. The same handling conditions apply to loadout as to storage. Air exposure is minimized by not allowing oil to splash as it enters the tank. The oil is sparged with nitrogen to remove air from the oil and to provide an airless headspace above the oil. Exposure to light is not a concern with metal rail cars and trucks. However, light may be a problem when the oil is bottled in clear glass. Most corn salad oils are presently bottled in clear glass or plastic, which may have a negative impact on shelf life unless the oil is properly stored. Most tank trucks are made of stainless steel.

C. Prevention of Oxidation

The shelf life of corn oil can be extended by the use of antioxidants, chelating agents, and "brush" (slight) hydrogenation (Erickson and List, 1985). Common antioxidants include BHA, BHT, TBHQ, propyl gallate, and tocopherol. Chelating agents include citric, phosphoric, and ascorbic acids. Antioxidants inhibit the autoxidative pathway, and chelating agents act as metal scavengers. Although the use of chelating agents and antioxidants can extend shelf life, their use cannot compensate for poor storage or handling conditions. Lightly hydrogenating the oil reduces the most highly unsaturated fatty acids, which aids in preventing oxidation.

VIII. NUTRITION

Before about 1940, the primary consumer cooking fats were lard and butter. Mazola corn oil had been introduced in 1911 by Corn Products Refining Company with the goal of establishing the brand name and building market share (Anonymous, 1985). By 1940, the essential role of fat in the diet was recognized, and the "health" image of Mazola Corn Oil's essential fatty acid content was being promoted. During the 1950s, corn oil was advertised as a substitute for saturated fats to reduce serum cholesterol (CRA, 1986). Promotion of corn oil subsequently emphasized the role of fat-modified diets on various heart disease risk factors. Consumer awareness of corn oil was noted in a study conducted by the American Soybean Association in 1978 (Erikson and Falb, 1979). Corn oil was ranked above both soybean and sunflower oils. The image created for corn oil carried through to the margarine industry.

Fat is, of course, an essential component of the diet, serving both physiological and biochemical functions (Harwood, 1978). In foods, it serves as an energy source, carrier of fat-soluble vitamins, and a source of polyunsaturated fatty acids (linoleic and linolenic acids). Corn oil is rich in polyunsaturated fatty acids. The typical linoleic acid content for U.S. production is 60.1%.

The role of essential fatty acids was demonstrated in 1929 for rats and in 1958 for human infants. The essential fatty acids linoleic and arachidonic are involved in fat transport and in the prevention of dermatitis (Alfin-Slater et al, 1980). The linoleic requirement is in the range of 1-3% of the dietary calories. Even a moderate dietary intake of 5-15 g of corn oil per day meets the essential fatty acid requirement of most people. The role of corn oil in lowering serum cholesterol was noted in about 1950 (Rathman, 1957). The favorable altering of serum lipid composition was later found. Combined, these changes were assumed to furnish protection against coronary heart disease.

Around 1980, the lowering of blood pressure by a change to a diet high in polyunsaturates and low in saturated fatty acids was observed (Vergroesen and Fleischman, 1978; Fleischman et al, 1979). The possible mechanism is through prostaglandins, a group of hormonelike lipid-derived substances. This is not well understood, however.

Corn oil is a source of various tocopherols that remain with the oil through the refining process. The tocopherols contribute to the stability of the oil, and several isomers possess vitamin E activity (see Chapter 10). The human requirement for vitamin E is proportionate to body size and polyunsaturated fat intake (Garrison, 1985). Corn oil can supply about 15% of the U.S. recommended daily allowance of vitamin E for an adult consuming 1 tablespoon (120 calories) per day.

IX. USES AND APPLICATIONS

Corn oil is considered a premium vegetable oil because of its flavor, color stability, and clarity at refrigerator temperatures. The nutritional benefits, primarily related to its polyunsaturated fatty acid and vitamin E contents, have made it a premium oil to the consumer (CRA, 1986). The two current major uses of corn oil are frying or salad applications and margarine formulations (Mounts and Anderson, 1983). About 50% of the refined oil is used for frying and salad oil, and 30-35% is used in margarine production. Currently, 100% corn oil margarines represent over 10% of the total margarine production (USDA, 1984). In the first two months of 1986, 28.4 million pounds of corn oil were used for margarine manufacture. This is approximately 170 million pounds per year (NAMM, 1986).

A. Margarines

A variety of margarine types are marketed in the United States. These include stick types, soft (tub) types, diet, imitation, liquid margarine, and vegetable oil spread. In most margarine formulations, liquid oil is blended with a partially hydrogenated component to give a fat having a sufficient solids content below the melting point to give the margarine its characteristic semisolid form. Both soft (tub) margarines and stick (hard) margarines are prepared commercially. In some instances, the liquid corn oil is blended with partially hydrogenated soybean or cottonseed oils. Regular margarines contain 80% fat and about 16-18% aqueous phase (Brekke, 1980). The aqueous phase may include an edible protein such as nonfat dry milk plus 2–3% salt, emulsifiers such as lecithin or monodiglycerides, preservatives, flavoring, and coloring. The coloring is primarily β-carotene. Fortification with 15,000 USP units of vitamin A is required. Vitamin D fortification, 2,000 USP units, is sometimes used (Crysam, 1979).

Other types of margarinelike spreads produced are 60% fat spreads and imitation margarines with 40% fat. The decreased-fat blends contain proportionately larger amounts of water. These have been developed to meet the consumer demands for lower calories or less fat. Butter blends containing corn oil have recently been introduced (Mounts and Anderson, 1983). The corn oil content varies but generally is in the 40% range. Table IV lists the typical fatty acid compositions of corn oil margarine and stick and soft margarines.

The basic steps in formulating a margarine are: blending the oil components, formulating the aqueous phase, preparing the emulsion, solidifying the emulsion, packaging, and tempering.

Formulation of the fat, cooling and working of the cooled emulsion, and tempering may be used to control the margarine's consistency. Mechanical working increases the plasticity of the emulsion, producing a spreadable product at refrigerator temperatures (Brekke, 1980).

In the preparation of margarine, the oil and aqueous phases are prepared separately and then blended in the proper proportions. The emulsion is then solidified, using tubular, scraped-surface heat exchangers. The emulsion temperature is decreased rapidly to 7–10° C. For stick margarines, the emulsion from the scraped-surface units is pumped to a stationary location to permit crystal growth, providing time for the product to become sufficiently firm for extruding, shaping, and wrapping (Chrysam, 1979). SFI data are used to control the process.

Soft tub margarines, containing as much as 80% liquid oil, are produced by mixing the chilled mass in large, agitated crystallizers. Working the chilled emulsion limits crystal growth and produces a flowable product. The product can be filled into tubs using liquid fillers. Tempering of the filled tubs at 5–10° C permits the development of a crystal structure that provides a desirable stabilized plasticlike product.

TABLE IV
Fatty Acid Composition for Typical Margarine Corn Oil
and Stick and Soft Margarines[a]

Fatty Acid	Percent of Total Fatty Acids in		
	Margarine Corn Oil	Stick Margarine	Soft Margarine
$C_{16:0}$	10.9	8.7	8.7
$C_{18:0}$	5.9	4.7	3.1
$C_{18:1}$	47.5	38.0	31.7
$C_{18:2}$	33.9	27.1	33.7
$C_{18:3}$	0.8	0.6	0.7
$C_{20:0}$	0.3	0.2	0.2

[a] Source: A. E. Staley Manufacturing Company.

B. Frying or Salad Oils

The oils used for salads must not solidify, cloud, or deposit a crystalline fraction at refrigerator temperatures (Brekke, 1980). Properly winterized corn oil easily passes a 5.5-hr winter test, as specified by the American Oil Chemist Society (AOCS, 1985).

When the oil is used in prepared salad dressings, the presence of a crystalline fraction may cause the breaking of the emulsion. The undesirable cloudy appearance of nonwinterized bottled oils is avoided by using winterized oils in salad dressings intended for the consumer market.

Fats in deep-frying applications aid in heat transfer and, because of absorption by the food, contribute to its nutritive value and flavor. Frying oils should have a bland or very weak flavor, a smoke point minimum of 204°C, and adequate resistance to deterioration during use. Acceptable flavor is attained by proper deodorization of the oil. Reduction in FFAs to 0.05% or less provides the increased smoke point. Silicone compounds (e.g., dimethyl polysiloxane) are added as antifoam compounds, resulting in increased fryer stability (Brekke, 1980).

In addition to deep frying, corn oil is used for pan or grill frying (Brekke, 1980; Crysam, 1979; Mounts and Anderson, 1983). Commercial pan-frying oils commonly contain lecithin as an antistick compound. The consumer's choice of oil for pan or grill frying is often based on price alone (Mounts and Anderson, 1983).

C. Other Uses

Many other uses exist for vegetable oils. Some of these include shortenings of various types (ranging from plastic to pourable products) for baked foods, icings and cream fillers, imitation dairy products, and confectionery coatings. Soybean oil is the principal oil used. Most of the soybean oil is modified through hydrogenation to give it the physical characteristics that meet the demands of each application.

Corn oil, although having a slightly lower iodine value and lower degree of unsaturation than soybean oil, can be manipulated in a similar way through hydrogenation. In most instances, the type or degree of hydrogenation is not presently known. Adjustment to fit intended applications is possible.

X. TRENDS

Use of corn oil to meet the needs of the marketplace will depend primarily on price and availability. Much of the future growth will probably depend on tax and trade policy. Should significant increases in alcohol production by wet milling occur, additional corn oil will probably be available. The recent trend has been to process whole corn for ethanol production. Corn oil's market position may change if corn wet-milling capacity is greatly increased or if sources of oils high in linoleic acid, particularly sunflower and safflower oils, are developed.

LITERATURE CITED

ALBRIGHT, L. F. 1970. Transfer and adsorption steps affecting partial hydrogenation of triglyceride oils. J. Am. Oil Chem. Soc. 47:490-493.

ALFIN-SLATER, R. B., and KRITCHEVSKY, D. 1980. Nutrition and the Adult. Plenum Press, New York.

ALLEN, R. R. 1982. Hydrogenation. Pages 1-96 in: Bailey's Industrial Oil and Fat Products, Vol. 2. D. Swern, ed. J. Wiley & Sons, New York.

ANONYMOUS. 1985. Corn oil. J. Am. Oil Chem. Soc. 62:1524-1531.

AOCS. 1985. Official Methods and Recommended Practices, 3rd ed. R. Walker, ed. Method Cc 11-53. Am. Oil Chem. Soc., Champaign, IL.

BREKKE, O. L. 1980. Food uses of soybean oil. Pages 393-438 in: Handbook of Soy Oil Processing and Utilization. D. R. Erickson, E. H. Pryde, O. L. Brekke, T. L. Mounts, and R. A. Falb, eds. Am. Soybean Assoc. and Am. Oil Chem. Soc., Champaign, IL.

CARR, R. A. 1976. Degumming and refining practices in the U.S. J. Am. Oil Chem. Soc. 53:347-352.

CHYSAM, M. M. 1979. Table spreads and shortenings. Pages 54-86 in: Bailey's Industrial Oil and Fat Products, Vol. 3. T. H. Applewhite, ed. J. Wiley & Sons, New York.

CRA. 1986. Corn Oil. Corn Refiners Assoc., Washington, DC.

ERICKSON, D. R., and FALB, R. A. 1979. Soy oil utilization—Current situation and potential. Pages 851-858 in: World Soybean Research Conference II: Proceedings. Am. Soybean Assoc., St. Louis, MO.

ERICKSON, D. R., and LIST, G. R. 1985. Storage, handling, and stabilization of edible fats and oils. Pages 273-310 in: Bailey's Industrial Oil and Fat Products, Vol. 3. T. H. Applewhite, ed. J. Wiley & Sons, New York.

FLEISCHMAN, A. I., BIERENBAUM, M., STIER, A., SOMOL, S. H., WATSON, P., and NASO, A. M. 1979. Hypotensive effect of increased dietary linoleic acid in mildly hypertensive humans. J. Med. Soc. N.J. 76:181-183.

GARRISON, R. H. 1985. The Nutrition Desk Reference. Keats Publishing, Inc., New Canaan, CT.

GAVIN, A. M. 1977. Edible oil deodorizing systems. J. Am. Oil Chem. Soc. 54:528-532.

HARWOOD, J. L. 1978. Nutritional aspects of oils and fats. Chem. Ind. 18:687-692.

HVOLBY, A. 1971. Removal of nonhydratable phosphatides from soybean oil. J. Am. Oil Chem. Soc. 48:503-509.

JACKSON, H. W. 1985. Oil flavor quality assessment. Pages 243-272 in: Bailey's Industrial Oil Fat Products, Vol. 3. T. H. Applewhite, ed. J. Wiley & Sons, New York.

LEO, D. A. 1985. Packaging of fats and oils. Pages 311-340 in: Bailey's Industrial Oil and Fat Products, Vol 3. T. H. Applewhite, ed. J. Wiley & Sons, New York.

MOUNTS, T. L., and ANDERSON, R. A. 1983. Corn oil production, processing and use. Pages 373-387 in: Lipids in Cereal Technology. P. J. Barnes, ed. Academic Press, New York.

NAMM. 1986. Margarine Statistics Report. (Jan.-Feb.). Natl. Assoc. of Margarine Manufacturers, Washington, DC.

NORRIS, F. A. 1982. Refining and bleaching. Pages 253-314 in: Bailey's Industrial Oil and Fat Products, Vol. 2. D. Swern, ed. J. Wiley & Sons, New York.

NORRIS, F. A. 1985. Deodorization. Pages 127-166 in: Bailey's Industrial Oil and Fat Products, Vol. 3. T. H. Applewhite, ed. J. Wiley & Sons, New York.

PATTERSON, H. B. W. 1976. Bleaching practices in Europe. J. Am. Oil Chem. Soc. 53:339-341.

PERKINS, E. G. 1984. Lipids. Pages 127-260 in: Food Chemistry and Nutrition. I. R. LaMontagne, ed. Univ. of Illinois, Champaign.

RATHMAN, D. M. 1957. Vegetable Oils in Nutrition. Canada Starch Co., Ltd., Montreal.

REINERS, R. A. 1978. Corn oil. Pages 18-23 in: Products of the Corn Refining Industry in Foods. Seminar Proceedings. Corn Refiners Association, Inc., Washington, DC.

REINERS, R. A., and GOODING, C. M., 1970. Corn oil. Pages 241-262 in: Corn: Culture, Processing, Products. G. E. Inglett, ed. Avi Publ. Co., Inc., Westport, CT.

SONNTAG, N. O. V. 1979a. Structure and composition of fats and oils. Pages 1-98 in: Bailey's Industrial Oil and Fat Products, Vol. 1. D. Swern, ed. J. Wiley & Sons, New York.

SONNTAG, N. O. V. 1979b. Composition and characteristics of individual fats and oils. Pages 289-478 in: Bailey's Industrial Oil and Fat Products, Vol. 1. D. Swern, ed. J. Wiley & Sons, New York.

SONNTAG, N. O. V. 1982. Analytical methods. Pages 407-526 in: Bailey's Industrial Oil and Fat Products, Vol 2. D. Swern, ed. J. Wiley & Sons, New York.

TAYLOR, F. 1975. Clay-heat refining process. U.S. patent 3,895,042.

USDA. 1984. Oil Crops: Outlook and Situation

Yearbook. OCS-8. U.S. Dept. Agric., Econ. Res. Serv., Washington, DC.
USDA. 1986. Oil Crops and Situation Yearbook. U.S. Dept. Agric., Econ. Res. Serv., Washington, DC. OCS-11. July. 34 pp.
USDC. 1985a. Grain mill products: 1982 census of manufacturers. MC 82-1-20D. Industry Series. U. S. Dept. Commerce, Washington, DC.
USDC. 1985b. Fats and oils: Production, consumption, and stock. Bureau of Census. Industry Division. U. S. Dept. Commerce, Washington, DC.
VERGROESEN, A. J., and FLEISCHMAN, A. I. 1978. Hypotensive effects of dietary prostaglandin precursor in hypertensive man. Prostaglandins 15:193-197.
WATSON, K. S., and MEIERHOEFER, C. H., 1976. Use or disposal of by-products and spent material from the vegetable oil processing industry in the U. S. J. Am. Oil Chem. Soc. 53:437-442.

CHAPTER 19

FERMENTATION PROCESSES AND PRODUCTS

WELDON F. MAISCH
Brown Forman Corporation
Louisville, Kentucky

I. INTRODUCTION

The recorded history of distilled spirits dates back to 1167 in the writings of Master Salerus, but oral tradition credits earlier Greek and Egyptian alchemists with the actual discovery of distilled beverages. The industry took shape in France with distillation of wine into brandy and developed its present-day form with the invention of the continuous still by Aeness Coffey in Ireland in 1830.

The first recorded beverage spirits in the United States were produced in 1640 using corn and rye as the substrates. Bourbon County, Kentucky, gave its name to the sour mash whiskey that became the bulwark of the American fermentation alcohol industry. During World War II, the United States developed a chemical alcohol industry that supplied 750 million gallons of 190° proof alcohol for the war program. Recent energy shortages have prompted the construction of new, large power (fuel) alcohol distilleries, with an accompanying new round of technical developments.

The utilization of corn substrates as fermentation media for the production of beverage and industrial grades of ethyl alcohol, as well as other chemical end products, has received renewed attention in the past 10 years. Previous reviews by Peppler and Perlman (1979), Rose (1977), Rehm and Reed (1983), and others have described much of this technology. Corn, cornmeal, starch, and corn glucose syrup can provide the carbon and energy source for virtually any fermentation. This chapter is limited to fermentations in which corn starch or glucose is used as a bioconversion building block for the production of large-volume commodity-type products, such as ethyl alcohol. In this process, yeast or other microorganisms convert glucose and/or other simple sugars into the end product.

The sugars are derived primarily from the starch polymers in the corn kernel. Milled corn is typically hydrated and gelatinized with heat to facilitate the enzymatic and/or chemical cleaving of the starch linkages, because the fermentation organisms themselves cannot readily utilize starch. Chapters 16 and 17 describe the principal methods of starch preparation and sugar production.

II. FERMENTATION SUBSTRATE

A. Starch-Bearing Substrates

The beverage alcohol distiller has traditionally used the simplest approach to obtaining a substrate. The starting point and the basic unit of measure has been the 56-lb bushel of shelled whole corn at 14% moisture, although any cereal grain can be used, including sorghum, wheat, barley, and rye. Table I presents averaged analytical data from actual sample receipts for corn offered as fermentation substrates by various suppliers of corn products. These data can be contrasted with the tables of corn composition listed by Watson (1984) and elsewhere in this book. The importance of knowing your substrate cannot be overemphasized.

At the typical distillery, ground whole corn is slurried, cooked to gel the starch, cooled, dosed with enzyme for starch saccharification, and sent to the fermentor for yeast inoculation. The mash bill lists the bushels of grain(s) and the gallons of water mixed to prepare the mash. "Beer gallonage" is a useful term referring to the gallons of mash that contain the equivalent of 56 lb of grain. Based upon the gallons of mash pumped to the fermentor, one calculates bushels of grain in the fermentor. The volume of alcohol produced is determined by physical analysis of the fermented beer. Yields usually are expressed as proof gallons of alcohol per bushel (PG/bu). A proof gallon is a 1-gal volume of 50% alcohol (v/v).

Since starch is the major component of corn that is convertible to ethanol, starch values are essential if the distiller is to determine true fermentation efficiency. Less starch in the grain will naturally translate into yields with lower PG/bu. Substrate analysis is also of value in determining coproduct credits. Only the carbohydrate portion of the grain is converted into product; all else can be recovered as a valuable coproduct called corn distillers dried grains with solubles, which is sold as a feedstuff. Protein and fat concentrations in this coproduct influence the selling price and can contribute significantly to profits.

Wet-milling and dry-milling processes are alternative methods of concentrating the starch component of corn. The dry-milling process (Chapter 11) can produce a substrate of higher starch content than whole corn, about 80%

TABLE I
Proximal Values of Samples Submitted for Fermentation Substrate Survey

Quality	Dry Milling Sample[a]			Distillery Corn[a]	Elevator Corn Screenings
	Whole Corn	Grits	Hominy Feed		
Carbohydrate, %	73.1	78.1	68.2	...	...
Starch, %	...	...	...	74.3	30–66
Protein, %	9.2	8.7	10.9	9.6	...
Fat, %	3.9	0.8	5.1	4.5[b]	...
Fiber, %	1.6	0.4	1.8	1.3[b]	...
Ash, %	1.2	0.4	1.4	1.3	...
Moisture, %	11.0	12.0	...	...	8–9
lb/bu (1.244 ft^3)	...	...	...	...	36

[a] Dry substance.
[b] Crude fat or fiber.

(Table I.) The wet-milling process (Chapter 12) produces a purified starch product almost completely convertible to ethanol. Both offer alternative substrate selections as well, especially for fuel alcohol fermentations. Hominy-grade feed fractions, for example, are moderately priced by-products that arise after the more valuable prime grit fractions have been removed during dry milling. A lower starch content results in lower alcohol yield, but the yield may be compensated for by its lower cost. The yield and composition of distillery by-products also vary in relation to the starch contents of the fermentation substrates. However, by mixing different fractions, one can make up for shortfalls that any one fraction may have in by-product value. Products out of specification or in temporary surplus can also become possible adjuncts for fermentation substrate.

When wet milling or dry milling precedes the fermentation, true PG/bu yields will always be lower because processing inefficiencies cause incomplete removal of the starch from the separated fractions. Purified starch fractions must be supplemented with nutrients for yeast growth. Light steep water can provide protein for yeast nutrition as well as supplement the carbohydrate level; it is often included in the fuel alcohol fermentation medium. The liquid generally requires pasteurization to control lactic acid organisms, and the levels of SO_2 in the steep water must be monitored. Yeast activities are affected above 100 ppm of free SO_2. Yeast sulfur metabolism is also a concern because H_2S production by the yeast complicates collection of the CO_2 by-product from the fermentors.

Crude corn fiber streams from the wet-milling process can contain a significant amount of starch (Table II). Although corn fiber is generally considered a feed ingredient, sending it to the feed recovery plant via the fermentor is worth consideration. Yield improvements from corn fermentations have been demonstrated when cellulase and pectinase preparations are added to corn substrates in the fermentor (Oksanen, et al, 1982; C. Roan, Rohm Enzyme, New York, personal communication). The mechanism for these higher yields requires further research. Table II shows the sugar composition that can be produced by acid hydrolysis of a crude corn fiber stream.

Corn screenings might also be a useful substrate for fuel alcohol if large quantities were collected near a fermentation facility (Table I). Bulk density and other properties that create handling problems would require equipment changes. The possibility of mycotoxins concentrating in the feed residues should also be addressed (Lillehoj et al, 1979).

TABLE II
Composition of Wet-Milled Corn Fiber as Recovered and After Mild Acid Hydrolysis[a]

	As Recovered (%)[b]		Acid-Hydrolyzed (%)[c]
Starch	15–30	Glucose	30–35
Protein	8–12	Xylose	18–25
Fat	1–2	Arabinose	10–15
Ash	0.5–1.5	Unidentified	20

[a] Data from Nagasuye and Assarson (1982).
[b] Dry substance.
[c] Of total sugars in hydrolysate.

B. Hydrolyzed Substrates

Acid hydrolysis of corn fiber could be another way to recover starch and possibly use the pentose sugars without complications from lignin (M. Ladisch, Purdue University, personal communication). Although pentose sugars are not directly fermentable to ethanol by distiller's yeasts, some *Candida* and *Pachyosolon* yeasts can accomplish the fermentation. *Saccharomyces* yeast ferments the keto-form pentose sugar, xylulose, which is the product of xylose isomerization by commercial glucose isomerase enzyme.

To meet the tremendous demand for high-fructose corn sweeteners, corn wet millers have developed plants and processes capable of producing large volumes of corn syrup. Wet millers can choose as fermentation media the partially converted streams arising early in their process or other streams up to the stream with >93% glucose content at the end of the saccharification lines.

Brewer's adjunct syrup is an intermediate conversion product of wet milling, often involving acid hydrolysis (Table III). Pomes (1980) demonstrated that the disaccharide fraction of adjunct syrup can be quite refractory to alcohol

TABLE III
Analysis of Carbohydrates in Fermentation: HPLC[a]

Carbohydrate	Amount (%) in Fraction			
	DP-1[b]	DP-2	DP-3	DP-4
Glucose[c]	30.9	1.10	0.19	0.80
Stillage	0.41	0.84	0.21	0.60
Brewer's adjunct syrup set[d]	34.4	24.0	13.3	...
Total used	98.4	94.3	100.0	...
14% Dextrin slurry				
10-DE[e]	25	18	51	...
30-DE	33	24	42	...
18% Dextrin slurry (90-DE)	92	4	1	3

[a] High-performance liquid chromatography.
[b] DP = degree of polymerization, where DP-1 is glucose.
[c] Enzyme-converted.
[d] Acid-converted.
[e] Dextrose equivalent.

TABLE IV
Analysis of Carbohydrates in Fermentation:
Gas-Liquid Chromatography of DP-2
Fraction from TABLE III[a]

Component[b]	Linkage Hydrolyzed	Percent in Substrate	Percent in Stillage
Maltose	1-4-α	7.5	none
Isomaltose	1-6-α	34.0	32.7
Cellobiose	1-4-β	5.3	4.5
Gentiobiose	1-6-β	27.4	24.4
Other		25.8	23.0

[a] Data from Nagasuye and Assarson (1982).
[b] From substrate converted by acid-enzyme process.

conversion. Based upon the American Society of Brewing Chemists' rapid method procedure (ASBC, 1981), glucose reduction averaged 98.4%, disaccharide reduction 94.3%, and trisaccharide reduction 100%.

C. Analytical Methods

The sugar components of syrup and other hydrolysates can be analyzed by several precise methods, including high-performance liquid chromatography (HPLC) and gas chromatography (Anonymous, 1982; Ivie, 1983).

HPLC separates compounds on the basis of their physical, chemical, or electrical interactions with a gel in a pressurized column (Table III). This instrument has greatly facilitated the analysis of carbohydrates in fermentation substrates and has produced increased understanding of beer components and carbohydrate utilization in ethanol fermentation. For example, the HPX87 column (BioRad, Cambridge, MA) can be used with direct injection of the fermenting beer. One can readily separate pentose and hexose sugars, organic acids, ethanol and glycerol via the refractive indexes of the compounds.

Gas chromatography (Table IV) can characterize the residue components of a particular fraction, such as one of the fractions from a substrate converted by acid-enzyme processing. Preparatory steps and the formation of trimethylsilyl-derivatives precede sample injection. The particular residue in Table IV was shown to contain mainly α- and β-1-6 linkages.

Combined, these data help to explain the ASBC determination of fermentables and strengthen recommendation of Pomes (1980) that percent fermentables should be described as:

$$\% \text{ fermentables} = (\% \text{ DP-1} \times 0.90) + (\% \text{ DP-2} \times 0.78) + (\% \text{ DP-3} \times 0.72) + 2.515,$$

where degree of polymerization (DP) is 1 for glucose, 2 for disaccharides, and 3 for trisaccharides.

III. PRODUCTS OF CONVERSION AND FERMENTATION

A. Ethanol

The classical equation for conversion of starch to ethanol is:

$$(C_6H_{10}O_5)_n + H_2O \xrightarrow[\text{or enzyme}]{H^+} C_6H_{12}O_6 \xrightarrow{\text{yeast}} 2C_2H_5OH + 2CO_2$$

starch (162) + water (18)　　　　glucose (180)　　　　ethanol (46) + carbon dioxide (44)

Table V presents the textbook chemical calculation for conversion of corn to ethanol. The fermentation is anaerobic and is classified as growth-associated; that is, cell growth and multiplication accompany the product formation. The fermentation yields 2 mol of ethanol and 2 mol of carbon dioxide from 1 mol of glucose. Note that nearly half of the substrate glucose weight can be wasted as CO_2 gas. Although this may be recovered and sold to improve profitability, the

TABLE V
Calculation of Yield Efficiency

Materials (weight in grams)		
Corn		2,503
H_2O		6,600
Malt		39
Precook	13	
Postcook	26	
Stillage		3,000
Enzyme liquid		50
Yeast		130
Top-off H_2O		678
Total		13,000

Theoretical yields expected (g)
   2,503 (corn)   × 0.70 (% starch)   = 1,752.1 starch
   1,752.1 (starch)   × 1.11 (hydrolysis gain) = 1,944.83 glucose
   1,944.83 (glucose) × 0.489 (CO_2 factor) = 951 CO_2
   1,944.83 (glucose) × 0.511 (EtOH factor) = 993.8 Ethanol

Experimental results
   CO_2 weight loss (g)   = 909
   Ethanol (g)
     (6.79% w/w × 13,000) = 883.8
   Glucose residue (%, w/v) = < 0.1

Efficiency
   909/951   = 95.5%
   883.8/993.8 = 88.93%

BTU value of the original starch is reduced. This is a key factor when calculating the new energy balance for production of power alcohol (product and processing energy in vs. fuel value out).

Table VI compares analytical data from two low-starch substrates through to product recovery. The recovered yields do not meet theoretical values for either alcohol or feedstuff, as the process efficiencies are not 100%. Any unutilized or unutilizable substrate should be accounted for as increased feedstuff recovery. Stoichiometric yields are not expected in the distillery. This is reflected in the calculation of fermentables as being only 90% of the starch value. Nevertheless, one can compare the costs of alcohol production based upon such data for substrate costs and performance.

On occasion, substrates of different origin are blended as shown in Table VI. To estimate coproduct credits for feedstuffs recovered from blended substrates, one multiplies the expected recovery for each substrate by the percentage of that substrate in the blend. If 60% of the blend is a substrate that yields 25% protein feed and 40% is a substrate that yields 22% protein, one can expect a protein value of 23.8% in the blend (Table VI). The substrate cost for the alcohol is the value of raw materials ($/100-wt) minus the by-product credit (wt in lb × ¢/lb).

Table VI contains data from a laboratory fermentation that can be used to determine fermentation efficiencies. Weight loss data show 95.5% of theoretical yield. The rapid evolution of CO_2 gas can strip water, ethanol, and other volatiles out of a fermentation substrate. Production units often include gas

TABLE VI
By-Product Credit Calculations Based upon Different Substrates (100-wt basis)

	Substrates	
	Corn Screening (CS)	Hominy Feed (HF)
Contents		
Protein	8.4	10.6
Fat	3.9	1.3
Fiber	1.1	1.7
Ash	1.1	2.2
Starch	65.1	52.7
Fermentables (starch × 0.9)	58.6	47.4
Moisture	12.4	11.4
Theoretical recovery, %		
Solids (13% moisture)	33.3	47.3
Protein	25.0	22.0
Fat	11.7	2.7
Alcohol (w/w)	10.06	8.14
Product recovered, %		
Alcohol (w/w), %	9.86	8.00
Feed (13% moisture), lb	32.1	43.2
Yield from blend of 60% CS and 40% HF		
Feed	32.1	43.2
	× .6	× .4
	19.3 +	17.3 = 36.6 lb
Protein	25.0	22.0
	× .6	× .4
	15.0 +	8.8 = 23.8%

collection systems with traps to prevent alcohol loss. Commercially, alcohol yields average 90% of the value one would predict from the table.

B. Other Fermentation Products

The classical equation for conversion of starch to lactic acid is:

$$(C_6H_{10}O_5)_n + H_2O \xrightarrow[\text{or enzyme}]{H^+} C_6H_{12}O_6 \xrightarrow{\text{bacteria}} 2C_3H_6O_3$$

    starch (162) + water (18)   glucose (180)   lactic acid (90)

The molar yield of lactic acid is 2 mol/mol of glucose when commercial strains of *Lactobacillus delbrueckii* are used in anaerobic fermentation. This organism is classified as a homofermentative lactic acid bacteria. Industrial data do not refer to the theoretical weight yields, however. An 85% yield of the DL-racemic acid mixture represents normal operations, because processing losses are significant. A portion of the substrate also goes toward cell metabolism.

A lactic acid fermentation can take four days (Schopmeyer, 1954), beginning with a substrate of 15% corn sugar, plus nutrients, all buffered to 6.0 pH with 10% calcium carbonate. The incubation temperature is 48.9°C (120°F), which helps keep down competition from contaminants. The reducing sugar content is monitored until a glucose residue of 0.10% is obtained.

Fermentation lactic acid is not produced in the United States at the present time. Recovery processing costs are too high to compete with the chemically synthesized products. Food-grade fermentation acid is imported from Europe and South America.

Citric acid and xanthan gum are two additional products derived from glucose by "fermentation." Both of these products are produced by aerobic processes and arise from overproduction by selected stimulated organisms. Most of the production details are considered proprietary information. The reaction for glucose to citric acid is:

$$C_6H_{12}O_6 + 1\frac{1}{2} O_2 \longrightarrow C_6H_8O_7 + 2 H_2O.$$

Lockwood (1979) reported that the maximum weight yield of citric acid is 88% of the weight of the hexose supplied. The substrate concentration may be approximately 20%, and the solutions historically are decationized because acid production is strongly influenced by metals. Clarified substrates are best suited for this production because isolation and purification steps are rather involved. Monitoring the process variables is complicated by the fact that cell growth and metabolic formation of the product do not occur at the same stages of the incubation.

The fermentation polysaccharide pullulan has also received interest recently. Partially hydrolyzed starch syrups are the preferred substrates.

C. Beer Production

Degerminated corn is frequently employed as an adjunct to malt in the preparation of fermentation wort for beer. The grits refined from corn starch are nearly pure carbohydrate and can yield 90% of their weight as extract. Brewer's syrups are also available from the wet-milling industry and can contain almost any sugar profile the brewer desires.

When corn grits are used in brewing, they are pregelatinized by boiling before they are added to the malt portions of the beer mash. The amount of adjunct that can be added to lager beers is approximately 50%, due to the limits imposed by the nitrogen requirements of the yeast, which only the malt portion of the mash will meet (Westermann and Huige, 1979). A further discussion of beer production from dry-milled corn products is in Chapter 11.

IV. SUBSTRATE PREPARATIONS

A. Cooking

Native corn starch consists of approximately 25% linear polymer glucose monomers linked α-1,4 (amylose) and 75% branched polymer glucose made up of smaller linear α-1,4 units joined through α-1,6 branch points (amylopectin).

Most industrial strains of microorganisms do not utilize starch efficiently, and the starch must be hydrolyzed with acid or enzymatically catalyzed to break the linkages and yield the glucose monomer. Chapter 17 gives details of the production of glucose syrups from corn starch.

Fiber-free starch streams are well suited to high-temperature, short-time jet cooking for hydrolysis, thus providing the liquefying and saccharifying enzymes ready access to the starch molecules. This procedure permits starch slurries of 30–40% solids and also yields syrups with high solids and high dextrose equivalent (DE). DE reflects the degree of hydrolysis in the starch stream by measuring the reducing sugar levels compared with the reducing power of pure glucose solutions, but it does not accurately reflect saccharide composition. Chromatographic procedures are required for this. However, during enzymatic saccharification with glucoamylase, low DE does mean less free glucose. High DE is important to distillers to achieve optimum alcohol yields.

Whole-grain distilleries generally employ hammer mills to grind their corn. The particle size influences cooking hydration and subsequent enzymatic conversion and thus the yield. Atmospheric cooking at 100°C (212°F) can be as effective as pressure cooking at 160°C (320°F) if fine grinds are used. However, downstream feed recovery and feed drying are influenced negatively when grinds are too fine. A typical screen analysis is 15–20% on No. 12 (1.52-mm), 35–50% on No. 20 (0.86-mm), and 50–75% on or through No. 60 (0.23-mm) screens.

When using conventional grinds, pressure-cooked mash yields more alcohol than atmospheric-cooked mash, by 0.1 PG/Bu or more. Presumably, this is due to successful hydration and gelling of the final 2–3% of the very refractory starch in the kernel. Analysis of starch hydrolysates and/or ferments by HPLC requires prior filtration. Thus, one analyzes the soluble portion of the stream only and will miss the insoluble starch resulting from poor cooking operations. This can cause confusion and substantially influence yield calculations.

Flavor considerations sometimes influence the choice of cooking temperatures used by beverage alcohol distillers. Some whole-grain distillers have joined wet millers and are using jet cookers at around 105°C for 5 min after a hot preliquefaction hold at 85°C with a high-temperature α-amylase enzyme.

B. Starch Conversion

The role of α-amylase enzymes in the initial steps of conversion is to decrease the viscosity of a gelatinized starch. This thinning and liquefaction activity produces only dextrins as the reaction products. Gelled amylose starch tends to retrograde upon cooling, and the retrograded product resists enzymatic hydrolysis. Both dilution and thinning help prevent the problem. Saccharifying enzymes hydrolyze the starch and dextrins to glucose or maltose. As shown in Table VII, barley malt contains both liquefying and saccharifying amylases, and maltose sugar is the principal product of their hydrolysis.

Malt enzymes have been the traditional choice of beverage alcohol distillers, and they remain so for whiskey production due to flavor considerations. Alcohol yeast contains the enzyme α-glucosidase, which converts the disaccharide into a single glucose monomer. Less expensive bacterial and fungal enzymes (Table VII) have replaced malt for production of spirits and power alcohol. The typical usage level for malt treated with gibberellin is 2%, based upon the weight of the grain, and 45,000 units per pound of malt is a normal unitage. Malt enzyme treatment cleaves 15–20% of the starch linkage during liquefaction. Malt enzymes are fairly heat- and pH-labile; for example, a temperature of 65.5° C (150° F) and a pH below 4.5 can inactivate them. Thus, they do not survive the cooking process. Splitting malt addition between pre- and postcooking stages is a common practice (Table V).

Bacterial α-amylases are also quite sensitive to low pH but are more heat-stable. The *Bacillus licheniformis* enzyme performs well in continuous 105° C liquefaction processes.

Originally, batch processing with *B. subtilis* α-amylase used 140° C heating cycles and required higher enzyme dosage. Maltulose formation by chemical isomerization has been reported during long holding times at higher temperatures. Maltulose is not hydrolyzed by glucoamylase. Reversion reactions also can occur during the production of glucose from starch via glucoamylase, after prolonged incubation. The resulting oligosaccharides are not fermentable by yeast.

Glucose yield from glucoamylase treatment is influenced by the concentration of solids. High concentrations of glucose shift the enzymatic equilibrium toward polymerization of glucoamylase, converting glucose into maltose and isomaltose. Therefore, exceeding a maximum glucose concentration at a given solids level or prolonging incubation can cause formation of reversion products and produce yield losses.

Very little HPLC data for mash conversion have been published by whole-grain distillers, but after 30 min of malt conversion at 62.8° C (145° F), a profile of 3–4% DP-1 (glucose), 40–42% DP-2 (maltose), and 53% dextrins may be typical. An additional 10 min of incubation with malt produces no DP-1 increase, but the DP-2 values could exceed 50%, DP-3 values become 18%, and dextrins fall toward 30%. Adding a saccharifying enzyme for a similar 10-min

TABLE VII
Enzymes Used in Starch Conversion

Enzyme	Type	Source	Amount	Activity
α-Amylase	Liquefying	*Bacillus subtilis*[a]	0.06% wt of starch	Decreases viscosity (Cleaves 1-4, pH 5.5, 70° C)
		B. licheniformis	0.06% wt of starch	Decrease viscosity (92° C)
		Barley malt	0.5–1.0 wt of grain	Decreases viscosity (60° C)
β-Amylase	Saccharifying	Barley malt[b]	2.0% wt of grain	Generates maltose (Cleaves 1-6, pH 5.5, 60° C)
Glucoamylase	Saccharifying	*Aspergillus niger*[b]	0.18% wt of starch (1.7 L/ton)	Generates glucose (Cleaves 1-6, pH 5.0, 60° C)
Pullulanase	Saccharifying	*B. acidopullulyticus*	0.2% wt of starch (2.0 kg/ton)	Cleaves 1-6, pH 5.0, 60° C

[a] Endoamylase.
[b] Exoamylase.

hold can increase glucose to 12%, maltose to 55%, and reduce dextrins below 30%.

Some glucoamylase ferments contain transglucoamylase activity, which synthesizes isomaltose. Whole-grain distillers have avoided these complications by letting yeast consumption of glucose prevent any buildups. Crude preparations of glucoamylase often contain amylase and protease enzymes, which can be beneficial to the whole-grain distiller.

Commercial pullulanases are a recent advancement in starch enzyme technology. By developing enzymes that are compatible with glucoamylase in pH and temperature, processors may be able to achieve better 1-6 bond cleavage and reduce glucoamylase usage while producing faster reaction rates (Jensen and Norman, 1984).

Cellulase preparations from *Trichoderma reesei* can potentially increase the alcohol yield from whole-corn fermentations. Cost reductions and process improvements have been reported by the Finnish State Alcohol Monopoly (Oksanen et al, 1982).

V. FERMENTATION PROCESS

A. Organisms

Published literature contains a wealth of information describing the response of *Saccharomyces* yeasts to the varying concentrations, constituents, and environmental circumstances of media presented to these baker's, brewer's, and distiller's strains. Most of the industrial yeasts have been selected by performance criteria, with consistency of performance weighing heavily in that selection.

Although genetic engineering techniques are being successfully applied to yeast genetics, an attribute such as alcohol tolerance remains ill-defined. Until the genes controlling the physiology of yeast response to alcohol concentration are properly mapped, progress toward the super yeast will be slow. Cell fusion, which physically blends the genomes of selected yeast strains, has produced improved strains. The stability of these new genetic materials must be better established. Full discussion is well beyond the scope of this work, but this chapter highlights some of the progress toward improved substrate utilization and efficiency.

Batch fermentation data following glucose consumption, weight loss, ethanol production, and cell growth normally show a general slowing as ethanol levels rise to 8% (v/v) (Fig. 1), although final ethanol levels may reach 13% and beyond when concessions are made for low temperature, special nutrition, and fermentation length. Jin et al (1981) improved alcohol tolerance and obtained higher yields by utilizing albumin and phosphatidylcholine supplementation with *S. cerevisiae* (sake strain). Viegas et al (1985) has reported increased alcoholic fermentation productivity via soy flour supplementation.

In the case of higher initial glucose levels, ethanol production can rise rapidly. The size and the history of a yeast inoculum can also have an effect upon such fermentation rates and upon yield efficiencies. For example, in batch experiments, Cysewski (1976) found specific growth rates of 0.46 hr^{-1} with air-saturated cultures but only 0.15 hr^{-1} under anaerobic conditions. Alcohol production paralleled growth until late in the anaerobic run, when growth stopped and alcohol production continued slowly until the glucose was completely fermented. Ethanol yields were identical. Front-loading of large inoculum or utilizing immobilized cells can conceivably circumvent cell growth and improve fermentation efficiency.

Recent experiments using high-gravity brewing worts and brewer's yeast, *S. uvarum*, have been reported. Slow fermentations and poor yeast viability generally are the results. Panchal (1981) showed that ethanol inside of the cells is more toxic than ethanol in the medium. High osmotic pressure appears to slow the escape of the alcohol. Casey et al (1984) reported 16.2% (v/v) alcohol and good yeast viability after finding that nutritional deficiencies in the wort could be overcome by the addition of nitrogen-containing nutrients, ergosterol, and oleic acid.

The chemical changes associated with fermentation of corn glucose syrup to ethanol release heat energy of 528 BTU/lb of alcohol. Ethanol coupled with high temperature is more toxic to the yeast, and cooling may be needed to sustain fermentation to completion. Fermentation temperature control is a critical concern during rapid fermentations. Beverage distillers control temperature for the production of their flavor distillates. Several research groups are working with specially developed strains described as ethanol-tolerant. This trait is sometimes associated with flocculating (self-settling) strains with altered cell wall and is reported with strains that release less heat of fermentation (D.

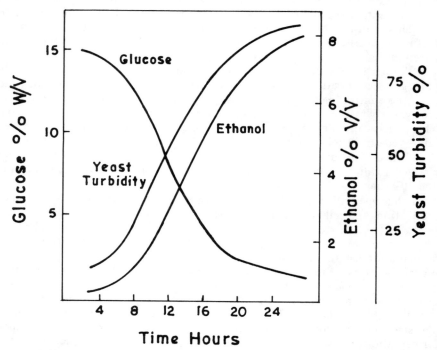

Fig. 1. Batch fermentation experiment using clear starch hydrolysate and no agitation: glucose, 15% initial concentration; time, 3–27 hr; 32°C.

Mattox, Ultraferm, Wheeling, IL, personal communication). This is associated with a reduction in the amounts of glycerol and fusel oil in the final ferment as well. Some yield increase could be expected as a result. Altered fermentation pathways have been suggested.

The bacterium *Zymomonas mobilis* does use a different pathway to produce ethanol from glucose. Distiller's yeast uses the Embden-Meyerhof pathway; the bacteria uses Entner-Doudoroff. Increased efficiency translates into higher alcohol yields and less by-product and biomass production. The specific rate of alcohol production can also be more rapid for bacterial fermentations. Alcohol toxicity has been a problem for these organisms, dictating dilute feedstocks. This requirement raises the cost of distillation and feed recovery (Esser and Schmidt, 1982; Esser and Karsch, 1984). A distillery strain carrying starch-digesting enzymes has been forecast by Lyons (1983).

B. Fermentation Mode

Substrate costs are the single most important factor determining the final cost of fermentation ethanol. Choosing a fermentation process that makes optimum use of the substrate becomes of paramount importance. Maiorella et al (1984) have recently published a plenary review of this topic.

At present, most alcohol producers rely upon batch fermentations (Fig. 1). Beverage alcohol producers do so out of consideration for flavor development. Power alcohol producers appreciate the flexibility and ease of operation that it offers. To achieve continuous operation, many have tied individual fermentors together, forming a forward-flowing, cascade, continuous system.

Figure 2 depicts data from an experimental 150-hr run using a cascade continuous fermentation. The medium consisted of 45% (v/v) low-DE liquefied corn starch (17% glucose equivalent), 19% (v/v) corn steep (5% solids), and 36% stillage backset. Stillage is dealcoholized fermentation beer that emerges from the base of the distillation column. This experiment was run with simultaneous saccharification and fermentation, using 100 units of glucoamylase per pound of starch at pH 4.5 and 32°C (90°F). Liquid residence time was 30 hr. A control batch fermentation set at the same time used an inoculum rate of 6 g/L and a 48-hr incubation. The 10.5% (v/v) alcohol (8.28% w/w) represents about 92% substrate utilization, compared with 96% for the batch data. (Data are not corrected for fermentables present in the steep and/or stillage backset.) However, biomass accumulated in the cascade fermentors. Large yeast populations facilitate glucoamylase saccharification via rapid utilization of freed glucose. Free glucose in all four vessels ran approximately 0.2%. By connecting the points indicated by the arrows, one obtains a graph similar to that of the batch data in Fig. 1. Note that fermentor No. 4 provides little more than a polishing step, even though fermentable substrate was available. The percent of viable cells was lowest in the final fermentor, also. Fermentation mashes of ground whole corn are more difficult to use in continuous systems, and generally, longer fermentation times are required.

Table VIII shows the material flow typical of batch fermentation of 1,000 lb of corn. Many considerations are omitted in the table, but it illustrates the material flow approach to distillery operations. The rather conventional high-temperature mashing of corn in Example A is done at 25% dry solids (ds). The innovative process (Example B) uses indirect heating, hydrochloric acid, and short cooking times at 40% ds (personal communication from Per Assarson, St. Lawrence Technology, Toronto, Canada). A new steel alloy has made it possible to construct the noncorrosive chamber that is central to this process. The manufacturer reports that one can virtually dial in the desired DE over a wide range of substrate mixtures without the formation of toxic by-products or the substrate loss associated with earlier attempts to acid-hydrolyze corn substrate. The two processes are presented with a balance in fermentable carbohydrates and in yield. Since the material flow points out differences in the processes better than a flow sheet of the conventional type, distillers can more readily perform cost analysis based upon the traditional cost per proof gallon.

Beverage alcohol distillers have long used $1/3$ CO_2, $1/3$ ethanol, and $1/3$ feed coproduct as their rule of thumb for distillery yields. Table VIII data show that 1,000 lb of corn was fed to each of the two processes, with 1,078 and 1,086 lb of product dry solids for the conventional and reactor mashing treatments, respectively. The sum of lines 5, 7, and 10 equals the total product. The yield efficiencies for alcohol and CO_2 can be compared to theoretical yields (51% and 49%, respectively, of the 778-lb glucose weight in line 4). The 357 lb of actual alcohol production vs. the 397-lb theoretical yield represents 90% efficiency

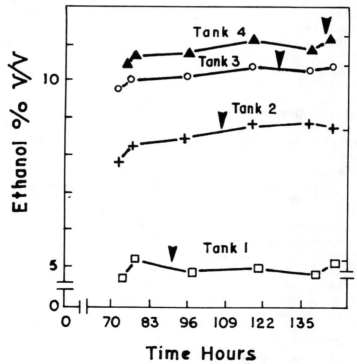

Fig. 2. Cascade continuous experiment using clear starch hydrolysate, 3.3%/hr feed rate, and 30-hr retention time.

calculated from alcohol yield. The CO_2 yield efficiency is 94% (359/381), which approximates the data for the example given in Table V.

The feed recovery from the process is 370 lb (ds), which exceeds the nonfermented total of 58 lb of solubles, 242 lb of insolubles (line 2), plus 62 lb of residue (line 4 minus lines 5 and 7). The material flow shows that both recycled solids, those returned to the fermentor as stillage from previous fermentations, and added chemicals can alter the final feed recovery weight. Eight pounds of extra feed credit equates to approximately $0.40/1,000 lb of original weight, based upon a coproduct price of $100/ton. However, the 40 lb of lost ethanol at 6.6 lb/gal equals $9.00, based upon an alcohol selling price of $1.50/gal.

Recycling thin stillage is a water conservation measure, and both processes have made provision for "back set" usage. By starting with a higher initial solids level in process B, the reactor process can take better advantage of the thin

TABLE VIII
Material Flow from 1,000 lb of Corn

Line No.	Process Stage	Components	Example A: Conventional Process (lb)	(total lb)	Example B: Innovative Process[a] (lb)	(total lb)
1	...	Condensate H_2O	2,837	2,837	1,324	1,324
2	Soak/slurry	Starch	700	4,000	700	2,500
		Solubles	58		58	
		Insolubles	242		242	
		H_2O	3,000		1,500	
3	Cook/react	Starch	700	3,960	700	2,172
		Solubles	58		58	
		Insolubles	242		242	
		Chemicals	...		7	
		H_2O	2,960		1,164	
4	Saccharification	Sugar	778	4,690	778	4,931
		Solubles	83		288	
		Insolubles	242		242	
		Chemicals	...		26	
		H_2O	3,587		3,587	
5	Fermentation	CO_2	359	−374	359	−374
		H_2O	15		15	
6	...	Alcohol	357	4,316	357	4,557
		Solubles	125		340	
		Insolubles	262		262	
		Chemicals	...		26	
		H_2O	3,572		3,572	
7	Distillation	Alcohol	357	−386	357	−386
		H_2O	29		29	
8	...	Solubles	125	3,930	340	4,171
		Insolubles	262		262	
		Chemicals	...		26	
		H_2O	3,543		3,543	
9	...	Recycle	705	705	2,760	2,760
10	Drying	Solubles	100	−402	100	−411
		Insolubles	262		262	
		Chemicals	...		7.7	
		H_2O	40		41	

[a]Data courtesy of St. Lawrence Technology, Toronto, Canada.

stillage recycle. Buildup of solids and/or toxic metabolites in the recycled stillage can occur, but much data are available on this topic (Wall et al, 1981).

Energy costs are a second major expense in alcohol production by fermentation. When clean, hot condensate is available for process use, energy costs can be reduced (line 1). When the grain solids concentration is high, there is less water to heat during the cooking process. This, too, can reduce energy requirements. The material flow for the reactor process provides for maximum utilization of these options. Ultimately, either equipment or yeast places a limit upon substrate solids. The review by Keim (1983) is an excellent reference in this area. Since these processes were designed with balanced carbohydrate levels, direct comparison can be made of conversion chemical costs, distillation costs, and drying cost for distiller's grains (Nagasuye and Assarson, 1982).

VI. FERMENTATION LOSSES

The use of whole grain complicates the analyses of fermentation yield losses. Beverage alcohol distillers traditionally rely upon final balling readings to determine whether soluble sugars remain in the finished beer, but the formation of the liquid product, ethanol, can cause a false negative balling reading. Positive readings show incomplete utilization. Balling is an expression of specific gravity of a solution in percent sugar compared to a standard sugar-in-water solution. Distillers also determine the final acid concentration to check on contamination by lactic-acid-producing organisms, which can take substrate from the yeast and may influence the performance of the enzymes that convert dextrins into fermentable soluble sugars (Maisch et al, 1979).

Yeast growth during alcohol fermentation requires energy and consumes substrate for biomass production. Microscopic cell counts are a rather unreliable method to measure biomass formation, but they are often employed in whole-grain fermentations. Viable cell determination using methylene blue decolorization is a practical aid for distillery monitoring (Lee et al, 1981). Centrifugation of the beer to determine the increase in yeast pellet size works well for clarified substrates.

Most alcohol fermentation media use recycled stillage (dealcoholized fermentation beer). In addition to water conservation, the recycling practice furnishes buffering and supplements yeast nutrients. However, intact dead yeast cells are normally present in recycled stillage, and this clouds biomass measurements. Beverage alcohol distillers have limited their stillage reuse to about 25% of the fermentation volume for flavor considerations. Power alcohol distillers often recycle to 50% volume or higher. Recycles of 100% through 10 or more fermentations with little negative impact have been reported, although the acetic acid concentrations would concern many yeast researchers.

Table IX shows final analyses of beer from several alcohol fermentations. In example A, 1.6 g of yeast was produced, accounting for 0.8% of the glucose utilized. The 183.3 g of product ethanol represents 95% of the glucose (glucose in grams $\times$ 0.51 = weight of ethanol), and glycerol accounts for an additional 3.2% of the glucose.

Carbon dioxide production can also be used to account for glucose utilization. CO_2 is seldom measured in large-volume fermentation. Weighing is impractical, and measuring gas evolution is complex, for a significant concentration can be retained in the solution.

The profile in Example A is for a glucose feed into and stillage residue remaining from a continuous cascade fermentation. Note that the fully saccharified starch contained residues that persisted through the fermentation. The efficiency of this fermentation may be based upon sugar consumed, but yields should be based upon the carbohydrate available. Little benefit is gained from carbohydrates passing through the process to become by-product feedstuff.

Example B in Table IX accounts for nearly 97% of the original glucose present in the fermentation beer and includes the CO_2 determination. Example C contains data from ALKO of Finland (Oura, 1977). Researchers there have distinguished between glycerol produced in association with cell growth and that associated with succinate formation. The latter results from cofactor regeneration. These data suggest that yield efficiency is limited, even if one achieves ethanol production independent of yeast growth. A recent patent has been issued to Spencer (1984) for ethanol production via isolated enzymatic reactors independent of whole cells. No experimental data are contained in the patent, however.

Example D is the analysis of fusel oil of a corn whiskey distillate. Fusel oil flavor components (congeners in whiskey distillates) can arise from amino acids in the mash via the Ehrlich mechanism involving α-keto acids in transamination and decarboxylation (Lewis, 1964). However, a carbohydrate pathway involving the synthesis of amino acids via similar α-keto acid building blocks seems to contribute the major concentration (fusel oil components are higher alcohols n-propyl, isobutyl, amyl, and isoamyl).

Fusel oil concentration can raise the values for alcohol yield, depending upon the method of yield determination. Fusel oil alcohols can be detected in fermentation beers at levels of 0.05–0.12% (w/w). Since fusel oils contain extra combustion energy, power alcohol distillers can blend the fusel oil draw from the still back into their anhydrous fuel alcohol. Thus, sugar diverted into fusel oil product may or may not constitute a yield loss.

The data in Table III showed that the majority of the DP-2 components found in starch hydrolysates are not fermented by distiller's yeast. Thus, the type of conversion process selected can have a real impact upon yield.

TABLE IX
Examples of Analysis of Fermentation Products

Example A: Products as Percent of Glucose Consumed			Example B: Products as Percent of Glucose Added		Example C: Products as Glucose Equivalents[a] (%)			Example D: Corn Whiskey Fusel Oil (g/100 L at 100 proof)	
Ethanol	95	(183.3 g)	Ethanol	45.5	Ethanol	94.75	(5,590 g)	Ethyl acetate	6.8
Glycerol	3.2	(6.2 g)	CO_2	45.4	Glycerol	2.24	(132 g)	n-Propyl	15.1
Yeast	0.8	(1.6 g)	Glycerol	4.26	Glycerol[b]	0.83	(49 g)	Isobutyl	35.7
By-products			Succinate	0.46	Succinate	0.37	(22 g)	Amyl	58.4
(difference)	1.0		Acetate	0.42	Yeast	0.80	(47 g)	Isoamyl	85.7
			Threalose	0.6				Acetate	0.7
			Total	96.64				Aldehyde	0.5

[a] 10,000 kg of added grain is equivalent to 5,900 kg of glucose.
[b] From yeast growth (Oura, 1977).

VII. PRODUCT FINISHING

A. Distillation

The final stage of ethanol production is the separation of the volatile alcohol (and flavor) components from the fermented mash (beer). Still designs vary to match the preselected type and quality of ethanol distillate. In whiskey production, the distillation unit is often a beer still equipped with a side doubler unit. Mash is fed to the still near midpoint, and steam is fed at the bottom. The downward-flowing mash is stripped of volatiles by the rising hot vapors from the boiling liquid in the still bottom. The inside of the column contains perforated plates. The product can be produced at 110–160° proof (Fig. 3). This distillate can be further refined by the doubler, a copper pot still. Provisions for refluxing and recycle can also be added, at the desire of the distiller.

Neutral spirits and power alcohol are typically produced on a multiple-column still. An aldehyde column is fed beer-still distillates and recycle streams. Its product is a low-proof alcohol cleaned of most aldehyde heads fractions. The rectifying column is next in line, removing intermediate- and high-boiling congeners (Fig. 4). The product is 190° proof. If a cleaner distillate is desired, the product can be withdrawn to permit a few percent of the ethanol to escape as vapor with the flavor components of the low-boiling heads.

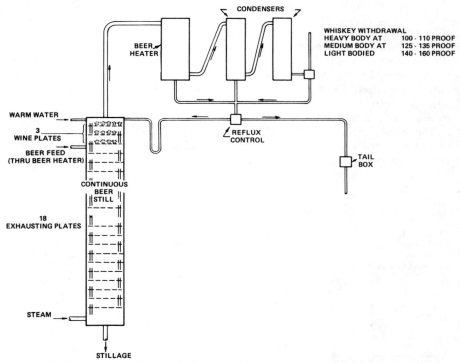

Fig. 3. Simple diagram of a whisky beer still. (Courtesy Hiram Walker & Sons)

B. Absolute Ethanol

At 95.6% (w/w) ethanol, water and ethanol mixtures form a constant boiling azeotrope. In order to break this azeotrope and obtain anhydrous ethanol, producers generally add a third component, such as benzene or diethyl ether. Ethanol can then be distilled from this mixture, leaving the other two components behind. This second distillation yields 99.9% ethanol but can add an additional 7,000–8,000 BTU to the distillation energy costs. Spirit distilleries produce 95% (w/w) ethanol consuming 20,000–25,000 BTU/gal.[1] Beverage alcohol distillers may spend 50% of their overall energy use on distillation.

Ladisch et al (1984) has developed an adsorptive-desorptive separation technique for dehydrating ethanol vapors at a lower energy cost. The process has been implemented at a large power alcohol plant.

C. Aging

Whiskey distillates are traditionally mellowed and developed for taste and aroma by an aging process. The distillate is reduced to a proof of 105–130° and placed into new or used charred oak barrels. Higher entry proof is permitted for

---

[1] R. Katzen, W. R. Ackley, G. D. Moon, U. R. Messick, B. F. Bruch, and D. F. Kavpisch. Low energy distillation systems. Presented at the 180th national meeting of the American Chemical Society, Las Vegas, 1980.

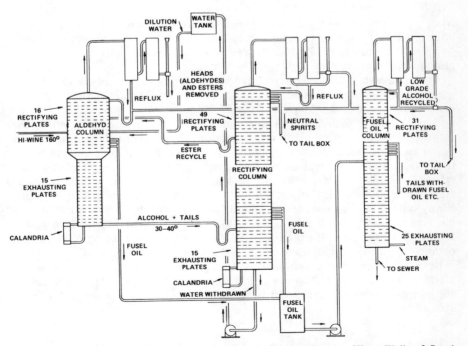

Fig. 4. Simple diagram of a spirit unit using extractive distillation. (Courtesy Hiram Walker & Sons)

certain classes of spirits. During the time the barrel is in storage, the rise and fall in temperature is accompanied by a small loss of product as the barrel breathes. By this process, the acids in the whiskey rise, and the ester and aldehyde concentrations also increase. The flavor and aroma of distilled spirits are often "defined" by the statutes of their country of origin. Distillates are withdrawn from the barrel after anywhere from two to 12 years, depending upon storage conditions and the types and flavor levels desired.

VIII. ECONOMIC IMPORTANCE

In 1983 in the United States, 58,500,000 bu of corn was used in the production of distilled spirits (Anonymous, 1984). The grind for brewer's adjuncts and for power alcohol exceeded 140,000,000 bu, although data are incomplete (Meyer, 1984). These data contrast with 32,100,000 bu for spirits in 1973 and approximately 10,000,000 bu for alcohol from wet-milling production in that year. In 1984, the corn wet-milling industry produced about 75% of the fuel ethanol production, but this level is expected to decline as more dry-milling plants come on stream (Gill and Allen, 1985).

The number of bushels of corn used for the production of fuel alcohol has increased 20-fold in the past 10 years. The projected use of corn for all alcohol production in 1985 is 280 million bushels (Anonymous, 1986)—40 million for beverage and 240 million for fuel. This figure could exceed 400 million by 1990.

The beverage alcohol industry produced 72,509,000 tax gallons (gallon volume of 100° proof ethanol) of whiskey, 15,395,000 tax gallons of gin, and 11,196,000 tax gallons of vodka in 1983. Power alcohol production in the United States was 330,000,000 gal (gallons of 200° proof ethanol). The dried grains coproduct from spirits totaled 696,100 short tons in 1983, compared with 440,300 short tons in 1973. Much power alcohol stillage is codried with the corn gluten feeds that are a by-product of wet milling.

As impressive as these data are, the economic impact on the nation is even greater when one considers tax revenues. For example, the 1983 federal taxes on distilled beverage alcohol were $3,756,636,000; state taxes were $2,985,713,000 and local taxes $249,079,000 (Anonymous, 1984). One may speculate that the future of fermentation alcohol production in the United States will be determined by governmental tax decisions. Power alcohol production has increased under the protection of a fuel tax break given to gasoline fortified with ethanol. This is $0.04 per gallon of gasoline, which translates into as much as $0.40 per gallon of alcohol, depending upon the amount added to gasoline. Many states add their own tax incentive. The status of the U.S. ethanol market has recently been reviewed by Gill and Allen (1985) in an article with that title. United States-produced ethanol, with subsidies, is price competitive as an octane-enhancing fuel additive.

Beverage distillers are anxiously awaiting the final impact of the new tax increase, which totals $12.50 per gallon (at 50% ethanol). This level of taxation will not be assessed on beer and wine alcohol. Initial reports suggest a 3–5% decline in sales.

LITERATURE CITED

ANONYMOUS. 1982. Pierce Chemical Handbook and General Catalog. Pierce Chemical Co., Rockford, IL.

ANONYMOUS. 1984. Annual Statistical Review. Distilled Spirits Council of the United States, Washington, DC.

ANONYMOUS. 1986. Feed and feed demand. Outlook and Situation Report, FdS 299. Mar. U.S. Dept. Agric., Econ. Res. Serv., Washington, DC.

ASBC. 1981. Methods of Analysis of the American Society of Brewing Chemists, 7th ed., revised. Adjunct Materials: Sugars and Syrups–7. The Society, St. Paul, MN.

CASEY, G. P., MAGNUS, C. A., and INGLEDEW, W. M. 1984. High-gravity brewing. Appl. Environ. Microb. 48:639-646.

CYSEWSKI, G. 1976. Fermentation kinetics of process economics for the production of ethanol. Ph.D. dissertation (77-4428), University of California, Berkeley.

ESSER, K., and KARSCH, T. 1984. Bacterial ethanol production: Advantages and disadvantages. Process Biochem. 19(3):116-121.

ESSER, K., and SCHMIDT, U. 1982. Alcohol production by biotechnology. Process Biochem. (May-June):46-49.

GILL, M, and ALLEN, L. 1985. Status of U.S. ethanol market. Pages 14-22 in: FEED. Outlook and Situation Report, Fd-297. U.S. Dept. Agric., Econ. Res. Serv., Washington, DC.

IVIE, K. F. 1983. High performance liquid chromatography in sugar analysis. Sugar Azucar (Feb.):44.

JENSEN, B. F., and NORMAN, B. E. 1984. *Bacillus acidopullulyticus* pullulanase: Application and regulatory aspects for use in the food industry. Process Biochem. (Aug.):129-134.

JIN, C. K., CHIANG, H. L., and WANG, S. S. 1981. Steady state analysis of the enhancement in ethanol production of a continuous fermentation process employing a protein-phospholipid complex as a protecting agent. Enzyme Microb. Technol. 3:249-257.

KEIM, C. R. 1983. Technology and economics of fermentation alcohol, an update. Enzyme Microb. Technol. 5:103-115.

LADISCH, M. R., VOLOCH, M., HONG J., BLENKOWSKI, P., and TSAO, G. T. 1984. Ethanol dehydration. Ind. Eng. Chem. Process Design Dev. 23:437-444.

LEE, S. S., ROBINSON, F. M., and WANG, H. Y. 1981. Rapid determination of yeast viability. Biotechnol. Bioeng. Symp., 11th. pp. 641-649.

LEWIS, M. J. 1964. Nitrogen metabolism of yeast. Tech. Q. Master Brew. Assoc. Am. 1(3):167-175.

LILLEHOJ, E. G., LAGODA, A., and MAISCH, W. F. 1979. The fate of aflatoxin in naturally contaminated corn during the ethanol fermentation. Can. J. Microbiol. 25:911-914.

LOCKWOOD, L. B. 1979. Production of organic acids by fermentation. Pages 355-387 in: Microbial Technology, Vol. 1. H. J. Peppler and D. Perlman, eds. Academic Press, New York.

LYONS, T. P. 1983. Alcohol fermentation in the United States. Pages 16-23 in: Advances in Fermentation. Chelsea College, London.

MAIORELLA, B. L., BLANCH, H. W., and WILKE, C. R. 1984. Economic evaluation of alternative ethanol fermentation processes. Biotechnol. Bioeng. 26:1003-1025.

MAISCH, W. F., SOBOLOV, M., and PETRICOLA, A. J. 1979. Distilled beverages. Pages 79-94 in: Microbial Technology, Vol. 1. H. J. Peppler and D. Perlman, eds. Academic Press, New York.

MEYER, P. A. 1984. Corn wet milling. Milling Baking News (Feb 28):19-20.

NAGASUYE, J., and ASSARSON, P. G. 1987. Preparation of acid hydrolysed starch substrates and their fermentability. Int. Carbohydrates Symp., 11th. Vancouver, B.C. Aug. 22, 1982. In press.

OKSANEN, J., PAARLAHTI, J., and ROERING, K. 1982. Use of cellulase preparations in grain alcohol fermentation. Pages 1-13 in: Esitelmä pidetty Finnish-Soviet Symp. Oct. 25–27. Tvärminne. ALKO, Finland.

OURA, E. 1977. Reaction productions of yeast fermentations. Process Biochem. 12(3):19-21.

PANCHAL, C. J., and STEWART, G. G. 1981. Regulatory factors in alcohol fermentation. Page 9-15 in: Current Developments in Yeast Research. G. G. Stewart and I. Russell, eds. Pergamon Press, Toronto.

PEPPLER, H. J., and PERLMAN, D., eds. 1979. Microbial Technology, Vols. 1 and 2. Academic Press, New York.

POMES, A. F. 1980. Corn syrup rapid method fermentables by HPLC. J. Am. Soc. Brew. Chem. 38:67-70.

REHM, H. F., and REED, G., eds. 1983. Biotechnology, Vol. 5. Verlag Chemie, Weinheim, West Germany.

ROSE, A. H., ed. 1977. Economic Microbiology, Vol. 1. Academic Press, New York.

SCHOPMEYER, H. H. 1954. Lactic acid.

Pages 391-419 in: Industrial Fermentation, Vol. 1. L. A. Underkofler and R. J. Hickey, eds. Chemical Publishing Co., Inc., New York.

SPENCER, D. B. 1984. Methods and apparatus for enzymatically producing ethanol. U.S. patent 4,451,566.

VIEGAS, C. A., SÁ-CORREIA, I., and NOVAIS, J. M. 1985. Nutrient-enhanced production of remarkably high concentrations of ethanol by *Saccharomyces bayanus* through soy flour supplementation. Appl. Environ. Microbiol. 50:1333-1335.

WALL, J. S., BOTHAST, R. J., LAGODA, A. A., SEXSON, K. R., and WU, Y. V. 1983. Effect of recycling distillery solubles on alcohol and feed production from corn fermentation. J. Agric. Food Chem. 31:770-775.

WATSON, S. A. 1984. Corn and sorghum starches: Production. Pages 417-468 in: Starch: Chemistry and Technology. R. L. Whistler, J. N. BeMiller, and E. F. Paschall, eds. Academic Press, Orlando, FL.

WESTERMANN, D. H., and HUIGE, N. J. 1979. Beer brewing. Pages 1-36 in: Microbial Technology, Vol. 1. H. J. Peppler and D. Perlman, eds. Academic Press, New York.

CHAPTER 20

BIOMASS USES AND CONVERSIONS

MARVIN O. BAGBY
Northern Regional Research Center
U.S. Department of Agriculture
Agricultural Research Service
Peoria, Illinois

NEIL W. WIDSTROM
Coastal Plains Experiment Station
U.S. Department of Agriculture
Agricultural Research Service
Tifton, Georgia

I. INTRODUCTION

Corn, or maize, is produced mainly for food and feed from the grain or for use of the entire plant as silage. As described in preceding chapters, the grain also is processed to provide a host of industrial products. However, substantial portions of the plant remain unused or underused. This chapter discusses the underutilized coproducts, corncobs and stalk residues, and gives some potential uses. Production of the corn plant as a source of extractable sugars for fermentation and other applications is also discussed.

II. CORNCOB RESIDUE

For every 100 kg of corn grain, approximately 18 kg of corncob is produced. Thus during the past five years, the annual corncob production worldwide averaged about 70 million tonnes. In 1979, about 1.1 million tonnes of cobs were used industrially. Nearly 0.7 million tonnes were converted into furfural and 0.3 million tonnes were used as granular products (Foley and Vander Hooven, 1981). An additional 0.1 million tonnes filled a variety of industrial applications, including production of xylose. An undetermined amount is consumed as animal feed. At $35 per tonne, the 1979 world value for industrial cobs was $38.5 million. The value increase from processing has been estimated as fivefold, thus equaling nearly $190 million worldwide.

The corncob consists of four distinctly different parts—pith, a woody ring, coarse chaff (tough woodlike flakes), and light chaff (Fig. 1). By weight, pith makes up about 1.9% of the cob, the woody ring about 60.3%, coarse chaff about 33.7%, and fine chaff about 4.1% (Clark and Lathrop, 1953). The ring and coarse chaff are hard, woody, and resistant to abrasion and granulation. These characteristics were the basis for their early industrial applications (Lathrop, 1947).

Foley (1978) compiled a comprehensive review of the chemical and physical properties of corncobs and of their uses, including proprietary products of The Andersons (Maumee, OH). Selected properties are shown in Table I.

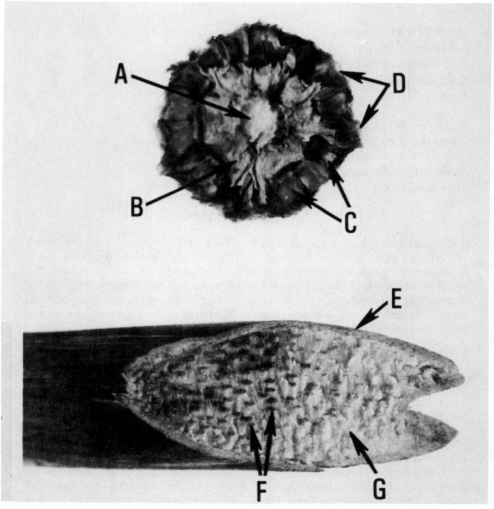

Fig. 1. Top: corncob cross section showing pith (A), woody ring (B), source of coarse chaff (C), and beeswing (D). Bottom: cornstalk cross section showing cortex (E), vascular fibrous bundles (F), and pith (G).

A. Cob Collection

Before the introduction of the picker-sheller and corn combine, cob collection accompanied grain harvest. Hudson (1984) discussed an industrial corncob refining operation. He reported that good weather and abundant harvest generally assured adequate supplies of cobs. However, if the corn price is high, the farmer becomes more interested in dealing with the grain. Collecting crop residues interferes with timely harvest of the commodity with the highest value. Thus, cobs and other crop residues must have value sufficient to pay their way before an enduring market can develop. Many researchers have tested collection procedures for retrieving cobs directly from the combine, either with the grain or separately.

Because hybrid corn seed is harvested with the cob, this unique operation offers the opportunity for cob utilization. Cobs have been used by a commercial seed producer as an alternative source of energy for drying seed corn (Peart et al, 1981). Claar et al (1981) report that, based on an energy and economic analysis, corncobs are a better fuel source than stover.

Because corncobs are available for only a short time, year-around supply necessitates storage. Dunning et al (1948) reviewed corncob-storage studies reported by grinding industries in the Midwest using various indoor and outdoor techniques. Representative samples removed from storage at 6-in. intervals, starting at the top of the pile and continuing to a depth of 42 in., were evaluated for contents of moisture, pentosan, and cellulose. Cobs stored indoors, with adequate ventilation, equilibrated at about 15% moisture and showed no change in composition during the initial five months. Pentosan and cellulose loss during outdoor storage was insignificant during the initial six months, but losses of both pentosan and cellulose were appreciable after 12 months of outdoor storage. Similarly, Smith et al (1985) evaluated material from commercial cob piles (1,000–2,000 t) stored outside for six and 18 months. Also, they studied smaller corncob piles (3–18 t) stored outside for 6, 9, and 18 months. The greatest loss of dry matter and available energy occurred in the outer 3 ft. The interior cobs from the large commercial piles did not deteriorate

TABLE I
Selected Chemical and Physical Properties of Corncob Fractions[a]

Property	Whole Cob	Fractions	
		Woody	Pith/Chaff
Cellulose, %	41.2	47.1	35.7
Hemicellulose, %	36.0	37.3	37.0
As pentosan, %	(34.6)	(36.5)	(34.7)
As xylan, %	(30.0)	(31.6)	(30.1)
Pectins, %	3.1	3.2	3.2
Lignin, %	6.1	6.8	5.4
Gross energy, Kcal/kg	3,998	4,113	4,157
Absorption[b]			
Oil, %	259	100	500
Oil (water-saturated cob), %	259	100	500
Water, %	369	133	727
Water (oil-soaked cob), %	208	75	409

[a] Source: Foley and Vander Hooven (1981); used by permission.
[b] Liquid absorbed as weight percent of dry corncob absorbing material.

significantly when moisture content remained below 12%. However, deterioration was appreciable when moisture content exceeded 20%. For example, 43% of the pentosan and 36% of the cellulose were lost after 18 months. Thus, corncobs are best stored with some appropriate shelter or in large, tall piles where the top 3–4 ft are considered protective cover.

B. Processing, Conversion, and Uses

As mentioned previously, the unique properties of cobs provide for many industrial uses. Most require grinding or milling. Much initial corncob processing research was conducted at the Northern Regional Research Center (Clark and Lathrop, 1952). The least complex grinding operations are for producing cattle feeds, mulch, and litter. Hammer-mill grinding usually satisfies the requirements, but screening to size and content for more efficient uses is recommended. To meet requirements for the myriad of uses, many techniques have been developed. Foley and Vander Hooven (1981) reported that the present practice involves a series of crushers and hammer mills, subsequent drying to 10% moisture or lower, and further size reduction by attrition or roller mills. Segregation and classification are then achieved by a variety of screening configurations and sieve sizes. To avoid machinery damage, attention must be given to the removal of stones, tramp metal, and other such materials. Throughout the milling, cobs are aspirated to separate pith and chaff from the woody portion. Special attention must be taken to avoid dust explosions and fires while these dry, finely divided, charged materials are being processed.

Clark and Lathrop (1953) summarized many initial uses for processed cobs. Early demands were for use as animal and poultry bedding, as mulches and soil conditioners, and fiber and roughage in animal feeds. Pith fractions provide an exceptional carrier of micronutrients in animal feed. Corncobs have served as a useful component of composts for growing mushrooms (Foley and Vander Hooven, 1981). Because of the absorption properties of corncob, fractions of suitable particle size have served as effective carriers of insecticides and other agricultural chemicals. Corncobs have served as fillers in many industrial applications (e.g., adhesives, roofing materials, caulking compounds, explosives, and plastics).

Because of the exceptional hardness of woody ring particles, corncobs have been widely used as material for finishing milled metal parts, plastics, hard rubber products, and glassware. During World War II, many ordnance materials were polished, deburred, and cleaned with ground corncobs. Ground cobs are used as scrubbing agents in powdered handsoaps, and cob fractions are used in the cleaning and dressing of furs and pelts (Clark and Ashbrook, 1953).

Most uses described above depend on the mechanical and physical characteristics of corncobs. However, corncobs are a major source of furfural (McKillip and Sherman, 1980). This aldehyde-substituted furan is readily produced by acid hydrolysis of the pentosans contained in corncobs, oat hulls, and other crop residues. The resulting pentose sugars are dehydrated during continued heating in the presence of strong mineral acids, such as sulfuric acid. The furfural is steam-distilled and collected for direct use or conversion to other chemicals. Those of significant commerce are the hydrogenated products, tetrahydrofuran and tetrahydrofurfuryl alcohol. Furfural serves as an intermediate for the manufacture of various other industrial chemicals, including hexamethylenediamine (used in the production of nylons). Tetrahydrofurfuryl nitrate as a 7–8% blend with ethanol has been substituted for diesel fuel in Brazil (Anonymous, 1983).

Corncobs could be fractionated into their major chemical constituents (cellulose, hemicellulose, and lignin) as well as their minor constituents. However, recent interest has emphasized the prior technology for conversion of the polysaccharides to their constituent simple sugars, mostly glucose and xylose (Dunning and Lathrop, 1945). These sugars can serve their usual market, including energy and carbon sources for fermentation media.

III. STALK RESIDUE

Corn harvest, worldwide, averaged 127 million hectares annually from 1979 through 1983. A conservative estimate for stalk residue of 1.5 t ha^{-1} indicates that 190 million tonnes were available for possible use. At \$30/t, the world market for stalk residue was \$5.7 billion. However, good soil and water management practices require that some crop residue be returned to the soil and hence dictate practical limits for collection (Larson, 1979). Further, existing technologies for industrial utilization of stalk residue are not sufficiently competitive in most economies to establish a sustained market.

Cornstalks consist of an outer cortex (containing fiber bundles) and vascular fibrous bundles surrounded by pith (parenchyma tissue) (Fig. 1). The pith contributes nearly 25% of the total dry stalk weight (Whittemore et al, 1935). Analyses for major chemical constituents of the different stalk fractions reveal only small differences. Selected, representative data are shown in Table II. Schultz et al (1984) and Sloneker (1976) reported cornstalk cellulose contents of 35 and 29.3%, respectively, and glucose contents, available after acid hydrolysis of stalk samples, of 40.2 and 37.7%, respectively. These glucose values are in reasonable agreement with the cellulose data; however, the glucose values

TABLE II
Selected Chemical Characteristics of Cornstalks at Harvest

Characteristics	Whole Stalk	Cortex	Vascular Bundles	Pith
Glucose,[a] %	38[b] (40)[c]	21[d]	21[d]	27[d]
Galactose, %	1[b] (NR)[c]	21[d]	21[d]	27[d]
Mannose, %	1[b] (3)[c]	1[d]	1[d]	1[d]
Xylose, %	16[b] (17)[c]	23[d]	21[d]	18[d]
Arabinose, %	2[b] (3)[c]	1[d]	1[d]	1[d]
Total sugar, %[a]	56.8[b] (NR)[c]	67.1[d]	65.3[d]	73.7[d]
Cellulose, %	38.4[e]	39.3[e]	37.1[e]	37.9[e]
Pentosan, %	27.6[e]	25.9[e]	26.4[e]	27.7[e]
Lignin, %	34.3[e]	24.2[d] (33.5)[e]	22.7[d] (35.2)[e]	14.6[d] (32.0)[e]

[a] Sugars of polysaccharide constituents analyzed following hydrolysis.
[b] Sloneker (1976).
[c] Schultz et al (1984); NR = not reported.
[d] McGovern (1982); monosaccharides adjusted to basis sugar.
[e] Webber (1929).

reported by McGovern (1982) are in poor agreement with the cellulose data. The data reported (Table II) for arabinose, xylose, and mannose are in close agreement; however, the galactose values are significantly different. The glucose and galactose data need further attention. Although it may be fortuitous, the sum of the two values reported by McGovern is in close agreement with cellulose data. McGovern (1982) commented on the low glucose and high galactose values obtained for his Wisconsin-produced corn but offered no explanation. None of the researchers reported fructose; however, low levels could have been present.

Lignin data reported by Schultz et al (1984) and Sloneker (1976), 21.2 and 3.1%, respectively, and the data in Table II are rather diverse. The range of values may result from differences in methods of hydrolysis and analysis and from differences among varieties, relative maturity at harvest, harvest date, and associated rainwater leaching. For comparison, consider the effect of maturity and field storage on the chemical composition of kenaf (Clark et al, 1967; Bagby et al, 1975). As the plants matured, structural carbohydrates and lignin contents increased, with corresponding declines in soluble substances. Similarly, during the two months standing in the field following a killing frost, soluble constituents decreased and relative percentages of cellulose, hemicellulose, and lignin increased (Bagby et al, 1975).

A. Stalk Collection and Storage

Cornstalk residue may be baled, cubed, stacked loose, chopped, or bundled and stored indoors or outdoors. Numerous approaches were discussed over 50 years ago (Sweeney and Arnold, 1930). Even more techniques are available today. Large balers and mechanical stackers, developed for silage and hay, are available to pick up plant residue directly from the field for transport to processing or storage sites (Richey et al, 1982). However, only about one half of the available residue is readily collected.

Because of low bulk density, handling can be improved by compression into bales, cubes, or pellets. End use, storage, and transportation needs may determine the most effective package. Miles and Miles (1980) state that if truck transport and storage are the determining factors, there is no need to exceed 16 lb/ft$_3$ (256 kg m^{-3}), because that density results in reaching the limits for both volume and weight.

As for all crop residues, year-around processing requires appropriate storage. The method for storage depends on how the material will be used and into what form it is packaged. Heid (1984) reported that storing crop residues outdoors may result in losses of 5–50%. Covered or indoor storage resulted in about a 5% loss during six months. However, that experience is mainly with relatively small stacks or large bales stored in single tiers. Sloneker (1976) noted that spoilage is greater if stacks or large bales are allowed to touch one another during outdoor storage. Large-scale use of stalk residue would probably require stacking to minimize storage area. No doubt further storage tests are necessary.

B. Processing, Conversion, and Uses

More than 50 years ago, Arnold (1933) listed 60 products from cornstalks, cornhusks, and cornstalk pith ranging from fodder to chemicals and materials such as α-cellulose, furfural, oxalic acid, paper, wall board, and door mats. In the late 1920s, an insulating-board mill at Dubuque, Iowa, used cornstalks as a raw material before switching to more economical materials. In 1927, the Cornstalks Products Company, Danville, Illinois, started production of cornstalk paper pulp (Wells and Steller, 1943). That mill produced about 45 t per day of pulp, but after a few years, it discontinued production for economic reasons associated largely with problems of continued supply and preservation of cornstalks. The pulp was used in blends with other fibers to form various grades of paper of acceptable market quality. In the later 1950s, the Israelis were partially successful in collecting and pulping cornstalks (Tall, 1959). However, this venture proved too expensive for continued operation. As with sugarcane bagasse, efficient pith removal improves the properties of the pulp and the efficiency of pulping chemicals.

Building panels from cornstalks were developed by Lewis et al (1960). The waxy coating on cornstalks interfered with resin bonding; however, flaming the stalks burned off the wax to provide a readily bonded surface. Although the panels were not of high strength, Lewis and co-workers concluded that their panels were sufficiently strong to use in one and two-story buildings without additional framing support. This technology has received little industrial attention.

Cornstalks have been processed to dissolving pulps in excess of 95% purity (Abou-State et al, 1983) by a process involving prehydrolysis hydration for 1 hr in boiling water. Subsequent hydrolysis with 0.75% sulfuric acid for 6 hr at 100°C and a 20:1 liquid-to-solids ratio followed by alkaline pulping using kraft, soda, or sulfite conditions for 5 hr at 100°C produced a satisfactory dissolving pulp. The degree of polymerization (970–1,140) compared favorably with commercial dissolving pulps from sugarcane bagasse and softwood having degrees of polymerization of 920 and 860, respectively. The residual pentosan contents ranged from 4.0 to 5.1%, a level similar to that of the commercial bagasse pulp (4.5%) but somewhat greater than that of the softwood pulp (3.2%).

As stated above, cornstalks are used as animal feed; however, low digestibility limits intake and nutritive benefit. Although ruminants possess the hydrolytic means to convert structural polysaccharides (i.e., cellulose and hemicellulose) to readily metabolized sugars, the encapsulating effect of lignin covering the fibers and polysaccharides is believed to limit their conversion. Other interrelated and limiting characteristics include cellulose crystallinity, hydrogen bonding, limited available surface area, and low hydration. The feed value of cornstalks has been improved by treatments known to diminish the above characteristics. For example, Oji et al (1977) reported results from several alkaline treatments (2% sodium hydroxide plus 2% calcium hydroxide, 3% ammonia, and 5% ammonia). Effects of these treatments were evaluated in feeding tests with eight wether lambs. Organic matter intakes increased by 45–51%, and gross energy digestibility improved by 12–14% in comparison with the control.

Various chemical and physical treatments have been applied to improve the susceptibility of lignocellulosics, such as cornstalks, to enzymatic conversion (Duckworth and Thompson, 1983; Ladisch et al, 1983). Renewed interest during recent years has focused on production of the constituent sugars and their subsequent fermentation to ethanol and other chemicals. As for corncob conversion, acid hydrolysis techniques also have been investigated extensively for the conversion of cornstalks to their constituent sugars (Dunning and Lathrop, 1945; Bhandari et al, 1984).

IV. SUGAR AND ETHANOL FROM THE CORN PLANT

The stalk portion of the corn plant has the capacity to store considerable energy. Stalk solids consist mainly of carbohydrates. Cellulose, hemicellulose, and soluble sugars are the major constituents of the mature plant. Other compounds include pectic substances, other soluble solids, insoluble lignin, and cutin (wax).

Limited consideration has been given to systems that utilize the entire corn plant. An example is the use of the whole plant for ruminant animal feed as fodder or ensilage, converting plant bulk to a more storable form. Annual worldwide silage production averaged 103.9 million tonnes during 1974–1983. Development of additional systems for single or multiple-use purposes may be useful to maximize the profitability of the crop, as well as to determine its ability to compete with other crops. Sugar production from high stalk-sugar varieties is one such possibility.

A. Background—Cornstalk Juice Sugar

Cornstalks have been considered a source of sugar for almost as long as the grain has been used in modern times. Blackshaw (1912) stated that sugar in the juice of cornstalks was known in the 1500s and that molasses was made from the juice in America during the 1700s. He noted that renewed interest in the manufacture of sugar from the juice of cornstalks resulted from a paper presented to the Linnean Society in 1843 by Prof. Croft. In 1850, a factory was established in France for sugar production from cornstalk juice, but it could not compete with the beet sugar industry.

Sucrose purity in soluble solids of cornstalk juice is lower than that of beets and cane, often resulting in poor crystallization. Collier (1884) recognized cornstalks as a promising source for sugar but viewed them as inferior to sorghum. Blackshaw's results (1913) on juice quantity and quality were similar to those of Collier, as were the findings of Clark (1913), but they stated that corn compared very favorably with sorghum as a sugar-producing plant. Stewart (1906, 1912) patented processes for making cornstalk sugar, even though he focused mainly on complete utilization of the plant for simultaneous production of sugar, cellulose, and alcohol.

Cornstalk juice for syrup production apparently was largely ignored by early investigators, although it was mentioned briefly by some. A detailed study of this alternative, using sweet-corn stalks, was begun in 1921 by the University of Minnesota in cooperation with the Minnesota State Canners' Association. This effort (Willaman et al, 1924) did not result in optimism about the development of a syrup-making operation in conjunction with the sweet-corn canning industry. All commercial corn syrups and sweeteners have been produced by hydrolysis of starch obtained from corn grain by the wet-milling industry beginning in the mid-1800s (Chapter 17). Stalk juice for making syrup has never been a competitive option, although sorghum syrup has been widely produced in many U.S. rural communities, including those of modern-day Amish farmers.

The carbohydrate content of the stalk juices from sweet-corn plants after harvest of the ears was reported by Gore (1947) in terms of alcohol yields. The estimated yields, all less than 950 L ha^{-1} (100 gal/acre), were comparable to conversion of approximately 2.5 t ha^{-1} (40 bu/acre) of grain yields at a conversion rate of 400 L of ethanol per tonne of grain. D'Ayala Valva et al (1980) estimated alcohol yields to be twice as large (2,000 L ha^{-1}) from juice sugar of dent hybrid cornstalks. Yields of this magnitude compare favorably with those of sweet sorghum, an unexpected result, considering that corn hybrids were developed for grain yield rather than for stalk sugar.

The energy crisis of the early 1970s stimulated renewed interest in the evaluation of corn as a carbohydrate source in the production of ethanol for fuel. Most interest centered around use of grain because the technology was already well developed. However, the sugar content of succulent green stalks of corn and other crops was investigated to determine its production and process efficiency for conversion to ethanol.

Inbred C103, identified by Singleton (1948) as having high stalk-sugar content, was suggested for use in silage hybrids to improve the feeding value. Smith (1963) later concluded that corns with high stalk sugar have no superiority over conventional corns in silage production, because free sugars are not found to improve silage palatability. Collings et al (1979) confirmed that soluble sugars decrease to less than 1% during the first 13 days of ensilage. Hybrids developed from Singleton's material (1948) were incorporated into the high-stalk-sugar corn developed by Blanco et al (1957). Cunningham et al (1980) found these lines to have stalk sugar contents (oven-dry basis) of 30–34%, about two to three and a half times greater than those of standard varieties. Blanco and co-workers developed their hybrids for the multiple purpose of producing grain and stalk products.

B. Plant-Sugar Relationships

The sugar contents of cornstalks have frequently been measured as a function of their relationship to other important agronomic traits. DeTurk et al (1939) reported that stalk sugars provided some protection against cold injury and proposed that this was probably attributable to a lowering of the freezing point of plant tissues. They also reported increased resistance to fungal attack by those plants having the highest stalk-sugar levels. A more detailed study by Craig and Hooker (1961) substantiated this early observation and also attributed senescence of pith tissue to low levels of sugar.

Sugar content in the normal cornstalk generally increases steadily throughout the life of the plant until the middle to late stages of the grain-filling period. At five to six weeks after pollination, depending on the hybrid, stalk-sugar content decreases substantially as carbohydrate reserves are transferred to the ear sink (Welton et al, 1930; Van Reen and Singleton, 1952). Stalk-sugar accumulation was reduced when leaves were removed during any stage of plant growth (Asanuma et al, 1967), but removal of ears or prevention of pollination had the opposite effect. Increases of 30–50% in sugar content or percent soluble solids occur in the stalks when plants are detasseled to prevent grain formation (Ertugrul et al, 1964), when earshoots are covered to prevent pollination (Sayre et al, 1931), or when plants are barren (Hume and Campbell, 1972). Male sterility, detasseling, or some other method of seed-development suppression is necessary if the intent is to harvest the stalks for the juice sugars alone. However, harvesting the corn plant for a single carbohydrate product may not be the best way to maximize yields.

Sugar concentration in the stalk tends to increase from the base (Welton et al, 1930; Van Reen and Singleton, 1952). Although the percentage of sugar in the juice increases from the base toward the top of the stalk, the total sugar content decreases because of the relative sizes and juice contents of the internodes. A typical distribution of soluble solids ranging from 13–16% in various internodes of more recently developed hybrids is shown in Fig. 2. One notable deviation from the findings of previous authors is the higher concentration of soluble solids in the two basal internodes in comparison to those higher on the stalk, near the ear. This increase might be attributable to recent selection to increase stalk rot resistance, a character known to correlate with sugar content. Such changes in sugar content are certainly not unique, in view of the success of Blanco et al (1957) in changing stalk sugar contents through selection. Van Reen and Singleton (1952) concluded that variable stalk-sugar contents found among dent inbreds could not be accounted for by differences due to grain production alone. More recent evaluations of hybrids and other cultivars (Widstrom et al, 1984) confirm that the available variation among genetic types is large enough to facilitate substantial progress in selection for high sugar in the stalks.

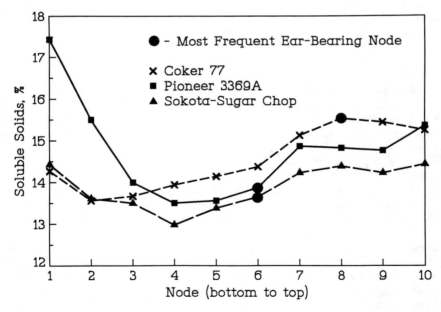

Fig. 2. Soluble solids (%) in cornstalk juice expressed six weeks after pollination from the bottom 10 nodes of three hybrids grown in 1981, 1982, and 1983. Each data point is the average of 180 individual samples.

C. Measurements for Comparisons

The sugar industry has developed various methods for determining sugar content or yield of plant juices. These methods usually give values for percent soluble solids, which are measured with a refractometer when quick relative values are desired. For more accurate measurement of sugars or of sucrose alone, polarimeter readings are used. For cornstalks, as for sugar cane, sugar contents of expressed or extracted juices may be readily determined using a refractometer for most breeding and field evaluation purposes (Betterton and Pappelis, 1964; Campbell and Hume, 1970). Refractometric readings measure percent soluble solids and are convertible to total sugar contents by referring to tables correlating sugar content and refractometer values.

Problems are encountered when comparisons are attempted among different crops. D'Ayala Valva et al (1980) suggested that easy comparisons could be made between cornstalks and sugarcane because the same processing facilities can be used for both to produce sugar, syrup, or alcohol. However, crops differ in their production requirements. For example, University of Michigan researchers have shown that corn silage and sugar beets require about the same production energy inputs, whereas potatoes require five to six times as much energy as either (Myers et al, 1981). Even the same crop may have varying requirements in terms of its output-input energy ratio. For corn produced in Mexico, this ratio can vary from 4.25 to 128.2, depending on whether corn grain is produced by hand-power or by the use of oxen (Pimentel and Burgess, 1980). The same ratio varies from 0.78 to 4.21 for irrigated corn in the United States. A similar energy analysis of sugarcane for alcohol production gave ratios ranging from 0.9 to 1.8, depending on whether crop residues or fossil fuels were used for processing energy (Hopkinson and Day, 1980). Regardless of the method used for comparing crops, most people devising systems now recognize the need for considering multiple uses, as did Blanco et al (1957). People have more awareness now of the need to consider all production opportunities that will eventually decrease the cost of energy delivered to the consumer (Lipinsky, 1978a), including conversion of the crop to ethanol.

D. Corn Compared with Other Crops as Ethanol Sources

The technology that will allow crop comparisons to be made on the basis of total carbohydrate production, reduced to yields of sugar or ethanol, will require improved processes for enzymatic hydrolysis of cellulose (Spano et al, 1980). A uniform system is needed to make easy comparisons among crops with high carbohydrate production potential such as pearl millet (Rao et al, 1982), sweet sorghum (Broadhead et al, 1978), sugarcane (Lipinsky, 1978b), and sugar beets and corn (Hills et al, 1981). Other crops like potatoes and Jerusalem artichoke, could be added to the list when production of total carbohydrates is considered. Some complex selection practices, such as those used for evaluating sugarcane varieties (Punia et al, 1982), could also be greatly simplified if the quantity of carbohydrate or energy measurement, rather than quality, received the main emphasis.

Breeding for high sugar production in the stalks of corn is an attainable goal (Widstrom et al, 1984). If that were done, cornstalks could be integrated into the existing processing systems for sugarcane and sugar beets, as suggested for sweet sorghum by Lipinsky et al (1979). Corn compares favorably with other crops in producing fermentable sugars or alcohol, even when only the grain production is considered (Hills et al, 1981, 1983) and has resulted in a large fuel ethanol industry in the United States (Chapter 19). Grain yields in the study by Hills et al (1981) were high at approximately 250 bu/acre (ethanol, 609 gal/acre) but were achievable under intensive cultural conditions and high energy input. The ethanol yield for corn grain was 74% of that for sugar beets and nearly equal to that of sweet sorghum. Work at Tifton, Georgia, indicates (Table III) that at grain yields of about 8 t ha^{-1} (about 300 bu per acre) at six weeks after pollination, over 40% additional ethanol yield can be added because of the remaining soluble sugar contributed by the stalk. Comparisons were made in ethanol yields per plant to provide a common denominator for combining stalk-sugar and grain production.

Additional factors must be considered if valid comparisons among crops are to be made. First, the adaptability of the crop to a wide growing area is an important consideration. Sugar beets and sugarcane both have a somewhat limited growing region in the United States, whereas corn and sorghum are widely grown and have a broad area of adaptation. In fact, two crops of corn can be grown per year over a rather wide area of the southeastern United States (Widstrom and Young, 1980). Although grain yields of a second crop are less than those of the first crop, dry-matter or stalk yields are large enough to make it a respectable forage option.

The length of time for a crop to mature for optimum carbohydrate production is another important consideration. This provision allows any crop to be compared with another on the basis of carbohydrate yield per unit area per unit time. Blanco and Blanco (1960) showed a substantial advantage for corn over sugar beets using this method. When adjusted to a net energy value (net energy equals total energy yield minus energy used in production) and to the same units of area and time, appropriate assessments of the crops can be made. Table IV illustrates the great differences among several crops in the time required for production. Its wide adaptation, short growing season, and the many options available for its use make corn a strong competitor with any crop when

TABLE III
Total Sugar and Starch Production (Reported as Potential Alcohol) of Stalks and Grain of Three Hybrids Grown in Georgia for Three Years and Harvested by Four Procedures

Hybrid	Ethanol Yields (ml/plant) of Harvesting Procedures[a]				
	Method 1	Method 2	Method 3	Method 4	Mean[b]
Coker 77	41	38	66	65	53 a
Pioneer 3369A	33	27	67	65	48 b
Sokota Sugar Chop	24	21	62	57	41 c
Mean[b]	33 y	29 z	65 x	63 x	

[a] Methods 1 and 2: harvest at six weeks postpollination of stalks for juice after covering ear (Method 1) or removing ear (Method 2) to prevent pollination. Method 3: simultaneous harvest at six weeks postpollination of grain and juice of pollinated plant. Method 4: harvest of mature grain (30% H_2O) at approximately six weeks postpollination and subsequent harvest of stalk juice two weeks later.
[b] Marginal means not followed by the same letter are significantly different at the 1% level of probability.

compared in terms of net energy produced per unit area over a specified length of time. This is apparently one reason why corn has become a dominant carbohydrate-producing crop in temperate climates.

V. SUMMARY

Corncobs satisfy numerous industrial applications, which result in the domestic processing of over a million tonnes annually. Furfural production consumes nearly 70% of the processed cobs, and granular products provide the next largest market. Since introduction of the picker-sheller and corn combine, corncobs are less readily available, and the grain no longer must support the collection and disposal of the cob. Consequently, corncob products must bear the cost of their separate collection, storage, and conversion before an enduring market can develop.

Cornstalk residues have been evaluated for a variety of products and have enjoyed short-term industrial markets as fiber resources for pulp, paper, and board products. However, the stalk residue was not economically competitive with other fiber resources. Techniques to improve digestibility and nutritional value of cornstalks are of continuing interest. Because of the recent petroleum shortages and the approaching petroleum dearth, renewed interest has emerged in the conversion of cornstalks to industrial feedstocks and chemicals through sugar production.

Cornstalk juices have been used to make molasses and other sweeteners since the 1700s. Cornstalk sugar has lower purity than that of sugarcane or sugar beets and is therefore used to make syrup. The sugar content of cornstalks can average as high as 30%, with the percentage increasing from the lower nodes toward the top. Some corns are much higher in stalk sugar than others, a trait that can be further increased because it is highly heritable.

Interest in maize as an energy-producing crop was renewed during the 1970s because of the energy crisis. For purposes of energy comparisons among crops, total carbohydrate production is an effective comparative measure. The comparisons should be based on net energy, with due regard for efficiency, and should also take into consideration the land area needed and the time required for production. Crops should be evaluated on a systems basis, that is, production options and versatility of product uses are important to maintain. Corn is very competitive as a high carbohydrate-yielding plant species when one considers 1) the short growing season it requires (two crops per year in many areas), 2) the multiple options available for its uses, 3) its broad area of adaptation, and 4) its efficient photosynthetic production of usable energy.

TABLE IV
Length of Growing Season and Area of Adaptation
for Several High Energy-Producing Crops

Crop	Length of Growing Season (mo)	Area of Adaptation in the United States
Corn	3–4	General
Sugar beets	At least 6	North
Sugarcane	12	Extreme south
Sweet sorghum	4–5½	South and central

LITERATURE CITED

ABOU-STATE, M. A., ABD EL-MEGEID, F. F., and NESSEEM, R. I. 1983. Ind. Eng. Chem. Prod. Res. Dev. 22:506-508.

ANONYMOUS 1983. A new substitute for diesel fuel. Chem. Week 133(7):40.

ARNOLD, L. K. 1933. Utilization of agricultural wastes and surpluses. Bull. 113. Iowa State College, Ames. 31 pp.

ASANUMA, K., NAKA, J., and TAMAKI, K. 1967. Effects of topping on the growth, the translocation and accumulation of carbohydrates in corn plants. Proc. Crop Sci. Soc. Jpn. 36:481-488.

BAGBY, M. O., ADAMSON, W. C., CLARK, T. F., and WHITE, G. A. 1975. Kenaf stem yield and composition: Influence of maturity and field storage. Pages 69-72 in: Tappi CA Rep. 58, Non-wood Plant Fiber Pulping Progress Rep. 6. Tech. Assoc. of the Pulp and Paper Industry, Atlanta, GA.

BETTERTON, H. O., and PAPPELIS, A. J. 1964. A method for rapid extraction of sugar from corn stalk tissue. Trans. Ill. Acad. Sci. 57:39-41.

BHANDARI, N., MacDONALD, D. G., and BAKHSHI, N. N. 1984. Kinetic studies of corn stover saccharification using sulfuric acid. Biotechnol. Bioeng. 26:320-327.

BLACKSHAW, G. N. 1912. The sugar content of maize stalks. S. Afr. J. Sci. 8:269-273.

BLACKSHAW, G. N. 1913. The sugar content of maize stalks. S. Afr. J. Sci. 9:42-48.

BLANCO, M., and BLANCO, J. L. 1960. Breeding for high sugar in the stalk at maturity and related subjects. Pages 25-28 in: Eucarpia. Rep. Congr. Eur. Assoc. Res. Plant Breeding. Mision Biologica de Galicia, Eontevedra, Spain.

BLANCO, M., BLANCO, J. L., and VEIGUINHA, A. S. 1957. Obtencion de hybridos de maiz de tallo azucarado, de doble aprovechamiento—grano y planta—y estudio comparativo de su valor industrial, agricola y economico. Genet. Iber. 9:1-102.

BROADHEAD, D. M., FREEMAN, K. C., and ZUMMO, N. 1978. "Wray"—A new variety of sweet sorghum for sugar production. Miss. Agric. For. Exp. Stn. Res. Rep. 3 pp.

CAMPBELL, D. K., and HUME, D. J. 1970. Evaluation of a rapid technique for measuring soluble solids in corn stalks. Crop Sci. 10:625-626.

CLAAR, P. W., II, CALVIN, T. S., and MARLEY, S. J. 1981. Economic and energy analysis of potential corn residues harvesting systems in agricultural energy. Pages 273-279 in: Agricultural Energy. Vol. 2, Biomass Energy Crop Production. Am. Soc. Agric. Eng., St. Joseph, MI.

CLARK, C. F. 1913. Preliminary report on sugar production from maize. Circ. 111. U.S. Dept. Agric., Bur. Plant. Ind., Washington, DC. 9 pp.

CLARK, T. F., and ASHBROOK, J. W. 1953. Why the emphasis on corn cobs? Chem. Dig. 12:9-10.

CLARK, T. F., and LATHROP, E. C. 1952. Dry grinding of agricultural residues, a new industrial enterprise. U.S. Dept. Agric., North. Reg. Res. Center, Peoria, IL.

CLARK, T. F., and LATHROP, E. C. 1953. Corncobs—Their composition, availability, agricultural and industrial uses. AIC 177 (rev.) U.S. Dept. Agric., North. Reg. Res. Center, Peoria, IL.

CLARK, T. F., UHR, S. C., and WOLFF, I. A. 1967. A search for new fiber crops. X. Effect of plant maturity and location of growth on kenaf composition and pulping characteristics. TAPPI 50(11):52A-56A.

COLLIER, P. 1884. Sorghum, Its Culture and Manufacture. Robert Clarke and Co., Cincinnati, OH.

COLLINGS, G. F., YOKOYAMA, M. T., COLLINGS, L. L., and WISELEY, S. L. 1979. Preliminary study of the soluble sugars available for fermentation during ensilement of whole chopped corn plant. Mich. Agric. Exp. Stn. Res. Rep. 388:113-115.

CRAIG, J., and HOOKER, A. L. 1961. Relation of sugar trends and pith density to Diplodia stalk rot in dent corn. Phytopathology 51:376-382.

CUNNINGHAM, R. L., BAGBY, M. O., and JUGENHEIMER, R. W. 1980. Sweet-stalked corn. Trans. Ill. Acad. Sci. 70:30-34.

D'AYALA VALVA, F., PATERNIANI, E., and DE OLIVEIRA, E. R. 1980. Evaluation of sugar content in corn stalks (*Zea mays* L.) for alcohol production. Maydica 25:185-197.

DeTURK, E. E., EARLEY, E. B., and HOLBERT, J. R. 1939. Chemistry, disease and cold injury of corn related. Pages 62-64 in: A Year's Progress in Solving Farm Problems of Illinois. F. J. Keilholz ed. Ill. Agric. Exp. Stn., Urbana.

DUCKWORTH, H. E., and THOMPSON, E. A., eds. 1983. Symposium on Ethanol from Biomass. Oct. 13–15, 1982. The Royal Society of Canada, Int., Winnipeg. 654 pp.

DUNNING, J. W., and LATHROP, E. C. 1945. The saccharification of agricultural residues. A continuous process. Ind. Eng. Chem. 37(1):24-29.

DUNNING, J. W., WINTER, P., and DALLAS, D. 1948. The storage of corncobs and other agricultural residues for industrial use. Agric. Eng. 29(1):11-13, 17.

ERTUGRUL, H., TOSUN, F., and CATALTAS, I. 1964. Sugar content of detasseled corn stalks. Seker 14:26-30.

FOLEY, K. M. 1978. Chemical properties, physical properties, and uses of the Anderson's corncob products and supplement. The Anderson's Cob Division Processing Group, Maumee, OH.

FOLEY, K. M., and VANDER HOOVEN, D. I. B. 1981. Properties and industrial uses of corncobs. Pages 523-543 in: Cereals: A Renewable Resource, Theory and Practice. Y. Pomeranz and Lars Munck, eds. Am. Assoc. Cereal Chem., St. Paul, MN.

GORE, H. C. 1947. Alcohol yielding power of succulent corn stalk juice. Fruit Prod. J. Am. Food Manuf. 24:46, 61.

HEID, W. G., Jr. 1984. Turning Great Plains crop residues and other products into energy. Agric. Econ. Rep. 523. U.S. Dept. Agric., Econ. Res. Serv., Washington, DC.

HILLS, F. J., JOHNSON, S. S., GENG, S., ABSHAHI, A., and PETERSON, G. R. 1981. Comparison of high-energy crops for alcohol production. Calif. Agric. 35(Nov.-Dec.):1-16.

HILLS, F. J., JOHNSON, S. S., GENG, S., ABSHAHI, A., and PETERSON, G. R. 1983. Comparison of four crops for alcohol yield. Calif. Agric. 37(Mar.-Apr.):17-19.

HOPKINSON, C. F., Jr., and DAY, J. W., Jr. 1980. Net energy analysis of alcohol production from sugarcane. Science 207:302-304.

HUDSON, W. J. 1984. Biomass energy and food—Conflicts? Pages 207-236 in: Food and Energy Resources. D. Pimentel and C. W. Hall, eds. Academic Press, Inc., New York.

HUME, D. J., and CAMPBELL, D. K. 1972. Accumulation and translocation of soluble solids in corn stalks. Can. J. Plant Sci. 52:363-368.

LADISCH, M. R., LIN, K. W., VOLOCH, M., and TSAO, G. T. 1983. Process considerations in the enzymatic hydrolysis of biomass. Enzyme Microb. Technol. 5(3):82-102

LARSON, W. E. 1979. Crop residues: Energy production or erosion control? J. Soil Water Conserv. 34(2):74-76.

LATHROP, E. C. 1947. Cobs enter industry. Pages 734-738 in: Yearbook of Agriculture. U.S. Dept. Agric., Washington, DC.

LEWIS, R. L., DALE, A. C., and WHISTLER, R. L. 1960. Building panels from cornstalks. Res. Bull. 690. Agric. Exp. Stn., Purdue University, Lafayette, IN.

LIPINSKY, E. S. 1978a. Fuels from biomass: Integration with food and materials systems. Science 199:644-651.

LIPINSKY, E. S. 1978b. Sugarcane versus corn versus ethylene as sources of ethanol for motor fuels and chemicals. Proc. Am. Soc. Sugarcane Technol. 7:152-162.

LIPINSKY, E. S., KRESOVICH, S., McCLURE, T. A., JACKSON, D. R., LAWHON, W. T., KALYONCU, A. A., and DANIELS, E. L. 1979. Pages 195-196 in: Sugar Crops as a Source of Fuels. Vol. 1, Agricultural Research. Battelle, Columbus, OH.

McGOVERN, J. N. 1982. Bonding in papyrus and papyrus-like material? TAPPI 65(5):159-162.

McKILLIP, W. J., and SHERMAN, E. 1980. Furan derivatives. Pages 499-527 in: Kirk-Othmer Encyclopedia of Chemical Technology, 3rd ed., Vol. 11. Martin Grayson ed. John Wiley & Sons, New York.

MILES, T. R., and MILES, T. R., Jr. 1980. Densification systems for agricultural residues in thermal conversion of solid wastes and biomass. Pages 179-193 in: Thermal Conversion of Solid Wastes and Biomass. J. L. Jones and S. B. Radding, eds. Am. Chem. Soc., Washington, DC.

MYERS, C. A., STOUT, B. A., BAAS, D., SURBROOK, S., NOTT, S., ROSENBERG, S., SCHAUER, W., SCHWOB, G., SIONAKIDES, G., HELSEL, Z., and MEINTS, V. 1981. Michigan farm energy audit completed. Agric. Energy 2:1-3.

OJI, U. I., MOWAT, D. N., and WINCH, J. E. 1977. Alkali treatment of corn stover to increase nutritive value. J. Anim. Sci. 44:798-802.

PEART, R. M., ZACK, H. R., and DOERING, O. C. 1981. Corn cob gasification for corn drying. Pages 338-341 in: Agricultural Energy. Vol. 2, Biomass Energy Crop Production. Am. Soc. Agric. Eng., St. Joseph, MI.

PIMENTEL, D., and BURGESS, M. 1980. Energy inputs in corn production. Pages 67-84 in: CRC Handbook of Energy Utilization in Agriculture. D. Pimentel, ed. Chem. Rubber Co., Cleveland, OH.

PUNIA, M. S., HOODA, R. S., and PARODA, R. S. 1982. Discriminant function analysis for sucrose content in sugarcane. Indian J. Genet. 42:293-295.

RAO, S. A., MENGESHA, M. H., and SUBRAMANIAN, V. 1982. Collection and preliminary evaluation of sweet-stalk pearl millet (*Pennisetum*). Econ. Bot. 36:286-290.

RICHEY, C. B., LILJEDAHL, J. B., and

LECHTENBERG, V. L. 1982. Corn stover harvest for energy production. Trans. ASAE 25:834-839, 844.

SAYRE, J. D., MORRIS, V. H., and RICHEY, F. D. 1931. The effect of preventing fruiting and of reducing the leaf area on the accumulation of sugars in the corn stem. J. Am. Soc. Agron. 23:751-753.

SCHULTZ, T. P., TEMPLETON, M. C., BIERMANN, C. J., and McGINNIS, G. D. 1984. Steam explosion of mixed hardwood chips, rice hulls, corn stalks, and sugar cane bagasse. J. Agric. Food Chem. 32:1166-1172.

SINGLETON, W. R. 1948. Sucrose in the stalks of maize inbreds. Science 107:174.

SLONEKER, J. H. 1976. Agricultural residues, including feedlot wastes. Biotechnol. Bioeng. Symp. 6:235-250.

SMITH, L. H. 1963. High-sugar corns show no promise as silage. Crops Soils 16:15.

SMITH, R. D., PEART, R. M., LILJEDAHL, J. B., BARRETT, J. R., and DOERING, O. C. 1985. Corncob property changes during outside storage. Trans. ASAE 28:937-942, 948.

SPANO, L., TASSINARI, T., RYU, D. D. Y., ALLEN, A., and MANDELS, M. 1980. Producing ethanol from cellulosic biomass. Pages 62-81 in: Biogas and Alcohol Fuels Production. J. Goldstein, ed. The J. G. Press, Emmaus, PA.

STEWART. F. L. 1906. Method of making maize-sugar. U.S. patent 811,523. Jan. 30.

STEWART, F. L. 1912. Process of preparing sugar-cane and maize juices. U.S. patent 1,018,994. Feb. 27.

SWEENEY, O. R., and ARNOLD, L. K. 1930. Cornstalks as an industrial raw material. Bull. 98. Iowa State College, Ames. 48 pp.

TALL, B. 1959. Paper from cornstalks. Sci. News Lett. March 7:149.

VAN REEN, R., and SINGLETON, W. R. 1952. Sucrose content in the stalks of maize inbreds. Agron. J. 44:610-614.

WEBBER, H. A. 1929. Cellulose from cornstalks. Ind. Eng. Chem. 21(3):270-275.

WELLS, S. D., and STELLER, R. L. 1943. Utilization of cornstalks in the manufacture of paper and paperboard. Paper Trade J. 116(15):45-51.

WELTON, F. A., MORRIS, V. H., and HARTZLER, A. J. 1930. Distribution of moisture, dry matter, and sugars in the maturing corn stem. Plant Physiol. 5:555-564.

WHITTEMORE, E. R., OVERMAN, C. B., and WINGFIELD, B. 1935. Separation of cornstalks into long fibers, pith, and fines. Misc. Publ. M 148. Natl. Bureau of Standards, Washington, DC. 8 pp.

WIDSTROM, N. W., and YOUNG, J. R. 1980. Double cropping corn on the coastal plain of the southeastern United States. Agron. J. 72:302-305.

WIDSTROM, N. W., BAGBY, M. O., PALMER, D. M., BLACK, L. T., and CARR, M. E. 1984. Relative stalk sugar yields among maize populations, cultivars, and hybrids. Crop Sci. 24:913-915.

WILLAMAN, J. J., BURR, G. O., and DAVISON, F. R. 1924. Corn-stalk Sirup Investigations. Bull. 207. Univ. Minn. Agric. Exp. Stn., St. Paul. 58 pp.

INDEX

Acetic anhydride, 491
Acid detergent fiber, 77
Adenine, 74
Adenosine, 74, 265, 266, 267, 268
Adhesives, from corn by-products
  corrugating, 368
  dextrins, 486
  general, 369
  plywood, 370
Adipic acid, 488
ADPG pyrophosphorylase, 265
Aeration, 115–117
Aflatoxin
  in by-products, 170
  destruction of, 166
  detection, 166–167
  in feeds, 172
  fungal species producing, 165–166
  government regulations concerning, 129, 166, 456
  in grain dust, 104
  incidence, 164
  isomers, 166
  in storage experiments, 161
  toxicological effects, 166, 456
Airflow, through corn
  fans, 99–100
  fines, effect of, 98–99
  resistance to, 97
  velocity, 97–98
Albumins, 73, 275, 289–290, 293
Alcohol industry, 214–215, 553, 572
Aleurone layer, 34, 65, 68
Alkaline-cooked products
  hominy, 418
  masa, 410–411, 414, 416
  nixtamalized corn flours, 416–417

  nutritional values, 413, 424–426
  other Mexican types of foods, 410–411
  tortillas, 410, 411, 413, 416, 417, 424, 426
Alkaline-cooking process, 422-423
Aluminum phosphide, 193, 194, 197
Amino acids
  analysis, 299–301
  in animal diets, 451
  content, in corn proteins, 284–287, 290, 291, 292, 293, 449–451
  in distiller's by-products, 469
  sequences, 287–288
  sulfur-bearing, 466
  in wet-milled feeds, 459, 463–464, 465
Ammonia
  in aflatoxin destruction, 166
  as mold inhibitor, 165
  treatment of corn stalks, 581
Amylopectin
  in corn starch, 480, 560
  structure, 261, 262, 479–480, 504
  in sweet corn, 433
Amyloplast, 262, 265
Amylose
  in corn starch, 264, 340, 480, 491, 560
  lipid complexes, 341
  structure, 262, 479, 504
  in sweet corn, 433
Amylose extender, *see* Mutants, endosperm
Amylose-amylopectin ratio, 45
Angle of repose, 114
Anthracnose, 39
Arabinoxylans, 260

Arepa, 406–407
Argentina, 132–133, 221, 233, 234, 235, 239, 242, 243
Ash, 72
Australia, 235

Bacterial species
  Acetobacter spp., 406
  Actinoplanes
    missouriensis, 514
  Aerobacter sp., 514
  Arthrobacter spp., 514, 520
  Bacillus spp., 506, 514
    acidopullulyticus, 511
    cereus, 520
    coagulans, 514
    licheniformis, 506, 509, 520, 562
    megaterium, 520
    stearothermophilus, 506, 509, 519
    subtilis, 506, 509, 562
    thuringiensis, 196
  Flavobacterium
    arborescens, 514
  Lactobacillus spp., 406, 559
  Pseudomonas stutzeri, 520
  Streptomyces spp., 514
  Zymomonas mobilis, 565
Bagasse, 581
Basal endosperm transfer cells, 255, 268
BCFM, *see* Broken corn and foreign material
Beef cattle feeding, 459–462, 465–466
  rumen by-pass, 459, 465, 466
Beer
  components, 560
  low-calorie, 360
  production, 358–359, 560
  traditional, 406

Beetles
  confused flour, 186, 197
  corn sap, 187, 190
  flat grain, 186, 197
  foreign grain, 187, 189, 190
  hairy fungus, 187, 190
  larger black flour, 187, 189
  minute, 187
  red flour, 186, 197
  red-horned grain, 187
  rusty grain, 186, 197
  sawtoothed grain, 186
  two-banded fungus, 187, 190
Beverage alcohol
  aging, 571–572
  corn quality for, 169
  distillation, 570
  volume, annual, 572
  whiskey, 367, 570, 571–572
Biological control, of insects, 196
Biomass, for drying, 95
Black layer
  as maturity indicator, 10, 34, 147
  structure, 69
Bleaching
  of crude oil, 541
  of kernels for grading, 158
  of starch, 486–487
Blending
  cause of spoilage, 92
  to equalize moisture, 92
  to make grade, 130, 134
  in marketing, 130–131
Blue-eye mold, 129, 156, 157
Bran
  in animal feeds, 370, 458, 461
  composition, 258–259
  definition, 68
  in foods, 77, 370
  production, 354
Branching enzyme, starch, 265, 266–267
Brazil, 243, 536
Breakage susceptibility
  breeding for improvement, 46–47
  from drying, 85, 96–97, 156
  in handling, 104
  measurement, 154–156
  reduction and prevention, 156
  from temperature change, 115
Breakfast cereals, ready-to-eat, 402–405
Breeding
  for kernel properties
    carbohydrate composition, 44–45
    lipid composition, 319–320, 325, 334, 336–337
    mold resistance, 165
    oil content, 314
    protein content and composition, 46, 301–340
    reduced breakage, 46–47, 156
    sweet corn, 432–437
  for plant properties
    disease resistance, 38–39, 165, 439–440
    high-sugar stalks, 585–586
    inbred lines, 43
    insect resistance, 39, 438
    recurrent selection, 40–42
    selection methods, 43–44
    systems, 39–43
Brewer's grits, 356
Brewing
  adjuncts, 213, 358, 359–360, 525, 526, 528, 556, 560
  process, 358–359
Broken corn and foreign material
  amount, at the farmstead, 85
  definition, 128, 134
  fermentation substrate, 555
  measurement, 134–135
  overestimation, in sampling, 127
  removal, 110
Building materials, from corn, 366, 370, 581 (see also Utilization, industrial)
Butylated hydroxyanisole, 538, 543, 546
Butylated hydroxytoluene, 538, 543, 546

Calcium
  in by-product feeds, 460
  for corn growing, 19, 20, 23
  in kernel, 455
  in lime-treated foods, 424, 426
Canada, 216, 235
Captan, 168, 441
Carbohydrates
  biosynthesis, 265–268
  cell walls, 68, 77–78, 259
  cellulose, 259, 577, 579
  di- and trisaccharides, 257–258
  distribution in kernel, 76–78, 256–257
  metabolic intermediates, 258
  monosaccharides, 257
  pentosans, 77, 259–260, 577, 578
  phytoglycogen, 260, 261, 433, 436
  starch, 261–269 (see also Starch)
  sugar alcohols, 258 (see also Sorbitol)
  transfer into kernel, 254–255
  water-soluble polysaccharides, 260–261
Carbon dioxide
  in corn storage losses, 160–161
  from ethanol production, 555, 568
  in inert gas storages, 195
  measurement in grain storage, 164
Carotene, 73–74, 333, 334, 453–454, 548
Carotenoids
  breeding for, 334
  in corn oil, 543, 545
  distribution in kernel, 333–334
  in feeds, 370, 453–454
  nutritional value, 453–454
Carryover stocks, 202, 245, 246, 248
Cell walls, 68, 70, 77–78, 259, 260
Cellulose
  in corncobs, 577
  in cornstalks, 579
  in kernel parts, 77, 259
  structure, 259
Cementing layer, of germ/endosperm, 62
Centrally planned economies, 238, 240
Centrifugation

Index / 593

basket centrifuge, 392
disk-nozzle centrifuge, 385, 392
hydroclones, 381, 385
in oil refining, 541
screen centrifuge, 384
solid-bowl type, 383
of starch, 385
Chicago Board of Trade, 241–242
China (PRC), 217, 238, 536
Chloropicrin, 193, 194, 195, 197
Cholesterol, 546, 547
Choline, 74, 455
Chromosomes, 31, 32, 33, 34
CIMMYT, 58
Citric acid
in corn oil refining, 543, 546
dry-milling product, 358, 367
by glucose fermentation, 560
Classes, of corn, 127–128
Cleaning, of corn
for dry milling, 353
effect of, for drying, 98
for wet milling, 110
Climates, influence of
on fatty acids, 319
on oil content, 313
on production, 11–13
Coarse-grain markets, 230–232
Cob
burning, 61, 577
collection, 575–578
colors, 423
processing, 578
production, 575
storage, 577
structure, 575–576
uses, 578–579
Color
carotenoids
in margarine, 548
in yellow corn, 333
discoloration of kernels, 97, 129, 157
genetics of, 34
in grading, 127–128, 129
of kernels, 58
Lovibond test, 544
removal
from crude oil, 541, 543
from starch hydrolysates, 511, 520

of snack foods, 169, 419, 423
Combustion
corn grain as fuel, 61
corncobs as fuel, 577
spontaneous, 60, 160
Commodity Credit Corporation, 240, 244, 245, 247, 248, 249
Common Agricultural Policy, 236, 237, 238, 239
Condensed fermented corn extractives, see Steep liquor
Confections, use of corn sweeteners in, 525, 527, 528
Consumption, world, 232 (see also Utilization)
Cooking properties, of corn, 423–424
Copra meal, 462
Corn Belt, 5, 6, 11, 12, 14–15, 16
Corn earworm, 39, 438
Corn flakes
flaking grits, 354, 356, 400–401, 402
process, 402, 404
Corn flour
composition, 356–357
in foods, 358, 362, 365, 401'
in industrial products, 358, 362, 366, 370
nixtamalized, see Nixtamalization
production, 353–355
Corn grits
in brewing, 359, 560
in foods, 358, 359, 365, 401, 402, 406, 407
production, 213, 355–356
Corn meal
in foods, 358, 362, 363, 401, 404, 406, 407, 419
partially cooked, 363, 364
production, 213, 355–356
Corn rootworm, 17, 39
Corn-soy-milk, 363, 364, 365
Cropping systems, 14–16
Cutin, 68, 332
Cystine, 73

Dairy cattle feeding, 462–463, 471, 474
Damaged kernels
determination of, 156–158
exterior damage, 152
in grading, 129, 156–157

in harvesting, 89
heat-damaged, 129, 151
insect-bored, 157
invisible damage, 152
mechanical damage, 152–153
mold damage, 129, 157
stress cracks, see Stress cracks
Debranching enzyme, in starch, 267
Degerminated products, 354, 358, 400–401 (see also specific products)
Degermination
equipment, 353, 354, 381
process, 62, 354–355, 380–381
Degree of polymerization, definition, 505
Degumming of oil, 539, 541
Density
bulk density, 59–60, 144–146
flotation test, 143
moisture, influence of, 144, 145, 146
specific density, 143–144
true density, 59, 143
Dent corn
description, 36, 55, 57, 58, 149, 386
origin, 3–4, 55
Deodorization of oil, 543
Deoxynivalenol, 167, 172
Deoxyribonucleic acid, 70, 287–288
Dewatering, in wet milling, 381, 383–384, 385, 389
Dextrins, 485–486
Dextrose
anhydrous, 513
applications, 525–527
crystalline, 513
liquid, 513
marketing, 217, 513–514
processing, 508–513
properties, 522–523
refining, 511–512
from starch hydrolysis, 509–511
USP grade, 513
Dextrose equivalent, definition, 505
Dietary fiber, see Fiber
Difolatan, 441
Discoloration, 97, 157
Diseases
animal and human, from

mycotoxins, 165–167, 456
management of, 27
plant
  bacterial, 27, 439
  breeding for resistance to, 38–39, 439
  ear rots, 167
  fungal, 27, 39, 439
  oil content, influence on, 313
  of seedling, 441
  virus, 27, 439
Distillation, of ethanol, 570, 571
Distiller's dried grains, 469, 470–471
Distillers Feed Research Council, 457
Distiller's solubles, 470–471, 472
Distilling industry, 467–472
Dried steep liquor concentrate, 464, 467
Dry milling
  corn quality for, 169
  history of, 351–352
  industry capacity and trends, 213, 352–353, 359–360, 371–372
  process, 353–355
  products, 68, 554 (see also Degerminated products)
Dryers
  batch-in-bin systems concurrent-countercurrent, 102
  continuous-flow cross-flow, 101–102
  cross-flow, 101
  dryeration, 102
  drum-dryers, starch, 485
  flash, 389
  rotary, 389
Drying
  of corn
    cultivar differences, 149
    deep-bed, 96
    with desiccants, 95
    of ear corn, 100
    economics of, 86, 221
    energy sources, 94–95
    mechanisms, 93–99
    quality, affected by, 96–97, 146, 153
    rates, 93–94, 137
    of seed corn, 48, 442
    shrink, 92–93, 137
    for storage, 131
  in storage, 100
  systems, 99–103
  of germ, 389
  of gluten feed, 389
  of starch, 392, 485
Dust, see Grain dust

Ear
  description, 31–32, 53
  development, 8–10
Eastern Europe, 217, 232
Egg quality factor, 467
Electronic moisture meters, 141–142
Elevators
  bucket, operation, 108
  country, 220, 224
  function in marketing, 220–223, 224–225, 226, 242
  terminal, 220, 224
  venting dust from legs, 108
Embryo (see also Germ)
  development, 70–71
  structure, 34, 62
Endosperm
  aleurone layer, 34, 65, 68
  carbohydrates, 257–265
  composition, 72–78
  development, 32, 69–71, 256–257, 265–267
  floury, 58
  genetics, 34, 36, 44, 258, 262, 431–437
  genotypes, see Mutants, endosperm
  horny/floury ratio, 55, 57, 149–150
  lipids, 72, 333, 336, 338
  protein matrix, 65
  proteins, 72–73, 296–298
  structure, 34, 63–68
  subaleurone, 65
  sweet corn, 431, 433–437
Endosperm dosage effects, 34, 36
Enzymes
  α-amylase, 78, 506, 509, 561, 562
  β-amylase, 507
  cellulase, 563
  glucoamylase, 507, 511, 562, 563
  glucose isomerase, 508
  glyoxysomal, 341
  immobilized, 511, 514–515, 516
  lipase, 342
  lipoxygenase, 342
  malt, 561
  pullulanase, 507, 563
  starch biosynthesizing, 265–268
Epichlorohydrin, 488, 489
Equilibrium relative humidity, 158, 159
Equine leucoencephalo-malacia, 167
Ergosterol, 162
Ethanol, fuel (power)
  absolute, 571
  azeotrope, 571
  heat energy content, 564
  processing, 467, 563–567, 582–587
  profitability calculations, 557–559
  from stalk, 583, 585
  from wet milling, 393
European corn borer, 39, 438
European (Economic) Community, 202, 216, 219, 231, 235, 236–237, 239, 240, 389, 536
Evaporation, 387–388, 396
Explosions, see Grain dust
Exports, 202, 216–219, 225, 240, 536 (see also World trade)
Extraction
  of corn oil, 538
  of juice from stalks, 582–583
Extrusion
  applications, 362–363
  of breakfast cereals, 403–404
  extrusion-cooking process, 360–362
  puffing, 404–405, 417–418
  snacks, 362, 403–404, 419–420

Fall armyworm, 39
Farming
  crop rotation, 14–15
  cultivar selection, 24
  dent corn, 13–27
  economics, 202–210
  government programs, 243–249
  industry size, U.S., 202–204
  management, 13–26
Fat, 72
Fatty acids
  in corn oil, 46, 316, 317–318, 535

essential, 316, 451, 546
free, 162, 538, 539, 540, 541, 544, 545, 549
genetic modification, 325, 327
2-hydroxy acids, 329
oxidation, 323–324
polyunsaturated, 316, 319, 535, 546, 547
saturated, 317, 338, 547
stereospecific distribution, 322
structure, 316
Feed manufacturing
corn in, 447–448
by-products in, 456–459
corn quality for, 171–173
linear programming, 172
Feeding value
of dent corn, 447–456
feed efficiency, 448
of fermentation by-products, 467–472
of high-oil corn, 315–316
of high-protein corn, 449–451
of low-test-weight corn, 148
of milling by-products, 457–467, 472–474
Feeds and feeding
amount consumed, of corn, 210–211, 447
animal feeding rates, 210–211
by-product ingredients, 370, 456–474
by-products, amount, 459
corn cobs in, 578
cornstalks in, 581, 583
from ethanol fermentation, 544, 567
high-oil corn in, 316
metabolizable energy in, 459, 463, 464, 465, 466, 471, 474
mycotoxins in, 165–167
pelleting of, 448
processing corn for, 387–389, 448
Fermentables
in corn syrups, 524, 528, 557
in hydrolysates, 561–562, 579–580
Fermentation
analytical methods in, 557
animal feed from, 467–472, 554, 567

batch, 563, 565
beverage ethanol, 570, 571–572
continuous, 565
economics and marketing, 214–215, 572
efficiency, 558, 568–569
energy costs, 568
fuel alcohol, 214, 393, 557–559, 569, 572
losses, 568–569
pharmaceutical products, 367
process, 563–568
substrate cooking and conversions, 560–563
substrates, 359–360, 402, 526, 554–557
Fertilization
of soil
influence on oil, 313, 318
sexual, in corn plant, 8–10, 32, 34
Ferulic acid, 77
Fiber
acid detergent, 77, 462, 463
and "bran," 69
dietary, 76, 258, 370, 371, 426
neutral detergent, 76, 77, 462, 463
separation, 381–384
Filtration
in brewing, 358–359
of crude oil, 539
in wet milling, 385, 392
Fine materials
in sampling, 130
distribution, 111, 130
in drying, 98–99
Fissures, endosperm, 152
Flaking (prime) grits
in corn flake manufacture, 354, 356, 400–401, 402
for corn quality test, 149
Flavor
of corn, 405, 439
of corn oil, 544, 549
of snacks, 419, 421
Flint corn, 3, 36, 55, 58, 128, 149, 386, 420
Flour (floury) corn, 36, 55, 149
Flowering habit, 31
Food uses, 399–427
for corn oil, 547–549
for corn sweeteners, 501, 525, 527, 528
for maltodextrins, 528

for milling products, 355–357, 360–365, 371, 390–392, 400–427
for sweet corn, 400, 431, 439, 443
trends, 211, 213–214
Foreign material, 128, 134
Fortified food products, 363–365
Free fatty acid test, for fungi, 162
Fructose (see also High-fructose corn syrup)
in corn, 255, 257, 433
crystalline, 517–518
by glucose isomerization, 508
isomeric forms, 523
Fumigants, 190–191, 193–195
criteria for success, 194
liquid, 194
Fungal species
Alternaria spp., 159
Aspergillus spp., 160, 165
candidus, 157
flavus, 104, 157, 158, 161, 164, 165, 166
glaucus, 157, 164
halophilicus, 91
niger, 507
oryzae, 506, 519
parasiticus, 165
Cephalosporium spp., 159
Cladosporium spp., 159
Colletotrichum graminicola, 39
Diplodia spp., 439
Erwinia stewartii, 439
Fusarium spp., 157, 159, 160, 167, 172, 439
moniliforme, 160, 167
Gibberella spp., 159, 167, 173, 439
zeae (*Fusarium graminearum*), 160, 439
Helminthosporium
maydis, 147, 439
turcicum, 39, 439
Nigrospora spp., 159, 160
Penicillium spp., 157, 158, 160, 165
Puccinia sorghi, 39, 439
Pythium spp., 439, 441
Rhizoctonia zeae, 441
Sphacelotheca reiliana, 439
Trichoderma reesei, 563
Trichothecium roseum, 167

Ustilago maydis, 439
Fungi
  growth, factors affecting, 158–159
  incidence in stored corn, 164
  inhibitors, *see* Fungistats; Mold inhibitors
  measurement of activity, 160–162
  and moisture, 159, 160
  types, 158
    field, 159–160
    storage, 160–161
Fungistats, 165, 168, 172, 441
Furfural, 575, 578–579, 581
Fusarium toxins, 167

Gas chromatography, 321, 335, 557
Genes (*see also* Mutants, endosperm)
  endosperm dosage effects, 34, 36
  for fatty acids, 319–320
  for leaf blight resistance, 439
  in sweet corn, 433–437
  for yellow pigments, 34
  for zein sequences, 287, 303–304
Genetics, *see* Breeding; Genes
Genotypes
  dent, 36, 55, 58, 128, 386
  flint, 36, 55, 58, 128, 386
  high-amylose, *see* High-amylose genotypes
  high-lysine, *see* High-lysine corn
  high-oil, 45, 313–314, 315–316
  mutants, *see* Mutants, endosperm
  open-pollinated, 37
  sweet, 37, 58, 433–437
  waxy, 44, 45, 50, 128, 262, 386
  white, *see* White corn
Germ (*see also* Embryo)
  amino acids, 293
  carbohydrates, 257–258
  composition, 72–73
  lipids, 72 (*see also* Lipids)
  lipoxygenase activity, 342
  minerals, 73
  proteins, 73, 293–296
  structure, 61–62

tocopherols, 336
Germ meal
  in feeds, 370, 390, 393
  in foods, 370–371
Germ processing
  dry milling, 355, 538
  oil expelling, 355, 390, 538
  oil refining, 539–543
  separation, *see* Degermination
  wet milling, 389–390, 538
Germination
  in planting, 7–8
  process, 62
  seed tests for, 49, 168
  in spoilage, 162
Globulins, 73, 275, 290–291, 293, 295
Glucose
  from cobs and stalks, 579
  in corn, 255, 257, 433
  isomeric forms, 522
  isomerization, 508, 514
  mutarotation, 522
Glutelins, 70, 275, 276, 291, 451
Gluten, corn, 305, 385, 389
Gluten feed, corn
  amino acid profile, 459
  for cattle, 459–463
  from ethanol production, 467, 469
  exporting, 237, 392, 457
  milk-producing property, 462
  pelleting, 389, 465
  for poultry, 464–465
  production, 217–219, 388–389
  for swine, 463–464
  wet, 458–459, 461
Gluten meal
  in cattle feeding, 465–466
  nutritional properties, 465
  in pet foods, 466
  in poultry feeds, 466
  production, 217–219, 305, 385
  rumenal bypass protein, 465–466
Glycerol
  in distiller's solubles, 469
  from fermentation, 568, 569
Glycosphingolipids, 329
Glycosylglycerides (glycolipids)
  in endosperm, 338
  galactosyldiglycerides, in

starch, 263
  in kernel, 327–328
Glyoxysomes, 341
Government (U.S.)
  marketing programs
    acreage diversion, 247
    Agricultural and Consumer Protection Act of 1973, 243, 244
    Agricultural Trade Development and Assistance Act of 1954 (P.L. 480), 363, 364, 365
    commodity disposal, 249
    commodity storage, 248
    emergency livestock feeding, 249
    Farm Facility Loan, 248
    Farmer-Owned Reserve, 223, 248. 250
    Payment-In-Kind, 201, 203, 223, 245, 248, 249, 250
    price-support, 243–246
    production adjustment, 246–248
    Soil Bank, 247
Grades and standards
  Argentine system, 132–133
  factors, 128, 134, 156–158
  FAQ system, 132
  Federal Grain Inspection Service, 127
  history, 126–127
  numerical, 128
  South African system, 132–133
  U.S. system, 127–128
    changes in, 126–127, 128
Grading
  accuracy, 129–130
  automation of tests, 130
  for damage, 156–158
  dividers, 128
  for export, 132–133
  for insects, 128, 185
  procedures, 128–129
  sampling for, *see* Sampling
  for toxic weed seeds, 128
Grain borer, lesser, 186, 190
Grain dust
  collection systems, 103–104
  control, 136
  description, 135
  explosions, 85, 103, 135, 578
  flow regulators, effect of, 111–112

oil addition, 136
source, 85
study of, 85–86, 135–136
suppressants, 104
utilization, 136
Grain exchange
organizations, 241–242
Grain moths
almond moth, 186
Angoumois grain moth, 186, 190
Indian meal moth, 186, 191, 192, 197
Grain stirrers, 100–101
Grain yields, U.S., 202–205
Gravity tables, 354, 355
Growing degree units, 12, 24, 147
Growing season, 11

Handling systems
belt conveyors, 109
bucket elevators, 108
chain and flight conveyors, 108
for ear corn, 104
equipment, 104, 112
grain spreaders, 110
gravity spouts, 110
mechanization, 85–86
pneumatic conveying, 110
power requirements, 106–107
screw conveyors, 106
for shelled corn, 104–112
Hardness, kernel
cultivar differences, 58
moisture influence, 151–152
protein relationship, 150
significance, 149–150
tests, 56–58, 150–152
Harvesters
field shellers, 83, 86, 87
pickers, 87
picker-shellers, 46, 88, 187
self-propelled combines, 89
for sweet corn, 442
Harvesting
accidents, 87
area harvested, U.S., 201, 203, 205, 579
dryers, effect of, on, 221
of ear corn, 86–87
ear corn shelling, 87
field shelling, 85, 86, 88–89
mechanization, 89
of seed corn, 48
shelling losses, 89

of sweet corn, 438, 442
systems, 83, 85
Heat energy
of corn kernels, 61
of corncobs, 577
of yeast fermentation, 564
Heat of vaporization, 93
Hemicelluloses, 62, 77, 259
(*see also* Pentosans)
Herbicides
in corn growing, 27
influence on oil content, 313
High-amylose genotypes
breeding, 44, 45, 50
starch granule properties, 262–264
wet milling, 386–387
High-fructose corn syrup
applications, 527
crystalline fructose, 517–518
growth of industry, 213, 214, 391
HFCS-42, 515–516
HFCS-55 and 90, 516–517
marketing, 391, 518
processing, 514–515
properties, 523–524
sweetness, 524
High-lysine corn (*see also* Opaque-2 corn)
assay methods, 299, 300
breeding and genetics, 45, 301–304
endosperm texture, 58, 150, 452
influence of nitrogen fertilizers, 301–302
properties, 58
nutritional, 45, 299–301, 426, 451–452
High-oil corn, 45, 313–316
High-performance liquid
chromatography, 167, 321, 335–336, 557
High-protein corn, 449–451
Hilar layer, 34, 69
Hominy feed
nutritional properties, 474
production, 355, 358, 370, 473–474
Horny-floury endosperm
ratio, 55, 57–58, 149–150
Husking, 87, 442
Husks, 39, 438
Hybrid corn (*see also*
Mutants, endosperm)
double-cross, 38

experimental, 43–44
high-oil genotypes, 313–316
history, 3–4, 37–38
performance trials, 44
protein trends in, 75
seeds, 215, 440–442
choice for planting, 24
production, 47–49
single-cross, 38
sweet corn, 433–437
test weight differences, 145
Hydrocarbons, 331
Hydroclones, 381, 385
Hydrogenation
and blending, 543
of corn oil, 324–325, 536–537, 542–543
of corn syrup, 528
solid fat index, 543, 544–545
Hydrol (corn molasses), 393
Hydroxyproline, 78, 260

Illinois oil strains
development, 313
oil bodies in, 312
oil content in, 45
proteins in, 303
triacylglycerols in, 321
Illinois protein strains, 46, 300, 303
Inbreeding, 40
Indoleacetic acid, 74
Inland Waterways Revenue
Act of 1978, 229
Insect species
Ahasverus advena, 187
Alphitophagus bifaciatus, 187
Carpophilus dimidiatus, 187
Cryptolestes spp., 187, 189, 197
ferrugineus, 186
pusillus, 186
Cynaeus angustus, 187
Diabrotica spp., 39
Diatraea grandiosella, 39
Ephestia cautella, 186
Heliothis zeae, 39, 438
Liposcelis spp., 187
Murmidius ovalis, 187
Oryzaephilus surinamensis, 186, 187
Ostrinia nubilalis, 39, 438
Platydema ruficorne, 187
Plodia spp., 197

interpunctella, 186, 187, 191, 192
Rhyzopertha dominica, 186
Sitophilus spp., 187, 189, 197
  granarius, 186
  oryzae, 186
  zeamais, 149, 186
Sitotroga cerealella, 186
Sporodoptera frugiperda, 39
Tenebrio molitor, 187
Tribolium spp., 187, 197
  castaneum, 186
  confusum, 186
Typhaea stercorea, 187
Insect trapping, 189
Insecticides
  aerosols, 193
  chloropicrin, 190
  chloropyriphos methyl, 193
  malathion, 191–192
  methoxychlor, 191
  pirimiphos methyl, 193
  pyrethrins, 191
Insects
  breeding for resistance to, 39, 438
  control
    biological, 196
    cleaning of storages, 190
    fumigation, 193–195
    grain protectant chemicals, 191
    insecticides, *see* Insecticides
    nonchemical methods, 195–196
    residual sprays, 191
    sampling for, 130, 189–190
  corn production, effect on, 27
  corn quality, effect on, 186–187
  development in stored grain
    factors affecting, 188–189
    stages, 188
  in exported grain, 187
  grade damaged by, 128, 129, 157, 185
  in grain storages
    detection and measurement, 189–190

incidence, 187
prevention, 191–193
species involved, *see* Insect species
sampling for, 130
in sweet corn, 438
trapping, 189
Inspection
  to assess grain condition, 189
  in marketing (grading), 128
International Center for Improvement of Maize and Wheat, 58
Iodine value, 544, 549
Irradiation, of insects, 196
Isomaltose, 510

Japan, 217, 230, 237–238

Kernel
  composition
    alteration by breeding, 45–47
    carbohydrates, 76–78, 260–265
    dietary fiber, 76
    lipids, 72, 75–76, 320–341
    minerals, 73
    proteins, 72, 74–75, 277–298
    starch, 70, 72
    sugar, 70, 72
    vitamins, 73–74
  damage, 89
  development, *see* Maturation, of kernel
  properties
    color, 58
    physical, 59–61
    size and shape, 55
  structure
    changes during cooking, 424
    description, 7–8, 34, 53–58, 61–69, 254–255

Lactic acid
  contamination of beverage alcohol, 568
  in corn steeping, 379
  from glucose fermentation, 559–560
  in steep liquor, 466, 467
Landry-Moureaux protein separations, 275–276
Latin America, 216

Leaf blights, 39, 147, 439
Lecithin, corn, 327, 539, 545, 548, 549
Lectins, 78
Lignin, 77, 259, 580, 581
Linoleic acid
  in corn oil, 46, 76, 316, 318, 319, 452, 537
  genetic control, 319–320
  nutritional value, 547
  preservation by tocopherols, 335
Linolenic acid, 327, 536, 546
Lipids
  classes, 320–337
  extraction, 337
  in kernel, 75–76, 295, 337–339
  nutritional value, 452–454
  and proteins, interaction, 295
  in starch granules, 262–264, 340–341
  triacylglycerols, 311, 321–325
Lysine
  analysis for, 299, 300
  in corn proteins, 291, 292, 293, 451
  in gluten feed, 463, 464
  in gluten meal, 465, 466
  in hybrids, 58
  nutritional importance, 452
  in o_2 and fl_2 mutants, 299, 300, 301
  synthetic, 452

Magnesium, 20, 23, 455
Maize, 1
Maize dwarf mosaic virus, 439
Malathion, 191–192
Maltodextrins, 521–522, 525, 528
Maltose
  in corn syrups, 519
  in developing kernel, 257–258
  in starch hydrolysates, 510
  in sweet corn, 433
Maltulose, 510, 562
Margarine, 547–548
Marketing, of corn (*see also* Transportation, of corn)
  of cob and stalk products, 575
  commercial, 224–225
  discounts, 130, 131, 137

economic system, 219–225
  for export, 132–133,
    230–232
  facilities, 220–221
  from farm, 223
  grades, 130–132
  identity-preserved
    contracts, 131–132
  moisture, accounting for
    in, 137–138
  pricing on quality, *see*
    Pricing systems
  of sweet corn, 444
  trends, 202, 250
Masa, *see* Nixtamalization
Mass selection, 39
Maturation, of kernel
  cultivar rating for, 24
  grain-filling period, 147
  growing season, effect of,
    11
  indexes, 10, 69, 146–147
  moisture, effect of, 12–13
  soft corn, 147
  structural, 9–10, 53–58,
    254–255
  temperature, effect of,
    11–12
  test weight, measure of,
    146–148
Mealworm, yellow, 187
Metabolic reactions in starch
  biosynthesis, 265–269
Metabolizable energy
  of corn, 447
  of distiller's by-products,
    470, 471
  of hominy feed, 474
  influence of fat absorption,
    452
  of lipids, 452
  for poultry, 465
  for swine, 463
  of wet-milled by-products,
    459, 465
Methionine, 73, 286, 451, 465
Methoprene, 196
Methyl bromide, 193, 194,
  195, 197
Mexico, 216
Milk line, 10, 147
Milling
  dry, *see* Dry milling
  evaluation factor, 170
  for fermentation, 561
  traditional, 406
  wet, *see* Wet milling
Mills
  attrition, 151, 578

Beall degerminator, 353,
  354
disk, 381, 382, 384
Entoleter, 382
hammer, 151, 561, 578
pin, 384
roller, 355, 578
Minerals
  in distiller's by-products,
    470–471
  in hominy feed, 472
  in kernels, 10, 73, 455
  required by plant, 19–20
Mixed corn, 127
Modal shares in grain
  marketing, 225–228,
    229–230
Modified atmosphere
  storage, 195
Moisture
  absorption, 60
  addition, 137
  for corn growing, 12–13
  effect on corn physical
    properties, 60, 136–137,
    144
  equilibrium, 90
  importance, 136–137
  and insect growth in
    storage, 188–189
  and marketing of corn,
    137–138
  measurement
    drying methods,
      138–140
    electronic, 141–142
    empirical methods, 140
    reference methods, 138
    in single kernels,
      142–143
    for sweet corn, 443
  migration, 115
  for safe storage, 90–92
Mold inhibitors
  ammonia, 165
  in animal feeds, 172
  propionic acid, 165, 172
  to reduce spoilage, 101
  sulfur dioxide, 101, 165
Mold invasion
  breeding resistance to, 165
  detection, 162–164
  loss of grade, 161
  and mechanical damage,
    161
  prevention, 164–165
  in storages, 92, 160
Monacyl lipids, 340
Mutants, endosperm

amylose extender, *ae*, 44,
  258, 262, 264, 436, 437
brittle, *bt*, 314, 435,
  436–437
dull, *du*, 436, 437
floury-2, fl_2, 45, 296, 298,
  303, 314, 451
multiple gene type, 437
for oil content, 319
opaque-2, o_2, *see* Opaque-2
  corn
opaque-7, o_7, 45, 298, 303
protein alteration,
  301–304
shrunken, *sh*, 435–436,
  437, 440, 441, 444
starch alteration, 44–45,
  257, 258, 264–265
sugary, *su*, 45, 257, 258,
  431, 433–435, 436, 437,
  438
sugary enhancer, *se*, 435,
  440, 441
waxy, *wx*, 44–45, 262, 264,
  436, 437
Mycotoxins
  aflatoxin, *see* Aflatoxin
  in animal feeds, 172
  in corn, 456
  deoxynivalinol, 167
  detection, 166–167
  mold species involved,
    165–167
  symptoms, 456
  tricothecenes, 167

National Sweet Corn
  Breeders Association, 433
Near-infrared reflectance
  methods
  for hardness measurement,
    151
  for moisture measurement,
    140–141
Neutral detergent fiber, 76
Niacin
  bound, 74, 455, 463
  content
    in by-product feeds, 463
    in corn, 74, 455
    in processed foods, 413,
      424
Nitrogen
  analysis for, 298
  cycle in soils, 21–22
  deficiency symptoms, 21
  fertilizer
    effects of, 15, 74, 295,
      301, 302

600 / Index

sources, 22
usage, 21, 205
gas in oil storage and
  packaging, 545, 546
nonprotein, 292–293, 295
protein conversion factor,
  298
Nixtamalization
  dry masa flour, 363, 416,
    417–418
  masa preparation, 411–424
Nuclear magnetic resonance,
  45, 140, 313
Nutrients, needed by corn
  plant, 19–22
Nutritional values
  of alkaline-cooked
    products, 424–426
  assays for protein, 299–301
  of by-products, 459–474
  of corn, 449–452
  of polyunsaturated fatty
    acids, 546–547
  vitamin E requirements,
    547

Oil
  absorbability, 452
  breeding for, 45, 46,
    313–316
  composition, 452, 536, 538
  content, 45, 75, 295, 312
  deposition in kernel, 71
  expelling, 355, 390, 538
  linoleic acid in, 452, 453
  market trends, 536, 549
  metabolizable energy, 452
  processing (see also
    Hydrogenation;
    Refining)
    of coproducts, 545
    of oil, 538–543
    volume, U.S. and world,
      535–536
  products
    lecithin, 327, 539, 545,
      548, 549
    margarine, 547–548
    refined oil, 539–543
    salad and frying oil, 547,
      549
    soapstock, 541, 545
    vegetable oil distillate,
      545
  properties
    autoxidation, 322, 323,
      538
    digestibility, 322
    fatty acids in, 316–318,

  536–538, 539, 540
  oxidative stability, 535,
    544, 546
  smoke point, 544, 545
  thermal oxidation, 323,
    538, 545
  thermal polymerization,
    538
  triglycerides, 311, 536,
    540
  quality, tests of, 543–545
  shipping, 546
  spherosomes (oil bodies),
    62, 312
  storage, 545–546
  volume of corn grown for,
    U.S., 311
Oil bodies, 62, 312
Oleic acid, 320, 537
Oligosaccharides
  in corn, 257–258
  as unfermentables, 562
Ontario heat unit system, 12,
  24
Opaque-2 corn, 45, 58, 150,
  273, 296, 299, 300, 301,
  302, 303, 304, 314, 406,
  426, 451
Origin and history of corn,
  3–4
Osborne protein classes, 275
  (see also individual classes)

Palmitic acid, 317, 318, 319
Paper manufacture
  with cornstalks, 581
  dry-milled products in, 368
  starches in, 485, 496–498
Parched corn, 422
Payment-In-Kind program,
  201, 203, 223, 245, 248,
  249, 250
Pedicel, 34, 53, 254
Pentosans
  as cell wall components,
    259–260
  content of
    in corn, 77, 260
    in corncobs, 577
  conversion to furfural, 578
  as fermentation substrate,
    556
  structure, 260
Pericarp
  damage, 152
  lipids in, 336, 339
  structure, 34, 53, 68–69,
    259, 260

  in sweet corn, 431, 439,
    440
Peroxide value, 544
Pest management, 27 (see
  also Insects, control)
Pet foods, 466
Phospholipids
  (phosphoglycerides)
  in crude corn oil, 539
  diacylphospholipids in
    starch, 263
  lysophospholipids in
    starch, 264
  minor types, 325
  modification by breeding,
    325
  nonhydratable, 541
  phosphatidylcholine, 325,
    327, 338
  phosphatidylethanolamine,
    325, 327, 338
  phosphatidylinositol, 325,
    327, 338
Phosphorus
  in by-product feeds, 460,
    470
  for corn production, 19,
    20, 21, 23
  in the kernel, 73, 455
  phytin phosphorus, 73, 470
Phytate, 258, 295, 455, 460
Phytin/phytic acid
  in distiller's grains, 470
  in the kernel, 73
  nutritional interactions,
    295, 455
Phytoglycogen, 260, 261,
  433, 435
PIK program, see Payment-
  In-Kind program
P.L. 480, see under
  Government (U.S.)
  marketing programs
Placento-chalazal tissue, 254
Plant
  climatic requirements,
    11–13
  composition, 10
  description, 1, 7, 8
  diseases, 27
  growing season, 11
  growth and development,
    1–2
  physiology of stalk sugars,
    583–584
  roots, 10
Planters, corn, 55, 441
Planting
  area planted, U.S.,

202–203, 232
  dent corn, 24–26
  rates, 26, 215
  sweet corn, 440–441
  time of, 24–25
Plasmodesmata, 255
Pod corn, 3, 36, 432
Pollination, 1, 8–9, 31, 32, 34
Pollution control
  air quality, 394–395
  grain dust, 86, 103–104
  waste water treatment, 393–394
Polyisoprenoid alcohols, 331–332
Popcorn, 55, 128, 131, 149, 202, 219, 420–421
Porosity (void volume), 59–60
Potassium
  for corn production, 19, 20, 21, 23
  in the kernel, 73, 455
  in steep liquor, 467
Poultry feeding, 464–465, 466, 467
Price support loans, 240, 244, 245, 250
Pricing systems
  of corn sweeteners, 513, 518, 521
  country elevators, 242–243
  grain exchanges, 241–242
  and quality differences, 131, 137
  standard bushel concept, 137–138
Production, of corn (*see also* Farming; Planting)
  acreage, 202–203
  compared to other crops, 6
  by contract, 50
  costs and returns, 205–210
  grain yields, 6, 13, 14, 203–205
  labor requirements, 85
  locations of, 5–6
  management, 13–26
  for silage and forage, 24, 201
  world, 4–6, 399
Production economics
  capital, 208
  cash receipts and expenses, 208
  costs, 205, 206, 207, 208
  drying economics, 221–223
  enterprise budget systems, 207–208

farm record systems, 205–206
  management, 205
  returns, 205, 207, 210
  trends, 202–210
Prolamins, 275, 277–278
Propionic acid
  as mold inhibitor, 165
  in poultry feeds, 172
  in rumen digestion, 471
  for storing high-moisture corn, 101
Propylene oxide, 489
Protein
  amino acid composition, 274
  assays of, 299–301
  biosynthesis in kernel, 70
  bodies, *see* Protein bodies
  content in kernel, 72, 74, 75, 449–451
  extraction methods, 275, 276
  fortification of foods with, 363, 371
  fractionation, 274–276
  industrial uses, 305
  matrix, 65–66, 70, 291, 296
  modification by breeding, 45, 46
  nutritional value, 449–452
  separation, from starch, 385
  separation methods, 277, 280–282
  in starch granules, 264
  storage proteins, 276, 291, 295
  yield, in corn crop, 273
Protein bodies
  composition, 66, 277–278
  development, 296–297
  in other grains, 297–298
  structure of, 277, 298
Psocids, 187
Puffing, 404–405, 417–418
Pullulan, 560
Purothionins, 291, 292
Pyridoxine, 74, 455

Quality, of kernels
  for alkaline-cooking process, 422–423
  effects of drying corn, 96–97, 136
  grades, 127–128, 132–133
  insects, effects of, 185–187
  microorganisms, effects of, 158–159

physical changes, 152–156
  for processing, 168–173
  sampling for, 127
  tests
    estimates of, 299–300
    for oils, 543–545
    for snack foods, 419
    for starches, 483
    for sweet corn, 443
  trends, 173–174
Quality protein maize, 58
  (*see also* High-lysine corn)

Recurrent selection, 40–42
Reduced soluble proteins, 280, 281, 282, 283, 286, 298
Refining
  caustic, 540
  of crude corn oil, 355, 390
  of high-fructose syrups, 516
  of regular corn syrups, 520
  of starch hydrolysates, 511–512
  steam, 541
Reproduction
  double fertilization, 33
  pollination, 8–9, 31, 32
Respiration
  accompanying mold invasion, 160, 161
  and dry matter loss, 160–161
Reverse osmosis, 396
Ribonucleic acid
  in kernel development, 70, 71
  zein messenger, 287
Rice, 1, 6, 239, 340, 359, 360
Rootworms, 17, 39
Rust disease, 39, 439

Sampling, 127, 128, 129, 130, 189
Scutellum, 34, 61, 62
Seed
  breeder, 49
  certified, 49
  chemical protectants, 441
  coat, 68
  coating with graphite, 441
  contents, 292
  foundation, 49
  marketing, 49, 131, 219
  for planting, 24, 215
  production
    classification, 49
    conditioning, 48

crossing techniques, 47–48
drying, 48, 168
harvesting, 48
isolation requirements, 48
of sweet corn, 441–442
quality, 24, 49, 168
registered, 49
size grading, 55, 441
for sweet corn, 440
testing, 49, 168
treatment with chemicals, 49
volume produced, 215
Selection in S_1 or S_2 progenies, 40
Selenium, 73, 455–456
Sieving
in grading, 134
in handling system, 110
sizes, U.S., 356
wedgewire screens, in wet milling, 383
Silk, 8–9, 32, 438
Smut, 439
Snack foods, 400 (see also Popcorn)
corn chips, 411, 418–419
corn quality for, 169–170
extruded snacks, 362, 419–420
parched corn, 422
tortilla chips, 410, 418–419, 423
Soft corn, 147, 406
Soil erosion, 15, 16
Solar energy in corn drying, 95
Solvent extraction, 390, 538
Sorbitol
from dextrose, 526
in sweet corn, 258, 495
Sorghum, 340, 411, 582, 585, 586
South Africa, Republic of, 132, 133, 217, 233, 234, 235, 239, 242, 536
South Korea, 217
Southwestern corn borer, 39
Soviet Union, 217, 232, 234, 238, 536
Soybean hulls, 461
Soybean meal, 447, 465
Soybeans, 6, 15, 202, 273, 311, 535
Spoilage
microorganisms, role of, 158–159

moisture, influence of, 91, 92, 159, 160
in storage, reduction of, 101
Spout lines, 110, 130, 189
Staggers Rail Act of 1980, 229
Stalks
amount produced, 579
composition, 579–580, 582
digestibility by animals, 581
for ethanol, 583, 585–586
genotypes, high stalk sugar, 583
juice, 582–583
processing, 581
products, 580–581
storage, 580
structure, 579
sugar in, 579–580, 583–584
measurement in juice, 585
for sweetener production, 582
Standard bushel pricing concept, 137–138
Starch
accumulation, in endosperm, 256
acid-thinned, 485–486
composition, 262–264
content, in kernel, 256, 447
cross-linked, 487–489
granules
biosynthesis, 265–268
gelatinization, 264, 481, 482
in kernel, 72
lipids, 262–264
physical properties, 264–265
structure, 262, 481
X-ray diffraction pattern, 262, 264
hydrolysis
acid, 505, 556
acid reversion products, 509
chemical gain factor, 505
for dextrose production, 509–511
enzyme, 506–507, 561–563
for fermentation, 561–563
marketing, 217
modification, 481–495

oxidized (bleached), 486–487
paste characteristics, 481
pregelled, 485
production, U.S., 479
proteins in, 264
substituted, 489
cationic, 494
esters, 491–492
hydroxyalkyl, 493–494
sweeteners from, see Sweeteners, corn
synthesis of, in kernel, 70
uses
food, 485, 487, 492, 496
nonfood, 485, 487, 492, 495
in paper manufacture, 496–498
for production of chemicals, 495
from waxy maize, 262, 264, 480, 521
wet milling of, 385–386, 392–393
Starch graft-polymers, 495
Starch phosphorylase, 265
Starch synthase, 264, 265, 266
Steam flaking, of whole corn, 448
Stearic acid, 317, 318, 319, 537
Steep liquor (condensed fermented corn extractives)
dried steep liquor concentrate, 464, 467
evaporation, 387–388
in feeds, 387
lactic acid in, 467
metabolizable energy, 464
nutrients in fermentation, 555
properties, 393, 466–467
unidentified growth factors, 472
Steeping
in alkaline processing, 413
index, 171
nutrient source for yeast, 555
in wet milling, 60, 62, 66–68, 377–380, 396
Stein breakage tester, 46, 47, 151, 154, 155
Stenvert hardness tester, 7, 151
Sterols, 329–331, 339

Stewart's wilt, 439
Stillage, 567, 568, 569
Storage, of grain
  aeration during, 115–117, 164
  in controlled environment, 117–118, 195
  of corn cobs, 577–578
  of corn gluten feed, 459
  of corn oil, 545–546
  of ear corn, 112
  future trends, 118–119
  moisture migration in, 115
  monitoring conditions, 118, 164
  reduction of insect damage in, 190–196
  safe moisture levels for, 90–92, 161
  structures
    bin and silo designs, 113–114
    bin failures, 113
    capacity, U.S., 223–225
    cribs, 112
    entrapment in, 115
    explosions in, 85–86, 103
    flat storage, 113
    temporary, 114–115
    underground, 112
Storage fungi, 160–161
Storage proteins, 295
Stover, 10
Stress cracks
  cause, 46, 97, 102, 136, 153–154
  in grading, 129
  hardness, effect on, 149
  measurement, 153–154
Sucrose
  content in kernel, 257–258
  in cornstalks, 582
  relative sweetness, 523, 525
  replacement by high-fructose corn syrup, 213, 527
  in sweet corn, 433, 434–435
  transfer into kernel, 255
Sugarcane, 581, 582, 585, 586
Sugars (see also Sweeteners, corn)
  in cell walls, 259
  content in kernel, 72, 76, 256, 257–258
  in green stalks, 582–584
  in kernel development, 70, 255, 256–257, 258
  in mature stalks, 579–580
  mutarotation, 522
  pentose, 578 (see also Pentosans)
  in sweet corn, 433–437
Sulfhydryl bonds, 280
Sulfur, 20, 23, 73
Sulfur amino acids, 73, 286, 292, 463, 466
Sulfur dioxide
  as mold inhibitor, 164, 165
  in starch recovery, 62, 66, 67, 377, 379
  thiamine, reduction of, 460
  yeast affected by, 555
Sweet corn (see also Mutants, endosperm)
  breeding, 58, 437–438
    for disease resistance, 439
    for eating quality, 439–440
    for pest resistance, 438
    for processing quality, 439
  standard types, 433
  super sweet types, 434–435
  for canning and freezing, 443
  composition, 257–258, 261, 433
  genotypes
    augmented-sugary, 434–435
    cultivar names, 433–434, 435, 436, 437
    elevated-sugar mutants, 432–437. 443
    multiple gene class, 437
    single gene classes, 435–437
    standard, 433–434
  grading, 128
  history, 432
  marketing, 132, 144
  production area, 431, 432, 443
  seed production, 440–442
  trypsin inhibitor in, 292
Sweeteners, corn (see also individual sweeteners)
  history, 502–504
  marketing, 513–514, 518, 521
  processing, 508–522
  utilization of corn for, 501
Swine feeding, 463–464
Syrups, corn
  brewing adjunct, 359–360
  as fermentation substrate, 556, 557
  high-conversion, 519
  high-fructose, see High-fructose corn syrup
  high-maltose, 519–520
  marketing, 520–521
  processing, 518, 520
  properties, 524–525
  specialty, 520
  sweetness, 525

Taiwan, 217, 230, 239
Temperature
  for corn growth, 11–12
  of drying, for wet milling, 171
  and fungal growth, 161–162
  and insect growth, 188
  for storage, 91–92, 118
  in stress crack formation, 153
Tempering, 102, 103, 353–354. 402
Teosinte, 3, 432
Test weight, 128, 129, 144–146
  and corn maturity, 146–148
  influence of, 148–149
Tetrazolium seed test, 168
Thailand, 233, 234, 235, 239
Thermal conductivity, 60, 94
Thiamine
  content in corn, 74, 455
  supplementation of feeds, 460
Thiram, 441
Tillage systems, 16–18
Tip cap, 53, 69
Tocols, 334–337
Tocopherols
  as antioxidants, 335, 546
  breeding for, 336–337
  content in corn, 335
  in oil distillate, 545
  and vitamin E activity, 337
Tocotrienols, 335, 336
Tortillas, 410–414, 416, 426
Traditional foods
  alcoholic beverages, 406
  alkaline-cooked products, 410–413
  arepa, 406–407
  porridges, 407–408
Transportation, of corn
  domestic movements, 226, 228

Index / 603

604 / Index

to export regions, 228–229
Inland Waterways Revenue Act of 1978, 229
mechanization, 85–86
modal shares, 225–226, 229–230
recent developments, 229–230
Staggers Rail Act of 1980, 229
Triacylglycerols (triglycerides)
  analytical methods for, 544
  fatty acid stereospecific positions, 322
  melting point, 544
  in oil, 311, 321–325, 536, 540
Trichothecenes, 167
Trisaccharides, 258, 505
Trypsin inhibitor, 78, 292, 456
Tryptophan
  analysis for, 300
  in by-product feeds, 463, 465
  in corn proteins, 426, 451
  and lysine, 452
  and niacin, 455

Unidentified growth factor, 467
United States
  carryover stocks, 202, 245, 246, 248
  grading system, *see* Grading
  marketing programs, *see* Government (U.S.) marketing programs
  production, 201–202, 203, 205, 311
  utilization, 201, 210, 216
Utilization
  economics of
    food and industrial uses, 210, 211–215
    livestock feeds, 210–211, 370
    seed, 215
  feed, *see* Feeds and feeding
  food, *see* Food uses
  industrial
    adhesives, 368, 369, 486
    charcoal briquets, 368
    corncobs, 578
    foundry binders, 367

    gypsum board binders, 366
    in oil well drilling, 369
    in ore refining, 369
    paper products, 368, 496–498
    plastics, 369–370, 495

Varieties, *see* Genotypes
Viscosity, 481, 483, 488
Vitamins
  A, 34, 73, 334, 454
  in by-products, 466, 472
  in corn, 73–74, 454–455
  E, 73, 337, 455, 456, 545
Vomitoxin, *see* Deoxynivalenol

Washing
  of alkaline-cooked corn, 414
  of raw corn, 353
  of starch, 383, 385–386
Water management
  in brewing, 555
  in corn growing, 18–19
  in wet milling, 379–380, 393–394, 395–396
Waxes, 332, 542
Waxy corn, 202
  breeding, 44, 45, 50
  grade, 128
  marketing, 131
  starch from, 480, 521
    granule properties, 262, 264, 420
  wet milling of, 386
Weed species
  Agrostemma githago, 128
  Amaranthus spp., 26
  Crotalaria spp., 128
  Cyperus spp., 26
  Echinochloa spp., 26
  Ricinus communis, 128
  Setaria spp., 26
Weeds
  in corn farming, 26–27
  toxic seeds in grading, 128
Weevils
  granary, 186, 188
  maize, 186, 197
  rice, 186, 190, 197
Western Europe, 217 (*see also* European [Economic] Community)
Wet milling (*see also* Feeds and feeding)
  automation, 395
  corn quality for, 170–171

  corn varieties processed, 386–387
  energy for, 395–396
  feed production, 387–389, 393, 456–467
  industry trends, 213, 390–392, 457, 459
  pollution control, 393–395
  process, 304, 305, 377–390
  products, 390–393, 508–528, 538, 543, 545
  separation of kernel components, 66–68, 380–386
  steeping, 60, 62, 66–68, 377–380
Wheat, 1, 6, 297
White corn, 36, 50, 127, 128, 131, 202, 235, 239, 405, 411, 418
Winterization, 539, 542, 549
Wisconsin breakage tester, 47, 155
World trade (*see also* Exports)
  export supply and demand, 232
  exporters, major, 233–236, 239
  grading systems, 132–133
  importers, major, 236–239
  markets, for U.S. corn, 216–217, 240
  policies of trading countries, 236–240
  trends, 202, 230
Wort, 359

Xanthan gum, 560
Xanthophylls, 333, 334, 454, 459, 466
  breeding for, 334
Xenia effect, 34, 37, 435, 437
Xylans, 77, 78, 259–260
Xylose
  in cell walls, 259–260
  in glucose isomerization, 508
  production from corn residue, 575, 579, 580

Yeast
  CO_2 production, 568
  energy release, 564
  ethanol production by, 563–565
  growth, 563–564, 568
  metabolic pathway, 565
  nutrient requirements, 563

role in quality
 deterioration, 158
Yeast species
 Candida spp., 556
 Saccharomyces spp., 556, 563
  cerevisiae, 563
  uvarum, 564
Yellow corn, 34, 127, 128

Yields, 6, 13, 14

Zearalenone, 167, 170
Zein
 composition, 274, 280, 285, 287–288, 299
 deposition in kernel, 70, 296–297
 in embryo, 293

 extraction, 278–282, 298, 301
 fractionation, 282–284
 genetics of synthesis, 303
 industrial uses, 305
 native, 280
 nutritional properties, 275
 and protein content, 75, 451